Advances in Modal Logic

Volume 8

Advances in Modal Logic

Volume 8

Edited by

Lev Beklemishev

Valentin Goranko

and

Valentin Shehtman

ISBN 978-1-84890-013-4

College Publications
Scientific Director: Dov Gabbay
Managing Director: Jane Spurr
Department of Computer Science
King's College London, Strand, London WC2R 2LS, UK

http://www.collegepublications.co.uk

Original cover design by Richard Fraser, Avalon Arts, UK
Printed by Lightning Source, Milton Keynes, UK

Contents

Preface vii

GERARD ALLWEIN AND WILLIAM L. HARRISON
Partially-ordered Modalities 1

MARTA BÍLKOVÁ, ONDREJ MAJER, MICHAL PELIŠ AND GREG RESTALL
Relevant Agents 22

KAI BRÜNNLER, REMO GOETSCHI AND ROMAN KUZNETS
A Syntactic Realization Theorem for Justification Logics 39

SERENELLA CERRITO AND MARTA CIALDEA MAYER
Nominal Substitution at Work with the Global and Converse Modalities 59

HANS VAN DITMARSCH, TIM FRENCH AND SOPHIE PINCHINAT
Future Event Logic — Axioms and Complexity 77

DAVID FERNÁNDEZ-DUQUE
Absolute Completeness of $\mathsf{S4}_u$ for Its Measure-Theoretic Semantics 100

SANTIAGO FIGUEIRA AND DANIEL GORÍN
On the Size of Shortest Modal Descriptions 120

AMÉLIE GHEERBRANT
Complete Axiomatization of the Stutter-invariant Fragment of the Linear Time μ-calculus 140

RAJEEV GORÉ, LINDA POSTNIECE AND ALWEN TIU
Cut-elimination and Proof Search for Bi-Intuitionistic Tense Logic 156

WESLEY H. HOLLIDAY AND THOMAS F. ICARD, III
Moorean Phenomena in Epistemic Logic 178

RYO KASHIMA
Completeness Proof by Semantic Diagrams for Transitive Closure of Accessibility Relation 200

STANISLAV KIKOT
Semantic Characterization of Kracht Formulas 218

CLEMENS KUPKE AND DIRK PATTINSON
On Modal Logics of Linear Inequalities 235

AGI KURUCZ
On the Complexity of Modal Axiomatisations over Many-dimensional Structures 256

AGI KURUCZ, FRANK WOLTER AND MICHAEL ZAKHARYASCHEV
Islands of Tractability for Relational Constraints: Towards Dichotomy Results for the Description Logic $\mathcal{EL}$ 271

ALEXANDER KURZ AND YDE VENEMA
Coalgebraic Lindström Theorems 292

MATI PENTUS
Complexity of the Lambek Calculus and Its Fragments 310

KATSUHIKO SANO AND MINGHUI MA
Goldblatt-Thomason-style Theorems for Graded Modal Language 330

LUIGI SANTOCANALE AND YDE VENEMA
Uniform Interpolation for Monotone Modal Logic 350

ILYA SHAPIROVSKY
Simulation of Two Dimensions in Unimodal Logics 371

DMITRY SKVORTSOV
A Remark on Propositional Kripke Frames Sound for Intuitionistic Logic 392

TOMOYUKI SUZUKI
Bi-approximation Semantics for Substructural Logic at Work 411

TINKO TINCHEV AND DIMITER VAKARELOV
Logics of Space with Connectedness Predicates: Complete Axiomatizations 434

SARA L. UCKELMAN AND SPENCER JOHNSTON
A Simple Semantics for Aristotelian Apodeictic Syllogistics 454

ERIC UFFERMAN, PEDRO ARTURO GÓNGORA AND FRANCISCO HERNÁNDEZ-QUIROZ
A Complete Proof System for a Dynamic Epistemic Logic Based upon Finite π-Calculus Processes 470

HEINRICH WANSING
Proofs, Disproofs, and Their Duals 483

Preface

Advances in Modal Logic (AiML) is an initiative founded in 1995 and aimed at presenting an up-to-date picture of the state of the art in modal logic and its many applications. It consists of a conference series together with volumes based on the conferences. The conference is the main international forum at which research on all aspects of modal logic is presented. The first one was held in 1996 in Berlin, Germany, and since then it has been organised biennially, with meetings in 1998 in Uppsala, Sweden; in 2000 in Leipzig, Germany (jointly with ICTL-2000); in 2002 in Toulouse, France; in 2004 in Manchester, UK; in 2006 in Noosa, Australia; and in 2008 in Nancy, France.

Information about AiML and related events, including conference proceedings, is available at the website `www.aiml.net`.

The eighth conference in the AiML series was organized by Steklov Mathematical Institute, Laboratoire J.-V.Poncelet and Moscow Center for Continuous Mathematical Education in Moscow, Russia. It was held on 24–27 August 2010; the conference web page can be found at `aiml10.mi.ras.ru`.

This volume contains invited and contributed papers from the conference. The conference included the following invited lectures.

- Alexandru Baltag (Oxford University): Iterating Model Transformers: cycles and fixed points, paradoxes and learning.
- Patrick Blackburn (LORIA Nancy): What are nominals?
- Dick de Jongh (University of Amsterdam): Some applications of universal models.
- Martin Otto (Darmstadt University of Technology): Expressive completeness.
- Mati Pentus (Moscow M.V. Lomonosov State University): Complexity of the Lambek Calculus and its fragments
- Heinrich Wansing (Dresden University of Technology): Proofs, disproofs, and their duals.

The Programme Committee received 43 regular paper submissions. Of them 24 were selected for this volume by a rigorous reviewing process in which each paper received at least 3 independent expert reviews. The volume includes papers on general problems in model theory, proof theory and algorithmic properties of modal logics, on systems for spatial, temporal and epistemic reasoning, on related kinds of logics - description, relevance, substructural, intuitionistic, and on related topics in algebraic logic. It is intended that these papers will be made available on the AiML website `www.aiml.net`.

The Steering Committee of AiML for 2008–2010 consisted of Carlos Areces (LORIA Nancy), Rob Goldblatt (Victoria University, Wellington), Ian Hodkinson (Imperial College, London), Mark Reynolds (University of Western Australia), Renate Schmidt

(University of Manchester), Yde Venema (University of Amsterdam), Valentin Shehtman (Moscow State University), Valentin Goranko (Technical University of Denmark), and Lev Beklemishev (Steklov Mathematical Institute, Moscow).

The Programme Committee for the conference was co-chaired by Valentin Shehtman and Valentin Goranko and also comprised: Carlos Areces (LORIA, Nancy, France), Philippe Balbiani (IRIT, Toulouse, France), Alexandru Baltag (University of Oxford, UK), Lev Beklemishev (Steklov Mathematical Institute, Moscow, Russia), Johan van Benthem (ILLC, University of Amsterdam, The Netherlands), Guram Bezhanishvili (New Mexico State University, USA), Patrick Blackburn (LORIA, Nancy, France), Torben Brauner (Roskilde University, Denmark), Balder ten Cate (LSV, ENS Cachan, France), Stéphane Demri (CNRS, Cachan, France), Melvin Fitting (City University of New York, USA), Silvio Ghilardi (University of Milano, Italy), Robert Goldblatt (Victoria University of Wellington, New Zealand), Rajeev Goré (The Australian National University, Australia), Andreas Herzig (IRIT, Toulouse, France), Ian Hodkinson (Imperial College London, UK), Wiebe van der Hoek (University of Liverpool, UK), Rosalie Iemhoff (University of Utrecht, The Netherlands), Alexander Kurz (University of Leicester, UK), Martin Lange (University of Kassel, Germany), Carsten Lutz (University of Bremen, Germany), Larisa Maksimova (Sobolev Institute of Mathematics, Novosibirsk, Russia), Edwin Mares (Victoria University of Wellington), Larry Moss (Indiana University, Bloomington, USA), Martin Otto (Technical University of Darmstadt, Germany), Dirk Pattinson (Imperial College London, UK), Mark Reynolds (University of Western Australia, Perth), Renate Schmidt (University of Manchester, UK), Nobu-Yuki Suzuki (Shizuoka University, Japan), Yde Venema (ILLC, University of Amsterdam, The Netherlands), Igor Walukiewicz (LABRI, Bordeaux, France), Frank Wolter (University of Liverpool, UK), and Michael Zakharyaschev (Birkbeck College, London, UK).

Many other people assisted with the reviewing process, including Guillaume Aucher, Meghyn Bienvenu, Jan Broersen, Kai Brünnler, Anna Bucalo, Serenella Cerrito, Giovanna D'Agostino, Guillaume Feuillade, Tim French, David Gabelaia, Torsten Hahmann, Jens Hansen, Chrysafis Hartonas, Koji Hasebe, Thomas Icard, Gerhard Jaeger, Ramon Jansana, Vladimir Krupski, Agi Kurucz, Dominique Larchey-Wendling, Minghui Ma, John McCabe-Dansted, George Metcalfe, Massimo Mugnai, Alessandra Palmigiano, Bryan Renne, Vladimir Rybakov, Joshua Sack, François Schwarzentruber, Anton Setzer, Kazushige Terui, Hans van Ditmarsch, Nikolas Vaporis, and others.

We would like to thank the members of the Programme Committee and all other reviewers for the time, professional effort and the expertise that they invested in ensuring the high scientific standards of the conference and its proceedings. Special thanks go to the authors for their excellent contributions, which are eloquent testimony for the thriving of the field of modal logic. We also thank Jane Spurr for bringing this volume to publication.

The Organizing Committee of AiML 2010 was chaired by Lev Beklemishev and included Andrei Kudinov (Institute for Information Transmission Problems), Mati Pentus (Moscow State University), Vladimir Podolsky (Steklov Mathematical Institute), Ilya Shapirovsky (Institute for Information Transmission Problems), Valentin Shehtman (Institute for Information Transmission Problems and Moscow State University), Alexei Talambutsa (Steklov Mathematical Institute), Michail Tsfasman (Laboratoire J.-V.Poncelet and Institute for Information Transmission Problems). We thank the organizers of the conference for their hard and dedicated work.

Finally, we are pleased to acknowledge the support that the conference received from the following organizations: Russian Academy of Sciences; Russian Foundation for Basic Research; Steklov Mathematical Institute, Moscow; M.V. Lomonosov Moscow State University; Moscow Center for Continuous Mathematical Education; EADS Foundation Chair in Mathematics; Poncelet Laboratory (UMI 2615 du CNRS).

July 2010

Lev Beklemishev, Valentin Goranko and Valentin Shehtman

Partially-ordered Modalities

Gerard Allwein

US Naval Research Laboratory, Code 5540, Washington, DC, USA

William L. Harrison

Department of CS, University of Missouri, Columbia, Missouri, USA

Abstract

Modal logic is extended by partially ordering the modalities. The modalities are normal, i.e., commute with either conjunctions or disjunctions and preserve either Truth or Falsity (respectively). The partial order does not conflict with type of modality (**K**, **S4**, etc.) although this paper will concentrate on **S4** since partially ordered **S4** systems appear to be numerous. The partially-ordered normal modal systems considered are both sound and complete. Hilbert and Gentzen systems are given. A cut-elimination theorem holds (for partially ordered **S4**), and the Hilbert and Gentzen systems present the same logic. The partial order induces a 2-category structure on a coalgebraic formulation of descriptive frames. Channel theory is used to 'move' modal logics when the source and target languages may be different. A particular partially ordered modal system is shown to be applicable to security properties.

Keywords: Modal logic, partial order, Hilbert, Gentzen, channel theory

1 Introduction

This paper presents modal logics with several modalities where the modalities are partially ordered. The partial order can be added to any normal modal logic, however individual partial orders derive from some particular (application oriented) domain of discourse. Propositional dynamic logic [9] places a fair amount of algebraic structure on modalities. A weakened form of this is had by replacing the algebraic structure with a partial order. The partial order typically arises from some application area where the modalities express abstract features of the area and the partial order expresses a relationship among the modalities.

Partially ordered modal systems have pleasant properties; the Hilbert-style axiomization is simple and, in the **S4** case (and we suspect others) a convenient Gentzen-style calculus which admits a cut elimination theorem. Using this theorem, it is easy to show that the Hilbert and Gentzen systems present the same logic. The simplicity of the logic is mirrored in the simplicity of the semantics. The collection of Kripke relations becomes partially ordered under the subset order. Soundness and completeness for partially ordered modal systems are also shown. One could mix modalities for the modal system **K** with **S4**, although a Gentzen system for such a logic might be a bit complicated. We suspect partial ordering modalities can be extended to other modal systems which are non-normal, leaving that extension to a later paper.

The semantics of a partially ordered system of modalities partially orders the relations of the Kripke semantics. The implications are more clearly seen when expressing that semantics in coalgebraic form. The coalgebra maps, which code the relations, are then partially ordered, and as such, form a category themselves. The result is the coalgebra maps are elements of 2-cells for a 2-category. The usual p-morphisms of modal logic are not changed except to enforce an additional requirement upon them that they respect the partial order of the coalgebras. This in effect makes them functors of the coalgebraic maps taken as the category of the underlying partial order. General frame morphisms are not effected by the partial order except for the same additional constraint imposed on the p-morphisms, i.e., that they respect the partial order.

The Vietoris topology usually given on (set) objects as the target of the Vietoris functor is not affected by the partial order on the coalgebras. So this too is independent of the partial order. Consequently, the Vietoris polynomials [10] are similarly unaffected.

A use of modal logic is in computer security. One wishes to 'move' theorems about a coarse grained security model to a fine grained system implementation model. However, the language for the security model and implementation model can be different. This paper shows way around this difficulty through the use of channel theory; a theorem about the security model can be 'moved' to the implementation model. The modalities require that the relation in the channel be a simulation relation.

This paper also presents a use of a partially ordered modal logic in the generalization of security properties which are second-order in nature [11]. Some second-order properties are expressible using modal logic and in this form, the properties are defined via certain functions on trace sets of data sequences in computer systems. These functions can be used to define closure operators; the closure operators have a natural partial order associated with them which is not a lattice order.

Section 2 presents the Hilbert and Gentzen systems. Section 3 presents the models for the partially ordered systems. Section 4 shows how to move modal logics using channel theory. Section 5 analyzes a particular generalization of security properties and shows how to connect them to a logic where the modalities represent those security properties in the logic.

2 Partially-ordered Modal Logics

The concentration will be on partially ordered **S4** modal logics since these are readily generated for computer systems by closing under functions on system behavior. They also have a nice Gentzen system that generalizes easily for the partially ordered modalities.

2.1 Hilbert-style Systems

A normal modal logic [13] is any set of formulae which contains the classical propositional tautologies and is closed under Modus Ponens and Substitution, and also contains the *normality formula* $\vdash \Box(A \to B) \to (\Box A \to \Box B)$ and closed under the rule: $\vdash A$ implies $\vdash \Box A$ ($\vdash$ is a provability turnstile here). These prescriptions can be suitably altered to include the **S4** nature of the logic and the partial order on the modalities.

Definition 2.1 The modal Hilbert system with partial order $(H, \geq)$ has the axioms of classical propositional logic and the axiom $[h](A \to B) \to ([h]\, A \to [h]\, B), h \in H$, and, in addition,

A1: $[k]\, A \to [h]\, A$ for $k \geq h$ and $k, h \in H$.

which clearly shows that the relationship between the necessity modalities and the partial order. One also has the rules for proofs from assumptions (repetition, modus ponens), and modal generalization:

$$\frac{A \in \Gamma}{\Gamma \vdash A}\, rep \qquad \frac{\Gamma \vdash A \qquad \Gamma \vdash A \to B}{\Gamma \vdash B}\, mp \qquad \frac{\vdash A}{\vdash [h]\, A}\, gen$$

and allowing that $\langle h \rangle\, A$ can be defined as $\neg\, [h]\, \neg A$. Here, Γ is a set of formulas and $\vdash$ is the provability relation.

To axiomitize **S4**, one adds the usual axioms:

A2: $[h]\, A \to A$.

A3: $[h]\, A \to [h]\, [h]\, A$.

Axioms A1 and A3 may be replaced with:

A3′: $[k]\, A \to [h]\, [k]\, A$, $k \geq h$.

The axiom A1 is the axiom that codes the partial order, it may also be expressed using possibility as:

A1′: $\langle k \rangle\, A \to \langle h \rangle\, A$ for $k \leq h$.

There are two derived rules for the Hilbert-system when proofs are allowed to have assumptions, the usual deduction theorem and an extension of *gen*.

Theorem 2.2 *The classical deduction theorem continues to hold and an expanded gen rule is a derived rule of the Hilbert-style system:*

$$[k_1]\, B_1, \ldots, [k_n]\, B_n \vdash A \text{ implies } [k_1]\, B_1, \ldots, [k_n]\, B_n \vdash [h]\, A, \;\; k_i \geq h.$$

2.2 *Gentzen System for Partially Ordered* **S4**

The rules for the classical propositional logic substrate of the modal system are Gentzen's original rules except that Permutation has been removed in favor of multisets. The context formulas in sequents are denoted with capital Greek letters.

Let the *active formula* in a premise of a rule be the instance of the formula which is altered and in a conclusion be the instance of the newly introduced formula. Let the *modal class* of a formula be either necessary, possible, or neutral depending upon whether the modal operator prefixing the formula is a necessity, possibility, or neither.

Definition 2.3 [Modal Condition (MC)] The Modal Condition is that all formulae on the same side of the $\vdash$ as the active formula must have the opposite modal class as the active formula, and all formulae on the opposite side of the $\vdash$ as the active formula must have the same modal class as the active formula.

The following rules define the modal partial order.

Definition 2.4 [Partially Ordered Modal Condition (NC)] Let NC be the condition

$$MC \text{ and } \forall C \in \Gamma \cup \Delta . c(C) \geq h$$

where $c(C)$ is the "closure" value of a formula using the modal partial order.

$$\frac{\Gamma, A \vdash \Delta \qquad NC}{\Gamma, \langle h \rangle A \vdash \Delta} \langle h \rangle \vdash \qquad\qquad \frac{\Gamma \vdash \Delta, B}{\Gamma \vdash \Delta, \langle h \rangle B} \vdash \langle h \rangle$$

$$\frac{\Gamma \vdash \Delta, A \qquad NC}{\Gamma \vdash \Delta, [h] A} \vdash [h] \qquad\qquad \frac{\Gamma, A \vdash \Delta}{\Gamma, [h] A \vdash \Delta} [h] \vdash$$

The cut rule can be eliminated as in the classical and the **S4** modal systems (see the appendix for the proof). The proof that the Hilbert system is translatable to the Gentzen system requires cut elimination. The proof that the Gentzen system is translatable to the Hilbert system uses the two derived Hilbert rules.

Theorem 2.5 *Cut is an admissible rule in the Gentzen system without cut.*

Theorem 2.6 *The Partial Order* **S4** *Gentzen system and the Partial Order* **S4** *Hilbert system are equivalent.*

3 Models

A *Kripke Frame* $(X, (\mathcal{R}, \geq))$ is a collection of points (worlds, states, etc.) and a partial order of binary relations $(\mathcal{R}, \geq)$. The relations of $\mathcal{R}$ will be indexed by the variables h and k in the presentation below. Hence $R_h \subseteq R_k$ is presented as $k \geq h$. The Kripke relations satisfy the following:

K1: Monotonicity: $R_h xy$ and $k \geq h$ implies $R_k xy$.

In addition, for **S4**, the following axioms are added

K2: Reflexivity: $R_h xx$

K3: Transitivity: $R_h zx$ and $R_h xy$ implies $R_h zy$.

One can also take, in place of K1 and K3, the following:

K3′: Transitivity + Monotonicity: for $k \geq h$, $R_k yz$ and $R_h xy$ implies $R_k xz$.

The modalities are evaluated using the usual prescription from modal logic using the following definition.

Definition 3.1

$$x \models \langle h \rangle P \text{ iff } \exists y.R_h xy \text{ and } y \models P$$
$$x \models [h]\, P \text{ iff } \forall y.R_h xy \text{ implies } y \models P.$$

It follows easily that: $[h] \neg P = \neg \langle h \rangle P$.

When the modal frame arises from a modal algebra (which is a Boolean lattice with modal operators), the modalities have the following canonical definition:

Definition 3.2 For A a set of maximal filters of the modal algebra,

$$[h]\, A = \{x \mid \forall y.R_h xy \text{ implies } y \in A\}, \quad \langle h \rangle A = \{x \mid \exists y.R_h xy \text{ and } y \in A\}$$

It is widely known that not all normal modal logics are complete with respect to Kripke frames. To obtain completeness, valuations must be added so that all frames and all valuations are considered. This is similar to regaining completeness for second-order logic by including an algebra of sets in a frame where the algebra is not the entire power set of elements in the domain.

Following [4] (originally [8]) but using [10], a general frame $(X, (\mathcal{R}, \geq), X_*)$ is Kripke frame $(X, (\mathcal{R}, \geq))$ and an Boolean algebra of sets X_* closed under derived modal operators using the prescriptions for $[h]\, A$ and $\langle h \rangle A$ in Definition 3.2. A frame is *differentiated* if for all $x, y \in X$ with $x \neq y$, there is a 'witness' $a \in X_*$ such that $x \in a$ and $y \notin a$; *tight* if whenever y is not an R_h-successor (for $R_h \in \mathcal{R}$) of x, there a 'witness' a such that $y \in a$ and $x \notin \langle R \rangle a$; and *compact* if for every $C \subseteq X_*$, if C has the finite intersection property, then $\bigcap C \neq \emptyset$.

Typically, X_* is thought of as the clopen basis for the Stone topology on the Kripke frame. The question arises as to the relationship between that topology and the "closed sets" of **S4** possibility operators. The clopen basis is a Boolean algebra and that algebra is closed under induced modal operators given by Definition 3.2.

It is possible to describe the clopen sets of the Boolean algebra as arising from the identity relation. The identity modal operator $[1_X]$ corresponds to the identity relation on X, and $[1_X]\, C = \langle 1_X \rangle C$ for all elements of X_* (or propositions) C. All partial orders of relations can be extended with this relation with little effect on the dual algebras.

Lemma 3.3 *For all* C, $[1_X]\, C = C = \langle 1_X \rangle C$.

The identity relation has the effect of making the lattice of sets of a general frame a modal algebra. Put another way, every Stone space has a modal dual, albeit a trivial one. Hence, every normal modal logic can be extended to a partially ordered modal logic by including the identity relation. If the modal logic is at least **T** meaning it satisfies at least the reflexivity axiom Rxx for all x, then the modal order is $1_X \subseteq R$ for R the modal relation for **T**.

Theorem 3.4 *The Gentzen rules for the system with the NC conditions are sound with respect to descriptive frames.*

The completeness argument is the usual algebraic argument using contraposition and a representation theorem. The modal representation theorem represents a modal algebra as an algebra of sets using the Kripke frame (Stone space) of the algebra. One defines the 1-1 homomorphism $\beta : A \longrightarrow \mathcal{P}(\mathcal{P}A)$ (where $\mathcal{P}$ is the powerset) from the modal algebra A to the double power set of A by:

$$\beta a = \{x \mid a \in x \text{ and } x \text{ is a maximal filter}\}.$$

It is not hard to show that $\beta\,[h]\,a = [h]\,\beta a$ and $\beta\,\langle h\rangle\,a = \langle h\rangle\,\beta a$. Set union, intersection, and complement interpret the classical logic logic connectives $\vee$, $\wedge$, and $\neg$. The Lindenbaum-Tarski modal algebra is generated via the logic by dividing out the word algebra of the logic by bi-implication and defining the operators via elements of the equivalence classes, i.e., $[P] \wedge [Q] \stackrel{def}{=} [P \wedge Q]$ where $[\ ]$ indicates bi-implication equivalence classes. To get a Kripke model requires that one take the (dual) Stone space containing all the maximal filters of the algebra and define the Kripke relations with:

$$R_h xy \text{ iff } [h]\,a \in x \text{ implies } a \in y.$$

Since $[h]$ and $\langle h\rangle$ are DeMorgan duals of each other, R_h admits an equivalent definition:

$$R_h xy \text{ iff } a \in y \text{ implies } \langle h\rangle\,a \in x.$$

The *canonical model* is the Kripke model generated by the Lindenbaum-Tarski algebra.

Lemma 3.5 *Monotonicity K1 holds in the canonical model. K2, and K3 hold if the frame is an* **S4** *frame.*

The following theorem holds via the usual contraposition argument.

Theorem 3.6 *The partially ordered, normal modal logics are complete with respect to descriptive frames.*

4 Moving Modal Logics

A tried and tested way to relate Kripke frames is via p-morphisms. These are also known as bounded morphisms, zig-zap maps, system maps, etc., and are the morphisms for the category of descriptive Kripke frames. The conditions guarantee that the power set algebras on the frames map properly when the inverse morphisms as inverse set maps are used.

An extrapolation of p-morphisms are bisimulation relations and their kin, simulation relations. These are inadequate when the modal formulas to be related come from different languages. This occurs when one is relating properties of a high-level model or specification to a low-level implementation. One way around this difficulty is to use channel theory.

4.1 p-morphisms for Partially Ordered Descriptive Frames

Kripke frames can be expressed in terms of coalgebras for the covariant power set functor $\mathcal{P}$. $\mathcal{P}$ takes X to the set of subsets of X and $f : X \to Y$ to the forward image of f, i.e., $\mathcal{P}(f)(A) = \{f(x) \mid x \in A\}$. The coalgebra for Kripke relation R in $\mathbb{X} = (X, (\mathcal{R}, \geq))$ is defined with:

$$R_h x = \{y \mid R_h xy\}$$

(where the symbol R_h is overloaded).

A p-morphism $p : \mathbb{X} \to \mathbb{Y}$ is then a system map which means the square commutes for all $R_h \in \mathcal{R}$ where pR_h is the relation in $\mathbb{Y}$ which is the target of the p-morphism for R_h. $\mathbb{Y}$ could well have many other Kripke relations. The commutation means that, as relations, (1) $R_h xy$ implies $(pR_h)(px)(py)$ and (2) $(pR_h)(px)y$ implies there is some z such that $R_h xz$ and $pz = y$. To form the category of all coalgebras on $\mathbb{X}$, partially order the relations. This partially orders the relations as coalgebra morphisms. Let $\mathsf{Coalg}(\mathbb{X})$ be the collection of coalgebra morphisms on $\mathbb{X}$. As a set, $\mathsf{Coalg}(\mathbb{X})$ then forms a simple category. A morphism of frames $p : \mathbb{X} \to \mathbb{Y}$ then can be expected to be p-morphism for all the relations of $\mathbb{X}$ with the additional constraint that it also be a functor $p : \mathsf{Coalg}(\mathbb{X}) \to \mathsf{Coalg}(\mathbb{Y})$.

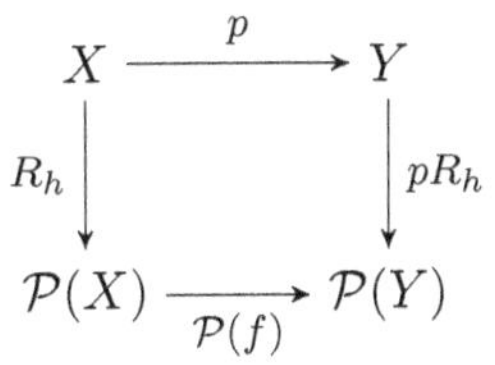

A morphism $p : \mathbb{X} = (X, (\mathcal{R}, \geq), X_*) \to \mathbb{Y} = (Y, (\mathcal{S}, \geq), Y_*)$ is a general frame morphism if it is a morphism for partially ordered frames and $p^{-1} : Y_* \to X_*$ is a modal homomorphism. General frame morphisms are also descriptive frame morphisms.

4.2 Channel Theory

The basic structures of channel theory [3,1] are deceptively simple. "Channel theory" is the colloquial term for Barwise and Seligman's term "information flow". Channel theory is both a qualitative information theory and a logic for distributed systems. The elements that are distributed are contexts called *classifications*. The classifications are connected by *infomorphisms*. Classifications and their morphisms appear in mathematics in various guises. What is unique about channel theory is that it uses a specific type of cocone called a "channel" as an organizing principle.

A classification contains two distinct collections of objects, tokens and types, connected with a binary relation. They could be anything that makes sense in using a classification as a model. However, most of modern language theory tends to use the term *types* in a different sense. The tokens will be model-theoretic entities such as states, theories, traces through a state space, etc. Bold-faced, slanted typefont always denote classifications.

Formally, the objects and morphisms are the same as Chu spaces [2]. However, in Chu spaces, no mention is made of the theory of a classification (see below) and most the work appears to be directed at their categorical structure. Here, the categorical structure, while used, is not of primary importance. Also, Scott in [14] uses similar structures but particularizes the formalism to talk about computation. There is also an extensive literature on institutions [6]; this reference integrates institutions with the work of Barwise and Seligman.

Definition 4.1 A *classification*, $\boldsymbol{X}$, is a pair of sets, $Tok(\boldsymbol{X})$, and $Typ(\boldsymbol{X})$, and a relation, $\models_{\boldsymbol{X}} \subseteq Tok(\boldsymbol{X}) \times Typ(\boldsymbol{X})$ written in infix, e.g., $x \models_{\boldsymbol{X}} A$.

Information in a classification is of the form "x being A"; x need not be a model for a logic. In Section 5, x would stand for an arbitrary trace in a security model. x is a carrier of information with A being some of the information x carries. We express this by saying "$x \models_{\boldsymbol{X}} A$" is an example of the basic unit of information in channel theory.

Channel theory has its own notion of morphism, called an *infomorphism.* It is similar to a pair of adjoint functors in that it is a pair of opposing arrows with a condition similar to the adjoint's bijection.

Definition 4.2 A morphism $f : \boldsymbol{X} \to \boldsymbol{Y}$ of classifications, sometimes called an **infomorphism**, is a pair of opposing maps, $\overrightarrow{f}$ and $\overleftarrow{f}$ such that $\overrightarrow{f} : Typ(\boldsymbol{X}) \to Typ(\boldsymbol{Y})$ and $\overleftarrow{f} : Tok(\boldsymbol{Y}) \to Tok(\boldsymbol{X})$, and for all x and A, the following condition is satisfied: $x^f \models_{\boldsymbol{X}} A$ iff $x \models_{\boldsymbol{Y}} A^f$. For ease of presentation, $\overleftarrow{f}(x)$ is displayed as x^f and $\overrightarrow{f}(A)$ as A^f.

General frame morphisms are instances of infomorphisms with $Tok(\boldsymbol{X})$ being the points X of a Kripke frame, $Typ(\boldsymbol{X})$ being the set algebra X_* and the $\models_{\boldsymbol{X}}$ relation being $\in$ relation.

A commuting *cocone* consists of a graph homomorphism G from a graph to the category of classifications, a vertex classification $\boldsymbol{C}$ called the channel's *core*, and a collection of arrows $g_i : G(i) \to \boldsymbol{C}$. It is required that for all $f : i \to j$, $g_i = g_j \circ G(f)$. The base of the cocone is the objects and arrows identified by G.

Definition 4.3 An *information channel* is a co-cone in the category of classifications and infomorphisms.

The smallest channel over a base is a colimit. Frequently, the smallest channel is not the most useful because a channel is used as a model. The smallest channel would simply connect the base with no additional modeling apparatus. A colimit in the category of classifications is a colimit on types and a limit on tokens.

Assuming a fixed classification $\boldsymbol{C}$, a *sequent* $\Gamma \vdash_{\boldsymbol{C}} \Delta$, is two sets of types connected by a relation $\vdash$. A *valid* sequent has the force of a meta-level implication of the form: for all tokens x, if $x \models_{\boldsymbol{C}} A$ for all of the types A in Γ, then $x \models_{\boldsymbol{C}} B$ for at least one type B in Δ. A classification's valid sequents are the classification's *theory*, also called the classification's *constraints.* A *channel's theory* refers to the theory in the core.

A channel $\boldsymbol{C}$ may connect a proximal classification $\boldsymbol{P}$, say a high level specification, with the distal classification $\boldsymbol{D}$, say a low level implementation [1]. Possibly

[1] The terms *proximal* and *distal* are merely convenient terms we use to refer to the two classifications

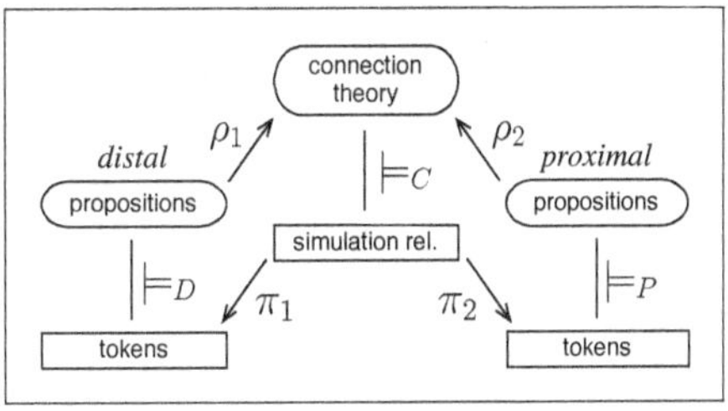

there are several design layers with every two adjacent layers connected by a channel, but the simple picture inset at the right will do for making the main argument apparent. In the diagram, the π_i are projections and ρ_i are injections into a disjoint sum. The rule for the morphisms $\langle \rho_i, \pi_i \rangle$ is: $\pi_i\langle x, y\rangle \models A$ iff $\langle x, y\rangle \models \rho_i A$. The proximal and distal languages are the type sets and are allowed to be different. The connection (channel) theory contains the rules for translation in the form of sequents.

A channel's sequents may be used to underwrite information flow through a channel where the pieces of information are tokens and the information they carry are properties. Using the channel in the diagram, let x be a token of $\boldsymbol{D}$, y a token of $\boldsymbol{P}$ and $\langle x, y\rangle$ a token of the channel $\boldsymbol{C}$. Further, let $\Gamma \subseteq Typ(\boldsymbol{D})$ and $\Delta \subseteq Typ(\boldsymbol{P})$ and Γ^{ρ_1} and Δ^{ρ_2} refer to the forward images of these sets under ρ_1 and ρ_2 respectively. If the sequent $\Gamma^{\rho_1} \vdash_C \Delta^{\rho_2}$ as a constraint in the channel, it will relate tokens from $\boldsymbol{D}$ to tokens from $\boldsymbol{P}$ using the following form of reasoning:

$x \models_{\boldsymbol{D}} \Gamma$ iff $\pi_1\langle x, y\rangle \models_{\boldsymbol{D}} \Gamma$	assumption
iff $\langle x, y\rangle \models_C \Gamma^{\rho_1}$	infomorphism condition
implies $\langle x, y\rangle \models_C \Delta^{\rho_2}$	channel constraint
iff $\pi_2\langle x, y\rangle \models_{\boldsymbol{P}} \Delta$	infomophism condition
iff $y \models_{\boldsymbol{P}} \Delta$	assumption

Our goal will be to transfer a theorem of the form $A' \vdash [h]\, B'$ at the proximal level to a theorem $A \vdash [h]\, B$ in the distal level. A system $\boldsymbol{P}$ simulates a system $\boldsymbol{D}$ with respect to $[h]$ just when there is channel $\boldsymbol{C}$ such that "if $\langle x, x'\rangle$ are in the simulation relation $Tok(\boldsymbol{C})$ and $\boldsymbol{D}$ transitions under the relation R_h from x to y, then $\boldsymbol{P}$ transitions under the relation $R_{h'}$ from x' to y' and $\langle x', y'\rangle \in Tok(\boldsymbol{C})$." For sequents of this form to transfer from proximal to distal, the following conditions must be met:

C1: The connection theory in $\boldsymbol{C}$ relates non-modal proximal and distal types.

C2: The projection π_1 is surjective, i.e., must cover $Tok(\boldsymbol{D})$.

C3: $\boldsymbol{P}$ simulates $\boldsymbol{D}$ via the channel tokens $Tok(\boldsymbol{C})$.

The proof of Theorem 4.4 is in the appendix.

Theorem 4.4 *For channel $\boldsymbol{C}$, if $\boldsymbol{P}$ simulates $\boldsymbol{D}$, $\rho_1 A \vdash_C \rho_2 A'$, and $\rho_2 B' \vdash_C \rho_1 B$:*

$$\left(A' \vdash_{\boldsymbol{P}} [h']\, B'\right) \text{ implies } \left(A \vdash_{\boldsymbol{D}} [h]\, B\right).$$

Bisimulations are extrapolations from p-morphisms and bisimulations in channel theory are extrapolations of general frame morphisms. Let $p : \mathbb{X} \to \mathbb{Y}$ be a general frame morphism. Treat the general frames $\mathbb{X}$ and $\mathbb{Y}$ as classifications with the tokens being the Kripke points (worlds), the types being the set algebras and $\models$ as the $\in$ relational between points and sets. A colimit with vertex $\mathbb{C}$ over the morphism $p : \mathbb{X} \to \mathbb{Y}$ is then a bisimulation the category of classifications.

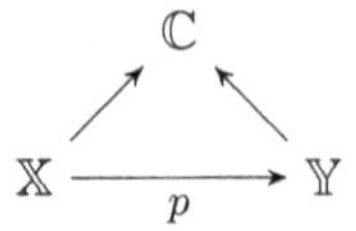

A simulation from distal to proximal uses the condition Rxy implies $(pR)(px)(py)$ whereas a simulation from proximal to distal uses the condition $(pR)(px)y$ implies there is some z such that $pz = y$ and Rxy. The (bi)simulation relation is the limit of the general frame morphism on the points of the domains and codomain of the general frame morphism.

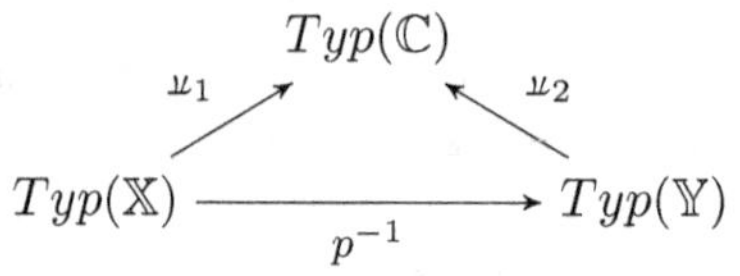

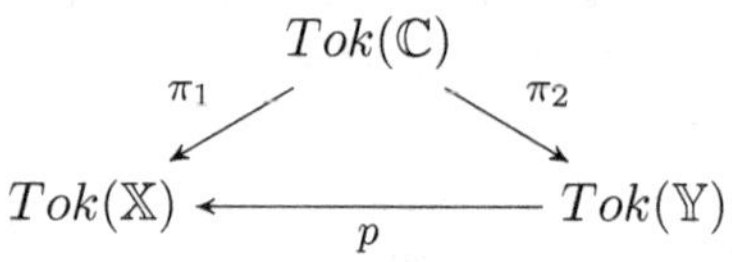

The colimit $p : \mathbb{X} \to \mathbb{Y}$ on the types identifies types in the source and target of p. The connection theory is empty since that work has now been taken over by the identification via p on the types.

Now if one drops the requirement that there be a morphism linking the two relations and simply use the (bi)simulation relation as the link, one has the definition of (bi)simulation. In place of the identification $pA' = A$ and $pB' = B$, one takes $\iota_1 A \vdash_{\mathbb{C}} \iota_2 A'$ and $\iota_2 B' \vdash_{\mathbb{C}} \iota_1 B$ for transferring $A' \models_{\mathbb{Y}} [h]\, B'$ in the classification $\mathbb{Y}$ to $A \models_{\mathbb{X}} [h]\, B$ in the classification $\mathbb{X}$.

4.3 High-Level Simulation

Partially ordering the modalities suggest that a "higher level" notion of simulation. Let $\leadsto$ be a relation on Kripke relations for both $\boldsymbol{P}$ and $\boldsymbol{D}$. Define

$$x \models \boxed{R}\, A \text{ iff } \forall S, \forall y. R \leadsto S \text{ and } Sxy \text{ implies } y \models A.$$

Definition 4.5 $\boldsymbol{P}$ simulates $\boldsymbol{D}$ with respect to R and R' when there is a channel $\boldsymbol{C}$ such that (where $C_{RR'}xy$ stands for a element of the $Tok(\boldsymbol{C})$)

$$R \leadsto S, Sxy, \text{and } C_{RR'}xx' \text{ implies } \exists S', \exists y'.R' \leadsto S', S'x'y' \text{ and } C_{RR'}yy'.$$

Theorem 4.6 *For channel $\boldsymbol{C}$, if $\boldsymbol{P}$ simulates $\boldsymbol{D}$ with respect to R and R', $\rho_1 A \vdash_C \rho_2 A'$, and $\rho_2 B' \vdash_C \rho_1 B$:*

$$\left(A' \vdash_{\boldsymbol{P}} \boxed{R}\, B'\right) \text{ implies } \left(A \vdash_{\boldsymbol{D}} \boxed{R}\, B\right).$$

Quantifying over all R properly paired with an R' yields a global, higher order necessity.

The proof is much like the proof for Theorem 4.4.

5 The Logic of Possibilistic Security

Separability, Generalized Noninterference, Noninference, and Generalized Noninference are the four possibilistic security properties handled by McLean [11]. Separability means that given a particular trace of high's behavior, any trace of low's behavior is possible, and vice versa; this relation is called *co-possibility.* Generalized Noninterference abstracts Goguen and Meseguer Noninterference [7]. The co-possibility relation becomes non-symmetric: any high-level trace is co-possible with any low-level trace, and *when only high-level input is considered* any low-level trace is co-possible with any high-level trace. Noninference "purges" high information from the input and output traces by overwriting that information with a constant value. This is weakened in Generalized Noninference, where only high input is purged. The relative strengths of these notions was captured by McLean a partial order; this order is reversed for the purposes of this paper in Fig. 1; the order indicates increasing restrictiveness from top to bottom. We have augmented by an additional element, Nothing, at the bottom for reasons that will become apparent later.

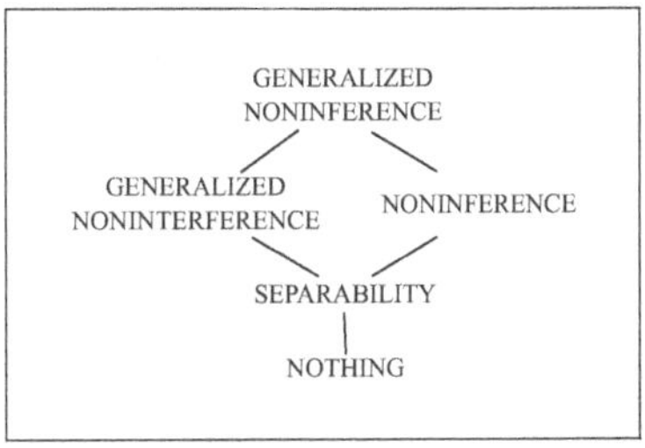

Figure 1

Relative Strengths of Possibilistic Models.

The diagram is somewhat misleading, because when viewed as a partial order of modalities, it turns out not to be a lattice. There is a Kripke relation for each element in the partial order. The partial order is the subset relation on the Kripke relations (as sets). However, Separability is not the set theoretic intersection of Generalized Noninterference and Noninference, and Generalized Noninference is not the set theoretic union. There are security properties which are the set theoretic intersection and union, but they are either unnamed or so far not put to any use. Also, since the relations involved here are at least transitive, the union of two transitive relations is not necessarily transitive.

McLean [11] uses state spaces (i.e., collections of system traces of input-output behavior) to show that the security properties do not attach themselves to traces but rather to sets of traces. He does this via an example showing that a particular property is not preserved through a reduction of system traces.

To formulate each possibilistic security property, McLean defines types of interleaving functions on traces of system behavior, where an *interleaving function* takes two traces and manufactures a third trace using some elements of the two traces. Each security property is then associated with a particular type of interleaving function. We summarize McLean's framework here for completeness.

Definition 5.1 [State Space] For non-negative integers m and n, let the sequences $\langle in_1, \ldots, in_m \rangle$ and $\langle out_1, \ldots, out_n \rangle$ be (respectively) tuples of distinct input and output variables such that the i-th input variable ranges over some alphabet I_i and the j-th output variable ranges over some alphabet O_j. A **state space** Σ is the set

$$\{\langle\langle in_1, \ldots, in_m\rangle, \langle out_1, \ldots, out_n\rangle\rangle \mid in_i \in I_i, out_j \in O_j\}.$$

An element of the state space is called a *system state.*

Definition 5.2 Assume that for some $1 \leq p < m$, $in_1, \ldots, in_p$ are inputs of high-level users, the rest are inputs of low-level users. Similarly, for some $1 \leq q < n$ $out_1, \ldots, out_q$ are the outputs of the respective high-level users and the rest the respective outputs of low-level users. The state notation is condensed to $\langle highin : lowin, highout : lowout\rangle$:

$$\langle\overbrace{\langle in_1, \ldots, in_p\rangle}^{highin} : \overbrace{\langle in_{p+1}, \ldots, in_m\rangle}^{lowin}, \overbrace{\langle out_1, \ldots, out_q\rangle}^{highout} : \overbrace{\langle out_{q+1}, \ldots, out_n\rangle}^{lowout}\rangle$$

A *trace* is a (possibly finite) sequence of states. Input $highin_i$ ($lowin_i$) refers to the high (low) inputs of the i-th state in the trace with outputs $highout_i$ and $lowout_i$ defined analogously. The concatenation $\langle highin_i : lowin_i\rangle$ refers to the sequence $\langle in_1, \ldots, in_p, in_{p+1}, \ldots, in_m\rangle$ in the i-th state. The set of all traces is denoted $\hat{\Sigma}$. A state space, Σ, partitioned into high and low, is called a *two level security state space.*

McLean's Example.

One form of confidentiality property is that one kind of behavior is unaffected by another kind of behavior. An example is that any legal low-level behavior, i.e., a trace of states restricted to low-level input and output, must be co-possible with any legal high-level behavior. However, McLean showed that expressing the co-possibility relation with traces is problematic for reasons we now summarize.

Let $P(t)$ be true for trace t just when every possible high-level input-output pair can be paired with t's input-output and the result still be an allowable sequence. Consider the property $Q(t) \equiv in(t) = out(t)$ (i.e., for all i, $in(t)_i = out(t)_i$ where the equality is over sequences of input elements and output elements). Hence $Q(t)$ is true of just those sequences where the high input is mapped directly to low output for all positions (no permutations). Let P and Q stand for their extensions. The maximal trace set $\hat{\Sigma}$ consisting of all possible traces has property P. This implies that all properties, as sets of traces, satisfy P. It follows that $Q \subseteq \hat{\Sigma} \subseteq P$. This is a contradiction since Q does not satisfy P. The intuitive appeal of the language does not match the extensional, first-order semantics of the language.

Definition 5.3 Let state space $\Sigma = \{\langle\langle in_1, \ldots, in_m\rangle, \langle out_1, \ldots, out_n\rangle\rangle \mid in_i \in I_i \wedge out_i \in O_i\}$, let $\mu \in \{0, 1, 2\}^m$ and let $\nu \in \{0, 1, 2\}^n$. A function $f : \hat{\Sigma} \times \hat{\Sigma} \to \hat{\Sigma}$ is a *selective interleaving function of type* $F_{\mu,\nu}$ if and only if $f(t_1, t_2) = t$ implies that for all i, j such that $1{\leq}i{\leq}m$ and $1{\leq}j{\leq}n$,

$$in[i](t) = in[i](t_1), \text{ if } \mu[i] = 1 \quad out[j](t) = out[j](t_1), \text{ if } \nu[j] = 1$$

$$in[i](t) = in[i](t_2), \text{ if } \mu[i] = 2 \quad out[j](t) = out[j](t_2), \text{ if } \nu[j] = 2$$

The interleaving class of type $F_{\langle 1^H:2^L\rangle,\langle 0^H:2^L\rangle}$ indicates $H+L$ inputs (from the notation $\langle 1^H : 2^L\rangle$) and $H+L$ outputs (from the notation $\langle 0^H : 2^L\rangle$). An *interleaving class* is a type of interleaving function. In this example, each function $f(t_1, t_2)$ in $F_{\langle 1^H:2^L\rangle,\langle 0^H:2^L\rangle}$ maps the high input of t_1 to the high input of the resulting trace, maps, the low input of t_2 to the low input of the resulting trace, does not care about high output, and maps the low output of t_2 to the low output of the resulting trace. The individual functions of $F_{\langle 1^H:2^L\rangle,\langle 0^H:2^L\rangle}$ may differ on how they set the high output of the resulting trace.

A *security class* is an interleaving class that corresponds to one of the security classes in the partial order. The following table summarizes the security classes with their interleaving class types:

Nothing	No functions
Separation	$F_{\langle 1^H:2^L\rangle,\langle 1^H:2^L\rangle}$
Generalized Noninterference	$F_{\langle 1^H:2^L\rangle,\langle 0^H:2^L\rangle}$
Noninference	$F_{\langle \lambda^H:2^L\rangle,\langle \lambda^H:2^L\rangle}$
Generalized Noninference	$F_{\langle \lambda^H:2^L\rangle,\langle 0^H:2^L\rangle}$

where λ is some fixed value to which the referenced input and output are set. Separability's lone interleaving function can be pictured as:

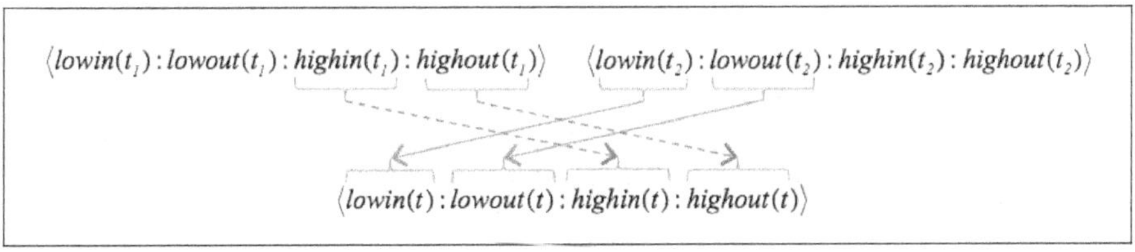

5.1 *Channel Theory and Possibilistic Security*

Example 5.4 [Trace Classification] Given a state space Σ, the trace classification $\boldsymbol{T}$ is

- $Tok(\boldsymbol{T}) = \{\langle s_1, s_2, \dots \rangle \mid s_i \in \Sigma\} \cup \{\langle s_1, \dots, s_n\rangle \mid s_i \in \Sigma\}$ $(= \hat{\Sigma})$;
- $Typ(\boldsymbol{T})$ are properties, i.e., open formulas of first-order logic with one free variable ranging over the tokens;
- $t \models_T A$ is the satisfaction relation (trace t satisfies property A).

A subset of the token set is called a *trace set*. A trace set U is a **reduction** of a trace set V if and only if $U \subseteq V$.

5.2 *Security Properties and Reductions.*

McLean and others use the term *refinement* for the term *reduction*. A reduction of a property P is a system S such that $S \subseteq P$ and so S is said to "refine" or "reduce" P. Consider a possible reduction infomorphism $r : \boldsymbol{T} \rightarrow \boldsymbol{T}'$ for $\boldsymbol{T}, \boldsymbol{T}'$ trace classifications. The channel types are first-order logic descriptions of traces.

If A is a description of a collection of traces, then $Tok(A)$ refers to all traces which satisfy A. Let $\overrightarrow{r} = 1_{Typ(\boldsymbol{T})}$ and $\overleftarrow{r} : Tok(\boldsymbol{T}') \rightarrowtail Tok(\boldsymbol{T})$ be an injection. The reduction r takes the property A into the property A^r and is an instance of the rule in the inset figure where Γ^r, Δ^r are r applied element-wise to the formulas in Γ, Δ. This rule preserves validity even when $\overleftarrow{r}$ is not an injection.

$$\frac{\Gamma \vdash_{\boldsymbol{T}} \Delta}{\Gamma^r \vdash'_{\boldsymbol{T}} \Delta^r}\ r\text{--Intro}$$

The notion of a possibilistic security property being a collection of traces is defective for the mere fact that these properties cannot be stated using it if traces are thrown out in going from $\boldsymbol{T}$ to $\boldsymbol{T}'$.

Continuing McLean's Example.

Recall, $P(t)$ is true just when every possible high-level input-output pair can be paired with t's input-output and the result still be an allowable sequence, i.e., still be a token in the trace classification. Let $\boldsymbol{T}$ be a trace set classification whose tokens are an entire state space, $Tok(\boldsymbol{T}) = \hat{\Sigma}$. As earlier, now in channel theoretic language, $Tok(\boldsymbol{T}) = Tok(P)$, and $Q(t) \equiv in(t) = out(t)$ i.e., for all i, $in(t)_i = out(t)_i$ where the equality is over sequences of input elements and output elements. Hence, $Tok(Q) \subseteq Tok(P)$ yet Q does not imply P.

The sequent $Q \vdash_{\boldsymbol{T}} P$ is a constraint of the classification $\boldsymbol{T}$. However, the property Q is being thought of as a classification that describes all of the tokens satisfying Q. Let $\boldsymbol{T}'$ refer to the classification with types Q and P but with only the tokens from $\boldsymbol{T}$ that satisfy Q. The obvious map is $r : \boldsymbol{T} \rightarrow \boldsymbol{T}'$ such that $\overleftarrow{r}$ is the injection induced by $Tok(\boldsymbol{T}') \subseteq Tok(\boldsymbol{T})$ and is the identity on types. The rule inset to the right should have produced a good constraint in $\boldsymbol{T}'$. It does not because r is not an infomorphism. In particular, $t^r \models_{\boldsymbol{T}} P$ iff $t \models_{\boldsymbol{T}'} P$ is false in the forward direction as long as $t \models_{\boldsymbol{T}'} P$ means $P(t)$. Put another way, the quantifiers defining $P(t)$ are not restricted to $Tok(\boldsymbol{T}')$. This is precisely the move (from $t \models_{\boldsymbol{T}'} P$ to $P(t)$) that one uses to conceive of modal logic as being a variant of second-order logic.

$$\frac{Q \vdash_{\boldsymbol{T}} P}{Q^r \vdash_{\boldsymbol{T}'} P^r}\ r\text{--Intro}$$

5.3 Formalizing the Intended Model.

Considering the partial order of security properties, any system closed under Separability is also closed under Generalized Noninterference. This would indicate that if Separability is interpreted using a modal closure operator $\langle s \rangle$ and Generalized Noninterference is interpreted using a modal closure operator $\langle g \rangle$, then $\langle s \rangle P \subseteq \langle g \rangle P$. However, let Noninference be interpreted by the modal closure operator $\langle n \rangle$. It would be expected that $\langle s \rangle P \subseteq \langle n \rangle P$. Closing under Noninference's purge function will not include the traces of $\langle s \rangle P$ if there is some trace $t \in P$ and $t \neq f(t_1, t_2)$ for f being Separability's interleaving function and $t \neq \langle \langle \lambda^{H+L}, \lambda^{H+L} \rangle, \ldots \rangle$.

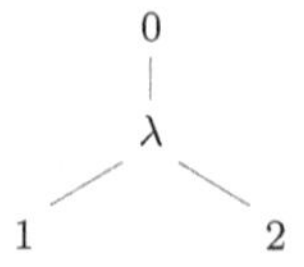

Extended Partial Order

In a similar vein, closing under one General Noninterference function will not necessarily imply closure under another. To use the partial order requires that closing be cumulative looking up the partial order and that it be inclusive at any one point in the order.

One way to achieve a cumulative and inclusive order is to use currying to turn the two-place functions into collections of single-place functions. To make the order cumulative, include functions from higher up in the order in a collection lower in the order. To make the order inclusive requires that it be closed under composition. To make the order work to define set closures requires that the identity function be an allowable function. The extended typing introduced for Noninference and Generalized Noninference can be further extended to correspond to the partial order in the Extended Partial Order.

Definition 5.5 Given functions $f_{t_1}t_2 \stackrel{def}{=} f(t_1, t_2)$ and $g_{t_2}t_1 \stackrel{def}{=} g(t_1, t_2)$. $F^{\circ}_{\mu,\nu}$ is the collection of curried interleaving functions generated by the two-place interleaving functions type $F_{\mu,\nu}$ and includes the identity function, $\mathrm{id}_{\hat{\Sigma}}$.

Definition 5.6 The *interleaving type partial order* is $F^{\circ}_{\mu_k,\nu_k} \geq F^{\circ}_{\mu_h,\nu_h}$ iff for all i, j, $\mu_k[i] \geq \mu_h[i]$ and $\nu_k[i] \geq \nu_h[i]$ where $\mu_h, \mu_k, \nu_h, \nu_k$ take values in the Extended Partial Order.

The typing structure is too restrictive for the interleaving type partial order to be cumulative. This can be remedied by taking the union of all the curried interleaving functions above and $\mathrm{id}_{\hat{\sigma}}$.

Definition 5.7 Define $k \geq h$ iff $h \subseteq k$, for $k \stackrel{def}{=} \bigcup \{F^{\circ}_{\mu_h,\nu_h} \mid F^{\circ}_{\mu_k,\nu_k} \geq F^{\circ}_{\mu_h,\nu_h}\}$.

Intuitively, an *interleaving class of operators* is the collection of interleaving operators generated via currying from an interleaving class of functions combined with the interleaving classes of operators further up the partial order. The Nothing interleaving class of operators is the empty set $\emptyset$, and the Separation class contains its lone interleaving operator and the identity operator on traces.

Theorem 5.8 *Let $k \geq h$ and $g \in k, f \in h$ for interleaving classes $F^{\circ}_{\mu_k,\nu_k}, F^{\circ}_{\mu_h,\nu_h}$ respectively, then $g \circ f, f \circ g \in k$ regardless of how they are internally curried, and h and k are semigroups.*

Let h be an interleaving class of operators. If $f, g \in h$, then $f \circ g \in h$. Since function composition is associative, h is a semigroup. If t is a trace and P is closed under a collection of interleaving operators, then t is still in P, hence the identity function, $\mathrm{id}_{\hat{\Sigma}}$ is a valid interleaving operator for all classes. This makes h a monoid. Also, if $k \geq h$ and $f \in h$, then $f \in k$.

To say that P is closed under an interleaving function f is to say that for all $t_1, t_2 \in P$, $f(t_1, t_2) = t$ and $t \in P$. There is a mismatch between the binary relations of modal logic and what appears to be a three place relation, namely $f(t_1, t_2) = t$. It will not work to simply apply g_{t_1} on the curried functions, since the condition is then rendered for all $t_2 \in P$, $g_{t_1}t_2 = t$ and there is no guarantee that $t_1 \in P$. The solution is to use pairs of traces. We define the *intended model* as the model of sequences and interleaving classes. Let h be an interleaving class, then (regardless of internal currying): $R_h\langle t_1, t_2\rangle\langle t_1, t\rangle$ iff $\exists f.f_{t_1} \in h$ and $f_{t_1}t_2 = t$. To finish the intended model, the R_h relations must be reversed, e.g., $R_h xy \mapsto \breve{R}_h yx$ which yields 'forward looking' relations for the $[h]$ and $\langle h\rangle$ operators.

Theorem 5.9 *The intended model satisfies axioms K1, K2, and K3.*

Theorem 5.10 *Separability is not the set theoretic intersection of Generalized Noninterference and Noninference, and Generalized Noninference is not the set theoretic union.*

6 Conclusion

The concentration in this paper was on **S4** since it has a pleasant Gentzen system and many applications can be found where properties are closures under some class of functions. The partial order is on the modalities themselves and represents a higher-order structure of the modalities reminiscent of dynamic logic. The difference between dynamic logic and partially ordered modal logics is the lack of algebraic structure on the modalities. The semantics of partial ordered modal logics reflects the partial order directly as the set theoretic subset relation on the Kripke relations. The partial order can be used to give a higher order notion of simulation for 'moving' logics among classifications in channel theory. This allows for expressing theorems of high level models or specification of system behavior to be transferred to low level implementations when the implementations satisfy several concrete criteria.

The partial order of the logic is reduced in the paper to a single axiom. This axiom can be added conservatively (i.e., preserving completeness) to any normal modal logic. The identity relation makes Boolean propositional logic a modal logic of the most simplest form. Any normal modal logic then becomes a partially ordered modal logic with the addition of the identity relation.

We would like to further explore the utility of partially ordering non-normal modalities as these have a number of Computer Science applications (e.g., they tend to crop up in linear logic). There are other aspects of modal logic we have not yet explored such as interactions with Sahlqvist formulas and correspondence theorems [4]. Substructural logics represent another avenue of research. Positive modal logic [5,12] weakens the Boolean propositional part of conventional modal logic to a logic without negation. It would be interesting to formulate a version of partially ordered modal logic for these conditions.

References

[1] Allwein, G., *A qualitative framework for Shannon information theories*, in: *Proceedings of the New Security Paradigms Workshop, 2004* (2005), pp. 23 – 31.

[2] Barr, M., *$*$-autonomous categories and linear logic*, Mathematical Structures Computer Science **1** (1991), pp. 159–178.

[3] Barwise, J. and J. Seligman, "Information Flow: The Logic of Distributed Systems," Cambridge University Press, 1997, cambridge Tracts in Theoretical Computer Science 44.

[4] Blackburn, P., M. de Rijke and Y. Venema, "Modal Logic," Cambridge University Press, 2001, cambridge Tracts in Theoretical Computer Science, No. 53.

[5] Dunn, J. M., *Positive modal logic*, Studia Logica **55** (1995), pp. 301–317.

[6] Goguen, J., *Information integration in institutions*, in: L. Moss, editor, *Thinking Logically: a Memorial Volume for Jon Barwise*, Indiana University Press, 200x pp. 1–48.

[7] Goguen, J. A. and J. Meseguer, *Security policies and security models*, in: *Proceedings of the 1982 IEEE Symposium on Security and Privacy* (1982), pp. 11–20.

[8] Goldblatt, R., *Metamathematics of modal logic*, Reports on Mathematical Logic **6** (1976), pp. 41–77.

[9] Harel, D., D. Kozen and J. Tiuryn, "Dynamic Logic," MIT Press, 2000.

[10] Kupke, C., A. Kurz and Y. Venema, *Stone coalgebras*, in: H. P. Gumm, editor, *Coalgebraic Methods in Computer Science, Electronic Notes in Theoretical Computer Science*, 1 **82**, 2003, pp. 170–190.

[11] McLean, J., *A general theory of composition for a class of "possibilistic" properties*, IEEE Transactions on Software Engineering **22** (1996), pp. 53–67.

[12] Palmigiano, A., *Coalgebraic semantics for positive modal logic*, in: H. P. Gumm, editor, *Coalgebraic Methods in Computer Science, Electronic Notes in Theoretical Computer Science*, 1 **82**, 2003, pp. 221–236.

[13] Sambin, G. and V. Vaccaro, *Topology and duality in modal logic*, Annals of Pure and Applied Logic **37** (1988), pp. 249–296.

[14] Scott, D. S., *Domains for denotational semantics*, in: *An extended version of the paper prepared for ICALP '82* (1982), pp. 1–47.

7 Appendix

7.1 Cut Elimination Proof

A *conclusion parameter* of an instance of a rule is an instance of a formula which is not newly introduced formula of a logical rule nor the formula introduced by thinning. A *premise parameter* is one that matches in an obvious way a conclusion parameter of an instance of a rule. To say a "formula is generated on the left" means that the bottom rule of the left subtree above the *mix* produces the formula. The formula may already exist in the conclusion of the rule, in that case another copy is generated. Similar remarks hold for generating on the right. Hence a formula is parametric just when it is not generated and generated when it is not parametric.

The proofs below are by examination of the rules using a double induction on (1) the rank, which is the sum of the distances along all the branches of the proof tree from the active cut formula to the leaves of the proof tree, and (2) the degree which is a count of the connective in the cut formula. The notation R_l refers to the rank of the left cut formula and R_r refers to the rank of the right cut formula.

Since cut elimination for the classical base of the logic is well known, only the cases involving the modal rules will be shown here. All uses of double lines in the proof will be to represent multiple uses of a rule. The locution $C \vdash C$ stand for $C \vdash$ and $\vdash C$. First we prove a lemma that will cut the number of cases that have to be independently considered down to a manageable number.

Lemma 7.1 *If the mix formula is parametric in the rule producing the left premise and that premise was not produced with the $\langle h \rangle \vdash$ or $\vdash [h]$ rules, then we can always reduce the rank of the cut formula in on the left subtree. A similar statement holds for right premises and right subtrees.*

Proof. $R_l > 1$.

$$\dfrac{\dfrac{\Gamma \vdash \Delta[A][B]}{\Gamma \vdash \Delta[A][\langle h \rangle B]}\ \vdash \langle h \rangle \qquad \Phi[A] \vdash \Psi}{\Gamma, \Phi \vdash \Delta[\langle h \rangle B]\Psi}\ cut$$

is transformed into

$$\dfrac{\dfrac{\Gamma \vdash \Delta[A][B] \qquad \Phi[A] \vdash \Psi}{\Gamma, \Phi \vdash \Delta[B], \Psi}\ cut}{\Gamma, \Phi \vdash \Delta[\langle h \rangle B], \Psi}\ \vdash \langle h \rangle$$

The case where the left rule above the *mix* is $[h] \vdash$ is similar. □

Theorem 7.2 *All uses of cut in a proof may be eliminiated.*

Proof.

The proof follows Gentzen's original proof, we only list cases relevant to the modalities.

Case 1: $R_l + R_r = 2$

Case 1.1: One premise is an axiom.

Case 1.2: $\vdash [h], [h] \vdash$

$$\dfrac{\dfrac{\Gamma \vdash \Delta[A]}{\Gamma \vdash \Delta[[h]\, A]}\ \vdash [h] \qquad \dfrac{\Phi[A] \vdash \Psi}{\Phi[[h]\, A] \vdash \Psi}\ [h] \vdash}{\Gamma, \Phi \vdash \Delta, \Psi}\ mix$$

which is transformed into

$$\dfrac{\Gamma \vdash \Delta[A] \qquad \Phi[A] \vdash \Psi}{\Gamma, \Phi \vdash \Delta, \Psi}\ mix$$

Case 1.3: $\vdash \langle h \rangle, \langle h \rangle \vdash$ This case is similar to Case 7.1.

Case 1.4: $\vdash K, \langle h \rangle \vdash$,

$$\dfrac{\dfrac{\Gamma \vdash \Delta[B]}{\Gamma \vdash \Delta[B, \langle h \rangle A]}\ \vdash K \qquad \dfrac{\Phi[A] \vdash \Psi}{\Phi[\langle h \rangle A] \vdash \Psi}\ \langle h \rangle \vdash}{\Gamma, \Phi \vdash \Delta[B], \Psi}\ mix$$

which is transformed into

$$\dfrac{\Gamma \vdash \Delta[B]}{\Gamma, \Phi \vdash \Delta[B], \Psi}\ K \vdash, \vdash K$$

This transform is justified since the rule producing the right hand premise assures us that we may thin in all of the elements needed to produce Φ and Ψ.

Case 1.5-7: ($\vdash K, [h] \vdash$), ($\vdash \langle h \rangle, K \vdash$), ($\vdash [h], K \vdash$) These are similar to Case 7.1.

Case 2: $\mathrm{R}_l > 1$

Case 2.1: The mix formula is parametric on the left.
Case 2.1.1: The mix formula is generated on the right.
Case 2.1.1.1: $\langle h \rangle \vdash$, any Logical Rule

$$\dfrac{\dfrac{\Gamma[B] \vdash \Delta[A] \quad NC}{\Gamma[\langle h \rangle B] \vdash \Delta[A]}\ \langle h \rangle \vdash \qquad \Phi[A] \vdash \Psi}{\Gamma[\langle h \rangle B], \Phi \vdash \Delta, \Psi}\ mix$$

is transformed to

$$\dfrac{\dfrac{\Gamma[B] \vdash \Delta[A] \quad \Phi[A] \vdash \Psi}{\Gamma[B], \Phi \vdash \Delta, \Psi}\ mix \qquad NC}{\Gamma[\langle h \rangle B], \Phi \vdash \Delta, \Psi}\ \langle h \rangle \vdash$$

From the NC condition of the left cut premise, A is of the form $\langle k \rangle C$ and $h \geq k$. Since the mix formula is generated on the right, $\langle k \rangle C$ was generated by a use $\langle k \rangle \vdash$. The NC condition on that use means $k \geq c(D)$ for all $D \in \Phi, \Psi$. Hence $h \geq c(D)$ for all $D \in \Gamma, \Delta, \Phi, \Psi$. Also, since the premise for the right hand cut sequent satisfies MC, and since Δ^* only differs from Δ by the elimination of A and similarly for Φ^*, the premise of the cut rule in the conclusion proof fragment also satisfies MC. Hence this premise satisfies NC and the use of the $\langle h \rangle$ is proper.

Case 2.1.1.2: $\vdash [h]$, any Logical Rule. This case is similar to the preceding case.

Case 2.1.1.3-4: ($[h] \vdash$, any non-modal Rule), ($\vdash \langle h \rangle$, any non-modal Rule). These cases are handled by the Lemma or, in the case of the right rule above the mix being $C \vdash$ or $K \vdash$, by Gentzen's original proof.

Case 2.1.2: The *mix* formula is parametric in the right.

Case 2.1.2.1: $\langle h \rangle \vdash, \vdash [k]$

$$\dfrac{\dfrac{\Gamma[A] \vdash \Delta[B] \quad NC}{\Gamma[\langle h \rangle A] \vdash \Delta[B]}\vdash \langle h \rangle \qquad \dfrac{\Phi[B] \vdash \Psi[C] \quad NC}{\Phi[B] \vdash \Psi[[k]\, C]}\vdash [h]}{\Gamma, \Phi^* \vdash \Delta^*, \Psi[[k]\, C]}\, mix$$

This case cannot happen since the NC condition on the right forces B to be of the form $[h']\, D$ and this contradicts the NC condition the left.

Case 2.1.2.2-4: ($\langle h \rangle \vdash, \langle k \rangle \vdash$), ($\vdash [h], \langle k \rangle \vdash$), ($\vdash [h], \vdash [h]$). These cases are similar to Case 2.1.2.1.

Case 2.1.2.3-4: ($\vdash [h]$, any Logical Rule), ($\vdash \langle h \rangle$, any Logical Rule). These cases are similar to Cases 2.1.1.3-4 and handled by the Lemma.

Case 2.2: The *mix* formula is generated on the left:

Case 2.2.1: The *mix* formula is generated on the right.

Case 2.2.1.1: $\vdash \langle h \rangle, \langle h \rangle \vdash$

$$\dfrac{\dfrac{\Gamma \vdash \Delta[A][\langle h \rangle A]}{\Gamma \vdash \Delta[\langle h \rangle A]}\vdash \langle h \rangle \qquad \dfrac{\Phi[A] \vdash \Psi \quad NC}{\Phi[\langle h \rangle A] \vdash \Psi}\langle h \rangle \vdash}{\Gamma, \Phi^* \vdash \Delta^*, \Psi}\, mix$$

is transformed into

$$\dfrac{\dfrac{\dfrac{\dfrac{\Gamma \vdash \Delta[A][\langle h \rangle A] \quad \Phi[\langle h \rangle A] \vdash \Psi}{\Gamma, \Phi^* \vdash \Delta^*[A], \Psi}\, mix}{\Gamma, \Phi^* \vdash \Delta^*[\langle h \rangle A], \Psi}\vdash \langle h \rangle \qquad \Phi[\langle h \rangle A] \vdash \Psi}{\Gamma, \Phi^*, \Phi^* \vdash \Delta^*, \Psi, \Psi}\, mix}{\Gamma, \Phi \vdash \Delta, \Psi}\, C \vdash C$$

The first mix reduces the rank of the mix formula, and the second reduces the rank on the left to 1.

Case 2.2.1.2: $\vdash [h], [h] \vdash$ This case mirrors the previous case. □

7.2 Proof of Simulation Theorem 4.4

Theorem 7.3 *For channel* $\boldsymbol{C}$*, if* $\boldsymbol{P}$ *simulates* $\boldsymbol{D}$*,* $\rho_1 A \vdash_{\boldsymbol{C}} \rho_2 A'$*, and* $\rho_2 B' \vdash_{\boldsymbol{C}} \rho_1 B$*:*

$$\left(A' \vdash_{\boldsymbol{P}} [h'] \, B'\right) \text{ implies } \left(A \vdash_{\boldsymbol{D}} [h] \, B\right).$$

Proof. Assume $x \models_{\boldsymbol{D}} A$ and that $R_h yx$ holds. Using C2, there is some tuple $\langle x, x'\rangle \in Tok(\boldsymbol{C})$ and $\pi_1\langle x, x'\rangle = x$. From the morphism condition on $\langle \pi_1, \rho_1\rangle$, $\langle x, x'\rangle \models_{\boldsymbol{C}} \rho_1 A$. C1 must include $\rho_1 A \vdash_{\boldsymbol{C}} \rho_2 A'$ in which case $\langle x, x'\rangle \models_{\boldsymbol{C}} \rho_2 A'$. The morphism condition on $\langle \pi_2, \rho_2\rangle$ implies $\pi_2\langle x, x'\rangle \models_{\boldsymbol{P}} A'$ and hence $x' \models_{\boldsymbol{P}} A'$. From the antecedent in the theorem, $x' \models_{\boldsymbol{P}} [h] \, B'$. Using C3, there is some y' such that $\langle y, y'\rangle \in Tok(\boldsymbol{C})$ and that $R'_h y' x'$ holds where R'_h is the modal relation corresponding to $[h]$. From the fact that $x' \models_{\boldsymbol{P}} [h] \, B'$, it follows that $y' \models B'$. Since $\pi_2\langle y, y'\rangle = y'$, $\pi_2\langle y, y'\rangle \models B'$ and from the morphism condition, $\langle y, y'\rangle \models_{\boldsymbol{C}} \rho_2 B'$. C1 must include $\rho_2 B' \vdash_{\boldsymbol{C}} \rho_1 B$ in which case $\langle y, y'\rangle \models_{\boldsymbol{C}} \rho_1 B$. From the morphism condition, $\pi_1\langle y, y'\rangle \models_{\boldsymbol{D}} B$ and hence $y \models_{\boldsymbol{D}} B$. The resulting conditions show that $x \models_{\boldsymbol{D}} [h] \, B$ and that x satisfies $A \vdash_{\boldsymbol{D}} [h] \, B$. □

Relevant Agents

Marta Bílková

Department of Logic, Charles University in Prague; Institute of Computer Science, Academy of Sciences of the Czech Republic, Pod Vodárenskou věží 2,18 2 07 Prague 8

Ondrej Majer

Institute of Philosophy, Academy of Sciences of the Czech Republic Jilská 1, 110 00 Prague 1

Michal Peliš

Institute of Philosophy, Academy of Sciences of the Czech Republic Jilská 1, 110 00 Prague 1

Greg Restall

Department of Philosophy, University of Melbourne Parkville 3010, Australia

Abstract

In [4], Majer and Peliš proposed a relevant logic for epistemic agents, providing a novel extension of the relevant logic R with a distinctive epistemic modality K, which is at the one and the same time factive ($K\varphi \to \varphi$ is a theorem) and an existential normal modal operator ($K(\varphi \vee \psi) \to (K\varphi \vee K\psi)$ is also a theorem). The intended interpretation is that $K\varphi$ holds (relative to a situation s) if there is a resource available at s, confirming φ. In this article we expand the class of models to the broader class of 'general epistemic frames'. With this generalisation we provide a sound and complete axiomatisation for the logic of general relevant epistemic frames. We also show, that each of the modal axioms characterises some natural subclasses of general frames.

Keywords: Modal Logic, Epistemic Logic, Relevant Logic, Substructural Logic, Frame Semantics.

---

[1] The work of the first and the third author was supported by the grant ICC/08/E018 of the Grant Agency of the Czech Republic (a part of ESF Eurocores-LogICCC project FP006). The work of the second author was supported by grant no. IAA900090703 of the Grant Agency of the Academy of Sciences of the Czech Republic.

1 Introduction

Representation of epistemic states of a rational agent and their changes has been for a long time an important issue in both logic and computer science. The majority solution represents the knowledge operator as a standard necessity-like modal operator; the standard modal axioms (K, T, 4, 5) then reflect epistemic properties (closure, truth, positive and negative introspection). The most popular representation (widely used also in computer science) is based on the epistemic version of S5, in which knowledge turns out to be an indistinguishability between epistemic states. The S5-representation has been extensively criticized (see, e.g., [3,2]) for being unrealistically strong. The agents represented here are 'too perfect'—they are, among other things, logically omniscient (they know all the logical truths) and fully introspective (they are explicitly aware of their both positive and negative knowledge).

Some of the strong properties can be avoided using modal systems weaker than S5. Other ones, like the logical omniscience, require a more essential change of framework—in recent literature we can find solutions based on the framework of dynamic epistemic logic. [2]

Our way to solve the problem of unrealistically strong properties is to employ a system weaker than that of a normal modal logic—the framework of distributive relevant logic. [3] The main reason for choosing relevant logic is that it fits very well our motivations. From a technical point of view we could use even weaker systems, see the section Conclusion. There is a number of ways to introduce modalities in the relevant framework (see [8] for a general overview). However we shall not add epistemic modality quite as an external independent operator; instead we define our knowledge operator using ingredients already contained in the relevant framework. The main reason is that we can provide an intuitively acceptable interpretation of the relational frames for distributive relevant logic and the definition of the epistemic operator naturally follows from this interpretation.

1.1 Rational agent

We assume that formulas of our epistemic language represent some collections of data. Data are typically incomplete and it can well happen they are inconsistent, but we still can work with them - we use them to make decisions and draw conclusions. Our prototypical agent is a scientist dealing with scientific data. Typically, she has to deal with data which are both theoretical and empirical and which are obtained from various sources (results of experiments, articles, books, technical reports etc.) Obviously different experiments might give results which contradict each other (due to an error of either equipment or experimenter) and even theories might explain some phenomena in a ways that are incompatible. Various data might be of a different quality. Obviously

[2] Duc in [2] provides a solution based on modifications of standard Kripke semantics (awareness and impossible worlds) as well as solutions based on a combination of temporal and epistemic logic and complexity approaches (algorithmic knowledge).

[3] The idea of combining epistemic and relevant frameworks has been used in the literature, however with a different aim—see, e.g., [1] and [9].

there is little use of inconsistent parts, but even consistent data are not on the same level - some of them might be results of experiments with a more reliable equipment, some might come from more respectable authorities etc. It is clear that our agent has to discriminate among the available data. Typically she prefers data which are confirmed.

1.2 Information states

Our basic entity will be an information state of an agent. It consists of local information—a collection of data immediately available to the agent (e.g. results of her experiments, observations...) and 'remote information'—collections of local data of another information states (e.g. data obtained by some other scientists or even by herself in past).

Local information consists of two basic kinds: experimental data ('facts')—inputs and outputs of experiments/observations and 'theories/laws'-generalizations extracted from the experimental data. If we consider these two kinds of data from the point of view of a logical framework, we can, with some simplification, see that basic data are typically represented by atoms, their conjunctions and disjunctions, while 'laws' are represented by conditionals (and their combinations).

The agent accepts data as knowledge if they are *confirmed* by some source (we require at least one confirmation, which makes our operator possibility-like).[4] As we shall see in the section 2.4, the relation of confirmation also deals with inconsistency of data as an inconsistent piece of data is never confirmed.

We assume that information states evolve. However, no information is lost—the evolution is in fact an accumulation of information—in this sense it reminds the persistence relation in intuitionistic logic.

2 Frame Semantics

There are more formal systems that can be called relevant logic. Our point of departure will be the distributive relevant logic R of Anderson and Belnap. We base our framework on the relational semantics for this logic, as developed by [5,6,7] and others. Before we give a formal definition, we discuss the elements of the relational semantics from the point of view of our epistemic motivation.

The universe of our semantics consists of *information states* (sometimes also called *situations*)—they play in our framework the same role as possible worlds in Kripke frames. We interpret a current state as data immediately available to our prototypical agent. Unlike possible worlds, states might be incomplete (neither φ nor $\neg\varphi$ is true in s) or inconsistent (both φ and $\neg\varphi$ are true in s).

As we said above, information states evolve. The relation tracing this process (or rather various ways the evolution can proceed) is traditionally called *involvement* and modelled by a partial order.

[4] It might seem that we use the term knowledge in rather specific sense, but let us point out, that there is no universal agreement about the criteria a logical representations of knowledge should obey. Various systems capture some features of knowledge while they leave some other unexplained—modal representation not being an exception.

Formulas are defined in the usual way in the language of relevant logic with a modality K:

$$\varphi ::= p \mid \top \mid t \mid \neg\varphi \mid \varphi \vee \varphi \mid \varphi \wedge \varphi \mid \varphi \rightarrow \varphi \mid K\varphi$$

Strong logical connectives $\otimes$ (group conjunction, fusion) and $\oplus$ (group disjunction) are definable by implication and negation, as well as the constants f and $\bot$:

$$(\varphi \oplus \psi) \equiv_{def} (\neg\varphi \rightarrow \psi) \quad (1)$$

$$(\varphi \otimes \psi) \equiv_{def} \neg(\varphi \rightarrow \neg\psi) \quad (2)$$

$$f \equiv_{def} \neg t \quad (3)$$

$$\bot \equiv_{def} \neg\top \quad (4)$$

Classical (weak, lattice) conjunction and disjunction correspond to the situation when the agent combines local data, i.e., data immediately available in the current state.

Implication is a modal connective in the sense that it depends on a neighborhood of a current state, which is given by a ternary *relevance* relation R. In fact it is analogous to the strong (necessary) implication in a standard Kripke frame except the neighborhood of a state s is given by pairs of states (y, z) such that $Rsyz$. The relation R reflects in our interpretation actual experimental setups. Let us call y, z the antecedent and the consequent state, respectively. Antecedent states correspond to some initial data (outcome of measurements or observations) of some experiment, while the related consequent states correspond to the corresponding resulting data of the experiment. Implication then corresponds to a (simple) kind of a rule: if I observe in my current state, that at every experiment (represented by a couple antecedent–consequent state) each observation of φ is followed by an observation of ψ, then I accept 'ψ follows φ' as a rule.

In Kripke models the *negation* of a formula φ is true at a world iff φ is not true there. As states can be incomplete and/or inconsistent, this is not an option any more if one deals with substructural logics. Negation becomes a modal connective and its meaning depends on the states related to the given state by *compatibility* relation C. Informally we can see the compatible states as collections of data our scientist wants to be consistent with (e.g. becuse of their reliabilty, impact etc.) Relevant negation does not correspond straightforwardly to 'necessary false'. We do not require that the negated formula in question is false in the neighborhood of the given state, we just require no state in the neighborhood contains a positive instance of this formula.

From the point of view of our motivation the interpretation of negation is rather straightforward—an agent can explicitly deny some information (a piece of data) only if its negation does not contradict any collection of data she wants to be consistent with. This condition also has a normative side: she has to be skeptical in the sense that she denies everything not positively supported by any of the compatible collection of data. If we want to have negative facts at the same basic level as positive facts, we can read the clause for the definition of compatibility relation in the other direction: the agent can relate her actual state just to the states which do not contradict her negative facts.

Properties of the compatibility relation obviously determine the kind of negation obtained, but as we shall see they moreover influence properties of the epistemic modality defined below. In standard relevant frameworks it is usually assumed that compatibility is symmetric, but it is in general neither reflexive (inconsistent states are not self-compatible) nor transitive.

One of the tricky points in the definition of relational semantics is the definition of truth in a relevant frame (model). If we take a hint from Kripke frames, we should equate truth in a frame with truth in every state. But this would gives us an extremely weak system with some very unpleasant properties [7]. For example the almost uniformly accepted identity axiom $(\alpha \to \alpha)$ and the Modus Ponens rule fail to hold in every state. Designers of relevant logics took a different route; instead of requiring truth in all states, they identify truth in a frame with truth in all logically 'well behaved' states. These states are called *logical.* In order to satisfy the 'good behavior' it is enough to require in a state l that all the information in any antecedent situation related to l is contained in the corresponding consequent situation as well. This is ensured by a half of the following condition: x is below y in the involvement relation if and only if there is a logical state l such that $Rlxy$. It is easy to see that logical situations validate both the identity axiom and (implicative) Modus Ponens.

We proceed to the formal definitions now.

Definition 2.1 A *relevant frame* is a tuple $F = (W, L, \leq, C, R)$, where W is a nonempty set of *situations* and

- $\leq$ is an *involvement* relation which is a partial order on W
- R is a ternary *relevance* relation on W satisfying the conditions[5] (where R^2xyzw means $(\exists s)(Rxys$ and $Rszw)$):

$$Rxyz \text{ and } x' \leq x, y' \leq y, z' \geq z \text{ implies } Rx'y'z' \tag{5}$$

$$Rxyz \text{ implies } Ryxz \tag{6}$$

$$Rxxx \tag{7}$$

$$R^2(xy)zw \text{ implies } R^2(xz)yw \tag{8}$$

- L, the set of *logical situations*, is a nonempty upwards closed subset of $(W, \leq)$, satisfying

$$x \leq y \text{ iff there is } z \in L \text{ such that } Rzxy \tag{9}$$

- C is a binary *compatibility* relation on W satisfying the conditions

$$Rxyz \text{ implies } (\forall z')z'Cz(\exists y')y'Cy \text{ where } Rxz'y' \tag{10}$$

$$xCy \text{ and } x' \leq x \text{ and } y' \leq y \text{ implies } x'Cy' \tag{11}$$

$$xCy \text{ implies } yCx \tag{12}$$

$$(\forall x)(\exists y)xCy \tag{13}$$

$$\forall x(\exists u)(xCu \text{ and } \forall z(xCz \text{ implies } z \leq u)) \tag{14}$$

Definition 2.2 A *relevant model* is $M = (F, V)$, where F is a relevant frame and $V : Prop \mapsto \mathcal{P}(W)$, such that each $V(p)$ is an upper subset of $(W, \leq)$, is a persistent

[5] We use the stronger versions of the relation conditions, called 'plump' in [8], because they correspond to structural axioms and we are interested in modal characterizability as well.

valuation of propositional formulas. The valuation generates the following satisfaction relation between states and formulas:

- $x \Vdash p$ iff $x \in V(p)$
- $x \Vdash t$ iff $x \in L$
- $x \Vdash \top$
- $x \Vdash \varphi \wedge \psi$ iff $x \Vdash \varphi$ and $x \Vdash \psi$
- $x \Vdash \varphi \vee \psi$ iff $x \Vdash \varphi$ or $x \Vdash \psi$
- $x \Vdash \neg\varphi$ iff for all y, xCy implies $y \nVdash \varphi$
- $x \Vdash \varphi \rightarrow \psi$ iff for all y, z, $Rxyz$ and $y \Vdash \varphi$ implies $z \Vdash \psi$

The relation between implication and fusion is that they form an adjoint pair (implication is the residuum of the fusion in algebraic terms):

$$\varphi \rightarrow (\psi \rightarrow \chi) \;\equiv\; \varphi \otimes \psi \rightarrow \chi$$

thus, semantically

$$x \Vdash \varphi \otimes \psi \;\text{ iff }\; \text{there are } y, z \text{ such that } Ryzx \text{ and } y \Vdash \varphi \text{ and } z \Vdash \psi.$$

The validity of formulas in a model, a frame, or in a class of models or frames is defined via validity in all *logical* states.

The results of persistency for all formulas, soundness and completeness of the semantics for the relevant logic R can be extracted as a special case from general definitions and proofs contained in [8].

The original motivation behind an epistemic modality in the framework of experimental data is that data can be accepted as knowledge only if they are *confirmed* by a source. As there are several possibilities what can be counted as a reliable source, we explicitly represent the relation of *being a source* by a new binary relation S on W and use it to define our epistemic modality K:

$$x \Vdash K\varphi \text{ iff for some s where } sSx, s \Vdash \varphi \tag{15}$$

$K\varphi$ thus meaning that φ is supported by at least one source.

Next we explore some possibilities how to define the S-relation and introduce corresponding classes of relevant frames.

2.1 Classic relevant frames

The S relation can be seen as determining which states are to be counted as reliable sources. The first attempt uses the ingredients already contained in a frame to define S. Following motivations presented in [4], we start with a requirement that a source state confirming data in a current state shall be compatible with the current state and it should (strictly) precede the current state in the involvement ordering (the second condition was supposed to make the confirmation 'independent'—to exclude the possibility that a state is a source for itself)

$$sSx \text{ iff } s < x \text{ and } sCx \tag{16}$$

Moreover, it is reasonable to require that modal formulas are persistent, this is guaranteed by the following condition:

$$sSx \text{ and } x \leq x' \text{ then } (\exists s')(s \leq s' \wedge s'Sx') \tag{17}$$

The class of frames defined this way however doesn't seem a good candidate to work with since it is an open question how to axiomatize the logic of classical frames in the modal language we have fixed. We conjecture it coincides with the logic of General frames defined below, and that the class of classical frames is not modally definable in the current language due to the presence of an anti-property $\neq$ contained in the definition of S in (16).

We call the class of frames satisfying conditions (16) and (17) the class of *Classic frames* and denote it $\mathcal{F}_c$.

Let us remind that validity in a class of frames is defined as a validity in all logical states in all models based on the frames from the class. Thus:

$$\mathcal{F}_c \Vdash \varphi \text{ iff } (\forall F \in \mathcal{F}_c)(\forall x \in L)(x \Vdash \varphi) \tag{18}$$

We can also weaken our requirements on available sources and admit that in some cases a state can be a source for itself, so we work with $\leq$ instead of $<$. We define a class of *Weak Classic frames* $\mathcal{F}_{wc}$ weakening the property (16) to:

$$sSx \text{ iff } s \leq x \text{ and } sCx \tag{19}$$

This class turns out to be distinguishable from the other two by the validity of the *introspection* axiom $K\varphi \rightarrow KK\varphi$. This was one of the axioms we criticised in the introduction and the validity of which we tried to avoid. We shall comment on this later.

2.2 General relevant frames

Providing an axiomatisation for the logic of classic (and weak classic) relevant epistemic frames is an open problem. Upon reflection, however, it is not clear that this class of epistemic frames is the natural target. To count *every* lower compatible state under the current state as a resource for use may be more restrictive than we need. For a more general class of frames, we allow for slightly more variation in the interpretation of K, by generalising the accessibility relation S. It is now required to satisfy the following two conditions: the condition (17) remains unchanged, while we replace 'iff' in the other condition with 'only if':

$$sSx \text{ then } s \leq x \text{ and } sCx \tag{20}$$

We denote the class of general frames $\mathcal{F}_g$ and provide in the next section an axiomatisation of this class. Validity of formulas in the class is defined:

$$\mathcal{F}_g \Vdash \varphi \text{ iff } (\forall F \in \mathcal{F}_g)(\forall x \in L)(x \Vdash \varphi) \tag{21}$$

2.3 Relating the classes of frames

We start with stating some basic properties of all the three classes defined above.

Lemma 2.3 (Persistency for formulas) *$x \Vdash \varphi$ and $x \leq y$ implies $y \Vdash \varphi$.*

Proof. We concentrate on the case of $K\varphi$ only. Suppose that $x \Vdash K\varphi$ and $x \leq y$. We show that $y \Vdash K\varphi$. Since $x \Vdash K\varphi$ there is some sSx satsifying φ. From the condition (17) $(\exists s')(s \leq s' \wedge s'Sy)$. From the induction hypothesis $s' \Vdash \varphi$, but then $y \Vdash K\varphi$ as desired. □

Lemma 2.4 (Selfcompatibility of sources) *sSx implies sCs.*

Proof. Suppose sSx. Then (in $\mathcal{F}_c, \mathcal{F}_{wc}$, and $\mathcal{F}_g$) $s \leq x$ and sCx. From the condition (11) it follows that sCs. □

Since every classic (weak classic) frame is a general frame, we have $\mathcal{F}_c \subseteq \mathcal{F}_g$ and $\mathcal{F}_{wc} \subseteq \mathcal{F}_g$, and thus, for the logics of the classes, the inclusions $L(\mathcal{F}_g) \subseteq L(\mathcal{F}_c)$ and $L(\mathcal{F}_g) \subseteq L(\mathcal{F}_{wc})$ hold. The other inclusions are open question.

We can however distinguish the class $\mathcal{F}_{wc}$ from $\mathcal{F}_g$ (and from $\mathcal{F}_c$ as well) with validity of the axiom of introspection $K\varphi \rightarrow KK\varphi$. Thus in particular $L(\mathcal{F}_g) \subset L(\mathcal{F}_{wc})$ is a proper inclusion.

Lemma 2.5 *$K\varphi \rightarrow KK\varphi$ fails in $\mathcal{F}_g$ and $\mathcal{F}_c$, while it holds in $\mathcal{F}_{wc}$.*

Proof. To show $K\varphi \rightarrow KK\varphi$ holds in $\mathcal{F}_{wc}$ consider $u \in L$, $Ruxy$ where $x \Vdash K\varphi$. We show $y \Vdash KK\varphi$. Since $x \Vdash K\varphi$, then there is s such that sSx and $s \Vdash \varphi$. Because of sCs (Lemma 2.4) and $s \leq s$, we obtain sSs. The situation s is a source for itself. From $s \Vdash K\varphi$ we get $x \Vdash KK\varphi$. By persistence (Lemma 2.3) $y \Vdash KK\varphi$.

To show $K\varphi \rightarrow KK\varphi$ fails in the class $\mathcal{F}_c$ and in the class $\mathcal{F}_g$, we consider the following counterexample: $W = L = \{x, s\}$ and sSx (thus in particular $s < x$ and sCx, and sCs by the Lemma 2.4. Moreover we have $Rsxx$ (and many others by the conditions for L and R to be satisfied). But we do not have sSs (as we would be forced in $\mathcal{F}_{wc}$ by $s \leq s$ and sCs). Now we put $V(p) = \{s, x\}$. Then $x \Vdash Kp$ but $x \nVdash KKp$, thus $Kp \rightarrow KKp$ fails in a logical state s. □

We are not interested much in the class $\mathcal{F}_{wc}$ itself, but the introspection axiom is interesting from another point of view– we show which class of relevant frames corresponds to the axiom of introspection in Section 4.

One more definition needed to state strong completeness is the *(local) consequence* over the class of frames $\mathcal{F}_g$. It is defined as follows (so that logical validity $\mathcal{F}_g \Vdash \varphi$ corresponds to $t \vDash_{\mathcal{F}_g} \varphi$):

$$\Gamma \vDash_{\mathcal{F}_g} \varphi \text{ iff } (\forall F \in \mathcal{F}_g)(\forall V)(\forall x)((\forall \gamma \in \Gamma)(F, V), x \Vdash \gamma \text{ implies } (F, V), x \Vdash \varphi) \qquad (22)$$

2.4 Basic properties of the modality K

The following examples of valid and invalid schemes hold for the validity in the class of general epistemic frames $\mathcal{F}_g$, as well as in the class of classic epistemic frames $\mathcal{F}_c$. Thus we use the symbol $\models$ and do not mention the class explicitly. We are so far not able to provide any example distinguishing these two classes.

Lemma 2.6 (Monotonicity of K) $\models \varphi \to \psi$ *implies* $\models K\varphi \to K\psi$

Proof. Consider $x \in L$, we show $x \Vdash K\varphi \to K\psi$. Consider any y, z, such that $Rxyz$ and $y \Vdash K\varphi$. We show that $z \Vdash K\psi$.

$y \Vdash K\varphi$ implies there is s such that (sSy and $s \Vdash \varphi$). $s \leq s$ implies there is $t \in L$ such that $Rtss$ and $Rsts$ (by the condition for L). Since in such $t \Vdash \varphi \to \psi$ and $s \Vdash \varphi$, we conclude $s \Vdash \psi$. Thus $y \Vdash K\psi$.

But $y \leq z$ (since $x \in L$ and $Rxyz$), and finally $z \Vdash K\psi$. □

It is to be expected that the proposed semantics based on the class of general frames blocks all the undesirable properties of both material and strict implication. Moreover, we ruled out (at least for some classes of frames) the validity of majority of the properties of standard epistemic logics that we have criticized, in particular, both positive and negative introspection, as well as some closure properties.

The two core modal principles valid in general frames (see Theorem 3.2) are *factivity*

$$K\varphi \to \varphi \tag{23}$$

and *strong factivity*

$$\neg\varphi \wedge K\varphi \to \bot \tag{24}$$

We call the second condition strong factivity, as it says not only that only information warranted here can be known, but that anything 'diswarranted' here is excluded from knowledge.

Factivity vs. strong factivity. Factivity does not follow from strong factivity. For factivity, we would need $\neg\varphi \otimes K\varphi \to \bot$ whence we could residuate to get $K\varphi \to (\neg\varphi \to \bot)$, and then contrapose the consequent $K\varphi \to (\top \to \varphi)$ and use the entailment from $(\top \to \varphi)$ to φ.

Of course, the claim that $\bot$ follows from the *fusion* of $\neg\varphi$ with $K\varphi$ is a stronger claim than it following from their conjunction. In the presence of weakening, however, factivity would follow from strong factivity.

Material factivity. It is worth observing that we also have validity of the weaker claim which we call *material factivity*:

$$\neg(\neg\varphi \wedge K\varphi), \text{ or equivalently, } \neg K\varphi \vee \varphi \tag{25}$$

Material factivity is a weaker condition than factivity since $\varphi \to \psi$ entails $\neg\varphi \vee \psi$ in R. It is also weaker then strong factivity since $(\varphi \to \bot) \to \neg\varphi$ is a theorem of R: $\neg\varphi \leftrightarrow (\varphi \to f)$ is a theorem, and we can prefix $\varphi \to$ on the theorem $\bot \to f$ in the usual manner. But in logics weaker than R, in which the law of excluded middle is rejected, material factivity remains a consequence of strong factivity, but no longer weak factivity.

K-axiom. K (in implicational form) would in fact correspond to a 'distribution of confirmation': if an implication is confirmed, then the confirmation of the antecedent implies the confirmation of the consequent. But the S relation is not connected to the R relation in any straightforward way, so existence of sources for $K(\alpha \to \beta)$ and for $K\alpha$ does not force existence of a source for $K\beta$. Hence

$$\not\models K(\alpha \to \beta) \to (K\alpha \to K\beta)$$

Introspection. We defined knowledge as independently confirmed data. In this reading the axioms **4** and **5** rather than to introspection correspond to a 'second order confirmation' (if α is confirmed then the confirmation of α is confirmed as well, similarly for the negative introspection). We showed in Lemma 2.5, that positive introspection fails in $\mathcal{F}_g$ and $\mathcal{F}_c$, while it holds in $\mathcal{F}_{wc}$. It is easy to see that negative introspection fails for all frames:

$$\not\models \neg K\alpha \to K\neg K\alpha$$

Necessity and negation. There is a difference between $s \nVdash K\varphi$ and $s \Vdash \neg K\varphi$. The former simply says that φ is not confirmed at the current situation s, while the latter is stronger (at least in the case of selfcompatible situations), it says that φ is not confirmed in any situation compatible with s. From this point of view it is uncontroversial that both $K\varphi$ (confirmation in the current situation) and $\neg K\varphi$ (the lack of confirmation in the compatible situations) might be true in some situation s (in this case s is not compatible with itself).

Closure properties. In the introduction we criticized too strong closure properties of the standard modal representations of knowledge. In fact the question how strong conditions shall be imposed on epistemic states to obtain an adequate representation is one of the crucial choices of the knowledge representation. It is also closely related to the problem of logical omniscience.

We can see the machinery of 'logical expansion' as having two basic ingredients. One is knowledge of all the tautologies of the logical system in question guaranteed by the Necessitation rule. The other is Modal Modus Ponens, which produces all consequences of any new piece of (non-logical) information. Our system turns out to be extremely weak and avoids both of these closure properties and some more. It can be seen as anti-logical and pragmatic—in a sense that our agent believes (accepts) just what is (or was) observed. Even the data corresponding to logical laws have to be confirmed.

Necessitation rule. The rule

$$\frac{\varphi}{K\varphi}$$

common to all normal epistemic logics, guarantees among other things that all the tautologies of the logical system in question are known. In our framework this would mean that all the logical truths are confirmed. This is in general not the case. Let us assume that φ is a valid formula (i.e., $l \Vdash \varphi$, for every logical situation l). The necessity rule would imply the validity of $K\varphi$. However, for $l \Vdash K\varphi$ we need a confirmation from a resource and the conditions for the relation S do not guarantee, that l has a logical situation as a resource (in fact it does not need to have a resource at all), so even if there is a source situation s for l, this situation does not have to be logical and hence the validity of ψ in s is not guaranteed.

Modal Modus Ponens. Closure of knowledge with respect to logical consequence, which is a part of logical omniscience (if an agent knows both φ and $\varphi \to \psi$, then she knows ψ as well) is forced by the validity of the modal Modus Ponens:

$$\frac{K\alpha \quad K(\alpha \to \beta)}{K\beta}$$

It is easy to see that it does not hold in our system. As we noted above, **K** is in fact a 'distribution of confirmation'. If both an implication and its antecedent are confirmed, there is no reason the consequent needs to be confirmed as well.

Let us note, that the weaker version of modal Modus Ponens holds

$$\frac{K\alpha \quad K(\alpha \to \beta)}{\beta}$$

however, it cannot be considered as any kind of omniscience. It just says that if both α and $(\alpha \to \beta)$ are confirmed, then β is a part of currently available data.

This rule holds not only in logical situations, but in all situations. If $K\alpha$ and $K(\alpha \to \beta)$ are true in an $s \in S$, then $s \Vdash \alpha$ and $s \Vdash \alpha \to \beta$ because of **T** axiom (*factivity*). It follows from the assumption $Rsss$ and the definition of implication, that $s \Vdash \beta$ as well.

Contradiction. Contradiction in relevant logic is non-explosive: φ and $\neg\varphi$ might hold in a contradictory situation, but it does not entail an arbitrary formula ψ. (This would require an R-connection to situation where ψ holds.) [6]

$$\not\models (\varphi \wedge \neg\varphi) \to \psi$$

As we noted above, a contradiction cannot be known (it is never confirmed).

$$\not\models K(\varphi \wedge \neg\varphi)$$

This has a trivial consequence, that knowledge of contradiction implies anything ($\models K(\varphi \wedge \neg\varphi) \to \psi$), so, in particular knowledge of contradiction implies knowledge of anything ($\models K(\varphi \wedge \neg\varphi) \to K(\psi)$). Nevertheless this does not lead to any kind of explosion as there is no such situation in which the antecedent is true. In standard models, $K(\varphi \wedge \neg\varphi)$ is never true either, but the reason is that $\varphi \wedge \neg\varphi$ is not true in any state (possible world). In our framework the situation is different: $\varphi \wedge \neg\varphi$ can be true in some situation (the agent obtained contradictory data), but $K(\varphi \wedge \neg\varphi)$ cannot.

Adjunction. Modal adjunction also does not hold—if $K\alpha$ and $K\beta$ are true in s, then obviously $(\alpha \wedge \beta)$ is true there, but $K(\alpha \wedge \beta)$ need not be. [7] Our agent is really careful here. Even if each of α and β are confirmed separately, their conjunction is not accepted as knowledge, unless there is a single resource confirming both of them (which in general does not need to be the case).

[6] The explosion does not occur even in the case of the strong conjunction; $(\varphi \otimes \neg\varphi) \to \psi$ does not hold.

[7] The same negative result holds also for strong conjunction. If $K\alpha$ and $K\beta$ are true in s, then $(\alpha \otimes \beta)$ is true in s (because of factivity and $Rsss$), but $K(\alpha \otimes \beta)$ need not be true in s.

Modal disjunction rule. In our system knowledge distributes with disjunction (see Theorem 3.2). It holds that

$$\frac{K(\alpha \vee \beta)}{K\alpha \vee K\beta}$$

One of the disjuncts has to be confirmed by the same source as the whole disjunction, because a disjunction is true at a source if at least one of the disjuncts is.

3 Axiomatics, Soundness, Completeness

We extend a standard Hilbert style axiomatisation of R in the language $\rightarrow, \wedge, \vee, \neg, \bot$, which can be adopted e.g. from [5], with aditional axioms for the constants t and $\top$ and the modality K. The remaining logical operators are definable as we noted in Section 2.

Definition 3.1 Calculus RK consists of axiom schemes

$$\varphi \rightarrow \varphi \qquad (\varphi \rightarrow \psi) \rightarrow ((\psi \rightarrow \chi) \rightarrow (\varphi \rightarrow \chi))$$
$$\varphi \rightarrow ((\varphi \rightarrow \psi) \rightarrow \psi) \qquad (\varphi \rightarrow (\varphi \rightarrow \psi)) \rightarrow (\varphi \rightarrow \psi)$$
$$(\varphi \wedge \psi) \rightarrow \varphi \qquad (\varphi \wedge \psi) \rightarrow \psi$$
$$\varphi \rightarrow (\varphi \vee \psi) \qquad \psi \rightarrow (\varphi \vee \psi)$$
$$((\varphi \rightarrow \psi) \wedge (\varphi \rightarrow \chi)) \rightarrow (\varphi \rightarrow (\psi \wedge \chi)) \qquad (\varphi \wedge (\psi \vee \chi)) \rightarrow ((\varphi \wedge \psi) \vee (\varphi \wedge \chi))$$
$$\neg\neg\varphi \rightarrow \varphi \qquad (\varphi \rightarrow \neg\psi) \rightarrow (\psi \rightarrow \neg\varphi)$$
$$(t \rightarrow \varphi) \leftrightarrow \varphi \qquad \varphi \rightarrow \top$$
$$K\varphi \rightarrow \varphi \qquad \neg\varphi \wedge K\varphi \rightarrow \bot$$
$$K(\varphi \vee \psi) \rightarrow K\varphi \vee K\psi$$

and the rules

$$\frac{\varphi \quad \psi}{\varphi \wedge \psi} \qquad \frac{\varphi \quad \varphi \rightarrow \psi}{\psi} \qquad \frac{\varphi \rightarrow \psi}{K\varphi \rightarrow K\psi}$$

Our first result is that this is an axiomatisation of the logic of general frames.

Theorem 3.2 (Soundness) *Any formula provable in* RK *is valid in all general frames.*

Proof. We spell out only soundness of the modal axioms, the soundness of the monotonicity rule has been established in Lemma 2.6.
(factivity) $K\varphi \rightarrow \varphi$ holds, since, whenever sSx and $s \Vdash \varphi$ then by the condition (20) $s \leq x$ and thus $x \Vdash \varphi$.
(strong factivity) $\neg\varphi \wedge K\varphi \rightarrow \bot$ (note that $x \Vdash \varphi \rightarrow \bot$ iff $x \nVdash \varphi$) holds in a logical state u, since if any $x \Vdash \neg\varphi \wedge K\varphi$, then for some s where sSx, $s \Vdash \varphi$, but from the definition of sSx we have sCx: so since $x \Vdash \neg\varphi$, we have $s \nVdash \varphi$, a contradiction. So there is no such x where $\neg\varphi \wedge K\varphi$ holds.
(K commuting with $\vee$) Suppose $x \in L$, $Rxyz$ and $y \Vdash K(\varphi \vee \psi)$ we show that $z \Vdash K\varphi \vee K\psi$. There is sSy satisfying $\varphi \vee \psi$, suppose, e.g., $s \Vdash \varphi$. From $Rxyz$ and $x \in L$ we know $y \leq z$. By the condition (17) there is s' such that $s \leq s'$ and $s'Sz$. Now $s' \Vdash \varphi$ and $z \Vdash K\varphi$. Then $z \Vdash K\varphi \vee K\psi$. □

We prove that the axiomatization RK is strongly complete with respect to the class of general frames, i.e. $\Gamma \nvdash \varphi$ implies $\Gamma \nvDash_{\mathcal{F}_g} \varphi$. We shall adopt the standard technique of *canonical model* construction.

Working with the logic R the natural concept of a theory (or a theory of a state in a model) is given by the notion of *prime theory.* A set of formulas Γ is a theory iff it is closed under provability: if $\varphi \vdash \psi$ and $\varphi \in \Gamma$ then $\psi \in \Gamma$, is closed under conjunction: if $\varphi \wedge \psi \in \Gamma$ then both $\varphi, \psi \in \Gamma$. Warning: in relevant logic, theories do not automatically contain all theorems. (If so, such theory is often called regular.) A theory is *prime* if it is moreover closed under disjunctions: if $\varphi \vee \psi \in \Gamma$ then $\varphi \in \Gamma$ or $\psi \in \Gamma$.

We use the set of all prime theories over RK as the canonical set of points. Note that the canonical model defined this way may contain points in which t is not included, thus it contains nonlogical situations. It is not surprising since e.g. to violate the weakening axiom we need some nonlogical situations included.

To show that any invalid consequence $\Gamma \nvdash \varphi$ can be falsified in a state of the canonical model we need to know that Γ can be expanded to a prime theory in such a way that φ is not included in the expansion. In R this is done extending simultaneously two sets—a pair $\langle \Gamma, \{\varphi\} \rangle$—keeping track on what should and what shouldn't be included simultaneously. (Showing φ is not a theorem corresponds to invalid consequence $t \nvdash \varphi$, thus we want to falsify φ in a logical state.)

Definition 3.3 Let Γ, Δ denote sets of formulas. $P = \langle \Gamma, \Delta \rangle$ is called a *pair* if no conjunction ($\wedge$) of formulas in Γ entails any disjunction ($\vee$) of formulas in Δ.

According to Pair Extension Theorem [8, p. 94] we can extend a pair P to a full pair $P' = \langle \Gamma', \Delta' \rangle$ which is maximal in the sense that $\Gamma' \cup \Delta'$ is the whole language. It follows, that in a full pair Γ' is always a prime theory [8, p. 93].

Theorem 3.4 (Strong Completeness) *The axiomatization* RK *is strongly complete with respect to the class $\mathcal{F}_g$ of general frames.*

Proof. We take the usual Henkin-style construction of a canonical model, from [8, Sections 11.3 and 11.4]. To this construction of canonical model for RK, we must pay attention to the behaviour of K, and the accessibility relation S defined on the canonical model, to verify that this model satisfies the frame conditions for S, and hence, is a general frame.

Following the proof in [8], we take the points W_r in the canonical frame to be all the prime theories of the logic RK, and define the canonical frame to be $F_r = (W_r, L_r, \leq_r, C_r, R_r)$, where canonical relations are defined:

- $L_r = \{x \mid t \in x\}$
- $R_r xyz$ iff for each $\varphi \to \psi \in x$ where $\varphi \in y; \psi \in z$
- $xC_r y$ iff for each $\neg\varphi \in x, \varphi \notin y$
- $xS_r y$ iff for each $\varphi \in x, K\varphi \in y$

Adding valuation V such that $V(p) = \{x \mid p \in x\}$ we get a canonical model M_r.

It is immediate that membership in the prime theory satisfies the modelling conditions for $\wedge$ and $\vee$, and half of the modelling conditions for the conditional, negation and for K. We have

- if $\varphi \to \psi \in x$, then if $R_r xyz$, and $\varphi \in y$, then $\psi \in z$

- if $\neg\varphi \in x$, then if $xC_r y$, then $\varphi \notin y$
- if $\varphi \in x$, then if $xS_r y$, then $K\varphi \in y$.

To ensure that we have the converse of these, we appeal to Pair Extension Theorem [8]. The proof for R_r and C_r is standard [8, pp. 256, 261], but the proof for S_r we reiterate here:

Lemma 3.5 (Extension of the valuation to formulas) $x \Vdash \varphi$ *iff* $\varphi \in x$

Proof. Everything is clear except of $x \Vdash K\varphi$ iff $K\varphi \in x$:

From left to right: $x \Vdash K\varphi$, thus there is $sS_r x$ satisfying φ. From the definition of S_r, $K\varphi \in x$.

From right to left: Let us assume $K\varphi \in x$, we need an $s \in W_r$ such that $sS_r x$ and $\varphi \in s$.

$P = \langle\{\varphi\}; \{\psi : K\psi \notin x\}\rangle$ is a pair. Because if not, then $\varphi \vdash \psi_1 \vee ... \vee \psi_n$, hence $K\varphi \vdash K(\psi_1 \vee ... \vee \psi_n)$ (monotonicity of K), and $K\varphi \vdash K\psi_1 \vee ... \vee K\psi_n$ (K commuting with $\vee$ axiom). But as $K\varphi \in x$, then $K\psi_1 \vee ... \vee K\psi_n \in x$ (as x is a theory), hence $K\psi_i \in x$ (as x is prime), which is a contradiction with the definition of P.
According to Pair Extension Theorem [8] we can extend P to a full pair $P' = \langle s, r\rangle$. It follows, that s is prime. It remains to show that $sS_r x$. Assume $\alpha \in s$. If $K\alpha \notin x$, then $\alpha \in r$ (definition of P), so $\alpha \notin s$. Contradiction. □

So, the canonical model satisfies the general conditions for a model. It remains to show the canonical frame falls within the class of general frames, checking the canonical relations satisfy the required conditions.

Lemma 3.6 $F_r \in \mathcal{F}_g$

Proof. It is almost immediate that R_r, C_r and S_r, so defined, satisfy the plump versions of conditions which we have used in this paper to define general frames. We skip these cases here and refer to [8] where they can be easily extracted from the general proof.

To make clear the situation of logical states we however show that L_r satisfy the required conditions. First observe that L_r is clearly an upper set w.r.t. inclusion. We show it satisfies the condition (9).

Suppose $R_r xyz$ and $x \in L_r$. To show $y \subseteq z$ suppose $\varphi \in y$. Since $t \in x$ and $t \vdash \varphi \to \varphi$, we have $\varphi \to \varphi \in x$. Since $R_r xyz$ and $\varphi \in y$, we obtain $\varphi \in z$ as desired.

For the converse suppose $y \subseteq z$. We have to find a prime theory x such that $t \in x$ and $R_r xyz$ holds. First observe that $P = \langle\{t\}, \{\varphi \to \psi \mid \varphi \in y, \psi \notin z\}\rangle$ is a pair: if not, $t \vdash \bigvee_{i\in I} (\varphi_i \to \psi_i)$ for some disjunction of implications from the left set in the pair P. Then $\bigwedge_{i\in I} \varphi_i \vdash \bigvee_{i\in I} \psi_i$. But since all $\varphi_i \in y$, we have $\bigvee_{i\in I} \psi_i \in y$ and thus some $\psi_i \in y$ – a contradiction.

According to Pair Extension Theorem [8] we can extend P to a full pair $P' = \langle x, u\rangle$ where x is a prime theory. Moreover, we have $t \in x$ and $x \in L_r$. $R_r xyz$ holds immediatelly from the definition of P: if $\varphi \in y$ and $\psi \notin z$ then, from the definition of P, $\varphi \to \psi \in u$ and thus $\varphi \to \psi \notin x$.

We show that S_r satisfies the modelling condition (20):

Suppose sS_rx. To show that $s \leq x$, reason as follows: If $\varphi \in s$, then $K\varphi \in x$ (by the definition of S_r) and by factivity (any prime theory containing $K\varphi$ is closed under factivity because $K\varphi \vdash \varphi$), $\varphi \in x$ too. Since inclusion on the canonical model is subsethood, we have s $\leq x$.

Suppose sS_rx. We know from the general proof in [8] that C_r is symmetric. It is therefore enough to show that xC_rs. Reason as follows: If $\neg\varphi \in x$, then suppose φ were in s. Then since sS_rx, we have $K\varphi \in x$. But we have $\neg\varphi \in x$. Thus, $\neg\varphi \wedge K\varphi \in x$, which by strong factivity is impossible (x is a prime theory, and no prime theories contain $\bot$). So, if $\neg\varphi \in x$, then $\varphi \notin s$, and hence, xC_rs. From symmetry sS_rx as desired.

Next we make sure that S_r satisfies the other condition (17). Suppose sS_rx and $x \subseteq x'$. We need a prime theory s' such that $s \subseteq s'$ and $s'S_rx'$. We claim we can take $s' = s$. We show sS_rx': suppose $\varphi \in s$, then by sS_rx and definition of S_r, $K\varphi \in x$. Since $x \subseteq x'$, we have $K\varphi \in x'$ as well. □

This finishes the proof of Theorem 3.4. □

4 Correspondence

We address a question of modal correspondence in this section. Not only are factivity and strong factivity consequences of the modelling conditions (20) and (17), we can strengthen this to a correspondence result. The modal axioms of factivity and strong factivity *characterize* the class of general frames $\mathcal{F}_g$ (they hold in a frame F if and only if $F \in \mathcal{F}_g$). We even obtain a more subtle view on the conditions we put on the relation S: factivity corresponds, on frames, to the first half of the condition (20): $sSx \rightarrow s \leq x$. Strong factivity corresponds to the other half of condition (20): $sSx \rightarrow sCx$. We may split the condition (20) into two conditions:

(S1) sSx implies $s \leq x$

(S2) sSx implies sCx

If we now define two classes of frames $\mathcal{F}_{S1}$ and $\mathcal{F}_{S2}$ (so that their intersection is $\mathcal{F}_g$) as those relevant frames with S added and satisfying (S1) and (17), (or (S2) and (17) respectivelly), we obtain immediatelly sound and complete axiomatizations of the two logics $L(\mathcal{F}_{S1})$ (using factivity) and L($\mathcal{F}_{S2}$) (using strong factivity).

Theorem 4.1 (Correspondence)

- $K\varphi \rightarrow \varphi$ *characterizes the class* $\mathcal{F}_{S1}$.
- $\neg\varphi \wedge K\varphi \rightarrow \bot$ *characterizes the class* $\mathcal{F}_{S2}$.

Proof. Soundness theorem 3.2 supplies one half of the correspondence result.

For the other half of the correspondence condition for factivity, suppose we have a frame in which the condition $sSx \rightarrow s \leq x$ fails. That is, we have a pair of points s, x where sSx and $s \not\leq x$. Let $V(p)$ be the set of all points in which s is included: $V(p) = \{u \mid s \leq u\}$. This is clearly an upper set. Then $s \in V(p)$ but $x \notin V(p)$. It follows that Kp holds at x, since sSx, but p does not. Thus, since by the condition (9) there is some $u \in L$ such that $Ruxx$, we have $u \nVdash Kp \to p$.

For strong factivity, suppose $sSx \to sCx$ fails. That is, we have a pair of points s, x where sSx and not sCx. Let $V(p)$ be the set of all points not compatible with x: $V(p) = \{u \mid \neg uCx\}$. This is upward closed from the conditions posed on C: if not uCx and $u \leq v$, then not vCx by (11). So, p is true at no point compatible with x, hence $\neg p$ holds at x. However, since s is not compatible with x, $s \in V(p)$, and hence Kp holds at x. Thus we have $x \Vdash \neg p \wedge Kp$ and since by the condition (9) there is some $u \in L$ such that $Ruxx$, strong factivity cannot hold in u. □

The following condition (implied by (19) using the fact $sSx \to sSs$) corresponds to the introspection axiom:

$$sSx \to (\exists y \geq s)(\exists t)(yStSx) \tag{26}$$

Theorem 4.2 *The class of relevant frames with S satisfying conditions (17) and (26) corresponds to $K\varphi \to KK\varphi$.*

Proof. Suppose (26) holds in a frame. Suppose $x \Vdash K\varphi$. Then there is sSx and $s \Vdash \varphi$. From (26) there is $y \geq s$, thus $y \Vdash \varphi$ from persistency. There is t such that $yStSx$. Thus $t \Vdash K\varphi$ and $x \Vdash KK\varphi$.

Suppose (26) doesn't hold in a frame. Then there is sSx and for no $y \geq s$ and no t we have $yStSx$. Define $V(p) = \{u \mid u \geq s\}$. Now $x \Vdash Kp$. No t such that tSx sees a p source via S (we have $\neg yStSx$ for each y), thus no such t satisfies Kp and x doesn't satisfy KKp. There must be a logical state u with $Ruxx$ and $u \nVdash Kp \to KKp$. □

5 Conclusion

Our aim was to provide an axiomatisation and completeness for relevant frames with an epistemic modality proposed in [4]. The motivation of the original operator allows for some natural generalisations and we provided an axiomatisation for the generalised operator (the logic of the class $\mathcal{F}_g$), while the axiomatisation for the original one (the logic of the class $\mathcal{F}_c$) is still an open problem. Moreover, we showed that the class of general frames is an intersection of the classes of frames satisfying factivity and the class of frames satisfying strong factivity. We also gave a modal characterisation of the class of epistemic relevant frames in which the introspection axiom holds.

There are several topics related to the subject we did not address here. We shall explore the proof theory of relevant epistemic modalities and define a display calculus for them.

Another thing we did not pay attention to is the modality I adjoint to K (i.e. $\vdash K\varphi \to \psi$ iff $\vdash \varphi \to I\psi$). It has a natural interpretation in our epistemic framework, it corresponds to what we can call *implicit knowledge* (a formula is implicit knowledge in a state iff it is true in all the states, for which the current state is a potential source.) We obtain $\varphi \to I\varphi$ as a theorem (everything explicitly waranted is implicitly known) and $\varphi \to IK\varphi$ (all that holds in a state is at least implicitly known there) as well as

$KI\varphi \to \varphi$ (nothing else can be known to be implicit then facts warranted in the state) are theorems.

Other modalities can of course be considered as e.g. another adjoint pair of natural semantical duals of diamond-like K and box-like I (i.e. a box-like modality acting backwards along S, and a diamond-like modality acting forward along S). To obtain persistent meanings of formulas we would have to add another half of condition (17). Such modalities do not have natural epistemic interpretation in our framework and as such are not a part of the current work, however we conjecture they are not definable from the old two and as such can increase the expressivity of the modal language. We conjecture we would be able e.g. to distinguish between the classes of classical frames, and the class of general frames using them, as well as extend our definability results.

Last we remark we could start with a logic weaker then R without loosing much from our original motivations. Obvious candidates would be logics with a weaker negation obtained, e.g., by weakening conditions for the compatibility relation. This would of course influence the properties the modality K. Another option is to give up the contraction or exchange (commutativity of fusion). A challenging option is to give up distributivity, but this would mean to use more complicated frame semantics.

References

[1] Cheng, J., *A Strong Relevant Logic Model of Epistemic Processes in Scientific Discovery*, in: E. Kawaguchi, H. Kangassalo, H. Jaakkola and I. Hamid, editors, *Information Modelling and Knowledge Bases XI*, IOS Press, 2000 pp. 136–159.

[2] Duc, H. N., "Resource-Bounded Reasoning about Knowledge," Ph.D. thesis, Faculty of Mathematics and Informatics, University of Leipzig (2001).

[3] Fagin, R., J. Halpern, Y. Moses and M. Vardi, "Reasoning about Knowledge," The MIT Press, 2003.

[4] Majer, O. and M. Peliš, *Epistemic Logic with Relevant Agents*, in: M. Peliš, editor, *The Logica Yearbook 2008*, College Publications, 2009 pp. 123–135.

[5] Mares, E. D., "Relevant Logic," Cambridge University Press, 2004.

[6] Paoli, F., "Substructural Logics: A Primer," Kluwer, 2002.

[7] Restall, G., *Negation in Relevant Logics: How I stopped worrying and learned to love the Routley Star*, in: D. Gabbay and H. Wansing, editors, *What is Negation?*, The Applied Logic Series **13**, Kluwer, 1999 pp. 53–76.

[8] Restall, G., "An Introduction to Substructural Logics," Routledge, 2000.

[9] Wansing, H., *Diamonds Are a Philosopher's Best Friends*, Journal of Philosophical Logic **31** (2002), pp. 591–612.

A Syntactic Realization Theorem for Justification Logics

Kai Brünnler, Remo Goetschi and Roman Kuznets

Institut für Informatik und angewandte Mathematik, Universität Bern
Neubrückstrasse 10, CH-3012 Bern, Switzerland

Abstract

Justification logics are refinements of modal logics where modalities are replaced by justification terms. They are connected to modal logics via so-called realization theorems. We present a syntactic proof of a single realization theorem that uniformly connects all the normal modal logics formed from the axioms d, t, b, 4, and 5 with their justification counterparts. The proof employs cut-free nested sequent systems together with Fitting's realization merging technique. We further strengthen the realization theorem for KB5 and S5 by showing that the positive introspection operator is superfluous.

Keywords: justification logic, modal logic, realization theorem, nested sequents, positive introspection

1 Introduction

Justification logic. The language of justification logic is a refinement of the language of modal logic. It replaces a single modality by a family of modalities, indexed by what are called *justification terms*. Given a modal formula such as $\Box A$, which can be read as *A is provable* or as *A is known*, a justification counterpart of this formula of the form $t : A$ can be read as *t is a proof of A* or as *A is known for reason t*.

The first justification logic, called the *Logic of Proofs* or LP, was introduced by Artemov [1,2] as a stepping stone for giving an arithmetical semantics for the modal logic S4. Justification logics are also interesting as epistemic logics. Justification terms have a structure and thus provide a measure of how hard it is to obtain knowledge of something. Because of that, justification logics avoid the well-known logical omniscience problem, as Artemov and Kuznets argue in [5].

The formal correspondence between S4 and LP is called a *realization theorem*. It has two directions. First, each provable formula of S4 can be turned into a provable formula of LP by realizing instances of modalities with justification terms. Second and vice versa, if all terms in a provable formula of LP are replaced with modalities, then the resulting modal formula is provable in S4.

Similar correspondences have been established for several other modal logics besides S4. An overview is given by Artemov in [3].

Methods for proving realization. There are two methods of establishing such correspondences: the syntactic method due to Artemov [1,2] and the semantic method due to Fitting [11]. The syntactic method makes use of cut-free Gentzen systems for modal logics, while the sematic method makes use of a Kripke-style semantics for justification logics. In contrast to the semantic method, the syntactic method is constructive. It provides an algorithm that, for each occurrence of a modality in a given provable modal formula, computes a justification term that realizes it.

The semantic method was used to prove several realization theorems: for S4, S5, K45, and KD45 [3,11,17]. Constructive realizations, via the syntactic method, are available for K, D, T, K4, D4, S4, and S5 [1,2,4,7,12,13]. In the case of S5, where no cut-free sequent system is available, two approaches have been used: first, a cut-free hypersequent system [4] and, second, an embedding of S5 into K45 [12]. This embedding also requires the use of a certain technique of *realization merging* developed by Fitting in [13]. However, neither approach applies to other modal logics that lack cut-free sequent systems, such as K5 and KB. The goal of this paper is to realize these logics and, in general, to provide a uniform constructive method of realizing all normal modal logics formed by the axioms d, t, b, 4, and 5.

Nested sequents. To that end, we use the cut-free proof systems given by Brünnler in [9], which are based on *nested sequents* and which capture all these modal logics. Nested sequents are essentially trees of sequents. They naturally generalize both sequents (which are nested sequents of depth zero) and hypersequents (which essentially are nested sequents of depth one). A crucial feature of these proof systems is *deep inference* [8,14], which is the ability to apply inference rules to formulas arbitrarily deep inside a nested sequent.

Outline. The paper is organized as follows. In Section 2 we introduce justification logics, in Section 3 we introduce nested sequent systems, and in Section 4 we recall Fitting's merging technique. We use them in Section 5 to prove our central result: the uniform realization theorem. In particular, this proves Pacuit's conjecture implicit in [16] that J5 is a justification counterpart of K5. It also creates justification counterparts for the modal logics D5, KB, DB, TB, and KB5, which, to our knowledge, did not have justification counterparts before. In Section 6 we go on to show that the operation of positive introspection is not necessary for the realization of KB5 and S5, which leads to new minimal realizations for them.

2 Justification Logic

Modal formulas. *Modal formulas* are given by the grammar

$$A ::= P_i \mid \neg P_i \mid (A \vee A) \mid (A \wedge A) \mid \Box A \mid \Diamond A \enspace ,$$

where i ranges over natural numbers, P_i denotes a *proposition*, and $\neg P_i$ denotes its *negation*. Negation of formulas is defined as usual by the De Morgan laws, with $\neg\neg P_i$ being P_i. Further, $A \to B$ denotes $\neg A \vee B$ and $\bot$ denotes $P_j \wedge \neg P_j$ for some fixed proposition P_j.

Justification formulas. *Justification terms*, or *terms* for short, are given by the grammar

$$t ::= c_i \mid x_i \mid (t \cdot t) \mid (t + t) \mid !t \mid ?t \quad .$$

The c_i are called *constants* and the x_i are called *variables*. The binary operators $\cdot$ and $+$ are called *application* and *sum* respectively. Application is left-associative. The unary operators ! and ? are called *positive introspection* (or *proof checker*) and *negative introspection* respectively. A sequence of n proof checker operators is denoted by $!^n$. Terms that do not contain variables are called *ground* and are denoted by p, p_1, p_2 and so on, whereas arbitrary terms are denoted by t, s, and q. We use the notation $t(x_1, \ldots, x_n)$ for terms that do not contain variables other than $x_1, \ldots, x_n$. *Justification formulas* are given by the grammar

$$A ::= P_i \mid \bot \mid (A \to A) \mid t : A \quad .$$

Negation, conjunction, and disjunction are defined as usual. Implication is right-associative and both conjunction and disjunction bind stronger than implication.

Axiom Systems. An axiom system for the modal logic K is assumed to be given. *Extensions* of system K are obtained by adding modal axioms from Figure 2 as described in Figure 3. The axiom system for the basic justification logic J consists of the axioms and rules given in Figure 1. The AN!-rule is called *axiom necessitation with embedded positive introspection*. *Extensions* of system J are obtained by adding justification axioms from Figure 2 as described in Figure 3. The justification axioms are mostly standard, except for jb, which is new. Observe that our choice of the jb-axiom does not increase the set of operations on terms but uses the well-known negative introspection operation. In Section 6 we will see that this is a natural choice. The reason why the zero-premise AN!-rule is defined as a rule and not as an axiom is to prevent it from referring to itself. We will often use the name of an axiom system to also denote its *logic*, which is its set of provable formulas.

From this point on by a *justification logic* we mean (the logic of) either system J or one of its extensions. Likewise, by a *modal logic* we mean either system K or one of its extensions. Each justification logic has a *corresponding* modal logic, and vice versa, as shown in Figure 3, with J corresponding to the modal logic K.

Remark 2.1 *Traditionally, the axiomatizations of justification logics that contain the* j4*-axiom had the following* axiom necessitation *rule, which is a simpler variant of the* AN!*-rule:*

$$\mathsf{AN}\ \frac{A \textit{ is an axiom instance}}{c_i : A} \quad .$$

Since in these systems the AN!*-rule is derivable, our axiomatizations produce the same logics.*

taut: A fixed complete set of propositional axioms

app: $s:(A \to B) \to t:A \to (s \cdot t):B$

sum: $s:A \to (s+t):A$ and $s:A \to (t+s):A$

$$\text{MP}\ \frac{A \quad A \to B}{B} \qquad \text{AN!}\ \frac{A \text{ is an axiom instance}}{!^n c_i : !^{n-1} c_i : \ldots : !!c_i : !c_i : c_i : A}$$

Fig. 1. The axiom system for the basic justification logic J

d: $\Box\bot \to \bot$ | t: $\Box A \to A$ | b: $A \to \Box\neg\Box\neg A$

jd: $t:\bot \to \bot$ | jt: $t:A \to A$ | jb: $A \to ?t:(\neg t:\neg A)$

4: $\Box A \to \Box\Box A$ | 5: $\neg\Box A \to \Box\neg\Box A$

j4: $t:A \to !t:t:A$ | j5: $\neg t:A \to ?t:(\neg t:A)$

Fig. 2. Modal axioms and their corresponding justification axioms

D	T	KB	K4	K5	DB	D4	D5	TB	K45	S4	KB5	D45	S5
d	t	b	4	5	d, b	d, 4	d, 5	t, b	4, 5	t, 4	b, 4, 5	d, 4, 5	t, 4, 5
JD	JT	JB	J4	J5	JDB	JD4	JD5	JTB	J45	LP	JB45	JD45	JT45
jd	jt	jb	j4	j5	jd, jb	jd, j4	jd, j5	jt, jb	j4, j5	jt, j4	jb, j4, j5	jd, j4, j5	jt, j4, j5

Fig. 3. Axiom systems of modal logic and of justification logic

Clearly, we can turn justification formulas into modal formulas by replacing terms with boxes, which is made formal in the next definition.

Definition 2.2 (Forgetful projection) *Given a justification formula A, its* forgetful projection *A° is defined as: $P_i^\circ := P_i$, $\bot^\circ := \bot$, $(A \to B)^\circ := A^\circ \to B^\circ$, and $(t{:}A)^\circ := \Box A^\circ$. The forgetful projection of a set of justification formulas is defined in the obvious way.*

An important fact about justification logics is that they can internalize their own proofs, i.e. if A is provable, then so is $t:A$ for some term t. This is formally stated in the lemma below, originally proved by Artemov [2] for LP. A proof for most of our justification logics can be found in [15]; the remaining cases are similar.

Lemma 2.3 (Internalization) *For any justification logic* JL*, if*

$$\mathsf{JL} \vdash A_1 \to \ldots \to A_n \to B \ ,$$

then there exists a term $t(x_1, \ldots, x_n)$ such that for all terms $s_1, \ldots, s_n$

$$\mathsf{JL} \vdash s_1 : A_1 \to \ldots \to s_n : A_n \to t(s_1, \ldots, s_n) : B \ .$$

Note that t is ground if $n = 0$.

3 The Nested Sequent Calculus

Nested sequents. *Nested sequents*, or *sequents* for short, are inductively defined as

$$\text{id}\,\frac{}{\Gamma\{P_i, \neg P_i\}} \qquad \vee\,\frac{\Gamma\{A, B\}}{\Gamma\{A \vee B\}} \qquad \wedge\,\frac{\Gamma\{A\} \quad \Gamma\{B\}}{\Gamma\{A \wedge B\}}$$

$$\text{ctr}\,\frac{\Gamma\{A, A\}}{\Gamma\{A\}} \qquad \text{exch}\,\frac{\Gamma\{\Delta, \Sigma\}}{\Gamma\{\Sigma, \Delta\}} \qquad \Box\,\frac{\Gamma\{[A]\}}{\Gamma\{\Box A\}} \qquad \text{k}\,\frac{\Gamma\{[A, \Delta]\}}{\Gamma\{\Diamond A, [\Delta]\}}$$

$$\text{d}\,\frac{\Gamma\{[A]\}}{\Gamma\{\Diamond A\}} \qquad \text{t}\,\frac{\Gamma\{A\}}{\Gamma\{\Diamond A\}} \qquad \text{b}\,\frac{\Gamma\{[\Delta], A\}}{\Gamma\{[\Delta, \Diamond A]\}} \qquad 4\,\frac{\Gamma\{[\Diamond A, \Delta]\}}{\Gamma\{\Diamond A, [\Delta]\}}$$

$$\text{5a}\,\frac{\Gamma\{[\Delta], \Diamond A\}}{\Gamma\{[\Delta, \Diamond A]\}} \qquad \text{5b}\,\frac{\Gamma\{[\Delta], [\Pi, \Diamond A]\}}{\Gamma\{[\Delta, \Diamond A], [\Pi]\}} \qquad \text{5c}\,\frac{\Gamma\{[\Delta, [\Pi, \Diamond A]]\}}{\Gamma\{[\Delta, \Diamond A, [\Pi]]\}}$$

Fig. 4. Rules of the nested sequent calculus

follows: the empty sequence $\varnothing$ is a nested sequent; if Σ and Δ are nested sequents and A is a formula, then Σ, A and $\Sigma, [\Delta]$ are nested sequents, where the comma denotes concatenation of sequences. The brackets in the expression $[\Delta]$ are called *structural box*. The *corresponding formula* of a sequent Γ, denoted $\underline{\Gamma}$, is inductively defined by $\underline{\varnothing} := \bot$, $\underline{\Sigma, A} := \underline{\Sigma} \vee A$, and $\underline{\Sigma, [\Delta]} := \underline{\Sigma} \vee \Box \underline{\Delta}$. For simplicity we often do not explicitly distinguish between a sequent and its corresponding formula. We use the letters Γ, Δ, Λ, Ω, Π, and Σ to denote sequents.

Sequent contexts. A *sequent context*, or *context* for short, is a sequent with (exactly) one occurrence of the symbol $\{\ \}$, called a *hole*, which does not occur inside formulas. Contexts are denoted by $\Gamma\{\ \}$. An inductive definition can be given as follows: $\{\ \}$ is a context and if $\Sigma\{\ \}$ is a context, then so are $[\Sigma\{\ \}]$ and $\Delta, \Sigma\{\ \}, \Pi$, where Δ and Π are sequents. The sequent $\Gamma\{\Delta\}$ is obtained by replacing the hole in $\Gamma\{\ \}$ with Δ. For example, if $\Gamma\{\ \} = A, [[B], \{\ \}]$ and $\Delta = C, [D]$, then $\Gamma\{\Delta\} = A, [[B], C, [D]]$.

Sequent systems. Consider the inference rules in Figure 4. *System* SK consists of the rules id, $\vee$, $\wedge$, ctr, exch, $\Box$, and k. *Extensions of system* SK are obtained by adding further rules from Figure 4 according to Figure 3, where 5 means that all three rules 5a, 5b, and 5c are added. Note that a name in the first row of Figure 3 now denotes both a (Hilbert-style) axiom system and a sequent system.

These sequent systems are essentially the same as the ones in [9], where their completeness is proved, so we have the following theorem.

Theorem 3.1 (Completeness) *System* SK *and its extensions are sound and complete with respect to their corresponding modal logics (as defined by the corresponding axiom systems).*

4 Annotations and Realizations

Our goal is to turn provable formulas of a given modal logic into provable formulas of the corresponding justification logic by replacing boxes with terms and diamonds with variables. In order to do so we use *annotations*, which are indices on modalities.

Annotations have no semantical meaning but allow us to keep track of occurrences of modal operators. We adopt Fitting's notation from [13].

Definition 4.1 (Annotations) Annotated modal formulas, *or* annotated formulas *for short, are built according to the grammar*

$$A ::= P_i \mid \neg P_i \mid (A \vee A) \mid (A \wedge A) \mid \Box_{2k+1} A \mid \Diamond_{2l} A \ ,$$

where i, k, and l range over natural numbers. An annotated sequent (context) *is a sequent (context) in which only annotated formulas occur and all structural boxes are annotated by odd indices. The* corresponding annotated formula *of an annotated sequent Γ is defined in the obvious way, with $\underline{\Sigma, [\Delta]_k} := \underline{\Sigma} \vee \Box_k \underline{\Delta}$.*

If A is a modal formula that is obtained from an annotated formula A' by dropping all indices on its modalities, then we call A' an annotated version *of A, and likewise for sequents. An annotated formula or sequent is called* properly annotated *if no index occurs twice in it. From now on we will always assume that an annotated formula or sequent is properly annotated, unless stated otherwise.*

Remark 4.2 *Since our modal formulas are in negation normal form, in contrast to [13] every subformula of a properly annotated formula is itself properly annotated.*

Definition 4.3 (Annotated rule instance) *An* annotated rule instance *is any instance of a rule in Figure 5 provided that its conclusion and each of its premises are properly annotated sequents and, in case of the* ctr*-rule, additionally A_1, A_2, and A_3 do not share indices and are annotated versions of the same modal formula. An* annotated proof *is built as usual from annotated rule instances.*

Remark 4.4 *Note that we do not define the negation of an annotated formula. The obvious definition, where $\neg\Box_k A$ is $\Diamond_k \neg A$, does not work because it does not produce an annotated formula. In particular, this prevents us from even formulating a cut-rule for annotated sequents.*

Lemma 4.5 (Annotating Proofs) *For each sequent calculus proof $\mathcal{P}$ there exists an annotated proof $\mathcal{P}'$ that is an* annotated version *of $\mathcal{P}$, meaning that $\mathcal{P}$ can be obtained from $\mathcal{P}'$ by dropping all annotations.*

Proof. We take $\mathcal{P}$, replace the endsequent with a properly annotated version of it, and straightforwardly propagate the annotations upwards. □

Now we can define realizations as functions from natural numbers to terms, with the restriction that even numbers are mapped to variables. This restriction is often called the *normality condition.*

Definition 4.6 (Realization function) *A* realization function *r is a partial mapping from natural numbers to terms such that if $r(2i)$ is defined, then $r(2i) = x_i$. A* realization function on *a given annotated formula (sequent) is one that is defined on all indices of that formula (sequent).*

Definition 4.7 (Realization) *If A is an annotated formula and r is a partial mapping*

$$\text{id}\ \frac{}{\Gamma\{P_i, \neg P_i\}} \qquad \vee\ \frac{\Gamma\{A, B\}}{\Gamma\{A \vee B\}} \qquad \wedge\ \frac{\Gamma\{A\} \quad \Gamma\{B\}}{\Gamma\{A \wedge B\}}$$

$$\text{ctr}\ \frac{\Gamma\{A_1, A_2\}}{\Gamma\{A_3\}} \qquad \text{exch}\ \frac{\Gamma\{\Delta, \Sigma\}}{\Gamma\{\Sigma, \Delta\}} \qquad \Box\ \frac{\Gamma\{[A]_k\}}{\Gamma\{\Box_k A\}} \qquad \text{k}\ \frac{\Gamma\{[A, \Delta]_k\}}{\Gamma\{\Diamond_{2m} A, [\Delta]_i\}}$$

$$\text{d}\ \frac{\Gamma\{[A]_k\}}{\Gamma\{\Diamond_{2m} A\}} \qquad \text{t}\ \frac{\Gamma\{A\}}{\Gamma\{\Diamond_{2m} A\}} \qquad \text{b}\ \frac{\Gamma\{[\Delta]_k, A\}}{\Gamma\{[\Delta, \Diamond_{2m} A]_i\}} \qquad \text{4}\ \frac{\Gamma\{[\Diamond_{2m} A, \Delta]_k\}}{\Gamma\{\Diamond_{2m} A, [\Delta]_i\}}$$

$$\text{5a}\ \frac{\Gamma\{[\Delta]_k, \Diamond_{2m} A\}}{\Gamma\{[\Delta, \Diamond_{2m} A]_i\}} \qquad \text{5b}\ \frac{\Gamma\{[\Delta]_k, [\Pi, \Diamond_{2m} A]_i\}}{\Gamma\{[\Delta, \Diamond_{2m} A]_l, [\Pi]_j\}} \qquad \text{5c}\ \frac{\Gamma\{[\Delta, [\Pi, \Diamond_{2m} A]_i]_k\}}{\Gamma\{[\Delta, \Diamond_{2m} A, [\Pi]_j]_l\}}$$

Fig. 5. Annotated rules of the nested sequent calculus

$$(P_i)^r := P_i \qquad (A \vee B)^r := A^r \vee B^r \qquad (\Diamond_{2l} A)^r := \neg r(2l) : \neg A^r$$

$$(\neg P_i)^r := \neg P_i \qquad (A \wedge B)^r := A^r \wedge B^r \qquad (\Box_{2k+1} A)^r := r(2k+1) : A^r$$

Fig. 6. Realization of a formula

from natural numbers to terms (not necessarily a realization function) that is defined on all indices of A, then the justification formula A^r is inductively defined as in Figure 6. Note that if r is a realization function, then $(\Diamond_{2l} A)^r = \neg x_l : \neg A^r$. Given an annotated sequent Γ, we define Γ^r as $(\underline{\Gamma})^r$.

We introduce some notation for stating restrictions on realization functions.

Definition 4.8 (diavars(A), $r \restriction A$) *Given an annotated formula A, we define*

$$\begin{aligned} \text{diavars}(A) &:= \{x_k \mid \Diamond_{2k} \textit{ occurs in } A\} \\ r \restriction A &:= r \restriction \{i \mid \Box_i \textit{ or } \Diamond_i \textit{ occurs in } A\} \end{aligned} \quad ,$$

where $f \restriction S$ is the restriction of the partial function f to the set S.

The next definition is mostly standard, see, e.g., Baader and Nipkow [6].

Definition 4.9 (Substitution) *A* substitution, *denoted by σ, is a total mapping from variables to terms. If σ is a substitution, then $\widetilde{\sigma}$ is the function that maps terms to terms and formulas to formulas by simultaneously replacing each occurrence of a variable x with the term $\sigma(x)$. The* domain of σ *is* dom(σ) $:= \{x \mid \sigma(x) \neq x\}$, *the* range of σ *is* range(σ) $:= \{\sigma(x) \mid x \in \text{dom}(\sigma)\}$, *and the* variable range of σ, *denoted by* vrange(σ), *is the set of variables that occur in terms in* range(σ). *We write $t\sigma$ and $A\sigma$ to denote $\widetilde{\sigma}(t)$ and $\widetilde{\sigma}(A)$ respectively. We also write $\sigma \circ r$ for $\widetilde{\sigma} \circ r$, where function composition is as usual, namely $(f_2 \circ f_1)(n) := f_2(f_1(n))$.*

The following lemma is standard, see, e.g., [15].

Lemma 4.10 (Substitution) *If $\mathsf{JL} \vdash A$ for a justification logic JL, then*

(i) $\mathsf{JL} \vdash A\sigma$ *for any substitution* σ *and*

(ii) $\mathsf{JL} \vdash A[P_i \mapsto B]$, *where* $A[P_i \mapsto B]$ *is the result of simultaneously replacing each occurrence of the proposition* P_i *in* A *with the formula* B.

The following immediate facts are used in many of the proofs that follow.

Lemma 4.11 (Facts about Substitutions and Realization Functions)

(i) $\sigma \circ r$ *is a realization function* *iff* $x_n \notin \mathrm{dom}(\sigma)$ *whenever* $r(2n)$ *is defined.*

(ii) $A^r\sigma = A^{\sigma \circ r}$.

(iii) *If* $\mathrm{dom}(r_1) \cap \mathrm{dom}(r_2) \subseteq \{n \mid n \textit{ is even}\}$, *then* $r_1 \cup r_2$ *is a realization function.*

(iv) *If* $\mathrm{dom}(\sigma_1) \cap \mathrm{dom}(\sigma_2) = \emptyset$, *then* $\sigma_1 \cup \sigma_2$ *is a substitution.*

(v) *If* $\sigma \circ r$ *is a realization function, then* $\mathrm{dom}(\sigma \circ r) = \mathrm{dom}(r)$.

(vi) *If* $r_1 \cup r_2$ *is a realization function, then* $\mathrm{dom}(r_1 \cup r_2) = \mathrm{dom}(r_1) \cup \mathrm{dom}(r_2)$.

(vii) *If* $\sigma_1 \cup \sigma_2$ *is a substitution, then* $\mathrm{dom}(\sigma_1 \cup \sigma_2) = \mathrm{dom}(\sigma_1) \cup \mathrm{dom}(\sigma_2)$.

(viii) $\mathrm{dom}(\sigma_2 \circ \sigma_1) \subseteq \mathrm{dom}(\sigma_1) \cup \mathrm{dom}(\sigma_2)$.

The proof of our main result, the realization theorem in the next section, is by induction on the depth of a given proof. For branching rules, we need to merge realizations. The following theorem allows us to do that. It is essentially Theorem 8.2 in Fitting [13]. There it is formulated for LP but the proof only makes use of the operations $+$ and $\cdot$ and the Internalization Lemma. Hence, the theorem also holds for all justification logics we consider.

Theorem 4.12 (Realization Merging) *Let* JL *be a justification logic,* A *be a properly annotated formula, and* r_1 *and* r_2 *be realization functions on* A. *Then there exists a realization function* r *on* A *and a substitution* σ *such that: 1) for any* x *the term* $\sigma(x)$ *contains no variables other than* x, *2)* $\mathrm{dom}(\sigma) \subseteq \mathrm{diavars}(A)$,

$$3)\quad \mathsf{JL} \vdash A^{r_1}\sigma \to A^r \ , \qquad \text{and} \qquad 4)\quad \mathsf{JL} \vdash A^{r_2}\sigma \to A^r \ .$$

(Note that it is not assumed that A^{r_1} *or* A^{r_2} *is provable.)*

The next lemma easily follows from the associativity of Boolean disjunction. It is needed because in general the formula $\underline{\Gamma, \Sigma}$ does not coincide with the formula $\underline{\Gamma} \vee \underline{\Sigma}$.

Lemma 4.13 (Associativity of Disjunction) *For any sequents* Γ *and* Σ *and for any realization function* r, *we have* $\mathsf{J} \vdash \underline{(\Sigma, \Gamma)}^r \leftrightarrow \underline{\Sigma}^r \vee \underline{\Gamma}^r$.

5 The Realization Theorem

We now prove the realization theorem. The argument is by induction on the depth of a given annotated proof. The following lemmas mostly correspond to the inductive cases for the various sequent calculus rules.

Lemma 5.1 (id-rule) *Given an annotated* id*-instance as in Figure 5, there exists a realization function* r *on its conclusion* Ω *such that* $\mathsf{J} \vdash \Omega^r$.

Proof. By induction on the structure of $\Gamma\{\,\}$.
Base case $\Gamma\{\,\} = \{\,\}$. The empty realization function suffices.
Induction step. By induction hypothesis, there exists a realization function r' on $\Sigma\{P_i, \neg P_i\}$ such that $\mathsf{J} \vdash \Sigma\{P_i, \neg P_i\}^{r'}$.
Case $\Gamma\{\,\} = [\Sigma\{\,\}]_k$. By the Internalization Lemma there exists a ground term p such that $\mathsf{J} \vdash p : \Sigma\{P_i, \neg P_i\}^{r'}$. Since the conclusion $\Omega = [\Sigma\{P_i, \neg P_i\}]_k$ is properly annotated, $r := (r' \restriction \Sigma\{P_i, \neg P_i\}) \cup \{(k, p)\}$ is a realization function on $[\Sigma\{P_i, \neg P_i\}]_k$ by Lemma 4.11. It follows that $\mathsf{J} \vdash ([\Sigma\{P_i, \neg P_i\}]_k)^r$.
Case $\Gamma\{\} = \Delta, \Sigma\{\}, \Pi$. Let r be a realization function on $\Delta, \Sigma\{P_i, \neg P_i\}, \Pi$ that extends $r' \restriction \Sigma\{P_i, \neg P_i\}$. Then $\mathsf{J} \vdash \Delta^r \vee \Sigma\{P_i, \neg P_i\}^r \vee \Pi^r$ follows by propositional reasoning. Therefore, by Lemma 4.13, $\mathsf{J} \vdash \left(\Delta, \Sigma\{P_i, \neg P_i\}, \Pi\right)^r$. □

Lemma 5.2 ($\wedge$-rule) *Given an annotated $\wedge$-instance as in Figure 5, let r_1 and r_2 be realization functions on its premises Λ_1 and Λ_2 respectively. Then there exists a substitution σ with* $\mathrm{dom}(\sigma) \subseteq \mathrm{diavars}(\Lambda_1) \cup \mathrm{diavars}(\Lambda_2) = \mathrm{diavars}(\Omega)$ *and a realization function r on the conclusion Ω such that* $\mathsf{J} \vdash (\Lambda_1)^{r_1}\sigma \to (\Lambda_2)^{r_2}\sigma \to \Omega^r$.

Proof. By induction on the structure of $\Gamma\{\,\}$.
Base case $\Gamma\{\,\} = \{\,\}$. Let $r := (r_1 \restriction A) \cup (r_2 \restriction B)$ and let σ be the identity substitution. The former is a realization function by Lemma 4.11 because $\Omega = A \wedge B$ is properly annotated. Thus, $A^{r_1} \wedge B^{r_2} = (A \wedge B)^r$ and $\mathsf{J} \vdash A^{r_1}\sigma \to B^{r_2}\sigma \to (A \wedge B)^r$ because it is a propositional tautology.
Induction step. By induction hypothesis there exists a substitution σ' with $\mathrm{dom}(\sigma') \subseteq \mathrm{diavars}(\Sigma\{A \wedge B\})$ and a realization function r' on $\Sigma\{A \wedge B\}$ such that

$$\mathsf{J} \vdash \Sigma\{A\}^{r_1}\sigma' \to \Sigma\{B\}^{r_2}\sigma' \to \Sigma\{A \wedge B\}^{r'} \quad . \tag{1}$$

Case $\Gamma\{\,\} = \left[\Sigma\{\,\}\right]_k$. By the Internalization Lemma,

$$\mathsf{J} \vdash r_1(k)\sigma' : (\Sigma\{A\}^{r_1}\sigma') \to r_2(k)\sigma' : (\Sigma\{B\}^{r_2}\sigma') \to\ t\big(r_1(k)\sigma', r_2(k)\sigma'\big) : \Sigma\{A \wedge B\}^{r'}$$

for some term $t(x, y)$. In other words,

$$\mathsf{J} \vdash \left([\Sigma\{A\}]_k\right)^{r_1}\sigma' \to \left([\Sigma\{B\}]_k\right)^{r_2}\sigma' \to \left([\Sigma\{A \wedge B\}]_k\right)^r$$

for $r := (r' \restriction \Sigma\{A \wedge B\}) \cup \{(k, t(r_1(k)\sigma', r_2(k)\sigma'))\}$, which by Lemma 4.11 is a realization function on the properly annotated sequent $\Omega = [\Sigma\{A \wedge B\}]_k$.
Case $\Gamma\{\,\} = \Delta, \Sigma\{\,\}, \Pi$. Since $\Omega = \Delta, \Sigma\{A \wedge B\}, \Pi$ is properly annotated, $\Sigma\{A \wedge B\}$ shares no indices with Δ, Π. Thus, by Lemma 4.11, both $\sigma' \circ (r_1 \restriction \Delta, \Pi)$ and $\sigma' \circ (r_2 \restriction \Delta, \Pi)$ are realization functions on Δ, Π. By Theorem 4.12 (Realization Merging) there exists a realization function r_m on Δ, Π and a substitution σ_m with $\mathrm{dom}(\sigma_m) \subseteq \mathrm{diavars}(\Delta, \Pi)$ such that

$$\mathsf{J} \vdash (\Delta, \Pi)^{\sigma' \circ (r_1 \restriction \Delta, \Pi)}\sigma_m \to (\Delta, \Pi)^{r_m} \quad , \tag{2}$$

$$\mathsf{J} \vdash (\Delta, \Pi)^{\sigma' \circ (r_2 \restriction \Delta, \Pi)}\sigma_m \to (\Delta, \Pi)^{r_m} \quad , \tag{3}$$

and x is the only variable in $\sigma_m(x)$, for any x. By Lemma 4.11, $(\Delta,\Pi)^{\sigma' \circ (r_1 \restriction \Delta,\Pi)}\sigma_m$ is $(\Delta,\Pi)^{r_1}\sigma'\sigma_m$ and $(\Delta,\Pi)^{\sigma' \circ (r_2 \restriction \Delta,\Pi)}\sigma_m$ is $(\Delta,\Pi)^{r_2}\sigma'\sigma_m$. Therefore, (2) and (3) are identical to

$$\mathsf{J} \vdash (\Delta,\Pi)^{r_1}\sigma'\sigma_m \to (\Delta,\Pi)^{r_m} \ , \tag{4}$$

$$\mathsf{J} \vdash (\Delta,\Pi)^{r_2}\sigma'\sigma_m \to (\Delta,\Pi)^{r_m} \ . \tag{5}$$

From (1) by Lemma 4.10 (Substitution) it follows that

$$\mathsf{J} \vdash \Sigma\{A\}^{r_1}\sigma'\sigma_m \to \Sigma\{B\}^{r_2}\sigma'\sigma_m \to \Sigma\{A \wedge B\}^{r'}\sigma_m \ .$$

From this, (4), and (5) it follows by propositional reasoning that

$$\mathsf{J} \vdash \Sigma\{A\}^{r_1}\sigma'\sigma_m \vee (\Delta,\Pi)^{r_1}\sigma'\sigma_m \ \to \ \Sigma\{B\}^{r_2}\sigma'\sigma_m \vee (\Delta,\Pi)^{r_2}\sigma'\sigma_m \\ \to \ \Sigma\{A \wedge B\}^{r'}\sigma_m \vee (\Delta,\Pi)^{r_m} \ . \tag{6}$$

Since $\mathrm{dom}(\sigma_m) \subseteq \mathrm{diavars}(\Delta,\Pi)$, it follows by Lemma 4.11 that $\sigma_m \circ (r' \restriction \Sigma\{A \wedge B\})$ is a realization function on $\Sigma\{A \wedge B\}$. Again by Lemma 4.11, we conclude that

$$r := (\sigma_m \circ (r' \restriction \Sigma\{A \wedge B\})) \cup (r_m \restriction \Delta,\Pi)$$

is a realization function on $\Delta,\Sigma\{A \wedge B\},\Pi$. And since

$$\Sigma\{A \wedge B\}^{r' \restriction \Sigma\{A \wedge B\}}\sigma_m = \Sigma\{A \wedge B\}^{\sigma_m \circ (r' \restriction \Sigma\{A \wedge B\})}$$

by Lemma 4.11, we can rewrite (6) as

$$\mathsf{J} \vdash \left(\Sigma\{A\} \vee (\Delta,\Pi)\right)^{r_1}\sigma \ \to \ \left(\Sigma\{B\} \vee (\Delta,\Pi)\right)^{r_2}\sigma \ \to \ \left(\Sigma\{A \wedge B\} \vee (\Delta,\Pi)\right)^{r} \tag{7}$$

for $\sigma := \sigma_m \circ \sigma'$ with $\mathrm{dom}(\sigma) \subseteq \mathrm{dom}(\sigma') \cup \mathrm{dom}(\sigma_m) \subseteq \mathrm{diavars}(\Delta,\Sigma\{A \wedge B\},\Pi)$. Finally, (7) is by Lemma 4.13 propositionally equivalent to

$$\mathsf{J} \vdash (\Delta,\Sigma\{A\},\Pi)^{r_1}\sigma \ \to \ (\Delta,\Sigma\{B\},\Pi)^{r_2}\sigma \ \to \ (\Delta,\Sigma\{A \wedge B\},\Pi)^{r} \ .$$

□

The proof of the following lemma is in Appendix A.

Lemma 5.3 **(ctr-rule)** *Given an annotated* ctr*-instance as in Figure 5, let r_1 be a realization function on its premise Λ. Then there exists 1) a realization function r on its conclusion Ω and 2) a substitution σ with* $\mathrm{dom}(\sigma) \subseteq \mathrm{diavars}(\Lambda)$ *such that* $\mathsf{J} \vdash \Lambda^{r_1}\sigma \to \Omega^{r}$.

Lemma 5.4 **($\vee$- and exch-rule)** *Given an annotated ρ-instance with $\rho \in \{\vee, \mathsf{exch}\}$ as in Figure 5, let r_1 be a realization function on its premise Λ. Then there exists a realization function r on its conclusion Ω such that* $\mathsf{J} \vdash \Lambda^{r_1} \to \Omega^{r}$.

Proof. By induction on the structure of $\Gamma\{\,\}$.
Base case $\Gamma\{\,\} = \{\,\}$. It suffices to take $r := r_1 \restriction \Omega$ for either rule. Indeed, we have $\underline{A, B} = A \vee B = \underline{A \vee B}$ for $\rho = \vee$. For $\rho = \mathsf{exch}$, the desired statement follows from Lemma 4.13.
Induction step. The arguments are the same as in the proof of Lemma 5.3, given in Appendix A, except that here the substitution is the identity substitution. □

Lemma 5.5 (k-rule) *Given an annotated* k*-instance as in Figure 5, let r_1 be a realization function on its premise Λ. Then there exists a realization function r on its conclusion Ω such that $\mathsf{J} \vdash \Lambda^{r_1} \to \Omega^r$.*

Proof. By induction on the structure of $\Gamma\{\,\}$.
Base case $\Gamma\{\,\} = \{\,\}$. For the propositional tautology $(A, \Delta)^{r_1} \to \neg A^{r_1} \to \Delta^{r_1}$, by the Internalization Lemma, $\mathsf{J} \vdash r_1(k) : (A, \Delta)^{r_1} \to x_m : \neg A^{r_1} \to t(r_1(k), x_m) : \Delta^{r_1}$ for some term $t(x, y)$. It follows by propositional reasoning that

$$\mathsf{J} \vdash r_1(k) : (A, \Delta)^{r_1} \to \neg x_m : \neg A^{r_1} \vee t(r_1(k), x_m) : \Delta^{r_1} \,, \quad \text{which is}$$
$$\mathsf{J} \vdash \quad \big([A, \Delta]_k\big)^{r_1} \to (\Diamond_{2m} A)^{r_1} \vee t(r_1(k), x_m) : \Delta^{r_1} \,.$$

For $r := (r_1 \restriction A, \Delta) \cup \{(i, t(r_1(k), x_m)), (2m, x_m)\}$ this is identical to

$$\mathsf{J} \vdash \big([A, \Delta]_k\big)^{r_1} \to (\Diamond_{2m} A, [\Delta]_i)^r \,.$$

Induction step. The arguments are the same as in Lemma 5.4. □

In order to realize the modal rules 5b and 5c, we will use realizations of theorems $\Box(\Box A \to A)$ and $\Box(\neg\Box\Box A \to \neg\Box A)$ of K5. They are provided by the following two auxiliary lemmas. We have to omit the proofs for space reasons.

Lemma 5.6 (Internalized Factivity) *There is a term $t(x)$ such that for any term s and any formula A we have that $\mathsf{J5} \vdash t(s) : (s : A \to A)$.*

Lemma 5.7 (Internalized Positive Introspection) *There are terms $t_1(x)$ and $t_2(x)$ such that $\mathsf{J5} \vdash t_1(t) : (\neg t_2(t) : t : A \to \neg t : A)$ for any term t and any formula A.*

The following lemma covers the remaining rules.

Lemma 5.8 (Modal Rules) *Given an annotated ρ-instance with $\rho \in \{$d, t, b, 4, 5a, 5b, 5c$\}$ as given in Figure 5, let r_1 be a realization function on its premise Λ. Then there is a realization function r on its conclusion Ω such that $\mathsf{J}\rho \vdash \Lambda^{r_1} \to \Omega^r$, where by* Jd *we mean* JD*, and so on, except for $\rho \in \{$5a, 5b, 5c$\}$ where we mean* J5.

Proof. By induction on the structure of $\Gamma\{\,\}$.
Base case $\Gamma\{\,\} = \{\,\}$. We need to consider each rule ρ in turn.
Subcases $\rho = \mathsf{d}, \mathsf{t}, 4$. The proof is similar to the k-rule and is omitted for space reasons.

Subcase $\rho = \mathsf{b}$. Since $\Delta^{r_1} \to \Delta^{r_1} \vee \neg x_m : \neg A^{r_1}$ is a propositional tautology, by the Internalization Lemma there exists a term $t_1(y)$ such that

$$\mathsf{JB} \vdash r_1(k) : \Delta^{r_1} \to t_1(r_1(k)) : (\Delta^{r_1} \vee \neg x_m : \neg A^{r_1}) \ . \tag{8}$$

Similarly, for a propositional tautology $\neg x_m : \neg A^{r_1} \to \Delta^{r_1} \vee \neg x_m : \neg A^{r_1}$, there exists a term $t_2(x)$ such that

$$\mathsf{JB} \vdash ?\, x_m : \neg x_m : \neg A^{r_1} \to t_2(?\, x_m) : (\Delta^{r_1} \vee \neg x_m : \neg A^{r_1}) \ . \tag{9}$$

It follows from (8) and (9) by axiom sum and propositional reasoning that

$$\mathsf{JB} \vdash r_1(k) : \Delta^{r_1} \vee ?\, x_m : \neg x_m : \neg A^{r_1} \to t : (\Delta^{r_1} \vee \neg x_m : \neg A^{r_1})$$

for $t := t_1(r_1(k)) + t_2(?\, x_m)$. Finally, from the instance $A^{r_1} \to ?\, x_m : \neg x_m : \neg A^{r_1}$ of axiom jb it follows that $\mathsf{JB} \vdash r_1(k) : \Delta^{r_1} \vee A^{r_1} \to t : (\Delta^{r_1} \vee \neg x_m : \neg A^{r_1})$. Hence the desired realization function is $r := (r_1 \upharpoonright \Delta, A) \cup \{(i, t), (2m, x_m)\}$.

Subcases $\rho = \mathsf{5a}, \mathsf{5c}$. The proof can be found in Appendix B.

Subcase $\rho = \mathsf{5b}$. By Lemma 5.7 there exist terms $t_1(x)$ and $t_2(x)$ that satisfy the condition $\mathsf{J5} \vdash t_1(x_m) : (\neg t_2(x_m) : x_m : \neg A^{r_1} \to \neg x_m : \neg A^{r_1})$. Thus, by app and MP,

$$\mathsf{J5} \vdash ?\, t_2(x_m) : \neg t_2(x_m) : x_m : \neg A^{r_1} \to (t_1(x_m) \cdot ?\, t_2(x_m)) : \neg x_m : \neg A^{r_1} \ .$$

From the instance $\neg t_2(x_m) : x_m : \neg A^{r_1} \to ?\, t_2(x_m) : \neg t_2(x_m) : x_m : \neg A^{r_1}$ of $\mathsf{j5}$ it follows:

$$\mathsf{J5} \vdash \neg t_2(x_m) : x_m : \neg A^{r_1} \to (t_1(x_m) \cdot ?\, t_2(x_m)) : \neg x_m : \neg A^{r_1} \ . \tag{10}$$

By a propositional tautology and the Internalization Lemma applied to it,

$$\mathsf{J5} \vdash p_1 : (x_m : \neg A^{r_1} \to \Pi^{r_1} \vee \neg x_m : \neg A^{r_1} \to \Pi^{r_1})$$

for some ground term p_1. Thus, by app and MP,

$$\mathsf{J5} \vdash t_2(x_m) : x_m : \neg A^{r_1} \to (p_1 \cdot t_2(x_m)) : (\Pi^{r_1} \vee \neg x_m : \neg A^{r_1} \to \Pi^{r_1}) \ .$$

Again by app and propositional reasoning,

$$\mathsf{J5} \vdash t_2(x_m) : x_m : \neg A^{r_1} \to r_1(i) : (\Pi^{r_1} \vee \neg x_m : \neg A^{r_1}) \to (p_1 \cdot t_2(x_m) \cdot r_1(i)) : \Pi^{r_1} \ ,$$

which is propositionally equivalent to

$$\mathsf{J5} \vdash r_1(i) : (\Pi^{r_1} \vee \neg x_m : \neg A^{r_1}) \to \neg t_2(x_m) : x_m : \neg A^{r_1} \vee s : \Pi^{r_1}$$

for $s := p_1 \cdot t_2(x_m) \cdot r_1(i)$. From this and (10) by propositional reasoning we obtain

$$\mathsf{J5} \vdash r_1(i) : (\Pi^{r_1} \vee \neg x_m : \neg A^{r_1}) \to (t_1(x_m) \cdot ?\, t_2(x_m)) : \neg x_m : \neg A^{r_1} \vee s : \Pi^{r_1} \ . \tag{11}$$

By the Internalization Lemma for the tautology $\neg x_m : \neg A^{r_1} \to \Delta^{r_1} \vee \neg x_m : \neg A^{r_1}$ and propositional reasoning, there is a term $t_3(x)$ such that

$$\mathsf{J5} \vdash r_1(i) : (\Pi^{r_1} \vee \neg x_m : \neg A^{r_1}) \;\to\; t_3(t_1(x_m) \cdot ?\, t_2(x_m)) : (\Delta^{r_1} \vee \neg x_m : \neg A^{r_1}) \vee s : \Pi^{r_1} . \quad (12)$$

Since $\Delta^{r_1} \to \Delta^{r_1} \vee \neg x_m : \neg A^{r_1}$ is a propositional tautology, by the Internalization Lemma there is a term $t_4(x)$ such that $\mathsf{J5} \vdash r_1(k) : \Delta^{r_1} \to t_4(r_1(k)) : (\Delta^{r_1} \vee \neg x_m : \neg A^{r_1})$. Therefore, by axiom sum,

$$\mathsf{J5} \vdash r_1(k) : \Delta^{r_1} \to t : (\Delta^{r_1} \vee \neg x_m : \neg A^{r_1}) \quad (13)$$

for $t := t_3(t_1(x_m) \cdot ?\, t_2(x_m)) + t_4(r_1(k))$. Similarly, by (12) and sum,

$$\mathsf{J5} \vdash r_1(i) : (\Pi^{r_1} \vee \neg x_m : \neg A^{r_1}) \;\to\; t : (\Delta^{r_1} \vee \neg x_m : \neg A^{r_1}) \vee s : \Pi^{r_1} \;. \quad (14)$$

Finally by propositional reasoning from (13) and (14),

$$\mathsf{J5} \vdash r_1(k) : \Delta^{r_1} \vee r_1(i) : (\Pi^{r_1} \vee \neg x_m : \neg A^{r_1}) \to t : (\Delta^{r_1} \vee \neg x_m : \neg A^{r_1}) \vee s : \Pi^{r_1} \;.$$

Hence the desired realization function is $r := (r_1 \restriction \Delta, \Diamond_{2m} A, \Pi) \cup \{(l, t), (j, s)\}$.

This completes the proof of the base case of the induction.

Induction step. The proof is the same as in Lemma 5.4. □

Now we are ready to prove our main result.

Theorem 5.9 (Realization) *For any modal logic* ML *and its corresponding justification logic* JL *we have that* $\mathsf{ML} = \mathsf{JL}^\circ$.

Proof. The inclusion $\mathsf{JL}^\circ \subseteq \mathsf{ML}$ is easy since forgetful projections of axioms and rules of any justification logic can easily be derived in the corresponding modal logic. So we now turn to the more interesting opposite inclusion. It follows from Theorem 3.1 (Completeness), Lemma 4.5 (Annotating Proofs), and the following

Claim. Let S be the sequent system for a modal logic ML and let $\mathcal{P}$ be an annotated proof with the endsequent Δ such that the unannotated version of $\mathcal{P}$ is a sequent calculus proof in S. Then there exists a realization function r on Δ such that $\mathsf{JL} \vdash \Delta^r$ for the justification logic JL that corresponds to ML.

We prove the claim by induction on the depth of $\mathcal{P}$ by case analysis on the lowermost rule.

Case id. The claim follows from Lemma 5.1.

Cases $\vee$- and exch-rules. The claim follows from the induction hypothesis and Lemma 5.4.

Case $\wedge \dfrac{\Gamma\{A\} \quad \Gamma\{B\}}{\Gamma\{A \wedge B\}}$**.** By induction hypothesis there exist realization functions r_1 and r_2 such that $\mathsf{JL} \vdash \Gamma\{A\}^{r_1}$ and $\mathsf{JL} \vdash \Gamma\{B\}^{r_2}$. By Lemma 5.2, there exists a realization function r on the conclusion $\Gamma\{A \wedge B\}$ and a substitution σ such that $\mathsf{J} \vdash \Gamma\{A\}^{r_1}\sigma \to \Gamma\{B\}^{r_2}\sigma \to \Gamma\{A \wedge B\}^r$. By Lemma 4.10, $\mathsf{JL} \vdash \Gamma\{A\}^{r_1}\sigma$ and $\mathsf{JL} \vdash \Gamma\{B\}^{r_2}\sigma$, hence, $\mathsf{JL} \vdash \Gamma\{A \wedge B\}^r$.

Case ctr-rule. The claim follows from the induction hypothesis, Lemma 5.3, and Lemma 4.10 (Substitution).
Case □-rule. The claim immediately follows from the induction hypothesis.
Case k-rule. The claim follows from the induction hypothesis and Lemma 5.5.
Cases for rules in {d, t, b, 4, 5a, 5b, 5c}. The claim follows from the induction hypothesis and Lemma 5.8. □

Remark 5.10 Fitting's Merging Theorem from [13] states a stronger result than used in this paper, namely that the proofs can be made *injective*. An injective proof uses each constant for only one axiom instance. We are confident that the results of this paper can also be extended to injective proofs.

6 A Strengthened Realization Theorem for S5 and KB5

We now introduce two new justification logics: JT5 and JB5. The axiom systems for them are obtained from the axiom systems for JT45 and JB45 respectively by removing the operator ! from the language and, therefore, dropping j4 and replacing AN! with AN from Remark 2.1. Note that, although S5 = KT5 = KT45 and KB5 = KB45, it is obvious that JT5 ≠ JT45 and JB5 ≠ JB45 simply because the languages are different. The proof of the Internalization Lemma relies on the AN!-rule, which is not admissible in either JT5 or JB5. Thus, we need to show the existence of a term $\mathrm{dpi}(x)$ that plays the role of $!x$ for these two logics, where dpi stands for *derived positive introspection.*

Lemma 6.1 (Positive Introspection in JB5 and JT5) *There is a term* $\mathrm{dpi}(x)$ *such that for any term t and any formula A*

$$\mathsf{JB5} \vdash t : A \rightarrow \mathrm{dpi}(t) : t : A \qquad \text{and} \qquad \mathsf{JT5} \vdash t : A \rightarrow \mathrm{dpi}(t) : t : A \ .$$

Proof. Since j5 is an axiom of JB5, by AN there exists a constant c_i such that

$$\mathsf{JB5} \vdash c_i : \big(\neg y : P \rightarrow ?y : \neg y : P\big) \tag{15}$$

for some proposition P and variable y. It can be shown using AN, app, and propositional reasoning that there exists a ground term p such that

$$\mathsf{JB5} \vdash p : \big((\neg y : P \rightarrow ?y : \neg y : P) \rightarrow \neg ?y : \neg y : P \rightarrow y : P\big) \ .$$

From this and (15) by app and MP, we have $\mathsf{JB5} \vdash (p \cdot c_i) : (\neg ?y : \neg y : P \rightarrow y : P)$. Again by app and MP, we have $\mathsf{JB5} \vdash ??y : \neg ?y : \neg y : P \rightarrow (p \cdot c_i \cdot ??y) : y : P$. Since

$$y : P \rightarrow ??y : \neg ?y : \neg y : P \tag{16}$$

is an instance of axiom jb, by propositional reasoning

$$\mathsf{JB5} \vdash y : P \rightarrow \mathrm{dpi}(y) : y : P \tag{17}$$

for $\mathrm{dpi}(y) := p \cdot c_i \cdot ??y$. We now show that (16) is provable in JT5. Indeed, formula $\neg ?y : \neg y : P \rightarrow ??y : \neg ?y : \neg y : P$ is an instance of j5. Hence, (16) follows by syllogism with $y : P \rightarrow \neg ?y : \neg y : P$, which is a contraposition of an instance of jt. Thus, (17) also holds if JB5 is replaced with JT5. The statement of the lemma for either logic now follows from (17) by the Substitution Lemma, which also holds for these logics. □

Because of Lemma 6.1, using $\mathrm{dpi}(t)$ instead of $!t$ we can adapt the standard proof of the Internalization Lemma to JT5 and JB5. As a consequence, versions of Theorem 4.12, as well as of Lemmas 5.1, 5.2, 5.3, 5.4, 5.5, and 5.8, for JT5 and JB5 also hold. The proofs apply literally except that in the case of the 4-rule in Lemma 5.8, we use Lemma 6.1 instead of axiom j4.

It follows from the Realization Theorem for JT45 and JB45 that $\mathsf{JT5}^\circ \subseteq \mathsf{S5}$ and $\mathsf{JB5}^\circ \subseteq \mathsf{KB5}$. The opposite inclusions can be shown by literally repeating the proof of the Realization Theorem.

Theorem 6.2 (Strengthened Realization) $\mathsf{S5} = \mathsf{JT5}^\circ$ *and* $\mathsf{KB5} = \mathsf{JB5}^\circ$.

7 Conclusion

We have used cut-free nested sequent systems to constructively realize each of our 15 modal logics. In doing so, we have reproved in a uniform way several known realization theorems and have realized logics that did not have justification counterparts before. For two logics, we have also shown that the positive introspection operation is superfluous.

For now we have realized these *logics*. However, some of them have more than one *axiomatization*. Justification counterparts of different axiomatizations of the same modal logic can be different, e.g., JT5 and JT45 are both justification counterparts of S5 but are based on different axiomatizations of it. Thus, it is a natural next step for us to try to obtain realizations for all the 32 different axiomatizations of these 15 logics. We believe that nested sequent systems with *structural modal rules* [9,10], which are *modular* in a certain sense, will allow us to do this.

Another direction for future research is to look for cut-free proof systems for all our justification logics. Currently many justification logics lack such proof systems, and the problems in obtaining them seem to be the same as for modal logics. Nested sequents have provided cut-free proof systems for all our modal logics, and thus we believe they can also provide cut-free proof systems for justification logics.

Acknowledgement

We thank Samuel Bucheli, Melvin Fitting, Meghdad Ghari, Richard McKinley, and the anonymous referees for helpful comments. Goetschi and Kuznets are supported by Swiss National Science Foundation grant 200021–117699.

References

[1] Artemov, S. N., *Operational modal logic*, Technical Report MSI 95–29, Cornell University (1995).
URL http://www.cs.gc.cuny.edu/~sartemov/publications/MSI95-29.ps

[2] Artemov, S. N., *Explicit provability and constructive semantics*, Bulletin of Symbolic Logic **7** (2001), pp. 1–36.
URL http://www.math.ucla.edu/~asl/bsl/0701/0701-001.ps

[3] Artemov, S. N., *The logic of justification*, The Review of Symbolic Logic **1** (2008), pp. 477–513.
URL http://dx.doi.org/10.1017/S1755020308090060

[4] Artemov, S. N., E. L. Kazakov and D. Shapiro, *Logic of knowledge with justifications*, Technical Report CFIS 99–12, Cornell University (1999).
URL http://www.cs.gc.cuny.edu/~sartemov/publications/S5LP.ps

[5] Artemov, S. N. and R. Kuznets, *Logical omniscience as a computational complexity problem*, in: A. Heifetz, editor, *Proceedings of TARK 2009* (2009), pp. 14–23.
URL http://sites.google.com/site/kuznets/TARK09.pdf

[6] Baader, F. and T. Nipkow, "Term *Re*writing and *All That*," 1998.

[7] Brezhnev, V. N., *On explicit counterparts of modal logics*, Technical Report CFIS 2000–05, Cornell University (2000).

[8] Brünnler, K., "Deep Inference and Symmetry in Classical Proofs," Ph.D. thesis, Technische Universität Dresden (2003).
URL http://www.iam.unibe.ch/~kai/Papers/phd.pdf

[9] Brünnler, K., *Deep sequent systems for modal logic*, Archive for Mathematical Logic **48** (2009), pp. 551–577.
URL http://www.iam.unibe.ch/til/publications/pubitems/pdfs/bru09.pdf

[10] Brünnler, K. and L. Straßburger, *Modular sequent systems for modal logic*, in: M. Giese and A. Waaler, editors, *Proceedings of TABLEAUX 2009*, Lecture Notes in Artificial Intelligence **5607**, Springer, 2009 pp. 152–166.
URL http://www.iam.unibe.ch/til/publications/pubitems/pdfs/bstr09.pdf

[11] Fitting, M., *The logic of proofs, semantically*, Annals of Pure and Applied Logic **132** (2005), pp. 1–25.
URL http://comet.lehman.cuny.edu/fitting/bookspapers/pdf/papers/LPSemantics.pdf

[12] Fitting, M., *The realization theorem for* S5*, a simple, constructive proof* (2008), forthcoming in *Proceedings of the Second Indian Conference on Logic and Its Relationship with Other Disciplines.*
URL http://comet.lehman.cuny.edu/fitting/bookspapers/pdf/papers/ForRohit.pdf

[13] Fitting, M., *Realizations and LP*, Annals of Pure and Applied Logic **161** (2009), pp. 368–387.
URL http://comet.lehman.cuny.edu/fitting/bookspapers/pdf/papers/RealizeTheorem.pdf

[14] Guglielmi, A., *A system of interaction and structure*, ACM Transactions on Computational Logic **8** (2007).
URL http://cs.bath.ac.uk/ag/p/SystIntStr.pdf

[15] Kuznets, R., "Complexity Issues in Justification Logic," Ph.D. thesis, CUNY Graduate Center (2008).
URL http://sites.google.com/site/kuznets/PhD.pdf

[16] Pacuit, E., *A note on some explicit modal logics*, in: *Proceedings of the 5th Panhellenic Logic Symposium* (2005), pp. 117–125.
URL http://www.illc.uva.nl/Publications/ResearchReports/PP-2006-29.text.pdf

[17] Rubtsova, N. M., *Evidence reconstruction of epistemic modal logic* S5, in: D. Grigoriev, J. Harrison and E. A. Hirsch, editors, *Proceedings of CSR 2006*, Lecture Notes in Computer Science **3967**, Springer, 2006 pp. 313–321.
URL http://dx.doi.org/10.1007/11753728_32

A Proof of Lemma 5.3 (Contraction)

Proof. By induction on the structure of $\Gamma\{\,\}$.
Base case $\Gamma\{\,\} = \{\,\}$. In order to demonstrate the statement, a subinduction on the structure of the common unannotated version A of formulas A_1, A_2, and A_3 is employed. The statement proved by subinduction is the same as in the main induction with an extra restriction on σ, namely that $\mathrm{vrange}(\sigma) \subseteq \mathrm{diavars}(\Omega)$.
Subinduction base: $A = P_i$ or $A = \neg P_i$. The identity substitution σ and $r := \emptyset$ suffice.
Subinduction step. The following cases have to be considered:
Subinduction case $A = B \vee C$. The annotated formulas $A_1 = B_1 \vee C_1$, $A_2 = B_2 \vee C_2$, and $A_3 = B_3 \vee C_3$ do not share indices. By subinduction hypothesis, there exist realization functions r'_B on B_3 and r'_C on C_3, as well as substitutions σ_B with $\mathrm{dom}(\sigma_B) \subseteq \mathrm{diavars}(B_1 \vee B_2)$ and σ_C with $\mathrm{dom}(\sigma_C) \subseteq \mathrm{diavars}(C_1 \vee C_2)$ such that

$$\mathsf{J} \vdash (B_1 \vee B_2)^{r_1}\sigma_B \to (B_3)^{r'_B} \qquad \text{and} \qquad \mathsf{J} \vdash (C_1 \vee C_2)^{r_1}\sigma_C \to (C_3)^{r'_C} \ .$$

Also, we have that $\mathrm{vrange}(\sigma_B) \subseteq \mathrm{diavars}(B_3)$ and $\mathrm{vrange}(\sigma_C) \subseteq \mathrm{diavars}(C_3)$. By Lemma 4.11, $\sigma := \sigma_B \cup \sigma_C$ is a substitution with $\mathrm{dom}(\sigma) \subseteq \mathrm{diavars}(\Lambda)$. In addition, for restrictions $r_B := r'_B \restriction B_3$ and $r_C := r'_C \restriction C_3$, both $\sigma_C \circ r_B$ and $\sigma_B \circ r_C$ are realization functions on B_3 and C_3 respectively. Since $(B_3)^{r'_B} = (B_3)^{r_B}$ and $(C_3)^{r'_C} = (C_3)^{r_C}$, by Lemma 4.10

$$\mathsf{J} \vdash (B_1 \vee B_2)^{r_1}\sigma_B\sigma_C \to (B_3)^{r_B}\sigma_C \qquad \text{and} \qquad \mathsf{J} \vdash (C_1 \vee C_2)^{r_1}\sigma_C\sigma_B \to (C_3)^{r_C}\sigma_B \ .$$

Note that σ_C has no effect on any term $\sigma_B(x) \in \mathrm{range}(\sigma_B)$ because $\sigma_B(x)$ only contains variables from $\mathrm{diavars}(B_3)$, which is disjoint from $\mathrm{diavars}(C_1 \vee C_2) \supseteq \mathrm{dom}(\sigma_C)$. Thus, $(B_1 \vee B_2)^{r_1}\sigma_B\sigma_C = (B_1 \vee B_2)^{r_1}\sigma$. Similarly, $(C_1 \vee C_2)^{r_1}\sigma_C\sigma_B = (C_1 \vee C_2)^{r_1}\sigma$. From this and Lemma 4.11 it follows that

$$\mathsf{J} \vdash (B_1 \vee B_2)^{r_1}\sigma \to (B_3)^{\sigma_C \circ r_B} \qquad \text{and} \qquad \mathsf{J} \vdash (C_1 \vee C_2)^{r_1}\sigma \to (C_3)^{\sigma_B \circ r_C} \ .$$

Finally, by propositional reasoning,

$$\mathsf{J} \vdash \big((B_1 \vee C_1) \vee (B_2 \vee C_2)\big)^{r_1}\sigma \ \to\ (B_3)^{\sigma_C \circ r_B} \vee (C_3)^{\sigma_B \circ r_C} \ .$$

In other words, $\mathsf{J} \vdash \Lambda^{r_1}\sigma \to \Omega^r$ for $r := (\sigma_C \circ r_B) \cup (\sigma_B \circ r_C)$, which by Lemma 4.11 is a realization function on $\Omega = B_3 \vee C_3$.
Subinduction case $A = B \wedge C$. It is analogous to $B \vee C$.
Subinduction case $A = \Diamond B$. The annotated formulas $A_1 = \Diamond_{2k}B_1$, $A_2 = \Diamond_{2m}B_2$, $A_3 = \Diamond_{2n}B_3$ do not share indices. By induction hypothesis, there are a realization function r'_B on B_3 and a substitution σ_B with $\mathrm{dom}(\sigma_B) \subseteq \mathrm{diavars}(B_1 \vee B_2)$ such that $\mathsf{J} \vdash (B_1 \vee B_2)^{r_1}\sigma_B \to (B_3)^{r'_B}$. In addition, $\mathrm{vrange}(\sigma_B) \subseteq \mathrm{diavars}(B_3)$. By propositional reasoning,

$$\mathsf{J} \vdash \neg(B_3)^{r'_B} \ \to\ \neg(B_1)^{r_1}\sigma_B \qquad \text{and} \qquad \mathsf{J} \vdash \neg(B_3)^{r'_B} \ \to\ \neg(B_2)^{r_1}\sigma_B \ .$$

By the Internalization Lemma, there exist terms $t_1(y)$ and $t_2(y)$ such that

$$\mathsf{J} \vdash x_n : \neg(B_3)^{r'_B} \to t_1(x_n) : \neg(B_1)^{r_1}\sigma_B \quad \text{and} \quad \mathsf{J} \vdash x_n : \neg(B_3)^{r'_B} \to t_2(x_n) : \neg(B_2)^{r_1}\sigma_B \ .$$

It then follows by propositional reasoning that

$$\mathsf{J} \vdash \neg t_1(x_n) : \neg(B_1)^{r_1}\sigma_B \vee \neg t_2(x_n) : \neg(B_2)^{r_1}\sigma_B \to \neg x_n : \neg(B_3)^{r'_B} \ . \tag{A.1}$$

Since $\mathrm{dom}(\sigma_B) \subseteq \mathrm{diavars}(B_1 \vee B_2) \not\ni x_n$, the substitution σ_B does not affect x_n and, hence, (A.1) is identical to

$$\mathsf{J} \vdash \left(\neg t_1(x_n) : \neg(B_1)^{r_1} \vee \neg t_2(x_n) : \neg(B_2)^{r_1}\right)\sigma_B \to \neg x_n : \neg(B_3)^{r'_B} \ .$$

Let $\sigma' := \{(x_k, t_1(x_n)), (x_m, t_2(x_n))\} \cup \{(x_i, x_i) \mid i \notin \{k, m\}\}$. By Lemma 4.10,

$$\mathsf{J} \vdash \left(\neg t_1(x_n) : \neg(B_1)^{r_1} \vee \neg t_2(x_n) : \neg(B_2)^{r_1}\right)\sigma_B\sigma' \to \neg x_n : \neg(B_3)^{r'_B}\sigma' \ . \tag{A.2}$$

Since $2k$ and $2m$ do not occur in B_3, $\sigma' \circ (r'_B \restriction B_3)$ is a realization function on B_3 by Lemma 4.11. Let $r := (\sigma' \circ (r'_B \restriction B_3)) \cup \{(2n, x_n)\}$. Clearly, it is a realization function on $\Diamond_{2n}B_3$. Since σ_B affects neither x_k nor x_m, (A.2) becomes

$$\mathsf{J} \vdash (\Diamond_{2k}B_1 \vee \Diamond_{2m}B_2)^{r_1}\sigma \to (\Diamond_{2n}B_3)^r$$

for $\sigma := \sigma' \circ \sigma_B$. In other words, $\mathsf{J} \vdash \Lambda^{r_1}\sigma \to \Omega^r$. It remains to note that, by Lemma 4.11, $\mathrm{dom}(\sigma) \subseteq \mathrm{dom}(\sigma_B) \cup \mathrm{dom}(\sigma') \subseteq \mathrm{diavars}(\Diamond_{2k}B_1 \vee \Diamond_{2m}B_2)$ and, in addition, we also have $\mathrm{vrange}(\sigma) \subseteq \mathrm{diavars}(B_3) \cup \{x_n\} = \mathrm{diavars}(\Diamond_{2n}B_3)$.

Subinduction case $A = \Box B$. The annotated formulas $A_1 = \Box_k B_1$, $A_2 = \Box_l B_2$, $A_3 = \Box_m B_3$ do not share indices. By induction hypothesis, there exists a realization function r_B on B_3 and a substitution σ with $\mathrm{dom}(\sigma) \subseteq \mathrm{diavars}(B_1 \vee B_2)$ such that $\mathsf{J} \vdash (B_1 \vee B_2)^{r_1}\sigma \to (B_3)^{r_B}$. In addition, $\mathrm{vrange}(\sigma) \subseteq \mathrm{diavars}(B_3)$. By propositional reasoning and the Internalization Lemma, there exist terms $t_1(y)$ and $t_2(y)$ such that

$$\mathsf{J} \vdash r_1(k)\sigma : (B_1)^{r_1}\sigma \to t_1(r_1(k)\sigma) : (B_3)^{r_B} \ ,$$

$$\mathsf{J} \vdash r_1(l)\sigma : (B_2)^{r_1}\sigma \to t_2(r_1(l)\sigma) : (B_3)^{r_B} \ .$$

By axiom sum, for $s := t_1(r_1(k)\sigma) + t_2(r_1(l)\sigma)$,

$$\mathsf{J} \vdash r_1(k)\sigma : (B_1)^{r_1}\sigma \to s : (B_3)^{r_B} \quad \text{and} \quad \mathsf{J} \vdash r_1(l)\sigma : (B_2)^{r_1}\sigma \to s : (B_3)^{r_B} \ .$$

Thus, by propositional reasoning,

$$\mathsf{J} \vdash (\Box_k B_1 \vee \Box_l B_2)^{r_1}\sigma \to (\Box_m B_3)^r$$

for $r := (r_B \restriction B_3) \cup \{(m, s)\}$. It is clear that r is a realization function on $\Box_m B_3$.

This completes the proof by subinduction of the base case $\Gamma\{\ \} = \{\ \}$.

Induction step. By induction hypothesis, there exists a realization function r' on $\Sigma\{A_3\}$ and a substitution σ with $\mathrm{dom}(\sigma) \subseteq \mathrm{diavars}(\Sigma\{A_1, A_2\})$ such that

$$\mathsf{J} \vdash \Sigma\{A_1, A_2\}^{r_1}\sigma \to \Sigma\{A_3\}^{r'} \ .$$

Case $\Gamma\{\ \} = [\Sigma\{\ \}]_k$. By the Internalization Lemma,

$$\mathsf{J} \vdash r_1(k)\sigma : \left(\Sigma\{A_1, A_2\}^{r_1}\sigma\right) \ \to\ t(r_1(k)\sigma) : \Sigma\{A_3\}^{r'}$$

for some term $t(x)$. In other words, the desired result

$$\mathsf{J} \vdash \left([\Sigma\{A_1, A_2\}]_k\right)^{r_1}\sigma \ \to\ \left([\Sigma\{A_3\}]_k\right)^{r} \ ,$$

is achieved for a realization function $r := (r' \upharpoonright \Sigma\{A_3\}) \cup \{(k, t(r_1(k)\sigma))\}$ and the same substitution σ.

Case $\Gamma\{\ \} = \Delta, \Sigma\{\ \}, \Pi$. By propositional reasoning,

$$\mathsf{J} \vdash \Delta^{r_1}\sigma \vee \Sigma\{A_1, A_2\}^{r_1}\sigma \vee \Pi^{r_1}\sigma \ \to\ \Delta^{r_1}\sigma \vee \Sigma\{A_3\}^{r'} \vee \Pi^{r_1}\sigma \ .$$

Since $\mathrm{dom}(\sigma) \subseteq \mathrm{diavars}(\Sigma\{A_1, A_2\})$, by Lemma 4.11, $\sigma \circ (r_1 \upharpoonright \Delta, \Pi)$ is a realization function on Δ, Π. Then for $r := \left(\sigma \circ (r_1 \upharpoonright \Delta, \Pi)\right) \cup (r' \upharpoonright \Sigma\{A_3\})$,

$$\mathsf{J} \vdash (\Delta \vee \Sigma\{A_1, A_2\} \vee \Pi)^{r_1}\sigma \ \to\ (\Delta \vee \Sigma\{A_3\} \vee \Pi)^{r} \ .$$

It remains to apply Lemma 4.13 to obtain the desired result

$$\mathsf{J} \vdash (\Delta, \Sigma\{A_1, A_2\}, \Pi)^{r_1}\sigma \ \to\ (\Delta, \Sigma\{A_3\}, \Pi)^{r}$$

for the realization function r and the same substitution σ.

Note that induction steps never alter σ. □

B Cases for Rules 5a and 5c in Lemma 5.8

Proof. Subcase $\rho = \mathsf{5a}$. By a propositional tautology $\Delta^{r_1} \to \Delta^{r_1} \vee \neg x_m : \neg A^{r_1}$ and the Internalization Lemma there exists a term $t_1(x)$ such that

$$\mathsf{J5} \vdash r_1(k) : \Delta^{r_1} \to t_1(r_1(k)) : \left(\Delta^{r_1} \vee \neg x_m : \neg A^{r_1}\right) \ . \tag{B.1}$$

Similarly, for a tautology $\neg x_m : \neg A^{r_1} \to \Delta^{r_1} \vee \neg x_m : \neg A^{r_1}$ there is $t_2(y)$ such that

$$\mathsf{J5} \vdash ?\,x_m : \neg x_m : \neg A^{r_1} \to t_2(?\,x_m) : (\Delta^{r_1} \vee \neg x_m : \neg A^{r_1}) \ .$$

From the instance $\neg x_m : \neg A^{r_1} \to ?\,x_m : \neg x_m : \neg A^{r_1}$ of $\mathsf{j5}$ by propositional reasoning

$$\mathsf{J5} \vdash \neg x_m : \neg A^{r_1} \to t_2(?\,x_m) : (\Delta^{r_1} \vee \neg x_m : \neg A^{r_1}) \ . \tag{B.2}$$

It follows from (B.1) and (B.2) by axiom sum and propositional reasoning that

$$\mathsf{J5} \vdash r_1(k) : \Delta^{r_1} \vee \neg x_m : \neg A^{r_1} \rightarrow t : (\Delta^{r_1} \vee \neg x_m : \neg A^{r_1}).$$

for $t := t_1(r_1(k)) + t_2(?\, x_m)$. In other words,

$$\mathsf{J5} \vdash ([\Delta]_k, \Diamond_{2m} A)^{r_1} \rightarrow ([\Delta, \Diamond_{2m} A]_i)^r$$

for $r := (r_1 \restriction \Delta, \Diamond_{2m} A) \cup \{(i, t)\}$.

Subcase $\rho = \mathsf{5c}$. The existence of terms $t_1(x_m)$, $t_2(x_m)$, and s that satisfy (11) follows as in the subcase of $\rho = \mathsf{5b}$. Thus, by propositional reasoning,

$$\mathsf{J5} \vdash \Delta^{r_1} \vee r_1(i) : (\Pi^{r_1} \vee \neg x_m : \neg A^{r_1}) \rightarrow \Delta^{r_1} \vee (t_1(x_m) \cdot ?\, t_2(x_m)) : \neg x_m : \neg A^{r_1} \vee s : \Pi^{r_1} \ .$$

By the Internalization Lemma there exists a term $s_1(x)$ such that

$$\mathsf{J5} \vdash r_1(k) : \left(\Delta^{r_1} \vee r_1(i) : (\Pi^{r_1} \vee \neg x_m : \neg A^{r_1})\right) \rightarrow s_1(r_1(k)) : \left(\Delta^{r_1} \vee q_1 : \neg x_m : \neg A^{r_1} \vee s : \Pi^{r_1}\right) \ , \tag{B.3}$$

where $q_1 := t_1(x_m) \cdot ?\, t_2(x_m)$ in the above formula. By Lemma 5.6 there exists a term $t(x)$ such that

$$\mathsf{J5} \vdash t(q_1) : \left(q_1 : \neg x_m : \neg A^{r_1} \rightarrow \neg x_m : \neg A^{r_1}\right) \ . \tag{B.4}$$

By a propositional tautology and the Internalization Lemma applied to it,

$$\mathsf{J5} \vdash p_2 : \Big(\left(q_1 : \neg x_m : \neg A^{r_1} \rightarrow \neg x_m : \neg A^{r_1}\right) \rightarrow \Delta^{r_1} \vee q_1 : \neg x_m : \neg A^{r_1} \vee s : \Pi^{r_1} \rightarrow \Delta^{r_1} \vee \neg x_m : \neg A^{r_1} \vee s : \Pi^{r_1}\Big)$$

for some ground term p_2. From this and (B.4) by app and MP it follows that

$$\mathsf{J5} \vdash (p_2 \cdot t(q_1)) : \left(\Delta^{r_1} \vee q_1 : \neg x_m : \neg A^{r_1} \vee s : \Pi^{r_1} \rightarrow \Delta^{r_1} \vee \neg x_m : \neg A^{r_1} \vee s : \Pi^{r_1}\right) \ .$$

It follows by app and MP that

$$\mathsf{J5} \vdash s_1(r_1(k)) : (\Delta^{r_1} \vee q_1 : \neg x_m : \neg A^{r_1} \vee s : \Pi^{r_1}) \rightarrow q_3 : (\Delta^{r_1} \vee \neg x_m : \neg A^{r_1} \vee s : \Pi^{r_1})$$

for $q_3 := p_2 \cdot t(q_1) \cdot s_1(r_1(k))$. By propositional reasoning with (B.3) it follows that

$$\mathsf{J5} \vdash r_1(k) : \left(\Delta^{r_1} \vee r_1(i) : (\Pi^{r_1} \vee \neg x_m : \neg A^{r_1})\right) \rightarrow q_3 : \left(\Delta^{r_1} \vee \neg x_m : \neg A^{r_1} \vee s : \Pi^{r_1}\right) \ .$$

Hence the desired realization function is $r := (r_1 \restriction \Delta, \Diamond_{2m} A, \Pi) \cup \{(j, s), (l, q_3)\}$. □

Nominal Substitution at Work with the Global and Converse Modalities

Serenella Cerrito

IBISC, Université d'Evry Val d'Essonne,
Tour Evry 2, 523 Place des Terrasses de l'Agora,
91000 Evry Cedex, France

Marta Cialdea Mayer

Dipartimento di Informatica e Automazione
Università di Roma Tre
Via della Vasca Navale 79, 00146 Roma, Italy

Abstract

This paper represents a continuation of a previous work, where a practical approach to the treatment of nominal equalities in tableaux for basic Hybrid Logic $HL(@)$ was proposed. Its peculiarity is a substitution rule accompanied by nominal deletion. The main advantage of such a rule, compared with other approaches, is its efficiency, that has been experimentally verified for the $HL(@)$ fragment.
The integration of substitution and nominal deletion with more expressive languages is not a trivial task. In this work the previously proposed tableaux calculus for $HL(@)$ is extended to hybrid logic with the global and converse modalities, taking into account also practical considerations. Though termination, in this case, relies on loop checks, the computational advantages of the substitution rule persist in this richer framework.

Keywords: Tableaux, Modal Logic, Hybrid Logic, Automated Deduction

1 Introduction

Kripke structures, providing the semantics of modal logics, can be seen as labelled graphs, or else, transition systems. As such, they are widely used in computer science for modelisation purposes. *Nominal equalities*, which are assertions saying that a and b are different names for the same vertex (state), are typical of hybrid logic. In fact, the hybrid syntax extends the modal one by allowing states to be named by means of *nominals*, i.e. atomic formulae which hold at exactly one state, and it is possible to express that a formula F - possibly a nominal b itself - holds at a state named a by use of the *satisfaction* operator ($@_a F$). In particular, an equality $@_a b$ states that the nominals a and b are synonymous.

The treatment of nominal equalities in proof systems and provers may raise many redundancies, because when a formula F holds in the world named by synonymous nominals $a_1, ..., a_n$, it can potentially be treated n times. A previous work [6], further revised and extended in [5], presented a tableaux calculus for $HL(@)$ (basic hybrid logic), called H, whose characterizing feature is the treatment of nominal equalities by means of a special substitution rule, that expands an equality $@_a b$ by replacing a with b while deleting the chain of nominals generated by a, thus allowing for a reduction of the number of redundancies possibly induced by nominal equalities. Though embedded in a different context, such a rule is essentially the same as the "merge and prune" method proposed by [14] for the description logic $\mathcal{SHOIQ}$.

The computational advantages of such an approach with respect to expanding $@_a b$ by copying formulae from a to b, like in [2], have been experimentally verified [9,8]. The gain in efficiency is due to the *nominal deletion* mechanism embedded in the substitution rule.

Like other calculi for basic hybrid logic, H enjoys strong termination (every tableau in H is finite, independently of the rule application order), and termination does not need loop checks. Moreover, the system does not use any extra-logical notation (like prefixes), i.e. it is an *internalized* calculus. Tableau calculi for modal logics can in fact be given either in the explicit style, by means of *prefixes*, or in the implicit style (internalized calculi). In the first case, tableau nodes are labelled by *prefixed formulae* of the form $\sigma : F$, where σ is a symbol of the meta-language and F a formula, in the second one by language-level formulae. Prefixes are useful either when they are complex expressions encoding the relation between states, or when there is no internal (object-language) mechanism to name worlds.

In the case of hybrid logic, both styles have been used. A prefixed calculus is presented for instance in [2], which constitutes the theoretical background of the prover HTab [12]. However, the use of (simple) prefixes in the case of hybrid logics seems a useless burden, since nominals and the satisfaction operator can play the same role. In fact, prefixes may sometimes make both meta-proofs and implementations more complicated. Beyond this fact, internalized hybrid calculi have the advantage that the addition of pure axioms automatically yields complete systems for the class of frames they define, although termination may in some cases become a non trivial issue [3].

In this work we show that the approach to nominal equalities proposed in [6,5] can be safely extended to $HL(@, \Diamond^-, \mathsf{E})$, i.e. the language containing, beyond the satisfaction operator, the global modality E (and its dual A) and the converse modalities $\Diamond^-$ and $\Box^-$, preserving the computational advantages of substitution (with nominal deletion). The overall main feature of the proposed calculus is its practical approach, aiming in fact at being as close as possible a theoretical basis of an implemented system. For this reason, a particular attention is paid to efficiency aspects.

In the extended calculus, termination is achieved by means of a loop checking mechanism, partially like in [2,3], i.e. using ancestor equality blocking. The paper also shows that such a mechanism coupled with substitution without nominal deletion does not ensure termination.

An additional gain in efficiency is obtained by a different treatment of the existential modality: no loop checks are performed when expanding a formula of the form $@_a \mathsf{E} F$, still preserving termination.

The presence of substitution, and its interplay with the nominal generating rules, raise specific technical subtleties in the termination and completeness proofs, which are given here with a fair amount of details.

1.1 Preliminaries

We conclude this section by briefly recalling the syntax and semantics of (multi-modal) hybrid logic with the global and converse modalities, $HL(@, \Diamond^-, \mathsf{E})$.

Given two disjoint sets of atoms, NOM (nominals) and PROP (ordinary atoms), and a set of relation labels REL, formulae are built up from atoms using the classical connectives, the satisfaction operator @, the unary modal operators $\Diamond_\tau$, $\Box_\tau$ and their converses $\Diamond^-_\tau$, $\Box^-_\tau$ (where $\tau \in \mathsf{REL}$), and the global modalities E and A. Formulae are defined by the following grammar, where $p \in \mathsf{PROP}$, $a \in \mathsf{NOM}$, $\tau \in \mathsf{REL}$:

$$F := p \mid a \mid \neg F \mid F \wedge F \mid F \vee F \mid @_a F \mid \Diamond_\tau F \mid \Box_\tau F \mid$$
$$\Diamond^-_\tau F \mid \Box^-_\tau F \mid \mathsf{A}\, F \mid \mathsf{E}\, F$$

An *interpretation* $\mathcal{M}$ is a quadruple $\langle W, \{R_\tau \mid \tau \in \mathsf{REL}\}, N, I\rangle$ where W is a non-empty set (whose elements are the *states* of the interpretation), each $R_\tau \subseteq W \times W$ is a binary relation on W (the *accessibility relations*), N is a function $\mathsf{NOM} \to W$ and I a function $W \to 2^{\mathsf{PROP}}$. We shall write $wR_\tau w'$ as a shorthand for $\langle w, w'\rangle \in R_\tau$.

If $\mathcal{M} = \langle W, \{R_\tau \mid \tau \in \mathsf{REL}\}, N, I\rangle$ is an interpretation, $w \in W$ and F a formula, the relation $\mathcal{M}_w \models F$ is defined by:

(i) $\mathcal{M}_w \models p$ if $p \in I(w)$, for $p \in \mathsf{PROP}$.

(ii) $\mathcal{M}_w \models a$ if $N(a) = w$, for $a \in \mathsf{NOM}$.

(iii) $\mathcal{M}_w \models \neg F$ if $\mathcal{M}_w \not\models F$.

(iv) $\mathcal{M}_w \models F \wedge G$ if $\mathcal{M}_w \models F$ and $\mathcal{M}_w \models G$.

(v) $\mathcal{M}_w \models F \vee G$ if either $\mathcal{M}_w \models F$ or $\mathcal{M}_w \models G$.

(vi) $\mathcal{M}_w \models @_a F$ if $\mathcal{M}_{N(a)} \models F$.

(vii) $\mathcal{M}_w \models \Box_\tau F$ if for each w' such that $wR_\tau w'$, $\mathcal{M}_{w'} \models F$.

(viii) $\mathcal{M}_w \models \Diamond_\tau F$ if there exists w' such that $wR_\tau w'$ and $\mathcal{M}_{w'} \models F$.

(ix) $\mathcal{M}_w \models \Box^-_\tau F$ if for each w' such that $w'R_\tau w$, $\mathcal{M}_{w'} \models F$.

(x) $\mathcal{M}_w \models \Diamond^-_\tau F$ if there exists w' such that $w'R_\tau w$ and $\mathcal{M}_{w'} \models F$.

(xi) $\mathcal{M}_w \models \mathsf{A}F$ if for each $w' \in W$, $\mathcal{M}_{w'} \models F$.

(xii) $\mathcal{M}_w \models \mathsf{E}F$ if there exists $w' \in W$ such that $\mathcal{M}_{w'} \models F$.

A formula F is *satisfiable* if there exist an interpretation $\mathcal{M}$ and a state w of $\mathcal{M}$, such that $\mathcal{M}_w \models F$. Two formulae F and G are logically equivalent ($F \equiv G$) when, for every interpretation $\mathcal{M}$ and state w of $\mathcal{M}$, $\mathcal{M}_w \models F$ if and only if $\mathcal{M}_w \models G$.

It is worth pointing out that, for any nominal a and formula F:

$$\neg @_a F \equiv @_a \neg F \qquad \neg \Diamond_\tau F \equiv \Box_\tau \neg F \qquad \neg \Box_\tau F \equiv \Diamond_\tau \neg F$$

$$\neg \Diamond_\tau^- F \equiv \Box_\tau^- \neg F \qquad \neg \Box_\tau^- F \equiv \Diamond_\tau^- \neg F \qquad \neg \mathsf{A} F \equiv \mathsf{E} \neg F \qquad \neg \mathsf{E} F \equiv \mathsf{A} \neg F$$

This allows one to restrict attention to formulae in negation normal form (where negation dominates only atoms), without loss of generality.

2 The tableau calculus $\mathbf{H}^+$

In the calculus H and its extension, that will be named H^+, tableau nodes are labelled by sets of *satisfaction formulae*, i.e. assertions of the form $@_a F$ written as comma separated sequences of formulae. A formula of the form $@_a F$ will be called *labelled by a*. A formula of the form $@_a \Diamond_\tau b$, where b is a nominal, is a *relational formula.*

The initial tableau for a set S of formulae is a node labelled by $S_a = \{@_a F \mid F \in S\}$, where a is a new nominal. Without loss of generality, formulae are assumed to be in negation normal form. The set S_a is called the *initial set.* Nominals occurring in S_a are called *native nominals* (in the tableau).

Table 1 contains the *logical* rules, i.e. all rules but substitution. A tableau node is closed if it contains either $@_a p$ and $@_a \neg p$ for some nominal a and atom p, or $@_a \neg a$ for some nominal a (otherwise it is open). A tableau branch is open if all its nodes are open, otherwise it is closed. A tableau is closed if all its branches are closed, otherwise it is open.

Note that all the rules of Table 1 are conservative, i.e. they do not "consume" their premises. The $\Diamond_\tau$, $\Diamond_\tau^-$ and E rules are called *nominal generating rules*, and formulae of the form $\Diamond_\tau F$, $\Diamond_\tau^- F$ and $\mathsf{E}F$ *nominal generating formulae.* It is worth pointing out that, contrarily to the $\Diamond_\tau$-rule, the table gives no restriction on the applicability of the $\Diamond_\tau^-$-rule; in fact, it is necessary to expand formulae of the form $@_a \Diamond_\tau^- c$ where $c \in \mathsf{NOM}$, in order to obtain possible premises for the $\Box_\tau$ and $\Box_\tau^-$ rules, of the form $@_c \Diamond_\tau a$.

A formula occurring in a tableau T is called *native* (in T) if and only if it is in the language of the initial set, i.e. it does not contain any non-native nominal. A formula occurring in a tableau node is an *accessibility formula* if it is a relational formula introduced by application of the $\Diamond_\tau$ or $\Diamond_\tau^-$ rule. Accessibility formulae are obviously not native. It is worth pointing out that only the direct modalities $\Diamond_\tau$ occur in accessibility formulae, and not the converse ones.

In order to define the last rule of the system, the substitution rule, the definition of father and children of a nominal, given in [6,5] for H, has to be extended.

Definition 2.1 Let Θ be a tableau branch. If either the $\Diamond_\tau$ or $\Diamond_\tau^-$ rule has been applied in Θ to a formula labelled by a, generating a new nominal b, then $a \prec_\Theta b$ (and we say that b is a child of a, and a is the father of b).

Boolean Rules	Label Rule
$$\frac{@_a(F \wedge G), S}{@_aF, @_aG, @_a(F \wedge G), S} \; (\wedge)$$	$$\frac{@_a@_bF, S}{@_bF, @_a@_bF, S} \; (@)$$
$$\frac{@_a(F \vee G), S}{@_aF, @_a(F \vee G), S \quad @_aG, @_a(F \vee G), S} \; (\vee)$$	

Rules for the direct and converse modalities

$$\frac{@_a\Box_\tau F, @_a\Diamond_\tau b, S}{@_bF, @_a\Box_\tau F, @_a\Diamond_\tau b, S} \; (\Box_\tau)$$

$$\frac{@_a\Diamond_\tau F, S}{@_a\Diamond_\tau b, @_bF, @_a\Diamond_\tau F, S} \; (\Diamond_\tau)$$
where b is a new nominal
(not applicable if F is a nominal)

$$\frac{@_a\Box_\tau^- F, @_b\Diamond_\tau a, S}{@_bF, @_a\Box_\tau^- F, @_b\Diamond_\tau a, S} \; (\Box_\tau^-)$$

$$\frac{@_a\Diamond_\tau^- F, S}{@_b\Diamond_\tau a, @_bF, @_a\Diamond_\tau^- F, S} \; (\Diamond_\tau^-)$$
where b is a new nominal

Rules for the global modalities

$$\frac{@_a\mathsf{A}F, S}{@_cF, @_a\mathsf{A}F, S} \; (\mathsf{A})$$
where c occurs in the premise

$$\frac{@_a\mathsf{E}F, S}{@_bF, @_a\mathsf{E}F, S} \; (\mathsf{E})$$
where b is a new nominal

Table 1
Logical rules of the tableau system

The relation $\prec_\Theta^+$ is the transitive closure of $\prec_\Theta$ and $\prec_\Theta^*$ the reflexive and transitive closure of $\prec_\Theta$. If $a \prec_\Theta^+ b$ we say that b is a descendant of a and a an ancestor of b in the branch Θ.

Nominals with no fathers are called *root nominals*.

Note that if a nominal b is newly introduced in a branch by expanding a formula or the form $@_a\mathsf{E}F$, then b is not a child of a.

The substitution rule, which is applicable only if $a \neq b$, is formulated as follows:

$$\frac{@_ab, S}{S^\#[a \mapsto b]} \; (Sub)$$

where $S^\#[a \mapsto b]$ is obtained from S by:

(i) replacing every occurrence of a with b;

(ii) deleting every formula containing a descendant of a.

When the substitution rule is applied, a is said to be *replaced in the branch* and the descendants of a are called *deleted in the branch*.

It is worth pointing out that, differently from the cases of the label and boolean rules which could also be formulated as "destructive", [1] the $\Diamond_\tau$ and $\Diamond_\tau^-$ rules must conserve their premises (though they are single-premise rules). In fact, when the descendants of a replaced nominal are deleted, these formulae have to be reused, with the new label, in order to keep completeness.

In order to ensure termination, first of all, trivial re-applications of rules must be ruled out. So, we establish that:

R1. rules that would not change the node (their expansion being already contained in the upper node) are not applicable.

Obviously, in practice, since substitution affects the whole node, such heavy membership tests can be avoided and replaced by a marking mechanism – which can also allow for the boolean and label rules being implemented destructively (see [9] or [8]).

Secondly, useless re-application of nominal generating rules must be avoided. To this aim, the expansion of existential formulae is restricted similarly to [16]. A memory of existential formulae is associated to each tableau node. When a formula of the form $@_a\mathsf{E}F$ is expanded, the "body" of the formula, $\mathsf{E}F$, is saved in the memory of the resulting node. When the substitution rule is applied, it affects also memorized formulae (which are not deleted), and the other rules keep the memory unchanged. The E rule is then subject to the following restriction:

R2. a formula of the form $@_a\mathsf{E}F$ is not expanded in a node S if $\mathsf{E}F$ belongs to the memory of S.

Moreover, when one of the rules $\Diamond_\tau$ or $\Diamond_\tau^-$ is applied, the expanded formula is marked as *inactive* in the lower node. When the substitution rule is fired by an equality $@_ab$, it affects (only) the markings of formulae labelled by a, as follows: every formula of the form $@_a\Diamond_\tau F$ or $@_a\Diamond_\tau^- F$ becomes active again, before replacing a with b and deleting its descendants. This is because, as already remarked, the children of a are deleted, and the formulae resulting from the substitution, $@_b\Diamond_\tau F'$ and $@_b\Diamond_\tau^- F'$ ($F' = F[a \mapsto b]$), need to be expanded again in order to generate the corresponding children of b.

It is necessary to specify how substitution interacts with the active/inactive marking mechanism, i.e. what happens when two differently marked formulae collapse by the effect of substitution. In such cases, the inactive marking takes precedence. In particular, when *Sub* is applied to a node containing an active formula F_1 and an inactive one F_2, then, if the application of the substitution to F_1 and F_2 produces the same formula G, G is marked as inactive in the conclusion of the rule. The same mechanism applies when the *Sub* rule replaces a with b in a node containing both $@_a\Diamond_\tau F_1$ (or $@_a\Diamond_\tau^- F_1$) and an inactive $@_b\Diamond_\tau F_2$ (or $@_b\Diamond_\tau^- F_2$), with $F_1[a \mapsto b] = F_2[a \mapsto b] = G$: the inactive marking takes precedence again in the conclusion of the rule. I.e. substitution does not produce two occurrences of $@_b\Diamond_\tau G$ (or $@_b\Diamond_\tau^- G$), an active occurrence and an inactive one, but a single inactive one.

The $\Diamond_\tau$ and $\Diamond_\tau^-$ rules are then restricted as follows:

[1] The destructive versions of the label and boolean rules would in fact be more faithful to a possible implementation. Though such a choice would affect only technical details, the completeness and termination proofs would however become more fastidious.

R3. the $\Diamond_\tau$ and $\Diamond^-_\tau$ rules are not applicable to inactive formulae.

Note that the simpler restriction that nominal generating formulae are never expanded more than once on a branch (R1) would be weaker than R2 and R3. In fact, in that case, when an already expanded formula of the form $@_a\Diamond_\tau F$, $@_a\Diamond^-_\tau F$ or $@_a\mathsf{E}F$ is changed by a substitution, resulting in a never expanded new formula, it would become expandable again (even if the substitution does not affect a). Moreover, restriction R2 cannot be replaced by marking an expanded $@_a\mathsf{E}F$ as inactive and establishing that $@_b\mathsf{E}F$ is not expanded whenever the node contains an inactive $@_a\mathsf{E}F$ (similarly to restriction R3); in fact an existential inactive formula could be deleted by the *Sub* rule, without its witness being deleted. Nor could restriction R2 be replaced by simply requiring that $@_a\mathsf{E}F$ is not expandable whenever the node contains some $@_bF$, for any nominal b (this would not be enough for the completeness proof given in Section 3.2).

Finally termination needs a blocking mechanism relying on loop checks, which exploits the notion of *twin nominals* (like in [3]):

Definition 2.2 Let T be a tableau, Θ a branch of T and S a node of Θ. Then:

- If a is a nominal occurring in S then

$$Forms_S(a) = \{F \mid @_aF \text{ occurs in } S \text{ and } F \text{ is native in } T\}$$

  i.e. $Forms_S(a)$ contains all the native formulae labelled by a in S.
- Two nominals a and b are said to be twins in S if $Forms_S(a) = Forms_S(b)$.
- A nominal a occurring in S is *directly blocked* by a nominal b in S if $b \prec^+_\Theta a$ and a and b are twins in S. It is *indirectly blocked* by b in S when $b \prec^+_\Theta a$ and b is directly blocked in S. Finally, a is *blocked* in S when it is either directly or indirectly blocked in S.

Note that a nominal a is non-blocked in S if and only if there is no pair of distinct twins b, c occurring in S such that $b, c \prec^*_\Theta a$.

Termination is ensured by the following restriction:

R4. *The rules $\Diamond_\tau$ and $\Diamond^-_\tau$ are not applicable to a formula labelled by a nominal a in a node S if a is blocked in S.*

A tableau node (respectively a tableau branch) is *complete* if no rule can be applied to expand it (possibly because of restrictions R1–R4). A tableau is complete when all its branches are complete.

The blocking mechanism is similar to the dynamic loop checking[2] introduced in [13] for description logics, then adapted to hybrid logic in [2]. However, existential formulae are treated differently, more in the style of [16]. A loop checking mechanism closer to that used in [2,3] – and adopted in the first proposal to extend H to the global and converse modalities, that is [7] – would treat E like the $\Diamond_\tau$ and $\Diamond^-_\tau$ operators: if a new

[2] If blocks on nominals are never undone then blocking is static. Otherwise it is called dynamic. In H^+ blocking is dynamic because the "twin" relation may be destroyed when new formulae are added to the branch.

nominal b is generated by expansion of an existential formula labelled by a, then $a \prec_\Theta b$, and it is established that a formula of the form $@_a \mathsf{E}F$ cannot be expanded when a is blocked. The main difference between the two methods is that, in H^+, any existential formula is expanded at most once, independently of its label, thus avoiding possible redundant computations.

We end up this section pointing out that nominal deletion not only allows one to avoid useless expansions, but is necessary in order to ensure termination, as shown by the infinite tableau in Figure 1, where nominals are replaced without deleting their descendants. For the sake or readability, when a rule without side effects is applied, the lower node shows only the newly added formulae. Note that, as an effect of substitution without nominal deletion, each nominal b_{i+1} becomes orphan of its father b_i that, otherwise, would have blocked it.

$$
\begin{array}{cl}
@_a \mathsf{A}\Diamond_\top(p \wedge a) & \\
\hline
@_a \Diamond_\top(p \wedge a) & (\mathsf{A}) \\
\hline
@_a \Diamond_\top b_0, @_{b_0}(p \wedge a) & (\Diamond_\top) \\
\hline
@_{b_0} p, @_{b_0} a & (\wedge) \\
\hline
@_{b_0} \Diamond_\top(p \wedge a) & (\mathsf{A}) \\
\hline
@_{b_0} \Diamond_\top b_1, @_{b_1}(p \wedge a) & (\Diamond_\top) \\
\hline
@_{b_1} p, @_{b_1} a & (\wedge) \\
\hline
@_a \mathsf{A}\Diamond_\top(p \wedge a), @_a \Diamond_\top(p \wedge a), @_a \Diamond_\top a, @_a(p \wedge a), @_a p, @_a a, @_a \Diamond_\top b_1, @_{b_1}(p \wedge a), @_{b_1} p, @_{b_1} a & (b_0 \mapsto a) \\
\hline
@_{b_1} \Diamond_\top(p \wedge a) & (\mathsf{A}) \\
\hline
@_{b_1} \Diamond_\top b_2, @_{b_2}(p \wedge a) & (\Diamond_\top) \\
\hline
@_{b_2} p, @_{b_2} a & (\wedge) \\
\hline
@_a \mathsf{A}\Diamond_\top(p \wedge a), @_a \Diamond_\top(p \wedge a), @_a \Diamond_\top a, @_a(p \wedge a), @_a p, @_a a, @_a \Diamond_\top b_2, @_{b_2}(p \wedge a), @_{b_2} p, @_{b_2} a & (b_1 \mapsto a) \\
\vdots &
\end{array}
$$

Fig. 1. Substitution without nominal deletion

As a final remark, one can observe that the tableau in Figure 1 would terminate (even without nominal deletion) if *subset blocking* were adopted, i.e. if a nominal were blocked whenever it labels a subset of the formulae labelled by some of its ancestors. In fact, $@_{b_1} \Diamond_\top(p \wedge a)$ could not be expanded since b_1 labels, at that point, a subset of the formulae labelled by a. Subset blocking is sufficient for converse-free formulae, [3] but does not guarantee completeness for $HL(@, \Diamond^-, \mathsf{E})$.

[3] One of the tableau decision procedures presented in [4] for $HL(@, \mathsf{E})$ uses in fact substitution – without nominal deletion – and (anywhere) subset blocking.

3 Properties of $\mathbf{H}^+$

3.1 *Termination*

In order to show that H^+ terminates with loop-checks (under any rule application strategy), one can use an argument similar to the one given in [2] for the prefixed calculus. However, there are obvious differences, and some more delicate points, due to the presence of the substitution rule and the different treatment of existential formulae.

In order to state the key property of the system, we need the following definition:

Definition 3.1 If T is a tableau rooted at S_0, then:

$$S_0^* = \{F \mid F = G[b_1 \mapsto c_1, ..., b_n \mapsto c_n] \text{ for some subformula } G \text{ of some formula in } S_0, \text{ and native nominals } c_1, ..., c_n\}$$

In other terms, S_0^* contains every formula that can be obtained from a subformula of some formula in the initial set, by replacing nominals with native nominals. Note that S_0^* is necessarily finite and closed with respect to subformulae.

The following result can easily been proved by induction on tableaux.

Lemma 3.2 (Quasi-subformula property) *Let Θ be a branch in a tableau rooted at S_0. Then for every nominal a, the set of native formulae labelled by a in Θ is a subset of the finite set*

$$S_0^* \cup \{\Diamond_\tau b \mid b \textit{ is a native nominal and } \tau \textit{ is in the language of } S_0\}$$

Moreover, any formula $@_a F$ occurring in Θ with F non-native is a relational formula (i.e. F has the form $\Diamond_\tau b$).

The following properties are direct consequences of Lemma 3.2. If T is a tableau rooted at S_0, then:

(i) If $@_a b$ occurs in a node of T, then b is a native nominal. Therefore, in the applications of the substitution rule, nominals are always replaced by native nominals.

(ii) A non-native nominal b may occur in tableau nodes only in relational formulae of the form $@_a \Diamond_\tau b$ or $@_b \Diamond_\tau a$, or as the label of a formula $@_b F$, where $F \in S_0^*$ is native.

(iii) In particular, b is native in any formula of the form $@_a \Diamond_\tau^- b$, since such formulae are not relational. Therefore, a given nominal a may label only a finite number of formulae of the form $\Diamond_\tau^- b$.

(iv) If there is only a finite number of nominals occurring in a tableau branch Θ, then the set of formulae occurring in Θ and labelled by any fixed nominal a is finite (if there is a finite number of nominals, a can label only a finite number of relational formulae).

Since a given nominal a may label only a finite number of formulae which can be expanded by means of the $\Diamond_\tau$ or $\Diamond_\tau^-$ rule (Lemma 3.2, and its consequence iii), the restrictions on the applicability of such rules allow us to prove:

Lemma 3.3 *If Θ is a tableau branch and a any nominal occurring in Θ, then $\{b \mid a \prec_\Theta b\}$ is finite.*

As a consequence, also the following result holds:

Lemma 3.4 *Let Θ be a tableau branch. If Θ is infinite then there is an infinite chain of nominals*

$$b_1 \prec_\Theta b_2 \prec_\Theta b_3 \ldots \,.$$

Proof. The presence of the substitution rule makes the argument a little more complicated than the corresponding one given in [2].

First of all we prove that if Θ is infinite, then there is an infinite number of nominals occurring in Θ. If there were only a finite number of nominals, in fact, each of them would label a finite number of formulae (by the consequence iv of Lemma 3.2). Now, since formulae are never added to nodes where they already occur, there should be at least one formula F occurring in a node S_i of Θ which disappears from the branch and then reappears in a node S_j below S_i. F can disappear only because some nominal occurring in it is either replaced or deleted. But when a nominal is replaced or deleted, it can never occur again in the branch below the application of *Sub* which replaces/deletes it.

The infinite number of nominals occurring in Θ can be arranged by $\prec_\Theta$ in a forest of trees rooted at root nominals, and there are finitely many such trees. In fact, only native nominals and nominals generated by the E rule have no fathers, and their number is finite (only a finite number of – necessarily native – formulae of the form $\mathsf{E}F$ may occur in the branch, and each of them is expanded at most once, independently of its label).

Moreover, each tree is finitely branching because any nominal can generate only a finite number of new ones, by Lemma 3.3. By König's Lemma, if one of such trees is infinite, it has an infinite branch, i.e. there is an infinite chain of nominals $b_1 \prec_\Theta b_2 \prec_\Theta b_3 \ldots$. □

Theorem 3.5 (Termination) *Every tableau is finite.*

Proof. By Lemma 3.4, if an infinite branch Θ exists, then there is an infinite chain of nominals

$$b_1 \prec_\Theta b_2 \prec_\Theta b_3 \ldots \,.$$

By Lemma 3.2, if $@_a F$ occurs in Θ and F is native, then F is an element of the finite set $\Sigma = S_0^* \cup \{\Diamond_\tau b \mid b$ is a native nominal and τ is in the language of $S_0\}$, where S_0 is the initial set.

Let n be the cardinality of Σ and consider the initial sub-chain:

$$b_1 \prec_\Theta b_2 \prec_\Theta \ldots \prec_\Theta b_{2^n+1} \prec_\Theta b_{2^n+2}$$

Let Θ' be the initial segment of Θ up to, but not including, the nominal-generating inference ($\Diamond_\tau$ or $\Diamond_\tau^-$) producing b_{2^n+2}, and let S_k be the last node of Θ'.

Since b_{2^n+1} occurs in S_k, all its ancestors occur in S_k, too (if some of them had been either replaced or deleted above S_k, b_{2^n+1} would have been deleted by the same application of the substitution rule). Since b_{2^n+1} is the father of b_{2^n+2} in Θ, and it generates b_{2^n+2} by expanding S_k, then b_{2^n+1} is not blocked in S_k, i.e. S_k does not contain two distinct twins $b_i, b_j \prec^* b_{2^n+1}$.

Because of the choice of n, however, at least two nominals b_i and b_j among $b_1, ..., b_{2^n+1}$ must be twins in S_k (i.e. they must label the same set of native formulae).

□

3.2 *Completeness*

In this section we prove that if Θ is a complete and open branch of an H^+ tableau rooted at S_0, then S_0 is satisfiable. The overall structure of the completeness proof is the same as the corresponding one for H [5], and exploits the termination theorem: we first consider the set labelling the last node of Θ, that is *downward saturated* (in some sense), and we show that any such set has a model. Then the model existence property is propagated upward to the root node. However, in the presence of substitution, the model existence argument is technically subtler, because of the interplay between *Sub* and the nominal generating rules, accompanied by the blocking mechanism.

The following notion of nominal representatives is used in order to define saturation.

Definition 3.6 Let Θ be a tableau branch, S a node of Θ and b a nominal occurring in S. The *representative* of b in S, written $\rho_S(b)$, is the nominal $a \prec_\Theta^* b$ such that a is a twin of b and a is not blocked, if it exists (undefined otherwise).

Note that $\rho_S(b)$ may be undefined. Consider in fact a situation where $a_1 \prec_\Theta^+ a_2 \prec_\Theta^+ b$ and a_1 and a_2 become twins after the generation of b (by effect of the converse rules). It may happen that, in the chain leading to b, there is no ancestor of b that is a twin of b (because of different choices in the expansion of disjunctive formulae). In such cases, b is blocked and $\rho_S(b)$ does not exist either. It is worth pointing out also that there is at most one non-blocked nominal $a \prec_\Theta^* b$ that is a twin of b. In fact, if a_1 and a_2 are distinct nominals such that $a_1 \prec_\Theta^* b$, $a_2 \prec_\Theta^* b$ and both a_1, a_2 are twins of b, then a_1 is also a twin of a_2, hence at least one among a_1, a_2 has a twin ancestor and is blocked.

The following result establishes useful properties of nominal representatives.

Lemma 3.7 *Let Θ be a tableau branch and S a node of Θ. Then:*

(i) *For any nominal a occurring in S, a is non-blocked in S if and only if $\rho_S(a) = a$. In particular, if a is a root nominal, $\rho_S(a) = a$.*

(ii) *Let a be a non-blocked nominal in S and b a nominal occurring in S. If either $a \prec_\Theta b$ or $b \prec_\Theta a$, then $\rho_S(b)$ is defined.*

(iii) *If $\rho_S(a) = b$ and F is native, then $@_a F \in S$ if and only if $@_b F \in S$.*

Here finally follows the notion of saturation, that is relative to a tableau node, since clauses ix–xi make reference to non-blocked nominals and nominal representatives. Such notions, in turn, depend on the branch, since the relation $\prec^*_\Theta$ is branch-dependent.

Definition 3.8 A node S of a tableau branch Θ is *downward saturated* if and only if the following conditions hold:

(i) S contains neither a formula of the form $@_a\neg a$, nor two formulae of the form $@_a p$ and $@_a\neg p$ for some atom p.

(ii) If $@_a(F \wedge G) \in S$, then $@_a F, @_a G \in S$.

(iii) If $@_a(F \vee G) \in S$, then either $@_a F \in S$ or $@_a G \in S$.

(iv) If $@_a @_b F \in S$, then $@_b F \in S$.

(v) If $@_a b \in S$ then $a = b$.

(vi) If $@_a \Diamond_\tau b, @_a \Box_\tau F \in S$, then $@_b F \in S$.

(vii) If $@_b \Diamond_\tau a, @_a \Box_\tau^- F \in S$, then $@_b F \in S$.

(viii) If $@_a \mathsf{A} F \in S$, then for all nominals b occurring in S, $@_b F \in S$.

(ix) If $@_a \Diamond_\tau F \in S$, F is not a nominal and a is not blocked in S, then there is a nominal b such that $\rho_S(b)$ is defined and $@_a \Diamond_\tau b, @_b F \in S$.

(x) If $@_a \Diamond_\tau^- F \in S$ and a is not blocked in S, then there is a nominal b such that $\rho_S(b)$ is defined and $@_b \Diamond_\tau a, @_b F \in S$.

(xi) If $@_a \mathsf{E} F \in S$, then there is a nominal b such that $\rho_S(b) = b$ and $@_b F \in S$.

The following lemma proves the adequacy of the set of expansion rules.

Lemma 3.9 *Any complete and open tableau node is downward saturated.*

Proof. The delicate cases of the proofs are the items concerning the nominal generating formulae.

ix. Let us assume that $@_a \Diamond_\tau F \in S$, F is not a nominal, and a is not blocked in S. Then, since S is complete, $@_a \Diamond_\tau F$ is inactive in S. This means that $@_a \Diamond_\tau F$ is obtained by a number of substitutions (possibly none) from some $@_a \Diamond_\tau G$ which has been expanded above S,[4] i.e. $F = G[c_1 \mapsto d_1, ..., c_n \mapsto d_n]$, $n \geq 0$. The expansion of $@_a \Diamond_\tau G$ has generated some $@_a \Diamond_\tau b$ and $@_b G$. Since a still occurs in S, it has not been replaced or deleted, and consequently b has not been deleted. Since the same substitutions modifying $\Diamond_\tau G$ also affect b and G, S contains $@_a \Diamond_\tau b[c_1 \mapsto d_1, ..., c_n \mapsto d_n]$ and $@_b G[c_1 \mapsto d_1, ..., c_n \mapsto d_n]$. If b has not been replaced, then $@_a \Diamond_\tau b, @_b F \in S$, and, by Lemma 3.7.ii, $\rho_S(b)$ is defined. If $b = c_j$ for some j, then S contains $@_a \Diamond_\tau c_j$ and $@_{c_j} F$; since c_j is a native nominal, $\rho_S(b)$ is defined by Lemma 3.7.i.
Case x is treated similarly.

xi. Let us assume that $@_a \mathsf{E} F \in S$. Then, since S is complete, the memory of S contains $\mathsf{E} F$, and this means that some $@_c \mathsf{E} G$ has been expanded in a node S' above S, where

[4] Note that, here, $\Diamond_\tau G$ is necessarily labelled by a, because of the of the interaction between substitution and the active/inactive markings.

F is obtained from G by means of a number of substitutions (those applied in the sub-branch leading from S' to S); i.e. $F = G[c_1 \mapsto d_1, ..., c_n \mapsto d_n]$, $n \geq 0$. The expansion of $@_c\mathsf{E}G$ has generated some $@_dG$. Now, since d is a root nominal, it cannot be deleted (though it can be replaced), therefore S contains $(@_dG)[c_1 \mapsto d_1, ..., c_n \mapsto d_n] = @_bF$ for some $b \in \{d, d_1, ..., d_n\}$. Since b is in any case a root nominal (either $b = d$ or b is native), $\rho_S(b) = b$, by Lemma 3.7.i.

□

The following lemma defines a model (in some sense) of any open and complete tableau node S. Note that S necessarily contains at least one root nominal a_0, which is non-blocked in it, and, by Lemma 3.7.i, $\rho_S(a_0) = a_0$.

Lemma 3.10 *Let S be an open and complete tableau node and a_0 any root nominal occurring in S. Let $\mathcal{M}^*$ be the interpretation defined as follows:*

$$W = \{a \mid a \textit{ is a non-blocked nominal occurring in } S\};$$

$$R_\tau = \{(\rho_S(a), \rho_S(b)) \mid @_a\Diamond_\tau b \in S \textit{ and both } \rho_S(a), \rho_S(b) \textit{ are defined}\};$$

$$\textit{For every nominal } a \textit{ occurring in } S: N^*(a) = \begin{cases} \rho_S(a) & \textit{if } \rho_S(a) \textit{ is defined} \\ a_0 & \textit{otherwise} \end{cases}$$

$$I(a) = \{p \mid @_ap \in S\} \textit{ for all } a \in W$$

If $a \in W$, $@_aF \in S$ and F has not the form $\Diamond_\tau b$ for some b such that $\rho_S(b)$ is undefined, then $\mathcal{M}^*_a \models F$.

Proof. We remark beforehand that since S is open and complete, it is downward saturated by Lemma 3.9. Let us assume that $a \in W$ and $@_aF \in S$, for $F \neq \Diamond_\tau b$ with $\rho_S(b)$ undefined. Since a is not blocked, $\rho_S(a)$ is defined, and $\rho_S(a) = a$ (by Lemma 3.7.i). Therefore $N^*(a) = a$. The proof that ${\mathcal{M}^*}_a \models F$ is by induction on F.

Base We distinguish three cases.

(i) F is a literal. If F is a propositional letter or its negation, then the result is true by construction. In fact, if $@_ap \in S$, for $p \in \mathsf{PROP}$, $\mathcal{M}^*_a \models p$ by definition, because $N^*(a) = a$. If $@_a\neg p \in S$, since S is open, $@_ap \notin S$ and again $\mathcal{M}^*_a \models \neg p$ by construction.

(ii) F is a nominal b. Then necessarily $b = a$, since S is saturated, and the result is trivial, since $N^*(a) = a$.

(iii) F is $\neg b$, for some nominal b. Since S is open, $b \neq a$. By Lemma 3.2, b is a native nominal, hence it is non-blocked and belongs to W, so that $N^*(b) = b \neq a$. Therefore $\mathcal{M}^*_a \models \neg b$.

Induction Step Several cases have to be considered, according to the form of F. Here follows the treatment of some of them, the others being either very simple or similar to those shown below.

(i) $F = \mathsf{E}G$. If $@_a\mathsf{E}G \in S$, since S is saturated, there is a nominal b such that $\rho_S(b) = b$ and $@_bG \in S$. By Lemma 3.7.i, b is not blocked in S, so that $b \in W$.

By the inductive hypothesis, $\mathcal{M}_b^* \models G$, thus $\mathcal{M}_a^* \models \mathsf{E}G$.

(ii) F is $\Diamond_\tau G$. We distinguish two cases.

(a) G is a nominal b. By hypothesis, $\rho_S(b)$ is defined and belongs to W. So, let $\rho_S(b) = c$. By construction of $\mathcal{M}^*$, we have: $N^*(a) = a$, $N^*(b) = \rho_S(b) = c$ and $aR_\tau c$. Hence $\mathcal{M}_a^* \models \Diamond_\tau b$.

(b) G is not a nominal, and therefore it is native. Since S is saturated and $a \in W$ is not blocked, there is a nominal b such that $\rho_S(b)$ is defined and $@_a\Diamond_\tau b, @_b G \in S$. Since $\rho_S(b)$ is a twin of b and G is native, if $c = \rho_S(b)$ we have $@_c G \in S$ by Lemma 3.7.iii. By construction of $\mathcal{M}^*$, $aR_\tau c$, and, by the inductive hypothesis, $\mathcal{M}_c^* \models G$. Hence $\mathcal{M}_a^* \models \Diamond_\tau G$.

(iii) F is $\Diamond_\tau^- G$. Observe that, differently from the above case, G is necessarily native (Lemma 3.2), because $@_a\Diamond_\tau^- c$ is not a relational formula, and this case is treated similarly to the second item of case ii.

(iv) $F = \Box_\tau^- G$. Let b be any element of W such that $bR_\tau a$. By definition, there are two nominals c and d such that $a = \rho_S(c)$, $b = \rho_S(d)$ and $@_d\Diamond_\tau c \in S$. Since a and c are twins and $\Box_\tau^- G$ is native, $@_c\Box_\tau^- G \in S$ by Lemma 3.7.iii. And since S is saturated, $@_d G \in S$, so that also $@_b G \in S$ because b and d are twins and G is native (Lemma 3.7.iii again). By the induction hypothesis, then $\mathcal{M}_b^* \models G$. Therefore $\mathcal{M}_a^* \models \Box_\tau^- G$.

□

Completeness is proved by lifting the model existence property upwards to the root set.

Theorem 3.11 (Completeness) *If S is unsatisfiable, then every complete tableau for S is closed.*

Proof. The proof consists in showing that if Θ is a complete and open branch of a tableau rooted at S_0, then S_0 is satisfiable.

Since Θ is finite by Theorem 3.5, $\Theta = S_0, S_1, ..., S_k$ for some k, where S_k is a complete and open node.

Let $\mathcal{M}^* = \langle W, \{R_\tau \mid \tau \in \mathsf{REL}\}, N^*, I\rangle$ be the model of S_k given by Lemma 3.10. Since N^* is undefined for nominals that do not occur in S_k, we can safely extend it to interpret all the nominals occurring in Θ. In order to do it, we define an equivalence relation on nominals (with respect to the branch Θ) as follows: $a \sim b$ if some node of Θ contains $@_a b$. The relation $\approx$ is the reflexive, symmetric and transitive closure of $\sim$. If a_0 is any native nominal occurring in S_k, then N is the extension of N^* such that for all nominals c occurring in Θ:

$$N(c) = \begin{cases} N^*(c) & \text{if } c \in W, \text{ i.e. } c \text{ occurs in } S_k \\ N^*(d) & \text{if for some } d \in W,\ c \approx d \\ a_0 & \text{otherwise} \end{cases}$$

It is clear that if some node of Θ contains $@_a b$, then $N(a) = N(b)$.

If $\mathcal{M} = \langle W, \{R_\tau \mid \tau \in \mathsf{REL}\}, N, I\rangle$, obviously, it still holds that for every $@_aF \in S_k$, if $a \in W$ and F has not the form $\Diamond_\tau b$ for some b with $\rho_{S_k}(b)$ undefined, then $\mathcal{M}_{N(a)} \models F$. We now prove that the satisfaction property propagates upwards, restricting our attention to nominals that are not deleted in Θ. Let us say that a formula $@_aF$ is *relevant* (w.r.t. Θ) if and only if either F is native, or both the following conditions hold:

- $@_aF$ contains only nominals that are never deleted in Θ, and
- F has not the form $\Diamond_\tau b$ for some b with $\rho_{S_k}(b)$ undefined.

Let us say that an interpretation $\mathcal{M}$ is a Θ-model of a node S of Θ if for every relevant formula $@_aF \in S$, $\mathcal{M}_{N(a)} \models F$. Obviously the specific interpretation $\mathcal{M}$ defined above is a Θ-model of S_k. We show that, for every $i = 0, ..., k-1$:

(•) if $\mathcal{M}$ is a Θ-model of S_{i+1}, then $\mathcal{M}$ is a Θ-model of S_i.

When $i = 0$ this is what we want, because the initial set obviously contains only native (hence relevant) formulae.

In order to prove (•), the cases where S_{i+1} is obtained from S_i by applying any rule but *Sub* are trivial, since $S_i \subseteq S_{i+1}$. So the only non-trivial case is the substitution rule, where:

$$S_i = @_ab, S'$$
$$S_{i+1} = S'^{\#}[a \mapsto b]$$

Since $N(a) = N(b)$ (by definition), $\mathcal{M}_{N(a)} \models b$, therefore $\mathcal{M}_{N(a)} \models @_ab$.

Let now $@_cF$ be any relevant formula in $S' = S_i \setminus \{@_ab\}$ such that $@_cF \neq (@_cF)[a \mapsto b] \in S_{i+1}$. If $@_cF$ is relevant then also $(@_cF)[a \mapsto b]$ is relevant. In fact:

- if $@_cF$ is native, then also $(@_cF)[a \mapsto b]$ is native because b is a native nominal (Lemma 3.2).
- If $@_cF$ contains only nominals that are never deleted in Θ, then the same holds for $(@_cF)[a \mapsto b]$. In fact, the only nominal possibly occurring in $(@_cF)[a \mapsto b]$ and not in $@_cF$ is b, and b is native (Lemma 3.2), so it cannot be deleted.
- If $@_cF$ is not a relational formula, obviously $(@_cF)[a \mapsto b]$ is not a relational formula either. So let us assume that $F = \Diamond_\tau d$ where $\rho_{S_k}(d)$ is defined. If $d \neq a$ there is nothing to prove (in fact, $(@_c\Diamond_\tau d)[a \mapsto b] = @_{c[a\mapsto b]}\Diamond_\tau d$ and $\rho_{S_k}(d)$ is defined). If $d = a$, then $(@_c\Diamond_\tau d)[a \mapsto b] = @_{c[a\mapsto b]}\Diamond_\tau b$. Since $\Diamond_\tau b$ is native, $(@_cF)[a \mapsto b]$ is relevant.

Therefore by the inductive hypothesis

$$\mathcal{M}_{N(c[a\mapsto b])} \models F[a \mapsto b]$$

where $c[a \mapsto b] = b$ if $c = a$, and $c[a \mapsto b] = c$ otherwise. Since $N(a) = N(b)$, $\mathcal{M}_{N(c[a\mapsto b])} \models F$. If $c[a \mapsto b] = c$, we are done. Otherwise, if $c[a \mapsto b] = b$ then $c = a$, so $N(c[a \mapsto b]) = N(b) = N(a) = N(c)$. Hence, also in this case $\mathcal{M}_{N(c)} \models F$. □

4 Concluding remarks

In this work we have extended the treatment of nominal equalities by means of substitution and nominal deletion, proposed in [6,5] for $HL(@)$, to the global and converse modalities. The approach followed in this work is *procedural*, in that the formal system embodies algorithmic choices. As a consequence, the proposed calculus, proved to be sound, complete and terminating, can be almost directly "synthesized" into a running prover, with a minimal need of checking that the good properties of the system are preserved.

The features of the calculus defined in this paper can be summarized into the following main points:

(i) No use of prefixes: the calculus is *internalized.*

(ii) A practical approach to the treatment of nominal equalities: they are handled by means of substitution, accompanied by the deletion of the chain of nominals generated by the replaced one.

(iii) Ancestor equality blocking for the $\Diamond_\tau$ and $\Diamond_\tau^-$ rules: formulae of the form $@_a\Diamond_\tau F$ or $@_a\Diamond_\tau^- F$ are not expanded when either there exists an ancestor b of a which labels the same set of native formulae as a (a is directly blocked) or the father of a is blocked (a is indirectly blocked).

(iv) A mechanism to prevent useless re-applications of the E-rule: a formula of the form $@_a\mathsf{E}F$ is not expanded whenever any $@_b\mathsf{E}F$ has already been expanded.

As regard to point iii, indirect blocking and equality loop checks (as opposed to subset blocking) are a necessity in the presence of the converse modalities. The paper shows that, in such a context, a simple substitution rule with no side effects yields a non-terminating calculus. And in fact, although substitution-based decision procedures have been defined for $HL(@,\mathsf{E})$, to our knowledge none of them has been extended to deal with the converse modalities.

Point ii is quite similar to the "merge and prune" mechanism used in [14], where a tableau calculus for description logic with nominals, number restrictions and transitive and inverse roles is defined. The overall context is however quite different: there, a tableau branch is a graph of nodes, labelled by sets of formulae and describing worlds in the model. Nodes containing a nominal in their label are called *nominal nodes*, the others are *blockable nodes*. When two nodes contain the same nominal, they are merged into one of them and the blockable successors of the other node (which is deleted) are "pruned". The blocking mechanism used there only affects blockable nodes, and requires *pairwise blocking* (pairwise blocking seems to be a necessity when both inverse roles and number restrictions are present). The possibility and usefulness of exporting some of the ideas in our approach to the context of description logics have to be explored.

The first tableau-based decision procedures for $HL(@,\mathsf{E})$ were proposed in [4], where systems defined in [17,18,1] are reformulated, and the global modalities are added. The first two systems (reformulations of [17] and [18]) use prefixes, while the third one is internalized (like in point i of our approach). The treatment of nominal equalities in the three systems is different: either by copying formulae from a "world" to another, or by means of substitution (similarly to point ii, but without nominal deletion), or else by means of a set of quite "natural" rules for equality. The procedures involve loop-checks based on subset blocking (which is sufficient in the absence of the converse modalities).

The first tableau-based decision procedure for hybrid logic including both the global and converse modalities was given in [2]. Ancestor equality blocking (point iii above) is the same blocking mechanism used *for every nominal generating rule* in the prefixed calculus defined there (and the internalized one given in [3]). The main difference between H and H^+ and the internalized calculi for the corresponding logics in [2,3] is represented by ii. In fact, in the latter systems, an equality $@_a b$ is handled by copying formulae labelled by a to b, except for accessibility formulae. When the copied formula generates a new nominal, two copies of such a nominal are therefore expanded, which can in turn generate copies of the same nominals, and so on. The redundancy of such a treatment of equalities and the computational gain of substitution with nominal deletion has been experimentally verified. The results of the comparison are reported in [9,8], which also compares the implementation of H with HTab [12], the more mature prover based on the prefixed calculus defined in [2].

As already remarked, the treatment of existential formulae in H^+ (point iv above) recalls the mechanism used in [16], which defines another procedure including the converse and difference modalities (which subsume the global ones). A similar mechanism (called *pattern-based blocking*) is used to block the expansions of $\Diamond$-formulae for the sublogic with no converse operators. In fact, pattern-based blocking yields a terminating system only in the absence of the converse modalities, and provided that applications of the $\Box$-rule are prioritized. When the converse modalities are present, chain-based blocking (with indirect blocking) has to be used. In [16], equalities are dealt with in an abstract and declarative way, constituting a still different approach w.r.t. point ii above. Algorithmic choices about the concrete treatment of such formulae are left open. A corresponding procedural approach is presented in [15], where a substitution rule, without nominal deletion, is used, in the context of the difference modality, but no converse operators. Differently from H^+ (point iii), pattern-based blocking is applied there to block the application of any nominal generating rule. Spartacus [11,10], the prover implementing tableaux for $HL(@, \mathsf{E})$ on the basis of the works in question, processes nominal equalities by merging the content of the corresponding "nodes", and electing one of them as the representative of both. Again, nominal generating formulae are there treated by pattern-based blocking.

Guidelines for future work include, on the practical side, the extension of the already mentioned prover [9,8], which at present implements only the restricted calculus H, and its refinement so as to include some basic optimization techniques. This would allow for an experimental verification of the fact that the extended calculus still benefits of the computational advantages of substitution with nominal deletion. On the theoretical side, the integration of the substitution mechanism used in this work with still more expressive languages can be studied.

ACKNOWLEDGEMENTS. The authors thank the anonymous referees of this work, who, with their useful comments and suggestions, gave the opportunity to make the presentation clearer and more accurate.

References

[1] Blackburn, P. and M. Marx, *Tableaux for quantified hybrid logic*, in: U. Egly and C. Fermüller, editors, *Automated Reasoning with Analytic Tableaux and Related Methods (TABLEAUX 2002)*, LNAI **2381** (2002), pp. 38–52.

[2] Bolander, T. and P. Blackburn, *Termination for hybrid tableaus*, Journal of Logic and Computation **17** (2007), pp. 517–554.

[3] Bolander, T. and P. Blackburn, *Terminating tableau calculi for hybrid logics extending K*, Electronic Notes in Theoretical Computer Science **231** (2009), pp. 21–39, proceedings of the 5th Workshop on Methods for Modalities (M4M-5), 2007.

[4] Bolander, T. and T. Braüner, *Tableau-based decision procedures for hybrid logic*, Journal of Logic and Computation **16** (2006), pp. 737–763.

[5] Cerrito, S. and M. Cialdea Mayer, *An efficient approach to nominal equalities in hybrid logic tableaux*, Journal of Applied Non-classical Logics (To appear).

[6] Cerrito, S. and M. Cialdea Mayer, *Terminating tableaux for HL(@) without loop-checking*, Technical Report IBISC-RR-2007-07, Ibisc Lab., Université d'Evry Val d'Essonne (2007), (`http://www.ibisc.univ-evry.fr/Vie/TR/2007/IBISC-RR2007-07.pdf`).

[7] Cerrito, S. and M. Cialdea Mayer, *Tableaux with substitution for hybrid logic with the global and converse modalities*, Technical Report RT-DIA-155-2009, Dipartimento di Informatica e Automazione, Università di Roma Tre (2009), (`http://web.dia.uniroma3.it/ricerca/rapporti/rt/2009-155.pdf`).

[8] Cialdea Mayer, M. and S. Cerrito, *Herod and Pilate: two tableau provers for basic hybrid logic*, in: J. Giesl and R. Hähnle, editors, *Proceedings of IJCAR 2010*, LNAI **6173** (2010), pp. 255–262.

[9] Cialdea Mayer, M., S. Cerrito, E. Benassi, F. Giammarinaro and C. Varani, *Two tableau provers for basic hybrid logic*, Technical Report RT-DIA-145-2009, Dipartimento di Informatica e Automazione, Università di Roma Tre (2009), (`http://web.dia.uniroma3.it/ricerca/rapporti/rt/2009-145.pdf`).

[10] Götzmann, D., "Spartacus: A Tableau Prover for Hybrid Logic," Master's thesis, Saarland University (2009).

[11] Götzmann, D., M. Kaminski and G. Smolka, *Spartacus: A tableau prover for hybrid logic*, in: *M4M6*, number 128 in Computer Science Research Reports, Roskilde University, 2009, pp. 201–212.

[12] Hoffmann, G. and C. Areces, *HTab: A terminating tableaux system for hybrid logic*, Electronic Notes in Theoretical Computer Science **231** (2009), pp. 3–19, proceedings of the 5th Workshop on Methods for Modalities (M4M-5), 2007.

[13] Horrocks, I. and U. Sattler, *A description logic with transitive and inverse roles and role hierarchies*, Journal of Logic and Computation **9** (1999), pp. 385–410.

[14] Horrocks, I. and U. Sattler, *A tableau decision procedure for $\mathcal{SHOIQ}$*, Journal of Automated Reasoning **39** (2007), pp. 249–276.

[15] Kaminski, M. and G. Smolka, *Hybrid tableaux for the difference modality*, Electronic Notes in Theoretical Computer Science **231** (2009), pp. 241–257, proceedings of the 5th Workshop on Methods for Modalities (M4M-5), 2007.

[16] Kaminski, M. and G. Smolka, *Terminating tableau systems for hybrid logic with difference and converse*, Journal of Logic, Language and Information **18** (2009), pp. 437–464.

[17] Tzakova, M., *Tableau calculi for hybrid logics*, in: N. Murray, editor, *Automated Reasoning with Analytic Tableaux and Related Methods (TABLEAUX 1999)*, LNAI **1617** (1999), pp. 278–292.

[18] van Eijck, J., *Constraint tableaux for hybrid logics* (2002), manuscript, CWI, Amsterdam.

Future Event Logic – Axioms and Complexity

Hans van Ditmarsch

Logic, University of Sevilla, Spain
hvd@us.es

Tim French

Computer Science and Software Engineering, The University of Western Australia
tim@csse.uwa.edu.au

Sophie Pinchinat

IRISA, University of Rennes
Sophie.Pinchinat@irisa.fr

Abstract

In this paper we present a sound and complete axiomatization of future event logic. Future event logic is a logic that generalizes a number of dynamic epistemic logics, by using a new operator $\triangleright$ that acts as a quantifier over the set of all refinements of a given model. (A refinement is like a bisimulation except that from the three relational requirements only 'atoms' and 'back' need to be satisfied.) Thus the logic combines the simplicity of modal logic with some powers of monadic second order quantification. We prove the axiomatization is sound and complete and discuss some extensions to the result.

Keywords: Bisimulation Quantifier, Modal Logic, Temporal Epistemic Logic, Multi-Agent System

1 Introduction

Modal logic is frequently used for modelling knowledge in multi-agent systems. The semantics of modal logic uses the notion of "possible worlds", between which an agent is unable to distinguish. In dynamic systems agents acquire new knowledge (say by an announcement, or the execution of some action) that allows to distinguish between worlds they previously could not separate. From the agents point of view, what were "possible worlds" become inconceivable. Thus, a future informative event may be modelled by a reduction in the agent's accessibility relation. In [21] the *future event logic* is introduced. It augments the multi-agent logic of knowledge with (only) an operation $\blacktriangleright\phi$ that stands for "ϕ holds after all informative events" — the diamond version $\triangleright\phi$ stands for "there is an informative event after which ϕ." The semantics of informative events axiomatized in this paper was presented in [21]; it encompasses action model execution à la Baltag et al [4]: on finite models, it can be easily shown that a model resulting from action model execution is a refinement of the initial model, and for a given refinement of a model we can construct an action model such that its execution is bisimilar to that refinement. Here we examine the important questions that arise for a new logic: expressivity; axiomatizations; and complexity. We visit these questions in both the context of modal logic, and the modal μ-calculus.

Previous works [10,15] have modelled informative events using a notion of model refinement. In [15] it was shown that model restrictions were not sufficient to simulate informative events, and they introduced *refinement trees* for this purpose—a precursor of the semantics of dynamic epistemic logics developed later [22]. We incorporate implicit quantification over informative events directly into the language using a similar notion of *refinement*; in our case a refinement is the inverse of simulation [1]. This work is also closely related to some recent work on bisimulation quantified modal logics [9,11]. The future event operators are weaker operators than bisimulation quantifiers [21], as they are only based on simulations rather than bisimulations, and do not allow us to vary the interpretation of propositional atoms. Bisimulation quantified modal logic has previously been axiomatized by providing a provably correct translation to the modal μ-calculus [8] (albeit a very complicated one).

Thus we may consider *refinement quantification* to be a generalization of future event operators [21] to other modal logics. This is significant in that it motivates the application of the new operator in many different settings: In logics for games [17,2] or in control theory [18,20], it may correspond to a player discarding some moves; for program logics [12] it may correspond to operational refinement [16]; and for topologics it may correspond to sub-space projections.

This paper will present the definitions for refinement quantification in the general settings of modal logic and the modal μ-calculus, and seek to motivate their use in a range of applied logics. We will then address the questions of expressivity, complexity and axiomatization. Specifically: sound and complete axiomatizations will be provided for both modal logic and the modal μ-calculus augmented with refinement quantification; we provide a double exponential upper-bound for each logic; and we show the use of refinement quantification does not change the expressive power of the logics, although they do make each logic exponentially more succinct.

2 Technical preliminaries

Structural notions

Assume a finite set of agents A and a countably infinite set of atoms P.

Definition 2.1 [Structures] A *model* $M = (S, R, V)$ consists of a *domain* S of (factual) *states* (or *worlds*), *accessibility* $R : A \to \mathcal{P}(S \times S)$, and a *valuation* $V : P \to \mathcal{P}(S)$. For $s \in S$, (M, s) is a *state* (also known as a pointed Kripke model).

For $R(a)$ we write R_a; accessibility R can be seen as a set of relations R_a, and V as a set of valuations $V(p)$. Given two states s, s' in the domain, $R_a(s, s')$ means that in state s agent a considers s' a possibility. As we will be often required to discuss several models at once, we will use the convention that $M = (S^M, R^M, V^M)$, $N = (S^N, R^N, V^N)$ etc. Also, given $s \in S^M$, we let M_s refer to the pair (M, s) or the *pointed model.*

In the first instance we will assume that there are no further restrictions on the *models.* That is, the underlying modal logic is $\mathcal{L}$ whose system of axioms is **K**. In future work we will consider how our results may be extended to epistemic logics, such as **S5** and **KD45**.

Definition 2.2 [Bisimulation, simulation, refinement] Let two models $M = (S, R, V)$ and $M' = (S', R', V')$ be given. A non-empty relation $\mathfrak{R} \subseteq S \times S'$ is a bisimulation, iff for all $s \in S$ and $s' \in S'$ with $(s, s') \in \mathfrak{R}$, for all $a \in A$:

atoms $s \in V(p)$ iff $s' \in V'(p)$ for all $p \in P$

forth-a for all $t \in S$, if $R_a(s, t)$, then there is a $t' \in S'$ such that $R'_a(s', t')$ and $(t, t') \in \mathfrak{R}$

back-a for all $t' \in S'$, if $R'_a(s', t')$, then there is a $t \in S$ such that $R_a(s, t)$ and $(t, t') \in \mathfrak{R}$

We write $M_s \leftrightarrows M'_{s'}$, iff there is a bisimulation between M and M' linking s and s'. Then we call M_s and $M'_{s'}$ bisimilar.

A relation that satisfies **atoms** and **forth-a** for every $a \in A$ is a *simulation*, and in that case $M'_{s'}$ is a *simulation* of M_s, and M_s is a *refinement* of $M'_{s'}$, and we write $M_s \preceq M'_{s'}$ (or $M'_{s'} \succeq M_s$).

A relation that satisfies **atoms** and **forth-b** for every $b \in A$, *as well as* **back-b** for every $b \in A - \{a\}$ is an *a-simulation*, and in that case $M'_{s'}$ is an *a-simulation* of M_s, and M_s is an *a-refinement* of $M'_{s'}$, and we write $M_s \preceq_a M'_{s'}$ (or $M'_{s'} \succeq_a M_s$).

We note that the definition of simulation and refinement above varies slightly to the one given by Blackburn et al [6]. Here we ensure that simulations and refinements preserve the interpretations of atoms, whereas [6], has them only preserving the truth of atoms. We take this approach as we feel it suits the epistemic domain we aspire to. It is also important to note that in an epistemic setting a refinement corresponds to the *diminishing uncertainty* of agents[1]. This means that there is a potential *decrease* in the number of states and transitions in a model. This is perhaps contrary to the concept of program refinement [16] where detail is added to a specification. However, in program refinement the added detail requires a more detailed state space (i.e. extra atoms) and as such is more the domain of bisimulation quantifiers, rather than refinement quantification. It is interesting to note the consequence of program refinement is a more deterministic system which agrees with the notion of diminishing uncertainty.

We give the following lemma for the properties of the relation $\succeq_a$.

[1] At least, with respect to formulas in which knowledge operators appear within the scope of an even number of negations. It is possible that in a refinement one agent may be less certain about what another agent does not know.

Lemma 2.3 *The relation $\succeq_a$ is reflexive and transitive (a pre-order), and satisfies the Church-Rosser property.*

Proof. Reflexivity follows from the observation that the identity relation satisfies **atoms**, and **back-**a and **forth-**a for all agents a, and therefore also the weaker requirement for refinement. Similarly, given two a-simulations B_1, and B_2, we can see that their composition, $\{(x, z) \mid \exists y,\ (x, y) \in B_1,\ (y, z) \in B_2\}$ is also an a-simulation. This is sufficient to demonstrate transitivity. The Church-Rosser property states that if $N_t \succeq_a M_s$ and $N_t \succeq_a M'_{s'}$, then there is some model $N'_{t'}$ such that $M_s \succeq_a N'_{t'}$ and $M'_{s'} \succeq_a N'_{t'}$. From Definition 2.2 it follows that M_s and $M'_{s'}$ must be bisimilar to one another with respect to $A - \{a\}$. We may therefore construct such a model $N'_{t'}$ by taking M_s (or $M'_{s'}$) and setting $R_a^{N'} = \emptyset$ and $R_b^{N'} = R_b^M$ for all $b \in A - \{a\}$. It can be seen that $N'_{t'} = (S^M, R^{N'}, V^M, s)$ satisfies the required properties. □

Finally, note that if $N_t \succeq_a M_s$ and $M_s \succeq_a N_t$ it is not necessarily the case that $M_s \leftrightarrows N_t$.

For example, consider the one agent models M and N where:

- $S^M = \{1, 2, 3\}$, $R_a^M = \{(1, 2), (2, 3)\}$ and $V^M(p) = \emptyset$ for all $p \in P$; and
- $S^N = \{a, b, c, d\}$, $R_a^N = \{(a, b), (b, c), (a, d)\}$ and $V^M(p) = \emptyset$ for all $p \in P$.

These two models are clearly not bisimilar, although $M_1 \preceq N_a$ via $\{(1, a), (2, b), (3, c)\}$ and $N_a \preceq M_1$ via $\{(a, 1), (b, 2), (c, 3), (d, 2)\}$.

3 Syntax and semantics

Assuming an interpretation where different $\Box_a$ operators stand for different epistemic operators (each describing what an agent knows), *future event logic* is able express what informative events are consistent with a given information state. The syntax and the semantics of future event logic are as follows.

Definition 3.1 [Language of $\mathcal{L}_{\triangleright}$] Given a finite set of agents A and a set of propositional atoms P, the language of $\mathcal{L}_{\triangleright}$ is inductively defined as

$$\phi ::= p \mid \neg\phi \mid (\phi \wedge \phi) \mid \Box_a\phi \mid \blacktriangleright_a\phi$$

where $a \in A$ and $p \in P$.

Standard abbreviations include: $\phi \vee \psi$ iff $\neg(\neg\phi \wedge \neg\psi)$; $\phi \to \psi$ iff $\neg\phi \vee \psi$; $\Diamond_a\phi$ iff $\neg\Box_a\neg\phi$. We write $\triangleright_a\phi$ for $\neg\blacktriangleright_a\neg\phi$. We propose a dynamic modal way to interpret the refinement quantification. This means that our future is the *computable future*: $\triangleright_a\phi$ is true now, iff there is an (unspecified) informative event for agent a, or a-refinement, after which ϕ is true.

Definition 3.2 [Semantics of future event logic] Assume an epistemic model $M =$

(S, R, V). The interpretation of $\phi \in \mathcal{L}_{\triangleright}$ is defined by induction.

$$M_s \models p \text{ iff } s \in V_p$$

$$M_s \models \neg\phi \text{ iff } M_s \not\models \phi$$

$$M_s \models \phi \wedge \psi \text{ iff } M_s \models \phi \text{ and } M_s \models \psi$$

$$M_s \models \Box_a \phi \text{ iff for all } t \in S : (s,t) \in R_a \text{ implies } M_t \models \phi$$

$$M_s \models \blacktriangleright_a \phi \text{ iff for all } M'_{s'} : M_s \succeq_a M'_{s'} \text{ implies } M'_{s'} \models \phi$$

The logic without the refinement quantifier $\blacktriangleright_a$ is the logic $\mathcal{L}$ of multi-agent epistemic logic.

In other words, $\blacktriangleright_a \phi$ is true in an epistemic state iff ϕ is true in all of its *a-refinements*. Note the inverse direction in the definition: the future epistemic state refines the current epistemic state. Typical model operations that produce an a-refinement are: blowing up the model (to a bisimilar model) such as adding copies that are indistinguishable from the current model and one another, removing states accessible only by agent a, and removing pairs of the accessibility relation for the agent a. Validity in a model, and validity, are defined as usual. For an extended discussion of these semantics and a comparison to related logics see [21].

Lemma 3.3 *The logic $\mathcal{L}_{\triangleright}$ is bisimulation invariant.*

Proof. This is straightforward, noting $\Box_a$ is bisimulation invariant, and the new operator $\blacktriangleright_a$ is clearly bisimulation invariant since a-simulation is transitive and bisimulation is just a specific type of simulation. Therefore, if $M_s \leftrightarroweq N_t$, and O_u is any model such that $O_u \preceq_a M_s$ then $M_s \preceq_a N_t$, so by Lemma 2.3, we have $O_u \preceq_a N_t$. Thus, $N_t \models \blacktriangleright_a \phi$ implies $M_s \models \blacktriangleright_a \phi$. The reverse direction is symmetric. □

Additionally, we may define $\mathcal{L}^{\mu}_{\triangleright}$, by including the fixed-point operators μ and ν. Specifically:

Definition 3.4 [Language of $\mathcal{L}^{\mu}_{\triangleright}$] Given a finite set of agents A and a set of propositional atoms P, the language of $\mathcal{L}^{\mu}_{\triangleright}$ is inductively defined as

$$\phi ::= p \mid \neg\phi \mid (\phi \wedge \phi) \mid \Box_a \phi \mid \blacktriangleright_a \phi \mid \mu x.\phi$$

where $a \in A$, $p \in P$, and the atom x only occurs positively (i.e. in the scope of an even number of negations) in the formula ϕ. We will refer to such an atom x as a *fixed-point variable*. The formula $\nu x.\phi$ is an abbreviation for $\neg\mu x.\neg\phi[\neg x \backslash x]$.

The restriction of this logic to the fragment without refinement quantifiers (the modal μ-calculus) will be referred to as $\mathcal{L}^{\mu}$. An important technical definition we require is that of a *disjunctive formula*. Let Γ be a finite set of $\mathcal{L}^{\mu}_{\triangleright}$ formulas. We let the *cover* operator $\nabla_a \Gamma$ be an abbreviation for $\Box_a \bigvee_{\gamma \in \Gamma} \gamma \wedge \bigwedge_{\gamma \in \Gamma} \Diamond_a \gamma$. (To avoid ambiguity, we note $\bigvee_{\gamma \in \emptyset} \gamma$

is always false, whilst $\bigwedge_{\gamma\in\emptyset}\gamma$ is always true). This operator has previously been used in the definition of disjunctive formulae [8], and has recently been axiomatized [5]. We also note its dual may be written $\triangle_a\Gamma$ as an abbreviation for $\Diamond_a \bigwedge_{\gamma\in\Gamma}\gamma \vee \bigvee_{\gamma\in\Gamma}\Box_a\gamma$.

Definition 3.5 [Disjunctive formula] A *disjunctive formula (df)* is specified by the following abstract syntax:

$$\alpha ::= x \mid \alpha \vee \alpha \mid \mu x.\alpha \mid \nu x.\alpha \mid \pi \wedge \overline{\nabla\Gamma} \mid \blacktriangleright_a\alpha \mid \triangleright_a\alpha$$

where π is a conjunction of free literals (atoms or negated atoms, but not fixed-point variables), and $\overline{\nabla\Gamma}$ is an abbreviation for $\nabla_{a_1}\Gamma_{a_1} \wedge ... \wedge \nabla_{a_n}\Gamma_{a_n}$ such that $a_1, ..., a_n$ are distinct elements of the set A, and each Γ_{a_i} is a finite set of disjunctive formulas. To avoid ambiguity we may refer to the disjunctive formulas of $\mathcal{L}^\mu$ (the ones without $\blacktriangleright_a$ or $\triangleright_a$ operators) as μ-disjunctive formulas.

Proposition 3.6 *Every formula ϕ of $\mathcal{L}^\mu$ is equivalent to a μ-disjunctive formula,*

This is shown in [13].

Example 3.7 [Knowledge and belief] Given are two agents that are uncertain about the value of a fact p, and where this is common knowledge, and where p is true. We assume that both accessibility relations are equivalence relations, and that the epistemic operators model the agents' knowledge. An informative event is possible after which a knows that p but b does not know that; this is expressed by

$$\triangleright(\Box_a p \wedge \neg\Box_b\Box_a p)$$

In Figure 1, the structure is on the left, and its refinement validating the post-condition is on the right. In this visualization, the actual state is the (bottom) right one, and states that are indistinguishable for an agent are linked and labeled with the name of that agent, and transitivity and reflexivity are assumed (so on the right, all three states are indistinguishable for agent b). Note that on the left, the formula $\triangleright(\Box_a p \wedge \neg\Box_b\Box_a p)$ is true, because $\Box_a p \wedge \neg\Box_b\Box_a p$ is true in the right structure: in the actual state there is no alternative for agent a, so $\Box_a p$ is true, whereas agent b considers it possible that the top-state is the actual state, and in that state agent a considers it possible that p is false. Therefore, $\neg\Box_b\Box_a p$ is also true in the bottom right state.

Example 3.8 [Controller synthesis] Consider a discrete-event system S to be controlled, with two possible actions c and u. Given a control objective ϕ expressed in, say the μ-calculus, the following formulas express respectively the well-known verification/synthesis problems:

- Controller synthesis: Assume action c is controllable as opposed to u. The system S is *controllable for ϕ* if and only if, $S \models \triangleright_c\phi$, as a c-refinement of S denotes the result of applying some control acting on action c.
- Module checking [14]: The system S is interpreted as an *open system* where action c is internal and action u comes from the environment. The system S satisfies ϕ whatever

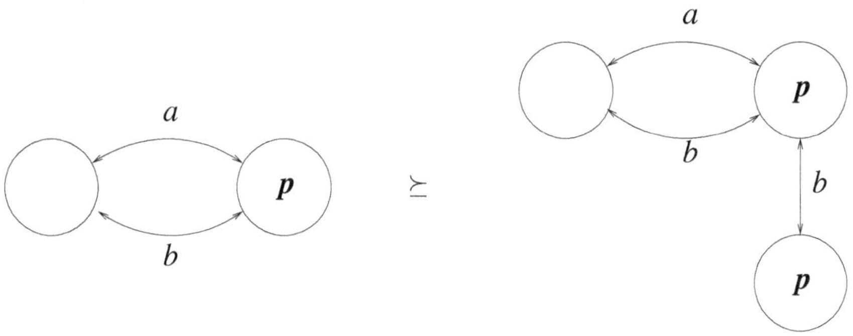

Fig. 1. The refinement in Example 3.7.

the environment does but blocking if, and only if, $S \models \blacktriangleright_u(\texttt{NonBlocking} \Rightarrow \phi)$, where $\texttt{NonBlocking} \equiv \nu x.\Diamond_u \top \wedge \Box x$ is an invariant telling that there always exists an environment reaction. The u-refinements with hypothesis NonBlocking denotes all possible non-blocking environments.

- Advanced controller synthesis: We combine the two cases above. We reconsider the control problem for ϕ where S is interpreted as an open system. The open system S can be controlled to achieve property ϕ if, and only if,

$$S \models \triangleright_c \blacktriangleright_u(\texttt{NonBlocking} \Rightarrow \phi)$$

Example 3.9 [Program logic] Consider a specification, MUTEX, of a mutual exclusion protocol and some property ϕ of this protocol specified in CTL. Now we may ask if we can find a refinement of MUTEX that satisfies ϕ but also such that if Process i is in the critical section ($cs(i)$) at time $n+1$, then this is known at time n. This is expressed as

$$\texttt{MUTEX} \models \triangleright(AG[EXcs(i) \Rightarrow AXcs(i)] \wedge \phi)$$

The refinement consists in moving the nondeterministic choices forward, so that a fork at time n becomes a fork at time $n-1$ with each branch having a single successor at time n.

We also note that Section 6.2 presents an application of the refinement quantification to two-player asynchronous games.

4 Axiomatization: $\mathcal{L}_\triangleright$

Here we present a series of axioms for the logic $\mathcal{L}_\triangleright$. We will derive a number of validities, show the axioms to be sound, and discuss a general strategy for showing their completeness. For simplicity, we will present the axiomatization in the single agent case (and hence the $\Box_a$ operator will simply be referred to as $\Box$), although the axiomatization and proofs easily generalize to the multi-agent case. We will also use the relation R simply as a set of pairs $\subseteq S^M \times S^M$, and use the abbreviation $sR^M = \{u \in S^M \mid (s,u) \in R^M\}$.

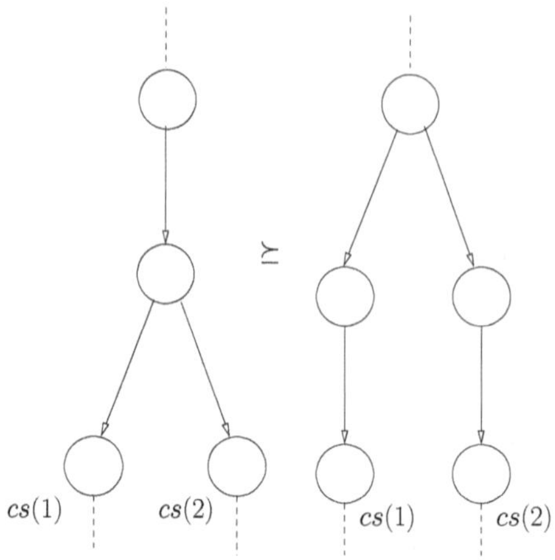

Fig. 2. The refinement in Example 3.9.

The axiomatization presented is a substitution schema, since the substitution rule itself is not valid. Note that for all atomic propositions p, $p \to \blacktriangleright p$, but the same is not true for an arbitrary formula (substitute $\Diamond\top$ for p in the formula of Example 3.7). This propositional case itself is presented as axiom **G1** and prevents the logic **FEL** from being a normal modal logic.

Definition 4.1 The axiomatization **FEL** is such that the axioms are all substitution instances of the following:

P All tautologies of propositional logic

K $\Box(\phi \to \psi) \to \Box\phi \to \Box\psi$

G0 $\blacktriangleright(\phi \to \psi) \to \blacktriangleright\phi \to \blacktriangleright\psi$

G1 $\alpha \leftrightarrow \blacktriangleright\alpha$ where α is a propositional formula

GK $\bigwedge_{\gamma\in\Gamma} \Diamond\triangleright\gamma \leftrightarrow \triangleright\nabla\Gamma$

along with the rules:

MP From $\vdash \phi \to \psi$ and $\vdash \phi$ infer $\vdash \psi$

Nec1 From $\vdash \phi$ infer $\vdash \Box\phi$

Nec2 From $\vdash \phi$ infer $\vdash \blacktriangleright\phi$

The axiomatization **K** for the logic $\mathcal{L}$ consists of the axioms **P**, **K**, and the rules **MP** and **Nec1**.

The axiomatization is surprisingly simple given the complexity of the semantic definition of the refinement quantification, $\blacktriangleright$. We note that while refinement is known to be reflexive, transitive and satisfies the Church-Rosser property (Lemma 2.3), the corresponding modal axioms are not required. Rather, these properties may be inferred from the axioms presented above.

4.1 *Example of derivation*

We present a simple derivation of $\Diamond\top \rightarrow \triangleright(\nabla\{p\} \vee \nabla\{\neg p\})$. In some cases several deductions have been combined into single statements, but this is restricted to cases of well known modal theorems.

$$
\begin{aligned}
1.\mathbf{P},\mathbf{Nec1},\mathbf{K} &\vdash \Diamond\top \leftrightarrow \Diamond(p \vee \neg p)\\
2.\mathbf{P},\mathbf{Nec1},\mathbf{K} &\vdash \Diamond(p \vee \neg p) \leftrightarrow (\Diamond p \vee \Diamond\neg p)\\
3.\text{See below} &\vdash \Diamond p \rightarrow \triangleright\nabla\{p\}\\
4.\text{See below} &\vdash \Diamond\neg p \rightarrow \triangleright\nabla\{\neg p\}\\
5.\mathbf{P},\mathbf{Nec2},\mathbf{G0} &\vdash \triangleright\Box p \rightarrow \triangleright(\nabla\{p\} \vee \nabla\{\neg p\})\\
6.\mathbf{P},\mathbf{Nec2},\mathbf{G0} &\vdash \triangleright\Box\neg p \rightarrow \triangleright(\nabla\{p\} \vee \nabla\{\neg p\})\\
7.\mathbf{P},\mathbf{MP} &\vdash \Diamond\top \rightarrow \triangleright(\nabla\{p\} \vee \nabla\{\neg p\})
\end{aligned}
$$

Lines 3 and 4 above require the following deduction, which is true for all propositional formula α:

$$
\begin{aligned}
1.\mathbf{G1} &\vdash \alpha \leftrightarrow \triangleright\alpha\\
2.\mathbf{P},\mathbf{Nec1},\mathbf{K} &\vdash \Diamond\alpha \leftrightarrow \Diamond\triangleright\alpha\\
3.\mathbf{GK}[\Gamma = \{\alpha\}] &\vdash \Diamond\alpha \leftrightarrow \triangleright\nabla\{\alpha\}
\end{aligned}
$$

4.2 *Soundness*

For notational convenience, given a finite set of $\mathcal{L}_\triangleright$ formulas, $\Gamma = \{\phi_1, \ldots, \phi_n\}$, we let $\triangleright\Gamma = \{\triangleright\phi \mid \phi \in \Gamma\}$ (and likewise for other unary operators).

Theorem 4.2 *The axiomatization* **FEL** *is sound for* $\mathcal{L}_\triangleright$.

Proof. As all models of $\mathcal{L}_\triangleright$ are models of $\mathcal{L}$, the schemas **P**, **K** and the rule **MP** and **Nec1** are all sound. We deal with the remaining schemas and rules below:

G0 Suppose that M_s is a model such that $M_s \models \blacktriangleright(\phi \rightarrow \psi)$. Then for every N_t, where $N_t \preceq M_s$, we have $N_t \models \phi \rightarrow \psi$. Therefore if it is also the case that for every N_t where $N_t \preceq M_s$, we have $N_t \models \phi$, then it follows that every such model also satisfies $\blacktriangleright\psi$.

G1 Suppose that α is a propositional formula. By Definition 2.2 for every model $N_t \preceq M_s$, for every propositional atom p, we have $s \in V^M(p)$ if and only if $t \in V^N(p)$. As the interpretation of α depends solely on the the valuation of propositions at s, then $M_s \models \alpha$ if and only if $N_t \models \alpha$ for every $N_t \preceq M_s$.

GK Suppose M_s is a model such that for some set Γ, $M_s \models \bigwedge_{\gamma\in\Gamma} \Diamond\triangleright\gamma$. Therefore for every $\gamma \in \Gamma$ there is some $t^\gamma \in sR^M$ such that $M_{t^\gamma} \models \triangleright\gamma$. Thus, for each $\gamma \in \Gamma$, there is some model $N^{\gamma}_{u^\gamma} \preceq M_{t^\gamma}$ such that $N^{\gamma}_{u^\gamma} \models \gamma$. Without loss of generality, we may assume that for each $\gamma \in \Gamma$ the models N^γ are disjoint. We construct the model M^Γ such that $S^{M^\Gamma} = S^M \cup \bigcup_{\gamma\in\Gamma} S^{N^\gamma}$, $R^{M^\Gamma} = \{(s, u^\gamma) \mid \gamma \in \Gamma\} \cup \bigcup_{\gamma\in\Gamma} R^{N^\gamma}$, and for all $p \in P$, $V^{M^\Gamma}(p) = V^M(p) \cup \bigcup V^{N^\Gamma}(p)$.

We can see that $M^\Gamma_s \preceq M_s$, via the relation $\mathcal{R}^\Gamma = \{(s,s)\} \cup \bigcup_{\gamma\in\Gamma} \mathcal{R}^\gamma$ where $\mathcal{R}^\gamma$ is the refinement relation corresponding to $N^\gamma_{u^\gamma} \preceq M_{t^\gamma}$. Furthermore, for each $t \in sR^{M^\Gamma}$ it is clear that $M^\Gamma_t \leftrightarroweq N^\gamma_{u^\gamma}$ for some γ, and thus $M^\Gamma_t \models \gamma$. Therefore $M^\Gamma_s \models \Box \bigvee_{\gamma\in\Gamma} \gamma$. Finally, for each $\gamma \in \Gamma$ there is some $u^\gamma \in sR^{M^\Gamma}$ where $M^\Gamma_{u^\gamma} \models \gamma$ so $M^\Gamma_s \models \bigwedge_{\gamma\in\Gamma} \Diamond\gamma$. Therefore $M^\Gamma_s \models \nabla\Gamma$, so $M_s \models \rhd\nabla\Gamma$.

Conversely, suppose that $M_s \models \rhd\nabla\Gamma$. Therefore, there is a model, $N_t \preceq M_s$ such that $N_t \models \nabla\Gamma$. Expanding the definitions, we have, for every $\gamma \in \Gamma$ there is some $u \in tR^N$ such that $N_u \models \gamma$, and for every $u \in tR^N$ there is some $v \in sR^M$ such that $N_u \preceq, M_v$. Combining these statements we have, for every $\gamma \in \Gamma$ there is some $v \in sR^M$ such that $M_v \models \rhd\gamma$, and thus $M_s \models \bigwedge_{\gamma\in\Gamma} \Diamond\rhd\gamma$.

Nec2 If ϕ is a validity, then it is satisfied by every model, so for any model M_s, ϕ is satisfied by every model $N_t \preceq M_s$, and hence every model satisfies $\blacktriangleright\phi$.

□

4.3 Completeness

We first show that every $\mathcal{L}_\rhd$ formula is logically equivalent to a $\mathcal{L}$ formula. We then show that if the latter is a theorem in **K**, the former is also a theorem, in **FEL**.

Lemma 4.3 *Every formula of $\mathcal{L}_\rhd$ is logically equivalent to a formula of $\mathcal{L}$.*

Proof. As the axiom **GK** is formulated in terms of the cover operator, it is convenient to prove this equivalence by means of an equally expressive version of the modal logic $\mathcal{L}$ that is also formulated with the cover operator [5].[2] (A direct proof in our own setting is quite possible, but considerably longer.) Consider the syntax of cover logic

$$\phi ::= \bot \mid \top \mid \phi \vee \phi \mid p \wedge \phi \mid \neg p \wedge \phi \mid \nabla\Gamma.$$

The semantics of $\nabla\Gamma$ is the obvious one if we recall our introduction by abbreviation of the cover operator: $M_s \models \nabla\Gamma$ iff for all $\phi \in \Gamma$ there is a $t \in R(s)$ such that $M_t \models \phi$, and for all $t \in R(s)$ there is a $\phi \in \Gamma$ such that $M_t \models \phi$. The modal box and diamond are definable as: $\Box\phi$ iff $\nabla\emptyset \vee \nabla\{\phi\}$, and $\Diamond\phi$ iff $\nabla\{\phi, \top\}$

Now consider the extension of cover logic with the refinement quantification $\rhd$. By the definition of $\Diamond$ in cover logic, axiom **GK** now takes shape $\bigwedge_{\gamma\in\Gamma} \nabla\{\rhd\gamma, \top\} \leftrightarrow \rhd\nabla\Gamma$. (And this is clearly also sound.) Given a formula ψ in cover logic with refinement, we prove by induction on the number of the occurrences of $\rhd$ in ψ that it is equivalent to an $\rhd$-free formula, and therefore to a formula in the modal logic $\mathcal{L}$. The base is trivial. Now assume ψ contains $n+1$ $\rhd$-operators. Choose a subformula of type $\rhd\phi$ of our given formula ψ, where ϕ is $\rhd$-free (i.e. choose an innermost $\rhd$). We prove by induction on the structure of ϕ that $\rhd\phi$ is logically equivalent to a formula χ without $\rhd$.

[2] We thank Yde Venema for suggesting this proof.

- $\triangleright\bot$ iff $\bot$.
- $\triangleright\top$ iff $\top$.
- $\triangleright(p \wedge \phi)$ iff $p \wedge \triangleright\phi$ (refinements do not affect atoms); IH.
- $\triangleright(\neg p \wedge \phi)$ iff $\neg p \wedge \triangleright\phi$ (refinements do not affect atoms); IH.
- $\triangleright(\phi \vee \psi)$ iff $\triangleright\phi \vee \triangleright\psi$ (directly from the semantics of $\triangleright$); IH.
- $\triangleright\nabla\Gamma$ iff $\bigwedge_{\gamma\in\Gamma} \nabla\{\triangleright\gamma, \top\}$; IH. (By induction, each $\triangleright\gamma$ is equivalent to an $\triangleright$-free formula ψ, and the resulting $\bigwedge_{\psi} \nabla\{\psi, \top\}$ is also $\triangleright$-free.)

Thus we are able to push the refinement operators deeper into the formula until they eventually reach $\top$ or $\bot$, at which point they disappear and we are left with χ (which does not contain $\triangleright$ and is equivalent to $\triangleright\phi$). Replacing $\triangleright\phi$ by χ in ψ gives a result with at least one less $\triangleright$-operator, to which the (original) induction hypothesis applies. □

Lemma 4.4 *Let $\phi \in \mathcal{L}_\triangleright$ be given and $\psi \in \mathcal{L}$ be equivalent to ϕ. If ψ is a theorem in* **K**, *then ϕ is a theorem in* **FEL**.

Proof. Given a $\phi \in \mathcal{L}_\triangleright$, Lemma 4.3 gives us an equivalent $\psi \in \mathcal{L}$. Assume that ψ is a theorem in **K**. We can extend the derivation of ψ to a derivation of ϕ by observing that the first five of the six itemized reduction steps in Lemma 4.3 are all provable equivalences, and that the last item is of course the axiom **GK**. (Where we also need to observe that the system **FEL** satisfies the substitution of equivalents: if ϕ_1 is equivalent to ϕ_2 and ϕ_1 is a subformula of ϕ_3, and ϕ_3 is a theorem, then $\phi_3[\phi_1\backslash\phi_2]$ is also a theorem.) □

Theorem 4.5 *The axiom schema* **FEL** *is sound and complete for the logic $\mathcal{L}_\triangleright$.*

Proof. The soundness proof is given in Theorem 4.2, so we are left to show completeness. Suppose that ϕ is valid: $\models \phi$. Applying Lemma 4.3 we know that there is some equivalent formula ψ not containing any refinement quantification. As ϕ is valid, from that and the validity $\phi \leftrightarrow \psi$ it follows that ψ is also valid in future event logic, and therefore also valid in the logic $\mathcal{L}$ (note that the model class is the same!). From the completeness of **K** it follows that ψ is derivable, i.e. it is a theorem. From Lemma 4.4 it follows that ϕ is a theorem. □

5 Axiomatization: $\mathcal{L}_\triangleright^\mu$

The axiomatization for $\mathcal{L}_\triangleright^\mu$ extends the axiomatization for $\mathcal{L}_\triangleright$ with the extra axiom and rule of Kozen's axiomatization of the modal μ-calculus (**F1** and **F2**), and two new interaction axioms (**G3** and **G4**). The axiomatization $\mathbf{FEL}_\mu$ is a substitution schema of the axioms and rules for $\mathcal{L}_\triangleright$, **FEL** (see Section 4), along with the axiom and rule for the modal μ-calculus:

F1 $\phi[\mu x.\phi\backslash x] \rightarrow \mu x.\phi$

F2 From $\phi[\psi\backslash x] \rightarrow \psi$ infer $\mu x\phi \rightarrow \psi$

and two new interaction axioms:

G3 $\blacktriangleright \mu x.\phi \leftrightarrow \mu x.\blacktriangleright \phi$ where $\mu x.\phi$ is a *df* (Def. 3.5)

G4 $\blacktriangleright \nu x.\phi \leftrightarrow \nu x.\blacktriangleright \phi$ where $\nu x.\phi$ is a *df*

These interaction axioms have an important associated condition: the refinement quantification will only commute with a fixed-point operator if the fixed-point formula is a disjunctive formula.

5.1 Soundness

The soundness proofs of Section 4.2 still apply and the soundness of **F1** and **F2** are well known [3], so we are left to show that **G3** and **G4** are sound.

Theorem 5.1 *The axioms* **G3** *and* **G4** *are sound.*

Proof. In this proof we will find it convenient to use the bisimulation quantifiers characterization of both fixed-point operators and refinement quantification. We recall what bisimulation quantifier is: Given an atom x and a formula ϕ, the expression $\exists x\phi$ means that there exists x such that ϕ, and it is interpreted as $M_s \models \exists x\phi$ if, and only if, for some N_t bisimilar to M_s *except for* x—for which we will write $N_t \underline{\leftrightarrow}_x M_s$— we have $N_t \models \phi$. We let $\forall x\phi$ abbreviate $\neg\exists x\neg\phi$, and a deeper technical discussion of the properties of bisimulation quantifiers may be found in [8].

(i) $\mu x.\phi$ is equivalent to $\forall x(\blacksquare(\phi \to x) \to x)$ [11] (where $\blacksquare$ is the universal modality which quantifies over all states in the model).

(ii) $\nu x.\phi$ is equivalent to $\exists x(\blacksquare(x \to \phi) \wedge x)$ [11].

(iii) $\blacktriangleright\phi$ is equivalent to $\forall r\phi^r$, where ϕ^r is the relativization of ϕ to the atom r, which may be computed recursively by replacing every occurrence of $\Box\psi$ in ϕ with $\Box(r \to \psi^r)$ [21].

(iv) $\triangleright\phi$ is equivalent to $\exists r\phi^r$.

Note that from [7] we know that bisimulation quantifiers are expressible in the modal μ-calculus, and thus the equivalences (i) and (ii) hold in the modal μ-calculus. Furthermore, in [21], the equivalences (iii) and (iv) are shown to hold for all logics that are closed under bisimulation and announcement. As the modal μ-calculus is such a logic, all four equivalences hold in the modal μ-calculus, and they may be reasonably applied in the proofs given below:

G3 It is more convenient in this proof to reason about the axiom in its contrapositive form: $\rhd \nu x.\phi \leftrightarrow \nu x.\rhd\phi$. Using the equivalent transformations above we have:

$$\begin{aligned}
\rhd\nu x.\phi &\leftrightarrow \exists r \exists x(\blacksquare(x \rightarrow \phi) \wedge x)^r \\
&\leftrightarrow \exists x \exists r(\blacksquare(x \rightarrow \phi^r) \wedge x) \\
&\leftrightarrow \exists x(\exists r \blacksquare(x \rightarrow \phi^r) \wedge x) \\
&\rightarrow \exists x(\blacksquare \exists r(x \rightarrow \phi^r) \wedge x) \\
&\leftrightarrow \exists x(\blacksquare(x \rightarrow \exists r \phi^r) \wedge x) \\
&\leftrightarrow \nu x.\rhd\phi
\end{aligned}$$

This proof simply applies known validities of bisimulation quantifiers. Note that the fourth line is not an equivalence in the general case. However, we may show that where ϕ is a disjunctive formula, the equivalence does hold. To do this, suppose M_s is any countable model such that $M_s \models \exists x(\blacksquare \exists r(x \rightarrow \alpha^r) \wedge x)$, where α is a disjunctive formula. As the μ-calculus enjoys the tree-model property, we may assume that there is some tree-like model $N_t \underline{\leftrightarrow}_x M_s$ such that $N_t \models x \wedge \blacksquare \exists r(x \rightarrow \alpha^r)$. We inductively build a series of models $N^i_t \underline{\leftrightarrow}_{r,x} N_t$ where $N^i = (S^N, R^N, V_i)$. We set $V_0(x) = \{t\}$, $V_0(r) = \emptyset$ and $V_0(p) = V^N(p)$ for all $p \notin \{r, x\}$. As $N_t \models \exists r \alpha^r$ and $\nu x.\alpha$ is a disjunctive formula, the only case where the atom x may influence the interpretation of $\exists r \alpha^r$ is at a set of states such that all states beyond that set of states are irrelevant to the interpretation of $\exists r \alpha^r$ at t (this set of states forms a *frontier*). This is because from Definition 3.5, if x is a sub-formula of α, then if x appears in the scope of a conjunction, it appears within the scope of a modality within that conjunction. Thus, there is a set of states $\{u_0, u_1, ...\} \in V^N(x)$ such that $N'_t \models \alpha^r$, where $N' = (S^N, R', V')$ for $V'(x) = \{t, u_0, u_1, ...\}$, $V'(y) = V^N(y)$ for $y \notin \{x, r\}$ and $R' = R^N \backslash \{(u_i, s) | s \in S^N, i = 0, 1, ...\}$. Consequently the valuation of r maybe restricted to states that are not reachable from any state, $\{u_0, u_1, ...\}$ Let $S_0 \subset S^N$ be the set of states reachable from t, but not reachable from u_i for any i. We define N^1 by setting $V_1(x) = V'(x)$, $V_1(r) = V'(r) \cap S_0$ and $V_1(y) = V^N(y)$ for $y \notin \{x, r\}$. As $u_0, u_1, ... \in V^N(x)$, we have $N_{u_i} \models x \wedge \blacksquare(\exists r(x \rightarrow \alpha^r)$ for all i.

As M_s is a countable model, we may assume an enumeration of the worlds (or states) in that model. The induction proceeds by taking the first state u_0 on the frontier and repeating the process (i.e. finding a valuation V' such that V' make x true on a frontier $\{v_0, v_1, ...\}$, agrees with V^N on the interpretation of all atoms except x and r, makes $N'_{u_0} \models \alpha^r$ and makes $N_{v_i} \models x \wedge \blacksquare(\exists r(x \rightarrow \alpha^r))$ for all i). We define V_2 by taking the union of $V_2(x) = V'(x)$ and $V_2(r) = V_1(r) \cup (V'(r) \cap S_1$ where S_1 is the set of states reachable from u_0, but not from v_i for and i, and all other atoms have their valuations unchanged. The states $\{v_0, v_1, ...\}$ are added to the set of frontier states and the induction continues. As the sets $V_i(x)$ and $V_i(r)$ are strictly increasing with i, this process is well defined, and its limit N^*_t will satisfy $\exists x \exists r(\blacksquare(x \rightarrow \alpha^r) \wedge x)$, as required. The construction is represented in Figure 3.

G4 We also use the contrapositive form of the axiom: $\rhd \mu x.\phi \leftrightarrow \mu x.\rhd\phi$.

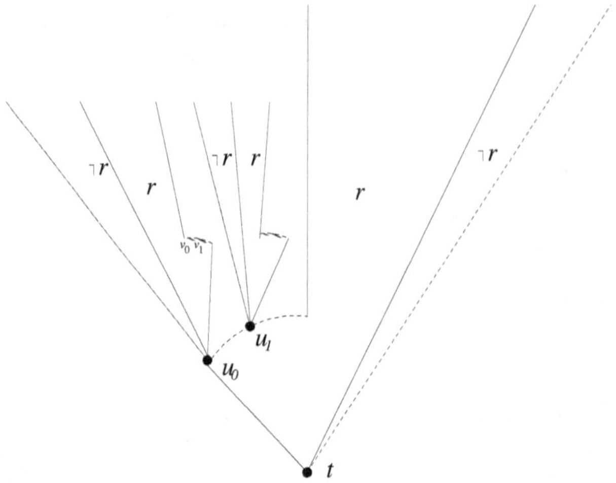

Fig. 3. The inductive step for the construction of N_t^*. The formula α^r is independent of any state where r is not true, or any state beyond the frontier defined by $u_0, u_1, \ldots$.

$$
\begin{aligned}
\triangleright\mu x.\phi &\leftrightarrow \exists r(\forall x(\blacksquare(\phi \rightarrow x) \rightarrow x))^r \\
&\rightarrow \forall x \exists r(\blacksquare(\phi \rightarrow x) \rightarrow x)^r \\
&\leftrightarrow \forall x \exists r(\blacklozenge(\phi \wedge \neg x) \vee x)^r \\
&\leftrightarrow \forall x(\exists r \blacklozenge(\phi \wedge \neg x)^r \vee x) \\
&\leftrightarrow \forall x(\blacklozenge \exists r(\phi \wedge \neg x)^r \vee x) \\
&\leftrightarrow \forall x(\blacklozenge(\exists r \phi^r \wedge \neg x) \vee x) \\
&\leftrightarrow \forall x(\blacksquare(\exists r \phi^r \rightarrow x) \rightarrow x) \\
&\leftrightarrow \mu x.\triangleright\phi
\end{aligned}
$$

While the forward implication is generally true, the right to left implication again relies on the fact that $\mu x\phi$ is a disjunctive formula. For this, write the formula $\mu x\alpha$ to emphasize it is a disjunctive formula. We use the inductive characterization of $\mu x.\triangleright\alpha$: if for $x \in P$ and $S \subseteq S^M$, $M^{[x \mapsto S]} = (S^M, R^M, V)$ such that $V(x) = S$ and $V(y) = V^M(y)$ for $y \neq x$, then we may inductively define $\|\triangleright\alpha\|_0 = \emptyset$, and $\|F\alpha\|_i = \{s \in S^M \mid M_s^{[x \mapsto \bigcup_{j<i} \|\triangleright\alpha\|_j]} \models \triangleright\alpha\}$. Then $M_s \models \mu x.\triangleright\alpha$ if and only if $s \in \|\triangleright\alpha\|_\tau$, where τ is an ordinal [3].

Now suppose $M_s \models \mu x.\triangleright\alpha$. Without loss of generality we may suppose that M is a countable tree-like model. As M_s satisfies $\mu x.\triangleright\alpha$, there must be some least ordinal τ whereby $s \in \|\triangleright\alpha\|_\tau$. We give a proof by induction, and the base case where $\tau = 0$ is trivial. Let $M^\tau = M^{[x \mapsto \bigcup_{j<\tau} \|\triangleright\alpha\|_j]}$, and then $M_s^\tau \models \triangleright\alpha$. As $\mu x\alpha$ is a disjunctive formula, we are again in the case where there is a refinement of M^τ and a frontier such that x may only be true at s or on the frontier, and no point beyond the frontier affects the interpretation of α. Formally, there is a set of states $\{u_0, u_1, \ldots\} \in V^{M^\tau}(x)$ such that $M'_s \models \exists r\alpha^r$, where $M' = (S', R', V')$ such that

- $S' \subseteq S^{M^\tau}$ is the set of states reachable from s, but not from any u_i;
- $V'(x) = \{t, u_0, u_1, \ldots\}$, $V'(y) = V^{M^\tau}(y)$ for $y \neq x$; and

- $R' = R^{M^\tau} \backslash \{(u_i, t) | t \in S^{M^\tau}, i = 0, 1, ...\}$.

We note that M'_s is a refinement of M^τ_s. Now as for each i, $u_i \in \|\triangleright\alpha\|_j$ for some $j < \tau$, by the inductive hypothesis we may assume there is some model $N^i = (S^i, R^i, V^i)$ where $N^i_{v_i} \preceq M^\tau_{u_i}$ and $N^i_{u_i} \models \mu x.\alpha$. We may append these models to M' to define $M^* = (S^*, R^*, V^*)$ where $S^* = S' \cup \bigcup_i S^i$, $R^* = R' \cup \bigcup_i R^i \cup \{(t, v_i) \mid (t, u_i) \in R'\}$, and $V^*(y) = V'(y) \cup \bigcup_i V^i(y)$ for all $y \in P$. It is clear that M^*_s is a refinement of M_s, and by the axiom **F1** we can see $M^*_s \models \mu x.\alpha$ as required.

□

We note that the general form of **G3** is not sound, for example, take $\phi = \mu z.\Diamond(y \rightarrow z) \rightarrow \Diamond(\neg y \rightarrow x)$. Then $\blacktriangleright\mu x.\phi$ is true if y is true at every immediate successor of the current state, whereas $\mu x.\blacktriangleright\phi$ is only true at states with no successor. Likewise **G4** is not true in the general case, as can be seen by taking $\phi = p \wedge \Box(\Diamond\top \rightarrow x)$. Then $\nu x.\blacktriangleright\phi$ is true if and only if p is true at every reachable state, and $\blacktriangleright\nu x.\phi$ is true only if p is true at every state within one step.

5.2 Completeness

The completeness proof of $\mathbf{FEL}_\mu$ proceeds exactly as for Theorem 4.5, replacing the formulas in cover logic with disjunctive formulas to get a statement similar to the one of Lemma 4.3.

Lemma 5.2 *Every formula of $\mathcal{L}^\mu_\triangleright$ is equivalent in $\mathbf{FEL}_\mu$ to a formula of the modal μ-calculus $\mathcal{L}^\mu$.*

Proof. Given a formula ψ, we prove by induction on the number of the occurrences of $\triangleright$ in ψ that it is equivalent to an $\triangleright$-free formula, and therefore to a formula in the modal μ-calculus $\mathcal{L}^\mu$. The base is trivial. Now assume ψ contains $n+1$ $\triangleright$-operators. Choose a subformula of type $\triangleright\phi$ of our given formula ψ, where ϕ is $\triangleright$-free (i.e. choose an innermost $\triangleright$). As ϕ is $\triangleright$-free, it follows from Proposition 3.6 that ϕ is semantically equivalent to a formula in disjunctive normal form, and by the completeness of Kozen's axiom system [23] this equivalence is provable in $\mathbf{FEL}_\mu$. By **Nec2** and **G0** it follows that $\triangleright\phi$ is provably equivalent to some formula $\triangleright\psi$ where ψ is a disjunctive formula. Thus without loss of generalization, we may assume in the following that ϕ is in disjunctive normal form. We may now proceed by induction over the complexity of ϕ, and conclude that $\triangleright\phi$ is logically equivalent to a formula χ without $\triangleright$. All cases of this induction are as before, we only show the final two, different cases:

- $\triangleright\mu x.\phi$ iff $\mu x.\triangleright\phi$ (by **G4** noting that all subformulas of a disjunctive formula are themselves disjunctive); IH.
- $\triangleright\nu x.\phi$ iff $\nu x.\triangleright\phi$ (by **G3**); IH.

Replacing $\triangleright\phi$ by χ in ψ gives a result with one less $\triangleright$-operator, to which the (original) induction hypothesis applies. □

Theorem 5.3 *The axiom schema $\mathbf{FEL}_\mu$ is sound and complete for the logic $\mathcal{L}^\mu_\triangleright$*

Proof. Soundness follows from Theorem 5.1 and Theorem 4.2. To see $\mathbf{FEL}_\mu$ is complete, suppose ϕ is a valid formula. Then by Lemma 5.2, ϕ is provably equivalent to some valid formula ψ of $\mathcal{L}^\mu$. As ψ is valid, it must be provable since **P**, **K**, **F1**, **F2**, **Nec1**, and **MP** give a sound and complete proof system for the modal μ-calculus [23]. A proof of ϕ follows by **MP**. □

6 Complexity

Both axiomatizations demonstrated the expressivity and decidability of $\mathcal{L}_\triangleright$ (expressively equivalent to **K**) and $\mathcal{L}^\mu_\triangleright$ (expressively equivalent to $\mathcal{L}^\mu$). Decidability for both follows from the fact that a computable translation is given in the completeness proofs. Note that as the translations given are recursive and involve translating formulas to disjunctive normal form, the translation is non-elementary in the size of of the original formula. In this section we examine the complexity of $\mathcal{L}_\triangleright$, providing both an elementary upper bound and a succinctness proof.

6.1 Upper-Bound

A decision procedure for $\mathcal{L}_\triangleright$ is given via a tableau. Given any $\mathcal{L}_\triangleright$ formula ϕ, we describe a tableau construction that either constructs a model for ϕ, or reports that ϕ is not satisfiable.

Definition 6.1 A formula is in positive normal form if it is built from the following abstract syntax.

$$\alpha ::= \top \mid \bot \mid p \mid \neg p \mid \alpha \wedge \alpha \mid \alpha \vee \alpha \mid \Box\alpha \mid \Diamond\alpha \mid \blacktriangleright\alpha \mid \triangleright\alpha$$

We note every $\mathcal{L}_\triangleright$ formula may be converted into positive normal form with linear change to the size of formula.

Tableau Definition:

Let ϕ be a formula in positive normal form. Suppose that each subformula of ϕ is uniquely indexed as ϕ_i for $i = 0, ...m$ (thus, two identical subformulas appearing in different places in ϕ would be indexed differently). Let $I = \{1, ..., m\}$, let $\subset$ be the subformula relation over these nodes (so $j \subset i$ if and only if ϕ_j is a subformula of ϕ_i), and given $\sigma \subseteq I$, let σ^+ be the set $\{j \in I \mid \exists i \in \sigma,\ i \subset j\}$. Suppose also that $\phi = \phi_0$. The initial tableau, $T_0 \in \wp(\wp(I))$ consists of the set of nodes, σ, each of which is a subset of I satisfying the following conditions:

- if $i \in \sigma$ and $\phi_i = \phi_j \wedge \phi_k$ then $j, k \in \sigma$;
- if $i \in \sigma$ and $\phi_i = \phi_j \vee \phi_k$ then either $j \in \sigma$ or $k \in \sigma$;
- if $i \in \sigma$ and $\phi_i = \triangleright\phi_j$ or $\phi_i = \blacktriangleright\phi_j$ then $j \in \sigma$;
- if $i, j \in \sigma$ then if $\phi_i = p$, then $\phi_j \neq \neg p$.

The tableau, T_n is then successively pruned according to a game for each node:

Definition 6.2 Let the two players be $\mathfrak{E}$ and $\mathfrak{A}$, and σ be some node in T_n. We define the *pruning game* $\mathcal{G}(T_n, \sigma)$ where each game position is a tuple (Θ, i) where $\Theta \subset T_n$ and $i \in I$. For two sets $\Theta_1, \Theta_2 \subset T_n$ we define $\Theta_1 \sqsubseteq \Theta_2$ if and only if for every $\theta \in \Theta_1$ there is some $\Theta' \subset \Theta_1$ and some $\lambda \in \Theta_2$ where $\theta \cup \bigcup_{\rho \in \Theta'} \rho = \lambda$.

Init Player $\mathfrak{E}$ selects some $\Theta \subseteq T_n$, and then the initial state is $(\Theta, 0)$.

Move given the state (Θ, i):

(i) if $\phi_i = \blacktriangleright\alpha_j$, $\mathfrak{A}$ selects some $\Theta' \sqsubseteq \Theta$, and the new game position is (Θ', j),

(ii) else if $\phi_i = \triangleright\alpha_j$, $\mathfrak{E}$ selects some $\Theta' \sqsubseteq \Theta$, and the new game position is (Θ', j),

(iii) else if $\phi_i = \phi_j \wedge \phi_k$, $\mathfrak{A}$ selects $\ell \in \{j, k\}$ and the new game position is (Θ, ℓ),

(iv) else if $\phi_i = \phi_j \vee \phi_k$, $\mathfrak{E}$ selects $\ell \in \{j, k\} \cap \sigma^+$ and the new state is (Θ, ℓ).

(v) else if $\phi_i \notin \sigma$ and $\phi_i = \Box\phi_j$ or $\phi_i = \Diamond\phi_j$, the new state is (Θ, j).

Wins The game proceeds until no further move can be made. For such a game position (Θ, i):

(i) if $\phi_i = p$, $\neg p$, $\top$ or $i \notin \sigma$, player $\mathfrak{E}$ wins,

(ii) else if $\phi_i = \bot$, player $\mathfrak{A}$ wins,

(iii) else if $\phi_i = \Box\phi_j$, then if for all $\theta \in \Theta$, $j \in \theta$, then $\mathfrak{E}$ wins and otherwise $\mathfrak{A}$ wins,

(iv) else if $\phi_i = \Diamond\phi_j$, then if for some $\theta \in \Theta$, $j \in \theta$, then $\mathfrak{E}$ wins and otherwise $\mathfrak{A}$ wins.

The next tableau is then $T_{n+1} = \{\sigma \in T_n \mid \mathfrak{E}$ has a winning strategy in $\mathcal{G}(T_n, \sigma)\}$. We note that each game is easily determined since because the subformula ϕ_i is strictly decreasing, there is a maximum of m moves in any game. Furthermore, T_0 is finite and $T_{n+1} \subseteq T_n$, so a fixed point T^* is eventually reached. If for some $\sigma \in T^*$ we have $0 \in \sigma$ the tableau reports that ϕ is satisfiable, and otherwise it reports ϕ is unsatisfiable. We let $\mathcal{G}(\sigma)$ abbreviate $\mathcal{G}(T^*, \sigma)$.

The intuition behind this tableau is that each node represents a state in the model, and records which parts of the formula ϕ are satisfied at that state. The semantics of the $\triangleright$ and $\blacktriangleright$ operator are captured by a game that is played at each state in the model (i.e. successors may be kept, pruned or split). To take a global view, the players are playing a game over T_n where $\mathfrak{E}$ is trying to show a model for ϕ exists, and $\mathfrak{A}$ is trying to show that whichever model $\mathfrak{E}$ builds does not satisfy ϕ. Every time they get to a new state (i.e. they reach a game position (Θ, i) where $i = \Box\phi_j$ or $\Diamond\phi_j$) they replay the series of moves that brought them to that state, so that each player may select, in turn, refinements of the set of successors for the new state.

Lemma 6.3 *If the tableau reports that ϕ is satisfiable, then ϕ has a model.*

Proof. Suppose that T^* is the final tableau, and $\sigma \in T^*$ and $\phi \in \sigma$. We build a model $M = (S, R, V)$ from T^* where $S = T^*$, for all $\theta \in S$, if $\theta \in V(p)$ if and only if $p \in \theta$, and for each $\theta \in S$, $\{\xi \mid (\theta, \xi) \in R\}$ is the first move of player $\mathfrak{E}$'s winning strategy in the game $\mathcal{G}(\theta)$. By induction over ϕ we may see that $M_\sigma \models \phi$. For our inductive hypothesis we assume if $\mathfrak{E}$ has a winning strategy for the game position $(\{\xi \mid (\theta, \xi) \in R^M\}, i)$, and $i \in \theta$, then $M_\theta \models \phi_i$. Let θM be the set of successors of θ in the model M.

(i) If $i \in \theta$ where $\phi_i \in \{p, \neg p, \top\}$ then $M_\theta \models \phi_i$.

(ii) If $i \in \theta$ where $\phi_i = \phi_j \wedge \phi_k$ then $\mathfrak{E}$ must have a winning strategy for the game position $(\theta M, i)$ in $\mathcal{G}(\theta)$, so $\mathfrak{E}$ must also have a winning strategy for $(\theta M, j)$ and a winning strategy $(\theta M, k)$, so by the inductive hypothesis we have $M_\theta \models \phi_i$.

(iii) If $i \in \theta$ where $\phi_i = \phi_j \vee \phi_k$ then $\mathfrak{E}$ must have a winning strategy for the game position $(\theta M, i)$ in $\mathcal{G}(\theta)$, so $\mathfrak{E}$ must also have a winning strategy for $(\theta M, j)$ or a winning strategy for $(\theta M, k)$, so by the inductive hypothesis we have $M_\theta \models \phi_i$.

(iv) If $i \in \theta$ where $\phi_i = \Box\phi_j$ then every $\xi \in \theta M$ must have $j \in \xi$. Therefore $\mathfrak{E}$ has a winning strategy from $(\xi M, j)$ in $\mathcal{G}(\xi)$, so every successor of θ satisfies ϕ_j.

(v) If $i \in \theta$ where $\phi_i = \Diamond\phi_j$ then some $\xi \in \theta M$ must have $j \in \xi$. Therefore $\mathfrak{E}$ has a winning strategy from $(\xi M, j)$ in $\mathcal{G}(\xi)$, so some successor of θ satisfies ϕ_j.

(vi) If $i \in \theta$ where $\phi_i = \blacktriangleright\phi_j$, then $\mathfrak{E}$ has a winning strategy in the game $\mathcal{G}(\theta)$ from the game position $(\theta M, i)$, so for every $A \sqsubseteq \theta M$, player $\mathfrak{E}$ has a winning strategy from the game position (A, j). Every refinement M'_θ of M_θ may be represented by the restrictions $A^\xi \sqsubseteq \xi M$ for all ξ reachable from θ. As $\mathfrak{E}$ has a winning strategy for all such game positions starting from (A^θ, j) we have $M'_\theta \models \phi_j$ and the result follows.

(vii) If $i \in \theta$ where $\phi_i = \triangleright\phi_j$, then $\mathfrak{E}$ has a winning strategy to select restrictions $A^\xi \sqsubseteq \xi M$ for all ξ reachable from θ, in the game $\mathcal{G}(\xi)$ so that she has a winning strategy from the game position (A^ξ, j). Collecting these restrictions together we are able to define a single refinement M'_θ of M_θ for which $\mathfrak{E}$ has a winning strategy from $(\theta M', j)$, and thus $M'_\theta \models \phi_j$.

By induction it follows that since $\mathfrak{E}$ has a winning strategy for $\mathcal{G}(\sigma)$, $M_\sigma \models \phi$. □

Lemma 6.4 *If ϕ is satisfiable, then the tableau reports that ϕ is satisfiable.*

Proof. If ϕ is satisfiable, then ϕ has some model, M_s. Seeing as ϕ is equivalent to a formula of $\mathcal{L}$ we may assume that M_s contains no infinite paths [6]. We use M_s to build a set of nodes in the tableau and define a winning strategy for $\mathfrak{E}$ in each, ensuring they are never pruned. The construction of the tableau mirrors the semantics of $\mathcal{L}_\triangleright$. Given the set of all refinements of M modulo bisimulation, $\mathcal{M}$, and the states in S^M, we index a set of nodes as n_t^N where $t \in S^M$ and $N \in \mathcal{M}$. We ensure $0 \in n_s^M$ and build up the nodes as follows:

(i) if $i \in n_t^N$ and $i = \phi_j \wedge \phi_k$, then $j, k \in n_t^N$,

(ii) if $i \in n_t^N$ and $i = \phi_j \vee \phi_k$, then $j \in n_t^N$ if and only if $N_t \models \phi_j$, and $k \in n_t^N$ if and only if $N_t \models \phi_k$,

(iii) if $i \in n_t^N$ and $i = \blacktriangleright\phi_j$, then for all $N'_t \preceq N_t$, $j \in n_t^{N'}$ [3],

(iv) if $i \in n_t^N$ and $i = \triangleright\phi_j$, then for all $N'_t \preceq N_t$, $j \in n_t^{N'}$ if and only if $N'_t \models \phi_j$,

(v) if $i \in n_t^N$ and $i = \Box\phi_j$, then for all u where $(t, u) \in R^N$, $j \in n_u^N$,

(vi) if $i \in n_t^N$ and $i = \Diamond\phi_j$, then for all u where $(t, u) \in R^N$, $j \in n_u^N$ if and only if $N_u \models \phi_j$.

[3] We assume, without loss of generality that every state in every refinement is associated with a single state from S^M

It is clear from the semantics of $\mathcal{L}_\triangleright$ that that for all $i \in n_t^N$, $N_t \models \phi_i$. Let $\mathcal{T}$ be the set of nodes $\{n_t \mid n_t = \bigcup_{N_t \preceq M_t} n_t^N\}$. Now $\mathfrak{E}$'s moves may be dictated by the model M_s. For the node n_t, the game $\mathcal{G}(n_t)$ simulates the set of formulas that must be true along the path leading to t (whose index is not in n_t in the model M), and the sets of formula true at M_t (whose index is in n_t). Throughout the play, player $\mathfrak{E}$ records a tuple (N, u) of the current refinement and state that is being used to evaluate the formula ϕ_i, where the game position is (Θ, i). Furthermore, at each step she may ensure that Θ represents the set of nodes $\{n_v \mid (t, v) \in R^N\}$. As $\mathfrak{E}$ is guided by the semantic interpretation of ϕ in M_s, her recorded tuple (N, u) for the game position (Θ, i) is such that $N_u \models \phi_i$. If at the end of play, $u \neq t$, then it will be the case that $i \notin n_t$, so player $\mathfrak{E}$ wins. Otherwise, we will have the game position (Θ, i) and either:

(i) $\phi_i \in \{\top, p, \neg p\}$ in which case $\mathfrak{E}$ wins ($\bot$ is not an option since $N_t \not\models \bot$); or

(ii) $\phi_i = \Box\phi_j$ so $N_t \models \Box\phi_j$ and thus for all u where $(t, u) \in R^N$, $N_u \models \phi_j$ so $j \in n_u = \Theta$ and $\mathfrak{E}$ wins; or

(iii) $\phi_i = \Diamond\phi_j$ so $N_t \models \Diamond\phi_j$ and thus for some u where $(t, u) \in R^N$, $N_u \models \phi_j$ so $j \in n_u = \Theta$ and $\mathfrak{E}$ wins.

Therefore no node $n_t \in \mathcal{T}$ is pruned from the tableau, and as $0 \in n_s$ the tableau reports that ϕ is satisfiable. □

Corollary 6.5 *The satisfiability problem for $\mathcal{L}_\triangleright$ can be determined in 2EXP time.*

Proof. This follows directly from the tableau description. If ϕ is a formula of size m, then there at most 2^m nodes in the initial tableau. To do the pruning steps we must search all possible strategies for $\mathfrak{E}$ to see if any are winning strategies. As the players moves involve sets of nodes, this takes time 2^{2^m}. For each step we must examine $\mathcal{G}_n(\sigma)$ for every node σ, and as the tableau are strictly decreasing, there are at most 2^m steps. Thus the overall complexity is $2^{O(2^m)}$. □

It is not yet known whether this complexity bound is optimal. However, below we show that $\mathcal{L}_\triangleright$ is exponentially more succinct than $\mathcal{L}^\mu$, which suggests that the 2EXP bound may be optimal.

6.2 Succinctness

Here we use the refinement quantification to show that $\mathcal{L}_\triangleright$ is able to express the property that two binary trees are n-bisimilar, with a formula of size $O(n^2)$. We will then show that neither **K**, nor $\mathcal{L}^\mu$ are able to express this property in size less than $2^{O(n)}$.

The basic idea of this construction is to encode a pebble game (or bisimulation game) for showing n-bisimilarity, using the refinement quantification to encode players moves. We restrict our attention to complete binary trees labelled by a single atom, a that marks a prefix closed subtree, and consider the property: "The left subtree marked by a is n-bisimilar to the right subtree marked by a". To enforce the binary nature of the tree we suppose that there is an atom ℓ that labels each left successor, and we suppose that r is an abbreviation for $\neg\ell$. We may then refer to the left successor using the modal abbreviations $\langle\ell\rangle\phi$ for $\Diamond(\ell \wedge \phi)$, and likewise for the right successor. Note orientation (left or right) of a successor does not affect whether two subtrees are bisimilar. They are just used to ensure that the rules of the game are followed.

Our intent is to encode a pebble game played by a *Spoiler* and a *Duplicator*. Each player takes turns at selecting a successor (or moving a pebble) in either subtree. Spoiler goes first, selecting a successor in either subtree (left or right), where a is true, and then Duplicator must select a successor in the other subtree where a is true. If Spoiler is ever unable to move Duplicator wins, and if Duplicator is unable to move, Spoiler wins. If Duplicator has a strategy to survive at least n moves, then the left and right sub-trees must be n-bisimilar [19].

In $\mathcal{L}_{\triangleright}$ we simulate "selecting a successor" by taking a refinement that leaves only the left or right successor, but otherwise leave the tree intact. To do this we introduce the abbreviation $trunk_m$ to represent a (1-2)-tree where nodes of height less than m have only a left successor, or only a right successor and nodes of height greater than or equal to m, but less than n have a left successor and a right successor:

$$trunk_m = \bigwedge_{i=1}^{m}(\Box^i(\ell \wedge a) \vee \Box^i(r \wedge a) \wedge \bigwedge_{i=m}^{n-1} \Box^i(\Diamond\ell \wedge \Diamond r)$$

We can present the definition for n-bisimilarity recursively, where:

$$\mathcal{B}_i^n = \blacktriangleright\left[\begin{pmatrix}\langle\ell\rangle trunk_i \wedge \langle r\rangle trunk_{i-1}\\ \vee \\ \langle r\rangle trunk_i \wedge \langle\ell\rangle trunk_{i-1}\end{pmatrix} \longrightarrow \triangleright\left[\langle\ell\rangle trunk_i \wedge \langle r\rangle trunk_i \wedge \mathcal{B}_{i+1}^n\right]\right]$$

and $\mathcal{B}_n^n = \langle\ell\rangle trunk_n \wedge \langle r\rangle trunk_n$. Then the property of n-bisimilarity is just equivalent to $\mathcal{B}_1^n$.

A game scenario in presented in Figure 4.

Lemma 6.6 *Let M_s be a complete binary tree. Then $M_s \models \mathcal{B}_1^n$ if and only if the subtree of the left node that is labelled by a is n-bisimilar to the subtree of the right node that is labelled by a.*

Proof. The proof follows the semantic encoding of a pebble game. If the left and right a-marked subtree are n-bisimilar, then Duplicator has a winning strategy in the game. Thus if $M_s \models \mathcal{B}_1^n$, any move Spoiler makes corresponds to a refinement that makes $trunk_1$ true at one branch, and $trunk_2$ true at the other. But for any move that Spoiler makes, Duplicator may find a move (a refinement that makes $trunk_2$ true for both subtrees) where the remaining subtrees are $(n-1)$-bisimilar (and thus $\mathcal{B}_2^n$ is true for the refined binary tree). Applying the argument inductively, it follows that if the left and right a-marked subtrees are n-bisimilar, then $M_s \models \mathcal{B}_1^n$.
Conversely, if $M_s \models \mathcal{B}_1^n$, then we may extract a winning strategy for Duplicator in the pebble game. Any move that Spoiler may make in the game will correspond to a refinement, M_s^1 that makes $trunk_1$ true at one subtree and $trunk_2$ true at the other. As $M_s \models \mathcal{B}_1^n$, for every such refinement, there is a further refinement, M_s^2 that has both

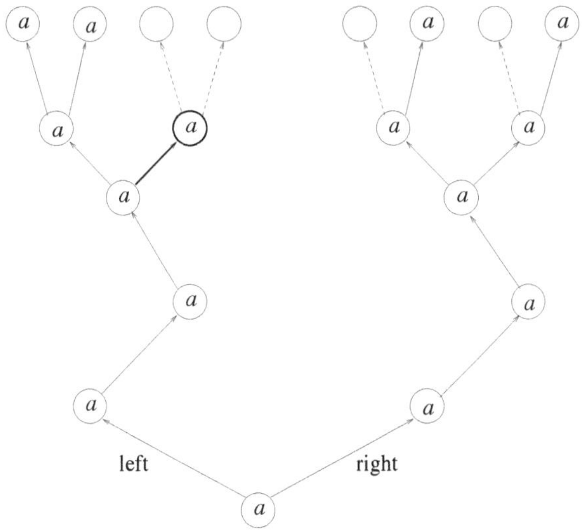

Fig. 4. The state of the game after four moves. The winning move for Spoiler to play is to select the bold successor. Then no matter what successor Duplicator picks, Spoiler may select one of it's successors, leaving Duplicator with no move to make

a-marked subtrees satisfying $trunk_2$, and furthermore, $M_s^2 \models \mathcal{B}_2^n$. Therefore, Duplicator may chose to move according to this bisimulation (i.e. by selecting which ever successor was preserved in the refinement). As $M_s \models \mathcal{B}_1^n$, Duplicator's strategy is guaranteed to last at least n moves and thus the left and right a-marked subtrees are n-bisimilar. □

Lemma 6.7 *No formula of the modal μ-calculus can express the property of $(n+1)$-bisimilarity in size less than $2^n/4$.*

Proof. We note that there are roughly 2^{2^n} non-bisimilar (1,2)-trees of height $n+1$. The actual number is specified by the recurrence $f(n) = (f(n-1)^2 + f(n-1))/2$ where $f(0) = 2$ and a simple induction will show that $f(n) > 2^{2^n/4}$. As every formula of the μ-calculus is expressively equivalent to an alternating automaton of equal size, if a formula of size less than $2^n/4$ were able to express $n+1$-bisimilarity then an alternating automaton of size less than $2^n/4$ would be able to accept all pairs of subtrees that are $n+1$-bisimilar. Given the 2-player parity game that results from applying the alternating automaton to any we may associate with every distinct sub-tree (up to bisimulation) the set of automaton states for which the automaton player has a winning strategy from that state in the the subtree. As there are more than $2^{2^n/4}$ non-bisimilar subtrees and less than $2^n/4$ states, there must be two non-bisimilar subtrees, T_1 and T_2 for which an automaton winning strategy exists for exactly the same set of states. The alternating automaton would not be able to distinguish the case where T_1 is the left successor and T_2 is the right successor (which it cannot accept), and the case where T_2 is the left and right successor (which it does accept). Therefore, any formula that expresses $n+1$-bisimilarity for $(1,2)$-trees must have at least $2^n/4$ subformulas. □

Corollary 6.8 *$\mathcal{L}_\triangleright$ is exponentially more succinct than $\mathcal{L}$.*

Proof. From the lemmas above, $\mathcal{L}_\triangleright$ is able to express n-bisimilarity with a formula of size $O(n^2)$, while $\mathcal{L}$ requires a formula of size $2^n/4$ at least. The quadratic growth of the $\mathcal{L}_\triangleright$ formula is cancelled out by by a constant in the exponent of the growth rate of the $\mathcal{L}$ formula, so we may find a family of formulas in $\mathcal{L}_\triangleright$, such that for a formula of size n, the smallest equivalent $\mathcal{L}$ formula is bound below by $2^{O(n)}$. □

We note this proof applies without change to show $\mathcal{L}^\mu_\triangleright$ is exponentially more succinct than $\mathcal{L}^\mu$.

7 Discussion and perspectives

The logic $\mathcal{L}_\triangleright$ is presented with respect to the class of all epistemic models. By restricting the class of models the logic is interpreted over we may associate different meanings with the modalities. For example, the epistemic logic **S5** is interpreted over all models where the accessibility relation is reflexive, transitive and symmetric (we will denote this class $S5$), and the logic **K4** is interpreted over all models with a transitive accessibility relation (denoted $K4$). Given any class of models $\mathcal{C}$, we define the logic $\mathcal{L}^{\mathcal{C}}_\triangleright$ to be as in Section 3 except:

(i) The interpretation is restricted to models in the class $\mathcal{C}$

(ii) The semantic interpretation of ▶ is given by:

$$M_s \models \blacktriangleright_a \phi \text{ iff for all } M'_{s'} \in \mathcal{C} : \ M_s \succeq_a M'_{s'} \text{ implies } M'_{s'} \models \phi.$$

A study of how various classes of models affect the properties of bisimulation quantified logics is given in [11]. For the effect of varying classes of models on the axiomatization given, we note that while the schema **FEL** is sound for $\mathcal{L}_\triangleright$, it is not the case that the axiom **GK** is sound for restricted classes of models. For example in the class of reflexive, transitive and symmetric models (i.e. $S5$ frames) we have $\Diamond\triangleright\Box p \wedge \Diamond\triangleright\neg\Box p$ is consistent, but $\triangleright\nabla(\Box p, \neg\Box p)$ is not. In future work we will examine axiomatizations and complexity for refinement quantifiers in logics such as **S5**, **KD45** and **K4**.

References

[1] Aczel, P., "Non-Well-Founded Sets," CSLI Publications, Stanford, CA, 1988, cSLI Lecture Notes 14.

[2] Alur, R., T. A. Henzinger and O. Kupferman, *Alternating-time temporal logic*, Journal of the ACM **49** (2002), pp. 672–713.

[3] Arnold, A. and D. Niwinski, "Rudiments of μ-calculus," North Holland, 2001.

[4] Baltag, A., L. Moss and S. Solecki, *The logic of public announcements, common knowledge, and private suspicions*, in: I. Gilboa, editor, *Proceedings of the 7th Conference on Theoretical Aspects of Rationality and Knowledge (TARK 98)*, 1998, pp. 43–56.

[5] Bilkova, M., A. Palmigiano and Y. Venema, *Proof systems for the coalgebraic cover modality.*, in: C. Areces and R. Goldblatt, editors, *Advances in Modal Logic* (2008), pp. 1–21.

[6] Blackburn, P., M. de Rijke and Y. Venema, "Modal Logic," Cambridge University Press, Cambridge, 2001, cambridge Tracts in Theoretical Computer Science 53.

[7] D'Agostino, G. and M. Hollenberg, *Logical questions concerning the mu-calculus: Interpolation, Lyndon and Los-Tarski*, Journal of Symbolic Logic **65** (2000), pp. 310–332.

[8] D'Agostino, G. and G. Lenzi, *An axiomatization of bisimulation quantifiers via the μ-calculus*, Theor. Comput. Sci. **338** (2005), pp. 64–95.

[9] D'Agostino, G. and G. Lenzi, *A note on bisimulation quantifiers and fixed points over transitive frames.*, J. Log. Comput. **18** (2008), pp. 601–614.

[10] Fagin, R., J. Halpern, Y. Moses and M. Vardi, "Reasoning about Knowledge," MIT Press, Cambridge MA, 1995.

[11] French, T., "Bisimulation quantifiers for modal logic," Ph.D. thesis, University of Western Australia (2006).

[12] Harel, D., D. Kozen and J. Tiuryn, "Dynamic Logic," MIT Press, 2000.

[13] Janin, D. and I. Walukiewicz, *Results on the expressive completeness of the propositional mu-calculus with respect to monadic second order logic*, in: *Concurrency Theory, 7th International Conference*, LNCS **1119** (1996), pp. 263–277.

[14] Kupferman, O., M. Vardi and P. Wolper, *Module checking*, Information and Computation **164** (2001), pp. 322–344.

[15] Lomuscio, A. and M. Ryan, *An algorithmic approach to knowledge evolution*, Artificial Intelligence for Engineering Design, Analysis and Manufacturing (AIEDAM) **13(2)** (1998).

[16] Morgan, C., "Programming from Specifications: Second Edition," Prentice Hall International, Hempstead, UK, 1994.
URL `http://web.comlab.ox.ac.uk/oucl/publications/books/PfS/`

[17] Pauly, M., "Logic for social software," Ph.D. thesis, University of Amsterdam (2001), iLLC Dissertation Series DS-2001-10.

[18] Ramadge, P. and W. Wonham, *On the supervisory control of discrete event systems*, in: *Proceedings of the IEEE*, 1989, pp. 81–98.

[19] Stirling, C., *The joys of bisimulation*, in: *MFCS*, 1998, pp. 142–151.

[20] Tsitsoklis, J., *On the control of discrete event dynamical systems*, Mathematics of Control Signals and Systems **2** (1989), pp. 95–107.

[21] van Ditmarsch, H. and T. French, *Simulation and information*, in: J. Broersen and J.-J. Meyer, editors, *Proceedings of KRAMAS, Sydney*, LNAI **5605** (2009), pp. 51–65.

[22] van Ditmarsch, H., W. van der Hoek and B. Kooi, "Dynamic Epistemic Logic," Synthese Library **337**, Springer, 2007.

[23] Walukiewicz, I., *Completeness of Kozen's axiomatisation of the propositional mu-calculus*, INFCTRL: Information and Computation (formerly Information and Control) **157** (2000).

Absolute Completeness of $\mathsf{S4}_u$ for Its Measure-Theoretic Semantics

David Fernández-Duque

Group for Logic, Language and Information
Universidad de Sevilla
dfduque@us.es

Abstract

Given a measure space $\langle X, \mu \rangle$, we define its *measure algebra* $\mathbb{A}_\mu$ as the quotient of the algebra of all measurable subsets of X modulo the relation $X \overset{\mu}{\sim} Y$ if $\mu(X \triangle Y) = 0$. If further X is endowed with a topology $\mathcal{T}$, we can define an interior operator on $\mathbb{A}_\mu$ analogous to the interior operator on $\mathcal{P}(X)$. Formulas of $\mathsf{S4}_u$ (the modal logic $\mathsf{S4}$ with a universal modality $\forall$ added) can then be assigned elements of $\mathbb{A}_\mu$ by interpreting $\Box$ as the aforementioned interior operator.
In this paper we prove a general completeness result which implies the following two facts:

(i) the logic $\mathsf{S4}_u$ is complete for interpretations on any subset of Euclidean space of positive Lebesgue measure;

(ii) the logic $\mathsf{S4}_u$ is complete for interpretations on the Cantor set equipped with its appropriate fractal measure.

Further, our result implies in both cases that given $\varepsilon > 0$, a satisfiable formula can be satisfied everywhere except in a region of measure at most ε.

Keywords: Modal logic, topological semantics, measure theory

1 Introduction

One of the primary appeals of modal logic is the flexibility in its interpretation. Since $\Box$ could be taken to have many different meanings, the same modal logic can often be used in several seemingly unrelated contexts.

The logic $\mathsf{S4}$ is a particularly good example of this, because along with its relational many-worlds semantics, it can be given a topological interpretation, as was already known by McKinsey and Tarski before 1940. With these semantics, modal logic can be used for reasoning about space, a perspective which has proven to be very fruitful.[1] Perhaps the most famous theorem in this field is McKinsey and Tarski's result that S4 is complete for topological interpretations on the real line and, more generally, for any separable metric space without isolated points [11]. More recently, this result has been followed by new proofs and strengthenings for the real line [4,13], as well as the Cantor set [12].

The result is modified slightly when we consider the universal modality from [9], giving rise to the logic $\mathsf{S4}_u$. We once again have completeness of $\mathsf{S4}_u$ for the class of finite topological spaces, but in general these must be disconnected [3]. In Euclidean spaces (and any other connected, separable metric spaces without isolated points) [14] proves that the logic we obtain is $\mathsf{S4}_u$ + Conn, where Conn denotes the connectedness axiom $\forall(\Box p \vee \Box q) \rightarrow \exists(\Box p \wedge \Box q)$.

This shows that a well-understood and -behaved modal logic can be used without trouble to reason about topological spaces, despite their richness and complexity. But why stop at topology? We can interpret S4 over spaces which have even deeper structure. The real line, for example, about which much work on S4 has focused, admits not only a natural metric (which is used to interpret the modal operator $\Box$) but also a natural measure. Thus in addition to the question *Can we satisfy a given formula* φ *on a model based on the real line?* we can ask *Can we satisfy a formula* φ *with a high probability on a model based on the real line?*

Formulas of S4 can be interpreted over subsets of Euclidean space "up to measure zero"; that is, over the algebra of measurable sets modulo all null sets. This intepretation was called to my attention in a lecture given by Dana Scott in the conference Topology, Algebra and Categories in Logic, 2009. I immediately became interested in the question of finding an analogue to McKinsey and Tarski's theorem.

Here we should remark that topological completeness of S4 does not *a priori* imply its measure-theoretic completeness, or vice-versa. It is true that every model of S4 based on Euclidean space gives rise to a measure-theoretic model (provided that all valuations of propositional variables are measurable); simply take the original valuation modulo null sets. However, the resulting model does not satisfy the same set of formulae. Indeed, many sets which are topologically "large", such as the set of rational numbers which is dense in the real line (or even a dense G_δ, which is topologically large in a more precise sense) can have measure zero and hence "disappear" under our measure-theoretic interpretation. Because of this, even a formula that was topologically satisfied by every point may no longer be satisfied after doing away with null sets.

As an example, consider the formula $\forall(\Diamond p \wedge \Diamond\neg p)$. This formula can be satisfied topologically on the real line by interpreting p as the set of rational numbers. Since both the interpretation of p and its complement are dense, it follows that every point satisfies $\Diamond p \wedge \Diamond\neg p$ and hence $\forall(\Diamond p \wedge \Diamond\neg p)$.

Meanwhile, if we were to translate this directly into a measure-theoretic model, we would be interpreting p as a null set because the set of rationals has measure zero.

[1] Although the basic modal language is not too expressive over the class of topological spaces, there are polymodal systems which turn out to be surprisingly powerful, such as the polymodal Gödel-Löb logic GLP [2] and Dynamic Topological Logic [1,10].

Therefore, every point would satisfy $\neg\Diamond p$, and our original formula would be false everywhere.

In order to give a measure-theoretic model, we would need to interpret p as a set such that every open set U intersects both p and its complement with positive measure. Such a set exists, but the reader unfamiliar with how to construct it may find doing so quite challenging! We will not give an explicit solution, but one can be extracted from our more general completeness proof.

Along these lines, existing proofs of topological completeness cannot serve as proofs of measure-theoretic completeness simply because it is not clear which of the sets that are generated have positive measure and which do not. This of course does not rule out modifying these proofs, taking care of the technical issues arising in the measure-theoretic setting: precisely what we shall set out to do.

On the other hand, working with measure-theoretic semantics has some advantages which might inspire us to think that S4 and related systems are sometimes more likely to be measure-theoretically complete than topologically complete. The reason for this is that there are several extensions of S4 which are incomplete for topological interpretations on Euclidean spaces precisely because said spaces are topologically connected; two examples of this are $\mathsf{S4}_u$, as mentioned above, and *Dynamic Topological Logic*, which can be shown to be incomplete for the plane due to local connectedness[2] [7]. However, measure-theoretically, Euclidean space is quite disconnected. Recall that a topological space is disconnected if it contains proper subsets which are both open and closed. Well, open balls in Euclidean space are both open and closed up to measure zero, because their boundaries carry no measure.

It is the author's opinion that there should be many more measure-theoretic completeness results to be found where topological completeness fails, but here we shall limit our discussion to $\mathsf{S4}_u$. Our main results are that $\mathsf{S4}_u$ is complete for interpretations on the measure algebra of any subset of $\mathbb{R}^N$ which has positive measure (the real line and the unit interval are examples of this, but this class of sets is much more general) and for interpretations on the measure algebra of the Cantor set, where we must take the Hausdorff measure of appropriate fractal dimension (in this case, $\ln(2)/\ln(3)$; see Appendix A). Further, in all of the above cases, if we take any $\varepsilon > 0$, a satisfiable formula φ can be satisfied everywhere except for a set of measure at most ε; in the case that the set we began with was a probability space (such as the unit interval), this means that every satisfiable formula can be satisfied with probability arbitrarily close to one.

2 Syntax and semantics

We will work in a bimodal language $\mathcal{L}$ consisting of propositional variables $p \in PV$ with the Boolean connectives $\neg$ and $\wedge$ (other Booleans are defined in the standard way) and two modal operators, $\Box$ and $\forall$.

[2] Dynamic Topological Logic is also incomplete for the real line but this can be shown using a formula which is not valid on all locally connected spaces [15].

The logic $\mathsf{S4}_u$ is that obtained by all $\mathsf{S4}$-axioms for $\Box$:

$$\Box\varphi \wedge \Box\psi \to \Box(\varphi \wedge \psi),$$
$$\Box\varphi \to \varphi,$$
$$\Box\varphi \to \Box\Box\varphi;$$

all **S5**-axioms for $\forall$ ($\mathsf{S4}$ with the additional axiom $\exists\varphi \to \forall\exists\varphi$) and the 'bridge' axiom $\forall\varphi \to \Box\varphi$, together with propositional tautologies, necessitation for both operators and modus ponens.

We wish to define semantics for $\mathsf{S4}_u$ on topological measure spaces, which we define below[3]:

Definition 2.1 [Measure algebra] Let $\mathfrak{X} = \langle X, \mathcal{A}, \mu \rangle$ be a measure space. We define the *measure algebra* of $\mathfrak{X}$, which we will denote $\mathbb{A}_\mu$, to be the set of equivalence classes of $\mathcal{A}$ under the relation $\overset{\mu}{\sim}$ given by $E \overset{\mu}{\sim} F$ if and only if $\mu(E \triangle F) = 0$.

In this paper we will refer to elements of $\mathbb{A}_\mu$ as *regions.*

Denote the equivalence class of $S \in \mathcal{A}$ by $[S]_\mu$. Boolean operations can be defined on $\mathbb{A}_\mu$ in the obvious way; $[E]_\mu \sqcap [F]_\mu = [E \cap F]_\mu$, $[E]_\mu - [F]_\mu = [E \setminus F]_\mu$. We can also define $[E]_\mu \sqsubseteq [F]_\mu$ by $\mu(E \setminus F) = 0$. In general we will use 'square' symbols for notation of the measure algebra and 'round' symbols for set notation in order to avoid confusion. As a slight abuse of notation, if $o \in \mathbb{A}_\mu$ and $o = [S]_\mu$ we may write $\mu(o)$ instead of $\mu(S)$; note that this is well-defined, independently of our choice of $S \in o$.

In order to interpret our modal operators, we need to consider measure spaces which also have a topological structure:

Definition 2.2 [topological measure space] A *topological measure space* is a triple $\langle X, \mathcal{T}, \mu \rangle$ where X is a set, $\mathcal{T}$ a topology on X and μ a σ-finite measure such that every open set is μ-measurable.

A set $S \subseteq X$ is *almost open* if $S \overset{\mu}{\sim} U$ for some $U \in \mathcal{T}$. The region $[S]_\mu$ is *open* if S is almost open.

Equivalently, we can say $o \in \mathbb{A}_\mu$ is open if $o = [U]_\mu$ for some open set U.

Given a σ-finite measure space $\langle X, \mu \rangle$ and $\mathcal{O} \subseteq \mathbb{A}_\mu$, the supremum of $\mathcal{O}$, which we will denote $\bigsqcup \mathcal{O}$, always exists; see Appendix B for details. With this operation we can define an interior operator on any measure algebra:

Definition 2.3 [interior] Let $\langle X, \mathcal{T}, \mu \rangle$ be a topological measure space and $o \in \mathbb{A}_\mu$. We define the *interior* of o by $o^\Box = \bigsqcup \{[U]_\mu \sqsubseteq o : U \in \mathcal{T}\}$.

Proposition 2.4 *If $\langle X, \mathcal{T}, \mu \rangle$ is a topological measure space and $o \in \mathbb{A}_\mu$,*

(i) $o^\Box$ *is open,*

(ii) $o^\Box \sqsubseteq o$,

---

[3] For a brief review of measure spaces, see Appendix A.

(iii) $(o^{\Box})^{\Box} = o^{\Box}$.

Proof. See Appendix B. $\Box$

We are now ready to define our semantics:

Definition 2.5 [Measure-theoretic semantics] If $\langle X, \mathcal{A}, \mu\rangle$ is a topological measure space, a *measurable valuation* on X is a function $[\![\cdot]\!] : \mathcal{L} \to \mathbb{A}_\mu$ satisfying

$$\begin{aligned}
[\![\alpha \wedge \beta]\!] &= [\![\alpha]\!] \sqcap [\![\beta]\!] \\
[\![\neg\alpha]\!] &= [X]_\mu - [\![\alpha]\!] \\
[\![\Box\alpha]\!] &= [\![\alpha]\!]^{\Box} \\
[\![\forall\alpha]\!] &= \begin{cases} [X]_\mu & \text{if } [\![\alpha]\!] = [X]_\mu \\ [\varnothing]_\mu & \text{otherwise.} \end{cases}
\end{aligned}$$

A *topological measure model* is a topological measure space equipped with a measurable valuation.

The system $\mathsf{S4}_u$ is sound for our semantics:

Theorem 2.6 (soundness) *Let $\langle X, \mathcal{T}, \mu, [\![\cdot]\!]\rangle$ be a topological measure model. Then, for every formula φ which is derivable in $\mathsf{S4}_u$, $[\![\varphi]\!] = [X]_\mu$.*

Proof. This follows from the fact that all axioms are valid and all rules preserve validity; note that the S4 axioms for $\Box$ are a direct consequence of Proposition 2.4. $\Box$

3 μ-Bisimulations

Our completeness proof depends on a well-known result that $\mathsf{S4}_u$ is complete for the class of finite Kripke frames where the accessibility relation is a preorder (that is, reflexive and transitive).

Definition 3.1 [Kripke frame; Kripke model] A (transitive, reflexive) *Kripke frame* is a preordered set $\langle W, \preccurlyeq\rangle$.

A *Kripke model* is a Kripke frame equipped with a valuation $(\!|\cdot|\!) : \mathcal{L} \to 2^W$ satisfying the standard clauses for Boolean operators,

$$w \in (\!|\Box\varphi|\!) \Leftrightarrow \forall v \preccurlyeq w, v \in (\!|\varphi|\!)$$

and

$$w \in (\!|\forall\varphi|\!) \Leftrightarrow \forall v \in W, v \in (\!|\varphi|\!).$$

The following well-known result can be found, for example, in [3]:

Theorem 3.2 *$\mathsf{S4}_u$ is complete with respect to the class of all finite, transitive, reflexive Kripke models.*

In order to prove our main result, we shall construct a type of bisimulation between a topological measure space and a given Kripke frame. For this we need to define the proper notion of bisimulation. In what follows, $\downarrow w = \{v : v \preccurlyeq w\}$ and a set $U \subseteq W$ is *open* if, for all $w \in U$, $\downarrow w \subseteq U$.

Definition 3.3 [almost continuous, strongly open] Let $\langle X, \mathcal{T}, \mu, [\![\cdot]\!]\rangle$ be a topological measure model and $\langle W, \preccurlyeq, (\!|\cdot|\!)\rangle$ a Kripke model.

Given a partial function[4] $\beta : X \to W$ and $S \subseteq X$, define $\beta[S]_\mu$ to be the set of all $w \in W$ such that $\beta^{-1}(w) \cap S$ has positive measure.

A partial function $\beta : X \to W$ is *almost continuous* if $\beta^{-1}(\downarrow w)$ is almost open for all $w \in W$. It is *strongly open* if whenever S is almost open, $\beta[S]_\mu$ is open, and *strongly surjective* if $\beta^{-1}(w)$ has positive measure for all $w \in W$.

Definition 3.4 [μ-Bisimulation]

With notation as above, a *μ-bisimulation* is a partial function $\beta : X \to W$ which is

(i) almost continuous,

(ii) strongly open,

(iii) defined almost everywhere,

(iv) strongly surjective and

(v) satisfies $[\![p]\!] = [\beta^{-1}(\!|p|\!)]_\mu$ for all $p \in PV$.

μ-Bisimulations preserve valuations of formulae. Before proving this fact we need a preliminary lemma.

Lemma 3.5 *If $\langle X, \mu\rangle$ is a measure space, W a finite set and $\beta : X \to W$ a partial function defined almost everywhere, then for every measurable $S \subseteq X$, $[S]_\mu \sqsubseteq [\beta^{-1}\beta[S]_\mu]_\mu$.*

Proof. Clearly

$$[S]_\mu = \bigsqcup_{w \in W} \left([\beta^{-1}(w)]_\mu \sqcap [S]_\mu\right),$$

since β is defined almost everywhere and W is finite. Now,

$$[\beta^{-1}(w)]_\mu \sqcap [S]_\mu = [\varnothing]_\mu$$

unless $w \in \beta[S]_\mu$, so we can write

$$\begin{aligned}
\textstyle\bigsqcup_{w \in W} \left([\beta^{-1}(w)]_\mu \sqcap [S]_\mu\right) &= \textstyle\bigsqcup_{w \in \beta[S]_\mu} [\beta^{-1}(w)]_\mu \sqcap [S]_\mu \\
\text{(Lemma B.2)} \qquad &= \left[\textstyle\bigcup_{w \in \beta[S]_\mu} \left(\beta^{-1}(w) \cap S\right)\right]_\mu \\
&\sqsubseteq \left[\textstyle\bigcup_{w \in \beta[S]_\mu} \left(\beta^{-1}(w)\right)\right]_\mu \\
&= \left[\beta^{-1}\beta[S]_\mu\right].
\end{aligned}$$

□

[4] That is, a function whose domain is a subset of X and possibly all of X.

Theorem 3.6 *Suppose that $\langle X, \mathcal{T}, \mu, [\![\cdot]\!]\rangle$ is a topological measure model, $\langle W, \preccurlyeq, (\!|\cdot|\!)\rangle$ a finite Kripke model and $\beta : X \to W$ a μ-bisimulation. Then, for every formula φ, $[\![\varphi]\!] = [\beta^{-1}(\!|\varphi|\!)]_\mu$.*

Proof. The proof follows by a simple induction on the build of formulas, with only the case for $\Box\varphi$ and $\forall\varphi$ being non-standard.

For $\Box\varphi$, note that $(\!|\Box\varphi|\!) \subseteq (\!|\varphi|\!)$ and by induction hypothesis $[\![\varphi]\!] = [\beta^{-1}(\!|\varphi|\!)]_\mu$. Now, $(\!|\Box\varphi|\!)$ is open in W and since β is almost continuous,

$$[\beta^{-1}(\!|\Box\varphi|\!)]_\mu = \left[\bigcup_{w\in(\!|\Box\varphi|\!)} \beta^{-1}(\downarrow w)\right]_\mu$$

is open, because each $\beta^{-1}(\downarrow w)$ is almost open. But

$$\left[\beta^{-1}(\!|\Box\varphi|\!)\right]_\mu \sqsubseteq [\beta^{-1}(\!|\varphi|\!)]_\mu = [\![\varphi]\!],$$

so $[\beta^{-1}(\!|\Box\varphi|\!)]_\mu \sqsubseteq [\![\varphi]\!]^\Box$ (recall that $[\![\varphi]\!]^\Box$ is the supremum over all open $o \sqsubseteq [\![\varphi]\!]$) and hence $[\beta^{-1}(\!|\Box\varphi|\!)]_\mu \sqsubseteq [\![\Box\varphi]\!]$.

For the other direction, consider $[\![\Box\varphi]\!]$. This is an open region and hence if $w \in \beta[\![\Box\varphi]\!]$, it follows that $\downarrow w \subseteq \beta[\![\Box\varphi]\!]$, because β is strongly open. But $\beta[\![\Box\varphi]\!] \subseteq \beta[\![\varphi]\!]$ and from our induction hypothesis we can see that $\beta[\![\varphi]\!] \subseteq (\!|\varphi|\!)$, so $\downarrow w \subseteq (\!|\varphi|\!)$. This implies that $w \in (\!|\Box\varphi|\!)$ and, given that w was arbitrary, $\beta[\![\Box\varphi]\!] \subseteq (\!|\Box\varphi|\!)$. Applying β^{-1} to both sides and using Lemma 3.5 we conclude that $[\![\Box\varphi]\!] \sqsubseteq [\beta^{-1}(\!|\Box\varphi|\!)]_\mu$, as desired.

The case of $\forall\varphi$ is simpler and uses the fact that β is strongly surjective and defined almost everywhere; we will skip the details. □

4 Provinces

We will focus much of our discussion on what we shall call *provinces*; these are an abstract class of spaces which have the basic properties we need of bounded subsets of Euclidean space, but are more general and include other familiar spaces (such as the Cantor set with its appropriate Hausdorff measure).

Definition 4.1 [Province] A *province* is a triple $\langle X, d, \mu\rangle$ where X is a set, d a metric and μ a measure on X satisfying

(i) every open set is μ-measurable;

(ii) every non-empty open set has finite positive measure;

(iii) for every $\varepsilon > 0$ there exists $\delta > 0$ such that, given $x \in X$, $\mu(B_\delta(x)) < \varepsilon$;

(iv) the boundary of every open ball has measure zero;

(v) X is totally bounded.

Lemma 4.2 *If i and $j_0, ..., j_{N-1}$ are open balls in a province $\langle X, d, \mu\rangle$, then*

$$i \setminus \bigcup_{n<N} j_n \overset{\mu}{\sim} \left(i \setminus \bigcup_{n<N} j_n \right)^{\circ}.$$

Proof. For every $n < N$, $\mu\left(\overline{j}_n \setminus j_n\right) = 0$ because the boundaries of open balls have measure zero in a province. Hence $\overline{j} \overset{\mu}{\sim} j$ and

$$i \setminus \bigcup_{n<N} j_n \overset{\mu}{\sim} i \setminus \bigcup_{n<N} \overline{j}_n.$$

But $\bigcup_{n<N} \overline{j}_n$ is closed (it is a finite union of closed sets) so $i \setminus \bigcup_{n<N} \overline{j}_n$ is an open subset of $i \setminus \bigcup_{n<N} j_n$ (in fact, its interior) and the result follows. □

Lemma 4.3 *If $\langle X, d, \mu\rangle$ is a province and $U \subseteq X$ is non-empty and open, then U is infinite.*

Proof. Suppose towards a contradiction that U was a finite, non-empty, open set and $x \in U$. Then, for δ small enough we have that $B_\delta(x) = \{x\}$ (take δ to be less than the distance between x and any point in U). Now, $B_\delta(x)$ is non-empty and open, so by the definition of province it has positive measure. Then there must exist $\varepsilon \in (0, \delta)$ such that $\mu(B_\varepsilon(x)) < \frac{\mu(B_\delta(x))}{2}$ (once again by the definition of province); but this cannot be, since $B_\varepsilon(x) = B_\delta(x) = \{x\}$, so they must have the same measure. Thus we have reached a contradiction and conclude that U must be infinite. □

5 Constructing μ-bisimulations

In this section we will construct continuous, open maps from an arbitrary province to a preorder. This, along with the Bisimulation Theorem, will establish the completeness of $\mathsf{S4}_u$ for these spaces and, more generally, for countable unions of provinces.

Definition 5.1 [Graded partition] Let $\langle X, d, \mu\rangle$ be a province. A *graded partition*[5] on X is a set $\mathfrak{p}$ of open balls in X such that

(i) $X \in \mathfrak{p}$ and

(ii) if $i, j \in \mathfrak{p}$ and $i \cap j \neq \varnothing$, then either $i \subseteq j$ or $j \subseteq i$.

Definition 5.2 [Placement function] Let $\mathfrak{p}$ be a graded partition on a province $\langle X, d, \mu\rangle$. The *placement function* of $\mathfrak{p}$ is the (possibly partial) function $\mathrm{pl}_\mathfrak{p} : X \to \mathfrak{p}$ assigning to each $x \in I$ the least $i \in \mathfrak{p}$ such that $x \in i$.

We will write $\mathrm{gr}_\mathfrak{p}$ instead of $\mathrm{pl}_\mathfrak{p}^{-1}$ and call $\mathrm{gr}_\mathfrak{p}(i)$ the *ground* of i.

[5] Compare to the *open ball trees* from [7].

Definition 5.3 [Fine graded partition] Let $\eta \in (0,1)$. A graded partition $\mathfrak{p}$ is *η-fine* if for every $i \in \mathfrak{p}$ we have that

$$\mu\left(\bigcup\{j \in \mathfrak{p} : j \subsetneq i\}\right) < \eta \cdot \mu(i).$$

Note that if $\mathfrak{p}$ is an η-fine graded partition and $i \in \mathfrak{p}$, it follows that $\mu(\mathrm{gr}_{\mathfrak{p}}(i)) > 0$; specifically, $\mu(\mathrm{gr}_{\mathfrak{p}}(i)) > (1-\eta)\mu(i)$.

Lemma 5.4 *Let $\eta \in (0,1)$ and $\mathfrak{p}$ be an η-fine graded partition. Then, given any $i \in \mathfrak{p}$, there exist only finitely many $j \in \mathfrak{p}$ such that $i \subseteq j$.*

Proof. Note that if $i \subsetneq j$, then $\mu(i) < \eta \cdot \mu(j)$, because $\mathfrak{p}$ is η-fine.

It follows that if

$$i = i_0 \subsetneq i_1 \subsetneq \ldots \subsetneq i_n$$

is a chain of elements of $\mathfrak{p}$, then $\mu(i_n) \geq (1/\eta)^n \mu(i)$; since $\mu(X)$ is finite and

$$\lim_{n\to\infty} (1/\eta)^n = \infty,$$

n must be bounded by some fixed $N < \omega$. But the elements of $\mathfrak{p}$ properly containing i are totally ordered, so there can be at most N of these. □

In view of the previous lemma we can give the following definition:

Definition 5.5 [Height] Given an η-fine graded partition $\mathfrak{p}$ and $i \in \mathfrak{p}$, we can define the *height* of i, denoted $\mathrm{hgt}(i)$, as the largest n such that there exists a sequence

$$i = i_0 \subsetneq i_1 \subsetneq \ldots \subsetneq i_n$$

in $\mathfrak{p}$.

We define $\mathfrak{p}[n]$ to be the set of all elements of $\mathfrak{p}$ of height n.

Lemma 5.6 *Let $\mathfrak{p}$ be an η-fine graded partition on a province $\langle X, d, \mu\rangle$. Then,*

$$\mu\left(\bigcup \mathfrak{p}[n]\right) < \eta^n \mu(X).$$

Proof. We prove this by induction on n. Clearly, every $i \in \mathfrak{p}[n+1]$ is contained in a unique $j \in \mathfrak{p}[n]$; since $\mathfrak{p}$ is η-fine, it follows that, for a fixed $j \in \mathfrak{p}[n]$,

$$\mu\left(\bigcup\{i \in \mathfrak{p}[n+1] : i \in j\}\right) < \eta \cdot \mu(j).$$

Now, writing

$$\mathfrak{p}[n+1] = \bigcup_{j \in \mathfrak{p}[n]} \{i \in \mathfrak{p}[n+1] : i \subseteq j\}$$

we have that

$$\mu(\mathfrak{p}[n+1]) = \mu\left(\bigcup_{j\in\mathfrak{p}[n]}\bigcup\{i\in\mathfrak{p}[n+1] : i\in j\}\right)$$

$$= \sum_{j\in\mathfrak{p}[n]} \mu\left(\bigcup\{i\in\mathfrak{p}[n+1] : i\in j\}\right)$$

$$< \eta\sum_{j\in\mathfrak{p}[n]}\mu(j)$$

$$\overset{\mathrm{IH}}{<} \eta^{n+1}\mu(X),$$

as desired (note that we are using the fact that elements of $\mathfrak{p}[n]$ are disjoint, a consequence of Definition 5.1.ii; this allows us to commute unions and sums). □

Corollary 5.7 *If $\mathfrak{p}$ is an η-fine graded partition on a province $\langle X, d, \mu\rangle$, then $\mathrm{pl}_{\mathfrak{p}}$ is defined almost everywhere on X.*

Proof. We know that $X \in \mathfrak{p}$ by definition, so $\mathrm{pl}_{\mathfrak{p}}(x)$ can only be undefined if x lies on some $i \in \mathfrak{p}$ but there is no minimal such i. If that is the case, we have an infinite sequence

$$i_0 \supsetneq i_1 \supsetneq \ldots \supsetneq i_n \supsetneq \ldots$$

of elements of $\mathfrak{p}$ containing x; therefore $x \in \bigcup\mathfrak{p}[n]$ for all $n < \omega$.

However, by Lemma 5.6, $\mu\left(\bigcup\mathfrak{p}[n]\right) < \eta^n\mu(X)$, so

$$\mu\left(\bigcap_{n<\omega}\bigcup\mathfrak{p}[n]\right) < \eta^k\mu(X)$$

for all k and must equal zero. Since $\mathrm{pl}_{\mathfrak{p}}$ is defined on the complement of this set in X, it follows that $\mathrm{pl}_{\mathfrak{p}}$ is defined almost everywhere. □

Definition 5.8 [Naming function] Let $\mathfrak{p}$ be a graded partition. A *naming function* on $\mathfrak{p}$ is a function $\nu : \mathfrak{p} \to W$, where W is a preordered set, such that $\nu(i) \preccurlyeq \nu(j)$ whenever $i \subseteq j$.

Given a named graded partition $\langle\mathfrak{p}, \nu\rangle$ on X, we define a partial function $\dot{\nu} : X \to W$ by $\dot{\nu} = \nu \circ \mathrm{pl}_{\mathfrak{p}}$.

Lemma 5.9 *If $\eta \in (0, 1)$ and $\langle\mathfrak{p}, \nu\rangle$ is a named η-fine graded partition on a province $\langle X, d, \mu\rangle$, then $\dot{\nu}$ is defined almost everywhere, almost continuous and, for all $i \in \mathfrak{p}$,*

$$\mu(\{x \in i : \dot{\nu}(x) = \nu(i)\}) > 0.$$

Proof. That $\dot{\nu}$ is defined almost everywhere is an immediate consequence of Lemma 5.7.

To see that $\dot{\nu}$ is continuous, pick $w \in W$ and consider $\downarrow w$. Suppose that $x \in \dot{\nu}^{-1}(w)$, so that $\nu(\mathrm{pl}_{\mathfrak{p}}(x)) = w$. Note that for all $i \in \mathfrak{p}$ such that $i \subseteq \mathrm{pl}_{\mathfrak{p}}(x)$, we have that $\nu(i) \in \downarrow w$ (by the definition of a naming function). Hence whenever $y \in \mathrm{pl}_{\mathfrak{p}}(x)$ and $\dot{\nu}(y)$ is defined, we have that $\dot{\nu}(y) \in \downarrow w$; since $\dot{\nu}$ is defined almost everywhere, we have that almost every point of $\mathrm{pl}_{\mathfrak{p}}(x)$ lies in $\dot{\nu}^{-1}(\downarrow w)$. Since $\mathrm{pl}_{\mathfrak{p}}(x)$ is an open neighborhood of x we conclude that $\dot{\nu}$ is almost continuous.

The last claim follows from the fact that

$$\mu(\mathrm{gr}_{\mathfrak{p}}(i)) > (1-\eta)\mu(i).$$

□

Definition 5.10 [Refinement; ε-refinement] Let $\langle \mathfrak{p}, \nu\rangle$ and $\langle \mathfrak{p}', \nu'\rangle$ be named graded partitions on a province $\langle X, d, \mu\rangle$. We say $\mathfrak{p}'$ is a *refinement* of $\mathfrak{p}$ if

(i) $\mathfrak{p} \subseteq \mathfrak{p}'$,

(ii) $\nu = \nu' \upharpoonright \mathfrak{p}$ and

(iii) if $i \in \mathfrak{p}$ and $j \in \mathfrak{p}'$ are such that $i \subseteq j$, it follows that $j \in \mathfrak{p}$.

Further, we say $\mathfrak{p}'$ is an *ε-refinement* of $\mathfrak{p}$ if, for almost every $x \in X$ and every $v \preccurlyeq \dot{\nu}(x)$, there exists $i \in \mathfrak{p}'$ such that $i \subseteq B_\varepsilon(x)$ and $\nu'(i) = v$.

Lemma 5.11 *Given $\eta \in (0,1)$ and $\varepsilon > 0$, every finite η-fine graded partition $\mathfrak{p}$ admits a finite ε-refinement which is also η-fine.*

Proof. Let $i \in \mathfrak{p}$, and consider $(\mathrm{gr}_{\mathfrak{p}}(i))^\circ \overset{\mu}{\sim} \mathrm{gr}_{\mathfrak{p}}(i)$ (Lemma 4.2). Because X is totally bounded, there exists a finite set

$$\{x_0, ..., x_{N-1}\} \subseteq (\mathrm{gr}_{\mathfrak{p}}(i))^\circ$$

which is $\varepsilon/2$-dense in $(\mathrm{gr}_{\mathfrak{p}}(i))^\circ$.

Pick $\delta \in (0,\varepsilon)$ such that, for all $x \in X$, $\mu(B_\delta(x)) < \eta/N$, $B_\delta(x_n) \subseteq (\mathrm{gr}_{\mathfrak{p}}(i))^\circ$ whenever $n < N$ and $\delta < d(x_n, x_m)$ whenever $n < m < N$.

Let $M = \# \downarrow \nu(i)$. By Lemma 4.3, $B_\delta(x_n)$ is infinite for each $n < N$, so we can find $y_{0,n}, ..., y_{M,n} \in B_\delta(x_n)$. Then, taking $\iota > 0$ small enough, we can ensure that the balls $B_\iota(y_{m,n})$ are mutually disjoint, contained in $B_\delta(x_n)$, and

$$\mu(B_\iota(x)) < \frac{\eta\mu(i) - \mu(i \setminus \mathrm{gr}_{\mathfrak{p}}(i))}{MN} \tag{1}$$

for all x.

Define $b(i,n,m) = B_\iota(y_{mn})$, $N(i) = N$ and $M(i) = M$. Fix a numbering $w_0^i, ..., w_{M-1}^i$ of the elements of $\downarrow \nu(i)$.

Now, let $\mathfrak{p}' = \mathfrak{p} \cup \{b(i,n,m) : i \in \mathfrak{p}, m < M, n < N(i)\}$ and

$$\nu'(i) = \begin{cases} \nu(i) & \text{if } i \in \mathfrak{p}, \\ \\ w^j_m & \text{if } i = b(j,n,m). \end{cases}$$

It is not hard to see that $\langle \mathfrak{p}', \nu' \rangle$ satisfies the desired properties. In particular, condition 1 guarantees that it remains η-fine. □

Definition 5.12 [Fine sequence of graded partitions] Let $\eta \in (0,1)$. A sequence of named graded partitions $\langle \mathfrak{p}_n, \nu_n \rangle_{n<\omega}$ is *η-fine* if

(i) each $\mathfrak{p}_n$ is η-fine and

(ii) each $\mathfrak{p}_{n+1}$ is an η^n-refinement of $\mathfrak{p}_n$.

We set $\mathfrak{p}_\omega = \bigcup_{n<\omega} \mathfrak{p}_n$ and define $\nu_\omega = \bigcup_{n<\omega} \nu_n$.

Lemma 5.13 *Given any province $\langle X, d, \mu \rangle$, $\eta \in (0,1)$, a preordered set $\langle W, \preccurlyeq \rangle$ and $w \in W$, there exists an η-fine sequence of graded partitions $\langle \mathfrak{p}_n, \nu_n \rangle_{n<\omega}$ with $\mathfrak{p}_0 = \{X\}$ and $\nu_\omega(X) = \nu_0(X) = w$.*

Proof. This follows by applying Lemma 5.11 ω times. □

Lemma 5.14 *If $\langle \mathfrak{p}_n, \nu_n \rangle_{n<\omega}$ is an $\eta/2$-fine sequence of graded partitions, then $\mathfrak{p}_\omega$ is η-fine.*

Proof. Pick $i \in \mathfrak{p}_\omega$, and let N be large enough so that $i \in \mathfrak{p}_N$.

For $N \leq n \leq \omega$ let $S_n = \bigcup \{j \in \mathfrak{p}_n : j \subsetneq i\}$.

Note that $\mu(S_n) < \eta/2 \mu(i)$ whenever $N \leq n \leq \omega$, because $\mathfrak{p}_n$ is η-fine.

Further, $S_n \subseteq S_m$ whenever $n \leq m$ and $S_\omega = \bigcup_{N<n<\omega} S_n$, from which it follows that

$$\mu(S_\omega) = \lim_{n\to\infty} \mu(S_n) \leq \eta/2,$$

the limit holding because μ is a measure.

In particular, $\mu(S_\omega) < \eta$, as desired. □

Lemma 5.15 *If $\langle \mathfrak{p}_n, \nu_n \rangle_{n<\omega}$ is an η-fine sequence of graded partitions, then $\dot{\nu}_\omega$ is strongly open.*

Proof. Let $U \subseteq X$ be a non-empty, open set and $w \in \dot{\nu}_\omega[U]_\mu$. This means that there exists $i \in \mathfrak{p}_\omega$ such that $\nu_\omega(i) = w$ and $\mu(i \cap U) > 0$.

Pick N large enough so that $i \in \mathfrak{p}_N$.

Now, for all $n \geq N$ we have that $\mathrm{gr}_{\mathfrak{p}_n}(i)$ is almost open, hence

$$\mathrm{gr}_{\mathfrak{p}_\omega}(i) = \bigcap_{n \geq N} \mathrm{gr}_{\mathfrak{p}_n}(i)$$

$$\overset{\mu}{\sim} \bigcap_{n \geq N} (\mathrm{gr}_{\mathfrak{p}_n}(i))^\circ;$$

the last step is valid by Lemma 4.2.

Pick $x \in \bigcap_{n \geq N} (\mathrm{gr}_{\mathfrak{p}_n}(i))^\circ$, $\varepsilon > 0$ small enough so that $B_\varepsilon(x) \subseteq U$ and M large enough so that $\eta^M < \varepsilon$.

Then, for every $v \preccurlyeq w$ there exists $j \in \mathfrak{p}_{M+1}$ such that $\nu_{M+1}(j) = v$ and $j \subseteq B_\varepsilon(x)$, because $\mathfrak{p}_{M+1}$ is an η^M-refinement of $\mathfrak{p}_M$.

But $\mu(\mathrm{gr}_{\mathfrak{p}_\omega}(j)) > 0$, and for any $y \in \mathrm{gr}_{\mathfrak{p}_\omega}(j)$ we have $\dot{\nu}_\omega(y) = v$.

Since v was arbitrary, we conclude that $\downarrow w \subseteq \dot{\nu}_\omega[U]_\mu$, and since U was arbitrary $\dot{\nu}_\omega$ is strongly open, as desired. □

Proposition 5.16 *Let $\langle X, d, \mu \rangle$ be a province, $\langle W, \preccurlyeq \rangle$ a finite preordered set and $w \in W$. Then, given $\varepsilon > 0$ there exists an almost continuous, strongly open map*

$$\beta : X \to W$$

such that

$$\mu(\beta^{-1}(W \setminus \{w\})) < \varepsilon.$$

Proof. Let $\eta = \varepsilon/\mu(X)$. By Lemma 5.13, there is an $\eta/2$-fine sequence of named graded partitions $\langle \mathfrak{p}_n, \nu_n \rangle_{n<\omega}$ such that $\nu_\omega(X) = w$. Then we can take $\beta = \dot{\nu}_\omega$.

Since $\mathfrak{p}_\omega$ is η-fine and $\nu_\omega(X) = w$, we have that

$$X \setminus \bigcup \{i \in \mathfrak{p}_\omega : i \subsetneq X\} \subseteq \beta^{-1}(w)$$

and

$$\mu\left(\bigcup \{i \in \mathfrak{p}_\omega : i \subsetneq X\}\right) < \eta \cdot \mu(X).$$

It follows that $\mu(\beta^{-1}(W \setminus \{w\})) < \varepsilon$, as desired. □

Corollary 5.17 *Let $\langle X, d, \mu \rangle$ be a province, $\langle W, \preccurlyeq \rangle$ a finite preordered set and $w \in W$. Then, for every $\varepsilon > 0$ there exists an almost continuous, strongly open surjection $\beta : X \to W$ such that $\mu(\beta^{-1}(W \setminus \{w\})) < \varepsilon$.*

Proof. Let $N = \#W$. By Lemma 4.3, X is infinite so we can find points $x_1, ..., x_{N-1}$ and $\delta > 0$ such that $B_\delta(x_i) \cap B_\delta(x_j) = \varnothing$ whenever $i \neq j$. Taking δ small enough, we can ensure that

$$\mu\left(X \setminus \bigcup_{n<N} B_\delta(x_n)\right) > 0$$

and

$$\mu\left(\bigcup_{n<N} B_\delta(x_n)\right) < {}^{\varepsilon}/2.$$

Define

$$X_0 = X \setminus \bigcup_{n<N} B_\delta(x_n)$$

and $X_n = B_\delta(x_n)$ for $1 \leq n < N$. By Lemma 4.2, X_0 is almost open, as are the rest of the X_n.

Write

$$W = \{w = w_0, w_1, ..., w_{N-1}\}.$$

Clearly each X_n is a province, so by Proposition 5.16 there exist almost continuous, strongly open maps $\beta_n : X_n \to W$ with $\mu(\beta_n^{-1}(w_n)) \neq 0$ and $\mu(X_0 \setminus \beta^{-1}(w)) < {}^{\varepsilon}/2$. Then we can set $\beta = \bigcup_{n<N} \beta_n$; it is not hard to see that β has all the desired properties. □

These results can be generalized to structures we will call *territories*:

Definition 5.18 [Territory] A *territory* is a triple $\langle X, d, \mu \rangle$ such that $X = \bigcup_{n<N} X_n$, where $N \leq \omega$, the sets X_n are disjoint, almost open subsets of X and

$$\langle X_n, d \upharpoonright X_n, \mu \upharpoonright X_n \rangle$$

is a province for all $n < N$.

Note that in the above definition, if N is finite then $\langle X, d, \mu \rangle$ is itself a province.

Corollary 5.19 *Let $\langle X, d, \mu \rangle$ be a territory.*

Then, for every $w \in W$ and $\varepsilon > 0$ there exists a strongly surjective, almost continuous and strongly open map $\beta : X \to W$ such that

$$\mu\left(X \setminus \beta^{-1}(w)\right) < \varepsilon.$$

Proof. Write $X = \bigcup_{n<N} X_n$ as in Definition 5.18. By Corollary 5.17, for all $n < N$ there is a surjective, continuous, open map $\beta_n : X_n \to W$ such that

$$\mu\left(X_n \setminus \beta^{-1}(w)\right) < {}^{\varepsilon}/2^{n+1},$$

so that

$$\begin{aligned}\mu\left(\textstyle\bigcup_{n<\omega} X_n \setminus \beta^{-1}(w)\right) &= \textstyle\sum_{n<\omega} \mu(X_n \setminus \beta^{-1}(w)) \\ &< \textstyle\sum_{n<\omega} {}^{\varepsilon}/2^{n+1} \\ &= \varepsilon.\end{aligned}$$

We can then take

$$\beta = \bigcup_{n<N} \beta_n.$$

□

6 Completeness results

In this section we will prove completeness of $\mathsf{S4}_u$ for its measure-theoretic semantics on subsets of Euclidean space as well as the Cantor set with its fractal measure. In fact we will use a stronger notion of completeness:

Definition 6.1 [Absolute completeness] Let $\mathfrak{X} = \langle X, \mathcal{T}, \mu \rangle$ be a topological measure space with $\mu(X) > 0$ and φ a formula of $\mathcal{L}$.

We say φ is *absolutely satisfiable* on $\mathfrak{X}$ if, for every $\varepsilon > 0$ there exists a valuation $[\![\cdot]\!]$ on $\mathfrak{X}$ such that $\mu([\![\neg\varphi]\!]) < \varepsilon$.

A logic Λ is *absolutely complete* for $\mathfrak{X}$ if for every $\varphi \in \mathcal{L}$, $\neg\varphi \notin \Lambda$ implies that φ is absolutely satisfiable on $\mathfrak{X}$.

Note that absolute satisfiability implies satisfiability and absolute completeness implies completeness.

Our main results are direct consequences of the following more general theorem:

Theorem 6.2 *Let φ be a formula of $\mathsf{S4}_u$ such that $\neg\varphi \notin \mathsf{S4}_u$. Suppose that $\langle X, d, \mu \rangle$ is a territory.*

Then, φ is absolutely satisfiable on $\langle X, d, \mu \rangle$.

Proof. Suppose φ is a satisfiable formula. Then, by Theorem 3.2 there exists a finite Kripke model $\langle W, \preccurlyeq, (\!|\cdot|\!) \rangle$ with some $w_* \in (\!|\varphi|\!)$.

By Corollary 5.19, there exists an almost continuous, open surjection $\beta : X \to W$ such that $\mu\left(X \setminus \beta^{-1}(w_*)\right) < \varepsilon$.

Setting $[\![\cdot]\!] = \beta^{-1}(\!|\cdot|\!)$, β becomes a μ-bisimulation, and by Theorem 3.4, $[\![\neg\varphi]\!] \sqsubseteq [X \setminus \beta^{-1}(w_*)]_\mu$, from which the result follows. □

As we will see, our main completeness results, for subsets of Euclidean space and the Cantor set, are special cases of Theorem 6.2.

In what follows, $|\cdot|$ denotes the N-dimensional Lebesgue measure.

Lemma 6.3 *Let X be a Lebesgue-measurable subset of $\mathbb{R}^N$ of positive measure. Then, up to a set of measure zero, X is a territory.*

Proof. First assume X is bounded. We will show that it is already a province.

X is totally bounded (as is every bounded subset of Euclidean space), and clearly for every $\varepsilon > 0$ there is $\delta > 0$ such that $|B_\delta(x) \cap X| < \varepsilon$ for all $x \in X$ (use the same δ that works for all of $\mathbb{R}^N$).

Now, let Y be the set of all Lebesgue points of density of X (see Appendix A). $|X \triangle Y| = 0$, and if $U \subseteq Y$ is open in Y and $x \in U$, then for ε small enough we have $B_\varepsilon(x) \cap Y \subseteq U$ and

$$|B_\varepsilon(x) \cap Y| > \frac{|B_\varepsilon(x)|}{2};$$

this implies that $|U| > 0$.

Finally, the boundary of every open ball in Y has measure zero because the boundary of any open ball in $\mathbb{R}^N$ does.

We conclude that Y is a province, and $Y \sim X$, as desired.

Now, if X is not bounded, write

$$X = \bigcup_{\mathbf{k} \in \mathbb{Z}^N} \{\mathbf{x} \in X : x_n \in [k_n, k_{n+1}) \text{ for all } n < N\}.$$

This is a countable, disjoint union of bounded subsets of $\mathbb{R}^N$ and hence each component which has positive measure can be written as a province, up to a set of measure zero. Clearly, X is equivalent to the union of these components. □

Corollary 6.4 *Given any Lebesgue-measurable $X \subseteq \mathbb{R}^N$ of positive measure, $\mathsf{S4}_u$ is absolutely complete for X.*

Proof. Immediate from Theorem 6.2 and Lemma 6.3. □

Corollary 6.5 *$\mathsf{S4}_u$ is absolutely complete for the Cantor set under the* $\ln(2)/\ln(3)$-*Hausdorff measure*[6].

Proof. Immediate from Proposition 5.16 and the fact that the Cantor set is a province under this measure. □

A Notions from measure theory

Here we review some notions from measure theory that are used throughout the text; we will assume basic familiarity with metric and topological spaces. All of the background we need should be covered in any standard text on real analysis and measure theory, such as [8].

If $\langle X, \mathcal{T} \rangle$ is a topological space and $S \subseteq X$, we will use S° to denote the topological interior of S and $\overline{S}$ to denote its closure.

If $\langle X, d \rangle$ is a metric space, $x \in X$ and $\varepsilon > 0$, then $B_\varepsilon(x)$ denotes the open ball around x with radius ε; that is, the set of all $y \in X$ such that $d(x, y) < \varepsilon$. Every metric space natuarlly acquires a topology given by $U \subseteq X$ being open if and only if, whenever $x \in U$, there exists $\varepsilon > 0$ such that $B_\varepsilon(x) \subseteq U$. All metric spaces will be assumed to be endowed with this topology.

A metric space $\langle X, d \rangle$ is *totally bounded* if for all $\varepsilon > 0$ there exist finitely many elements $x_0, ..., x_{N-1} \in X$ such that for every $y \in X$ there is $n < N$ with $d(y, x_n) < \varepsilon$. Every bounded subset of Euclidean space is totally bounded.

A *measure space* is a triple $\langle X, \mathcal{A}, \mu \rangle$ where X is a set, $\mathcal{A} \subseteq 2^X$ is a σ-algebra (that is, a collection of sets containing $\varnothing$ and X which is closed under set difference and countable unions) and $\mu : \mathcal{A} \to [0, \infty]$ (the non-negative reals with a maximal element ∞ added) satisfying

(i) $\mu(\varnothing) = 0$

(ii) $\mu(A \setminus B) = \mu(A) - \mu(B)$ if $B \subseteq A$ and

[6] See Appendix A.

(iii) if $\langle A_n \rangle_{n<\omega}$ is an increasing sequence of elements of $\mathcal{A}$,

$$\mu\left(\bigcup_{n<\omega} A_n\right) = \lim_{n\to\infty} \mu(A_n).$$

Elements of $\mathcal{A}$ will be called *μ-measurable.* Note that $\mathcal{A}$ can be reconstructed from μ, since it is the domain of μ; because of this we will often omit explicit mention of $\mathcal{A}$ and speak of measure spaces as pairs $\langle X, \mu \rangle$. We say μ is *σ-finite* if there are countably many sets $S_n \subseteq X$ such that $\mu(S_n)$ is finite for all $n < \omega$ and $X = \bigcup_{n<\omega} S_n$. Measure spaces which are σ-finite cannot contain an uncountable collection of disjoint sets of positive measure.

We always assume that Euclidean space $\mathbb{R}^N$ is equipped with the standard Euclidean metric and Lebesgue measure; the latter will be denoted $|\cdot|$.

Given a set $S \subseteq \mathbb{R}^N$ and $x \in \mathbb{R}^N$, x is a *Lebesgue point of density* of S if

$$\lim_{\varepsilon\to 0} \frac{|B_\varepsilon(x) \cap S|}{|B_\varepsilon(x)|} = 1.$$

It is a famous theorem of Lebesgue that for every measurable set $S \subseteq \mathbb{R}^N$, almost every $x \in S$ is a point of density of S; that is, the set of elements of S which are not Lebesgue points of S has measure zero. For a proof and further details see, for example, [5].

The Cantor set has Lebesgue measure zero. However, it has measure one under the $\ln(2)/\ln(3)$-dimensional Hausdorff measure. Hausdorff measures can be used to measure fractals (sets of non-integer dimension) and provide a generalization of Lebesgue measure; a full definition is beyond our scope, but it is known that the Cantor set is a province under this measure (although the terminology is our own and this would be stated differently elsewhere). A thorough treatment of Hausdorff measures can be found in [6].

B Properties of the interior operator

Here we will develop some of the theory needed to establish the properties we use of the interior operator on a measure algebra.

We first note that the relation $\sqsubseteq$ is well-behaved under taking countable unions:

Lemma B.1 *Suppose that $\langle X, \mu \rangle$ is a measure space, $E \subseteq X$ and $\langle S_n \rangle_{n<\omega}$ is a sequence of subsets of S such that $[S_n]_\mu \sqsubseteq [E]_\mu$ for all $n < \omega$.*

Then,

$$\left[\bigcup_{n<\omega} S_n\right] \sqsubseteq [E]_\mu.$$

Proof. We have that

$$\begin{aligned}\mu\left(\left(\bigcup_{n<\omega} S_n\right) \setminus E\right) &\le \mu\left(\bigcup_{n<\omega}(S_n \setminus E)\right)\\ &\le \sum_{n<\omega} \mu(S_n \setminus E)\\ &= \sum_{n<\omega} 0\\ &= 0.\end{aligned}$$

But this implies that $\left[\bigcup_{n<\omega} S_n\right] \sqsubseteq [E]_\mu$, as desired. □

If $\langle X, \mu\rangle$ is a measure space and $\mathcal{O} \subseteq \mathbb{A}_\mu$, $\bigsqcup \mathcal{O}$ denotes the *supremum* of $\mathcal{O}$, that is, the least $u \in \mathbb{A}_\mu$ such that $o \sqsubseteq u$ for all $o \in \mathcal{O}$. As we show below, if μ is σ-finite, $\bigsqcup \mathcal{O}$ is always defined [7].

Lemma B.2 *If $\langle X, \mu\rangle$ is a measure space and $\langle S_n\rangle_{n<\omega}$ is a sequence of subsets of X, then $\bigsqcup \langle [S_n]_\mu\rangle_{n<\omega}$ is defined and equals $\left[\bigcup_{n<\omega} S_n\right]$.*

Proof. Clearly $\left[\bigcup_{n<\omega} S_n\right]$ is an upper bound for $\bigsqcup \langle [S_n]_\mu\rangle_{n<\omega}$; Lemma B.1 guarantees that it is the least upper bound, since any element of $\mathbb{A}_\mu$ which is greater than all $[S_n]_\mu$ is also greater than their union. □

All operations on the measure algebra are essentially countable in the following sense:

Lemma B.3 *If $\langle X, \mathcal{A}, \mu\rangle$ is a σ-finite measure space and $\mathcal{O} \subseteq \mathbb{A}_\mu$, then $\bigsqcup \mathcal{O}$ (the supremum of $\mathcal{O}$) is defined and there is a sequence $\langle e_n\rangle_{n<\omega}$ of elements of $\mathcal{O}$ such that*

$$\bigsqcup \mathcal{O} = \bigsqcup_{n<\omega} e_n.$$

Proof. Suppose $\mathcal{O} = \langle o_\xi\rangle_{\xi<\gamma}$, where γ is a possibly uncountable cardinal.

By cardinal induction, we can assume that $O_\xi = \bigsqcup_{\zeta<\xi} o_\xi$ is defined for all $\xi < \gamma$.

Let I be the set of all $\xi < \gamma$ such that $\mu(O_\zeta) < \mu(O_\xi)$ for all $\zeta < \xi$. Write I as an increasing sequence $I = \langle \iota_\xi\rangle_{\xi<\lambda}$.

One can see that for all $\chi \neq \zeta < \lambda$,

$$(O_{\iota_{\xi+1}} - O_{\iota_\xi}) \sqcap (O_{\iota_{\zeta+1}} - O_{\iota_\zeta}) = [\varnothing]_\mu$$

and each $O_{\iota_{\xi+1}} - O_{\iota_\xi}$ is of positive measure; since μ is σ-finite, it follows that there can be only countably many of them, and therefore I must be countable.

We claim that $\bigsqcup_{\xi\in I} O_\xi = \bigsqcup \mathcal{O}$. Note that this is sufficient to establish our result; by induction hypothesis, for each $\xi < \gamma$ we can write $O_\xi = \bigsqcup_{n<\omega} e^\xi_n$ with $e^\xi_n \in \mathcal{O}$, so that $\bigsqcup_{\xi\in I} O_\xi = \bigsqcup_{\xi\in I} \bigsqcup_{n<\omega} e^\xi_n$. Since I is countable the latter is a supremum over a countable set, as desired.

---

[7] Indeed, $\mathbb{A}_\mu$ is a complete Boolean algebra, a fact which was proven by Tarski in [16]. The same paper indicates that Jaskowski proved the result in the special case of the Lebesgue measure algebra in 1931, but did not publish the proof at the time.

Clearly $\bigsqcup_{\xi\in I} O_\xi \sqsubseteq \bigsqcup \mathcal{O}$, so to show that $\bigsqcup_{\xi\in I} O_\xi = \bigsqcup \mathcal{O}$ it suffices to prove that $\bigsqcup_{\xi\in I} O_\xi$ is an upper bound for $\mathcal{O}$.

Pick $\zeta < \gamma$ and consider the least ordinal ϑ such that $\mu(O_\vartheta) = \mu(O_\zeta)$. By the way we defined I we have that $\vartheta \in I$; but then $O_\zeta \sqsubseteq O_\vartheta$, and hence $O_\zeta \sqsubseteq \bigsqcup_{\xi\in I} O_\xi$. Since $o_\zeta \sqsubseteq O_\zeta$ and ζ was arbitrary, the claim follows. □

We are now ready to prove Proposition 2.4.

Proof. [Proof of Proposition 2.4] Let $\langle X, \mathcal{T}, \mu\rangle$ be a topological measure space and o an element of its measure algebra.

By Lemma B.3 there are countably many open sets U_n with $[U_n]_\mu \sqsubseteq o$ such that $o^\Box = \bigsqcup_{n<\omega} [U_n]_\mu$; by Lemma B.2, $\bigsqcup_{n<\omega} [U_n]_\mu = \left[\bigcup_{n<\omega} U_n\right]_\mu$. Then,

(i) $o^\Box$ is open, since $\bigcup_{n<\omega} U_n$ is an open set;

(ii) follows from the definition of the interior operator;

(iii) follows immediately from the fact that $o^\Box$ is open.

□

Acknowledgements

I would like to thank Theodora Bourni for her insightful comments about the Hausdorff measure of the Cantor set, as well as Dana Scott for kindly answering my questions about measure algebras.

References

[1] S.N. Artemov, J.M. Davoren, and A. Nerode. Modal logics and topological semantics for hybrid systems. *Technical Report MSI 97-05*, 1997.

[2] L.D. Beklemishev, G. Bezhanishvili, and T. Icard. On topological models of GLP. 2009.

[3] B. Bennett and I. Düntsch. Axioms, algebras and topology. In M. Aiello, I. Pratt-Harman, and J. van Benthem, editors, *Handbook of Spatial Logics*. 2007.

[4] G. Bezhanishvili and M. Gehrke. Completeness of S4 with respect to the real line: revisited. *Annals of Pure and Applied Logic*, 131:287–301, 2005.

[5] D. Cohn. *Measure Theory.* Boston, 1980.

[6] H. Federer. *Geometric Measure Theory.* Springer-Verlag, Berlin, Heidelberg, New York, 1969.

[7] D. Fernández-Duque. Dynamic topological completeness for $\mathbb{R}^2$. *Logic Journal of IGPL*, 2007. doi: 10.1093/jigpal/jzl036.

[8] G. Folland. *Real Analysis: Modern Techniques and Their Applications.* Wiley-Interscience, 1999.

[9] V. Goranko and S. Passy. Using the universal modality: gains and questions. *J. Logic Comput*, 2:5–30, 1992.

[10] P. Kremer and G. Mints. Dynamic topological logic. *Annals of Pure and Applied Logic*, 131:133–158, 2005.

[11] J.C.C. McKinsey and A. Tarski. The algebra of topology. *Ann. of Math.*, 2:141–191, 1944.

[12] G. Mints. A completeness proof for propositional S4 in cantor space. In E. Orlowska, editor, *Logic at Work: Essays Dedicated to the Memory of Helena Rasiowa*. Physica-Verlag, Heidelberg, 1998.

[13] G. Mints and T. Zhang. A proof of topological completeness for S4 in (0,1). *Annals of Pure and Applied Logic*, 133:231–245, 2005.

[14] V. Shehtman. "Everywhere" and "here". *J. Appl. Non-Classical Logics*, 9:369–379, 1999.

[15] S. Slavnov. Two counterexamples in the logic of dynamic topological systems. *Technical Report TR-2003015*, 2003.

[16] A. Tarski. Über additive und multiplikative mengenkörper und mengenfunktionen. *C. R. Soc. Sci. Lett. Varsovie Cl. III Sci. Math-Phys. 30*, pages 151–181, 1937.

On the Size of Shortest Modal Descriptions

Santiago Figueira [1]

Departamento de Computación, FCEyN, Universidad de Buenos Aires and CONICET
Pabellón I, Ciudad Universitaria
(C1428EGA), Buenos Aires, Argentina

Daniel Gorín [2]

Departamento de Computación, FCEyN, Universidad de Buenos Aires
Pabellón I, Ciudad Universitaria
(C1428EGA), Buenos Aires, Argentina

Abstract

We address the problems of *separation* and *description* in some fragments of modal logics. The former consists in finding a formula that is true in some given subset of the domain and false in another. The latter is a special case when one separates a singleton from the rest. We are interested in the shortest *size* of both separations and descriptions. This is motivated by applications in computational linguistics. Lower bounds are given by considering the minimum size of Spoiler's strategies in the classical Ehrenfeucht-Fraïssé game. This allows us to show that the size of such formulas is not polynomially bounded (with respect to the size of the finite input model). Upper bounds for these problems are also studied. Finally we give a fine hierarchy of succinctness for separation over the studied logics.

Keywords: Modal logic, referring expression, shortest formula size, lower bound, Ehrenfeucht-Fraïssé, succinctness.

1 Introduction

We informally say that a formula φ *describes* an element e in the domain of some model $\mathcal{M}$ whenever φ is true when evaluated at e and false when evaluated at every other point in the domain of $\mathcal{M}$. One can then define the *description problem* as that of finding a description for a given e, if such description exists. [3]

[1] S. Figueira was partially supported by CONICET (grant PIP 370), ANPCyT (grant PICT 2067) and UBA (grant UBACyT X615).

[2] D. Gorín was partially supported by ANPCyT (grant PICT 2067).

[3] We are being deliberately unspecific about the logic in question here, since one can in principle define a description problem for any logic with suitable semantics.

This a fundamental problem in the *Generation of Referring Expressions* (GRE), a key task in the field of Natural Language Generation with continuous active research (see [6,7,8,16,17] among others). GRE is the generation of *noun-phrases* that refer unequivocally to certain objects in the context of conversation. The description problem amounts to finding the relevant features that identify an object (e.g., $kid \wedge \langle carries \rangle (ball \wedge red)$) the outcome of which can be given to a *surface realization* module that deals with the generation of an equivalent expression in natural language (e.g., "the kid carrying a red ball").

In this paper we focus on the description problem for the basic modal logic $\mathcal{ML}$ and some of its syntactical fragments. Multi-modal versions of these languages have been previously considered in the context of GRE because of their combination of expressiveness and good computational behavior [3].

In particular, we are interested in the computational complexity of the description problem for modal languages. This question was initially addressed in [3] where it is shown that a standard algorithm for computing bisimulation minimization [15,12] can be adapted to compute an $\mathcal{ML}$-description for every equivalence class in the minimized model (we revisit this idea in Section 5). Since bisimulation minimization can be done in polynomial time, if new formulas are built by combining, in constant-time, formulas that were computed in previous iterations, the resulting algorithm will run in polynomial time too.

Can we conclude that the description problem for $\mathcal{ML}$ can be solved in polynomial time? One must be careful here. In order to implement formula constructors (such as $\wedge$, $\Box$, etc.) as constant-time operations one needs to resort to pointers or similar mechanisms based on aliasing; the upshot of this is that we will be computing $\mathcal{ML}$-descriptions that are *compactly represented* as direct acyclic graphs (DAG). The size of these DAGs is, by construction, bounded by a polynomial in the size of the model, but it is not clear, in principle, that such a bound exists with respect to the expansion of these DAGs to full-blown trees.

It is shown in [2] that for certain class of models this algorithm can lead to DAGs whose expansions cannot be bounded by a polynomial (cf. Section 5). However, every element in that class of models has a description of linear size. That example only proves that this algorithm may compute very degenerate solutions, but already shows that one has to be careful about complexity claims for this problem. [4]

The description problem is a particular instance of a more general problem: given a model $\mathcal{M}$ and two non-empty sets C, D (of the domain of $\mathcal{M}$), find a formula that is true at every element in C and false at every element in D. We call this the *separation problem*. In this article we show that no polynomial can bound the size of the solutions to the separation and description problems. More precisely, we give exponential lower bounds for the worst-case size of solutions for the separation problem for C and D singleton sets. We show similar lower bounds when weak fragments of $\mathcal{ML}$ (such as the one without negation) are used.

The article is structured as follows. We begin in Section 2 by introducing the notation

[4] Surface realization algorithms do not exploit subformula sharing, but will produce a noun-phrase that is proportional (typically, linear) in length to the size of the formula.

we will use throughout the paper. In Section 3 we present the tool we will use to establish lower bound results: *uniform strategy trees.* These formalize a strategy for Spoiler in an Ehrenfeucht-Fraïssé game in a way such that the size of a minimum winning uniform strategy tree corresponds to the size of the minimum formula that separates the elements in the initial position of the game. Using these, we give, in Section 4, exponential lower bounds for the separation problem using different fragments of $\mathcal{ML}$ formulas. In Section 5 we give an upper bound for this problem that is slightly higher than the lower bound of Section 4. Though the upper and lower bounds are almost tight, the question of which are the optimal bounds remains open. Finally, in Section 6 we relate these results with the standard notion of succinctness, and use them to form a hierarchy for the studied fragments of $\mathcal{ML}$ in terms of it. Conclusions and future work are presented in Section 7.

2 Preliminaries

We will work on the basic modal language, presented for convenience in negation normal form. Results can be trivially extended to multi-modal languages.

Definition 2.1 [Syntax] The language of the basic modal logic $\mathcal{ML}$ is given by the following grammar:

$$\varphi ::= \top \mid \bot \mid p \mid \neg p \mid \varphi \wedge \varphi \mid \varphi \vee \varphi \mid \Diamond\varphi \mid \Box\varphi$$

where $p, q, r \ldots$ are propositional symbols. A *literal* is formula of the form $\top$, $\bot$, p or $\neg p$. $\mathcal{ML}^{\Diamond\wedge\neg}$ is the fragment of $\mathcal{ML}$ with no occurrences of $\Box$ and $\vee$. $\mathcal{ML}^{\Diamond\wedge}$ is the fragment of $\mathcal{ML}^{\Diamond\wedge\neg}$ with no literals of the form $\neg p$.

For $\varphi \in \mathcal{ML}$, we use $\overline{\varphi}$ to denote the negation of φ: $\overline{\top} = \bot$, $\overline{p} = \neg p$, $\overline{\varphi \wedge \psi} = \overline{\varphi} \vee \overline{\psi}$, $\overline{\Diamond\varphi} = \Box\overline{\varphi}$, etc. We use $d(\varphi)$ for the *modal depth* of φ, i.e., the maximum number of nested modalities occurring in φ.

To measure *formula size*, we define $|\varphi|$ as the number of literals (counting repetitions) that occur in φ; therefore, we have $|\varphi \vee \psi| = |\varphi \wedge \psi| = |\varphi| + |\psi|$; $|\neg\varphi| = |\varphi|$; $|\Diamond\varphi| = |\Box\varphi| = |\varphi|$; $|p| = 1$ for propositional symbols p; and $|\top| = |\bot| = 1$. Of course there are other reasonable notions of formula size, for instance those counting modal or boolean operators, or even parenthesis. But any reasonable measure of size $||\cdot||$ should satisfy $||\Diamond\varphi|| \geq ||\varphi||, ||\varphi \wedge \psi|| \geq ||\varphi|| + ||\psi||$, etc. and, therefore, will satisfy $||\varphi|| \geq |\varphi|$. Hence all the lower bounds presented in this work will hold for any reasonable definition of size.

As usual, formulas are interpreted using Kripke models $\mathcal{M} = \langle W, R, V\rangle$ where W is a non-empty carrier set, R is a binary relation on W and V maps proposition symbols to subsets of W. We use $sucs_{\mathcal{M}}(w)$ for $\{w' \mid (w, w') \in R\}$ (or simply $sucs(w)$ if $\mathcal{M}$ is clear from context). The size of the finite model $\mathcal{M} = \langle W, R, V\rangle$ (with finite domain W and finite valuation V), denoted $|\mathcal{M}|$, is taken to be $|W| + |R| + |V|$. Here $|V|$ denotes the size of the set of all pairs (w, p) where where p ranges over a finite set of propositional symbols and $w \in V(p)$, while $|R|$ is the number of pairs $(w, v) \in R$.

Definition 2.2 [Semantics] Given $\mathcal{M} = \langle W, R, V\rangle$, the satisfaction relation $\models$ is inductively defined as:

$$\begin{array}{ll}
\mathcal{M}, w \models \top & \\
\mathcal{M}, w \models p & \text{iff } w \in V(p) \\
\mathcal{M}, w \models \neg p & \text{iff } w \notin V(p) \\
\mathcal{M}, w \models \varphi \wedge \psi & \text{iff } \mathcal{M}, w \models \varphi \text{ and } \mathcal{M}, w \models \psi \\
\mathcal{M}, w \models \varphi \vee \psi & \text{iff } \mathcal{M}, w \models \varphi \text{ or } \mathcal{M}, w \models \psi \\
\mathcal{M}, w \models \Diamond\varphi & \text{iff } \mathcal{M}, w' \models \varphi \text{ for some } w' \in \mathit{sucs}_{\mathcal{M}}(w) \\
\mathcal{M}, w \models \Box\varphi & \text{iff } \mathcal{M}, w' \models \varphi \text{ for every } w' \in \mathit{sucs}_{\mathcal{M}}(w)
\end{array}$$

If $C \subseteq W$, we write $\mathcal{M}, C \models \varphi$ when $\mathcal{M}, w \models \varphi$ for every $w \in C$. When $\mathcal{M}$ is fixed or clear from context, we shall use the shorter versions $w \models \varphi$ and $C \models \varphi$.

The *description problem* can be seen as a particular case of the more general problem of finding a formula that *separates* two arbitrary sets.

Definition 2.3 Let $\mathcal{M} = \langle W, R, V\rangle$ and let $C, D \subseteq W$ be two non-empty sets. We say that *φ separates C and D in $\mathcal{M}$* whenever $\mathcal{M}, C \models \varphi$ and $\mathcal{M}, D \models \overline{\varphi}$. When φ separates $\{w\}$ and $W \setminus \{w\}$ in $\mathcal{M}$, we say that *φ is a description for w*. For $c, d \in W$, by 'φ separates c and d' we mean 'φ separates $\{c\}$ and $\{d\}$'.

3 Games, strategies and shortest description size

The standard way of establishing lower bounds on formula size is using Adler-Immerman games [1] (for other techniques and logics see [9,14,18]). In these games, one of the players tries to build a tree that induces a formula of the same size separating two models (or points in a model), while an opponent tries to prevent it. The latter has an optimal strategy in these games, so one only needs to show that the former has a strategy that beats it. This is essentially the technique we will employ.

In order to make this paper self-contained, we will define in this section all the machinery needed. But we shall do it with a slight twist. The trees constructed during Adler-Immerman games can be reinterpreted as decision trees that act as winning strategies for Spoiler in classical Ehrenfeucht-Fraïssé games. This means that while existence of a winning strategy for Duplicator in these games can be used to give lower bounds on the number of nested modalities needed for some task, the minimum size of (certain formalization of) a strategy for Spoiler can be used to give lower bounds on the size of a formula.

We start then defining the classical n-turn Ehrenfeucht-Fraïssé game for $\mathcal{ML}$. Instead of being played on two Kripke models, we find it convenient to define it on two elements of the same Kripke model. Since modal truth is invariant under disjoint unions, no generality is lost.

Definition 3.1 The n-turn Ehrenfeucht-Fraïssé game over model $\mathcal{M} = \langle W, R, V \rangle$ starting on $(w, v) \in W^2$ (notation: $\mathcal{G}_\mathcal{M}(w, v, n)$) is played between two players, Spoiler and Duplicator. The rules of the games are:

p_l: Spoiler picks a p such that $w \in V(p)$ and $v \notin V(p)$ and wins.

p_r: Spoiler picks a p such that $v \in V(p)$ and $w \notin V(p)$ and wins.

R_l: If $n > 0$, Spoiler may pick a $w' \in sucs(w)$; then Duplicator must respond choosing a $v' \in sucs(v)$ or otherwise loses. In the first case, the turn ends and they continue to play $\mathcal{G}_\mathcal{M}(w', v', n-1)$.

R_r: If $n > 0$, Spoiler may pick a $v' \in sucs(v)$; then Duplicator must respond choosing a $w' \in sucs(w)$ or otherwise loses. In the first case, the turn ends and they continue to play $\mathcal{G}_\mathcal{M}(w', v', n-1)$.

Duplicator wins whenever Spoiler cannot play. We write $\mathcal{G}_\mathcal{M}^{\Diamond\wedge\neg}(w, v, n)$ for the variation of $\mathcal{G}_\mathcal{M}(w, v, n)$ without rule R_r, and $\mathcal{G}_\mathcal{M}^{\Diamond\wedge}(w, v, n)$ for the one that additionally drops rule p_r. We will write $\mathcal{G}(w, v, n)$ when the model is clear from context.

Rules p_l and p_r are not typically part of presentations of Ehrenfeucht-Fraïssé games for $\mathcal{ML}$; they include, instead, the additional constraint on rules R_l and R_r that w' and v' must agree propositionally. Our formulation is clearly equivalent.

Informally, a *strategy* for the game $\mathcal{G}_\mathcal{M}(w, v, n)$ is a way of playing in which a player's moves are determined by the previous ones. It is a *winning* strategy for player P when P, following the commands of the strategy, wins the game independently of the opponent's moves. Before turning into the formal definition of *strategy* and *winning strategy* let us mention some is well-known results (see [5, Chapter 3] for more details) regarding winning strategies, bisimilarity and modal equivalence. The following are equivalent:

- Duplicator has a winning strategy for $\mathcal{G}_\mathcal{M}(w, v, n)$;
- For every formula φ of $\mathcal{ML}$ with modal depth n, $\mathcal{M}, w \models \varphi$ iff $\mathcal{M}, v \models \varphi$

Hence, w and v are modally equivalent in $\mathcal{M}$ if and only if, for every n, Duplicator has a winning strategy for $\mathcal{G}_\mathcal{M}(w, v, n)$. If we drop the restriction of n-rounds and allow for infinite games, then a winning strategy for player P denotes a way of playing in such a way that P can always answer to his opponent's move. Let $\mathcal{G}_\mathcal{M}(w, v)$ denote this infinite game with no limit in the number of rounds. Then Duplicator has a winning strategy for $\mathcal{G}_\mathcal{M}(w, v)$ (that is, one that prevents Spoiler from reaching any of his winning states) if and only if u and v are bisimilar in $\mathcal{M}$. If u and v are bisimilar in $\mathcal{M}$ then u and v are modally equivalent in $\mathcal{M}$. The converse is not true for arbitrary $\mathcal{M}$ but it holds when $\mathcal{M}$ is finitely branching (this is known as the Hennessy-Milner Theorem [4,10]).

Any strategy for $\mathcal{G}_\mathcal{M}(w, v, n)$ can be formalized in a more or less straightforward way using a simple lookup table. However, we are interested in a formalization that will ultimately allow us to correlate strategy and formula size. We therefore formalize Spoiler's strategies using a form of decision tree, which we call *uniform strategy trees.*

Definition 3.2 Let $\mathcal{M} = \langle W, R, V \rangle$ be some fixed model. A *uniform strategy tree* for Spoiler is an annotated tree.We write $x \to y$ to mean that nodes x and y are linked by an edge and use $x \xrightarrow{C} y$ when the edge is annotated with a non-empty $C \subseteq W$. Nodes can be of six different types: those of type 1 and 2 are annotated with a non-empty set $C \subseteq W$ and are denoted $\langle C \rangle$ and $[C]$, respectively; those of type 3 and 4 are annotated with a proposition symbol p and denoted (p) and $(\overline{p})$; finally, those of type 5 and 6, denoted $(\wedge)$ and $(\vee)$, are not annotated (that is, they do not have any other information apart from the type itself).

Let $C, D \subseteq W$ be non-empty sets; we say that a uniform strategy tree with root x is *winning for* $\mathcal{G}_{\mathcal{M}}(C, D)$ whenever these inductive conditions hold:

(i) If $x = \langle E \rangle$, then we must have $E \cap \mathit{sucs}(\{w\}) \neq \emptyset$ for all $w \in C$, and if $\mathit{sucs}(D) \neq \emptyset$, then $x \to y$ for some y that is winning for $\mathcal{G}_{\mathcal{M}}(E, \mathit{sucs}(D))$.

(ii) If $x = [E]$, then we must have $E \cap \mathit{sucs}(\{w\}) \neq \emptyset$ for all $w \in D$, and if $\mathit{sucs}(C) \neq \emptyset$, then $x \to y$ for some y that is winning for $\mathcal{G}_{\mathcal{M}}(\mathit{sucs}(C), E)$.

(iii) If $x = (p)$, then we must have $C \cap V(p) = C$ and $D \cap V(p) = \emptyset$.

(iv) If $x = (\overline{p})$, then we must have $C \cap V(p) = \emptyset$ and $D \cap V(p) = D$.

(v) If $x = (\wedge)$, then $D = \bigcup\{A \mid \exists y, x \xrightarrow{A} y \text{ and } y \text{ is winning for } \mathcal{G}_{\mathcal{M}}(C, A)\}$.

(vi) If $x = (\vee)$, then $C = \bigcup\{A \mid \exists y, x \xrightarrow{A} y \text{ and } y \text{ is winning for } \mathcal{G}_{\mathcal{M}}(A, D)\}$.

When a uniform strategy tree that is winning for $\mathcal{G}_{\mathcal{M}}(C, D)$ has no nodes of type 2 nor 6 we say that it is winning for $\mathcal{G}^{\diamond\wedge\neg}_{\mathcal{M}}(C, D)$; if it doesn't have nodes of type 4 either, we say that it is also winning for $\mathcal{G}^{\diamond\wedge}_{\mathcal{M}}(C, D)$. Again, we will drop the model when clear from context and say, e.g., that a strategy is *winning for* $\mathcal{G}(C, D)$.

The size $|s|$ of a uniform strategy tree s is the number of *leaf nodes* in s; its depth $d(s)$ is the maximum number of nested nodes of type 1 or 2.

Notice that every uniform strategy tree s has, by definition, a finite height. Therefore, it has a finite size if and only if every node is finitely branching.

Readers familiar with Adler-Immerman games may recognize in conditions (i)–(vi) the rules of the modal version of these games (for formulas in negation normal form). Since in Adler-Immerman games Duplicator has an optimal strategy, it is not surprising that we can give a static characterization of them.

The first thing we need to show is that winning uniform strategy trees indeed constitute winning strategies.

Theorem 3.3 *If there exists a uniform strategy tree for Spoiler with $d(s) \leq n$ that is winning for $\mathcal{G}^{\star}_{\mathcal{M}}(C, D)$, then Spoiler wins every game $\mathcal{G}^{\star}_{\mathcal{M}}(w, v, n)$ with $w \in C$, $v \in D$ (for $\mathcal{G}^{\star} \in \{\mathcal{G}, \mathcal{G}^{\diamond\wedge\neg}, \mathcal{G}^{\diamond\wedge}\}$).*

Proof. We proceed by induction on the tree, so let x be its root. If $x = (p)$ or $x = (\overline{p})$, then by definition, Spoiler can play p according to rule p_l or p_r, respectively, and win immediately. If $x = \langle E \rangle$, then Spoiler may play according to rule R_l, picking (non-deterministically) some $w' \in E$ that is an R-successor of w (observe that since $E \cap \mathit{sucs}(\{w\}) \neq \emptyset$, some such successor exists); if Duplicator answers with some $v' \in \mathit{sucs}(D)$, then $\mathit{sucs}(D) \neq \emptyset$ and there must exist some $x \to y$ such that y is winning for $\mathcal{G}(E, \mathit{sucs}(D))$, and by inductive hypothesis, Spoiler wins every instance of $G(w', v', n-$

1). If $x = [E]$, Spoiler may play according to rule R_r and we reason analogously. Suppose now $x = (\wedge)$; for some A with $v \in A$ we must have $x \xrightarrow{A} y$ and since y is winning for $\mathcal{G}(C, A)$, we conclude that Spoiler wins every instance of $\mathcal{G}(w, v, n)$. The case for $x = (\vee)$ is analogous. □

We will now prove again the well-known Ehrenfeucht-Fraïssé Theorem for $\mathcal{ML}$ but paying attention not only to the modal depth of formulas but also to their sizes. For the rest of this section, we assume a fixed but otherwise arbitrary model $\mathcal{M} = \langle W, R, V\rangle$.

Lemma 3.4 *Let $C, D \subseteq W$ be non-empty. If $\varphi \in \mathcal{L}^\star$ separates C and D in $\mathcal{M}$, then Spoiler has a uniform strategy tree s that is winning for $\mathcal{G}^\star_{\mathcal{M}}(C, D)$, with $|s| \leq |\varphi|$ and $d(s) \leq d(\varphi)$ (for $(\mathcal{L}^\star, \mathcal{G}^\star) \in \{(\mathcal{ML}, \mathcal{G}), (\mathcal{ML}^{\Diamond\wedge\neg}, \mathcal{G}^{\Diamond\wedge\neg}), (\mathcal{ML}^{\Diamond\wedge}, \mathcal{G}^{\Diamond\wedge})\}$).*

Proof. We proceed by induction on φ. Since C and D are non-empty, we cannot have $\varphi = \top$ nor $\varphi = \bot$. If $\varphi = p$, then a leaf-node (p) suffices while $(\overline{p})$ works in case $\varphi = \neg p$. If $\varphi = \Diamond\psi$ then we know there exists a $E \subseteq sucs(C)$ such that $E \cap sucs(\{w\}) \neq \emptyset$ for all $w \in C$ and $E \models \psi$. In case $sucs(D) = \emptyset$, we can use $\langle E\rangle$ (or $\langle sucs(C)\rangle$) as strategy tree. Otherwise, by inductive hypothesis, there is a strategy with root y that is winning for $\mathcal{G}(E, sucs(D))$, $|y| \leq |\psi|$ and $d(y) \leq d(\psi)$. Therefore, the uniform strategy tree with root $x = \langle E\rangle$ such that $x \to y$ must be winning for $\mathcal{G}(C, D)$, $|x| = |y| \leq |\psi| = |\Diamond\psi|$ and $d(x) = 1 + d(y) \leq 1 + d(\psi) = d(\Diamond\psi)$. The case for $\varphi = \Box\psi$ is analogous. Suppose now that $\varphi = \psi_1 \wedge \cdots \wedge \psi_k$ and let $F_i = \{v \in D \mid v \models \overline{\psi_i}\}$. Observe that $D = \bigcup_{i=1}^k F_i$, so for some i, $F_i \neq \emptyset$. For each $1 \leq i \leq k$, if $F_i \neq \emptyset$ then there exists, by inductive hypothesis, a uniform strategy tree y_i that is winning for $\mathcal{G}(C, F_i)$. Therefore, the uniform strategy tree whose root x is $(\wedge)$ and such that $x \xrightarrow{F_i} y_i$ for every $F_i \neq \emptyset$, is winning for $\mathcal{G}(C, D)$. Observe also that, by inductive hypothesis, $|x| \leq \sum_{i=i}^k |\psi_i| = |\varphi|$ and, similarly, $d(x) \leq d(y)$. The case for $\varphi = \Box(\psi_1 \vee \cdots \vee \psi_k)$ is analogous. □

Theorem 3.5 (Ehrenfeucht-Fraïssé Theorem) *If Duplicator has some strategy that is winning for $\mathcal{G}^\star_{\mathcal{M}}(w, v, n)$, then for every $\varphi \in \mathcal{L}^\star$ with $d(\varphi) \leq n$, $\mathcal{M}, w \models \varphi$ implies $\mathcal{M}, v \models \varphi$ ($(\mathcal{L}^\star, \mathcal{G}^\star) \in \{(\mathcal{ML}, \mathcal{G}), (\mathcal{ML}^{\Diamond\wedge\neg}, \mathcal{G}^{\Diamond\wedge\neg}), (\mathcal{ML}^{\Diamond\wedge}, \mathcal{G}^{\Diamond\wedge})\}$).*

Proof. Suppose $w \models \varphi$ and $v \not\models \varphi$. This means that φ separates w and v and, by Lemma 3.4, Spoiler has a uniform strategy tree s with $d(s) \leq n$ that is winning for $\mathcal{G}^\star(\{w\}, \{v\})$. Therefore, by Theorem 3.3, Spoiler wins every instance of $\mathcal{G}^\star(w, v, n)$, so Duplicator cannot have a winning strategy. □

Observe that the condition "$\mathcal{M}, w \models \varphi$ implies $\mathcal{M}, v \models \varphi$" is equivalent to "$\mathcal{M}, w \models \varphi$ iff $\mathcal{M}, v \models \varphi$" in $\mathcal{ML}$, but not in $\mathcal{ML}^{\Diamond\wedge}$ nor $\mathcal{ML}^{\Diamond\wedge\neg}$, since they are not closed under negation.

For the converse of Lemma 3.4 we need the additional requirement that s is finitely branching.

Lemma 3.6 *If Spoiler has a uniform strategy tree s* of finite size *that is winning for $\mathcal{G}^\star(C, D)$, then there exists a $\varphi \in \mathcal{ML}^\star$ such that $|\varphi| \leq |s|$ and $d(\varphi) \leq d(s)$ that separates C and D (for $(\mathcal{L}^\star, \mathcal{G}^\star) \in \{(\mathcal{ML}, \mathcal{G}), (\mathcal{ML}^{\Diamond\wedge\neg}, \mathcal{G}^{\Diamond\wedge\neg}), (\mathcal{ML}^{\Diamond\wedge}, \mathcal{G}^{\Diamond\wedge})\}$).*

Proof. We proceed by induction on s, and let x be its root. If $x = (p)$, then p trivially satisfies $C \models p$ and $D \models \overline{p}$. Similarly, $\neg p$ can handle the case $x = (\overline{p})$. In case $x = \langle E \rangle$, then either $sucs(D) = \emptyset$ and $\Diamond\top$ is the formula we need or else there exists an y that is winning for $\mathcal{G}(E, sucs(D))$ and, by inductive hypothesis for some ψ we have $E \models \psi$, $sucs(D) \models \overline{\psi}$, $|\psi| \leq |y|$ and $(\psi) \leq d(y)$. Clearly, $D \models \Box\overline{\psi}$, $|\Diamond\psi| \leq |x|$, $d(\Diamond\psi) \leq d(x)$ and, because $E \cap sucs(\{w\}) = \emptyset$ for every $w \in C$, we can also conclude $C \models \Diamond\psi$. The case for for $x = [E]$ is analogous. Suppose now that $x = (\wedge)$. Since s has finite size, there can be only finitely many y such that $x \xrightarrow{A} y$ (there is at least one y since D is non-empty). For every such y there exists, by inductive hypothesis, a formula φ_y such that $C \models \varphi_y$ and $A \models \overline{\varphi_y}$; by taking φ to be the conjunction of all such φ_y, we get $C \models \varphi$ and $D \models \overline{\varphi}$. The case for $x = (\vee)$ is symmetrical. □

We are now ready to give the main result of this section.

Definition 3.7 We say that a uniform strategy tree s that is winning for $\mathcal{G}^\star_{\mathcal{M}}(C, D)$ is *minimum* whenever for any other uniform strategy tree s' winning for $\mathcal{G}^\star_{\mathcal{M}}(C, D)$, $|s| \leq |s'|$ (for $\mathcal{G}^\star \in \{\mathcal{G}, \mathcal{G}^{\Diamond\wedge\neg}, \mathcal{G}^{\Diamond\wedge}\}$). Similarly, a formula $\varphi \in \mathcal{ML}^\star$ that separates C and D in $\mathcal{M}$ is *minimum* whenever for any $\psi \in \mathcal{ML}^\star$ that separates C and D in $\mathcal{M}$, $|\varphi| \leq |\psi|$ (for $\mathcal{ML}^\star \in \{\mathcal{ML}, \mathcal{ML}^{\Diamond\wedge\neg}, \mathcal{ML}^{\Diamond\wedge}\}$).

Theorem 3.8 *If s is a minimum uniform strategy tree winning for $\mathcal{G}^\star_{\mathcal{M}}(C, D)$ and $\varphi \in \mathcal{ML}^\star$ is a minimum formula that separates C and D in $\mathcal{M}$, then $|s| = |\varphi|$ (for $(\mathcal{L}^\star, \mathcal{G}^\star) \in \{(\mathcal{ML}, \mathcal{G}), (\mathcal{ML}^{\Diamond\wedge\neg}, \mathcal{G}^{\Diamond\wedge\neg}), (\mathcal{ML}^{\Diamond\wedge}, \mathcal{G}^{\Diamond\wedge})\}$).*

Proof. By Lemma 3.6, there exists a ψ that separates C and D such that $|\psi| \leq |s|$, and since φ is minimum, we know $|\varphi| \leq |\psi| \leq |s|$. Now, by Lemma 3.4, there exists an s' that is winning for $\mathcal{G}^\star(C, D)$ with $|s'| \leq |\varphi|$, and since s is minimum we conclude $|s| \leq |s'| \leq |\varphi| \leq |\psi| \leq |s|$. □

A simple inspection of Definition 3.2 shows that if a uniform tree strategy is winning for $\mathcal{G}(C, D)$, then it is also winning for $\mathcal{G}(C', D')$ for every non-empty $C' \subseteq C$ and $D' \subseteq D$. This shows that in order to give a lower bound for the description problem for w it suffices to guarantee that w has a description and give a lower bound for the size of a formula that separates $\{w\}$ and D for some D with $w \notin D$. This will be pursued in Section 4 and the following results will be useful.

Proposition 3.9 *If s is a uniform strategy tree winning for $\mathcal{G}^\star(C, D)$ whose root is of the form $\langle E \rangle$ (resp. $[E]$), then there exists a uniform strategy tree s' with root $\langle E' \rangle$ (resp. $[E']$) that is winning for $\mathcal{G}^\star(C, D)$ and such that $|s| = |s'|$ and $E' \subseteq sucs(C)$ (resp. $E' \subseteq sucs(D)$), for $\mathcal{G}^\star \in \{\mathcal{ML}, \mathcal{ML}^{\Diamond\wedge\neg}, \mathcal{ML}^{\Diamond\wedge}\}$.*

Proof. Let $x = \langle E \rangle$ be the root of s; define $x' = \langle E \cap sucs(C) \rangle$ and set $x' \to y$ for every y such that $x \to y$. Since C is not empty and for every $w \in C$, $E \cap sucs(w) \neq \emptyset$, we know $E \cap sucs(C)$ is not empty. If $x \to y$ then y is winning for $\mathcal{G}(E, sucs(D))$ and, by the observation above, y is also winning for $\mathcal{G}(E \cap sucs(C), sucs(D))$ and therefore x' is winning for $\mathcal{G}(C, D)$. The case for $[E]$ is analogous. □

Proposition 3.10 *If s is a uniform strategy tree that is winning for $\mathcal{G}^\star(C,D)$ ($\mathcal{G}^\star \in \{\mathcal{G}, \mathcal{G}^{\Diamond\wedge\neg}, \mathcal{G}^{\Diamond\wedge}\}$), then $C \cap D = \emptyset$.*

Proposition 3.11 *Let $\mathcal{G}^\star \in \{\mathcal{G}, \mathcal{G}^{\Diamond\wedge\neg}, \mathcal{G}^{\Diamond\wedge}\}$ and let s be a winning strategy for $\mathcal{G}^\star(C,D)$ of minimum size. If D is singleton (resp. C is singleton) and the root of s is of type $(\wedge)$ (resp. of type $(\vee)$) then there is a winning strategy s' for $\mathcal{G}^\star(C,D)$ such that $|s| = |s'|$ and the root of s' is not of type $(\wedge)$ (resp. of type $(\vee)$).*

Proof. If s is of type $(\wedge)$ and D is singleton then $s \xrightarrow{D} s_1$ is the only possible beginning of s (here s_1 is the only child of s because s is minimum). Like s, the subtree s_1 is winning for $\mathcal{G}^\star(C,D)$ and $|s| = |s_1|$. Let n be the least such that $s \xrightarrow{D} s_1 \xrightarrow{D} s_2 \ldots s_{n-1} \xrightarrow{D} s_n$ and s_n is not of type $(\wedge)$. By a simple induction one can show that the subtree s_n of s is winning for $\mathcal{G}^\star(C,D)$ and $|s| = |s_n|$. The case for C singleton is analogous. □

Proposition 3.12 *If s is a uniform strategy tree winning for $\mathcal{G}(C,D)$, then there exists a winning strategy tree s' for $\mathcal{G}(D,C)$ and $|s| = |s'|$.*

Proof. We obtain s' from s by applying on each node of s substitution σ, where $\sigma = [\langle E\rangle \mapsto [E], [E] \mapsto \langle E\rangle, (p) \mapsto (\overline{p}), (\overline{p}) \mapsto (p), (\wedge) \mapsto (\vee), (\vee) \mapsto (\wedge)]$. By a trivial induction, s is winning for $\mathcal{G}(C,D)$ iff $s' = \sigma(s)$ is winning for $\mathcal{G}(D,C)$. □

4 Lower bound for the size of modal descriptions

We say that, for a modal logic $\mathcal{L}$, the size of the $\mathcal{L}$-separation problem is bounded by f if for all finite models $\mathcal{M} = \langle W, R, V\rangle$ and non-empty $C, D \subseteq W$ if there is an $\mathcal{L}$-formula that separates C and D then there is one such formula of size at most $f(|\mathcal{M}|)$. We say it is polynomially bounded when it is bounded by some polynomial. Similarly, we say that f is a lower bound for the size of the $\mathcal{L}$-separation problem when there are infinitely many models $\mathcal{M} = \langle W, R, V\rangle$ and non-empty $C, D \subseteq W$ such that an $\mathcal{L}$-formula $\varphi_\mathcal{M}$ separates C and D, and all such formulas have size at least $f(|\mathcal{M}|)$. We say that the size of the $\mathcal{L}$-separation problem has an exponential lower bound when there is a fixed $b > 1$ such that b^x is a lower bound for the $\mathcal{L}$-separation problem. The notions are analogously defined for the $\mathcal{L}$-description problem.

In general one cannot conclude that an a in the domain of $\mathcal{M}$ has exclusively $\mathcal{L}$-descriptions of size at least $f(\mathcal{M})$ from the fact that a is $\mathcal{L}$-separable from some b exclusively by formulas of size at least $f(|\mathcal{M}|)$ (a could have no $\mathcal{L}$-descriptions at all). However, the implication is true when the a in question does have an $\mathcal{L}$-description.

In this section we show that, for $\mathcal{L} \in \{\mathcal{ML}, \mathcal{ML}^{\Diamond\wedge\neg}, \mathcal{ML}^{\Diamond\wedge}\}$, the size of the $\mathcal{L}$-separation and $\mathcal{L}$-description problems has an exponential lower bound and therefore it cannot be polynomially bounded. We use the machinery introduced in Section 3 to show lower bounds on Spoiler's winning uniform strategy trees and hence the size of the corresponding formulas.

Theorem 4.1 *There is a recursive family of acyclic finite models with two distinguished points $(\mathcal{M}_n, a_n, b_n)_{n\in\mathbb{N}}$ such that $|\mathcal{M}_n| \in O(n)$ and the size of the shortest $\mathcal{ML}$-formula φ_n that separates a_n from b_n in $\mathcal{M}_n$ is exponential in n. Furthermore, there exists an $\mathcal{ML}$-description of a_n in $\mathcal{M}_n$.*

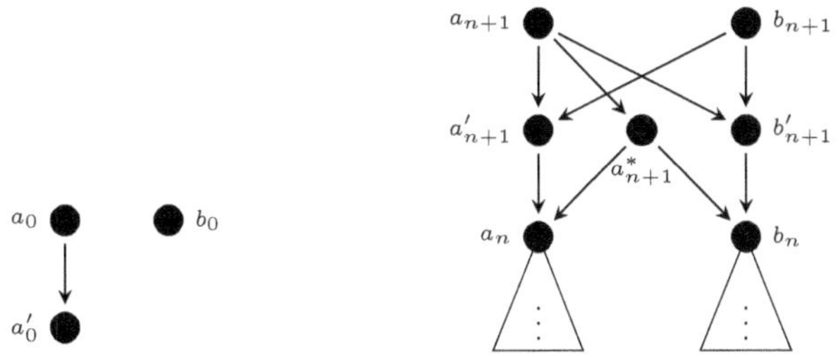

a) $(\mathcal{M}_0, a_0, b_0)$ b) $(\mathcal{M}_{n+1}, a_{n+1}, b_{n+1})$

Fig. 1. Recursive family of models $(\mathcal{M}_n, a_n, b_n)_{n\in\mathbb{N}}$

Proof. The definition of $(\mathcal{M}_n, a_n, b_n)_{n\in\mathbb{N}}$ is shown in Figure 1. It is not hard to see that for all n, $\mathcal{M}_n$ is acyclic and $|\mathcal{M}_n| \in O(n)$. We now show by induction on n that for all n there exists a minimum uniform strategy tree s_n such that $|s_n| = 2^n$. Since $|\mathcal{M}_n| \in O(n)$, we conclude from this that $|s_n|$ is exponential in $|\mathcal{M}|$ and by Theorem 3.8 the minimum formula φ_n that separates a_n from b_n is exponential in $|\mathcal{M}|$ too.

For $n = 0$, we have that $s_0 = \langle\{a'_0\}\rangle$ is clearly winning for $\mathcal{G}(\{a_0\},\{b_0\})$ and since $|s_0| = 1$, s_0 is minimum. Now assume s_n is a minimum uniform strategy tree that is winning for $\mathcal{G}(\{a_n\},\{b_n\})$ and let s_{n+1} be a minimum winning strategy tree for $\mathcal{G}(\{a_n\},\{b_n\})$. We do a case analysis of s_{n+1} to rule out possibilities and ensure that s_{n+1} is unique (up-to redundant occurrences of nodes of type $(\wedge)$, see below) and exponentially large.

In what follows, we avoid subscript $n+1$ for convenience (e.g., we write a' for a'_{n+1}). The reader can track the name of the nodes we use and the shape of the resulting strategy in Figure 2. We use the convention for nodes of type $\langle\cdot\rangle$ and $[\cdot]$ guaranteed by Proposition 3.9.

The first thing to observe is that since $\mathcal{M}_n \models \neg p$, for all p, no nodes of type (p) or $(\overline{p})$ can occur in s_{n+1} at all. Secondly, observe that using Proposition 3.11, we can assume without loss of generality that the root of s_{n+1} is not of type $(\wedge)$ or $(\vee)$. Next we can rule out also the case $s_{n+1} = [E] \to x$, for some $E \subseteq \{a', b'\}$, for that would imply that x is winning for $\mathcal{G}(\{a', b'\}, E)$, which contradicts Proposition 3.10.

We can assume, therefore that $s_{n+1} = \langle E\rangle \to x$, for some non-empty $E \subseteq \{a', b', a^*\}$ and that x is winning for $\mathcal{G}(E, \{a', b'\})$. But by Proposition 3.10, we may conclude $a' \notin E$ and $b' \notin E$. Hence, we must have $E = \{a^*\}$.

We now perform a similar case analysis on x. By Proposition 3.11, we may assume that x is not of type $(\vee)$. If $x = \langle F\rangle \to y$, for some non-empty $F \subseteq \{a_n, b_n\}$, then y would have to be winning for $\mathcal{G}(F, \{a_n, b_n\})$ which contradicts Proposition 3.10. Similarly, we can see that we cannot have $x = [F] \to y$.

Therefore, we can assume that x is of type $(\wedge)$, winning for $\mathcal{G}(\{a^*\}, \{a', b'\})$ and minimum. Notice that we can ignore, without loss of generality, the case where x has only one successor y with $x \overset{\{a',b'\}}{\to} y$ (for in that case y would also be a minimum uniform strategy tree for $\mathcal{G}(\{a^*\}, \{a', b'\})$). We conclude, then, that x has two children y_1 and y_2 such that $x \overset{\{a'\}}{\to} y_1$ and $x \overset{\{b'\}}{\to} y_2$. Furthermore, y_1 is winning for $\mathcal{G}(\{a^*\}, \{a'\})$ and y_2 is winning for $\mathcal{G}(\{a^*\}, \{b'\})$.

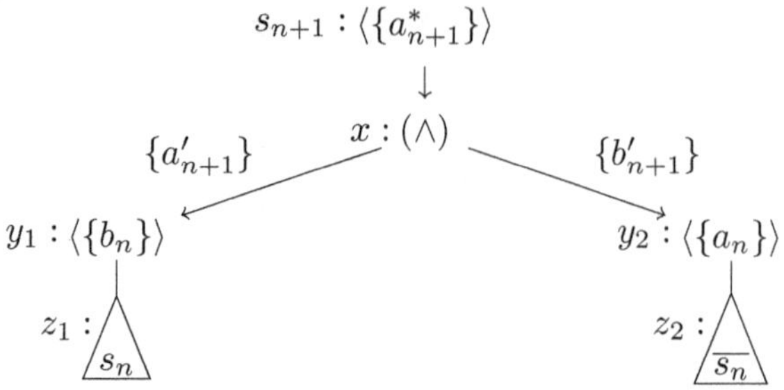

Fig. 2. s_{n+1}, minimum uniform strategy tree winning for $\mathcal{G}_{\mathcal{M}_{n+1}}(\{a_{n+1}\}, \{b_{n+1}\})$.

Using again Proposition 3.11, we conclude that y_1 and y_2 are not of type $(\wedge)$ nor $(\vee)$. If $y_1 = [G_1] \to z_1$, the only possibility for G_1 is $\{a_n\}$. So z_1 would have to be winning for $\mathcal{G}(\{a_n, b_n\}, \{a_n\})$ contradicting Proposition 3.10.

Hence we must have $y_1 = \langle E_1\rangle \to z_1$ and $y_2 = \langle E_1\rangle \to z_2$ with $E_i \subseteq \{a_n, b_n\}$; z_1 must be winning for $\mathcal{G}(\{E_1\}, \{a_n\})$ and z_2 for $\mathcal{G}(\{E_2\}, \{b_n\})$. By Proposition 3.10, $a_n \notin E_1$ and $b_n \notin E_2$, so $E_1 = \{b_n\}$ and $E_2 = \{a_n\}$.

For s to be minimum, both z_1 and z_2 have to be minimum strategies winning for $\mathcal{G}(\{b_n\}, \{a_n\})$ and $\mathcal{G}(\{a_n\}, \{b_n\})$ respectively. We can therefore assume that $z_1 = s_n$ and, by inductive hypothesis, $|z_1| = 2^n$. Using Proposition 3.12, we conclude $|z_2| = 2^n$ and since $|s_{n+1}| = |z_1| + |z_2|$, we obtain $|s_{n+1}| = 2^{n+1}$.

By Lemma 3.6, there exists a separating formula φ_n associated to each s_n. It is not hard to see that they are $\varphi_0 := \Diamond\top$ and $\varphi_{n+1} := \Diamond(\Diamond\varphi_n \wedge \Diamond\overline{\varphi_n})$. But observe that φ_n is stronger than $\Diamond^{2n+1}\top$ and since clearly $\mathcal{M}_n, w \not\models \Diamond^{2n+1}\top$ for all w other than a_n and b_n, we have that φ_n is a description for a_n. □

Corollary 4.2 *The size of the $\mathcal{ML}$-separation and $\mathcal{ML}$-description problems has an exponential lower bound and therefore it is not polynomially bounded.*

An inspection of the proof of Theorem 4.1 reveals that any $\mathcal{ML}$-formula that separates a_n and b_n in $\mathcal{M}_n$ (with $n > 1$) must use $\Box$ and $\vee$. This already implies that one cannot separate a_n and b_n in $\mathcal{M}_n$ using the logics $\mathcal{ML}^{\Diamond\wedge}$ or $\mathcal{ML}^{\Diamond\wedge\neg}$ (one could alternatively show that consistently playing a'_n constitutes a winning strategy for Duplicator).

In order to show that the size of the $\mathcal{L}$-separation and $\mathcal{L}$-description problem for $\mathcal{L} \in \{\mathcal{ML}^{\Diamond\wedge}, \mathcal{ML}^{\Diamond\wedge\neg}\}$ has an exponential lower bounds, we need to find another family of models. The models in this case turned out to be somehow more complex.

Theorem 4.3 *There is a recursive family of acyclic finite models with two distinguished points $(\mathcal{N}_n, a_n, b_n)_{n\in\mathbb{N}}$ such that $|\mathcal{N}_n| \in O(n)$ and the size of the shortest $\mathcal{ML}^{\Diamond\wedge\neg}$-formula ψ_n that separates a_n and b_n in $\mathcal{N}_n$ is exponential in n. Furthermore, there exists an $\mathcal{ML}^{\Diamond\wedge\neg}$-description of a_n in $\mathcal{N}_n$.*

Proof. The definition of $(\mathcal{N}_n, a_n, b_n)_{n\in\mathbb{N}}$ is given in Figure 3. Notice that now the models interpret a propositional variable p. It is not hard to see that for all n, $\mathcal{N}_n$ is acyclic and $|\mathcal{N}_n| \in O(n)$. One proceeds as in the proof of Theorem 4.1, and shows by induction on n that the minimum uniform strategy tree winning for $\mathcal{G}^{\Diamond\wedge\neg}(\{a_n\}, \{b_n\})$ has size, in this case, $2^n 3 - 2$, which is the closed form of $|s_0| = 1; |s_{n+1}| = 2|s_n| + 2$.

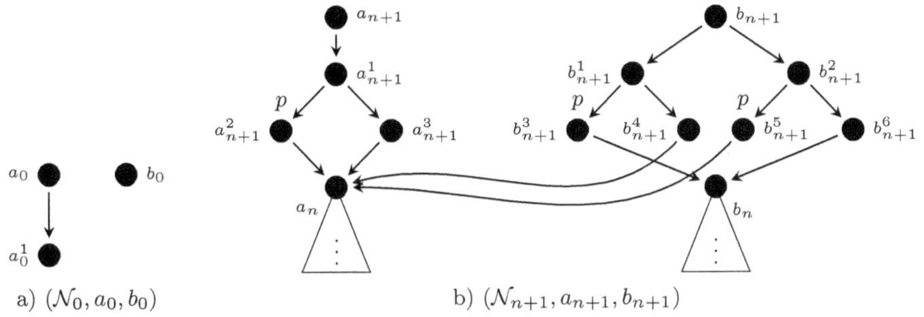

Fig. 3. Recursive family of models $(\mathcal{N}_n, a_n, b_n)_{n\in\mathbb{N}}$

Details can be found in Appendix A. □

Corollary 4.4 *The size of the $\mathcal{ML}^{\Diamond\wedge\neg}$-separation and $\mathcal{ML}^{\Diamond\wedge\neg}$-description problems has an exponential lower bound and therefore it is not polynomially bounded.*

The proof of Theorem 4.3 shows that atomic negation is necessary to separate a_n and b_n in $\mathcal{N}_n$. Therefore, there is no $\mathcal{ML}^{\Diamond\wedge}$-formula separating a_n and b_n in $\mathcal{N}_n$. However a simple modification of the models $(\mathcal{N}_n)_{n\in\mathbb{N}}$ in the proof of Theorem 4.3 shows the same result for $\mathcal{ML}^{\Diamond\wedge}$ instead of $\mathcal{ML}^{\Diamond\wedge\neg}$.

Theorem 4.5 *There is a recursive family of acyclic finite models with two distinguished points $(\mathcal{N}'_n, a_n, b_n)_{n\in\mathbb{N}}$ such that $|\mathcal{N}'_n| \in O(n)$ and the size of the shortest $\mathcal{ML}^{\Diamond\wedge}$-formula ψ' that separates a_n and b_n in $\mathcal{N}'_n$ is exponential in n. Furthermore, there exists an $\mathcal{ML}^{\Diamond\wedge}$-description of a_n in $\mathcal{N}_n$.*

Proof. Define $\mathcal{N}'_n$ in the same way as $\mathcal{N}_n$ in the proof of Theorem 4.3, but introduce a second propositional variable q and set $V(q) = \{a^3_n, b^4_n, b^6_n\}$ in $\mathcal{N}'_n$; that is, make q true in all the nodes of the third level of $\mathcal{N}_n$ where p was false. The proof is completely analogous. □

Corollary 4.6 *The size of the $\mathcal{ML}^{\Diamond\wedge}$-separation and $\mathcal{ML}^{\Diamond\wedge}$-description problems has an exponential lower bound and therefore it is not polynomially bounded.*

5 Upper bound for the size of modal descriptions

In the previous section we found an exponential lower bound for the size of a modal formula that describes an element of the domain, more precisely, $O(b^{|\mathcal{M}|})$ for $b \in (1, 2]$. We will now analyze the complexity of a simple algorithm that computes such formulas in order to find an upper bound for this problem. We will see that its complexity is $O(2^{|\mathcal{M}|^2} \cdot |\mathcal{M}|)$, so, while the lower bound is not tight it is still reasonable. We expect that tighter upper bounds can be obtained by considering better algorithms.

Assume a fixed *finite* model $\mathcal{M} = \langle W, R, V\rangle$. We define now a simple procedure which maintains, at each step s, a relation $\sim_s \subseteq W \times W$ and a map $f_s : W \to \mathcal{ML}$ that satisfy the following invariant: "if $w \not\sim_s v$, then Spoiler has a winning strategy for the game $\mathcal{G}(w, v, d(f_s(w))$; witnessed by the fact that $w \models f_s(w)$ and $v \not\models f_s(w)$". The algorithm computes the largest such $\sim_s$, which, of course, corresponds to the maximum autobisimulation on $\mathcal{M}$ (in fact, it can be seen as a variation of Hopcroft's bisimulation algorithm [11]). The procedure goes as follows:

- *Step* 0: Let $P(v) := \{p \mid v \in V(p)\}$ and $\overline{P}(v) := \{\neg p \mid v \notin V(p)\}$. Define

$$f_0(w) := \bigwedge(P(w) \cup \overline{P}(w)) \qquad \text{for all } w \in W$$
$$\sim_0 := \{(u,v) \mid P(u) = P(v)\}$$

- *Step* $s+1$: Pick any two $u, v \in W$ with $u \sim_s v$ that satisfy Condition 1 or 2 below and proceed accordingly. If no such elements exist, stop.
  - *Condition 1:* For some $u' \in \mathit{sucs}(u)$, there is no element $v' \in \mathit{sucs}(v)$ such that $u' \sim_s v'$. In that case, set

$$f_{s+1}(u) := f_s(u) \wedge \Diamond f_s(v')$$
$$f_{s+1}(v) := f_s(v) \wedge \Box\overline{f_s(v')}$$
$$f_{s+1}(x) := f_s(x) \qquad \text{for any } x \notin \{u,v\}$$
$$\sim_{s+1} := \sim_s \setminus\{(u,v),(v,u)\}$$

  - *Condition 2:* For some $v' \in \mathit{sucs}(v)$, there is no element $u' \in \mathit{sucs}(u)$ such that $u' \sim_s v'$. In that case, set

$$f_{s+1}(u) := f_s(u) \wedge \Box\overline{f_s(v')}$$
$$f_{s+1}(v) := f_s(v) \wedge \Diamond f_s(v')$$
$$f_{s+1}(x) := f_s(x) \qquad \text{for any } x \notin \{u,v\}$$
$$\sim_{s+1} := \sim_s \setminus\{(u,v),(v,u)\}$$

Clearly the invariant holds after *Step 0*, and assuming it holds at the beginning of *Step* $s+1$, it is straightforward to see that whenever Condition 1 holds, then Spoiler wins any $\mathcal{G}(u,v,n)$ (for certain n) by playing first u' according to rule R_l, so the invariant is maintained (similar for Condition 2). The procedure is guaranteed to terminate and, if it does by stage k then, because of the invariant, $f_k(u)$ is a description for u whenever $u \not\sim_k v$ for $u \neq v$.

Notice that this procedure does not compute a *minimum* description for u (this problem appears to be harder). Even more, it is shown in [2] that for the class of converse well-founded, linear models, there exist executions of this procedure[5] that lead to formulas whose size cannot be bound by a polynomial. However, every element in a model in this class has a modal formula description of linear size.

In any case, the analysis of this simple procedure will give us an upper bound for the size of the *minimum* modal description of an element.

The first thing to observe is that the procedure terminates at most by stage $\frac{1}{2}(|W|^2 - |W|)$. This is because at each step s we have $|\sim_{s+1}| = |\sim_s| - 2$, $|\sim_0| \leq |W|^2$ and for every $w \in W$ and s, $w \sim_s w$.

Let $M(s) = \max\{|f_s(v)| \mid v \in W\}$. It is clear that $M(0) \in O(|V|)$ and (since the witnesses u and v are distinct) we have $M(s+1) \leq 2 \cdot M(s)$. We conclude that

[5] The procedure is non-deterministic on the choice of u and v on *Step* $s+1$.

$M(s) \in O(2^s \cdot |V|)$. Therefore, if $v \in W$ has a modal description, the one computed by this procedure has size at most $M(\frac{1}{2}(|W|^2 - |W|))$.

Theorem 5.1 *Let $\mathcal{M} = \langle W, R, V \rangle$ and $v \in W$. If $\varphi \in \mathcal{ML}$ is a minimum description of v in $\mathcal{M}$, then $|\varphi| \in O(2^{\frac{1}{2}|W|^2} \cdot |V|)$.*

Obtaining an upper bound for $\mathcal{ML}^{\Diamond\wedge}$ and $\mathcal{ML}^{\Diamond\wedge\neg}$ is not difficult. The main difference is that the simulation notion for these two logics is no longer symmetric. Hence we have to treat (u, v) and (v, u) separately. For $\mathcal{ML}^{\Diamond\wedge\neg}$ the same procedure applies with the following three modifications: a) since Spoiler cannot play according to rule R_r, only *Condition 1* is considered; b) only the value for u has to be updated in f_{s+1} (i.e., $f_{s+1}(v) = f_s(v)$); c) $\sim_{s+1}$ has to be defined as $\sim_s \setminus \{(u, v)\}$. For $\mathcal{ML}^{\Diamond\wedge}$ the same three modifications have to be made, plus d) replace $=$ by $\subseteq$ in the initialization of $\sim_0$.

The procedure for $\mathcal{ML}^{\Diamond\wedge}$ or $\mathcal{ML}^{\Diamond\wedge\neg}$ terminates at most by step $|W|^2 - |W|$ and the same analysis as the one explained for $\mathcal{ML}$ applies in this case. Hence the upper bound for the size of minimum descriptions in $\mathcal{ML}^{\Diamond\wedge}$ or $\mathcal{ML}^{\Diamond\wedge\neg}$ is $O(|W^2| - |W|)$.

Theorem 5.2 *Let $\mathcal{M} = \langle W, R, V \rangle$ and $v \in W$. If $\varphi \in \mathcal{ML}^{\Diamond\wedge\neg}$ (or $\varphi \in \mathcal{ML}^{\Diamond\wedge}$) is a minimum description of v in $\mathcal{M}$, then $|\varphi| \in O(2^{|W|^2} \cdot |V|)$.*

6 Succinctness for separation

The bounds established in the previous sections resemble classical succinctness results. It is therefore interesting to analyze in which way they differ.

Succinctness deals with how short a formula can be found to express a given property. It is especially important when studying two equally expressive logics $\mathcal{L}$ and $\mathcal{L}'$. In these situations, succinctness is the foremost quantitative measure to distinguish $\mathcal{L}$ and $\mathcal{L}'$. Informally speaking, if we find a an infinite collection of properties Φ_n, each expressible in $\mathcal{L}$ with a formula φ_n of size $f(n)$ and all the formulas of $\mathcal{L}'$ expressing the same property Φ_n (that is, all the $\mathcal{L}'$-formulas semantically equivalent to φ_n) are much larger than $f(n)$, then we say that $\mathcal{L}$ is more succinct than $\mathcal{L}'$.

In this section we will see that the results of Section 4 can also be used to distinguish the three logics considered (and, more generally, other logics) by means of a quantitative measure. Informally, if an $\mathcal{L}$-formula φ_n of size $f(n)$ separates C_n and D_n in $\mathcal{M}_n$ (for an infinite collection of models $\mathcal{M}_n$) and C_n and D_n are also separable in $\mathcal{L}'$, but only with $\mathcal{L}'$-formulas much larger than $f(n)$ then we say that $\mathcal{L}$ is more *succinct for separation* than $\mathcal{L}'$. It is interesting to observe, though, that this notion is not a form of succinctness as described above: the short $\mathcal{L}$-formula φ_n need not be semantically equivalent to none of the exponentially larger ones of $\mathcal{L}$.

We will ultimately show that $\mathcal{ML}$ is more succinct for separation than $\mathcal{ML}^{\Diamond\wedge\neg}$, which in turn is more succinct for separation than $\mathcal{ML}^{\Diamond\wedge}$. We shall do this in a rather formal (arguably, too formal) way, but this will allow us to close this section drawing some promising connections with the field of Algorithmic Information Theory.

Let us turn to the formal analysis. We say that $\mathcal{L} \geq_p \mathcal{L}'$ if there is a truth-preserving translation T mapping $\mathcal{L}'$-formulas into $\mathcal{L}$-formulas and p is a polynomial such that $|T(\varphi)| \leq p(|\varphi|)$ for all $\mathcal{L}'$-formula φ.

Knowing $\mathcal{L} \geq_p \mathcal{L}'$ tells us, for example, that every two sets separated by a formula φ of $\mathcal{L}'$ can be separated by a formula of $\mathcal{L}$ which is not much larger than φ. But $\mathcal{L}$ might in principle separate sets in a much shorter way.

Definition 6.1 Given a modal logic $\mathcal{L}$ and a suitable Kripke model $\mathcal{H} = \langle W, R, V \rangle$, let $S^{\mathcal{L}}_{\mathcal{H}} : \mathcal{P}(W)^2 \to \mathbb{N}$ be the *separation complexity* of $\mathcal{L}$ within $\mathcal{H}$. For $C, D \subseteq W$, $S^{\mathcal{L}}_{\mathcal{H}}(C, D)$ is defined as the size of the shortest $\mathcal{L}$-formula which separates C and D in $\mathcal{H}$ in case there is such separator or ∞ otherwise.

Clearly if $\mathcal{L} \geq_p \mathcal{L}'$ then the following hold:

$$S^{\mathcal{L}}_{\mathcal{H}} \leq p \circ S^{\mathcal{L}'}_{\mathcal{H}} \tag{1}$$

$$S^{\mathcal{L}}_{\mathcal{H}}(C, D) = S^{\mathcal{L}}_{\mathcal{H}}(D, C) \tag{2}$$

$$S^{\mathcal{L}}_{\mathcal{H}}(C', D') \leq S^{\mathcal{L}}_{\mathcal{H}}(C, D) \text{ (whenever } C' \subseteq C \text{ and } D' \subseteq D). \tag{3}$$

The notion of size of the $\mathcal{L}$-separation problem bounded by f introduced in Section 4 can be restated in the following way: the size of the $\mathcal{L}$-separation problem is bounded by f if for all finite model $\mathcal{H} = \langle W, R, V \rangle$ and $C, D \subseteq W$ if $S^{\mathcal{L}}_{\mathcal{H}}(C, D) < \infty$ then $S^{\mathcal{L}}_{\mathcal{H}}(C, D) \leq f(|\mathcal{H}|)$.

The argument we have used in Section 4 to show that the size of the $\mathcal{L}$-separation problem (for $\mathcal{L} \in \{\mathcal{ML}, \mathcal{ML}^{\Diamond\wedge\neg}, \mathcal{ML}^{\Diamond\wedge}\}$) has an exponential lower bound is to exhibit a sequence $(\mathcal{H}_n, a_n, b_n)_{n\in\mathbb{N}}$ where a_n and b_n are elements of $\mathcal{H}_n$ such that there is $b > 1$ such that for all n, $\infty > S^{\mathcal{L}}_{\mathcal{H}_n}(\{a_n\}, \{b_n\}) > b^{|\mathcal{H}_n|}$. On the other hand, results of Section 5 show that for all $\mathcal{H} = \langle W, R, V \rangle$ and $a \in W$, $S^{\mathcal{ML}}_{\mathcal{H}}(\{a\}, W \setminus \{a\}) \in O(|V| \cdot 2^{\frac{1}{2}|W|^2})$ and $S^{\mathcal{L}'}_{\mathcal{H}}(\{a\}, W \setminus \{a\}) \in O(|V| \cdot 2^{|W|^2})$, for $\mathcal{L}' \in \{\mathcal{ML}^{\Diamond\wedge}, \mathcal{ML}^{\Diamond\wedge\neg}\}$.

We next introduce formally our notion of succinctness for separation that may be applied to logics which do not necessarily have the same expressive power.

Definition 6.2 Let $\mathcal{L} \geq \mathcal{L}'$. We say that $\mathcal{L}$ is *f-more succinct for separation* than $\mathcal{L}'$ if there is a sequence

$$(\mathcal{H}_n = \langle W_n, R_n, V_n \rangle, C_n \subseteq W_n, D_n \subseteq W_n)_{n\in\mathbb{N}} \tag{4}$$

of finite models with two distinguished sets such that C_n, D_n is separable in $\mathcal{L}'$ and there is a polynomial p such that for almost all n (that is for all n except finitely many), $S^{\mathcal{L}'}_{\mathcal{H}_n}(C_n, D_n) - S^{\mathcal{L}}_{\mathcal{H}_n}(C_n, D_n) > f(|\mathcal{H}_n|)$.

The idea is that $\mathcal{L}$ is f-more succinct for separation than $\mathcal{L}'$ when there is a sequence of examples (4) showing that the difference between $S^{\mathcal{L}'}_{\mathcal{H}_n}$ and $S^{\mathcal{L}}_{\mathcal{H}_n}$ grows faster than $f(|\mathcal{H}_n|)$. When f is of the form b^x, for a fixed $b > 1$, we simply say that $\mathcal{L}$ is exponentially more succinct for separation than $\mathcal{L}'$.

We ask if the additional expressive power of $\mathcal{ML}$ over $\mathcal{ML}^{\Diamond\wedge\neg}$ is enough to be exponentially more succinct for separation than $\mathcal{ML}^{\Diamond\wedge\neg}$, and if the additional expressive power of $\mathcal{ML}^{\Diamond\wedge\neg}$ is enough to be exponentially more succinct for separation than $\mathcal{ML}^{\Diamond\wedge}$. As we have anticipated, in both cases the answer is yes.

Theorem 6.3 *$\mathcal{ML}$ is exponentially more succinct for separation than $\mathcal{ML}^{\Diamond\wedge\neg}$.*

Proof. Recall $(\mathcal{N}_n, a_n, b_n)_{n\in\mathbb{N}}$ from the proof of Theorem 4.3. For each n, a_n and b_n can be separated in $\mathcal{N}_n$ by χ_n, where $\chi_0 := \Diamond\top$ and $\chi_{n+1} := \Diamond\Box\Diamond\chi_n$. Clearly $|\chi_n| \in O(n)$, so $\mathcal{ML}$ is exponentially more succinct for separation than $\mathcal{ML}^{\Diamond\wedge\neg}$. In fact, $\mathcal{ML}^{\Diamond\wedge\neg}$ plus $\Box$ is already exponentially more succinct for separation than $\mathcal{ML}^{\Diamond\wedge\neg}$. Surprisingly χ_n does not use $\wedge$ or $\neg$. □

Theorem 6.4 *$\mathcal{ML}^{\Diamond\wedge\neg}$ is exponentially more succinct for separation than $\mathcal{ML}^{\Diamond\wedge}$.*

Proof. Recall the proof of Theorem 4.5. Let $\mathcal{N}'_n = \langle W_n, R_n, V_n\rangle$. For a new propositional symbol r, define $\mathcal{N}''_n = \langle W_n, R_n, V_n[r \to \{b_n\}]\rangle$. It is easy to verify that the proof of Theorem 4.5 goes through with $\mathcal{N}''_n$ instead of $\mathcal{N}'_n$. Now elements a_n and b_n can be separated in $\mathcal{N}''_n$ by the constant-size formula $\theta = \neg r$ of $\mathcal{ML}^{\Diamond\wedge\neg}$. Then $\mathcal{ML}^{\Diamond\wedge\neg}$ is exponentially more succinct for separation than $\mathcal{ML}^{\Diamond\wedge}$. □

We close this section with a short digression. For a fixed and suitable model of computation M, the Kolmogorov complexity of a string σ relative to M, denoted $K_M(\sigma)$, is defined as the length of the shortest program which computes σ in M, or ∞ if there is no such program. (For more details on Kolmogorov Complexity Theory, see [13].) This underlying model of computation M may range from finite automata to Turing machines relativized to oracles. Here the *meaning* of a program is seen as the output it produces in the fixed model of computation M. Hence programs are seen as descriptors of strings, and the Kolmogorov complexity of a given string σ is just the length of the shortest description of σ within M. Informally speaking, stronger models of computation yield smaller Kolmogorov complexity. That is, if M is more powerful than M' then $K_M \leq K_{M'}$ (up to additive constant).

The notion of separation complexity S given in Definition 6.1 has some similarities with the classical Kolmogorov Complexity K. First, S needs some underlying language $\mathcal{L}$ and a suitable model $\mathcal{H}$. Second, in the context of logic, the *meaning* of a formula φ is given by its extension, that is by the set of points of $\mathcal{H}$ where φ is true. Hence formulas are seen as descriptors of elements of $\mathcal{H}$. As with K, if $\mathcal{L}$ is more expressive than $\mathcal{L}'$, in the sense of $\mathcal{L} \geq \mathcal{L}'$, then $S_{\mathcal{L}}$ is 'smaller' than $S_{\mathcal{L}'}$ in the sense of equation (1). But unlike programs which are simply executed in M to produce some output, for formulas evaluated in a fixed model $\mathcal{H}$ we may conceive different 'semantic tasks': here separation was analyzed, but one can conceive many others as well.

It is not the purpose of this article to study the resemblance of the *algorithmic* Kolmogorov Complexity with other *logical* description complexities. Although a fine analysis is needed, we want to point out that some results from the algorithmic side and the logical side may be somehow harmonized in a natural way.

7 Conclusions and future work

The line of research that motivated this work comes from the study of the computational complexity of the description problem for modal languages. We seek for efficient algorithms to compute modal descriptions, for various languages –including sub-boolean ones. Is it true that the problem of finding an $\mathcal{L}$-description for a given element is computable in polynomial time? The answer depends in the way the output is represented. If one allows the output formula to be representable as a DAG then the answer is 'yes' [2]. But if we stick to the standard complexity computational model of Turing machines where 'compute a formula' means, literally, to write it down in the output tape then the answer is 'no': we have shown that the length of the output formula may be exponentially larger than the input model.

We have employed *classical* Ehrenfeucht-Fraïssé games as a theoretical tool for proving lower bounds on formula size. In this respect, our work is close to Adler and Immerman's [1], who propose a new kind of Ehrenfeucht-Fraïssé game to establish lower bounds for various kinds of logics. In their game, Spoiler can be seen as trying to construct what we have called a winning uniform strategy tree while Duplicator tries to identify deficiencies in it. The fact that Duplicator possesses an optimal strategy in this kind of games suggests, in our opinion, that the problem does not require a dynamic view in terms of games, but can be analyzed using the static notion of strategy over standard games.

We have only analyzed a few modal fragments, but the problem of the size of $\mathcal{L}$-descriptions is of course applicable to other logics. One can, for instance, study this problem for First Order Logic or Propositional Logic. These are two extremes, since $\mathcal{P} \leq \mathcal{ML}^{\Diamond\wedge} \leq \mathcal{ML}^{\Diamond\wedge\neg} \leq \mathcal{ML} \leq \mathcal{FO}^{=}$.

Consider $\mathcal{FO}^{=}$, the first-order logic with equality (over the modal correspondence language). It is well-known that one can characterize up-to-isomorphism any finite model $\mathcal{H}$ with domain $\{a_1 \dots a_n\}$ using a sentence $\varphi \in \mathcal{FO}^{=}$ that is polynomial in the size of $\mathcal{H}$. Taking this as a basis, one can define for each a_i a formula of $\varphi_i(x) \in \mathcal{FO}^{=}$, with one free variable x, polynomial in $\mathcal{H}$ (that is there is a polynomial such that for all such $\mathcal{H}$, $|\varphi_i(x)| \leq p(|\mathcal{H}|)$), such that if a_i is $\mathcal{FO}^{=}$-describable then φ_i is a suitable description. In fact, this polynomial formula can be constructed in polynomial time.

Proposition 7.1 *The size of the $\mathcal{FO}^{=}$-description problem is polynomially bounded.*

We now go the the other extreme and regard Propositional Logic $\mathcal{P}$ as a fragment of $\mathcal{ML}$. For any finite Kripke model $\mathcal{H} = \langle W, R, V\rangle$ with $W = \{a_1, \dots, a_n\}$ and Dom $V = \{p_1, \dots, p_m\}$, we define, $\psi_k := \bigwedge_{a_k \in V(p_j)} p_j \wedge \bigwedge_{a_k \notin V(p_j)} \neg p_j$. Now if a_k is $\mathcal{P}$-describable in $\mathcal{H}$ then ψ_k is one such $\mathcal{P}$-description.

Proposition 7.2 *The size of the $\mathcal{P}$-description problem is polynomially bounded.*

Propositions 7.1 and 7.2 are clearly opposed to Corollaries 4.2, 4.4 and 4.6. While the modal fragments studied in this article ($\mathcal{ML}$, $\mathcal{ML}^{\Diamond\wedge\neg}$ and $\mathcal{ML}^{\Diamond\wedge}$) do not have polynomially bounded descriptions problems two extreme logics in terms of expressivity do.

It is interesting to study the size of the description problem for other fragments not addressed in this article such that $\mathcal{ML}^{\Diamond\wedge\neg}$ plus $\Box$ but without $\vee$, or others with restrictions in the shape of nestings of $\wedge$ and $\vee$. Even for the logics considered here, it would be interesting to have a better understanding of the computational complexity of their description and separation problems. In particular, one would like to close the gap between lower and upper bounds and determine the complexity of finding a minimum description or separation.

References

[1] Adler, M. and N. Immerman, *An n! lower bound on formula size*, ACM Trans. Comput. Logic **4** (2003), pp. 296–314.

[2] Areces, C., S. Figueira and D. Gorín, *The question of expressiveness in the generation of referring expressions*, Technical report, University of Buenos Aires (2010). URL `http://arxiv.org/abs/1006.4621v1`

[3] Areces, C., A. Koller and K. Striegnitz, *Referring expressions as formulas of description logic*, in: *Proc. of the 5th INLG*, Salt Fork, OH, USA, 2008, pp. 42–49.

[4] Blackburn, P., M. de Rijke and Y. Venema, "Modal Logic," Cambridge University Press, 2001.

[5] Blackburn, P., J. van Benthem and F. Wolter, "Handbook of Modal Logic, Volume 3 (Studies in Logic and Practical Reasoning)," Elsevier Science Inc., New York, NY, USA, 2006.

[6] Dale, R., *Cooking up referring expressions*, in: *Proc. of the 27th ACL*, 1989, pp. 68–75.

[7] Dale, R. and N. Haddock, *Generating referring expressions involving relations*, in: *Proc. of the 5th EACL*, 1991, pp. 161–166.

[8] Dale, R. and E. Reiter, *Computational interpretations of the Gricean maxims in the generation of referring expressions*, Cognitive Science **19** (1995).

[9] Dawar, A., M. Grohe, S. Kreutzer and N. Schweikardt, *Model theory makes formulas large*, in: *In Proceedings of the 34th International Colloquium on Automata, Languages and Programming*, 2007.

[10] Hennessy, M. and R. Milner, *Algebraic laws for indeterminism and concurrency*, Journal of the ACM **32** (1985), pp. 137–162.

[11] Hopcroft, J., "An $nlog(n)$ algorithm for minimizing states in a finite automaton," In Z. Kohave, editor, *Theory of Machines and Computations*, Academic Press, 1971.

[12] Kanellakis, P. and S. Smolka, *CCS expressions finite state processes, and three problems of equivalence*, Inf. Comput. **86** (1990), pp. 43–68.

[13] Li, M. and P. Vitányi, "An introduction to Kolmogorov complexity and its applications," Springer, 1997, 2nd edition.

[14] Markey, N., *Temporal logic with past is exponentially more succinct*, in: *Bulletin of the EATCS 79*, 2003, pp. 122–128.

[15] Paige, R. and R. Tarjan, *Three partition refinement algorithms*, SIAM J. Comput. **16** (1987), pp. 973–989.

[16] Stone, M., *On identifying sets*, in: *Proc. of the 1st INLG*, 2000, pp. 116–123.

[17] van Deemter, K., *Generating referring expressions: Boolean extensions of the incremental algorithm*, Computational Linguistics **28** (2002), pp. 37–52.

[18] Wilke, T., *CTL^+ is exponentially more succinct than CTL*, in: *FSTTCS*, 1999, pp. 110–121.

A Proof of Theorem 4.3

The definition of $(\mathcal{N}_n, a_n, b_n)_{n\in\mathbb{N}}$ is given in Figure 3 (Section 4). Notice that now the models interpret a propositional variable p. It is not hard to see that for all n, $\mathcal{N}_n$ is acyclic and $|\mathcal{N}_n| \in O(n)$.

We proceed as in the proof of Theorem 4.1, and show by induction on n that the minimum uniform strategy tree winning for $\mathcal{G}^{\Diamond\wedge\neg}(\{a_n\},\{g_n\})$ has size, in this case, $2^n 3 - 2$, which is the closed form of $|s_0| = 1; |s_{n+1}| = 2|s_n| + 2$.

For $n = 0$, we take $s_0 = \langle a_0^1 \rangle$ and it is clearly a minimum uniform strategy tree winning for $\mathcal{G}^{\Diamond\wedge\neg}(\{a_0\},\{b_0\})$. Assume now that s_n is the minimum uniform strategy tree winning for $\mathcal{G}^{\Diamond\wedge\neg}(\{a_n\},\{b_n\})$. We perform again a case analysis of s_n ruling out possibilities, but recall that uniform strategy trees for $\mathcal{G}^{\Diamond\wedge\neg}$ comprise only nodes of type $\langle\cdot\rangle$, $(\wedge)$, (p) or $(\overline{p})$. The reader can track the name of the nodes and the general shape of the strategy in Figure A.1. Again, we avoid subscript $n+1$ and write, for instance a^1 for a^1_{n+1}. We use the convention for nodes of type $\langle\cdot\rangle$ guaranteed by Proposition 3.9. We will write $\langle a\rangle$ for $\langle\{a\}\rangle$.

Since there are no propositional symbols true at a or b, the root of s_n is not of type (p) nor $(\overline{p})$. By Proposition 3.11, the only possibility is then $s_n = \langle a^1\rangle$; so let x be the child of s, that is $\langle a^1\rangle \to x$, where x is winning for $\mathcal{G}^{\Diamond\wedge\neg}(\{a^1\},\{b^1,b^2\})$. Again, x is not of type (p) nor $(\overline{p})$. Suppose then $x = \langle E\rangle \to y$, for a non-empty $E \subseteq \{a^2,a^3\}$ and a y winning for $\mathcal{G}^{\Diamond\wedge\neg}(E,\{b^3,b^4,b^5,b^6\})$. This would imply that y is winning for $\mathcal{G}^{\Diamond\wedge\neg}(\{a^2\},\{b^5\})$ or for $\mathcal{G}^{\Diamond\wedge\neg}(\{a^3\},\{b^4\})$, which is absurd, so this possibility is discarded.

We conclude that x is of type $(\wedge)$. Again, we can ignore without loss of generality the case $x \overset{\{b^1,b^2\}}{\to} y$, and assume that x has two children y_1 and y_2, such that $x \overset{\{b^1\}}{\to} y_1$ and $x \overset{\{b^2\}}{\to} y_2$. Furthermore, y_1 is winning for $\mathcal{G}^{\Diamond\wedge\neg}(\{a^1\},\{b^1\})$ and y_2 is winning for $\mathcal{G}^{\Diamond\wedge\neg}\{a^1\},\{b^2\})$.

Again, we observe that y_1 and y_2 can only be of type $\langle\cdot\rangle$. Suppose $y_1 = \langle E_1\rangle \to z_1$, for a non-empty $E_1 \subseteq \{a^2,a^3\}$ and a z_1 that is winning for $\mathcal{G}^{\Diamond\wedge\neg}(E_1,\{b^3,b^4\})$. Since clearly there cannot be a uniform strategy tree winning for $\mathcal{G}^{\Diamond\wedge\neg}(\{a^3\},\{b^4\})$, we have $a^3 \notin E_1$ and, then $E_1 = \{a^2\}$ and z_1 is winning for $\mathcal{G}^{\Diamond\wedge\neg}(\{a^2\},\{b^3,b^4\})$. In a similar way, we conclude that $y_2 = \langle a^3\rangle \to z_2$ with z_2 winning for $\mathcal{G}^{\Diamond\wedge\neg}(\{a^3\},\{b^5,b^6\})$.

Since p is true in b^3 and false in b^4, z_1 is not of type (p) nor $(\overline{p})$. The same can be said about z_2. Suppose $z_1 = \langle F\rangle \to h_1$; then necessarily $F = \{a_n\}$ and h_1 has to be winning for $\mathcal{G}^{\Diamond\wedge\neg}(\{a_n\},\{a_n,b_n\})$ contradicting Proposition 3.10. Similarly, z_2 cannot be of type $\langle\cdot\rangle$ either.

Therefore, z_1 and z_2 must be of type $(\wedge)$ and, once again, we can assume without loss of generality that they have both two children each. Suppose $z_1 \overset{\{b^3\}}{\to} h_1$, $z_1 \overset{\{b^4\}}{\to} h_2$, $z_2 \overset{\{b^5\}}{\to} h_3$ and $z_2 \overset{\{b^6\}}{\to} h_4$ where h_1 is winning for $\mathcal{G}^{\Diamond\wedge\neg}(\{a^2\},\{b^3\})$, h_2 is winning for $\mathcal{G}^{\Diamond\wedge\neg}(\{a^2\},\{b^4\})$, h_3 is winning for $\mathcal{G}^{\Diamond\wedge\neg}(\{a^3\},\{b^5\})$ and h_4 is winning for $\mathcal{G}^{\Diamond\wedge\neg}(\{a^3\},\{b^6\})$. Furthermore, we assume all such strategies are minimum, so h_2 and h_3 are necessarily of type (p) and $(\overline{p})$ respectively. Since no propositional variable distinguishes a^2 from a^3, h_1 is not of type (p) nor $(\overline{p})$. And by Proposition 3.11 h_1 is not of type $(\wedge)$ either. The same can be said about h_4.

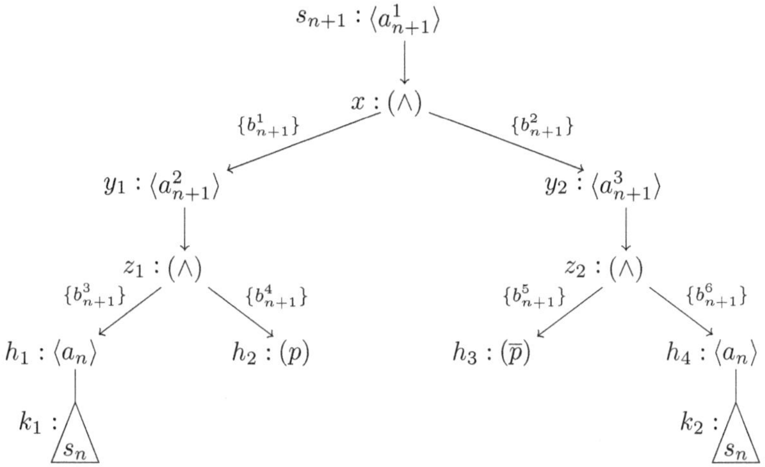

Fig. A.1. s_{n+1}, minimum uniform strategy tree winning for $\mathcal{G}^{\Diamond\wedge\neg}_{\mathcal{N}_{n+1}}(\{a_{n+1}\}, \{b_{n+1}\})$.

Hence $h_1 = \langle a_n \rangle \to k_1$ and $h_4 = \langle a_n \rangle \to k_2$, where k_1 and k_2 are minimum strategies winning for $\mathcal{G}^{\Diamond\wedge\neg}(\{a_n\}, \{b_n\})$, that is $k_1 = k_2 = s_n$. Therefore, we have that $|s_{n+1}| = 2|s_n| + 2$, so these uniform strategy trees cannot be polynomially bounded. As in the proof of Theorem 4.1, it is easy to see that the associated formulas $\psi_0 := \Diamond\top$ and $\psi_{n+1} := \Diamond(\Diamond(p \wedge \Diamond\psi_n) \wedge \Diamond(\neg p \wedge \Diamond\psi_n))$ describe a_n in $\mathcal{N}_n$.

Complete Axiomatization of the Stutter-invariant Fragment of the Linear Time μ-calculus

Amélie Gheerbrant

ILLC
Universiteit van Amsterdam
P.O. Box 94242
1090 GE AMSTERDAM

Abstract

The logic $\mu\mathsf{TL}(\mathsf{U})$ is the fixpoint extension of the "Until"-only fragment of linear-time temporal logic. It also happens to be the stutter-invariant fragment of linear-time μ-calculus $\mu\mathsf{TL}$. We provide complete axiomatizations of $\mu\mathsf{TL}(\mathsf{U})$ on the class of finite words and on the class of ω-words. We introduce for this end another logic, which we call $\mu\mathsf{TL}(\Diamond_\Gamma)$, and which is a variation of $\mu\mathsf{TL}$ where the Next time operator is replaced by the family of its stutter-invariant counterparts. This logic has exactly the same expressive power as $\mu\mathsf{TL}(\mathsf{U})$. Using already known results for $\mu\mathsf{TL}$, we first prove completeness for $\mu\mathsf{TL}(\Diamond_\Gamma)$, which finally allows us to obtain completeness for $\mu\mathsf{TL}(\mathsf{U})$.

Keywords: Completeness, Modal Fixed-Point Logics, μ-Calculus, Linear Temporal Logic, Stutter-Invariance

1 Introduction

Stutter-invariance is a property that is argued by some authors (see [9]) to be natural and desirable for a temporal logic, especially in the context of concurrent systems. Roughly, a temporal logic is stutter-invariant if it cannot detect the addition of identical copies of a state. The stutter-invariant fragment of linear-time temporal logic LTL is known to be its "Until"-only fragment $\mathsf{LTL}(\mathsf{U})$ and is obtained by disallowing the use of the "Next" operator (see [11]). It has been extensively studied and it is widely used as a

[1] I am grateful to Alexandru Baltag, Balder ten Cate, Gaëlle Fontaine, Johan van Benthem and to the anonymous referees for helpful comments on earlier drafts. This work was supported by a GLoRiClass fellowship of the European Commission (Research Training Fellowship MEST-CT-2005-020841).

specification language. Nevertheless, it has been pointed out (see in particular [5]) that $\mathsf{LTL(U)}$ fails to characterize the class of stutter-invariant ω-regular languages. In order to extend the expressive power of this framework, while retaining stutter-invariance, some ways of extending it have been proposed. In [5], Kousha Etessami proposed for instance the logic $\mathsf{SI\text{-}EQLTL}$, which extends $\mathsf{LTL(U)}$ by means of a certain restricted type of quantification over proposition letters. He showed that $\mathsf{SI\text{-}EQLTL}$ characterizes exactly stutter-invariant ω-regular languages.

In this paper, we focus on another logic, which we call $\mu\mathsf{TL(U)}$ and which is the fixpoint extension of the "Until"-only fragment of linear-time temporal logic. $\mu\mathsf{TL(U)}$ has exactly the same expressive power as $\mathsf{SI\text{-}EQLTL}$ (this follows from results in [6]), which implies that it also characterizes exactly stutter-invariant ω-regular languages. It is also known that it satisfies uniform interpolation (see [6]), which is a sign that $\mu\mathsf{TL(U)}$ is a well-behaved logic. Additionally, it is known that $\mathsf{LTL(U)}$ is PSPACE complete both for model checking and for satisfiability (c.f. [4]). It is also know that $\mu\mathsf{TL}$ is PSPACE complete both for model checking and for satisfiability (c.f. [15]). So PSPACE completeness follows for $\mu\mathsf{TL(U)}$ in both cases. This is another argument in favor of $\mu\mathsf{TL(U)}$: while much more expressive than $\mathsf{LTL(U)}$, it has the same complexity. Here we further contribute to the study of the logical properties of $\mu\mathsf{TL(U)}$ by completely axiomatizing it over the class of ω-words and over the class of finite words. We introduce for this end another logic, which we call $\mu\mathsf{TL}(\Diamond_\Gamma)$, and which is a variation of $\mu\mathsf{TL}$ where the Next time operator is replaced by the family of its stutter-invariant counterparts. We use this logic as a technical tool to show completeness results for $\mu\mathsf{TL(U)}$.

Outline of the paper: In Section 2, we recall basic facts and notions about linear-time μ-calculus $\mu\mathsf{TL}$. We also give a precise definition of the notion of stutter-invariance and introduce $\mu\mathsf{TL(U)}$, the stutter-invariant fragment of $\mu\mathsf{TL}$. In Section 3, we introduce the logic $\mu\mathsf{TL}(\Diamond_\Gamma)$ and show that $\mu\mathsf{TL(U)}$ and $\mu\mathsf{TL}(\Diamond_\Gamma)$ have exactly the same expressive power on finite and ω-words. In section 4, we give axiomatizations of $\mu\mathsf{TL}(\Diamond_\Gamma)$ that we respectively show to be complete on these two classes of structures. Finally, these results are put to use in Section 5, where we show similar completeness results for $\mu\mathsf{TL(U)}$.

2 Preliminaries

In this section, we introduce the syntax and semantics of linear time μ-calculus $\mu\mathsf{TL}$. We also recall its axiomatization on some interesting classes of linear orders and introduce the notion of stutter-invariance.

2.1 Linear time μ-calculus

By a *propositional vocabulary* we mean a countable (possibly finite) non-empty set of propositional letters $\sigma = \{p_i \mid i \in I\}$.

Definition 2.1 [Syntax of $\mu\mathsf{TL}$] Let σ be a propositional vocabulary, and let $\mathcal{V} = \{x_1, x_2, \ldots\}$ be a disjoint countably infinite set of *propositional variables.* We inductively define the set of μTL-formulas in vocabulary σ as follows:

$$\varphi, \psi, \xi := At \mid \top \mid \neg\varphi \mid \varphi \wedge \psi \mid \varphi \vee \psi \mid \Diamond\varphi \mid \mu x_i.\xi$$

where $At \in \sigma \cup \mathcal{V}$ and, in the last clause, x_i occurs only positively in ξ (i.e., within the scope of an even number of negations). We will use $\varphi \rightarrow \psi$, $\nu x_i.\xi$, $\Box\varphi$, $\varphi \mathsf{U} \psi$, $\mathsf{F}\varphi$ as shorthand for, respectively, $\neg(\varphi \wedge \neg\psi$, $\neg\mu x_i.\neg\xi(\neg x_i)$, $\neg\Diamond\neg\varphi$, $\mu y.(\psi \vee (\varphi \wedge \Diamond y))$ and $\mu y.(\varphi \vee \Diamond y)$. We will also use $\mathsf{G}\varphi$ as shorthand for $\neg\mathsf{F}\neg\varphi$.

A *linear flow of time* is a structure $\mathcal{L} = (W, <)$, where W is a non-empty set of points and $<$ is a linear order on W. A *linear time σ-structure* is a structure $\mathfrak{M} = (\mathcal{L}, V)$ where $\mathcal{L} = (W, R)$ is a linear flow of time and $V : \sigma \rightarrow \wp(W)$ a valuation. Whenever $w \in W$ is a point, we call $\mathfrak{M}, w$ a *pointed σ-structure.* Linear time μ-calculus is usually considered over restricted classes of linear orders. In this paper, we will only consider it over the following classes of linear flows of time:

- $\mathbf{L}_\omega$, the class of linear orders of order type ω, i.e., flows of time $(W, <)$ that are isomorphic to $(\mathbb{N}, <)$, where $\mathbb{N}$ is the set of natural numbers with the natural ordering,
- $\mathbf{L}_{fin}$, the class of finite linear orders,
- the union $\mathbf{L}_\omega \cup \mathbf{L}_{fin}$ of these two classes

We will often refer to structures based on $\mathbf{L}_\omega$ as *ω-words* or $\mathbf{L}_\omega$-structures, to structures based on $\mathbf{L}_{fin}$ as *finite words* or $\mathbf{L}_{fin}$-structures and more generally, to structures based on $\mathbf{L}$ as $\mathbf{L}$-structures.

Definition 2.2 [Semantics of μTL] Given a μTL-formula φ, a structure $\mathfrak{M} = ((W, <), V)$ and an assignment $g : \mathcal{V} \rightarrow \wp(W)$, we define a subset $[\![\varphi]\!]_{\mathfrak{M},g}$ of $\mathfrak{M}$ that is interpreted as the set of points at which φ is true. This subset is defined by induction in the usual way. Let ImSuc(w), be the set of direct successors of the point w with respect to $<$, we only recall:

$$[\![\Diamond\varphi]\!]_{\mathfrak{M},g} = \{w \in W : [\![\varphi]\!]_{\mathfrak{M},g} \cap \text{ImSuc}(w) \neq \emptyset\}$$
$$[\![\mu x.\varphi]\!]_{\mathfrak{M},g} = \bigcap\{A \subseteq W : [\![\varphi]\!]_{\mathfrak{M},g[x/A]} \subseteq A\}$$

where $g[x/A]$ is the assignment defined by $g[x/A](x) = A$ and $g[x/A](y) = g(y)$ for all $y \neq x$. If $w \in [\![\varphi]\!]_{\mathfrak{M},g}$, we write $\mathfrak{M}, w \models_g \varphi$ and we say that φ is true at $w \in \mathfrak{M}$ under the assignment g. If φ is a sentence, or if $\mathfrak{M}, w \models_g \phi$ holds for every valuation g, we simply write $\mathfrak{M}, w \models \varphi$.

In order to understand the semantics of the μ-operator, consider a μTL-formula $\varphi(x)$ and a structure $\mathfrak{M} = ((W, <), V)$ together with a valuation g. This formula induces an operator F^φ taking a set $A \subseteq W$ to the set $\{v : (\mathfrak{M}, v) \models_{g[x/A]} \varphi(x)\}$. μTL is concerned with least fixpoints of such operators. If $\varphi(x)$ is positive in x, the operator F^φ is monotone, i.e., $x \subseteq y$ implies $\mathrm{F}^\varphi(x) \subseteq \mathrm{F}^\varphi(y)$. Monotone operators F^φ always have a least fixpoint, defined as the intersection of all their prefixpoints: $\bigcap\{A \subseteq W : \{v : (\mathfrak{M}, v) \models_{g[x/A]} \varphi(x)\} \subseteq A\}$. The formula $\mu x.\varphi(x)$ denotes this least fixpoint.

Note that the $\Diamond$ operator is interpreted as the "Next" operator of temporal logic and that the temporal operators U and F that we defined as shorthand have their usual meaning:

- $(\mathcal{L}, V, w) \models_g \mathsf{F}\varphi$ iff there exists w' such that $w \leq w'$ and $(\mathcal{L}, V, w') \models_g \varphi$
- $(\mathcal{L}, V, w) \models_g \varphi \mathsf{U} \psi$ iff there exists w' such that $w \leq w'$, $(\mathcal{L}, V, w') \models_g \psi$ and for all w'' such that $w \leq w'' < w'$, $(\mathcal{L}, V, w'') \models_g \varphi$

Before we give the complete axiomatization of $\mu\mathsf{TL}$ on $\mathbf{L}_\omega$, $\mathbf{L}_{fin}$ and $\mathbf{L}_\omega \cup \mathbf{L}_{fin}$, let us first recall the axiomatization of the *μ-calculus.* In the μ-calculus, instead of considering a linear order $<$, we consider an arbitrary binary relation R on W. In this more general context, (W, R) can be an arbitrary graph and we call it a *frame.* [2] The corresponding structures are called *Kripke structures.* Let $\mathrm{RSuc}(w) = \{w' : (w, w') \in R\}$, the semantics of $\Diamond$ is now as follows:

$$[\![\Diamond\varphi]\!]_{\mathfrak{M},g} = \{w \in W : [\![\varphi]\!]_{\mathfrak{M},g} \cap \mathrm{RSuc}(w) \neq \emptyset\}$$

Definition 2.3 Let σ be a finite propositional vocabulary and $\varphi, \psi \in \mu\mathsf{TL}$ arbitrary formulas. We call $\mathrm{BV}(\varphi)$ and $\mathrm{FV}(\varphi)$ respectively, the set of bound variables in φ and the set of free variables in φ. The Kozen system K_μ consists of the Modus Ponens, the Substitution rule, the Necessitation rule and the following axioms and rules:

A1 propositional tautologies,

A2 $\vdash \Box\varphi \leftrightarrow \neg\Diamond\neg\varphi$ (dual),

A3 $\vdash \Box(\varphi \to \psi) \to (\Box\varphi \to \Box\psi)$ (K),

A4 $\vdash \varphi[x/\mu x.\varphi] \to \mu x.\varphi$ (fixpoint axiom),

FR If $\vdash \varphi[x/\psi] \to \psi$, then $\vdash \mu x.\varphi \to \psi$ (fixpoint rule)

where x does not belong to $\mathrm{BV}(\varphi)$ and $\mathrm{FV}(\psi) \cap \mathrm{BV}(\varphi) = \emptyset$.

Theorem 2.4 *If φ is a $\mu\mathsf{TL}$-formula, let $K_\mu + \varphi$ be the smallest set which contains both K_μ and φ and is closed for the Modus Ponens, Substitution, fixpoint and Necessitation rules. The following holds:*

(i) *K_μ is complete with respect to the class of Kripke structures.*

(ii) *$K_\mu + \Diamond\varphi \leftrightarrow \Box\varphi$ is complete with respect to the class of ω-words.*

(iii) *$K_\mu + \Diamond\varphi \to \Box\varphi + \mu x.\Box x$ is complete with respect to the class of finite words.*

(iv) *$K_\mu + \Diamond\varphi \to \Box\varphi$ is complete with respect to the class of finite and ω-words.*

Proof. (i) was shown in [17] and the three other completeness results might actually be derivable from it. But direct (and simpler) proofs for (ii) and (iii) can be found respectively in [8] and [13]. In order to establish (iv), we will rely on (i), (ii) and (iii), using an argument from Johan van Benthem and Balder ten Cate (private communication). We first show the following claim:

[2] Note that on arbitrary graphs, we do not introduce $\mathsf{F}\varphi$, $\mathsf{G}\varphi$ and $\varphi\mathsf{U}\psi$ as shorthands for $\mu\mathsf{TL}$-formulas anymore: as we consider frames instead of linear flows of time, this would not really map the usual meaning of these temporal operators.

- *Claim: Let K_x be a system extending K_μ with a finite set of axioms and closed under Substitution, Modus Ponens and the fixpoint and Necessitation rules. Let θ be a closed formula with $K_x \vdash \theta \to \Box\theta$. For all formulas ξ, if $K_x + \theta \vdash \xi$, then $K_x \vdash \theta \to \xi$.* The proof goes by induction on the length of K_x-derivations. The only difficult case is whenever the last line in the proof is obtained via the application of the fixpoint rule. So assume the property holds for all derivations of length n and $K_x + \theta \vdash \mu y.\varphi \to \psi$ is the last line of a derivation of length $n+1$. We want to show that $K_x \vdash \theta \to (\mu y.\varphi \to \psi)$. By induction hypothesis, $K_x \vdash \theta \to (\varphi[x/\psi] \to \psi)$. So by propositional tautologies also $K_x \vdash (\theta \wedge \varphi[\psi/x]) \to \psi$. By the fixpoint rule, $K_x \vdash \mu x.(\theta \wedge \varphi) \to \psi$. Now from $K_x \vdash \theta \to \Box\theta$ it follows by propositional tautologies that $K_x \vdash (\Diamond\neg\theta \vee \neg\theta) \to \neg\theta$ and by the fixpoint rule $K_x \vdash \mu x.(\neg\theta \vee \Diamond x) \to \neg\theta$, so $K_x \vdash \theta \to \neg\mu x.(\neg\theta \vee \Diamond x)$. Now it is valid in the μ-calculus that $\neg\mu x.(\neg\theta \vee \Diamond x) \to (\mu x.(\theta \wedge \varphi) \leftrightarrow \mu x.\varphi)$, so it is also derivable in K_x. It follows that $K_x \vdash \theta \to (\mu x.\varphi \to \psi)$.

Now assume that ξ is valid on finite and ω-words. As it is valid on finite words, by (iii), $K_\mu + \Diamond\varphi \to \Box\varphi + \mu x.\Box x \vdash \xi$. As $\mu x.\Box x$ satisfies the condition of the claim we get:

$$\text{(a) } K_\mu + \Diamond\varphi \to \Box\varphi \vdash \mu x.\Box x \to \xi$$

ξ is also valid on ω-words, and hence by (ii), $K_\mu + \Diamond\varphi \to \Box\varphi + \Box\varphi \to \Diamond\varphi \vdash \xi$. Note that $\Diamond\top$ can equivalently be substituted for $\Box\varphi \to \Diamond\varphi$ there. As $K_\mu + \Diamond\varphi \to \Box\varphi \vdash \neg\mu x.\Box x \to \Diamond\top$, we can also take θ to be $\neg\mu x.\Box x$, which also satisfies the condition of the claim. Indeed by the $\Diamond\varphi \to \Box\varphi$ axiom, it is enough to prove $\neg\mu x.\Box x \to \Diamond\neg\mu x.\Box x$. But this is equivalent to $\Box\mu x.\Box x \to \mu x.\Box x$, which is derivable in K_μ (since $\mu x.\Box x \leftrightarrow \Box(\mu x.\Box x)$). It follows that:

$$\text{(b) } K_\mu + \Diamond\varphi \to \Box\varphi \vdash \neg(\mu x.\Box x) \to \xi$$

$K_\mu + \Diamond\varphi \to \Box\varphi \vdash \xi$ follows from (a) and (b), which proves (iv).

□

2.2 Stutter-invariance

We will now introduce $\mu\mathsf{TL}(\mathsf{U})$, a semantic fragment of $\mu\mathsf{TL}$ extending $\mathsf{LTL}(\mathsf{U})$ (linear-time temporal logic without the Next operator, see [6]). We also provide a precise definition for the notion of stutter-invariance and recall that, in terms of expressive power, $\mu\mathsf{TL}(\mathsf{U})$ is exactly the stutter-invariant fragment of $\mu\mathsf{TL}$ (which was shown in [6]).

Definition 2.5 [Syntax of $\mu\mathsf{TL}(\mathsf{U})$] Let σ be a propositional vocabulary, and let $\mathcal{V} = \{x_1, x_2, \ldots\}$ be a disjoint countably infinite set of *propositional variables.* We inductively define the set of $\mu\mathsf{TL}(\mathsf{U})$-formulas in vocabulary σ as follows:

$$\varphi, \psi, \xi := At \mid \top \mid \neg\varphi \mid \varphi \wedge \psi \mid \varphi \vee \psi \mid \varphi \mathsf{U} \psi \mid \mu x_i.\xi$$

where $At \in \sigma \cup \mathcal{V}$ and, in the last clause, x_i occurs only positively in ξ (i.e., within the scope of an even number of negations). We will use $\varphi \to \psi$, $\nu x_i.\xi$, $\mathsf{F}\varphi$ and $\mathsf{G}\varphi$ as shorthand for, respectively, $\neg(\varphi \wedge \neg\psi$, $\neg\mu x_i.\neg\xi(\neg x_i)$, $\top\mathsf{U}\varphi$ and $\neg(\top\mathsf{U}\neg\varphi)$.

Note that the temporal operators F and G defined as shorthand have their usual meaning. We interpret μTL(U)-formulas in the same type of structures as μTL-formulas, i.e., structures of the form $\mathfrak{M} = (\mathcal{L}, V)$ where $\mathcal{L} \in \mathbf{L}_{fin} \cup \mathbf{L}_{\omega}$.

Definition 2.6 [Semantics of μTL(U)] Given a μTL(U)-formula φ, a structure $\mathfrak{M} = ((W, <), V)$ and an assignment $g : \mathcal{V} \to \wp(W)$, we define a subset $[\![\varphi]\!]_{\mathfrak{M},g}$ of $\mathfrak{M}$ that is interpreted as the set of points at which φ is true. This subset is defined by induction in the usual way. We only recall:

$$[\![\varphi \mathsf{U} \psi]\!]_{\mathfrak{M},g} = \{w \in W : \exists w' \geq w, w' \in [\![\psi]\!]_{\mathfrak{M},g} \text{ and } \forall w \leq w'' < w', w'' \in [\![\varphi]\!]_{\mathfrak{M},g}\}$$

$$[\![\mu x.\varphi]\!]_{\mathfrak{M},g} = \bigcap \{A \subseteq W : [\![\varphi]\!]_{\mathfrak{M},g[x/A]} \subseteq A\}$$

where $g[x/A]$ is the assignment defined by $g[x/A](x) = A$ and $g[x/A](y) = g(y)$ for all $y \neq x$.

In the remaining, we always assume $\mathbf{L} \in \{\mathbf{L}_{\omega}, \mathbf{L}_{fin}, \mathbf{L}_{fin} \cup \mathbf{L}_{\omega}\}$.

Definition 2.7 [Stuttering] Let σ be a propositional signature, and $\mathfrak{M} = ((W, <), V, w)$, $\mathfrak{M}' = ((W', <), V', w')$ pointed **L**-structures in vocabulary σ. We say that $\mathfrak{M}'$ is a stuttering of $\mathfrak{M}$ if and only if there is a surjective function $s : W' \to W$ such that

(i) $s(w') = w$

(ii) for every $w_i, w_j \in W'$, $w_i < w_j$ implies $s(w_i) \leq s(w_j)$

(iii) for every $w_i \in W'$ and $p \in \sigma$, $w_i \in V'(p)$ iff $s(w_i) \in V(p)$

We say that an **L**-structure $\mathfrak{M}$ is *stutter-free relative to* **L** whenever for all $\mathfrak{M}'$ such that $\mathfrak{M}$ is a stuttering of $\mathfrak{M}'$, $\mathfrak{M}'$ is isomorphic to $\mathfrak{M}$.

Let for instance $\mathfrak{M}, w$ be an ω-word in vocabulary $\{p\}$ with $V(p) = W$. $\mathfrak{M}, w$ is stutter-free relative to $\mathbf{L}_{\omega}$, but it is not stutter-free relative to $\mathbf{L}_{fin} \cup \mathbf{L}_{\omega}$. Indeed, let $\mathfrak{M}', w'$ be a finite word in vocabulary $\{p\}$ containing one single point w'. Assume $V'(p) = \{w'\}$, then $\mathfrak{M}, w$ is a stuttering of $\mathfrak{M}', w'$ and relative to $\mathbf{L}_{fin} \cup \mathbf{L}_{\omega}$, $\mathfrak{M}', w'$ is stutter-free, while $\mathfrak{M}, w$ is not.

Definition 2.8 [Stutter-Invariant Class of Structures] Let σ be a propositional signature and **K** a class of **L**-structures in vocabulary σ. Then **K** is a stutter-invariant class iff for every **L**-structure $\mathfrak{M}$ in vocabulary σ and for every **L**-stuttering $\mathfrak{M}'$ of $\mathfrak{M}$, $\mathfrak{M} \in \mathbf{K} \Leftrightarrow \mathfrak{M}' \in \mathbf{K}$.

We say that a sentence φ is stutter-invariant relative to **L** whenever the class of **L**-structures in which φ is satisfied is stutter-invariant. Every μTL(U)-sentence is stutter-invariant relative to **L** (see [6]). To see that it is not possible in μTL(U) to define $\Diamond\varphi$, it is hence enough to observe that the sentence $\Diamond p$ is not stutter-invariant. Also, considering a **L**-structure $\mathfrak{M}, w$, there is always a unique (up to isomorphism) $\mathfrak{M}', w'$ which is stutter-free relative to **L** and such that $\mathfrak{M}, w$ is a stuttering of $\mathfrak{M}', w'$. Observe that it follows that if a μTL(U)-formula is satisfiable in some **L**-structure, it is also satisfiable in a **L**-structure which is stutter-free relative to **L**. Additionally, on **L**, we can show that μTL(U) is exactly the stutter-invariant fragment of μTL:

Theorem 2.9 *Let φ be a μTL-sentence which is stutter-invariant relative to* **L**. *Then, there exists a μTL(U)-sentence φ^* which is equivalent to φ on* **L**-*structures.*

Proof. The proof can be found in [6]. □

3 The logic $\mu\mathsf{TL}(\Diamond_\Gamma)$

In this Section, we introduce the logic $\mu\mathsf{TL}(\Diamond_\Gamma)$ and we show that, as far as expressivity is concerned, it is a fragment of μTL. More precisely, we show that $\mu\mathsf{TL}(\Diamond_\Gamma)$ has exactly the same expressive power as μTL(U). In the last Sections, we will see that $\mu\mathsf{TL}(\Diamond_\Gamma)$ can be used as a very convenient tool to show completeness results for μTL(U).

$\mu\mathsf{TL}(\Diamond_\Gamma)$ is a variation of μTL where instead of the regular $\Diamond$ modality, we consider the family of its stutter-invariant counterparts. For each finite set Γ of $\mu\mathsf{TL}(\Diamond_\Gamma)$-sentences, we consider a $\Diamond_\Gamma$ operator which intuitively means "at the next distinct point with respect to Γ" (i.e., distinct with respect to the values it assigns to the formulas in Γ). To design this operator, we took inspiration from [5], where a "next distinct" operator was mentioned in passing. This operator was interpreted in σ-structures as our $\Diamond_\sigma$ operator. In order to obtain a well-behaved operator, we relativize it here to any finite set Γ of sentences. This gives rise to a better-behaved logic, where we can define a natural notion of substitution and where the truth of σ-formulas in σ-structures is preserved in σ^+-expansions of these structures (with $\sigma^+ \supseteq \sigma$).

We interpret $\mu\mathsf{TL}(\Diamond_\Gamma)$-formulas in the same type of structures as μTL-formulas, i.e., structures of the form $\mathfrak{M} = (\mathcal{L}, V)$ where $\mathcal{L} \in \mathbf{L}_{fin} \cup \mathbf{L}_\omega$. For any finite set of $\mu\mathsf{TL}(\Diamond_\Gamma)$-formulas and for any points w, w', we write $w \equiv_\Gamma w'$ if w and w' satisfy the same formulas in Γ.

Definition 3.1 Let σ be a finite propositional signature, and let $\mathcal{V} = \{x_1, x_2, \ldots\}$ be a disjoint countably infinite stock of *propositional variables.* We inductively define the set of $\mu\mathsf{TL}(\Diamond_\Gamma)$-formulas as follows:

$$\varphi, \psi, \xi := At \mid \top \mid \neg\varphi \mid \varphi \wedge \psi \mid \varphi \vee \psi \mid \Diamond_\Gamma \varphi \mid \mu x_i.\xi$$

where $At \in \sigma \cup \mathcal{V}$, Γ is a finite set of $\mu\mathsf{TL}(\Diamond_\Gamma)$-formulas and, in the last clause, x_i occurs only positively in ξ (i.e., within the scope of an even number of negations). We use $\Box_\Gamma\varphi$, $\varphi \to \psi$ and $\nu x_i.\xi(x_i)$ as shorthand for $\neg\Diamond_\Gamma\neg\varphi$, $\neg(\varphi \wedge \neg\psi$ and $\neg x_i \mu \neg \xi(\neg x_i)$, respectively. We interpret $\mu\mathsf{TL}(\Diamond_\Gamma)$-formulas as μTL-formulas, except that:

$$(\mathcal{L}, V, w) \models \Diamond_\Gamma\varphi \text{ if } \exists w' > w \text{ such that } w \not\equiv_\Gamma w', \forall w'' \text{ with } w < w'' < w', w'' \equiv_\Gamma w \text{ and } (\mathcal{L}, V, w') \models \varphi$$

The resulting logic is stuttering invariant. We write $\mathrm{Voc}(\varphi)$ for the vocabulary of φ and $\mathrm{Voc}(\Gamma)$ for $\bigcup_{\varphi\in\Gamma} \mathrm{Voc}(\varphi)$. Note that we include in the vocabulary of a formula *all* the proposition letters occurring in it, including those which occur in the formulas contained in the sets Γ indexing its modalities. This remark particularly matters for the notion of substitution, as whenever a formula is to be uniformly substituted for a proposition letter, the operation has to be done everywhere, including in the formulas contained in the sets indexing the modalities. Otherwise, validity would not be preserved by uniform substitution. Consider for instance $(p \wedge \Diamond_{\{p\}}\top) \rightarrow \Diamond_{\{p\}}\neg p$. It is clear that this formula is valid and that for any $\mu\mathsf{TL}(\Diamond_\Gamma)$-formula φ, $\models (\varphi \wedge \Diamond_{\{\varphi\}}\top) \rightarrow \Diamond_{\{\varphi\}}\neg\varphi$ also holds. But it is also very clear that $\not\models (\varphi \wedge \Diamond_{\{p\}}\top) \rightarrow \Diamond_{\{p\}}\neg\varphi$.

We will now provide a way to compare $\mu\mathsf{TL}(\Diamond_\Gamma)$ and $\mu\mathsf{TL}(\mathsf{U})$, by defining two recursive procedures transforming each formula from one language into an equivalent formula from the other language.

Definition 3.2 Let $\Gamma = \{\varphi_0, \ldots, \varphi_{n-1}\}$ be a finite set of $\mu\mathsf{TL}(\Diamond_\Gamma)$-formulas. Whenever $\Gamma \neq \emptyset$, we define B_Γ as the set of all possible mappings $\Gamma \rightarrow \{\bot, \top\}$, and for each $g \in B_\Gamma$, we let β_g be the formula $\alpha_0 \wedge \ldots \wedge \alpha_{n-1}$ where $\alpha_j = \varphi_j$ if $g(\varphi_j) = \top$ and $\alpha_j = \neg\varphi_j$ if $g(\varphi_j) = \bot$. By convention, we set $B_\emptyset = \{\bot, \top\}$.[3]

Definition 3.3 [$\mu\mathsf{TL}(\mathsf{U})$-translation of a $\mu\mathsf{TL}(\Diamond_\Gamma)$-formula] Let φ be a $\mu\mathsf{TL}(\Diamond_\Gamma)$-formula, we recursively define its $\mu\mathsf{TL}(\mathsf{U})$-translation $\varphi_{\mu\mathsf{TL}(\mathsf{U})}$ via the following procedure. $At_{\mu\mathsf{TL}(\mathsf{U})} = At$, $(\neg\varphi)_{\mu\mathsf{TL}(\mathsf{U})} = \neg\varphi_{\mu\mathsf{TL}(\mathsf{U})}$, $(\varphi \wedge \psi)_{\mu\mathsf{TL}(\mathsf{U})} = \varphi_{\mu\mathsf{TL}(\mathsf{U})} \wedge \psi_{\mu\mathsf{TL}(\mathsf{U})}$, $(\mu x.\varphi)_{\mu\mathsf{TL}(\mathsf{U})} = \mu x.\varphi_{\mu\mathsf{TL}(\mathsf{U})}$, and $(\Diamond_\Gamma\varphi)_{\mu\mathsf{TL}(\mathsf{U})} = \bigvee_{g \in B_\Gamma}(\beta_g \wedge \beta_g \mathsf{U}(\neg\beta_g \wedge \varphi_{\mu\mathsf{TL}(\mathsf{U})}))$.

Proposition 3.4 *Let* $\mathbf{L} \in \{\mathbf{L}_\omega, \mathbf{L}_{fin}, \mathbf{L}_\omega \cup \mathbf{L}_{fin}\}$ *and* φ *be a* $\mu\mathsf{TL}(\Diamond_\Gamma)$*-formula,* φ *and* $\varphi_{\mu\mathsf{TL}(\mathsf{U})}$ *are equivalent on* $\mathbf{L}$*-structures.*

Proof. We show that a class of σ-structures based on $\mathbf{L}_{fin} \cup \mathbf{L}_\omega$, is definable by a $\mu\mathsf{TL}(\Diamond_\Gamma)$-formula if and only if it is definable by its $\mu\mathsf{TL}(\mathsf{U})$-translation. Let $((W,<), V, w)$ be a σ-structure (by induction hypothesis, we assume the property holds for φ, $\varphi_{\mu\mathsf{TL}(\mathsf{U})}$).

Assume $((W,<),V,w) \models \Diamond_\Gamma\varphi$, i.e., there exists $w' > w$ such that $w \not\equiv_\Gamma w'$ and $\forall w''$ with $w < w'' < w', w'' \equiv_\Gamma w$ and $((W,<),V,w') \models \varphi$. So there are $g \neq g' \in B_\Gamma$ such that $\mathfrak{M}, w \models \beta_g$ and there exists $w' > w$ with $((W,<),V,w') \models \beta_{g'} \wedge \varphi_{\mu\mathsf{TL}(\mathsf{U})}$ and for all w'' such that $w \leq w'' < w'$, $((W,<),V,w'') \models \beta_g$. By induction hypothesis, $((W,<),V,w) \models \bigvee_{g \in B_\Gamma}(\beta_g \wedge \beta_g \mathsf{U}(\neg\beta_g \wedge \varphi_{\mu\mathsf{TL}(\mathsf{U})}))$.

Assume $((W,<),V,w) \models \bigvee_{g \in B_\Gamma}(\beta_g \wedge \beta_g \mathsf{U}(\neg\beta_g \wedge \varphi_{\mu\mathsf{TL}(\mathsf{U})}))$. So there are $g \neq g'$ such that $\beta_g \mathsf{U}(\beta_{g'} \wedge \varphi_{\mu\mathsf{TL}(\mathsf{U})})$, i.e., there exists w' such that $w \leq w'$, $((W,<),V,w') \models \beta_{g'} \wedge \varphi_{\mu\mathsf{TL}(\mathsf{U})}$ and for all w'' such that $w \leq w'' < w'$, $((W,<),V,w'') \models \beta_g$. As $g \neq g'$, also $w \not\equiv_\Gamma w'$. By induction hypothesis, $((W,<),V,w) \models \Diamond_\Gamma\varphi$. □

Definition 3.5 [$\mu\mathsf{TL}(\Diamond_\Gamma)$-translation of a $\mu\mathsf{TL}(\mathsf{U})$-formula] Let φ be $\mu\mathsf{TL}(\mathsf{U})$-formula in vocabulary σ, we recursively define its $\mu\mathsf{TL}(\Diamond_\Gamma)$-translation $\varphi_{\mu\mathsf{TL}(\Diamond_\Gamma)}$ via the following procedure. $At_{\mu\mathsf{TL}(\Diamond_\Gamma)} = At$, $(\neg\varphi)_{\mu\mathsf{TL}(\Diamond_\Gamma)} = \neg\varphi_{\mu\mathsf{TL}(\Diamond_\Gamma)}$, $(\varphi \wedge \psi)_{\mu\mathsf{TL}(\Diamond_\Gamma)} = \varphi_{\mu\mathsf{TL}(\Diamond_\Gamma)} \wedge \psi_{\mu\mathsf{TL}(\Diamond_\Gamma)}$, $(\mu x.\varphi)_{\mu\mathsf{TL}(\Diamond_\Gamma)} = \mu x.\varphi_{\mu\mathsf{TL}(\Diamond_\Gamma)}$, and $(\varphi \mathsf{U} \psi)_{\mu\mathsf{TL}(\Diamond_\Gamma)} = \mu x.(\psi_{\mu\mathsf{TL}(\Diamond_\Gamma)} \vee (\varphi_{\mu\mathsf{TL}(\Diamond_\Gamma)} \wedge \Diamond_\sigma x))$.

Proposition 3.6 *Let* $\mathbf{L} \in \{\mathbf{L}_\omega, \mathbf{L}_{fin}, \mathbf{L}_\omega \cup \mathbf{L}_{fin}\}$ *and* φ *be a* $\mu\mathsf{TL}(\mathsf{U})$*-formula. Then* φ *and* $\varphi_{\mu\mathsf{TL}(\Diamond_\Gamma)}$ *are equivalent on* $\mathbf{L}$*-structures.*

---

[3] We adopt this convention because we allowed Γ to be empty (see the instantiation of Axiom $A6'$ where $\Gamma = \emptyset$, our convention will guaranty that $\Diamond_\emptyset\varphi$, which is not satisfiable, is also inconsistent), but we could also have required that $\Gamma \neq \emptyset$.

Proof. We show that a class of σ-structures based on $\mathbf{L}_{fin} \cup \mathbf{L}_\omega$, is definable by a $\mu\mathsf{TL}(\mathsf{U})$-formula if and only if it is definable by its $\mu\mathsf{TL}(\Diamond_\Gamma)$-translation. Let $((W,<),V,w)$ be a σ-structure (by induction hypothesis, we assume the property holds for φ, $\varphi_{\mu\mathsf{TL}(\Diamond_\Gamma)}$ and ψ, $\psi_{\mu\mathsf{TL}(\Diamond_\Gamma)}$ respectively).

Assume $((W,<),V,w) \models \varphi \mathsf{U} \psi$. This means either that w satisfies ψ, or w satisfies φ and it is separated from some subsequent w' satisfying ψ by a finite sequence of points which all satisfy φ. So, by induction hypothesis, $((W,<),V,w) \models \mu x.(\psi_{\mu\mathsf{TL}(\Diamond_\Gamma)} \vee (\varphi_{\mu\mathsf{TL}(\Diamond_\Gamma)} \wedge \Diamond_\sigma x))$, because $\mu x.(\psi_{\mu\mathsf{TL}(\Diamond_\Gamma)} \vee (\varphi_{\mu\mathsf{TL}(\Diamond_\Gamma)} \wedge \Diamond_\sigma x))$ states that the current state belongs to the least fixpoint which contains all the points satisfying $\psi_{\mu\mathsf{TL}(\Diamond_\Gamma)}$, together with all the points that satisfy $\varphi_{\mu\mathsf{TL}(\Diamond_\Gamma)}$ and which are immediate predecessors of a point which is already in the fixpoint.

Assume $\mu x.(\psi_{\mu\mathsf{TL}(\Diamond_\Gamma)} \vee (\varphi_{\mu\mathsf{TL}(\Diamond_\Gamma)} \wedge \Diamond_\sigma x))$, i.e., w belongs to the least fixpoint which contains all the points satisfying $\psi_{\mu\mathsf{TL}(\Diamond_\Gamma)}$, together with all the points that satisfy $\varphi_{\mu\mathsf{TL}(\Diamond_\Gamma)}$ and which are immediate predecessors of a point which is already in the fixpoint. This means that either w satisfies $\psi_{\mu\mathsf{TL}(\Diamond_\Gamma)}$, or it satisfies $\varphi_{\mu\mathsf{TL}(\Diamond_\Gamma)}$ and it is separated from some subsequent w' satisfying $\psi_{\mu\mathsf{TL}(\Diamond_\Gamma)}$ by a finite sequence of successor points which all satisfy $\varphi_{\mu\mathsf{TL}(\Diamond_\Gamma)}$ and by induction hypothesis, $((W,<),V,w) \models \varphi U \psi$. □

Corollary 3.7 *$\mu\mathsf{TL}(\mathsf{U})$ and $\mu\mathsf{TL}(\Diamond_\Gamma)$ have the same expressive power on the class of finite and ω-words.*

Proof. Follows from Propositions 3.4 and 3.6. □

Remark 3.8 It follows that $\Diamond_\Gamma$ can be used as shorthand either in $\mu\mathsf{TL}$ or in $\mu\mathsf{TL}(\mathsf{U})$, that U can be used as shorthand in $\mu\mathsf{TL}(\Diamond_\Gamma)$ and that $\mu\mathsf{TL}(\Diamond_\Gamma)$ is definable as a semantic fragment of $\mu\mathsf{TL}$. In the remainder of the paper, this will be assumed.

Now in order to see that both $\mu\mathsf{TL}(\Diamond_\Gamma)$ and $\mu\mathsf{TL}(\mathsf{U})$ strictly extend $\mathsf{LTL}(\mathsf{U})$, let us give an example of a class of finite words which is known to be not definable in $\mathsf{LTL}(\mathsf{U})$, while it is definable in $\mu\mathsf{TL}(\Diamond_\Gamma)$ and $\mu\mathsf{TL}(\mathsf{U})$. The following $\mu\mathsf{TL}(\Diamond_\Gamma)$-formula is satisfied at the root of a finite word in any vocabulary σ expanding $\{p\}$ exactly whenever this word contains an even number of sequences (of arbitrary length) of states satisfying p:

$$\Phi := \mu x.((\neg p \wedge \Box_p x) \vee (p \wedge \Diamond_p \mu y.((\neg p \wedge \Diamond_p y) \vee (p \wedge \Box_p x))))$$

By Proposition 3.4, $\Phi_{\mu\mathsf{TL}(\mathsf{U})} \in \mu\mathsf{TL}(\mathsf{U})$ is equivalent to Φ. Note also that by removing the subscripts in the modal operators in Φ, we obtain the following $\mu\mathsf{TL}$-formula:

$$\Phi' := \mu x.((\neg p \wedge \Box x) \vee (p \wedge \Diamond \mu y.((\neg p \wedge \Diamond y) \vee (p \wedge \Box x))))$$

which is satisfied at the root of a finite word in vocabulary σ exactly whenever this word contains an even number of p (i.e., whenever p is satisfied at an even number of states). Note also that the property defined via Φ is the closure under stuttering of the one defined via Φ'. This suggests a natural procedure - via indexing of the modalities in the formula - to characterize in $\mu\mathsf{TL}(\Diamond_\Gamma)$ the closure under stuttering of $\mu\mathsf{TL}$-properties, which illustrates a close connection between the syntax of $\mu\mathsf{TL}$ and $\mu\mathsf{TL}(\Diamond_\Gamma)$. It is admittedly difficult to write specifications in $\mu\mathsf{TL}$ (c.f. [16]), but the difficulty doesn't seem to be higher in the case of $\mu\mathsf{TL}(\Diamond_\Gamma)$.

4 Complete axiomatization of $\mu\mathsf{TL}(\Diamond_\Gamma)$

In this Section, we show some completeness results for the logic $\mu\mathsf{TL}(\Diamond_\Gamma)$. We will use them in the next Section as a tool to obtain similar results for the logic $\mu\mathsf{TL}(\mathsf{U})$.

Proposition 4.1 *Let φ be a $\mu\mathsf{TL}(\Diamond_\Gamma)$-formula in vocabulary σ containing no free occurrence of the variable x. On the class of finite and ω-words, the following formulas are equivalent:*

- $\bigvee_{g\in B_\sigma}(\beta_g \wedge \mu x.((\neg\beta_g \wedge \varphi) \vee (\beta_g \wedge \Diamond_\sigma x))$
- $\bigvee_{g\in B_\sigma}(\beta_g \wedge \mu x.((\neg\beta_g \wedge \varphi) \vee (\beta_g \wedge \Diamond x))$
- $\Diamond_\sigma\varphi$

Proof. Recall that U can be defined as shorthand in $\mu\mathsf{TL}(\Diamond_\Gamma)$. We already noted in Section 2 and in Proposition 3.6 that on linear orders, the formulas $\varphi\mathsf{U}\psi$, $\mu x.(\psi\vee(\varphi\wedge\Diamond x))$ and $\mu x.(\psi \vee (\varphi \wedge \Diamond_\sigma x))$ are equivalent. We also noted in Proposition 3.4 that in this context, the formulas $\Diamond_\sigma\varphi$ and $\bigvee_{g\in B_\sigma}(\beta_g\wedge\beta_g\mathsf{U}(\neg\beta_g\wedge\varphi))$ are equivalent. The Proposition follows. □

Definition 4.2 $K_{\mu\mathsf{TL}(\Diamond_\Gamma)}$ consists of the Modus Ponens, the Substitution rule, for each Γ, the corresponding Necessitation rule (i.e., if $\vdash \phi$, then $\vdash \Box_\Gamma\varphi$) and the following axioms and rules:

$A1'$ propositional tautologies,

$A2'$ $\vdash \Box_\Gamma\varphi \leftrightarrow \neg\Diamond_\Gamma\neg\varphi$ (dual),

$A3'$ $\vdash \Diamond_\Gamma\varphi \rightarrow \Box_\Gamma\varphi$ (linearity),

$A4'$ $\vdash \Box_\Gamma(\varphi\rightarrow\psi) \rightarrow (\Box_\Gamma\varphi \rightarrow \Box_\Gamma\psi)$ (K),

$A5'$ $\vdash \varphi[x/\mu x.\varphi] \rightarrow \mu x.\varphi$ (fixpoint axiom),

FR$'$ If $\vdash \varphi[x/\psi] \rightarrow \psi$, then $\vdash \mu x.\varphi \rightarrow \psi$(fixpoint rule),

$A6'$ $\vdash \Diamond_\Gamma\varphi \leftrightarrow \bigvee_{g\in B_\Gamma}(\beta_g\wedge\mu x.((\neg\beta_g\wedge\varphi)\vee(\beta_g\wedge\Diamond_\sigma x)))$, where $\mathrm{Voc}(\Diamond_\Gamma\varphi)\subseteq\sigma$ (inductive meaning of $\Diamond_\Gamma$),

for each finite set $\Gamma = \{\varphi_0,\ldots,\varphi_{n-1}\}$ of $\mu\mathsf{TL}(\Diamond_\Gamma)$-sentences and where in the three last Axioms, x does not belong to $\mathrm{BV}(\varphi)$ and $\mathrm{FV}(\psi)\cap\mathrm{BV}(\varphi)=\emptyset$.

Lemma 4.3 *Axiom A6$'$ is sound on the class of finite and ω-words.*

Proof. Let σ be a finite vocabulary, Γ a finite set of $\mu\mathsf{TL}(\Diamond_\Gamma)$-formulas and φ a $\mu\mathsf{TL}(\Diamond_\Gamma)$-formula with $\mathrm{Voc}(\Diamond_\Gamma\varphi)\subseteq\sigma$ and $x\notin\mathrm{FV}(\varphi)$. As $\mu\mathsf{TL}(\Diamond_\Gamma)$ define only stutter-invariant classes of structures, we can consider a stutter-free σ-model $\mathfrak{M}$ with $w\in\mathfrak{M}$ and it is enough to show that the following are equivalent:

(i) $\mathfrak{M}, w \models \Diamond_\Gamma\varphi$

(ii) $\mathfrak{M}, w \models \bigvee_{g\in B_\Gamma}(\beta_g \wedge \mu x.((\neg\beta_g\wedge\varphi)\vee(\beta_g\wedge\Diamond_\sigma x)))$

As for Proposition 4.1, this follows from what was observed in Section 2 and 3.

□

Theorem 4.4 $K_{\mu\mathsf{TL}(\Diamond_\Gamma)}$ *is complete for* $\mu\mathsf{TL}(\Diamond_\Gamma)$ *with respect to the class of* ω*-words and with respect to the class of finite and* ω*-words.*

Proof. Let φ be a $K_{\mu\mathsf{TL}(\Diamond_\Gamma)}$-consistent formula in vocabulary σ. By Axiom $A6'$, we can restrict our attention to σ-formulas containing only $\Diamond_\sigma$ modalities. Again by Axiom $A6'$, we can define a recursive procedure transforming φ into a $K_{\mu\mathsf{TL}(\Diamond_\Gamma)}$-equivalent formula φ'. We set $At' = At$, $(\neg\varphi)' = \neg\varphi'$, $(\varphi \wedge \psi)' = \varphi' \wedge \psi'$, $(\mu x.\varphi)' = \mu x.\varphi'$, and $(\Diamond_\sigma\varphi)' = \bigvee_{g\in B_\sigma}(\beta_g \wedge \mu y.((\neg\beta_g \wedge \varphi') \vee (\beta_g \wedge \Diamond_\sigma y)))$. Consider now the $\mu\mathsf{TL}$-formula φ'', which we define as the result of removing in φ' all the subscripts of the modalities. Notice that by Proposition 4.1, ϕ' and ϕ'' are equivalent. We claim that φ'' is $K_\mu + \Diamond\varphi \rightarrow \Box\varphi$-consistent. For suppose not. Then, there exists a proof of $\neg\varphi''$ using the axioms and rules of $K_\mu + \Diamond\varphi \rightarrow \Box\varphi$. Now, replace every occurrence of the operator $\Diamond$ by $\Diamond_\sigma$ in each axiom and rule used in the proof. The result is a correct $K_{\mu\mathsf{TL}(\Diamond_\Gamma)}$-proof, where only correct axioms and rules of $K_{\mu\mathsf{TL}(\Diamond_\Gamma)}$ are used (because the $K_\mu + \Diamond\varphi \rightarrow \Box\varphi$ axioms and rules can be obtained from the $K_{\mu\mathsf{TL}(\Diamond_\Gamma)}$ ones simply by removing the indexes of the modalities). Additionally, this is a proof of the formula $\neg\varphi'$ (as the original φ' can also be obtained from φ'' by adding the subscript σ to every $\Diamond$ in φ''). But this contradicts the fact that φ' was $K_{\mu\mathsf{TL}(\Diamond_\Gamma)}$-consistent. So φ'' is $K_\mu + \Diamond\varphi \rightarrow \Box\varphi$-consistent. By Theorem 2.4, there is an ω-word or a finite word $\mathfrak{M}$ such that $\mathfrak{M}, w \models \varphi''$ and it follows from Proposition 4.1 (by which ϕ' and ϕ'' are equivalent) that $\mathfrak{M}, w \models \varphi'$, i.e. (by Axiom $A6'$), $\mathfrak{M}, w \models \varphi$. Completeness with respect to the class of ω-words follows too, because every finite word has an ω-word stuttering. $\Box$

Theorem 4.5 $K_{\mu\mathsf{TL}(\Diamond_\Gamma)} + \mu x.\Box_\Gamma x$ *is complete for* $\mu\mathsf{TL}(\Diamond_\Gamma)$ *with respect to the class of finite words.*

Proof. We can apply the same reasoning as for the proof of Theorem 4.5, using completeness of $K_\mu + \Diamond\varphi \rightarrow \Box\varphi + \mu x.\Box x$ on finite words, instead of completeness of $K_\mu + \Diamond\varphi \rightarrow \Box\varphi$ on finite and ω-words. $\Box$

Let $\mathfrak{M}$ be an ω-word. We say that $\mathfrak{M}$ is a *pseudo-finite* word whenever there exists a finite word $\mathfrak{M}'$ such that $\mathfrak{M}$ is a stuttering of $\mathfrak{M}'$. Note that $K_{\mu\mathsf{TL}(\Diamond_\Gamma)} + \mu x.\Box_\Gamma x$ is also complete for $\mu\mathsf{TL}(\Diamond_\Gamma)$ with respect to the class of finite and pseudo-finite words, as every pseudo-finite word is the stuttering of a finite word.

Remark 4.6 Axiom $A6'$ *is not derivable* from the other axioms and rules. Otherwise, every $\Box_\Gamma$ would simply be interpreted as the regular $\Box$ operator of $\mu\mathsf{TL}$. Now, more precisely, let $K^{-A6'}_{\mu\mathsf{TL}(\Diamond_\Gamma)}$ be the smallest set of $\mu(\Diamond_\Gamma)$-formulas which is closed under all axioms and rules in $K_{\mu\mathsf{TL}(\Diamond_\Gamma)}$, except Axiom $A6'$. Suppose Axiom $A6'$ is derivable in $K^{-A6'}_{\mu\mathsf{TL}(\Diamond_\Gamma)}$. Then, $K^{-A6'}_{\mu\mathsf{TL}(\Diamond_\Gamma)}$ would be complete with respect to the class of ω-words. Therefore, as on ω-words $\models (p \wedge \Diamond_{\{p\}}\top) \rightarrow \Diamond_p\neg p$, also in $K^{-A6'}_{\mu\mathsf{TL}(\Diamond_\Gamma)}$, $\vdash (p \wedge \Diamond_{\{p\}}\top) \rightarrow \Diamond_p\neg p$ and there would exists a $K^{-A6'}_{\mu\mathsf{TL}(\Diamond_\Gamma)}$-proof of this formula. But now we could replace in that proof, every modal operator by the regular $\Diamond$ operator. This would be a correct $K_\mu + \Diamond\varphi \rightarrow \Box\varphi$-proof of $(p \wedge \Diamond\top) \rightarrow \Diamond\neg p$. But as on $\omega - words$, $\not\models (p \wedge \Diamond\top) \rightarrow \Diamond\neg p$, this contradicts the soundness of $K_{\mu\mathsf{TL}} + \Diamond\varphi \rightarrow \Box\varphi$. It follows that Axiom $A6'$ is not derivable in $K^{-A6'}_{\mu\mathsf{TL}(\Diamond_\Gamma)}$.

5 Complete axiomatization of $\mu\mathsf{TL}(\mathsf{U})$

Recall that $\mathsf{LTL}(\mathsf{U})$ is the fragment of $\mu\mathsf{TL}(\mathsf{U})$ where the μ-operator is disallowed. In [10], the authors propose an axiomatization of $\mathsf{LTL}(\mathsf{U})$ which is complete on the class of ω-words and finite words. In order to axiomatize $\mu\mathsf{TL}(\mathsf{U})$, we extend here the Axioms and rules in [10] with the usual fixpoint rule and Axiom, together with an additional axiom accounting for the way the Until operator and the μ-operator can interact together. Using the completeness result in [10] with the completeness of $K_{\mu\mathsf{TL}(\Diamond_\Gamma)}$, this allows us to derive a similar completeness Theorem for $\mu\mathsf{TL}(\mathsf{U})$. Recall that, in $\mu\mathsf{TL}(\mathsf{U})$, we use $\mathsf{G}\varphi$ as shorthand for $\neg(\top\mathsf{U}\neg\varphi)$ and $\Diamond_\tau\phi$ as shorthand for $\bigvee_{g\in B_\tau}\beta_g\wedge(\beta_g\mathsf{U}(\neg\beta_g\wedge\varphi))$.

Definition 5.1 The $K_{\mu\mathsf{TL}(\mathsf{U})}$ system consists of the Modus Ponens, the G Necessitation rule (i.e., if $\vdash\varphi$, then $\vdash\mathsf{G}\varphi$) the Substitution rule and the following axioms and rules (these rules, as well as Axioms *A1''* to *A9''*, are borrowed from [10]):

A1'' propositional tautologies,

A2'' The Until operator is non strict:
$\vdash\varphi\rightarrow\bot\mathsf{U}\varphi$,

A3'' For any consistent formula there exists a model that is a discrete linear order:
- $\vdash\mathsf{F}\varphi\rightarrow\neg\varphi\mathsf{U}\varphi$,
- $\vdash\varphi\wedge\mathsf{F}\psi\rightarrow\neg\psi\mathsf{U}(\varphi\wedge\varphi\mathsf{U}(\neg\varphi\mathsf{U}\psi))$,

A4'' Properties that hold throughout a computation hold at the initial state:
$\vdash G\varphi\rightarrow\varphi$,

A5'' Conventional logical deduction holds within individual states (K axiom):
- $\vdash(\mathsf{G}(\varphi\rightarrow\psi)\rightarrow(\varphi\mathsf{U}\xi\rightarrow\psi\mathsf{U}\xi))$
- $\vdash(\mathsf{G}(\varphi\rightarrow\psi)\rightarrow(\xi\mathsf{U}\varphi\rightarrow\xi\mathsf{U}\psi))$

A6'' Persistence of an Until formula until its second argument is satisfied:
$\vdash\varphi\mathsf{U}\psi\rightarrow(\varphi\mathsf{U}\psi)\mathsf{U}\psi$

A7'' Immediacy of satisfaction of an Until formula at the current state:
$\vdash\varphi\mathsf{U}(\varphi\mathsf{U}\psi)\rightarrow\varphi\mathsf{U}\psi$

A8'' States of the time line are not skipped over in evaluating an Until formula:
$\vdash\varphi\mathsf{U}\psi\wedge\neg(\xi\mathsf{U}\psi)\rightarrow\varphi\mathsf{U}(\varphi\wedge\neg\xi)$

A9'' Models are linearly ordered:
$\vdash\varphi U\psi\wedge\xi U\theta\rightarrow((\varphi\wedge\xi)\mathsf{U}(\psi\wedge\theta)\vee(\varphi\wedge\xi)\mathsf{U}(\psi\wedge\xi)\vee(\varphi\wedge\xi)\mathsf{U}(\varphi\wedge\theta))$

A10'' $\vdash\varphi[x/\mu x.\varphi]\rightarrow\mu x.\varphi$, (fixpoint axiom),

FR'' If $\vdash\varphi[x/\psi]\rightarrow\psi$, then $\vdash\mu x.\varphi\rightarrow\psi$ (fixpoint rule),

A11'' $\vdash\mu x.(\psi\vee(\varphi\wedge\Diamond_\sigma x))\leftrightarrow\varphi\mathsf{U}\psi$, where $\mathrm{Voc}(\varphi)\cup\mathrm{Voc}(\psi)\subseteq\sigma$ (inductive meaning of U),

where in the three last Axioms, x does not belong to $\mathrm{BV}(\varphi)\cup\mathrm{BV}(\psi)$ and $\mathrm{FV}(\psi)\cap\mathrm{BV}(\varphi)=\emptyset$.

Lemma 5.2 *Let* $\phi\in\mu\mathsf{TL}(\mathsf{U})$. *Then* $\phi\leftrightarrow(\phi_{\mu\mathsf{TL}(\Diamond_\Gamma)})_{\mu\mathsf{TL}(\mathsf{U})}$ *is derivable in* $K_{\mu\mathsf{TL}(\mathsf{U})}$.

Proof. By induction on the complexity of φ (number of Boolean, modal and fixed-point operators in φ). The base case is immediate. Assume the property holds for all formulas of complexity n. Let $\varphi \in \mu\mathsf{TL}(\mathsf{U})$ of complexity $n+1$ be of the form $\xi \mathsf{U} \psi$ for some $\xi, \psi \in \mu\mathsf{TL}(\mathsf{U})$ (otherwise, by induction hypothesis, the property follows immediately). We have:

$$(\xi \mathsf{U} \psi)_{\mu\mathsf{TL}(\Diamond_\Gamma)} := \mu x.(\psi_{\mu\mathsf{TL}(\Diamond_\Gamma)} \vee (\xi_{\mu\mathsf{TL}(\Diamond_\Gamma)} \wedge \Diamond_\sigma x))$$

and

$$((\xi \mathsf{U} \psi)_{\mu\mathsf{TL}(\Diamond_\Gamma)})_{\mu\mathsf{TL}(\mathsf{U})} := \mu x.((\psi_{\mu\mathsf{TL}(\Diamond_\Gamma)})_{\mu\mathsf{TL}(\mathsf{U})} \vee ((\xi_{\mu\mathsf{TL}(\Diamond_\Gamma)})_{\mu\mathsf{TL}(\mathsf{U})} \wedge \bigvee_{g \in B_\sigma} (\beta_g \wedge \beta_g \mathsf{U}(\neg \beta_g \wedge x))$$

By induction hypothesis, $((\xi \mathsf{U} \psi)_{\mu\mathsf{TL}(\Diamond_\Gamma)})_{\mu\mathsf{TL}(\mathsf{U})}$ is provably equivalent in $K_{\mu\mathsf{TL}(\mathsf{U})}$ to:

$$\mu x.(\psi \vee ((\xi \wedge \bigvee_{g \in B_\sigma} (\beta_g \wedge \beta_g \mathsf{U}(\neg \beta_g \wedge x))$$

By Axiom $A11''$ the following is derivable in $K_{\mu\mathsf{TL}(\mathsf{U})}$:

$$\mu x.(\psi \vee ((\xi \wedge \bigvee_{g \in B_\sigma} (\beta_g \wedge \beta_g \mathsf{U}(\neg \beta_g \wedge x)) \leftrightarrow \xi \mathsf{U} \psi$$

The property follows. □

Lemma 5.3 *The $\mu\mathsf{TL}(\mathsf{U})$-translations of the axioms and rules of $K_{\mu\mathsf{TL}(\Diamond_\Gamma)}$ are derivable in $K_{\mu\mathsf{TL}(\mathsf{U})}$.*

Proof. Except for the $\mu\mathsf{TL}(\mathsf{U})$-translation of the fixpoint Axiom and of the fixpoint rule (which are both trivially derivable from $K_{\mu\mathsf{TL}(\mathsf{U})}$, as they also belong to it), as well as Axiom $A6'$, there is no explicit occurrence of the μ-operator in the $\mu\mathsf{TL}(\mathsf{U})$-translation of the Axioms and rules of $K_{\mu\mathsf{TL}(\Diamond_\Gamma)}$. As they are sound on the class of ω-words and finite words, by the completeness Theorem in [10], together with Proposition 3.4, they are derivable in $\mathsf{LTL}(\mathsf{U})$. It follows that they are also derivable in $K_{\mu\mathsf{TL}(\mathsf{U})}$, because the Axioms and rules of $K_{\mu\mathsf{TL}(\mathsf{U})}$ simply extend those of $\mathsf{LTL}(\mathsf{U})$.

Now consider the $\mu\mathsf{TL}(\mathsf{U})$-translation of Axiom $A6'$:

$$\bigvee_{g \in B_\Gamma} (\beta_g \wedge \beta_g \mathsf{U}(\neg \beta_g \wedge \varphi))$$

$$\leftrightarrow$$

$$\bigvee_{g \in B_\Gamma} (\beta_g \wedge \mu y.((\neg \beta_g \wedge \varphi) \vee (\beta_g \wedge \bigvee_{g' \in B_\sigma} (\beta_{g'} \wedge \beta_{g'} \mathsf{U}(\neg \beta_{g'} \wedge y)))))$$

This formula is derivable from propositional tautologies, together with the substitution rule and Axiom $A11''$ of $K_{\mu\mathsf{TL}(\mathsf{U})} \vdash \mu y.(\psi \vee (\varphi \wedge \Diamond_\sigma x)) \leftrightarrow \varphi \mathsf{U} \psi$ (which is actually shorthand for $\vdash \mu y.(\psi \vee (\varphi \wedge \bigvee_{g \in B_\sigma} (\beta_g \wedge \beta_g \mathsf{U}(\neg \beta_g \wedge y)))) \leftrightarrow \varphi \mathsf{U} \psi$). Finally, let us point out that the restriction of our axioms and rules to $\mathsf{LTL}(\mathsf{U})$-formulas is actually slightly stronger than the axiomatization proposed in [10]. The authors chose to prefix all their modal axioms and rules by G and to allow the generalization rule only on propositional tautologies (our generalization rule is a derived rule in their framework). But our axioms and rule being sound, it is safe to use the completeness of their system as we do here.□

Proposition 5.4 *Let $\phi \in \mu\mathsf{TL}(\mathsf{U})$ be $K_{\mu\mathsf{TL}(\mathsf{U})}$-consistent, then its $\mu\mathsf{TL}(\Diamond_\Gamma)$-translation $\phi_{\mu\mathsf{TL}(\Diamond_\Gamma)}$ is $K_{\mu\mathsf{TL}(\Diamond_\Gamma)}$-consistent.*

Proof. Let $\phi \in \mu\mathsf{TL}(\mathsf{U})$ be $K_{\mu\mathsf{TL}(\mathsf{U})}$-consistent. Now suppose $\phi_{\mu\mathsf{TL}(\Diamond_\Gamma)}$ is not $K_{\mu\mathsf{TL}(\Diamond_\Gamma)}$-consistent. So there is a $K_{\mu\mathsf{TL}(\Diamond_\Gamma)}$-proof of $\neg\phi_{\mu\mathsf{TL}(\Diamond_\Gamma)}$. By Lemma 5.2 and 5.3, this entails that there is a $K_{\mu\mathsf{TL}(\mathsf{U})}$-proof of $\neg\phi$, which contradicts the $K_{\mu\mathsf{TL}(\mathsf{U})}$-consistency of ϕ. □

Corollary 5.5 *$K_{\mu\mathsf{TL}(\mathsf{U})}$ is complete for $\mu\mathsf{TL}(\mathsf{U})$ with respect to the class of ω-words.*

Proof. Let φ be a $K_{\mu\mathsf{TL}(\mathsf{U})}$-consistent formula. Now let φ' be the $\mu\mathsf{TL}(\Diamond_\Gamma)$-translation of φ. By Proposition 5.4, φ' is $K_{\mu\mathsf{TL}(\Diamond_\Gamma)}$-consistent and so, by Theorem 4.4, φ' is satisfied in some ω-word $\mathfrak{M}, w$. By Proposition 3.6, φ and φ' are equivalent on ω-words. Hence also $\mathfrak{M}, w \models \varphi$. □

Proposition 5.6 *Let $\phi \in \mu\mathsf{TL}(\mathsf{U})$ be $K_{\mu\mathsf{TL}(\mathsf{U})} + \mu y.\Box_\Gamma y$-consistent, then its $\mu\mathsf{TL}(\Diamond_\Gamma)$-translation $\phi_{\mu\mathsf{TL}(\Diamond_\Gamma)}$ is $K_{\mu\mathsf{TL}(\Diamond_\Gamma)} + \mu y.\Box_\Gamma y$-consistent.*

Proof. The proof is similar to the proof of Proposition 5.4. □

Corollary 5.7 *$K_{\mu\mathsf{TL}(\mathsf{U})} + \mu y.\Box_\Gamma y$ is complete for $\mu\mathsf{TL}(\mathsf{U})$ with respect to the class of finite words.*

Proof. Similarly follows from Proposition 3.6, Theorem 4.5 and Proposition 5.6. □

Remark 5.8 Let $K^{-A11''}_{\mu\mathsf{TL}(\mathsf{U})}$ be the smallest set of $\mu\mathsf{TL}(\mathsf{U})$-formulas which is closed under all axioms and rules in $K_{\mu\mathsf{TL}(\mathsf{U})}$ except Axiom $A11''$. Axiom $A11''$ *is not derivable* in $K^{-A11''}_{\mu\mathsf{TL}(\mathsf{U})}$. Observe that the $\mu\mathsf{TL}$-translation of every axiom and rule of $K^{-A11''}_{\mu\mathsf{TL}(\mathsf{U})}$6 is sound when instantiated by $\mu\mathsf{TL}$-formulas and that, by completeness of $\mu\mathsf{TL}$, their $\mu\mathsf{TL}$-translations are also derivable in $\mu\mathsf{TL}$. So if Axiom $A11''$ was derivable in $K^{-A11''}_{\mu\mathsf{TL}(\mathsf{U})}$, its $\mu\mathsf{TL}$-translation would be also derivable (and hence, valid) in $K_\mu + \Diamond\varphi \to \Box\varphi$. But let $\mathfrak{M}$ be a finite word in vocabulary $\{p\}$ with $W = \{w_0, w_1, w_2\}$, $w_i < w_{i+1}$ and $V(p) = w_2$. Obviously $\mathfrak{M}, w_0 \models \mu x.(p \vee (\Diamond\Diamond p \wedge \Diamond_{\{p\}} x))$, but $\mathfrak{M}, w_0 \not\models (\Diamond\Diamond p)\mathsf{U}p$, i.e., $\mathfrak{M}, w_0 \not\models \mu x.(p \vee (\Diamond\Diamond p \wedge \Diamond_{\{p\}} x)) \leftrightarrow (\Diamond\Diamond p)\mathsf{U}p$.

6 Conclusions and future works

In this paper, we studied the logic $\mu\mathsf{TL}(\mathsf{U})$. We introduced for that purpose the logic $\mu\mathsf{TL}(\Diamond_\Gamma)$ as a technical tool in order to easily obtain completeness results for $\mu\mathsf{TL}(\mathsf{U})$. In [6], we used an even simpler trick to show that $\mu\mathsf{TL}(\mathsf{U})$ satisfies uniform interpolation. A number of other interesting logical properties of $\mu\mathsf{TL}(\mathsf{U})$ remain to be investigated. In particular, we could examine counterparts of the Łos Tarski Theorem and of the Lyndon Theorem, which the μ-calculus was shown in [2] to satisfy. More generally, the logic $\mu\mathsf{TL}(\Diamond_\Gamma)$ could also be used as a tool in order to easily transfer results from

$\mu\mathsf{TL}$ to languages capturing exactly its stutter-invariant fragment (see for instance the frameworks in [5], [12], or [3]).

The method that we used here in order to show completeness results could also be reused in other contexts. It could for instance be applicable to the extension of $\mu\mathsf{TL}(\mathsf{U})$ with past tense operators or to the μ-calculus on trees (either finite or infinite). For a discussion of stuttering on trees, see [1] and [7]. It should be noted, though, that on (especially infinite) trees, there is still no general consensus on the appropriate notion of stuttering and that it is questionable whether the "Until only" fragment and the stutter-invariant fragment of the μ-calculus actually coincide . A further generalization would be to consider finite game trees, which actually carry a bit more structure than plain finite trees. In the context of game equivalence, the notion of stuttering could indeed constitute an interesting alternative to the notion of bisimulation (for a discussion see [14]).

References

[1] Browne, M. C., E. M. Clarke and O. Grumberg, *Characterizing Finite Kripke Structures in Propositional Temporal Logic*, Theor. Comput. Sci. **59** (1988), pp. 115–131.

[2] D'Agostino, G. and M. Hollenberg, *Logical Questions Concerning the μ-Calculus: Interpolation, Lyndon and Lös-Tarski*, Journal of Symbolic Logic **65** (2000), pp. 310–332.

[3] Dax, C., F. Klaedtke and S. Leue, *Specification Languages for Stutter-Invariant Regular Properties*, in: *ATVA*, 2009, pp. 244–254.

[4] Demri, S. and P. Schnoebelen, *The Complexity of Propositional Linear Temporal Logics in Simple Cases*, Information and Computation **174** (2002), pp. 84–103.

[5] Etessami, K., *Stutter-Invariant Languages, ω-Automata, and Temporal Logic*, in: N. Halbwachs and D. Peled, editors, *Proceedings of CAV* (1999), pp. 236–248.

[6] Gheerbrant, A. and B. ten Cate, *Craig Interpolation for Linear Temporal Languages*, in: *CSL*, 2009, pp. 287–301.

[7] Gross, R., "Invariance under Stuttering in Branching-Time Temporal Logic," Master's thesis, Israel Institute of Technology, Haifa (2008).

[8] Kaivola, R., "Using Automata to Characterise Fixed-Point Temporal Logics," Ph.D. thesis, University of Edinburgh (1997).

[9] Lamport, L., *What Good is Temporal Logic?*, in: R. E. A. Mason, editor, *Proceedings of the IFIP 9th World Computer Congress* (1983), pp. 657–668.

[10] Moser, L. E., P. M. Melliar-Smith, G. Kutty and Y. S. Ramakrishna, *Completeness and Soundness of Axiomatizations for Temporal Logics without Next*, Fundamenta Informatica **21** (1994), pp. 257–305.

[11] Peled, D. and T. Wilke, *Stutter-Invariant Temporal Properties are Expressible Without the Next-Time Operator*, Inf. Process. Lett. **63** (1997), pp. 243–246.

[12] Rabinovich, A. M., *Expressive Completeness of Temporal Logic of Action*, in: *MFCS*, 1998, pp. 229–238.

[13] ten Cate, B. and G. Fontaine, *An Easy Completeness Proof for the Modal μ-Calculus on Finite Trees*, in: *FOSSACS*, 2010, pp. 161–175.

[14] van Benthem, J., *Extensive Games as Process Models*, Journal of Logic, Language and Information **11** (2002), pp. 289–313.

[15] Vardi, M. Y., *A Temporal Fixpoint Calculus*, in: *Proceedings of POPL*, 1988, pp. 250–259.

[16] Vardi, M. Y., *From Philosophical to Industrial Logics*, in: *ICLA*, 2009, pp. 89–115.

[17] Walukiewicz, I., *A Note on the Completeness of Kozen's Axiomatization of the Propositional μ-Calculus*, The Bulletin of Symbolic Logic **2** (1996).

Cut-elimination and Proof Search for Bi-Intuitionistic Tense Logic

Rajeev Goré, Linda Postniece and Alwen Tiu

Logic and Computation Group, School of Computer Science
The Australian National University
{Rajeev.Gore,Linda.Postniece,Alwen.Tiu}@anu.edu.au

Abstract

We consider an extension of bi-intuitionistic logic with the traditional modalities $\Diamond$, $\Box$, $\blacklozenge$ and $\blacksquare$ from tense logic Kt. Proof theoretically, this extension is obtained simply by extending an existing sequent calculus for bi-intuitionistic logic with typical inference rules for the modalities used in display logics. As it turns out, the resulting calculus, **LBiKt**, seems to be more basic than most intuitionistic tense or modal logics considered in the literature, in particular, those studied by Ewald and Simpson, as it does not assume any *a priori* relationship between the modal operators $\Diamond$ and $\Box$. We recover Ewald's intuitionistic tense logic and Simpson's intuitionistic modal logic by modularly extending **LBiKt** with additional structural rules. The calculus **LBiKt** is formulated in a variant of display calculus, using a form of sequents called nested sequents. Cut elimination is proved for **LBiKt**, using a technique similar to that used in display calculi. As in display calculi, the inference rules of **LBiKt** are "shallow" rules, in the sense that they act on top-level formulae in a nested sequent. The calculus **LBiKt** is ill-suited for backward proof search due to the presence of certain structural rules called "display postulates" and the contraction rules on arbitrary structures. We show that these structural rules can be made redundant in another calculus, **DBiKt**, which uses deep inference, allowing one to apply inference rules at an arbitrary depth in a nested sequent. We prove the equivalence between **LBiKt** and **DBiKt** and outline a proof search strategy for **DBiKt**. We also give a Kripke semantics and prove that **LBiKt** is sound with respect to the semantics, but completeness is still an open problem. We then discuss various extensions of **LBiKt**.

Keywords: Intuitionistic logic, modal logic, intuitionistic modal logic, deep inference.

1 Introduction

Intuitionistic logic Int forms a rigorous foundation for many areas of Computer Science via its constructive interpretation and via the Curry-Howard isomorphism between natural deduction proofs and well-typed terms in the λ-calculus. Central to both concerns are syntactic proof calculi with cut-elimination and backwards proof-search for finding derivations automatically.

In traditional intuitionistic logic, the connectives $\rightarrow$ and $\wedge$ form an adjoint pair in that $(A \wedge B) \rightarrow C$ is valid iff $A \rightarrow (B \rightarrow C)$ is valid iff $B \rightarrow (A \rightarrow C)$ is valid. Rauszer [22] obtained BiInt by extending Int with a binary connective $\prec$ called "exclusion" which is adjoint to $\vee$ in that $A \rightarrow (B \vee C)$ is valid iff $(A \prec B) \rightarrow C$ is valid iff $(A \prec C) \rightarrow B$ is valid. Crolard [4] showed that BiInt has a computational interpretation in terms of continuation passing style semantics. Uustalu and Pinto recently showed that Rauszer's sequent calculus [21] and Crolard's extensions of it fail cut-elimination, but a nested sequent calculus with cut-elimination [9] and a labelled sequent calculus [18] with cut-free-completeness have been found for BiInt.

The literature on Intuitionistic Modal/Tense Logics (IM/TLs) is vast [6,24] and typically uses Hilbert calculi with algebraic, topological or relational semantics. We omit details since our interest is primarily proof-theoretic. Sequent and natural deduction calculi for IMLs are rarer [14,1,17,3,5,12,7]. Extending them with "converse" modalities like $\blacklozenge$ and $\blacksquare$ causes cut-elimination to fail as it does for classical modal logic S5 where $\Diamond$ is a self-converse. Labels [15,24,16] can help but are not purely proof-theoretic since they encode the Kripke semantics.

The closest to our work is that of Sadrzadeh and Dyckhoff [23] who give a cut-free sequent calculus using deep inference for a logic with an adjoint pair of modalities $(\blacklozenge, \Box)$ plus only $\wedge$, $\vee$, $\top$ and $\bot$. As all their connectives are "monotonic", cut-elimination presents no difficulties.

Let BiKt be the bi-intuitionistic tense logic obtained by extending BiInt with two pairs of adjoint modalities $(\Diamond, \blacksquare)$ and $(\blacklozenge, \Box)$, with no explicit relationship between the modalities of the same colour, namely, $(\Diamond, \Box)$ and $(\blacklozenge, \blacksquare)$. The modalities form an adjunction as follows: $A \rightarrow \Box B$ iff $\blacklozenge A \rightarrow B$ and $A \rightarrow \blacksquare B$ iff $\Diamond A \rightarrow B$.

Our shallow inference calculus **LBiKt** is a merger of two sub-calculi for BiInt and Kt derived from Belnap's inherently modular display logic. **LBiKt** has syntactic cut-elimination, but is ill-suited for backward proof search. Our deep inference calculus **DBiKt** is complete with respect to the cut-free fragment of **LBiKt** and is more amenable to proof search as it contains no display postulates and contraction rules. To complete the picture, we also give a Kripke semantics for BiKt based upon three relations $\leq$, $R_\Diamond$ and $R_\Box$. The logic BiKt enjoys various desirable properties:

* **Conservativity:** it is a conservative extension of intuitionistic logic Int, dual intuitionistic logic DInt, and bi-intuitionistic logic BiInt;
* **Classical Collapse:** it collapses to classical tense logic by the addition of four structural rules;
* **Disjunction Property:** If $A \vee B$ is a theorem not containing $\prec$ then A is a theorem or B is a theorem;
* **Dual Disjunction Property:** If $A \wedge B$ is a counter-theorem not containing $\rightarrow$ then so is A or B;
* **Independent $\Diamond$ and $\Box$:** there is no *a priori* relationship between these connectives.

The independence of $\Diamond$ and $\Box$ is a departure from traditional intuitionistic tense or modal logics, e.g., those considered by Ewald [6] and Simpson [24]. Both Ewald and Simpson allow a form of interdependency between $\Diamond$ and $\Box$, expressed as the axiom $(\Diamond A \rightarrow \Box B) \rightarrow \Box(A \rightarrow B)$, which is not derivable in **LBiKt**. However, we shall see in Section 7

$$
\begin{array}{llll}
\tau^-(A) & = A & \tau^+(A) & = A \\
\tau^-(X,Y) & = \tau^-(X) \wedge \tau^-(Y) & \tau^+(X,Y) & = \tau^+(X) \vee \tau^+(Y) \\
\tau^-(X \triangleright Y) & = \tau^-(X) \prec \tau^+(Y) & \tau^+(X \triangleright Y) & = \tau^-(X) \to \tau^+(Y) \\
\tau^-(\circ X) & = \Diamond \tau^-(X) & \tau^+(\circ X) & = \Box \tau^+(X) \\
\tau^-(\bullet X) & = \blacklozenge \tau^-(X) & \tau^+(\bullet X) & = \blacksquare \tau^+(X)
\end{array}
$$

$$\tau(X \triangleright Y) = \tau^-(X) \to \tau^+(Y)$$

Fig. 1. Formula Translation of Nested Sequents

that we can recover Ewald's intuitionistic tense logic and Simpson's intuitionistic modal logic by extending **LBiKt** with two structural rules.

Due to space limit, some proofs are omitted, but they can be found in an extended version of this paper [11].

2 Nested Sequents

The formulae of BiKt are built from a set *Atoms* of atomic formulae via the grammar below, with $p \in Atoms$:

$$A ::= p \mid \top \mid \bot \mid A \to A \mid A \prec A \mid A \wedge A \mid A \vee A \mid \Box A \mid \Diamond A \mid \blacksquare A \mid \blacklozenge A.$$

A structure is defined by the following grammar, where A is a BiKt formula:

$$X := \emptyset \mid A \mid (X, X) \mid X \triangleright X \mid \circ X \mid \bullet X.$$

The structural connective "," is associative and commutative and $\emptyset$ is its unit. We always consider structures modulo these equivalences. To reduce parentheses, we assume that "$\circ$" and "$\bullet$" bind tighter than "," which binds tighter than "$\triangleright$". Thus, we write $\bullet X, Y \triangleright Z$ to mean $(\bullet(X), Y) \triangleright Z$.

A *nested sequent* is a structure of the form $X \triangleright Y$. This notion of nested sequents generalises Kashima's nested sequents [13] for classical tense logics, Brünnler's nested sequents [2] and Poggiolesi's tree-hypersequents [19] for classical modal logics. Figure 1 shows the formula-translation of nested sequents. On both sides of the sequent, $\circ$ is interpreted as a white (modal) operator and $\bullet$ as a black (tense) operator. Note that however, on the lefthand side of the sequent, $\triangleright$ is interpreted as exclusion, while on the righthand side, it is interpreted as implication.

A *context* is a structure with a hole or a placeholder $[]$. Contexts are ranged over by $\Sigma[]$. We write $\Sigma[X]$ for the structure obtained by filling the hole $[]$ in the context $\Sigma[]$ with a structure X. A *simple* context is defined via:

$$\Sigma[] ::= [] \mid \Sigma[], (Y) \mid (Y), \Sigma[] \mid \circ\Sigma[] \mid \bullet\Sigma[]$$

Intuitively, the hole in a simple context is never under the scope of $\triangleright$. Positive and negative contexts are defined inductively as follows:

* If $\Sigma[]$ is a simple context then $\Sigma[] \rhd Y$ is a negative context and $Y \rhd \Sigma[]$ is a positive context.
* If $\Sigma[]$ is a positive/negative context then so are $(\Sigma[], Y)$, $(Y, \Sigma[])$, $\bullet(\Sigma[])$, $\circ(\Sigma[])$, $\Sigma[] \rhd Y$, and $Y \rhd \Sigma[]$.

We write $\Sigma^-[]$ to indicate that $\Sigma[]$ is a negative context and $\Sigma^+[]$ to indicate that it is a positive context. Intuitively, if one views a nested sequent as a tree (with structural connectives and formulae as nodes), then a hole in a context is negative (positive) if it appears to the left (right) of the closest ancestor node labelled with $\rhd$. As a consequence of the overloading of $\rhd$ as a structural proxy for both $\to$ and $-\!\!<$, further nesting of a positive/negative context within $\rhd$ does not change its polarity. This is different from the traditional notion of polarity which is defined in terms of either $\to$ or $-\!\!<$ alone, but not both. This aspect is different from display calculi and may cause confusion at first reading. Our statement of the display property in Lemma 3.2–Lemma 3.2 accounts for this difference.

The context $\Sigma[]$ is *strict* if it has any of the forms:

$$\Sigma'[X \rhd [\,]] \qquad \Sigma'[[\,] \rhd X] \qquad \Sigma'[\circ[\,]] \qquad \Sigma'[\bullet[\,]]$$

Intuitively, in the formation tree of a strict context, the hole must be an immediate child of $\rhd$ or $\circ$ or $\bullet$. This notion of strict contexts will be used in later in Section 3.

Example 2.1 The context $\bullet([], (X \rhd Y))$ is a simple context but $\bullet(([], X) \rhd Y)$ is not. Both $\bullet([], (X \rhd Y)) \rhd Z$ and $\bullet(([], X) \rhd Y) \rhd Z$ are negative contexts. The context $\bullet[] \rhd Z$ is a strict context but $\bullet(([], X) \rhd Y) \rhd Z$ is not.

3 Nested Sequent Calculi

We now present the two nested sequent calculi that we will use in the rest of the paper: a shallow inference calculus **LBiKt** and a deep inference calculus **DBiKt**. Fig. 2 gives the rules of the shallow inference calculus **LBiKt**. The inference rules of **LBiKt** can only be applied to formulae at the top level of nested sequents, and the structural rules s_L, s_R, $\rhd_L$, $\rhd_R$, $rp_\circ$ and $rp_\bullet$, also called the *residuation rules*, are used to bring the required substructures to the top level. These rules are similar to residuation postulates in display logic, are essential for the cut-elimination proof of **LBiKt**, but contain too much non-determinism for effective proof search. Another issue with proof search in **LBiKt** is the structural contraction rules, which allow contraction on arbitrary structures, not just formulae as in traditional sequent calculi. **LBiKt** is as a merger of two calculi: the LBiInt calculus [9,20] for the intuitionistic connectives, and the display calculus [8] for the tense connectives.

We use $\circ$ and $\bullet$ as structural proxies for the non-residuated pairs $(\Diamond, \Box)$ and $(\blacklozenge, \blacksquare)$ respectively, whereas Wansing [25] uses only $\bullet$ as a structural proxy for the residuated pair $(\blacklozenge, \Box)$ and recovers $(\Diamond, \blacksquare)$ via classical negation, while Goré [8] uses $\circ$ and $\bullet$ as structural proxies for the residuated pairs $(\Diamond, \blacksquare)$ and $(\blacklozenge, \Box)$ respectively. As we shall see later, our choice allows us to retain the modal fragment $(\Diamond, \Box)$ by simply eliding all

Identity and logical constants:

$$\frac{}{X, A \rhd A, Y}\ id \qquad \frac{}{X, \bot \rhd Y}\ \bot_L \qquad \frac{}{X \rhd \top, Y}\ \top_R$$

Structural rules:

$$\frac{X \rhd Z}{X, Y \rhd Z}\ w_L \qquad \frac{X \rhd Z}{X \rhd Y, Z}\ w_R \qquad \frac{X, Y, Y \rhd Z}{X, Y \rhd Z}\ c_L \qquad \frac{X \rhd Y, Y, Z}{X \rhd Y, Z}\ c_R$$

$$\frac{(X_1 \rhd Y_1), X_2 \rhd Y_2}{X_1, X_2 \rhd Y_1, Y_2}\ s_L \qquad \frac{X_1 \rhd Y_1, (X_2 \rhd Y_2)}{X_1, X_2 \rhd Y_1, Y_2}\ s_R \qquad \frac{X_2 \rhd Y_2, Y_1}{(X_2 \rhd Y_2) \rhd Y_1}\ \rhd_L \qquad \frac{X_1, X_2 \rhd Y_2}{X_1 \rhd (X_2 \rhd Y_2)}\ \rhd_R$$

$$\frac{\bullet X \rhd Y}{X \rhd \circ Y}\ rp_\circ \qquad \frac{\circ X \rhd Y}{X \rhd \bullet Y}\ rp_\bullet \qquad \frac{X_1 \rhd Y_1, A \qquad A, X_2 \rhd Y_2}{X_1, X_2 \rhd Y_1, Y_2}\ cut$$

Logical rules:

$$\frac{X, B_i \rhd Y}{X, B_1 \wedge B_2 \rhd Y}\ \wedge_L\ i \in \{1, 2\} \qquad \frac{X \rhd A, Y \qquad X \rhd B, Y}{X \rhd A \wedge B, Y}\ \wedge_R$$

$$\frac{X, A \rhd Y \qquad X, B \rhd Y}{X, A \vee B \rhd Y}\ \vee_L \qquad \frac{X \rhd B_i, Y}{X \rhd B_1 \vee B_2, Y}\ \vee_R\ i \in \{1, 2\}$$

$$\frac{X \rhd A, Y \qquad X, B \rhd Y}{X, A \to B \rhd Y}\ \to_L \qquad \frac{X, A \rhd B}{X \rhd A \to B, Y}\ \to_R$$

$$\frac{A \rhd B, Y}{X, A \prec B \rhd Y}\ \prec_L \qquad \frac{X \rhd A, Y \qquad X, B \rhd Y}{X \rhd A \prec B, Y}\ \prec_R$$

$$\frac{A \rhd X}{\Box A \rhd \circ X}\ \Box_L \qquad \frac{X \rhd \circ A}{X \rhd \Box A}\ \Box_R \qquad \frac{A \rhd X}{\blacksquare A \rhd \bullet X}\ \blacksquare_L \qquad \frac{X \rhd \bullet A}{X \rhd \blacksquare A}\ \blacksquare_R$$

$$\frac{\circ A \rhd X}{\Diamond A \rhd X}\ \Diamond_L \qquad \frac{X \rhd A}{\circ X \rhd \Diamond A}\ \Diamond_R \qquad \frac{\bullet A \rhd X}{\blacklozenge A \rhd X}\ \blacklozenge_L \qquad \frac{X \rhd A}{\bullet X \rhd \blacklozenge A}\ \blacklozenge_R$$

Fig. 2. **LBiKt**: a shallow inference system for BiKt

rules that contain "black" operators from our deep sequent calculus.

Fig. 3 gives the rules of the deep inference calculus **DBiKt**. Here the inference rules can be applied at any level of the nested sequent, indicated by the use of contexts. Notably, there are no residuation rules; indeed one of the goals of our paper is to show that the residuation rules of **LBiKt** can be simulated by deep inference and propagation rules in **DBiKt**. Another feature of **DBiKt** is the use of polarities in defining contexts to which rules are applicable. For example, the premise of the $\Box_{L1}$ rule denotes a negative context Σ which itself contains a formula A and a $\bullet$-structure, such that the $\bullet$-structure contains $\Box A$.

DBiKt achieves the goal of merging the DBiInt calculus [20] and a two-sided version of the DKt calculus [10]. While in the shallow inference case, a calculus for BiKt could be obtained relatively easily by merging shallow inference calculi for BiInt and tense logics, the combination of calculi is not so obvious in the deep inference case. Although the propagation rules for $\rhd$-structures remain the same as in the BiInt case [20], the

Identity and logical constants:

$$\frac{}{\Sigma[X, A \rhd A, Y]}\ id \qquad \frac{}{\Sigma[\bot, X \rhd Y]}\ \bot_L \qquad \frac{}{\Sigma[X \rhd \top, Y]}\ \top_R$$

Propagation rules:

$$\frac{\Sigma^-[A, (A, X \rhd Y)]}{\Sigma^-[A, X \rhd Y]}\ \rhd_{L1} \qquad \frac{\Sigma^+[(X \rhd Y, A), A]}{\Sigma^+[X \rhd Y, A]}\ \rhd_{R1}$$

$$\frac{\Sigma[X, A \rhd W, (A, Y \rhd Z)]}{\Sigma[X, A \rhd W, (Y \rhd Z)]}\ \rhd_{L2} \qquad \frac{\Sigma[(X \rhd Y, A), W \rhd A, Z]}{\Sigma[(X \rhd Y), W \rhd A, Z]}\ \rhd_{R2}$$

$$\frac{\Sigma^-[A, \bullet(\Box A, X)]}{\Sigma^-[\bullet(\Box A, X)]}\ \Box_{L1} \qquad \frac{\Sigma^+[A, \bullet(\Diamond A, X)]}{\Sigma^+[\bullet(\Diamond A, X)]}\ \Diamond_{R1} \qquad \frac{\Sigma^-[A, \circ(\blacksquare A, X)]}{\Sigma^-[\circ(\blacksquare A, X)]}\ \blacksquare_{L1} \qquad \frac{\Sigma^+[A, \circ(\blacklozenge A, X)]}{\Sigma^+[\circ(\blacklozenge A, X)]}\ \blacklozenge_{R1}$$

$$\frac{\Sigma[\blacksquare A, X \rhd \bullet(A \rhd Y), Z]}{\Sigma[\blacksquare A, X \rhd \bullet Y, Z]}\ \blacksquare_{L2} \qquad \frac{\Sigma[\circ(X \rhd A), Y \rhd Z, \Diamond A]}{\Sigma[\circ X, Y \rhd Z, \Diamond A]}\ \Diamond_{R2}$$

$$\frac{\Sigma[\Box A, X \rhd \circ(A \rhd Y), Z]}{\Sigma[\Box A, X \rhd \circ Y, Z]}\ \Box_{L2} \qquad \frac{\Sigma[\bullet(X \rhd A), Y \rhd Z, \blacklozenge A]}{\Sigma[\bullet X, Y \rhd Z, \blacklozenge A]}\ \blacklozenge_{R2}$$

Logical rules:

$$\frac{\Sigma^-[A \vee B, A] \qquad \Sigma^-[A \vee B, B]}{\Sigma^-[A \vee B]}\ \vee_L \qquad \frac{\Sigma^+[A \vee B, A, B]}{\Sigma^+[A \vee B]}\ \vee_R$$

$$\frac{\Sigma^-[A \wedge B, A, B]}{\Sigma^-[A \wedge B]}\ \wedge_L \qquad \frac{\Sigma^+[A \wedge B, A] \qquad \Sigma^+[A \wedge B, B]}{\Sigma^+[A \wedge B]}\ \wedge_R$$

$$\frac{\Sigma^-[A -\!\!< B, (A \rhd B)]}{\Sigma^-[A -\!\!< B]}\ -\!\!<_L \qquad \frac{\Sigma^+[A \to B, (A \rhd B)]}{\Sigma^+[A \to B]}\ \to_R$$

$$\frac{\Sigma^-[X, A \to B \rhd A] \qquad \Sigma^-[X, A \to B, B]}{\Sigma^-[X, A \to B]}\ \to_L \qquad \Sigma^-[]\ \text{is a strict context}$$

$$\frac{\Sigma^+[X, A -\!\!< B, A] \qquad \Sigma^+[B \rhd X, A -\!\!< B]}{\Sigma^+[X, A -\!\!< B]}\ -\!\!<_R \qquad \Sigma^+[]\ \text{is a strict context}$$

$$\frac{\Sigma^-[\Diamond A, \circ A]}{\Sigma^-[\Diamond A]}\ \Diamond_L \qquad \frac{\Sigma^+[\Box A, \circ A]}{\Sigma^+[\Box A]}\ \Box_R \qquad \frac{\Sigma^-[\blacklozenge A, \bullet A]}{\Sigma^-[\blacklozenge A]}\ \blacklozenge_L \qquad \frac{\Sigma^+[\blacksquare A, \bullet A]}{\Sigma^+[\blacksquare A]}\ \blacksquare_R$$

Fig. 3. **DBiKt**: a deep inference system for BiKt

propagation rules for $\circ$- and $\bullet$-structures are not as simple as in the DKt calculus [10]. Since we do not assume any direct relationship between $\Box$ and $\Diamond$, or $\blacksquare$ and $\blacklozenge$, propagation rules like $\blacksquare_{L2}$ need to involve the $\rhd$ structural connective so they can refer to both sides of the nested sequent.

Note that in the rules $\to_L$ and $-\!\!<_R$ in **DBiKt**, we require that the contexts in which the principal formulae reside are strict contexts. This is strictly speaking not necessary, i.e., we could remove the proviso without affecting the expressivity of the proof system. The proviso does, however, reduce the non-determinism in partitioning the contexts in $\to_L$ or $-\!\!<_R$. Consider, for example, the nested sequent $\circ(a, b \to c) \rhd d$. Without the requirement of strict contexts, there are two instances of $\to_L$ with that nested sequent as the conclusion:

$$\dfrac{\circ(a, (b \to c \rhd b)) \rhd d \quad \circ(a, (b \to c, c)) \rhd d}{\circ(a, b \to c) \rhd d} \to_L$$

$$\dfrac{\circ(a, b \to c \rhd b) \rhd d \quad \circ(a, b \to c, c) \rhd d}{\circ(a, b \to c) \rhd d} \to_L$$

In the first instance, the context is $\circ(a, [\,]) \rhd d$, which is not strict, whereas in the second instance, it is $\circ([\,]) \rhd d$, which is strict. In general, if there are n formulae connected to $b \to c$ via the comma structural connective, then there are 2^n possible instances of $\to_L$ without the strict context proviso.

We write $\vdash_{\mathbf{LBiKt}} \pi : X \rhd Y$ when π is a derivation of the shallow sequent $X \rhd Y$ in **LBiKt**, and $\vdash_{\mathbf{DBiKt}} \pi : X \rhd Y$ when π is a derivation of the sequent $X \rhd Y$ in **DBiKt**. In either calculus, the height $|\pi|$ of a derivation π is the number of sequents on the longest branch.

Example 3.1 Below we derive Ewald's axiom 9 for IK_t [6] in **LBiKt** and **DBiKt**. The **LBiKt**-derivation on the left read bottom-up brings the required sub-structure $\blacklozenge A$ to the top-level using the residuation rule $rp_\circ$ and applies $\blacklozenge_R$ backward. The **DBiKt**-derivation on the right instead applies $\Box_R$ deeply, and propagates the required formulae to the appropriate sub-structure using $\blacklozenge_{R1}$. Note that contraction is implicit in $\blacklozenge_{R1}$, and all propagation rules.

$$\dfrac{\dfrac{\dfrac{\dfrac{\dfrac{}{A \rhd A}\,id}{\bullet A \rhd \blacklozenge A}\,\blacklozenge_R}{A \rhd \circ\blacklozenge A}\,rp_\circ}{A \rhd \Box\blacklozenge A}\,\Box_R}{\emptyset \rhd A \to \Box\blacklozenge A}\,\to_R
\qquad\qquad
\dfrac{\dfrac{\dfrac{\dfrac{}{\emptyset \rhd (A \rhd A, \circ(\blacklozenge A))}\,id}{\emptyset \rhd (A \rhd \circ(\blacklozenge A))}\,\blacklozenge_{R1}}{\emptyset \rhd (A \rhd \Box\blacklozenge A)}\,\Box_R}{\emptyset \rhd A \to \Box\blacklozenge A}\,\to_R$$

Display property

A (deep or shallow) nested sequent can be seen as a tree of traditional sequents. The structural rules of **LBiKt** allows shuffling of structures to display/un-display a particular node in the tree, so inference rules can be applied to it. This is similar to the display property in traditional display calculi, where any substructure can be displayed and un-displayed. We state the display property of **LBiKt** more precisely in subsequent lemmas. We shall use two "display" rules which are easily derivable using s_L, s_R, $\rhd_L$ and $\rhd_R$:

$$\dfrac{(X_1 \rhd X_2) \rhd Y}{X_1 \rhd X_2, Y}\,rp^L_\rhd \qquad\qquad \dfrac{X_1 \rhd (X_2 \rhd Y)}{X_1, X_2 \rhd Y}\,rp^R_\rhd$$

Let $DP = \{rp^L_\rhd, rp^R_\rhd, rp_\circ, rp_\bullet\}$ and let DP-derivable mean "derivable using rules only from DP".

Lemma 3.2 (Display property for simple contexts) *Let $\Sigma[]$ be a simple context. Let X be a structure and p a propositional variable not occurring in X nor $\Sigma[]$. Then there exist structures Y and Z such that:*

(i) *$Y \triangleright p$ is DP-derivable from $X \triangleright \Sigma[p]$ and*

(ii) *$p \triangleright Z$ is DP-derivable from $\Sigma[p] \triangleright X$.*

Lemma 3.3 (Display property for positive contexts) *Let $\Sigma[]$ be a positive context. Let X be a structure and p a propositional variable not occurring in X nor $\Sigma[]$. Then there exist structures Y and Z such that:*

(i) *$Y \triangleright p$ is DP-derivable from $X \triangleright \Sigma[p]$, and*

(ii) *$Z \triangleright p$ is DP-derivable from $\Sigma[p] \triangleright X$.*

Lemma 3.4 (Display property for negative contexts) *Let $\Sigma[]$ be a negative context. Let X be a structure and p a propositional variable not occurring in X nor $\Sigma[]$. Then there exist structures Y and Z such that:*

(i) *$p \triangleright Y$ is DP-derivable from $X \triangleright \Sigma[p]$ and*

(ii) *$p \triangleright Z$ is DP-derivable from $\Sigma[p] \triangleright X$.*

Since the rules in *DP* are all invertible, the derivations constructed in the above lemmas are invertible derivations. That is, we can derive $Y \triangleright p$ from $X \triangleright \Sigma[p]$ and vice versa. Note also that since rules in the shallow system are closed under substitution, this also means $Y \triangleright Z$ is derivable from $X \triangleright \Sigma[Z]$, and vice versa, for any Z.

Cut elimination in LBiKt

Our cut-elimination proof is based on the method of proof-substitution presented in [9]. It is very similar to the general cut elimination method used in display calculi. The proof relies on the display property and the fact that inference rules in **LBiKt** are closed under substitutions. The proof is omitted here, but is available in the extended version of this paper [11]. We illustrate one case with an example.

Consider the derivation below ending with a cut on $\Diamond A$:

$$\dfrac{\begin{matrix}\Pi_1 \\ X_1 \triangleright Y_1, \Diamond A\end{matrix} \qquad \begin{matrix}\Pi_2 \\ \Diamond A, X_2 \triangleright Y_2\end{matrix}}{X_1, X_2 \triangleright Y_1, Y_2}\ cut$$

Instead of permuting the cut rule locally, we trace the cut formula $\Diamond A$ until it becomes principal in the derivations Π_1 and Π_2, and then apply cut on a smaller formula. Suppose that Π_1 and Π_2 are respectively the derivations (1) and (2) in Figure 4. We first transform Π_1 by substituting $(X_2 \triangleright Y_2)$ for $\Diamond A$ in Π_1 and obtain the sub-derivation with an open leaf as shown in Figure 4(3). We then prove the open leaf by uniformly substituting $\circ(X_1')$ for $\Diamond A$ in Π_2, and applying cut on a sub-formula A, as shown in Figure 4(4).

Theorem 3.5 *If $X \triangleright Y$ is* **LBiKt***-derivable then it is also* **LBiKt***-derivable without using cut.*

$$
\begin{array}{cccc}
\begin{array}{c}
\Psi_1 \\
\dfrac{X_1' \rhd A}{\circ(X_1') \rhd \Diamond A}\,\Diamond_R \\
\vdots \\
X_1 \rhd Y_1, \Diamond A
\end{array}
&
\begin{array}{c}
\Psi_2 \\
\dfrac{\circ A \rhd Y_2'}{\Diamond A \rhd Y_2'}\,\Diamond_L \\
\vdots \\
\Diamond A, X_2 \rhd Y_2
\end{array}
&
\begin{array}{c}
\circ X_1' \rhd (X_2 \rhd Y_2) \\
\vdots \\
\dfrac{X_1 \rhd Y_1, (X_2 \rhd Y_2)}{X_1, X_2 \rhd Y_1, Y_2}\,s_R
\end{array}
&
\begin{array}{c}
\dfrac{\dfrac{\begin{array}{c}\Psi_1 \\ X_1' \rhd A\end{array} \quad \dfrac{\begin{array}{c}\Psi_2 \\ \circ A \rhd Y_2'\end{array}}{A \rhd \bullet Y_2'}\,rp_\bullet}{X_1' \rhd \bullet Y_2'}\,cut}{\circ X_1' \rhd Y_2'}\,rp_\bullet \\
\vdots \\
\dfrac{\circ X_1', X_2 \rhd Y_2}{\circ X_1' \rhd (X_2 \rhd Y_2)}\,\rhd_R
\end{array}
\\
(1) & (2) & (3) & (4)
\end{array}
$$

Fig. 4. An example of cut reduction

4 Equivalence between LBiKt and DBiKt

We now show that **LBiKt** and **DBiKt** are equivalent. We first show that every derivation in **DBiKt** can be mimicked by a cut-free derivation in **LBiKt**. The interesting cases involve showing that the propagation rules of **DBiKt** are derivable in **LBiKt** using residuation. This is not surprising since the residuation rules in display calculi are used exactly for the purpose of displaying and un-displaying sub-structures so that inference rules can be applied to them.

Theorem 4.1 *For any X and Y, if $\vdash_{\mathbf{DBiKt}} \pi : X \rhd Y$ then $\vdash_{\mathbf{LBiKt}} \pi' : X \rhd Y$.*

Proof. *(Outline)* We show that each deep inference rule ρ is derivable in the shallow system by a case analysis of the context $\Sigma[\,]$ in which the deep rule ρ applies. Note that if a deep inference rule ρ is applicable to $X \rhd Y$, then the context $\Sigma[\,]$ in this case is either $[\,]$, a positive context or a negative context. In the first case, it is easy to show that each valid instance of ρ where $\Sigma[\,] = [\,]$ is derivable in the shallow system. For the case where $\Sigma[\,]$ is either positive or negative, we use the display property. We show here the case where ρ is a rule with a single premise; the other cases are similar. Suppose ρ is as shown below left. By the display properties, we need to show only that the rule on the right is derivable in the shallow system for some structure X':

$$
\frac{\Sigma^+[U]}{\Sigma^+[V]}\,\rho \qquad\qquad \frac{X' \rhd U}{X' \rhd V}
$$

A more detailed proof is given in the extended paper. $\dashv$

We now show that any *cut-free* **LBiKt**-derivation can be transformed into a cut-free **DBiKt**-derivation. This requires proving cut-free admissibility of various structural rules in **DBiKt**. The admissibility of general weakening and *formula* contraction (not general contraction, which we will show later) are straightforward by induction on the height of derivations.

Lemma 4.2 (Admissibility of general weakening) *For any structures X and Y: if $\vdash_{\mathbf{DBiKt}} \pi : \Sigma[X]$ then $\vdash_{\mathbf{DBiKt}} \pi' : \Sigma[X, Y]$ such that $|\pi'| \leq |\pi|$.*

Lemma 4.3 (Admissibility of formula contraction) *For any structure X and formula A: if $\vdash_{\mathbf{DBiKt}} \pi : \Sigma[X, A, A]$ then $\vdash_{\mathbf{DBiKt}} \pi' : \Sigma[X, A]$ such that $|\pi'| \leq |\pi|$.*

Once weakening and contraction are shown admissible, it remains to show that the residuation rules of **LBiKt** are also admissible. In contrast to the case with the deep inference system for bi-intuitionistic logic, the combination of modal and intuitionistic structural connectives complicates the proof of this admissibility. It seems crucial to first show "deep" admissibility of certain forms of residuation for $\triangleright$. We state the required lemmas below.

Lemma 4.4 (Deep admissibility of structural rules) *The following statements hold for **DBiKt**:*

(i) Deep admissibility of s_L. *If $\vdash_{\mathbf{DBiKt}} \pi : \Sigma[(X \triangleright Y), Z \triangleright W]$ then $\vdash_{\mathbf{DBiKt}} \pi' : \Sigma[X, Z \triangleright Y, W]$ such that $|\pi'| \leq |\pi|$.*

(ii) Deep admissibility of s_R. *If $\vdash_{\mathbf{DBiKt}} \pi : \Sigma[X \triangleright Y, (Z \triangleright W)]$ then $\vdash_{\mathbf{DBiKt}} \pi' : \Sigma[X, Z \triangleright Y, W]$ such that $|\pi'| \leq |\pi|$.*

(iii) Deep admissibility of $\triangleright_L$. *If $\vdash_{\mathbf{DBiKt}} \pi : \Sigma[X \triangleright Y, Z]$ and Σ is either the empty context $[\,]$ or a negative context $\Sigma_1^-[\,]$, then $\vdash_{\mathbf{DBiKt}} \pi' : \Sigma[(X \triangleright Y) \triangleright Z]$.*

(iv) Deep admissibility of $\triangleright_R$. *If $\vdash_{\mathbf{DBiKt}} \pi : \Sigma[X, Y \triangleright Z]$ and Σ is either the empty context $[\,]$ or a positive context $\Sigma_1^+[\,]$, then $\vdash_{\mathbf{DBiKt}} \pi' : \Sigma[X \triangleright (Y \triangleright Z)]$.*

We now show that the residuation rules of **LBiKt** for $\circ$- and $\bullet$-structures are admissible in **DBiKt**, i.e., they can be simulated by the propagation rules of **DBiKt**.

Lemma 4.5 (Admissibility of residuation) *The following statements hold in **DBiKt**:*

(i) Admissibility of $rp_\bullet$. *If $\vdash_{\mathbf{DBiKt}} \pi : X \triangleright \bullet Z$ then $\vdash_{\mathbf{DBiKt}} \pi' : \circ X \triangleright Z$.*

(ii) Admissibility of $rp_\bullet$. *If $\vdash_{\mathbf{DBiKt}} \pi : \circ X \triangleright Z$ then $\vdash_{\mathbf{DBiKt}} \pi' : X \triangleright \bullet Z$.*

(iii) Admissibility of $rp_\circ$. *If $\vdash_{\mathbf{DBiKt}} \pi : X \triangleright \circ Z$ then $\vdash_{\mathbf{DBiKt}} \pi' : \bullet X \triangleright Z$.*

(iv) Admissibility of $rp_\circ$. *If $\vdash_{\mathbf{DBiKt}} \pi : \bullet X \triangleright Z$ then $\vdash_{\mathbf{DBiKt}} \pi' : X \triangleright \circ Z$.*

The proof of admissibility of general contraction is more involved and requires proving several distribution properties among structural connectives. The proof can be found in Appendix A.

Lemma 4.6 (Admissibility of general contraction) *For any structure Y: if $\vdash_{\mathbf{DBiKt}} \pi : \Sigma[Y, Y]$ then $\vdash_{\mathbf{DBiKt}} \pi' : \Sigma[Y]$.*

Once all structural rules of **LBiKt** are shown admissible in **DBiKt**, it is easy to show that every derivation in **LBiKt** can be translated to a derivation in **DBiKt**.

Theorem 4.7 *For any X and Y, if $\vdash_{\mathbf{LBiKt}} \pi : X \triangleright Y$ then $\vdash_{\mathbf{DBiKt}} \pi' : X \triangleright Y$.*

Corollary 4.8 *For any X and Y, $\vdash_{\mathbf{LBiKt}} \pi : X \triangleright Y$ if and only if $\vdash_{\mathbf{DBiKt}} \pi' : X \triangleright Y$.*

Proof. By Theorems 4.1 and 4.7. $\dashv$

5 Proof Search

In this section we outline a proof search strategy for **DBiKt**, closely following the approaches presented in [20] and [10]. Here we emphasize the aspects that are new/different because of the interaction between the tense structures $\circ$ and $\bullet$ and the intuitionistic structure $\triangleright$.

Our backward proof search strategy proceeds in three stages: saturation, propagation and *realisation*. The saturation phase applies the "static rules" (i.e. those that do not create extra structural connectives) until further backward application do not lead to any progress. The propagation phase propagates formulaes across different structural connectives, while the realisation phase applies the "dynamic rules" (i.e., those that create new structural connectives, e.g., $\rightarrow_R$).

A context $\Sigma[\,]$ is said to be *headed by a structural connective* $\#$ if the topmost symbol in the formation tree of $\Sigma[\,]$ is $\#$. A context $\Sigma[\,]$ is said to be a *factor* of $\Sigma'[\,]$ if $\Sigma[\,]$ is a subcontext of $\Sigma'[\,]$ and $\Sigma[\,]$ is headed by either $\triangleright$, $\circ$ or $\bullet$. We write $\widehat{\Sigma}[\,]$ to denote the smallest factor of $\Sigma[\,]$. We write $\widehat{\Sigma}[X]$ to denote the structure $\Sigma_1[X]$, if $\Sigma_1[\,] = \widehat{\Sigma}[\,]$. We define the *top-level* formulae of a structure as:

$$\{\!|X|\!\} = \{A \mid X = (A, Y) \text{ for some } A \text{ and } Y\}.$$

For example, if $\Sigma[] = (A, B \triangleright C, \bullet(D, (E \triangleright F) \triangleright []))$, then $\widehat{\Sigma}[G] = (D, (E \triangleright F) \triangleright G)$, and $\{\!|D, (E \triangleright F)|\!\} = \{D\}$.

Let $-\!\!<_{L1}$ and $\rightarrow_{R1}$ denote two new derived rules (see [20] for their derivation):

$$\frac{\Sigma^-[A, A -\!\!< B]}{\Sigma^-[A -\!\!< B]} -\!\!<_{L1} \qquad \frac{\Sigma^+[A \rightarrow B, B]}{\Sigma^+[A \rightarrow B]} \rightarrow_{R1}$$

We now define a notion of a *saturated structure*, which is similar to that of a traditional sequent. Note that we need to define it for both structures headed by $\triangleright$ and those headed by $\circ$ or $\bullet$. A structure $X \triangleright Y$ is *saturated* if it satisfies the following:

(1) $\{\!|X|\!\} \cap \{\!|Y|\!\} = \emptyset$

(2) If $A \wedge B \in \{\!|X|\!\}$ then $A \in \{\!|X|\!\}$ and $B \in \{\!|X|\!\}$

(3) If $A \wedge B \in \{\!|Y|\!\}$ then $A \in \{\!|Y|\!\}$ or $B \in \{\!|Y|\!\}$

(4) If $A \vee B \in \{\!|X|\!\}$ then $A \in \{\!|X|\!\}$ or $B \in \{\!|X|\!\}$

(5) If $A \vee B \in \{\!|Y|\!\}$ then $A \in \{\!|Y|\!\}$ and $B \in \{\!|Y|\!\}$

(6) If $A \rightarrow B \in \{\!|X|\!\}$ then $A \in \{\!|Y|\!\}$ or $B \in \{\!|X|\!\}$

(7) If $A -\!\!< B \in \{\!|Y|\!\}$ then $A \in \{\!|Y|\!\}$ or $B \in \{\!|X|\!\}$

(8) If $A -\!\!< B \in \{\!|X|\!\}$ then $A \in \{\!|X|\!\}$

(9) If $A \to B \in \{\!|Y|\!\}$ then $B \in \{\!|Y|\!\}$

For structures of the form $\circ X$ or $\bullet X$, we need to define two notions of saturation, *left saturation* and *right saturation*. The former is used when $\circ X$ is nested in a negative context, and the latter when it is in a positive context. A structure $\circ X$ or $\bullet X$ is *left-saturated* if it satisfies (2), (4), (8) above, and

6' If $A \to B \in \{\!|X|\!\}$ then $B \in \{\!|X|\!\}$.

Dually, $\circ Y$ or $\bullet Y$ is *right-saturated* if it satisfies (3), (5), (9) above, and

7' If $A \prec B \in \{\!|Y|\!\}$ then $A \in \{\!|Y|\!\}$.

We define structure membership for any two structures X and Y as follows: $X \in Y$ iff $Y = X, X'$ for some X', modulo associativity and commutativity of comma. For example, $(A \triangleright B) \in (A, (A \triangleright B), \circ C)$. The *realisation of formulae* by a structure X is defined as follows:

* $A \to B$ ($A \prec B$, resp.) is right-realised (resp. left-realised) by X iff there exists $Z \triangleright W \in X$ such that $A \in \{\!|Z|\!\}$ and $B \in \{\!|W|\!\}$.
* $\Box A$ ($\Diamond A$ resp.) is right-realised (resp. left-realised) by X iff there exists $\circ(Z \triangleright W) \in X$ or $\circ W \in X$ (resp. $\circ(W \triangleright Z) \in X$ or $\circ W \in X$) such that $A \in \{\!|W|\!\}$.
* $\blacksquare A$ ($\blacklozenge A$ resp.) is right-realised (resp. left-realised) by X iff there exists $\bullet(W \triangleright Z) \in X$ or $\bullet Z \in X$ (resp. $\bullet(Z \triangleright W) \in X$ or $\bullet Z \in X$) such that $A \in \{\!|Z|\!\}$.

We say that a structure X is left-realised iff every formula in $\{\!|X|\!\}$ with top-level connective $\prec$, $\Diamond$ or $\blacklozenge$ is left-realised by X. Right-realisation of X is defined dually. We say that a structure occurrence X in $\Sigma[X]$ is *propagated* iff no propagation rules are (backwards) applicable to any formula occurrences in X. We define the super-set relation on structures as follows:

* $X_1 \triangleright Y_1 \supset X_0 \triangleright Y_0$ iff $\{\!|X_1|\!\} \supset \{\!|X_0|\!\}$ or $\{\!|Y_1|\!\} \supset \{\!|Y_0|\!\}$.
* $\circ X \supset \circ Y$ iff $\bullet X \supset \bullet Y$ iff $\{\!|X|\!\} \supset \{\!|Y|\!\}$.

To simplify presentation, we use the following terminology: Given a structure $\Sigma[A]$, we say that $\widehat{\Sigma}[A]$ is saturated if $\widehat{\Sigma}[A]$ is $X \triangleright Y$ and it is saturated; or $\widehat{\Sigma}[A]$ is either $\circ X$ or $\bullet X$ and it is either left- or right-saturated (depending on its position in $\Sigma[A]$). We say that $\widehat{\Sigma}[A]$ is propagated if its occurrence in $\Sigma[A]$ is propagated, and we say that A is realised by $\widehat{\Sigma}[A]$, if either

* $\widehat{\Sigma}[A] = (X \triangleright Y)$ and either $A \in \{\!|X|\!\}$ is left-realised by X, or $A \in \{\!|Y|\!\}$ is right realised by Y; or
* $\widehat{\Sigma}[A]$ is either $\circ X$ or $\bullet X$, and, depending on the polarity of $\Sigma[\ ]$, A is either left- or right-realised by X.

We now outline an approach to proof search in **DBiKt**. We approach this by modifying **DBiKt** to obtain a calculus $\mathbf{DBiKt_1}$ that is more amenable to proof search. Our approach follows that of our previous work on bi-intuitionistic logic [20] since we define syntactic restrictions on rules to enforce a search strategy. For example, we stipulate that a structure must be saturated and propagated before child structures can be created using the $\to_R$ rule (see condition ii of Definition 5.1). Additionally and more importantly, our proof search calculus addresses the issue that some modal propagation rules of **DBiKt**, e.g. $\Box_{L2}$, create $\triangleright$-structures during backward proof search. This property of **DBiKt** is undesirable and gives rise to non-termination if rules like $\Box_{L2}$ are applied naively.

Definition 5.1 Let **DBiKt**$_1$ be the system obtained from **DBiKt** with the following changes:

(i) Add the derived rules $\prec_{L1}$ and $\to_{R1}$.

(ii) Restrict rules $\prec_L$, $\to_R$ with the following condition: the rule is applicable only if $\widehat{\Sigma}[A\#B]$ is saturated and propagated, and $A\#B$ is not realised by $\widehat{\Sigma}[A\#B]$, for $\# \in \{\to, \prec\}$.

(iii) Replace rules $\triangleright_{L1}$ and $\triangleright_{R1}$ with the following:

$$\frac{\Sigma[A, (A, X \triangleright Y), W \triangleright Z]}{\Sigma[(A, X \triangleright Y), W \triangleright Z]}\triangleright_{L1} \qquad \frac{\Sigma[W \triangleright Z, (X \triangleright Y, A), A]}{\Sigma[W \triangleright Z, (X \triangleright Y, A)]}\triangleright_{R1}$$

(iv) Restrict rules $\triangleright_{L2}$ and $\triangleright_{R2}$ with the following condition: the rule is applicable only if $A \notin \{\!|Y|\!\}$.

(v) Replace rules $\Diamond_L$, $\Box_R$, $\blacklozenge_L$, $\blacksquare_R$ with the following, where the rule is applicable only if $\widehat{\Sigma}[\#A]$ is saturated and propagated and $\#A$ is not realised by$\widehat{\Sigma}[\#A]$, for $\# \in \{\Diamond, \Box, \blacklozenge, \blacksquare\}$:

$$\frac{\Sigma^-[\Diamond A, \circ(A \triangleright \emptyset)]}{\Sigma^-[\Diamond A]}\Diamond_L \qquad \frac{\Sigma^+[\Box A, \circ(\emptyset \triangleright A)]}{\Sigma^+[\Box A]}\Box_R$$

$$\frac{\Sigma^-[\blacklozenge A, \bullet(A \triangleright \emptyset)]}{\Sigma^-[\blacklozenge A]}\blacklozenge_L \qquad \frac{\Sigma^+[\blacksquare A, \bullet(\emptyset \triangleright A)]}{\Sigma^+[\blacksquare A]}\blacksquare_R$$

(vi) Replace rules $\blacksquare_{L2}, \Box_{L2}$ with the following, where $A \notin \{\!|Y_1|\!\}$:

$$\frac{\Sigma[\blacksquare A, X \triangleright \bullet(A, Y_1 \triangleright Y_2), Z]}{\Sigma[\blacksquare A, X \triangleright \bullet(Y_1 \triangleright Y_2), Z]}\blacksquare_{L2} \qquad \frac{\Sigma[\Box A, X \triangleright \circ(A, Y_1 \triangleright Y_2), Z]}{\Sigma[\Box A, X \triangleright \circ(Y_1 \triangleright Y_2), Z]}\Box_{L2}$$

(vii) Replace rules $\Diamond_{R2}, \blacklozenge_{R2}$ with the following, where $A \notin \{\!|X_2|\!\}$:

$$\frac{\Sigma[\circ(X_1 \triangleright X_2, A), Y \triangleright Z, \Diamond A]}{\Sigma[\circ(X_1 \triangleright X_2), Y \triangleright Z, \Diamond A]}\Diamond_{R2} \qquad \frac{\Sigma[\bullet(X_1 \triangleright X_2, A), Y \triangleright Z, \blacklozenge A]}{\Sigma[\bullet(X_1 \triangleright X_2), Y \triangleright Z, \blacklozenge A]}\blacklozenge_{R2}$$

(viii) Replace rules $\Box_{L1}, \Diamond_{R1}, \blacksquare_{L1}, \blacklozenge_{R1}$ with the following:

$$\frac{\Sigma^-[A, \bullet(\Box A, X \triangleright Y)]}{\Sigma^-[\bullet(\Box A, X \triangleright Y)]}\Box_{L1} \qquad \frac{\Sigma^+[A, \bullet(Y \triangleright \Diamond A, X)]}{\Sigma^+[\bullet(Y \triangleright \Diamond A, X)]}\Diamond_{R1}$$

$$\frac{\Sigma^-[A, \circ(\blacksquare A, X \triangleright Y)]}{\Sigma^-[\circ(\blacksquare A, X \triangleright Y)]}\blacksquare_{L1} \qquad \frac{\Sigma^+[A, \circ(Y \triangleright \blacklozenge A, X)]}{\Sigma^+[\circ(Y \triangleright \blacklozenge A, X)]}\blacklozenge_{R1}$$

(ix) Replace rules $\to_L$, $\prec_R$ with the following:

$$\frac{\Sigma[X, A \to B \rhd A, Y] \qquad \Sigma[X, A \to B, B \rhd Y]}{\Sigma[X, A \to B \rhd Y]} \to_L$$

$$\frac{\Sigma[X \rhd Y, A \prec B, A] \qquad \Sigma[X, B \rhd Y, A \prec B]}{\Sigma[X \rhd Y, A \prec B]} \prec_R$$

(x) Restrict rules $\to_L$, $\prec_R$, $\rhd_{L1}$, $\rhd_{R1}$, $\wedge_L$, $\wedge_R$, $\vee_L$, $\vee_R$ and all modal propagation rules to the following: Let $\Sigma[X_0]$ be the conclusion of the rule and let $\Sigma[X_1]$ (and $\Sigma[X_2]$) be the premise(s). The rule is applicable only if: $\widehat{\Sigma}[X_1] \supset \widehat{\Sigma}[X_0]$ and $\widehat{\Sigma}[X_2] \supset \widehat{\Sigma}[X_0]$.

We conjecture that **DBiKt** and **DBiKt$_1$** are equivalent and that backward proof search in **DBiKt** terminates. Note that by equivalence here we mean that **DBiKt** and **DBiKt** proves the same set of formulae, but not necessarily the same set of structures. This is because the propagation rules in **DBiKt$_1$** are more restricted so as to allow for easier termination checking. For example, the structure $A \rhd \bullet(\Diamond A)$ is derivable in **DBiKt** but not in **DBiKt$_1$**, although its formula translate is derivable in both systems. It is likely that a combination of the techniques from [20] and [10] can be used to prove termination of proof search in **DBiKt$_1$**, given its similarities to the deep inference systems used in those two works.

6 Semantics

We now give a Kripke-style semantics for BiKt and show that **LBiKt** is sound with respect to the semantics. Our semantics for BiKt extend Rauszer's [22] Kripke-style semantics for BiInt by clauses for the tense logic connectives. We use the classical first-order meta-level connectives &, "or", "not", $\Rightarrow$, $\forall$ and $\exists$ to state our semantics.

A Kripke *frame* is a tuple $\langle W, \leq, R_\Diamond, R_\Box \rangle$ where W is a non-empty set of worlds and $\leq \subseteq (W \times W)$ is a reflexive and transitive binary relation over W, and each of $R_\Diamond$ and $R_\Box$ are arbitrary binary relations over W with the following *frame conditions*:

F1$\Diamond$ if $x \leq y$ & $xR_\Diamond z$ then $\exists w.\ yR_\Diamond w$ & $z \leq w$

F2$\Box$ if $xR_\Box y$ & $y \leq z$ then $\exists w.\ x \leq w$ & $wR_\Box z$.

A Kripke *model* extends a Kripke frame with a mapping V from *Atoms* to 2^W obeying *persistence:* $\forall v \geq w.\ w \in V(p) \Rightarrow v \in V(p)$. Given a model $\langle W, \leq, R_\Diamond, R_\Box, V \rangle$, we say that $w \in W$ *satisfies* p if $w \in V(p)$, and write this as $w \Vdash p$. We write $w \not\Vdash p$ to mean $(not)(w \Vdash p)$; that is, $\exists v \geq w.\ v \notin V(p)$. The relation $\Vdash$ is then extended to formulae as given in Figure 5. A BiKt-formula A is *BiKt-valid* if it is satisfied by every world in every Kripke model. A nested sequent $X \rhd Y$ is BiKt-valid if its formula translation is BiKt-valid.

Our semantics differ from those of Simpson [24] and Ewald [6] because we use two modal accessibility relations instead of one. In our calculi, there is no direct relationship between $\Diamond$ and $\Box$ (or $\blacklozenge$ and $\blacksquare$), but $\Diamond$ and $\blacksquare$ are a residuated pair, as are $\blacklozenge$ and $\Box$. Semantically, this corresponds to $R_\blacklozenge = R_\Box^{-1}$ and $R_\blacksquare = R_\Diamond^{-1}$; therefore the clauses in

$w \Vdash \top$		for every $w \in W$	$w \Vdash \bot$		for no $w \in W$
$w \Vdash A \land B$	if	$w \Vdash A$ & $w \Vdash B$	$w \Vdash A \lor B$	if	$w \Vdash A$ or $w \Vdash B$
$w \Vdash A \to B$	if	$\forall v \geq w.\ v \Vdash A \Rightarrow v \Vdash B$	$w \Vdash \neg A$	if	$\forall v \geq w.\ v \nVdash A$
$w \Vdash A \prec B$	if	$\exists v \leq w.\ v \Vdash A$ & $v \nVdash B$	$w \Vdash \sim A$	if	$\exists v \leq w.\ v \nVdash A$
$w \Vdash \Diamond A$	if	$\exists v.\ wR_\Diamond v$ & $v \Vdash A$	$w \Vdash \Box A$	if	$\forall z.\forall v.\ w \leq z$ & $zR_\Box v \Rightarrow v \Vdash A$
$w \Vdash \blacklozenge A$	if	$\exists v.\ wR_\Box^{-1} v$ & $v \Vdash A$	$w \Vdash \blacksquare A$	if	$\forall z.\forall v.\ w \leq z$ & $zR_\Diamond^{-1} v \Rightarrow v \Vdash A$

Fig. 5. Semantics for BiKt

Figure 5 are couched in terms of $R_\Diamond$ and $R_\Box$ only. Our frame conditions F1$\Diamond$ and F2$\Box$ are also used by Simpson whose F2 captures the "persistence of being seen by" [24, page 51] while for us F2$\Box$ is simply the "persistence of $\blacklozenge$".

LBiKt is sound with respect to BiKt. The soundness proof is straightforward by the definition of the semantics and the inference rules.

Theorem 6.1 (Soundness) *If A is a BiKt-formula and $\emptyset \triangleright A$ is* **LBiKt***-derivable, then A is BiKt-valid.*

We conjecture that **DBiKt** is also complete w.r.t. the semantics; an outline of the proof will be given in the extended version of the paper.

7 Modularity, Extensions and Classicality

We first exhibit the modularity of our deep calculus **DBiKt** by showing that fragments of **DBiKt** obtained by restricting the language of formulae and structures also satisfy cut admissibility. We then show how we can obtain Ewald's intuitionistic tense logic IKt [6], Simpson's intuitionistic modal logic IK [24] and regain classical tense logic Kt. We also discuss extensions of **DBiKt** with axioms T, 4 and B but they do not correspond semantically to reflexivity, transitivity and symmetry [24].

Modularity

A nested sequent is *purely modal* if contains no occurrences of $\bullet$ nor its formula translates $\blacksquare$ and $\blacklozenge$. We write **DInt** for the sub-system of **DBiKt** containing only the rules *id*, the logical rules for intuitionistic connectives, and the propagation rules for $\triangleright$. The logical system **DIntK** is obtained by adding to **DInt** the deep introduction rules for $\Box$ and $\Diamond$, and the propagation rules $\Box_{L2}$ and $\Diamond_{R2}$. The logical system **DBInt** is obtained by adding to **DInt** the deep introduction rules for $\prec$. In the following, we say that a formula is an *IntK-formula* if it is composed from propositional variables, intuitionistic connectives, and $\Box$ and $\Diamond$. Observe that in **DBiKt**, the only rules that create $\bullet$ upwards are $\blacklozenge_L$ and $\blacksquare_R$. Thus in every **DBiKt**-derivation π of an IntK formula, the internal sequents in π are purely modal, and hence π is also a **DIntK**-derivation. This observation gives immediately the following modularity result.

Theorem 7.1 (Modularity) *Let A be an Int (resp. BiInt and IntK) formula. The nested sequent $\emptyset \triangleright A$ is* **DInt***-derivable (resp.* **DBInt**- *and* **DIntK***-derivable) iff $\emptyset \triangleright A$ is* **DBiKt***-derivable.*

A consequence of Theorem 3.5, Theorem 4.1, Theorem 4.7 and Theorem 7.1, is that the cut rule is admissible in **DInt**, **DBInt** and **DIntK**. As the semantics of **LBiKt** (hence, also **DBiKt**) is conservative w.r.t. to the semantics of both intuitionistic and bi-intuitionistic logic, the following completeness result holds.

Theorem 7.2 *An Int (resp. BiInt) formula A is valid in Int (resp. BiInt) iff $\emptyset \rhd A$ is derivable in* **DInt** *(resp.* **DBInt***).*

Obtaining Ewald's IKt

To obtain Ewald's IKt [6] we need to collapse $R_\Diamond$ and $R_\Box$ into one temporal relation R and leave out our semantic clauses for $\prec$ and $\sim$. That is, we need to add the following conditions to the basic semantics: $R_\Diamond \subseteq R_\Box$ and $R_\Box \subseteq R_\Diamond$. Proof theoretically, this is captured by extending **LBiKt** with the structural rules:

$$\frac{X \rhd \bullet Y \rhd \bullet Z}{X \rhd \bullet(Y \rhd Z)}\ \bullet\!\rhd_R \qquad \frac{X \rhd \circ Y \rhd \circ Z}{X \rhd \circ(Y \rhd Z)}\ \circ\!\rhd_R$$

We refer to the extension of **LBiKt** with these two structural rules as **LBiKtE**.

Simpson's intuitionistic modal logic IK [24] can then be obtained from Ewald's system by restricting the language to the modal fragment. Note that cut-elimination still holds for **LBiKtE** because these structural rules are closed under formula substitution and the cut-elimination proof for **LBiKt** still goes through when additional structural rules of this kind are added. We refer the reader to [10] for a discussion on how cut elimination can be proved for this kind of extension.

A BiKt-frame is an *E-frame* if $R_\Box = R_\Diamond$. A formula A is *E-valid* if it is true in all worlds of every E-model. An IKt formula A is a theorem of IKt iff it is E-valid [6]. The rules $\circ\!\rhd_R$ and $\bullet\!\rhd_R$ are sound for E-frames. The proofs of the following lemmas can be found in Appendix B.

Lemma 7.3 *Rule $\circ\!\rhd_R$ is sound iff $R_\Box \subseteq R_\Diamond$.*

Lemma 7.4 *Rule $\bullet\!\rhd_R$ is sound iff $R_\Diamond \subseteq R_\Box$.*

Theorem 7.5 *If A is derivable in* **LBiKtE** *then A is E-valid.*

Proof. Straightforward from the soundness of **LBiKt** w.r.t. BiKt-semantics (which subsumes Ewald's semantics) and Lemma 7.3 and Lemma 7.4. ⊣

Completeness of **LBiKtE** w.r.t. IKt and IK can be shown by deriving the axioms of IKt and IK. The completeness proof will be given in the extended version of the paper.

Theorem 7.6 *System* **LBiKtE** *is complete w.r.t. Ewald's IKt and Simpson's IK.*

Theorem 7.7 (Conservativity over IKt and IK) *If A is an IKt-formula (IK formula), then A is IKt-valid (IK-valid) iff $\emptyset \rhd A$ is derivable in* **LBiKtE**.

$$\frac{\Sigma^-[A, \Box A]}{\Sigma^-[\Box A]}\ T\Box \qquad \frac{\Sigma[\Box A, X \rhd \circ(\Box A \rhd Y), Z]}{\Sigma[\Box A, X \rhd \circ Y, Z]}\ 4\Box_L \qquad \frac{\Sigma^-[A, \circ(\Box A, X)]}{\Sigma^-[\circ(\Box A, X)]}\ B\Box_L$$

$$\dfrac{\dfrac{\dfrac{}{p, \Box p \rhd p}\ id}{\Box p \rhd p}\ T\Box}{\rhd \Box p \to p}\ \to_R \qquad \dfrac{\dfrac{\dfrac{\dfrac{}{\Box p \rhd \circ(\Box p \rhd \Box p)}\ id}{\Box p \rhd \circ \Box p}\ 4\Box_L}{\Box p \rhd \Box\Box p}\ \Box_R}{\rhd \Box p \to \Box\Box p}\ \to_R \qquad \dfrac{\dfrac{\dfrac{\dfrac{}{p, \circ\Box p \rhd p}\ id}{\circ \Box p \rhd p}\ B\Box_L}{\Diamond\Box p \rhd p}\ \Diamond_L}{\rhd \Diamond\Box p \to p}\ \to_R$$

Fig. 6. Some example propagation rules and the axioms they capture

Regaining classical tense logic Kt

To collapse BiKt to classical tense logic Kt we add the rules $\bullet\rhd_R$ and $\circ\rhd_L$, giving Ewald's IKt with $R_\Diamond = R_\Box$ via Lemmas 7.3-7.4, and then add following two rules:

$$\frac{X_1, X_2 \rhd Y_1, Y_2}{(X_1 \rhd Y_1), X_2 \rhd Y_2}\ s_L^{-1} \qquad \frac{X_1, X_2 \rhd Y_1, Y_2}{X_1 \rhd Y_1, (X_2 \rhd Y_2)}\ s_R^{-1}$$

The law of the excluded middle and the law of (dual-)contradiction can then be derived as shown below:

$$\dfrac{\dfrac{\dfrac{\dfrac{p \rhd p, \bot}{(\emptyset \rhd p), p \rhd \bot}\ s_L^{-1}}{(\emptyset \rhd p) \rhd (p \to \bot)}\ \to_R}{\emptyset \rhd p, (p \to \bot)}\ s_L}{\emptyset \rhd p \vee (p \to \bot)}\ \vee_L \qquad \dfrac{\dfrac{\dfrac{\dfrac{p, \top \rhd p}{\top \rhd p, (p \rhd \emptyset)}\ s_R^{-1}}{(\top \prec p) \rhd (p \rhd \emptyset)}\ \prec_L}{p, (\top \prec p) \rhd \emptyset}\ s_R}{p \wedge (\top \prec p) \rhd \emptyset}\ \wedge_R$$

Further extensions

Our previous work on deep inference systems for classical tense logic [10] shows that extensions of classical tense logic with some standard modal axioms can be formalised by adding numerous propagation rules to the deep inference system for classical tense logic given in that paper. We illustrate here with a few examples how such an approach to extensions with modal axioms can be applied to BiKt. Figure 6 shows the propagation rules that are needed to derive axiom T, 4 and B. For each rule, the derivation of the corresponding axiom is given below the rule. Other nesting combinations will be needed for full completeness. Dual rules allow derivations of $p \to \Diamond p$ and $\Diamond\Diamond p \to \Diamond p$. The complete treatement of these and other possible extensions of **LBiKt** is left for future work.

References

[1] G. Amati and F. Pirri. A uniform tableau method for intuitionistic modal logics i. *Studia Logica*, 53(1):29–60, 1994.

[2] K Brünnler and L Straßburger Modular Sequent Systems for Modal Logic In *Proc. TABLEAUX*, LNCS:5607;152-166. Springer, 2009.

[3] M J Collinson, B. Hilken and D. Rydeheard. Semantics and proof theory of an intuitionistic modal sequent calculus. Technical report, University of Manchester, UK, 1999.

[4] T. Crolard. A formulae-as-types interpretation of Subtractive Logic. *J. of Logic and Comput.*, 14(4):529–570, 2004.

[5] R. Davies and F. Pfenning. A modal analysis of staged computation. *J. ACM*, 48(3):555–604, 2001.
[6] W. B. Ewald. Intuitionistic tense and modal logic. *J. Symb. Log*, 51(1):166–179, 1986.
[7] D. Galmiche and Y. Salhi. Calculi for an intuitionistic hybrid modal logic. In *Proc. IMLA*, 2008.
[8] R. Goré. Substructural logics on display. *Log. J of Interest Group in Pure and Applied Logic*, 6(3):451–504, 1998.
[9] R. Goré, L. Postniece, and A. Tiu. Cut-elimination and proof-search for bi-intuitionistic logic using nested sequents. In *Proc. AiML* 7:43–66. College Publications, 2008.
[10] R. Goré, L. Postniece, and A. Tiu. Taming displayed tense logics using nested sequents with deep inference. In *Proc. TABLEAUX*, LNCS:5607;189–204. Springer, 2009.
[11] R. Goré, L. Postniece, and A. Tiu. Cut-elimination and proof search for bi-intuitionistic tense logic. *CoRR*, abs/1006.4793, 2010.
[12] Y. Kakutani. Calculi for intuitionistic normal modal logic. In *Proceedings of PPL 2007*.
[13] R. Kashima. Cut-free sequent calculi for some tense logics. *Studia Logica*, 53:119–135, 1994.
[14] A. Masini. 2-sequent calculus: Intuitionism and natural deduction. *J. Log. Comput.*, 3(5):533–562, 1993.
[15] G Mints. On some calculi of modal logic. Proc. Steklov Inst. of Mathematics, 98:97-122, 1971.
[16] T. Murphy VII, K. Crary, R. Harper, and F. Pfenning. A symmetric modal lambda calculus for distributed computing. In *LICS*, pages 286–295, 2004.
[17] F. Pfenning and H.-C. Wong. On a modal lambda calculus for S4. *Electr. Notes Theor. Comput. Sci.*, 1, 1995.
[18] L. Pinto and T. Uustalu. Proof search and counter-model construction for bi-intuitionistic propositional logic with labelled sequents. In *TABLEAUX*, pages 295–309, 2009.
[19] F Poggiolesi. The Tree-hypersequent Method for Modal Propositional Logic. Trends in Logic: Towards Mathematical Philosophy, pp 9–30, Springer, 2009.
[20] L. Postniece. Deep inference in bi-intuitionistic logic. In *Proc. WoLLIC*, LNCS 5514:320–334. Springer, 2009.
[21] C. Rauszer. A formalization of the propositional calculus of H-B logic. *Studia Logica*, 33:23–34, 1974.
[22] C. Rauszer. An algebraic and Kripke-style approach to a certain extension of intuitionistic logic. *Dissertationes Mathematicae*, 168, 1980.
[23] M. Sadrzadeh and R. Dyckhoff. Positive logic with adjoint modalities: Proof theory, semantics and reasoning about information. *Electr. Notes in TCS*, 249:451–470, 2009.
[24] A. K. Simpson. *The proof theory and semantics of intuitionistic modal logic*. PhD thesis, Univ. of Edinburgh, 1994.
[25] H. Wansing. Sequent calculi for normal modal proposisional logics. *J. Logic and Computation*, 4(2):125–142, Apr. 1994.

A Admissibility of general contraction

In the following, we label a dashed line with the lemma used to obtain the conclusion from the premise.

Lemma A.1 *If* $\vdash_{\mathbf{DBiKt}} \pi : \Sigma^+[\circ(X \rhd Y), \circ Y]$ *and contraction on structures is admissible for all derivations* π_1 *such that* $|\pi_1| \leq |\pi|$ *then* $\vdash_{\mathbf{DBiKt}} \pi' : \Sigma^+[\circ(X \rhd Y)]$.

Proof. By induction on the height of π. The interesting cases are when π ends with a propagation rule that moves a formula into either $\circ(X \rhd Y)$ or $\circ Y$:

* Suppose π ends as below left. Then by Lemma 4.4(ii), there is a derivation π_2 of $\Box A \rhd \circ(A, X \rhd Y), \circ Y$ such that $|\pi_2| \leq |\pi_1|$. Then we can apply the induction hypothesis to π_2 to obtain a derivation π_3 of $\Box A \rhd \circ(A, X \rhd Y)$. Then the derivation below right gives the required:

$$\dfrac{\begin{matrix}\pi_1\\ \Box A \rhd \circ(A \rhd (X \rhd Y)), \circ Y\end{matrix}}{\Box A \rhd \circ(X \rhd Y), \circ Y}\ \Box_{L2} \qquad \dfrac{\dfrac{\begin{matrix}\pi_3\\ \Box A \rhd \circ(A, X \rhd Y)\end{matrix}}{\Box A \rhd \circ(A \rhd (X \rhd Y))}\ \text{Lemma 4.4(iv)}}{\Box A \rhd \circ(X \rhd Y)}\ \Box_{L2}$$

* Suppose π ends as below left. Then applying Lemma 4.2 twice, we obtain a derivation π_2 of $\Box A \rhd \circ(A, X \rhd Y), \circ(A, X \rhd Y)$ such that $|\pi_2| \leq |\pi_1|$. Then we apply the assumption of this lemma to π_2 to obtain a derivation π_3 of $\Box A \rhd \circ(A, X \rhd Y)$. Then the derivation below right gives the required:

$$\dfrac{\begin{matrix}\pi_1 \\ \Box A \rhd \circ(X \rhd Y), \circ(A \rhd Y)\end{matrix}}{\Box A \rhd \circ(X \rhd Y), \circ Y}\ \Box_{L2} \qquad \dfrac{\dfrac{\begin{matrix}\pi_3 \\ \Box A \rhd \circ(A, X \rhd Y)\end{matrix}}{\Box A \rhd \circ(A \rhd (X \rhd Y))}\ \text{Lemma 4.4(iv)}}{\Box A \rhd \circ(X \rhd Y)}\ \Box_{L2}$$

⊣

Lemma A.2 *If* $\vdash_{\mathbf{DBiKt}} \pi : \Sigma^-[\circ(X \rhd Y), \circ X]$ *then* $\vdash_{\mathbf{DBiKt}} \pi' : \Sigma^-[\circ(X \rhd Y)]$.

Proof. By induction on the height of π. The interesting cases are when π ends with a propagation rule that moves a formula into either $\circ(X \rhd Y)$ or $\circ X$:

* Suppose π ends as below left. Then by Lemma 4.4(i), there is a derivation π_2 of $\circ(X \rhd Y, A), \circ X \rhd \Diamond A$ such that $|\pi_2| \leq |\pi_1|$. Then we can apply the induction hypothesis to π_2 to obtain a derivation π_3 of $\circ(X \rhd Y, A) \rhd \Diamond A$. Then the derivation below right gives the required:

$$\dfrac{\begin{matrix}\pi_1 \\ \circ((X \rhd Y) \rhd A), \circ X \rhd \Diamond A\end{matrix}}{\circ(X \rhd Y), \circ X \rhd \Diamond A}\ \Diamond_{R2} \qquad \dfrac{\dfrac{\begin{matrix}\pi_3 \\ \circ(X \rhd Y, A) \rhd \Diamond A\end{matrix}}{\circ((X \rhd Y) \rhd A) \rhd \Diamond A}\ \text{Lemma 4.4(iii)}}{\circ(X \rhd Y) \rhd \Diamond A}\ \Diamond_{R2}$$

* Suppose π ends as below left. Then applying Lemma 4.2 twice, we obtain a derivation π_2 of $\circ(X \rhd Y, A), \circ(X \rhd Y, A) \rhd \Diamond A$ such that $|\pi_2| \leq |\pi_1|$. Then we apply the assumption of this lemma to π_2 to obtain a derivation π_3 of $\circ(X \rhd Y, A) \rhd \Diamond A$. Then the derivation below right gives the required:

$$\dfrac{\begin{matrix}\pi_1 \\ \circ(X \rhd Y), \circ(X \rhd A) \rhd \Diamond A\end{matrix}}{\circ(X \rhd Y), \circ X \rhd \Diamond A}\ \Diamond_{R2} \qquad \dfrac{\dfrac{\begin{matrix}\pi_3 \\ \circ(X \rhd Y, A) \rhd \Diamond A\end{matrix}}{\circ((X \rhd Y) \rhd A) \rhd \Diamond A}\ \text{Lemma 4.4(iii)}}{\circ(X \rhd Y) \rhd \Diamond A}\ \Diamond_{R2}$$

⊣

Lemma A.3 *If* $\vdash_{\mathbf{DBiKt}} \pi : \Sigma^+[\bullet(X \rhd Y), \bullet(Y)]$ *then* $\vdash_{\mathbf{DBiKt}} \pi' : \Sigma^+[\bullet(X \rhd Y)]$.

Lemma A.4 *If* $\vdash_{\mathbf{DBiKt}} \pi : \Sigma^-[\bullet(X \rhd Y), \bullet(X)]$ *then* $\vdash_{\mathbf{DBiKt}} \pi' : \Sigma^-[\bullet(X \rhd Y)]$.

Lemma 4.6 (Admissibility of general contraction) For any structure Y: if $\vdash_{\mathbf{DBiKt}}$ $\pi : \Sigma[Y, Y]$ then $\vdash_{\mathbf{DBiKt}} \pi' : \Sigma[Y]$.

Proof. By induction on the size of Y, with a sub-induction on $|\pi|$.

* For the base case, use Lemma 4.3.

* For the case where Y is a $\triangleright$-structure, we show the sub-case where Y in a negative context, the other case is symmetric:

$$
\cfrac{\cfrac{\cfrac{\cfrac{\cfrac{\Sigma[(Y_1 \triangleright Y_2), (Y_1 \triangleright Y_2) \triangleright Z]}{\Sigma[Y_1, (Y_1 \triangleright Y_2) \triangleright Y_2, Z]}\ \text{Lemma 4.4(i)}}{\Sigma[Y_1, Y_1 \triangleright Y_2, Y_2, Z]}\ \text{Lemma 4.4(i)}}{\Sigma[Y_1 \triangleright Y_2, Y_2, Z]}\ \text{IH}}{\Sigma[Y_1 \triangleright Y_2, Z]}\ \text{IH}}{\Sigma[(Y_1 \triangleright Y_2) \triangleright Z]}\ \text{Lemma 4.4(iii)}
$$

* For the case where Y is a $\circ$– or $\bullet$-structure and π ends with a propagation rule applied to Y, there are three non-trivial sub-cases:
  · A formula is propagated into Y and Y is in a positive context, as below left. Then by Lemma A.1, there is a derivation π_1' of $\Box A, X \triangleright \circ(A \triangleright Z)$. Then the derivation below right gives the required:

$$
\cfrac{\begin{array}{c}\pi_1\\ \Box A, X \triangleright \circ(A \triangleright Z), \circ Z\end{array}}{\Box A, X \triangleright \circ Z, \circ Z}\ \Box_{L2}
\qquad
\cfrac{\begin{array}{c}\pi_1'\\ \Box A, X \triangleright \circ(A \triangleright Z)\end{array}}{\Box A, X \triangleright \circ Z}\ \Box_{L2}
$$

  · A formula is propagated into Y and Y is in a negative context, as below left. Then by Lemma A.2, there is a derivation π_1' of $\circ(Z \triangleright A) \triangleright X, \Diamond A$. Then the derivation below right gives the required:

$$
\cfrac{\begin{array}{c}\pi_1\\ \circ(Z \triangleright A), \circ Z \triangleright X, \Diamond A\end{array}}{\circ Z, \circ Z \triangleright X, \Diamond A}\ \Diamond_{R2}
\qquad
\cfrac{\begin{array}{c}\pi_1'\\ \circ(Z \triangleright A) \triangleright X, \Diamond A\end{array}}{\circ Z \triangleright X, \Diamond A}\ \Diamond_{R2}
$$

  · A formula is propagated out of Y, as below left. In this case we use the sub-induction hypothesis to obtain a derivation π_1' of $X \triangleright A, \circ(\blacklozenge A, Z)$. Then the derivation below right gives the required:

$$
\cfrac{\begin{array}{c}\pi_1\\ X \triangleright A, \circ(\blacklozenge A, Z), \circ(\blacklozenge A, Z)\end{array}}{X \triangleright \circ(\blacklozenge A, Z), \circ(\blacklozenge A, Z)}\ \blacklozenge_{R1}
\qquad
\cfrac{\begin{array}{c}\pi_1'\\ X \triangleright A, \circ(\blacklozenge A, Z)\end{array}}{X \triangleright \circ(\blacklozenge A, Z)}\ \blacklozenge_{R1}
$$

⊣

B Modularity, Extensions and Classicality

Theorem 7.6. System **LBiKtE** is complete w.r.t. Ewald's IKt and Simpson's IK.

Proof. We show the non-trivial cases; the rest are similar or easier. Derivations of Simpson's axiom 2 and Ewald's axiom 5 and 7 are given in Figure B.1, derivations of Simpson's axiom 5 and Ewald's axiom 10 and 11' are given in Figure B.2. ⊣

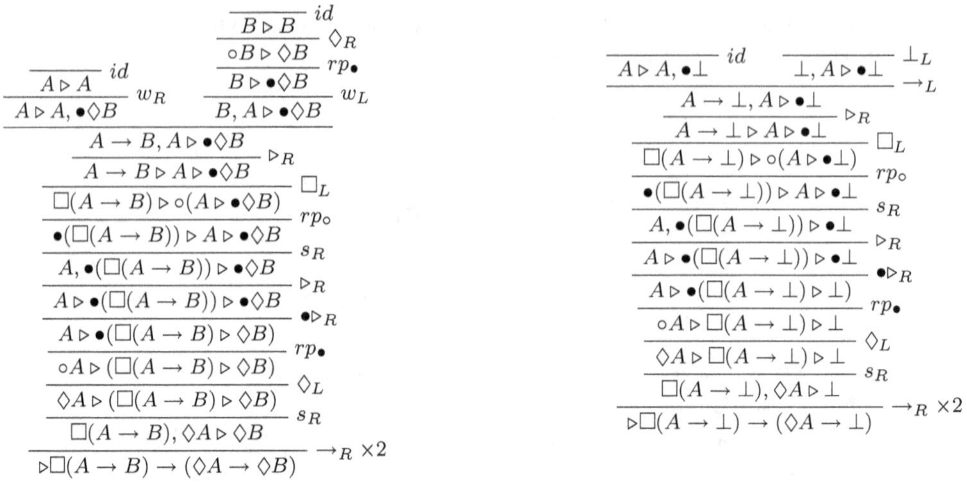

Fig. B.1. Derivations of Simpson's axiom 2 and Ewald's axiom 5 (left) and Ewald's axiom 7 (right)

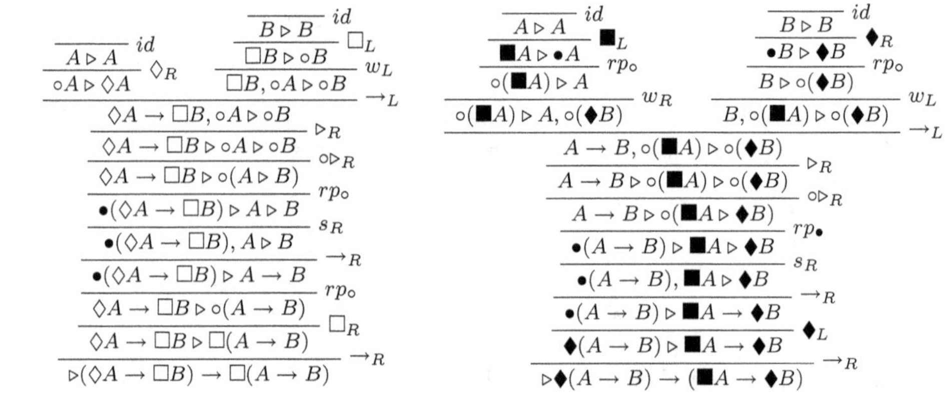

Fig. B.2. Derivations of Simpson's axiom 5 and Ewald's axiom 10 (left) and Ewald's axiom 11' (right)

Lemma 7.3. Rule $\circ\!\triangleright_R$ is sound iff $R_\Box \subseteq R_\Diamond$.

Proof. ($\Leftarrow$) We show that if the frame condition holds, then the rule is sound. We assume that: (1) $R_\Box \subseteq R_\Diamond$, and (2) that the formula translation $\Diamond A \to \Box B$ of the premise is valid. We then show that the formula translation $\Box(A \to B)$ of the conclusion is valid. For a contradiction, suppose that $\Box(A \to B)$ is not valid. That is, there exists a world u such that $u \not\Vdash \Box(p \to q)$. Then (4) there exist worlds x and y such that $u \leq x$ & $xR_\Box y$ and $y \not\Vdash p \to q$. Thus there exists z s.t. $z \geq y$ and $z \Vdash p$ and $z \not\Vdash q$. The pattern $xR_\Box y \leq z$ implies there is a world w with $x \leq wR_\Box z$ by F2$\Box$. The frame condition (1) then gives $wR_\Diamond z$ too, meaning that $w \Vdash \Diamond p$. From (2) we get $w \Vdash \Box q$, which gives us $z \Vdash q$, giving us the contradiction we seek. Therefore the premise $\Box(A \to B)$ is valid and the rule is sound.

($\Rightarrow$) We show that if the rule is sound, then the failure of the frame condition gives a contradiction. So suppose that the rule is sound. The rule implies that $\rhd(\Diamond A \to \Box B) \to \Box(A \to B)$ is derivable. For a contradiction, suppose we have a frame with $R_\Box \not\subseteq R_\Diamond$. That is, (5): there exist x and y such that $xR_\Box y$ but not $xR_\Diamond y$. Let $W = \{u, w, x, y, z\}$, let $<$ be the relation $\{(u, x), (x, w), (y, z)\}$ and let $\leq$ be the reflexive-transitive closure of $<$. Let $R_\Diamond \;= \{\}$, $R_\Box \;= \{(x, y), (w, z)\}$ and let $V(p) \;= \{z\}, V(q) = \{\}$. Then the model $\langle W, \leq, R_\Diamond, R_\Box, V\rangle$ satisfies (5), and has $u \Vdash \Diamond p \to \Box q$ but $u \nVdash \Box(p \to q)$. $\dashv$

Lemma 7.4. Rule $\bullet\rhd_R$ is sound iff $R_\Diamond \subseteq R_\Box$.

Proof. $R_\Diamond \subseteq R_\Box$ means $R_\blacksquare \subseteq R_\blacklozenge$; the rest of the proof is analogous to the proof of Lemma 7.3. $\dashv$

Moorean Phenomena in Epistemic Logic

Wesley H. Holliday and Thomas F. Icard, III

Department of Philosophy
Stanford University
Stanford, California, USA

Abstract

A well-known open problem in epistemic logic is to give a syntactic characterization of the *successful* formulas. Semantically, a formula is successful if and only if for any pointed model where it is true, it remains true after deleting all points where the formula was false. The classic example of a formula that is not successful in this sense is the "Moore sentence" $p \wedge \neg\Box p$, read as "p is true but you do not know p." Not only is the Moore sentence unsuccessful, it is *self-refuting*, for it never remains true as described. We show that in logics of knowledge and belief for a single agent (extended by **S5**), Moorean phenomena are the source of all self-refutation; moreover, in logics for an *introspective* agent (extending **KD45**), Moorean phenomena are the source of all unsuccessfulness as well. This is a distinctive feature of such logics, for with a non-introspective agent or multiple agents, non-Moorean unsuccessful formulas appear. We also consider how successful and self-refuting formulas relate to the *Cartesian* and *learnable* formulas, which have been discussed in connection with Fitch's "paradox of knowability." We show that the Cartesian formulas are exactly the formulas that are not *eventually* self-refuting and that not all learnable formulas are successful. In an appendix, we give syntactic characterizations of the successful and the self-refuting formulas.

Keywords: epistemic logic, successful formulas, Moore sentence

1 Introduction

According to the epistemic interpretation of modal logic, the points in a modal model represent ways the world might be, consistent with an agent's information. In this context, "learning" a formula amounts to eliminating those points in the model where the formula is false. The resulting submodel represents the agent's information state after learning has occurred. Some formulas—though not all—remain true whenever they are learned. A well-known open problem [12,4,5,6,7,2] in epistemic logic is to give a syntactic characterization of these *successful* formulas. Partial results have been obtained (Section 2), but a full solution has proven elusive.

The classic example of an *unsuccessful* formula is the Moore sentence $p \wedge \neg\Box p$, read as "p is true but you do not know p." This example is a second-person variation of

G.E. Moore's famous puzzle [16] involving the paradoxical first-person assertion, "p is true but I do not believe p." Hintikka devoted a chapter of his seminal 1962 monograph *Knowledge and Belief* [13] to an analysis of such sentences, including the second-person Moore sentence. Hintikka observed its unsuccessfulness as follows: "You may come to know that what I said *was* true, but saying it in so many words has the effect of making what is being said false" [13, p. 69]. Yet the formal question of unsuccessfulness did not arise for Hintikka. Only with the advent of Dynamic Epistemic Logic (see, e.g., [7]) and the idea of learning as model reduction have the Moorean phenomena and unsuccessfulness been formally related.

The reason the Moore sentence is unsuccessful is that it can only be true in a model if p is true at some point and false at some other point accessible from the first. When the agent learns the sentence, all points where p is false are eliminated from the model, including all witnesses for $\neg\Box p$, so the sentence becomes false. This shows that the Moore sentence is not only unsuccessful, it is *self-refuting*, for it always becomes false when learned. Related to the self-refuting property of the Moore sentence is the fact that the sentence cannot be known. Indeed, the Moore sentence is at the root of Fitch's famous "paradox of knowability" [11,3,9]: if there is an unknown truth, then there is an unknowable truth. For if p is true but unknown, then the Moore sentence $p \wedge \neg\Box p$ is true, but the Moore sentence cannot be known, because $\Box(p \wedge \neg\Box p)$ is inconsistent with standard assumptions about knowledge.

While the Moore sentence is conspicuously unsuccessful, other unsuccessful formulas are less conspicuous. The formula $\neg(p \vee q) \vee (p \wedge (\Box p \vee \Diamond q))$ is also unsuccessful, as we show in Example 5.12 below, but is the reason Moorean? While on the surface this formula looks unlike a Moore sentence, it is in fact possible to transform the formula to reveal its Moorean character. Indeed, we will prove that for a wide range of logics, such a transformation is possible for every unsuccessful formula. However well-disguised, their nature is always Moorean.

In Section 2 we establish notation, give the definitions of *successful*, *self-refuting*, etc., and review what is already known in the literature on successful formulas. In Section 3 we show that in logics of knowledge and belief for a single agent (extended by **S5**), Moorean phenomena are the source of all self-refutation; moreover, in logics for an *introspective* agent (extending **KD45**), Moorean phenomena are the source of all unsuccessfulness as well. This is a distinctive feature of such logics, for as we show in Section 4, in logics for a non-introspective agent or multiple agents, non-Moorean unsuccessful formulas appear. Finally, in Section 5 we relate successful and self-refuting formulas to the *Cartesian* and *learnable* formulas, which have been discussed in connection with Fitch's paradox, and to the *informative*, *eventually* self-refuting, *super-successful* formulas, which we introduce here. In Appendix A we give syntactic characterizations of the successful and the self-refuting formulas.

2 Preliminaries and Previous Results

Throughout we work with a fixed, unimodal language with $\Box$ and its dual $\Diamond$, and an infinite set Prop of propositional variables. The expression $\Diamond^+\varphi$ abbreviates $\Diamond\varphi \wedge \varphi$. We use the standard semantics of modal logic, where a model is a triple $\langle W, R, V\rangle$ with W any set of points, $R \subseteq W \times W$ any relation, and $V : \mathsf{Prop} \to \wp(W)$ a valuation function. A pointed model is a pair $\mathcal{M}, w$ with $\mathcal{M}$ a model and $w \in W$. The satisfaction relation $\vDash$ between pointed models and formulas is defined as usual.

It is typical to take **KD45** as a logic of belief and **S5** as a logic of knowledge. Apart from the assumptions that one does not believe or know anything inconsistent, and that what is known is true, these logics assume *positive* (axiom **4**) and *negative* (axiom **5**) *introspection*: if one believes something, then one believes that one believes it; if one does not believe something, then one believes that one does not believe it; and *mutatis mutandis* for knowledge. Most work on successful formulas has assumed **S5**. Since our results apply to a wider range of logics, we assume through Section 3 that we are working with at least **KD45**, so all of our models will be serial, transitive, and Euclidean. We call such models *quasi-partitions*.

Where $\mathcal{M}'$ is a submodel (not necessarily proper) of $\mathcal{M}$, we write $\mathcal{M}' \subseteq \mathcal{M}$. Rather than studying formulas preserved under arbitrary submodels, we will study formulas preserved under a special way of taking submodels, as in the following.

Definition 2.1 Given a model $\mathcal{M} = \langle W, R, V\rangle$, the *relativization of* $\mathcal{M}$ *to* φ is the (possibly empty) submodel $\mathcal{M}_{|\varphi} = \langle W_{|\varphi}, R_{|\varphi}, V_{|\varphi}\rangle$ of $\mathcal{M}$, where $W_{|\varphi} = \{w \in W : \mathcal{M}, w \vDash \varphi\}$, $R_{|\varphi}$ is R restricted to $W_{|\varphi}$, and $V_{|\varphi}(p) = V(p) \cap W_{|\varphi}$.

Definition 2.2 A formula φ is *successful* (in logic **L**) iff for every pointed model (of **L**), $\mathcal{M}, w \vDash \Diamond^+\varphi$ implies $\mathcal{M}_{|\varphi}, w \vDash \varphi$. A formula is *unsuccessful* (in **L**) iff it is not successful. A formula is *self-refuting* (in **L**) iff for every pointed model (of **L**), $\mathcal{M}, w \vDash \Diamond^+\varphi$ implies $\mathcal{M}_{|\varphi}, w \nvDash \varphi$.[1]

In the standard definitions of successful and self-refuting formulas, where **L** is assumed to be **S5**, the precondition only requires that φ be true at w. Since we are also working with **KD45**, we additionally require that φ be true at an accessible point, so that $\mathcal{M}_{|\varphi}$ is a quasi-partition provided $\mathcal{M}$ is. Our definition reduces to the standard one in the case of **S5**. In either case, unsatisfiable formulas are self-refuting and successful. In the case of **KD45**, a satisfiable formula such as $p \wedge \Box\neg p$, read as "p is true but you believe $\neg p$," is self-refuting and successful, since $\Diamond^+(p \wedge \Box\neg p)$ is unsatisfiable. While it may be more intuitive to require the satisfiability of $\Diamond^+\varphi$ for a successful φ, we will follow the standard definition in not requiring satisfiability.

The following lemma relates success and self-refutation across different logics.

Lemma 2.3 *Let* **L** *be a sublogic of* **L**$'$. *If* φ *is unsuccessful in* **L**$'$, *then* φ *is unsuccessful in* **L**, *and if* φ *is self-refuting in* **L**, *then* φ *is self-refuting in* **L**$'$.

Proof. Immediate from Definition 2.2, given that models of **L**$'$ are models of **L**. □

An obstacle to giving a simple syntactic characterization of the set of successful formulas is its lack of closure properties. Successful formulas are not closed under negation

---

[1] The term 'successful' is used by Gerbrandy [12], by analogy with the success postulate of belief revision, while the term 'self-refuting' is used by van Benthem [3]. Self-refuting formulas have also been called *strongly unsuccessful* [2].

(take $\neg p \vee \Box p$), conjunction (take p and $\neg\Box p$), or implication (use $\varphi \to \bot$) [6]. We show in Proposition 5.11 that they are also not closed under disjunction. Conversely, if a negated formula is successful, the unnegated formula may be unsuccessful (take $\neg(p \wedge \Diamond\neg p)$), if a conjunction is successful, some of the conjuncts may be unsuccessful (take $(p \wedge \Diamond\neg p) \wedge \neg p$), and if a disjunction is successful, some or even all of the disjuncts may be unsuccessful (Proposition 5.9 and Example 5.2).

By contrast, the formulas preserved under arbitrary submodels are well-behaved. The following result was proved independently by van Benthem and Visser [20,8]. A formula is *universal* iff it can be constructed using only literals, $\wedge$, $\vee$, and $\Box$.

Theorem 2.4 *A formula is preserved under submodels (of all relational models) iff it is equivalent (in* **K***) to a universal formula.*

Similarly, a formula is preserved under model extensions iff it is equivalent to an *existential* formula, constructed using only literals, $\wedge$, $\vee$, and $\Diamond$ [8].

Lemma 2.3 and the right-to-left direction of Theorem 2.4 give the following.

Corollary 2.5 *Universal formulas are successful in any normal modal logic.*

As noted by van Benthem [4], for any model $\mathcal{M}$ and formula φ, there is a universal formula φ' such that $\mathcal{M}_{|\varphi} = \mathcal{M}_{|\varphi'}$.[2] However, this result assumes the relation for $\mathcal{M}$ is at least a quasi-partition. For example, given the model in Figure 1, where the relation is not Euclidean, there is no universal formula ψ such that $\mathcal{M}_{|\Diamond p} = \mathcal{M}_{|\psi}$. This is symptomatic of the fact, established in Section 4, that in logics without both axioms **4** and **5**, there are non-Moorean sources of unsuccessfulness.

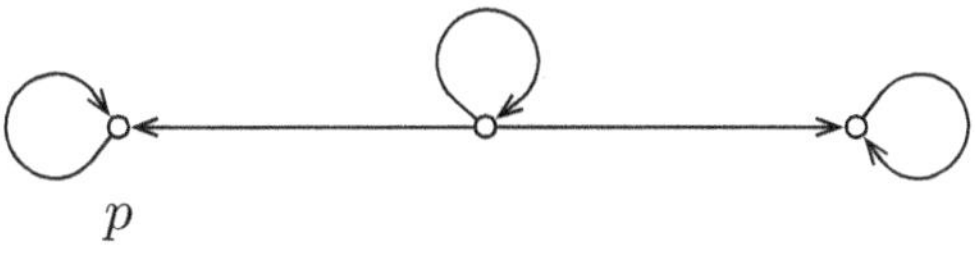

Fig. 1.

Gerbrandy [12] proved a proposition similar to Corollary 2.5, but in a slightly different formal context and with $\Diamond$ only as a defined operator. He showed that if all instances of $\Box$ are within the scope of an even number of negations, then the formula is successful. However, the converse of Corollary 2.5 does not hold, even in the case of formulas containing no nested modal operators. For example, the formula $\Diamond p \vee p$ is successful in **K**, but it is not equivalent to any universal formula. Yet other connections with universal formulas do hold. Qian [18] showed that a formula containing no nested modal operators is successful in **K** iff it can be transformed using a certain algorithm into a universal formula; moreover, a "homogeneous" formula, containing no nested modal operators and no propositional variables outside the scope of modal operators, is successful in **K** iff it is equivalent to a universal formula. The restriction to **K** is essential, for the formula $\Diamond p$

---

[2] The result in [4] is given for the case of multi-agent epistemic logic with a common knowledge operator, assuming $\mathcal{M}$ is finite. The proof for the single-agent case is trivial and requires no assumption of finiteness.

is successful in extensions of **KD45** (not in **K**), but it is not equivalent to any universal formula.

As far as we know, there are no other published results on the syntax of successful formulas in the basic modal language. However, successful formulas are often studied in the more general context of Dynamic Epistemic Logic (DEL) [1,12,7], where model changing operations are formalized in the logic itself by adding dynamic operators to the language. In the simplest fragment of DEL extending **S5**, known as Public Announcement Logic (PAL), there are sentences of the form $[\varphi]\psi$, read as "after the *announcement* of φ, ψ is true," for which satisfaction is defined:

$$\mathcal{M}, w \vDash [\varphi]\psi \text{ iff } \mathcal{M}, w \vDash \varphi \text{ implies } \mathcal{M}_{|\varphi}, w \vDash \psi.$$

An advantage of the PAL setting is that it allows for simple definitions of success and self-refutation [6]. Successful formulas are those for which $[\varphi]\varphi$ is valid (in **S5**), and self-refuting formulas are those for which $[\varphi]\neg\varphi$ is valid (in **S5**). While we will not deal directly with this language, our work will apply indirectly to PAL given the following result, due to Plaza [17], which relates PAL to the basic modal language.

Theorem 2.6 *Every formula in the language of* PAL *is equivalent (over partitions) to a formula in the basic modal language.*

Direct results on successful formulas in PAL with multiple knowledge modalities and the *common knowledge* operator have been obtained by van Ditmarsch and Kooi in [6], which also presents an analysis of the role of (un)successful formulas in a variety of scenarios involving information change.

Another benefit of the PAL definitions of successful and self-refuting is that they give an upper bound on the complexity of checking a formula for these properties. The following result including a lower bound for the success problem is due to Johan van Benthem in correspondence.

Theorem 2.7 *The success problem for* **S5** *is* coNP*-complete.*

Proof. For an upper bound, φ is successful iff $[\varphi]\,\varphi$ is valid, and the validity problem for single-agent PAL is coNP-complete [14]. For a lower bound, we use the following reduction of the validity problem for **S5**, which is coNP-complete, to the success problem for **S5**: φ is valid iff for a new variable p, $\psi \equiv p \wedge (\Diamond\neg p \vee \varphi)$ is successful. From left to right, if $\mathcal{M}, w \vDash \psi$ then $\mathcal{M}_{|\psi}, w \vDash p$, and since φ is valid, $\mathcal{M}_{|\psi}, w \vDash \varphi$, so $\mathcal{M}_{|\psi}, w \vDash \psi$. From right to left, take any pointed model $\mathcal{M}, w$. Extend the language with a new variable p and extend $\mathcal{M}$ to $\mathcal{M}'$ with one new point v, related to all other points, with the same valuation as w for all variables of the old language. Make p true everywhere except at v. Then since $\mathcal{M}', w \vDash p \wedge \Diamond\neg p$, we have $\mathcal{M}', w \vDash \psi$, and since ψ is successful, we have $\mathcal{M}'_{|\psi}, w \vDash \psi$. But v is eliminated in $\mathcal{M}'_{|\psi}$, so $\mathcal{M}'_{|\psi}, w \nvDash \Diamond\neg p$. Hence $\mathcal{M}'_{|\psi}, w \vDash \varphi$. We conclude that $\mathcal{M}, w \vDash \varphi$, for $\mathcal{M}'_{|\psi}$ and $\mathcal{M}$ differ only with respect to the valuation of p, which φ does not contain. □

A similar argument shows that φ is valid iff for a new variable p, $p \wedge (\Diamond\neg p \vee \neg\varphi)$ is self-refuting, so the self-refutation problem is coNP-complete given the upper bound from PAL. From left to right the reduction is obvious, while from right to left the same extension of $\mathcal{M}$ to $\mathcal{M}'$ works. Similar arguments also show that the success and self-refutation problems are PSPACE-complete for multimodal **S5**.

3 The Moorean Source of Unsuccessfulness

To show that Moorean phenomena are the source of all unsuccessfulness in logics for an introspective agent, we proceed via normal forms for such logics. The following normal form derives from Carnap [10], and Proposition 3.2 is standard.

Definition 3.1 A formula is in *normal form* iff it is a disjunction of conjunctions of the form $\delta \equiv \alpha \wedge \Box\beta_1 \wedge ... \wedge \Box\beta_n \wedge \Diamond\gamma_1 \wedge ... \wedge \Diamond\gamma_m$ where α and each γ_i are conjunctions of literals and each β_i is a disjunction of literals.

Proposition 3.2 *For every formula φ, there is a formula φ' in normal form such that φ and φ' are equivalent in* **K45**.

Theorem 1.7.6.4 of [15] gives the analogue of Proposition 3.2 for **S5**. Inspection of the proof shows that the necessary equivalences hold in **K45**.

We use the following notation and terminology in this section and Appendix A.

Definition 3.3 Given $\delta \equiv \alpha \wedge \Box\beta_1 \wedge ... \wedge \Box\beta_n \wedge \Diamond\gamma_1 \wedge ... \wedge \Diamond\gamma_m$ in normal form, we define $\delta^{\alpha} \equiv \alpha$, $\delta^{\alpha\Box} \equiv \alpha \wedge \Box\beta_1 \wedge ... \wedge \Box\beta_n$, and similarly for $\delta^{\alpha\Diamond}$, $\delta^{\Box}$, $\delta^{\Box\Diamond}$, and $\delta^{\Diamond}$.

Definition 3.4 Where χ is a conjunction or disjunction of literals, let $L(\chi)$ be the set of literals in χ. A set of literals is *open* iff no literal in the set is the negation of any of the others.

Definition 3.5 A conjunction $\delta \equiv \alpha \wedge \Box\beta_1 \wedge ... \wedge \Box\beta_n \wedge \Diamond\gamma_1 \wedge ... \wedge \Diamond\gamma_m$ in normal form is **KD45**-*clear* iff: (i) $L(\alpha)$ is open; (ii) there is an open set of literals $\{l_1, ..., l_n\}$ with $l_i \in L(\beta_i)$; and (iii) for every γ_k there is a set of literals $\{l_1, ..., l_n\}$ with $l_i \in L(\beta_i)$ such that $\{l_1, ..., l_n\} \cup L(\gamma_k)$ is open. A disjunction in normal form is **KD45**-clear iff at least one of its disjuncts is **KD45**-clear.

In the following, by "clear" and "satisfiable" we mean **KD45**-clear and satisfiable in a quasi-partition, respectively. In the case of **S5** clear, we must require in clause (ii) of Definition 3.5 that $\{l_1, ..., l_n\} \cup L(\alpha)$ is open, in which case (i) is unnecessary. In both cases, the following lemma holds with the appropriate definition of clarity.

Lemma 3.6 *A formula in normal form is satisfiable iff it is clear.*

Proof. [Sketch] Suppose φ in normal form is satisfiable. Then there is some disjunct δ of φ satisfied at a pointed model. Read off the appropriate open sets of literals from the current point for clarity condition (i), from some accessible point for (ii), and from witnesses for each $\Diamond\gamma_k$ in δ for (iii). In the other direction, suppose there are open sets of literals as described. Construct a model where δ^{α} is true at the root point w, which is possible by clarity condition (i). For each $\Diamond\gamma_k$, add a point v accessible from w and extend the valuation such that all conjuncts of γ_k and some disjunct of each β_i are true at v, which is possible by (iii). If there are no $\Diamond\gamma_k$ formulas in δ, add an accessible point v and extend the valuation such that some disjunct of each β_i is true at v, which is possible by (ii). Then φ is true at w. □

The following definition fixes the basic class of Moore conjunctions, as well as a wider class of Moorean conjunctions. The intuition is that a Moore conjunction simultaneously asserts a lack of information about another fact being asserted, and a Moorean conjunction is one that behaves like a Moore conjunction in some context.

We write $\sim\varphi$ for the negation of φ in negation normal form, with $\neg$ applying only to literals.

Definition 3.7 Let δ be a conjunction in normal form.

(i) δ is a *Moore conjunction* iff $\delta \wedge \Diamond\delta^\alpha$ is not clear or there is a $\Diamond\gamma_k$ conjunct in δ such that $\delta \wedge \Diamond(\delta^\alpha \wedge \gamma_k)$ is not clear.

(ii) δ is a *Moorean conjunction* iff there is a $\Diamond\gamma_k$ conjunct in δ such that $\delta \wedge \Diamond\delta^\alpha \wedge \Box(\sim\delta^\alpha \vee \sim\gamma_k)$ is clear.

Assuming **S5**, $\delta \wedge \Diamond\delta^\alpha$ may be replaced by δ in both (i) and (ii).

The definition of a Moore conjunction generalizes from the paradigmatic case of $p\wedge\Diamond\neg p$ to include formulas such as $p\wedge\Diamond q\wedge\Box(q \rightarrow \neg p)$. The formulas $p\wedge\neg p$ and $p\wedge\Box\neg p$ are also Moore conjunctions, since for these δ the formula $\delta\wedge\Diamond\delta^\alpha$ is not clear, but for the same reason they are not Moorean conjunctions. An example of a Moorean conjunction that is not a Moore conjunction is $p \wedge \Diamond q$. We consider this formula Moorean because in a context where the agent knows that q implies $\neg p$, so $\Box(q \rightarrow \neg p)$ holds, learning $p\wedge\Diamond q$ has the same effect as learning $p\wedge\Diamond\neg p$. By contrast, the formula $p\wedge\Diamond q\wedge\Diamond(p\wedge q)$ rules out the Moorean context with its last conjunct, and it is not Moorean according to Definition 3.7(ii).

In the proofs of Lemma 3.9 and Theorem 3.13 below, we will assume without loss of generality that all models considered are connected (i.e., $\forall w, v \in W : wRv \vee vRw$), in which case the following basic facts hold.

Lemma 3.8 *Where $\mathcal{M}$ is a connected quasi-partition and δ is a conjunction in normal form:*

(i) $\mathcal{M}, w \vDash \delta \Rightarrow \mathcal{M} \vDash \delta^{\Box\Diamond}$;

(ii) $\mathcal{M}, w \vDash \delta \Rightarrow \mathcal{M}_{|\delta} = \mathcal{M}_{|\delta^\alpha}$ *and* $\mathcal{M}_{|\delta} \vDash \delta^\alpha$;

(iii) $\mathcal{M}_{|\delta}, w \vDash \delta \Rightarrow \mathcal{M}_{|\delta} \vDash \delta$.

We now prove the main lemma used in the proof of Theorem 3.13.

Lemma 3.9 *Let δ be a conjunction in normal form. The following hold for both* **KD45** *and* **S5**.

(i) *δ is self-refuting if and only if it is a Moore conjunction.*

(ii) *δ is unsuccessful if and only if it is a Moorean conjunction.*

Proof. ($\Leftarrow$ *(i)*) Suppose δ is a Moore conjunction. Case 1: $\delta \wedge \Diamond\delta^\alpha$ is not clear. By

Lemmas 3.6 and 3.8(i), $\delta \wedge \Diamond\delta^{\alpha}$ is clear iff $\Diamond^{+}\delta$ is satisfiable, so in this case $\Diamond^{+}\delta$ is unsatisfiable and hence δ is self-refuting. Case 2: $\delta \wedge \Diamond\delta^{\alpha}$ is clear, so suppose $\mathcal{M}, w \vDash \Diamond^{+}\delta$. For the $\Diamond\gamma_k$ in δ such that $\delta \wedge \Diamond(\delta^{\alpha} \wedge \gamma_k)$ is not clear, $\neg\delta \vee \Box(\neg\delta^{\alpha} \vee \neg\gamma_k)$ is valid by Lemma 3.6, so $\mathcal{M}, w \vDash \Box(\neg\delta^{\alpha} \vee \neg\gamma_k)$ given $\mathcal{M}, w \vDash \delta$. Then since $\Box(\neg\delta^{\alpha} \vee \neg\gamma_k)$ is universal, it is preserved under submodels by Theorem 2.4, so $\mathcal{M}_{|\delta}, w \vDash \Box(\neg\delta^{\alpha} \vee \neg\gamma_k)$. From Lemma 3.8(ii), $\mathcal{M}_{|\delta}, w \vDash \Box\delta^{\alpha}$, so $\mathcal{M}_{|\delta}, w \vDash \Box\neg\gamma_k$. Since $\Diamond\gamma_k$ is a conjunct in δ, $\mathcal{M}_{|\delta}, w \nvDash \delta$. Since $\mathcal{M}$ was arbitrary, δ is self-refuting.

((*i*) $\Rightarrow$) We prove the contrapositive.[3] Suppose δ is not a Moore conjunction. Then $\delta \wedge \Diamond\delta^{\alpha}$ is clear, and for every $\Diamond\gamma_k$ in δ, $\delta \wedge \Diamond(\delta^{\alpha} \wedge \gamma_k)$ is clear $(*)$. If there are no $\Diamond\gamma_k$ conjuncts in δ, then δ is a universal formula with $\Diamond^{+}\delta$ satisfiable, so it is not self-refuting by Theorem 2.4. Suppose there are $\Diamond\gamma_k$ conjuncts in δ. We claim that $\delta' \equiv \delta \wedge \Box\beta_{n+1} \wedge ... \wedge \Box\beta_{n+j}$ is clear where $\{\beta_{n+1}, ..., \beta_{n+j}\} = L(\delta^{\alpha})$. Given assumption $(*)$ and clarity condition (iii) for each $\delta \wedge \Diamond(\delta^{\alpha} \wedge \gamma_k)$, we have that for all $\Diamond\gamma_k$ in δ there is a set $\{l_1, ..., l_n\}$ with $l_i \in L(\beta_i)$ such that $\{l_1, ..., l_n\} \cup L(\delta^{\alpha} \wedge \gamma_k)$ is open. Taking $\{l_{n+1}, ..., l_{n+j}\} = L(\delta^{\alpha})$, $\{l_1, ..., l_{n+j}\} \cup L(\gamma_k) = \{l_1, ..., l_n\} \cup L(\delta^{\alpha} \wedge \gamma_k)$ is open, which gives clarity conditions (i), (ii) and (iii) for δ'. Since δ' is clear, suppose $\mathcal{N}, w \vDash \delta'$. From the fact that $\vDash \delta' \leftrightarrow (\delta \wedge \Box\delta^{\alpha})$ we have $\mathcal{N}, w \vDash \delta \wedge \Box\delta^{\alpha}$. Given $\mathcal{N}, w \vDash \Box\delta^{\alpha}$ and the assumption that $\mathcal{N}$ is connected, $\mathcal{N} \vDash \delta^{\alpha}$; given $\mathcal{N}, w \vDash \delta$ and Lemma 3.8(i), $\mathcal{N} \vDash \delta^{\Box\Diamond}$. Hence $\mathcal{N} \vDash \delta$, in which case $\mathcal{N}, w \vDash \Diamond^{+}\delta$ and $\mathcal{N}_{|\delta} = \mathcal{N}$. It follows that $\mathcal{N}_{|\delta}, w \vDash \delta$, so δ is not self-refuting.

((*ii*) $\Leftarrow$) Suppose δ is a Moorean conjunction. Since for some $\Diamond\gamma_k$ in δ, $\chi \equiv \delta \wedge \Diamond\delta^{\alpha} \wedge \Box(\sim\delta^{\alpha} \vee \sim\gamma_k)$ is clear, there is a model with $\mathcal{M}, w \vDash \chi$. Given $\mathcal{M}, w \vDash \delta \wedge \Diamond\delta^{\alpha}$, we have $\mathcal{M}, w \vDash \Diamond^{+}\delta$ by Lemma 3.8(i). Given $\mathcal{M}, w \vDash \Box(\sim\delta^{\alpha} \vee \sim\gamma_k)$, by the same reasoning as in Case 2 of ((*i*) $\Leftarrow$), $\mathcal{M}_{|\delta}, w \nvDash \delta$. Therefore δ is unsuccessful.

((*ii*) $\Rightarrow$) We prove the contrapositive. Suppose $\mathcal{M}, w \vDash \Diamond^{+}\delta$ and δ is not a Moorean conjunction. Then for all $\Diamond\gamma_k$ in δ, $\chi_k \equiv \delta \wedge \Diamond\delta^{\alpha} \wedge \Box(\sim\delta^{\alpha} \vee \sim\gamma_k)$ is not clear. To show $\mathcal{M}_{|\delta}, w \vDash \delta$, it suffices to show $\mathcal{M}_{|\delta}, w \vDash \delta^{\Diamond}$, since $\delta^{\alpha\Box}$ is universal and therefore preserved under submodels. Consider some $\Diamond\gamma_k$ conjunct in δ. It follows from our assumption that χ_k is unsatisfiable, whence $(\delta \wedge \Diamond\delta^{\alpha}) \to \Diamond(\delta^{\alpha} \wedge \gamma_k)$ is valid. Then from $\mathcal{M}, w \vDash \Diamond^{+}\delta$ we obtain $\mathcal{M}, w \vDash \Diamond(\delta^{\alpha} \wedge \gamma_k)$, so there is a v with wRv and $\mathcal{M}, v \vDash (\delta^{\alpha} \wedge \gamma_k)$. By Lemma 3.8(ii), v is retained in $\mathcal{M}_{|\delta}$. Since γ_k is propositional, $\mathcal{M}_{|\delta}, v \vDash \gamma_k$ and hence $\mathcal{M}_{|\delta}, w \vDash \Diamond\gamma_k$. Since $\Diamond\gamma_k$ was arbitrary, $\mathcal{M}_{|\delta}, w \vDash \delta^{\Diamond}$ and hence $\mathcal{M}_{|\delta}, w \vDash \delta$. Since $\mathcal{M}$ was arbitrary, δ is successful. □

Lemma 3.9 gives necessary and sufficient conditions for the successfulness of a conjunction in normal form. We now introduce an apparently stronger notion.

Definition 3.10 A formula φ is *super-successful* (in **L**) iff for every pointed model (of **L**), $\mathcal{M}, w \vDash \Diamond^{+}\varphi$ implies $\mathcal{M}', w \vDash \varphi$ for every $\mathcal{M}'$ such that $\mathcal{M}_{|\varphi} \subseteq \mathcal{M}' \subseteq \mathcal{M}$.

If φ is super-successful and $\mathcal{M}, w \vDash \varphi$, then as points that are not in $\mathcal{M}_{|\varphi}$ are eliminated from $\mathcal{M}$, φ remains true at w. Since we take the elimination of points as an agent's acquisition of new information, this means that φ remains true as the agent approaches,

[3] The following argument establishes something stronger than we need for Lemma 3.9, but we use it to establish Corollary 5.3 below. Compare the ($\Rightarrow$) direction of the proof of Theorem A.3 in Appendix A.

by way of the incremental acquisition of new information, the epistemic state of $\mathcal{M}_{|\varphi}$ wherein the agent knows φ. Intuitively, we can say that a super-successful formula remains true while an agent is "on the way" to learning it.

We will use the next lemma in the proof of Theorem 3.13.

Lemma 3.11 *If δ is a successful conjunction in normal form, δ is super-successful.*

Proof. Suppose δ is not super-successful, so there is a pointed model such that $\mathcal{M}, w \models \Diamond^+\delta$ and an $\mathcal{M}'$ such that $\mathcal{M}_{|\varphi} \subseteq \mathcal{M}' \subseteq \mathcal{M}$ and $\mathcal{M}', w \not\models \delta$. Since $\mathcal{M}' \subseteq \mathcal{M}$ and $\delta^{\alpha\Box}$ is preserved under submodels, we must have $\mathcal{M}', w \not\models \delta^{\Diamond}$. But then $\mathcal{M}_{|\delta}, w \not\models \delta^{\Diamond}$ given that $\mathcal{M}_{|\delta} \subseteq \mathcal{M}'$ and $\delta^{\Diamond}$ is preserved under extensions. Hence $\mathcal{M}_{|\delta}, w \not\models \delta$, so δ is unsuccessful. □

We now lift the definition of Moore and Moorean to arbitrary formulas.

Definition 3.12 Let φ be an arbitrary formula.

(i) φ is a *Moore sentence* iff any normal form of φ is a disjunction of Moore conjunctions.

(ii) φ is a *Moorean sentence* iff any normal form of φ contains a Moorean conjunction as a disjunct.

The following theorem gives necessary conditions for self-refuting and unsuccessful formulas. In Appendix A we strengthen this result with conditions that are sufficient as well as necessary.

Theorem 3.13 *Let φ be an arbitrary formula.*

(i) *If φ is self-refuting in any sublogic of* **S5**, *then φ is a Moore sentence.*

(ii) *If φ is unsuccessful in any extension of* **KD45**, *then φ is a Moorean sentence.*

Proof. By Lemma 2.3 it suffices to show the consequent of (i) for φ that is self-refuting in **S5** and the consequent of (ii) for φ that is unsuccessful in **KD45**. Since φ is self-refuting (resp. unsuccessful) iff any equivalent normal form of φ is self-refuting (resp. unsuccessful), let us assume that φ is already in normal form.

(i) Suppose φ is not a Moore sentence, so by Definition 3.12 there is a disjunct δ of φ that is not a Moore conjunction. Then by Lemma 3.9, δ is not self-refuting, so there is a pointed model with $\mathcal{M}, w \models \Diamond^+\delta$ and $\mathcal{M}_{|\delta}, w \models \delta$. By Lemma 3.8(iii), $\mathcal{M}_{|\delta} \models \delta$. It follows that $(\mathcal{M}_{|\delta})_{|\varphi} = \mathcal{M}_{|\delta}$, since all points in $\mathcal{M}_{|\delta}$ satisfy one of the disjuncts of φ, namely δ. Then given $\mathcal{M}_{|\delta}, w \models \Diamond^+\delta$, we have $\mathcal{M}_{|\delta}, w \models \Diamond^+\varphi$ and hence $(\mathcal{M}_{|\delta})_{|\varphi}, w \models \varphi$. Therefore φ is not self-refuting.

(ii) We prove something stronger. Suppose φ is not a Moorean sentence, so by Definition 3.12 no disjunct of φ is a Moorean conjunction. Then each disjunct of φ is successful by Lemma 3.9. Consider a pointed model such that $\mathcal{M}, w \models \Diamond^+\varphi$, so for some disjunct δ of φ, we have $\mathcal{M}, w \models \delta$. Since φ is a disjunction, $\mathcal{M}_{|\delta} \subseteq \mathcal{M}_{|\varphi}$. By Lemma 3.11, δ is super-successful, so for any $\mathcal{M}'$ with $\mathcal{M}_{|\varphi} \subseteq \mathcal{M}' \subseteq \mathcal{M}$, we have $\mathcal{M}', w \models \delta$ and hence $\mathcal{M}', w \models \varphi$. Since $\mathcal{M}$ was arbitrary, φ is super-successful. □

4 Unsuccessfulness in Other Logics

We now consider the sources of unsuccessfulness in logics for an agent without introspection (logics without axioms **4** and **5**) and in logics for multiple agents.

From an epistemic perspective, the most interesting (normal) proper sublogics of **S5** are obtained by dropping axiom **5** and adding something weaker in its place. Indeed, logics such as **S4**, **S4.x** for $\mathbf{x} = \mathbf{2}, \mathbf{3}, \mathbf{4}$, etc., have been proposed as logics of knowledge. Call logics **L** and **L**′ *comparable* if **L** is a sublogic of **L**′ or *vice versa*.

Proposition 4.1 *For any normal, proper sublogic* **L** *of* **S5**, *comparable to* **S4.4**, *there is a formula (consistent with* **S5**) *that is unsuccessful in* **L** *but is not Moorean.*[4]

Proof. First, we claim that $\varphi \equiv \Diamond p \wedge \Diamond\neg p$ is unsuccessful in **S4.4** and hence in any sublogic of **S4.4** by Lemma 2.3. In the **S4.4** model $\mathcal{M}$ in Figure 2, φ is true at the left point, but in $\mathcal{M}_{|\varphi}$, the right point is eliminated, so φ becomes false at the left point. Note that the formula is already in normal form and is not Moorean.

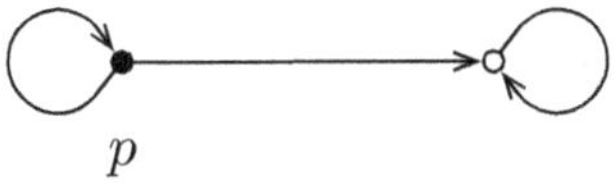

Fig. 2.

Next, we claim that $\mathcal{M}$ is a model of any logic **L** that is a proper extension of **S4.4** and a proper sublogic of **S5**. Suppose not, so there is a theorem φ of **L** with $\mathcal{M} \nvDash \varphi$. Change φ to φ' by substituting $(p \wedge \neg p)$ for any propositional variable q other than p. Since **L** is normal and therefore closed under substitution, φ' is also a theorem of **L** and therefore of **S5**. Moreover, since for all variables q other than p, $\mathcal{M} \vDash \neg q$, the substitution of $(p \wedge \neg p)$ for q preserves (un)satisfiability in $\mathcal{M}$, so $\mathcal{M} \nvDash \varphi'$. Hence φ' is not a theorem of **S4.4**. But by a result of Zeman [21], for any formula ψ, containing exactly one variable, that is a theorem of **S5** but not a theorem of **S4.4**, adding ψ to **S4.4** gives **S5**. Hence **L** is **S5**, a contradiction. Since $\mathcal{M}$ models any logic between **S4.4** and **S5**, φ is unsuccessful in these logics. □

Proposition 4.1 shows that **S5** is unique among the typical logics of knowledge insofar as all of its unsuccessful formulas are Moorean. The counterexample for the weaker logics shows that without negative introspection, one can come to know p by being truly told, "You do not know whether or not p," a surprising case of unsuccessfulness. The following proposition, although weaker than Proposition 4.1, shows that non-Moorean unsuccessful formulas appear if we weaken logics of knowledge and belief in other ways as well.

Proposition 4.2 *For any sublogic* **L** *of* **KTB** *or* **KD5**, *there is a formula (consistent with* **S5**) *that is unsuccessful in* **L** *but is not Moorean.*

Proof. For **KTB** consider $\varphi \equiv \Diamond p \wedge \Diamond q$ and the model $\mathcal{M}$ in Figure 3. In $\mathcal{M}_{|\varphi}$, the left and right points are eliminated, so φ becomes false at the center point.

[4] **S4.4** is **S4** plus $\varphi \rightarrow (\Diamond\Box\varphi \rightarrow \Box\varphi)$. It is a proper extension of **S4.3** and therefore also of **S4.2** [21].

Fig. 3.

For **KD5** consider $\psi \equiv \neg q \vee (\Box p \wedge \Diamond\Diamond q)$ and the model $\mathcal{M}$ in Figure 4. Only the right point is eliminated in $\mathcal{M}_{|\psi}$, so ψ becomes false at the left point.

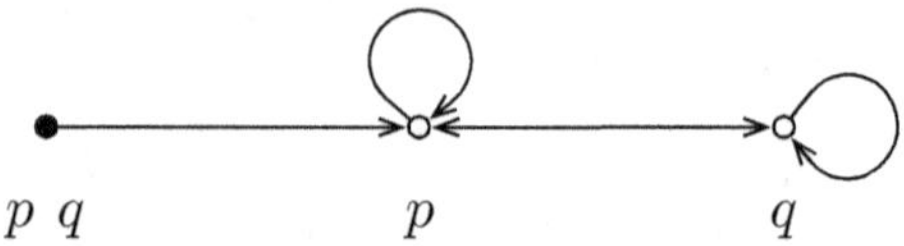

Fig. 4.

The formula ψ is equivalent in **KD45** to the normal form $\neg q \vee (\Box p \wedge \Diamond q)$, which is not Moorean, since neither disjunct is a Moorean conjunction.[5] □

There are other formulas that one may wish to categorize as Moorean, perhaps even as Moore sentences, but which are inconsistent with **KD45**.

Example 4.3 The formula $\Box p \wedge \Diamond\neg\Box p$ is a kind of "higher order" Moore sentence, which is satisfiable on intransitive frames but is also self-refuting over such frames. Similarly, the formula $\Diamond p \wedge \Diamond\neg\Diamond p$ is satisfiable yet self-refuting on non-Euclidean frames. The first formula says that the agent is not aware of what he believes, while the second says that he is not aware of what he does not believe. These formulas are the very witnesses of a failure to validate axioms **4** and **5**, respectively.

A natural question is whether there are non-Moorean sources of unsuccessfulness in languages more expressive than the basic modal language. Consider, for example, a language with multiple modalities, as in multi-agent epistemic logic. Without giving a formal definition of Moorean in the multi-agent case, it is nonetheless clear that there are more ways to be Moorean in the multi-agent case than in the single-agent case, even assuming quasi-partitions for each relation. For example, there are self-refuting, *indirect* Moore sentences such as $\Box_a p \wedge \Diamond_b \neg p$, which imply single-agent Moore sentences (in this case, $p \wedge \Diamond_b \neg p$) in multi-agent **S5**. There are also self-refuting, *higher order* Moore sentences such as $\Box_a p \wedge \Diamond_b \neg\Box_a\, p$ and $\Diamond_a p \wedge \Diamond_b \neg\Diamond_a\, p$, which resemble the higher order, single-agent Moore sentences consistent with logics weaker than **KD45**, noted in Example 4.3. However, not all unsuccessful formulas in the multi-agent context have a Moorean character.

Example 4.4 In the model $\mathcal{M}$ in Figure 5, the formula $\varphi \equiv \Diamond_a p \wedge \Diamond_a \Diamond_b \neg p$ is true at the left point but false at the right point, since $\Diamond_a p$ is false at the right point.

[5] Note that while the right disjunct is not a Moorean conjunction, in **S5** (but not in **KD45**) it implies the Moorean conjunction $p \wedge \Diamond q$. This is an instance of the more general fact that in **S5** some successful formulas imply unsuccessful formulas.

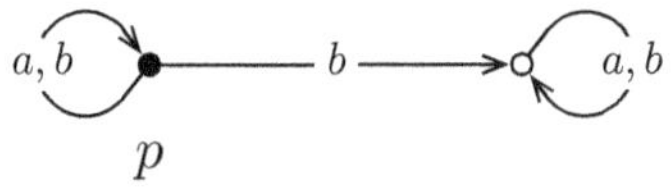

Fig. 5.

As a result, in the model $\mathcal{M}_{|\varphi}$ the right point is eliminated, in which case φ becomes false at the left point because $\Diamond_a \Diamond_b \neg p$ becomes false there.

The formula $\Diamond_a p \wedge \Diamond_a \Diamond_b \neg p$ does not resemble any single-agent Moorean formula, yet it is nonetheless unsuccessful. Given the connection that we have observed between introspection and Moorean phenomena in the single-agent case, we can see why there should be non-Moorean unsuccessful formulas in the multi-agent case: agents do not have introspective access to each other's knowledge or beliefs.

5 Related Classes of Formulas

Theorem 3.13 shows that in a wide range of logics, every self-refuting formula is a Moore sentence (i) and every unsuccessful formula is a Moorean sentence (ii). However, neither the converse of (i) nor the converse of (ii) holds in general. In this section, we relate the failures of the converses of (i) and (ii) to other classes of formulas. In Theorems A.3 and A.6 in Appendix A, we overcome these failures and give syntactic characterizations of the self-refuting and unsuccessful formulas as the *strong* Moore sentences and *strong* Moorean sentences, respectively.

For simplicity we assume in this section that we are working with **S5** only.

5.1 Informative, Cartesian, and eventually self-refuting formulas

Definition 5.1 A formula φ is (*potentially*) *informative* iff there is a pointed model such that $\mathcal{M}, w \vDash \varphi$ and $\mathcal{M}_{|\varphi} \neq \mathcal{M}$. Otherwise φ is *uninformative.* A formula φ is *always informative* iff for all pointed models such that $\mathcal{M}, w \vDash \varphi$, $\mathcal{M}_{|\varphi} \neq \mathcal{M}$.

Note that if a formula is not always informative, then it is not self-refuting, for there is a model such that $\mathcal{M}, w \vDash \varphi$ but $\mathcal{M}_{|\varphi} = \mathcal{M}$, so $\mathcal{M}_{|\varphi}, w \vDash \varphi$. This observation explains some of the counterexamples to the converse of Theorem 3.13(i), Moore sentences that are not self-refuting.

Example 5.2 The formula $(p \wedge \Diamond \neg p) \vee (q \wedge \Diamond \neg q)$ is a Moore sentence, but it is not always informative (take a model with only two connected points, one of which satisfies p but not q and the other of which satisfies q but not p) and hence not self-refuting. The formula $(p \wedge \Diamond \neg p) \vee (\neg p \wedge \Diamond p)$ is a Moore sentence, but it is uninformative and hence successful, but not self-refuting. These examples show that neither the self-refuting nor the unsuccessful formulas are closed under disjunction.

From our previous results, we have the following corollary.

Corollary 5.3 *A conjunction in normal form is always informative iff it is self-refuting.*

Proof. By inspection of the proof of Lemma 3.9(i). □

However, a Moore sentence that is always informative may not be self-refuting.

Example 5.4 The formula $\varphi \equiv (p \wedge \Diamond\neg p) \vee (p \wedge q \wedge \Diamond\neg q)$ is always informative, for if $\mathcal{M}, w \vDash (p \wedge \Diamond\neg p)$, then there is a witness to $\Diamond\neg p$ that is eliminated in $\mathcal{M}_{|\varphi}$, and if $\mathcal{M}, w \vDash (p \wedge q \wedge \Diamond\neg q)$, then either the witness to $\Diamond\neg q$ does not satisfy $(p \wedge \Diamond\neg p)$, in which case it does not satisfy φ and is eliminated in $\mathcal{M}_{|\varphi}$, or it does satisfy $(p \wedge \Diamond\neg p)$, in which case there is a witness to $\Diamond\neg p$ that is eliminated in $\mathcal{M}_{|\varphi}$. In either case, $\mathcal{M}_{|\varphi} \neq \mathcal{M}$, so φ is always informative. However, φ is not self-refuting, as shown by the partition model $\mathcal{M}$ with $W = \{w, v, u\}$, $V(p) = \{w, v\}$, and $V(q) = \{w, u\}$. We have $\mathcal{M}, w \vDash p \wedge q \wedge \Diamond\neg q$, and given $W_{|\varphi} = \{w, v\}$, also $\mathcal{M}_{|\varphi}, w \vDash p \wedge q \wedge \Diamond\neg q$.

Interestingly, the formula φ is *self-refuting within two steps*; given $\mathcal{M}, w \vDash \varphi$ and $\mathcal{M}_{|\varphi}, w \vDash \varphi$, we have $\left(\mathcal{M}_{|\varphi}\right)_{|\varphi}, w \nvDash \varphi$. Let $\delta_1 \equiv (p \wedge \Diamond\neg p)$ and $\delta_2 \equiv (p \wedge q \wedge \Diamond\neg q)$ be the disjuncts of φ. All $\neg p$-points are eliminated in $\mathcal{M}_{|\varphi}$, so $\mathcal{M}_{|\varphi} \vDash \neg\delta_1$. Hence $\left(\mathcal{M}_{|\varphi}\right)_{|\varphi} = \left(\mathcal{M}_{|\varphi}\right)_{|\delta_2}$. But since $\mathcal{M}_{|\varphi}, w \vDash \delta_2$ and δ_2 is self-refuting, $\left(\mathcal{M}_{|\varphi}\right)_{|\delta_2}, w \nvDash \delta_2$. Since δ_1 is existential, $\left(\mathcal{M}_{|\varphi}\right)_{|\delta_2}, w \nvDash \delta_1$, so we conclude that $\left(\mathcal{M}_{|\varphi}\right)_{|\varphi}, w \nvDash \varphi$.

Example 5.4 points to the interest of self-refutation "in the long run."

Definition 5.5 Given a model $\mathcal{M}$, we define $\mathcal{M}_{|^n\varphi}$ recursively by $\mathcal{M}_{|^0\varphi} = \mathcal{M}$, $\mathcal{M}_{|^{n+1}\varphi} = \left(\mathcal{M}_{|^n\varphi}\right)_{|\varphi}$. A formula φ is *self-refuting within* n *steps* iff there is an n such that for all pointed models, if $\mathcal{M}, w \vDash \varphi$, then $\mathcal{M}_{|^m}, w \nvDash \varphi$ with $m \leq n$; φ is *eventually self-refuting* iff for all pointed models, if $\mathcal{M}, w \vDash \varphi$, then there is an n such that $\mathcal{M}_{|^n}, w \nvDash \varphi$.

Using Definition 5.5, we can generalize Corollary 5.3. First, recall Fitch's paradox, mentioned in Section 1, which we can restate as: *if all truths are knowable, then all truths are* (*already*) *known.* One proposal for avoiding the paradox [19] is to restrict the claim that all truths are knowable to the claim that all *Cartesian* truths are knowable, in which case it does not follow that all truths are known.

Definition 5.6 φ is *Cartesian* iff $\Box\varphi$ is satisfiable.[6]

The class of formulas φ for which $\Box\varphi$ is satisfiable seems a natural object of study in its own right. The following proposition establishes a connection between these formulas, defined "statically," and the other formulas classes that we have defined "dynamically" in terms of model transformations. It also generalizes Corollary 5.3.

Proposition 5.7 *The following are equivalent:*

(i) *φ is always informative.*

(ii) *φ is not Cartesian.*

(iii) *φ is eventually self-refuting.*

---

[6] The term 'Cartesian' is due to Tennant [19], though his definition is in terms of consistency rather than satisfiability. The term 'knowable' seems more natural, but 'knowable' is used in a different sense in the literature [9], for what we call 'learnable' in Definition 5.13.

Proof. Suppose φ is always informative and $\mathcal{M}, w \vDash \varphi$. Without loss of generality, assume $\mathcal{M}$ is connected. Since φ is always informative, $\mathcal{M}_{|\varphi} \neq \mathcal{M}$. Hence there is a point v such that wRv and $\mathcal{M}, v \nvDash \varphi$, in which case $\mathcal{M}, w \nvDash \Box\varphi$. Since $\mathcal{M}$ was arbitrary, $\Box\varphi$ is unsatisfiable, so φ is not Cartesian.

Suppose φ is not Cartesian, and without loss of generality, assume φ is in normal form. We prove by induction that for all $n \geq 0$, if $\mathcal{M}_{|^n\varphi}, w \vDash \varphi$, then there are $\Diamond\gamma_1, ..., \Diamond\gamma_{n+1}$ distinct subformulas of φ with $\mathcal{M}_{|^{n+1}\varphi} \vDash \neg\Diamond\gamma_1 \wedge ... \wedge \neg\Diamond\gamma_{n+1}$ (*). It follows that since φ has some finite number n of distinct $\Diamond\gamma_k$ subformulas, we must have $\mathcal{M}_{|^m\varphi}, w \nvDash \varphi$ for some $m \leq n$. For the base case, assume $\mathcal{M}, w \vDash \varphi$. Since φ is not Cartesian, $\mathcal{M}_{|\varphi}, w \nvDash \Box\varphi$. Let v be such that wRv and $\mathcal{M}_{|\varphi}, v \nvDash \varphi$, and note that $\mathcal{M}, v \vDash \varphi$, for otherwise v would not have been retained in $\mathcal{M}_{|\varphi}$. From the fact that universal formulas are preserved under submodels, it follows that for some $\Diamond\gamma_1$ in φ, we have $\mathcal{M}, v \vDash \Diamond\gamma_1$ but $\mathcal{M}_{|\varphi}, v \nvDash \Diamond\gamma_1$. Hence $\mathcal{M}_{|\varphi} \vDash \neg\Diamond\gamma_1$ by Lemma 3.8(i). For the inductive step, assume that $\mathcal{M}_{|^{n+1}\varphi}, w \vDash \varphi$. Since w is retained in $\mathcal{M}_{|^{n+1}\varphi}$, we have $\mathcal{M}_{|^n\varphi}, w \vDash \varphi$, so by the induction hypothesis there are $\Diamond\gamma_1, ..., \Diamond\gamma_{n+1}$ distinct subformulas of φ for which (*) holds. By the same reasoning as before, since φ is Cartesian, $\mathcal{M}_{|^{n+2}\varphi}, w \nvDash \Box\varphi$, so there is some z with wRz and some $\Diamond\gamma_{n+2}$ in φ such that $\mathcal{M}_{|^{n+1}\varphi}, z \vDash \Diamond\gamma_{n+2}$ but $\mathcal{M}_{|^{n+2}\varphi} \vDash \neg\Diamond\gamma_{n+2}$. Moreover, since $\Diamond\gamma_k$ formulas are preserved under extensions, $\mathcal{M}_{|^{n+2}\varphi} \vDash \neg\Diamond\gamma_1 \wedge ... \wedge \neg\Diamond\gamma_{n+1}$ as well. Finally, given $\mathcal{M}_{|^{n+1}\varphi}, z \vDash \Diamond\gamma_{n+2}$ and (*), $\Diamond\gamma_{n+2}$ is distinct from $\Diamond\gamma_1, ..., \Diamond\gamma_{n+1}$.

Suppose φ is not always informative. Then there is a model with $\mathcal{M}, w \vDash \varphi$ and $\mathcal{M}_{|\varphi} = \mathcal{M}$, in which case $\mathcal{M}_{|^n\varphi} = \mathcal{M}$ for all n, so φ is not eventually self-refuting. □

Corollary 5.8 *φ is eventually self-refuting iff it is self-refuting within n steps, with n bounded by the number of distinct diamond formulas in a normal form of φ.*

Proof. By inspection of the proof of the previous proposition. □

5.2 Successful, super-successful, and learnable formulas

The converse of Theorem 3.13(ii) fails because a formula φ in normal form that contains a Moorean conjunction as a disjunct may be successful. For example, it is easy to see that if the disjunction of the non-Moorean conjunctions in φ is a consequence of the disjunction of the Moorean conjunctions in φ, then φ is not only successful but super-successful, given Lemma 3.11. Examples of such φ include $(p \wedge \Diamond\neg p) \vee p$ and $(p \wedge \Diamond\neg p) \vee \Diamond\neg p$. There are also successful formulas ψ that contain Moorean conjunctions as disjuncts, but do not meet the condition of φ. These formulas nevertheless manage to be successful by a kind of compensation: when one disjunct of ψ goes from true at $\mathcal{M}, w$ to false at $\mathcal{M}_{|\psi}, w$, another disjunct compensates by going from false at $\mathcal{M}, w$ to true at $\mathcal{M}_{|\psi}, w$. The formula $(p \wedge \Diamond\neg p) \vee \Box p$ exhibits this kind of compensation against a Moore conjunction, while the formula in the proof of the following proposition does so against a Moorean conjunction.

Proposition 5.9 *Not all successful formulas are super-successful.*

Proof. The formula $\varphi \equiv (p \wedge \Diamond q) \vee \Box p$ is successful but not super-successful. Suppose $\mathcal{M}, w \vDash \varphi$, and without loss of generality assume that $\mathcal{M}$ is connected. Case 1: $\mathcal{M}, w \vDash$

$\Box p$. Then $\mathcal{M}_{|\varphi} = \mathcal{M}$, so $\mathcal{M}_{|\varphi}, w \vDash \varphi$. Case 2: $\mathcal{M}, w \nvDash \Box p$. Then $\mathcal{M}_{|\varphi} = \mathcal{M}_{|p \wedge \Diamond q}$, and given $\mathcal{M}_{|p \wedge \Diamond q} \vDash \Box p$, we again have $\mathcal{M}_{|\varphi}, w \vDash \varphi$. Hence φ is successful. But φ is not super-successful, as shown by the partition model $\mathcal{N}$ with $W = \{w, v, u\}$, $V(p) = \{w\}$, and $V(q) = \{u\}$. We begin with $\mathcal{N}, w \vDash \varphi$, for while $\mathcal{N}, w \nvDash \Box p$, it holds that $\mathcal{N}, w \vDash p \wedge \Diamond q$. Moreover, given $W_{|\varphi} = \{w\}$, we still have $\mathcal{N}_{|\varphi}, w \vDash \varphi$, for while $\mathcal{N}_{|\varphi}, w \nvDash p \wedge \Diamond q$, it holds that $\mathcal{N}_{|\varphi}, w \vDash \Box p$. However, in the extension $\mathcal{N}'$ of $\mathcal{N}_{|\varphi}$ with $W' = \{w, v\}$, we now have $\mathcal{N}', w \nvDash \varphi$. The fact that $u \notin W'$ gives $\mathcal{N}', w \nvDash p \wedge \Diamond q$, and the fact that $v \in W'$ gives $\mathcal{N}', w \nvDash \Box p$. □

Proposition 5.10 *A formula φ is super-successful iff for a propositional variable p that does not occur in φ, $\varphi \vee p$ is successful.*

Proof. ($\Rightarrow$) Suppose φ is super-successful and $\mathcal{M}, w \vDash \varphi \vee p$, where p does not occur in φ. If $\mathcal{M}, w \vDash p$, then $\mathcal{M}_{|\varphi \vee p}, w \vDash p$, and if $\mathcal{M}, w \vDash \varphi$, then given $\mathcal{M}_{|\varphi} \subseteq \mathcal{M}_{|\varphi \vee p}$ and the assumption that φ is super-successful, $\mathcal{M}_{|\varphi \vee p}, w \vDash \varphi$. In either case, $\mathcal{M}_{|\varphi \vee p}, w \vDash \varphi \vee p$, so $\varphi \vee p$ is successful.

($\Leftarrow$) Suppose φ is not super-successful, so there is an $\mathcal{M} = \langle W, R, V \rangle$ with $w \in W$ such that $\mathcal{M}, w \vDash \varphi$, and an $\mathcal{M}' = \langle W', R', V' \rangle$ such that $\mathcal{M}_{|\varphi} \subseteq \mathcal{M}' \subseteq \mathcal{M}$ and $\mathcal{M}', w \nvDash \varphi$. Let $\mathcal{M}^* = \langle W, R, V^* \rangle$ be the same as $\mathcal{M}$ except that $V^*(p) = W' \setminus W_{|\varphi}$. Then since p does not occur in φ, we have $\mathcal{M}^*, w \vDash \varphi \vee p$ given $\mathcal{M}, w \vDash \varphi$. Moreover, since $W^*_{|\varphi \vee p} = W'$, we have $\mathcal{M}^*_{|\varphi \vee p}, w \nvDash \varphi$ given $\mathcal{M}', w \nvDash \varphi$. From the assumption that $\mathcal{M}, w \vDash \varphi$, it follows that $w \in W_{|\varphi}$, in which case $w \notin V^*(p)$ and hence $\mathcal{M}^*_{|\varphi \vee p}, w \nvDash p$. But then $\mathcal{M}^*_{|\varphi \vee p}, w \nvDash \varphi \vee p$, so $\varphi \vee p$ is unsuccessful. □

By Proposition 5.10, complexity and syntactic characterization results for successful formulas carry over immediately to super-successful formulas.

From the previous propositions we obtain a surprising failure of closure.

Corollary 5.11 *The set of successful formulas is not closed under disjunction.*

Proof. Immediate from Propositions 5.9 and 5.10. □

Example 5.12 The formula $\neg p \wedge \neg q$ is successful, and by the proof of Proposition 5.9, $(p \wedge \Diamond q) \vee \Box p$ is also successful. However, the disjunction of these formulas, $\chi \equiv (\neg p \wedge \neg q) \vee ((p \wedge \Diamond q) \vee \Box p)$ is unsuccessful, as shown by the model $\mathcal{N}$ in the proof of Proposition 5.9. We begin with $\mathcal{N}, w \vDash \chi$, since while $\mathcal{N}, w \nvDash \neg p \wedge \neg q$ (and hence $\mathcal{N}_{|\chi}, w \nvDash \neg p \wedge \neg q$), we have already seen that $\mathcal{N}, w \vDash (p \wedge \Diamond q) \vee \Box p$. However, $\mathcal{N}_{|\chi} = \mathcal{N}'$, and we have already seen that $\mathcal{N}', w \nvDash (p \wedge \Diamond q) \vee \Box p$.

By contrast, the set of super-successful formulas is closed under disjunction, since $\mathcal{M}_{|\varphi \vee \psi}$ is an extension of $\mathcal{M}_{|\varphi}$ and $\mathcal{M}_{|\psi}$.

Another consequence of the previous results concerns the relation between successful and *learnable* formulas. Like the Cartesian formulas, the learnable formulas have been discussed in connection with Fitch's paradox [3,9].

Definition 5.13 A formula φ is (*always*) *learnable* iff for all pointed models, if $\mathcal{M}, w \vDash$

φ, then there is some ψ such that $\mathcal{M}_{|\psi}, w \vDash \Box\varphi$.[7]

Formulas that are learnable (and satisfiable) are Cartesian according to Definition 5.6, but the converse does not hold [3]. For example, the formula $p \wedge \Diamond q$ is Cartesian, since $\Box(p \wedge \Diamond q)$ is satisfiable, but it is not (always) learnable, for if $\mathcal{M}, w \vDash \Box(p \to \neg q)$ and $\mathcal{M}' \subseteq \mathcal{M}$, then $\mathcal{M}', w \nvDash \Box(p \wedge \Diamond q)$.

All successful formulas are learnable [9], since if $\mathcal{M}, w \vDash \varphi$ and φ is successful, then not only $\mathcal{M}_{|\varphi}, w \vDash \varphi$ but also $\mathcal{M}_{|\varphi}, w \vDash \Box\varphi$. For if v is retained in $\mathcal{M}_{|\varphi}$, then $\mathcal{M}, v \vDash \varphi$, in which case $\mathcal{M}_{|\varphi}, v \vDash \varphi$ by the successfulness of φ. Hence for a successful φ, we can always take ψ in Definition 5.13 to be φ itself. On the other hand, it is natural to ask whether some unsuccessful formulas are learnable as well.

Corollary 5.14 *Not all learnable formulas are successful.*

Proof. Let $\delta_1 \vee \delta_2$ be an unsuccessful disjunction with δ_1 and δ_2 successful, as given by Corollary 5.11. If $\mathcal{M}, w \vDash \delta_1 \vee \delta_2$, then $\mathcal{M}, w \vDash \delta_i$ for $i = 1$ or $i = 2$. Since δ_i is successful, we have $\mathcal{M}_{|\delta_i}, w \vDash \Box\delta_i$, in which case $\mathcal{M}_{|\delta_i}, w \vDash \Box(\delta_1 \vee \delta_2)$. Since $\mathcal{M}$ was arbitrary, $\delta_1 \vee \delta_2$ is (always) learnable. □

6 Conclusion

From a technical point of view, we have studied the question of when a modal formula is preserved under relativizing models to the formula itself. In the epistemic interpretation of modal logic, this preservation question takes on a new significance: it concerns whether an agent retains knowledge of what is learned, which requires that what is learned remain true. We have shown (Theorem 3.13) that for an introspective agent, the only true sentences that may become false when learned are variants of the Moore sentence. For an agent without introspection or multiple agents without introspective access to each other's knowledge or beliefs, there are non-Moorean sources of unsuccessfulness (Propositions 4.1 and 4.2, Example 4.4).

In connection with our study of Moorean phenomena, we have observed a number of related results. We saw that the sentences that always provide information to an agent, no matter the agent's prior epistemic state, are exactly those sentences that cannot be known—and will eventually become false if repeated enough (Proposition 5.7); we saw that there are sentences that always remain true when they are learned, but whose truth value may oscillate while an agent is on the way to learning them (Proposition 5.9); and we saw that there are sentences that sometimes become false when learned directly, but which an agent can always come to know indirectly by learning something else (Corollary 5.14).

In Appendix A, we return to the problem with which we began. We give syntactic characterizations of the self-refuting and unsuccessful formulas as the *strong* Moore sentences and *strong* Moorean sentences, respectively.

[7] The term 'learnable' is used by van Benthem [3]. Balbiani et al. [9] use the term 'knowable'.

Acknowledgement

We wish to thank Hans van Ditmarsch and the anonymous referees for their helpful comments on an earlier draft of this paper, and Johan van Benthem for stimulating our interest in the topic of successful formulas.

References

[1] Baltag, A., L. Moss and S. Solecki, *The logic of public announcements, common knowledge and private suspicions*, in: I. Gilboa, editor, *Proceedings of the 7th Conference on Theoretical Aspects of Rationality and Knowledge (TARK 98)*, Morgan Kaufmann, 1998 pp. 43–56.

[2] Baltag, A., H. van Ditmarsch and L. Moss, *Epistemic logic and information update*, in: P. Adriaans and J. van Benthem, editors, *Philosophy of Information*, North-Holland, 2008 pp. 361–456.

[3] van Benthem, J., *What one may come to know*, Analysis **64** (2004), pp. 95–105.

[4] van Benthem, J., *One is a lonely number: on the logic of communication*, in: Z. Chatzidakis, P. Koepke and W. Pohlers, editors, *Logic Colloquium '02*, ASL & A.K. Peters, 2006 pp. 96–129.

[5] van Benthem, J., *Open problems in logical dynamics*, in: D. Gabbay, S. Goncharov and M. Zakharyashev, editors, *Mathematical Problems from Applied Logic I*, Springer, 2006 pp. 137–192.

[6] van Ditmarsch, H. and B. Kooi, *The secret of my success*, Synthese **151** (2006), pp. 201–232.

[7] van Ditmarsch, H., W. van der Hoek and B. Kooi, "Dynamic Epistemic Logic," Springer, 2008.

[8] Andréka, H., I. Németi and J. van Benthem, *Modal languages and bounded fragments of predicate logic*, Journal of Philosophical Logic **27** (1998), pp. 217–274.

[9] Balbiani, P., A. Baltag, H. van Ditmarsch, A. Herzig, T. Hoshi and T. de Lima, *'Knowable' as 'known after an announcement'*, The Review of Symbolic Logic **1** (2008), pp. 305–334.

[10] Carnap, R., *Modalities and quantification*, The Journal of Symbolic Logic **11** (1946), pp. 33–64.

[11] Fitch, F. B., *A logical analysis of some value concepts*, The Journal of Symbolic Logic **28** (1963), pp. 135–142.

[12] Gerbrandy, J., "Bisimulations on Planet Kripke," Ph.D. thesis, University of Amsterdam (1999), ILLC Dissertation Series DS-1999-01.

[13] Hintikka, J., "Knowledge and Belief: An Introduction to the Logic of the Two Notions," College Publications, 2005.

[14] Lutz, C., *Complexity and succinctness of public announcement logic*, in: P. Stone and G. Weiss, editors, *Proceedings of the Fifth International Joint Conference on Autonomous Agents and Multiagent Systems (AAMAS 06)* (2006), pp. 137–143.

[15] Meyer, J.-J. Ch. and W. van der Hoek, "Epistemic Logic for AI and Computer Science," Cambridge Tracts in Theoretical Computer Science **41**, Cambridge University Press, 1995.

[16] Moore, G. E., *A reply to my critics*, in: P. Schilpp, editor, *The Philosophy of G.E. Moore*, The Library of Living Philosophers **4**, Northwestern University, 1942 pp. 535–677.

[17] Plaza, J., *Logics of public communications*, in: M. Emrich, M. Pfeifer, M. Hadzikadic and Z. Ras, editors, *Proceedings of the 4th International Symposium on Methodologies for Intelligent Systems: Poster Session Program*, Oak Ridge National Laboratory, ORNL/DSRD-24, 1989 pp. 201–216.

[18] Qian, L., *Sentences true after being announced*, in: *Proceedings of the Student Session of the 1st North American Summer School in Logic, Language, and Information (NASSLLI)*, Stanford University, 2002.

[19] Tennant, N., "The Taming of The True," Oxford: Clarendon Press, 1997.

[20] Visser, A., J. van Benthem, D. de Jongh and G. R. R. de Lavalette, *NNIL, A study in intuitionistic propositional logic*, Logic Group Preprint Series 111, Philosophical Institute, Utrecht University (1994).

[21] Zeman, J. J., *A study of some systems in the neighborhood of S4.4*, Notre Dame Journal of Formal Logic **12** (1971), pp. 341–357.

A Appendix

In this appendix, we address the problem of giving syntactic characterizations of the successful and self-refuting formulas. Given Theorem 2.6, Lemma 3.6, and the PAL definitions of successful and self-refuting, there is a trivial characterization of both classes: φ is successful iff the result of reducing the PAL formula $\neg[\varphi]\varphi$ to normal form in the basic modal language is not clear; φ is self-refuting iff the result of reducing the PAL formula $\neg[\varphi]\neg\varphi$ to normal form in the basic modal language is not clear. Although by Proposition 2.7, the PAL definitions of successful and self-refuting lead to optimal methods for checking for these properties, they do not provide much insight beyond the semantic definitions of the formula classes. In Theorems A.3 and A.6 below, we give syntactic characterizations of the self-refuting and unsuccessful formulas that reveal more than the trivial characterization.

We state the following results for **S5** with the standard definitions of successful ($\forall \mathcal{M}, w : \mathcal{M}, w \vDash \varphi \Rightarrow \mathcal{M}_{|\varphi}, w \vDash \varphi$) and self-refuting ($\forall \mathcal{M}, w : \mathcal{M}, w \vDash \varphi \Rightarrow \mathcal{M}_{|\varphi}, w \nvDash \varphi$). With minor changes the results also hold for **KD45**, given the modified definitions using precondition $\Diamond^{+}\varphi$ instead of φ.

We will continue to use the abbreviations δ^{α}, $\delta^{\Box\Diamond}$, etc., from Definition 3.3. We will also use $\hat{\delta}^{\Box\Diamond}$ to denote a conjunct of $\delta^{\Box\Diamond}$, i.e., a $\Box\beta_i$ or $\Diamond\gamma_k$ in δ. As before, $\sim\varphi$ is the negation of φ in negation normal form, with $\neg$ applying only to literals.

Definition A.1 φ is a *strong Moore sentence* iff for any normal form φ^* of φ, no disjunct of φ^* is *compensated in* φ^*. A disjunct $\delta \equiv \alpha \wedge \Box\beta_1 \wedge ... \wedge \Box\beta_n \wedge \Diamond\gamma_1 \wedge ... \wedge \Diamond\gamma_m$ of φ^* is compensated in φ^* iff there is a disjunct δ' of φ^*, a subset S of the disjuncts of φ^*, a sequence of disjuncts $\sigma_{\gamma_1}, ..., \sigma_{\gamma_m}$ (not necessarily distinct) from S, and for every $\sigma \notin S$ a conjunct $\hat{\sigma}^{\Box\Diamond}$ of $\sigma^{\Box\Diamond}$, such that $\chi \equiv \delta' \wedge \delta^{\alpha} \wedge \chi_{\Box} \wedge \chi_{\Diamond}$ is clear, where

$$\chi_{\Box} \equiv \bigwedge_{\sigma \in S} \sigma^{\Box\Diamond} \wedge \bigwedge_{\sigma \notin S} \sim\hat{\sigma}^{\Box\Diamond} \wedge \bigwedge_{\sigma \in S, j \leq n} \Box(\sim\sigma^{\alpha} \vee \beta_j);$$

$$\chi_{\Diamond} \equiv \bigwedge_{k \leq m} \Diamond\left(\sigma^{\alpha}_{\gamma_k} \wedge \gamma_k\right).$$

Note that χ is in normal form, so the definition of clarity properly applies. The reason for the use of $\sim\hat{\sigma}^{\Box\Diamond}$ is that $\sim\sigma^{\Box\Diamond}$ may be a disjunction, in which case χ would not be in normal form. Hence we pull the disjunction out of the formula and add an existential quantifier to the condition, since $\sigma^{\Box\Diamond}$ is false iff there is some conjunct $\hat{\sigma}^{\Box\Diamond}$ of $\sigma^{\Box\Diamond}$ that is false.

As we will see in the proof of Theorem A.3, the existence of a compensated disjunct in φ is equivalent to there being a model in which φ has a "successful update," in the sense that $\mathcal{M}, w \vDash \varphi$ and $\mathcal{M}_{|\varphi}, w \vDash \varphi$. For φ to have a successful update in $\mathcal{M}$, there must be a disjunct δ' of φ true at a point w in $\mathcal{M}$ and a disjunct δ of φ (possibly distinct from δ') true at w in $\mathcal{M}_{|\varphi}$. What is required for δ to be true at w in $\mathcal{M}_{|\varphi}$ is that its propositional part δ^{α} is true at w in $\mathcal{M}$, that its universal part $\delta^{\Box}$ is already true at w in $\mathcal{M}$ or *becomes* true at w in $\mathcal{M}_{|\varphi}$, and that its existential part $\delta^{\Diamond}$ remains true at w in $\mathcal{M}_{|\varphi}$. This is exactly what χ captures.

Proposition A.2 *Every strong Moore sentence is a Moore sentence.*

Proof. By Definitions A.1 and 3.12, it suffices to show that for φ in normal form, if a disjunct δ of φ is not compensated in φ, then δ is a Moore conjunction. To prove the contrapositive, suppose δ is not a Moore conjunction. Then by Definition 3.7, $\delta \wedge \Diamond\delta^{\alpha}$ is clear and $\delta \wedge \Diamond(\delta^{\alpha} \wedge \gamma_k)$ is clear for every $\Diamond\gamma_k$ conjunct in δ. It follows that $\delta \wedge \bigwedge_{k \leq m} \Diamond(\delta^{\alpha} \wedge \gamma_k)$ is clear and hence satisfiable. Given $\mathcal{M}, w \vDash \delta \wedge \bigwedge_{k \leq m} \Diamond(\delta^{\alpha} \wedge \gamma_k)$, let S be the set of disjuncts in φ such that $\mathcal{M}, w \vDash \bigwedge_{\sigma \in S} \sigma^{\Box\Diamond} \wedge \bigwedge_{\sigma \notin S} \neg\sigma^{\Box\Diamond}$. For every $\sigma \notin S$, pick a conjunct $\hat{\sigma}^{\Box\Diamond}$ of $\sigma^{\Box\Diamond}$ such that $\mathcal{M}, w \vDash \bigwedge_{\sigma \in S} \sigma^{\Box\Diamond} \wedge \bigwedge_{\sigma \notin S} \sim \hat{\sigma}^{\Box\Diamond}$. Since $\delta \in S$, let the sequence $\sigma_{\gamma_1}, ..., \sigma_{\gamma_m}$ of disjuncts from S be such that $\sigma_{\gamma_i} \equiv \delta$ for $1 \leq i \leq m$. We claim that for the χ as in Definition A.1 based on these choices, $\mathcal{M}, w \vDash \chi$. From the fact that $\mathcal{M}, w \vDash \bigwedge_{k \leq m} \Diamond(\delta^{\alpha} \wedge \gamma_k)$ and our choice of $\sigma_{\gamma_1}, ..., \sigma_{\gamma_m}$, we have $\mathcal{M}, w \vDash \chi_{\Diamond}$. Since $\mathcal{M}, w \vDash \delta^{\Box}$, it is immediate that $\mathcal{M}, w \vDash \bigwedge_{\sigma \in S, j \leq n} \Box(\sim\sigma \vee \beta_j)$. Together with $\mathcal{M}, w \vDash \bigwedge_{\sigma \in S} \sigma^{\Box\Diamond} \wedge \bigwedge_{\sigma \notin S} \sim \hat{\sigma}^{\Box\Diamond}$, this gives $\mathcal{M}, w \vDash \chi_{\Box}$. Finally, setting $\delta' \equiv \delta$, we have $\mathcal{M}, w \vDash \chi$. Hence χ is clear, so δ is compensated in φ. □

We can now generalize Theorem 3.13(i).

Theorem A.3 *φ is self-refuting if and only if φ is a strong Moore sentence.*

Proof. φ is self-refuting iff any equivalent normal form of φ is self-refuting, so let us assume φ is already in normal form.

($\Leftarrow$) We prove the contrapositive. Suppose φ is not self-refuting, so there is a pointed model such that $\mathcal{M}, w \vDash \varphi$ and $\mathcal{M}_{|\varphi}, w \vDash \varphi$. Given $\mathcal{M}_{|\varphi}, w \vDash \varphi$, there is a disjunct δ of φ such that $\mathcal{M}_{|\varphi}, w \vDash \delta$. We claim that δ is compensated in φ, so φ is not a strong Moore sentence. It suffices to show the satisfiability of an appropriate χ as in Definition A.1.

Since $\mathcal{M}, w \vDash \varphi$, there is a disjunct δ' of φ such that $\mathcal{M}, w \vDash \delta'$. This gives the first conjunct of χ. Since $\mathcal{M}_{|\varphi}, w \vDash \delta$, we have $\mathcal{M}, w \vDash \delta^{\alpha}$ because δ^{α} is propositional and hence preserved under extensions. This gives the second conjunct of χ.

Next we claim $\mathcal{M}, w \vDash \chi_{\Box}$. Let S be the set of the disjuncts of φ such that $\mathcal{M}, w \vDash \bigwedge_{\sigma \in S} \sigma^{\Box\Diamond} \wedge \bigwedge_{\sigma \notin S} \neg\sigma^{\Box\Diamond}$, and let $\hat{\sigma}^{\Box\Diamond}$ be a false conjunct of each $\sigma^{\Box\Diamond}$ for $\sigma \notin S$. For *reductio*, suppose $\mathcal{M}, w \nvDash \bigwedge_{\sigma \in S, j \leq n} \Box(\sim\sigma^{\alpha} \vee \beta_j)$, where $\Box\beta_1 \wedge ... \wedge \Box\beta_n \equiv \delta^{\Box}$. Then there is some v with wRv and $\mathcal{M}, v \vDash \sigma^{\alpha} \wedge \neg\beta_j$ for some $\sigma \in S$ and $j \leq n$. Since $\sigma \in S$, we have $\mathcal{M}, w \vDash \sigma^{\Box\Diamond}$. It follows by Lemma 3.8(i) that $\mathcal{M}, v \vDash \sigma^{\Box\Diamond}$, in which case $\mathcal{M}, v \vDash \sigma$

given $\mathcal{M}, v \vDash \sigma^{\alpha}$. Hence v is retained in $\mathcal{M}_{|\varphi}$. But then given $\mathcal{M}_{|\varphi}, v \nvDash \beta_j$, we have $\mathcal{M}_{|\varphi}, w \nvDash \Box\beta_j$, which contradicts the assumption that $\mathcal{M}_{|\varphi}, w \vDash \delta$. We conclude that $\mathcal{M}, w \vDash \chi_{\Box}$, which gives the third conjunct of χ.

Finally, we claim $\mathcal{M}, w \vDash \chi_{\Diamond}$. Given $\mathcal{M}_{|\varphi}, w \vDash \delta^{\Diamond}$, take an arbitrary $\Diamond\gamma_k$ in δ, and let v be such that wRv and $\mathcal{M}_{|\varphi}, v \vDash \gamma_k$. It follows that $\mathcal{M}, v \vDash \sigma$ for some disjunct σ of φ, which we label as σ_{γ_k}, for otherwise v would not be retained in $\mathcal{M}_{|\varphi}$. Since γ_k is propositional, $\mathcal{M}, v \vDash \gamma_k$ given $\mathcal{M}_{|\varphi}, v \vDash \gamma_k$. Therefore $\mathcal{M}, w \vDash \Diamond\left(\sigma^{\alpha}_{\gamma_k} \wedge \gamma_k\right)$. Then from the fact that $\mathcal{M}, v \vDash \sigma^{\Box\Diamond}_{\gamma_k}$, we have $\mathcal{M}, w \vDash \sigma^{\Box\Diamond}_{\gamma_k}$ by Lemma 3.8(i), so $\sigma_{\gamma_k} \in S$. Since $\Diamond\gamma_k$ was arbitrary, $\mathcal{M}, w \vDash \chi_{\Diamond}$, which gives the final conjunct of χ.

($\Rightarrow$) Again we prove the contrapositive. Suppose δ is not a strong Moore sentence, so there is some δ that is compensated in φ, for which an appropriate χ as in Definition A.1 is clear. Where $\mathcal{M}, w \vDash \chi$, we claim that $\mathcal{M}_{|\varphi}, w \vDash \delta$. We will show $\mathcal{M}_{|\varphi}, w \vDash \delta^{\alpha}$, $\mathcal{M}_{|\varphi}, w \vDash \delta^{\Box}$, and $\mathcal{M}_{|\varphi}, w \vDash \delta^{\Diamond}$ separately.

Given $\mathcal{M}, w \vDash \delta$, we have $\mathcal{M}_{|\varphi}, w \vDash \delta^{\alpha}$ since δ^{α} is propositional.

Next, for any v retained in $\mathcal{M}_{|\varphi}$, we have $\mathcal{M}, v \vDash \sigma^{\alpha}$ for some disjunct σ of φ. It must be that $\sigma \in S$, for otherwise $\mathcal{M} \vDash \neg\sigma$ given $\mathcal{M}, w \vDash \chi_{\Box}$ and Lemma 3.8(i). Then from the fact that $\mathcal{M}, w \vDash \bigwedge_{\sigma \in S, j \leq n} \Box(\sim \sigma^{\alpha} \vee \beta_j)$, we have $\mathcal{M}_{|\varphi}, v \vDash \beta_j$ for all $j \leq n$. Since v was arbitrary, $\mathcal{M}_{|\varphi}, w \vDash \delta^{\Box}$.

Finally, given $\mathcal{M}, w \vDash \chi_{\Diamond}$, for any $\Diamond\gamma_k$ in δ we have $\mathcal{M}, w \vDash \Diamond(\sigma_{\gamma_k} \wedge \gamma_k)$ with $\sigma_{\gamma_k} \in S$. Let v be such that wRv and $\mathcal{M}, v \vDash \sigma^{\alpha}_{\gamma_k} \wedge \gamma_k$. Since $\sigma_{\gamma_k} \in S$, $\mathcal{M}, w \vDash \sigma^{\Box\Diamond}_{\gamma_k}$. It follows by Lemma 3.8(i) that $\mathcal{M}, v \vDash \sigma^{\Box\Diamond}_{\gamma_k}$, in which case $\mathcal{M}, v \vDash \sigma_{\gamma_k}$ given $\mathcal{M}, v \vDash \sigma^{\alpha}_{\gamma_k}$. Hence v is retained in $\mathcal{M}_{|\varphi}$. Since γ_k is propositional, we have $\mathcal{M}_{|\varphi}, v \vDash \gamma_k$ given $\mathcal{M}, v \vDash \gamma_k$, whence $\mathcal{M}_{|\varphi}, w \vDash \Diamond\gamma_k$. Since γ_k was arbitrary, $\mathcal{M}_{|\varphi}, w \vDash \delta^{\Diamond}$.

We conclude that $\mathcal{M}_{|\varphi}, w \vDash \delta$, in which case $\mathcal{M}_{|\varphi}, w \vDash \varphi$. Given $\mathcal{M}, w \vDash \chi$, we have $\mathcal{M}, w \vDash \delta$ and hence $\mathcal{M}, w \vDash \varphi$, so φ is not self-refuting. $\Box$

Finally, we will prove an analogous generalization of Theorem 3.13(ii).

Definition A.4 φ is a *strong Moorean sentence* iff for any normal form φ^* of φ, there is a disjunct δ and non-empty sets S and T of disjuncts of φ^*, with for every $\theta \in T$, a $\Diamond\gamma_\theta$ in θ, such that $\chi \equiv \delta \wedge \chi_1 \wedge \chi_2 \wedge \chi_3$ is clear, where

$$t(\theta) \equiv \theta^{\alpha\Diamond} \wedge \bigwedge_{\sigma \in S,\, \Box\beta \text{ in } \theta} \Box(\sim \sigma^{\alpha} \vee \beta);$$

$$\chi_1 \equiv \bigwedge_{\theta \in T} t(\theta) \wedge \bigwedge_{\theta \notin T} \neg t(\theta);\ \chi_2 \equiv \bigwedge_{\sigma \in S} \sigma^{\Box\Diamond} \wedge \bigwedge_{\sigma \notin S} \neg\sigma^{\Box\Diamond};$$

$$\chi_3 \equiv \bigwedge_{\sigma \in S, \theta \in T} \Box(\sim \sigma^{\alpha} \vee \sim \gamma_\theta).$$

The formula χ is not yet in normal form, due to the $\neg t(\theta)$ conjuncts in χ_1 and the $\neg\sigma^{\Box\Diamond}$ conjuncts in χ_2, so strictly the definition of clarity does not apply. However, it is straightforward to put χ into normal form using the same method involving $\hat{\sigma}^{\Box\Diamond}$ as in Definition A.1, together with some distribution of $\wedge$ and $\vee$. Since in this case the necessary modifications add four existential quantifiers to the definition, for simplicity we do not write them out. When we say that χ is clear, strictly we mean that the modified formula is clear.

As we will see in the proof of Theorem A.6, the clarity of χ is equivalent to there being a model in which φ has an unsuccessful update. For φ to have an unsuccessful update in $\mathcal{M}$, there must be some disjunct δ in φ such that $\mathcal{M}, w \vDash \delta$, but no disjunct δ' such that $\mathcal{M}_{|\varphi}, w \vDash \delta'$. To ensure that there are no such δ', we need only keep track of those disjuncts θ of φ whose propositional part θ^{α} and existential part $\theta^{\Diamond}$ are true in $\mathcal{M}$ and whose universal part $\theta^{\Box}$ was already true in $\mathcal{M}$ or *becomes* true in $\mathcal{M}_{|\varphi}$, since all other disjuncts will be false in $\mathcal{M}_{|\varphi}$. This is the purpose of χ_1. For each such θ, we must ensure that there is a diamond formula $\Diamond\gamma_{\theta}$ in θ that becomes false in $\mathcal{M}_{|\varphi}$, because none of its witnesses satisfy any of the disjuncts that are satisfied somewhere in $\mathcal{M}$. This is the purpose of χ_3 and χ_2.

Proposition A.5 *Every strong Moorean sentence is a Moorean sentence.*

Proof. Assume φ is a strong Moorean sentence, so an appropriate χ as in Definition A.4 is clear. Note that the distinguished disjunct δ of φ^* must be a member of both S and T. Then given that $\delta \wedge \chi_3$ is clear, we have that $\delta \wedge \Box(\sim \delta^{\alpha} \vee \sim \gamma_{\delta})$ is clear, where $\Diamond\gamma_{\delta}$ is a conjunct of δ. Hence δ is a Moorean conjunction by Definition 3.7, in which case φ is a Moorean sentence by Definition 3.12. $\Box$

Theorem A.6 *φ is unsuccessful if and only if φ is a strong Moorean sentence.*

Proof. As before, let us assume that φ is already in normal form.

($\Leftarrow$) Suppose φ is a strong Moorean sentence, so an appropriate χ as in Definition A.4 is clear. Consider a pointed model such that $\mathcal{M}, w \vDash \chi$. For *reductio*, assume $\mathcal{M}_{|\varphi}, w \vDash \varphi$. Then there exists a disjunct θ in φ such that $\mathcal{M}_{|\varphi}, w \vDash \theta$.

We claim that $\theta \in T$. For if $\theta \notin T$, then given $\mathcal{M}, w \vDash \chi_1$ there are two cases. Case 1: $\mathcal{M}, w \nvDash \theta^{\alpha\Diamond}$. Then $\mathcal{M}_{|\varphi}, w \nvDash \theta^{\alpha\Diamond}$ since $\theta^{\alpha\Diamond}$ is existential. Case 2: $\mathcal{M}, w \vDash \Diamond(\sigma^{\alpha} \wedge \neg\beta)$ for some $\sigma \in S$ and $\Box\beta$ in θ. Let v be such that wRv and $\mathcal{M}, v \vDash \sigma^{\alpha} \wedge \neg\beta$. Since $\sigma \in S$, we have $\mathcal{M}, w \vDash \sigma^{\Box\Diamond}$, in which case $\mathcal{M}, v \vDash \sigma$ given Lemma 3.8(i) and $\mathcal{M}, v \vDash \sigma^{\alpha}$. Hence v is retained in $\mathcal{M}_{|\varphi}$, and since β is propositional, $\mathcal{M}_{|\varphi}, v \nvDash \beta$, so $\mathcal{M}_{|\varphi}, w \nvDash \Box\beta$ and $\mathcal{M}_{|\varphi}, w \nvDash \theta^{\Box}$. In both cases, $\mathcal{M}_{|\varphi}, w \nvDash \theta$, a contradiction. Therefore $\theta \in T$.

Given $\mathcal{M}_{|\varphi}, w \vDash \theta$, for every $\Diamond\gamma$ in θ there is a v with wRv and $\mathcal{M}_{|\varphi}, v \vDash \gamma$. Since v was retained in $\mathcal{M}_{|\varphi}$, we have $\mathcal{M}, v \vDash \sigma$ for some $\sigma \in S$. Then given $\mathcal{M}, w \vDash \chi_3$ and $\theta \in T$, we have $\mathcal{M}, v \vDash \neg\gamma$. Since γ is propositional, $\mathcal{M}_{|\varphi}, v \vDash \neg\gamma$, a contradiction. We conclude that $\mathcal{M}_{|\varphi}, w \nvDash \varphi$, so φ is unsuccessful.

($\Rightarrow$) Suppose φ is unsuccessful, so there is a pointed model with $\mathcal{M}, w \vDash \varphi$ but $\mathcal{M}_{|\varphi}, w \nvDash \varphi$. To show that φ is a strong Moorean sentence, it suffices to show that an appropriate χ as in Definition A.4 is satisfiable. Given $\mathcal{M}, w \vDash \varphi$, we can read off from w the disjunct δ and sets S and T such that $\mathcal{M}, w \vDash \delta \wedge \chi_1 \wedge \chi_2$. It only remains to show $\mathcal{M}, w \vDash \chi_3$. For *reductio*, suppose there is a $\theta \in T$ such that for all $\Diamond\gamma$ in θ, $\mathcal{M}, w \nvDash \bigwedge_{\sigma \in S} \Box(\sim \sigma^{\alpha} \vee \sim \gamma)$, i.e., $\mathcal{M}, w \vDash \Diamond(\sigma^{\alpha} \wedge \gamma)$ for some $\sigma \in S$. Then we claim that $\mathcal{M}_{|\varphi}, w \vDash \theta$. As before, we take the three parts of θ separately.

For $\theta^{\Diamond}$, consider any $\Diamond\gamma$ in θ and let v be such that wRv and $\mathcal{M}, v \vDash \sigma^{\alpha} \wedge \gamma$. Given $\mathcal{M}, w \vDash \chi_2$ and the fact that $\sigma \in S$, we have $\mathcal{M}, w \vDash \sigma^{\Box\Diamond}$, so $\mathcal{M}, v \vDash \sigma$ by Lemma 3.8(i). Hence v is retained in $\mathcal{M}_{|\varphi}$. Since $\mathcal{M}, v \vDash \gamma$ and γ is propositional, $\mathcal{M}_{|\varphi}, v \vDash \gamma$ and therefore $\mathcal{M}_{|\varphi}, w \vDash \Diamond\gamma$. Since $\Diamond\gamma$ was arbitrary, $\mathcal{M}_{|\varphi}, w \vDash \theta^{\Diamond}$.

For $\theta^{\Box}$, take any v retained in $\mathcal{M}_{|\varphi}$ such that wRv, and we have $\mathcal{M}, v \vDash \sigma^{\alpha}$ for some $\sigma \in S$. It follows that for any $\Box\beta$ in θ, we have $\mathcal{M}, v \vDash \beta$ given $\mathcal{M}, w \vDash \chi_1$. Since β is propositional, $\mathcal{M}_{|\varphi}, v \vDash \beta$. Since v and β were arbitrary, $\mathcal{M}_{|\varphi}, w \vDash \delta^{\Box}$.

Finally, for θ^{α}, since by assumption $\theta \in T$ and $\mathcal{M}, w \vDash \chi_1$, we have $\mathcal{M}, w \vDash \theta^{\alpha}$ and hence $\mathcal{M}_{|\varphi}, w \vDash \theta^{\alpha}$.

We have shown $\mathcal{M}_{|\varphi}, w \vDash \theta$ and hence $\mathcal{M}_{|\varphi}, w \vDash \varphi$, which contradicts our initial assumption. It follows that for every $\theta \in T$ there is a $\Diamond\gamma_{\theta}$ in θ such that $\mathcal{M}, w \vDash \chi_3$. $\Box$

Completeness Proof by Semantic Diagrams for Transitive Closure of Accessibility Relation

Ryo Kashima

Department of Mathematical and Computing Sciences, Tokyo Institute of Technology. Ookayama, Meguro, Tokyo 152-8552, Japan.
kashima@is.titech.ac.jp

Abstract

We treat the smallest normal modal propositional logic with two modal operators $\Box$ and $\Box^+$. While $\Box$ is interpreted in Kripke models by the accessibility relation R, $\Box^+$ is interpreted by the transitive closure of R. Intuitively the formula $\Box^+\varphi$ means the infinite conjunction $\Box\varphi \wedge \Box\Box\varphi \wedge \Box\Box\Box\varphi \wedge \cdots$. There is a Hilbert style axiomatization of this logic (a characteristic axiom is $\Box\varphi \wedge \Box^+(\varphi \to \Box\varphi) \to \Box^+\varphi$, called "induction axiom"), and its completeness with respect to finite models was shown by the canonical model method. This paper gives an alternative proof of this completeness. We use the method of "semantic diagram", which is a variant of semantic tableaux, as follows. Given an unprovable formula φ, we first make a small model (consisting of one world that forces φ to be false); then we add worlds step by step using the Hilbert system as an oracle, and finally we get a finite countermodel for φ. The point is how to handle $\Box^+$ in this construction.

Keywords: completeness of modal logic, transitive closure of accessibility relation, semantic diagram

1 Introduction

In Kripke models, the modal operator $\Box$ is interpreted as

$$w \models \Box\varphi \iff x \models \varphi \text{ for any } x \text{ such that } wRx$$

where w and x are possible worlds and R is the accessibility relation. Then we introduce a new modal operator $\Box^+$ by

$$w \models \Box^+\varphi \iff x \models \varphi \text{ for any } x \text{ such that } wR^+x$$

where R^+ is the transitive closure of R. Intuitively $\Box^+\varphi$ means the infinite conjunction as follows:

$$\Box^+\varphi \quad \leftrightarrow \quad \Box\varphi \wedge \Box\Box\varphi \wedge \Box\Box\Box\varphi \wedge \cdots .$$

This paper treats the smallest normal modal propositional logic with the operators $\Box$ and $\Box^+$ as above. This logic will be called K^+.

The relationship between $\Box$ and $\Box^+$ in K^+ is equal to that between the operators E ("everyone knows") and C ("common knowledge") in the common knowledge logic, since

$$C\varphi \quad \leftrightarrow \quad E\varphi \wedge EE\varphi \wedge EEE\varphi \wedge \cdots .$$

Moreover the relationship is similar to that between the operators X ("next time") and G ("globally") in temporal logic, since

$$G\varphi \quad \leftrightarrow \quad \varphi \wedge X\varphi \wedge XX\varphi \wedge \cdots .$$

There are Hilbert style systems for the common knowledge logic and the temporal logic, and the completeness with respect to finite models (i.e., a formula is provable in a system if it is true in every finite model) was proved by using canonical models and filtrations (see, e.g., [2] and [4]). Of course the argument can be applied to K^+ — there is a Hilbert system, which we will call HK^+ (a characteristic axiom is the *induction axiom*: $\Box\varphi \wedge \Box^+(\varphi \to \Box\varphi) \to \Box^+\varphi$), and the completeness with respect to finite models can be shown by using canonical models and filtrations.

The purpose of this paper is to give a new proof for the completeness of HK^+. We use the method of "semantic diagram", which is a variant of semantic tableaux, as follows. Given an unprovable formula α_0, we first make a small model (consisting of one world that forces α_0 to be false); then we add worlds step by step using HK^+ as an oracle, and finally we get a finite countermodel for α_0.

Here we give an informal explanation of the point of our method. It is well known that the finite set $\mathrm{Sub}^{\pm}(\alpha_0) = \{\varphi, \neg\varphi \mid \varphi \text{ is a subformula of } \alpha_0\}$ is sufficient for the construction of a countermodel for α_0. Then the point of our method is how to make the witness of $\Diamond^+\varphi$. If $\Diamond^+\varphi \in \Gamma$ and a world $\boxed{\Gamma}$ (this means all the elements of Γ are true at this world) is in a Kripke model, then we may consider a path to the witness $\boxed{\varphi}$ to be of the form

$$\boxed{\Gamma} \xrightarrow{R} \boxed{\Gamma'} \xrightarrow{R} \cdots \xrightarrow{R} \boxed{\Gamma''} \xrightarrow{R} \boxed{\varphi} \tag{1}$$

where $\Gamma, \Gamma', \ldots, \Gamma''$ are *mutually distinct* subsets of $\mathrm{Sub}^{\pm}(\alpha_0)$. For example, suppose imaginarily that the powerset $\mathcal{P}(\mathrm{Sub}^{\pm}(\alpha_0))$ consists of just three sets Γ, Δ and Λ; then the candidates of paths to the witness can be limited to the five paths:

$$\begin{array}{l} \boxed{\Gamma} \xrightarrow{R} \boxed{\varphi} \\ \boxed{\Gamma} \xrightarrow{R} \boxed{\Delta} \xrightarrow{R} \boxed{\varphi} \\ \boxed{\Gamma} \xrightarrow{R} \boxed{\Lambda} \xrightarrow{R} \boxed{\varphi} \\ \boxed{\Gamma} \xrightarrow{R} \boxed{\Delta} \xrightarrow{R} \boxed{\Lambda} \xrightarrow{R} \boxed{\varphi} \\ \boxed{\Gamma} \xrightarrow{R} \boxed{\Lambda} \xrightarrow{R} \boxed{\Delta} \xrightarrow{R} \boxed{\varphi} \end{array} \tag{2}$$

This limitation is justified by the following argument. Given a long path from Σ_1 $(=\Gamma)$ to φ as

$$\boxed{\Sigma_1} \xrightarrow{R} \boxed{\Sigma_2} \xrightarrow{R} \cdots \xrightarrow{R} \boxed{\Sigma_k} \xrightarrow{R} \boxed{\varphi}, \tag{3}$$

we can extract a *skipping path* $(\Sigma_{a_1}, \Sigma_{a_2}, \ldots, \Sigma_{a_m})$ such that

- Σ_{a_1} is *the last* Σ_1 *before* φ; that is, Σ_{a_1} is the same set as Σ_1, and none of $\Sigma_{a_1+1}, \Sigma_{a_1+2}, \ldots, \Sigma_k$ are the same set as Σ_1;
- Σ_{a_2} is the last Σ_{a_1+1} before φ;

  ⋮

- $\Sigma_{a_m} = \Sigma_k$ is the last $\Sigma_{a_{m-1}+1}$ before φ.

Then

$$\boxed{\Sigma_{a_1}} \xrightarrow{R} \boxed{\Sigma_{a_2}} \xrightarrow{R} \cdots \xrightarrow{R} \boxed{\Sigma_{a_m}} \xrightarrow{R} \boxed{\varphi}$$

is the very path denoted by (1), of length $\leq |\mathcal{P}(\mathrm{Sub}^{\pm}(\alpha_0))|$.

This principle of extraction (of length-limited paths from unlimited paths) is the core of our method. While such a principle was used in Brünnler and Lange [1] and Gaintzarain et al. [3] for temporal logics, the originality of this paper is that our method *does not need the until operator*. If a binary operator U' is available as

$$w \models \alpha U' \beta \iff \exists w_1, \ldots, \exists w_n \Big(w R w_1 R \cdots R w_n, w_i \models \alpha \text{ for } i < n, \text{ and } w_n \models \beta \Big),$$

then the condition "Σ_{a_1} is the last Σ_1 before φ" can be easily described by putting

$$\Sigma_{a_1} = \Sigma_1 \cup \{(\neg \Sigma_1) U' \varphi\}.$$

(Brünnler and Lange [1] and Gaintzarain et al. [3] introduced similar description as an inference rule of sequent calculi, and proved the completeness of the calculi.) However our K^+ does not have the until operator; hence we realize the extraction by *explicit enumeration of all the possible candidates of paths to the witness*, like (2) above.

2 Axiomatization

Formulas are constructed from the following symbols: propositional variables (the set of propositional variables is called **Prop**); logical connectives $\wedge$ and $\neg$; and modal operators $\Box$ and $\Box^+$. We will use letters $p, q, \dots$ to denote propositional variables, and letters $\alpha, \beta, \dots \varphi, \psi, \dots$ to denote formulas. Other symbols $(\bot, \top, \to, \vee, \Diamond, \Diamond^+, \dots)$ are defined by the usual abbreviations. Parentheses are omitted by the convention that the unary operators $\neg$, $\Box$, $\Box^+$, $\Diamond$, and $\Diamond^+$ bind stronger than other connectives, $\wedge$ and $\vee$ bind stronger than $\to$, and that $\alpha_1 \to \alpha_2 \to \cdots \to \alpha_n = \alpha_1 \to (\alpha_2 \to (\cdots \to (\alpha_{n-1} \to \alpha_n) \cdots))$. For example, the axiom scheme (A2) below is $(\Box(\alpha \to \beta)) \to ((\Box\alpha) \to \Box\beta)$, and $\neg\alpha \wedge \beta \to \Box^+\gamma \vee \delta = ((\neg\alpha) \wedge \beta) \to ((\Box^+\gamma) \vee \delta)$.

A *Kripke model* is a triple $M = \langle W, R, V \rangle$ where W is a nonempty set (the set of *possible worlds*), R is a binary relation on W (the *accessibility relation*), and V is a function from $W \times$ **Prop** to $\{\mathsf{True}, \mathsf{False}\}$. M is said to be *finite* if W is a finite set. The transitive closure of R is denoted by R^+; that is, xR^+y holds if and only if $x = a_0 R a_1 R \cdots R a_n = y$ for some $a_0, a_1, \dots, a_n$ $(n \geq 1)$. The notion "a formula φ is true at a world w in M", written by "$M, w \models \varphi$" (or "$w \models \varphi$" for short), is defined as usual: $w \models p \iff V(w,p) = \mathsf{True}$; $w \models \alpha \wedge \beta \iff w \models \alpha$ and $w \models \beta$; $w \models \neg\alpha \iff w \not\models \alpha$; $w \models \Box\alpha \iff x \models \alpha$ for any x such that wRx; and $w \models \Box^+\alpha \iff x \models \alpha$ for any x such that wR^+x. We say that a formula φ is *valid in M* if and only if $M, x \models \varphi$ for any world x.

The system HK$^+$ is defined as follows (cf. the axiomatization of linear temporal logic in [4, §9]). The axiom schemata are

(A1) instances of classical tautologies,

(A2) $\Box(\alpha \to \beta) \to \Box\alpha \to \Box\beta$ ('K axiom' for $\Box$),

(A3) $\Box^+(\alpha \to \beta) \to \Box^+\alpha \to \Box^+\beta$ ('K axiom' for $\Box^+$),

(A4) $\Box^+\alpha \to \Box\alpha \wedge \Box\Box^+\alpha$, and

(A5) $\Box\alpha \wedge \Box^+(\alpha \to \Box\alpha) \to \Box^+\alpha$ (induction axiom)

and the inference rules are

(R1) $\dfrac{\alpha \to \beta \quad \alpha}{\beta}$ (modus ponens), and

(R2) $\dfrac{\alpha}{\Box^+\alpha}$ (generalization for $\Box^+$).

Note that the 'transitive axiom' $\Box^+\alpha \to \Box^+\Box^+\alpha$ is derivable using (A4) and the instance $\Box\Box^+\alpha \wedge \Box^+(\Box^+\alpha \to \Box\Box^+\alpha) \to \Box^+\Box^+\alpha$ of induction axiom. The generalization rule for $\Box$ is also derivable using (R2) and (A4).

By "$\vdash \varphi$", we mean "φ is provable in HK$^+$". The purpose of this paper is to give a new proof of the completeness of HK$^+$ with respect to finite models, which states "if α_0 is valid in any finite model, then $\vdash \alpha_0$" or equivalently "if $\nvdash \alpha_0$, then there is a finite countermodel for α_0". The soundness (converse of the completeness) can be easily shown as usual.

3 Special formulas

In this section, we show provability of certain formulas which will be used in the next section.

Two formulas α and β are said to be *provably equivalent* when $\vdash (\alpha \to \beta) \wedge (\beta \to \alpha)$. If $\Gamma = \{\gamma_1, \gamma_2, \ldots, \gamma_n\}$ is a finite set of formulas, then "$\vdash \Gamma \Rightarrow \varphi$" means "$\vdash (\gamma_1 \wedge \gamma_2 \wedge \cdots \wedge \gamma_n) \to \varphi$". Note that we do not mind permutations or duplications in Γ because, for example, $((\gamma_1 \wedge \gamma_2) \wedge \gamma_3) \to \varphi$ and $((\gamma_2 \wedge \gamma_1) \wedge (\gamma_3 \wedge \gamma_1)) \to \varphi$ are provably equivalent.

Lemma 3.1 *(1) If $\vdash \{\varphi_1, \varphi_2, \ldots, \varphi_n\} \Rightarrow \psi$, then $\vdash \{\varphi_1 \vee \rho, \varphi_2 \vee \rho, \ldots, \varphi_n \vee \rho\} \Rightarrow \psi \vee \rho$.*

(2) If $\vdash \{\varphi_1, \varphi_2, \ldots, \varphi_n\} \Rightarrow \psi$, then $\vdash \{\rho \to \varphi_1, \rho \to \varphi_2, \ldots, \rho \to \varphi_n\} \Rightarrow \rho \to \psi$.

(3) If $\vdash \{\varphi_1, \varphi_2, \ldots, \varphi_n\} \Rightarrow \psi$, then $\vdash \{\Box\varphi_1, \Box\varphi_2, \ldots, \Box\varphi_n\} \Rightarrow \Box\psi$.

(4) If $\vdash \{\varphi_1, \varphi_2, \ldots, \varphi_n\} \Rightarrow \psi$, then $\vdash \{\Box^+\varphi_1, \Box^+\varphi_2, \ldots, \Box^+\varphi_n\} \Rightarrow \Box^+\psi$.

Proof. (1) and (2) are properties of classical logic. (3) and (4) are properties of normal modal logics. □

Lemma 3.2 *If formulas $\sigma, \sigma', \tau, \tau'$ and ω satisfy the conditions (a) $\vdash \sigma \to \Box\tau$, (b) $\vdash \sigma' \to \Box\tau'$, and (c) $\vdash \neg\sigma' \to \Box\tau$; then we have $\vdash \{\sigma \to \Box^+(\tau \to \omega),\ \sigma \to \Box^+(\tau \to \sigma' \to \Box^+(\tau' \to \omega))\} \Rightarrow \sigma \to \Box^+\omega$.*

Proof. See Appendix A. □

In the rest of this section, a natural number $N \geq 2$ and formulas ω, σ_i, τ_i ($i = 1, 2, \ldots, N$) are fixed. A formula is called *special* if and only if it is of the form

$$\sigma_{f(1)} \to \Box^+\Big(\tau_{f(1)} \to \sigma_{f(2)} \to \Box^+\Big(\tau_{f(2)} \to \cdots \to \sigma_{f(m)} \to \Box^+(\tau_{f(m)} \to \omega) \cdots\Big)\Big)$$

for some natural number m and some function f that satisfy the following conditions.

- $1 \leq m \leq N$.
- f is an injection (one-to-one) from $\{1, 2, \ldots, m\}$ to $\{1, 2, \ldots, N\}$.
- $f(1) = 1$.

The set of special formulas is called **SP**, which is a finite set. For example, if $N = 3$, then

$$\begin{aligned} \mathbf{SP} = \{\ & \sigma_1 \to \Box^+(\tau_1 \to \omega), \\ & \sigma_1 \to \Box^+(\tau_1 \to \sigma_2 \to \Box^+(\tau_2 \to \omega)), \\ & \sigma_1 \to \Box^+(\tau_1 \to \sigma_3 \to \Box^+(\tau_3 \to \omega)), \\ & \sigma_1 \to \Box^+(\tau_1 \to \sigma_2 \to \Box^+(\tau_2 \to \sigma_3 \to \Box^+(\tau_3 \to \omega))), \\ & \sigma_1 \to \Box^+(\tau_1 \to \sigma_3 \to \Box^+(\tau_3 \to \sigma_2 \to \Box^+(\tau_2 \to \omega)))\ \}. \end{aligned} \tag{4}$$

Note that the shapes of these formulas are same as the paths (2) in Section 1. If $N = 4$, then **SP** consists of sixteen formulas.

Theorem 3.3 (Main theorem on special formulas) *Suppose that*

Fig. 1.

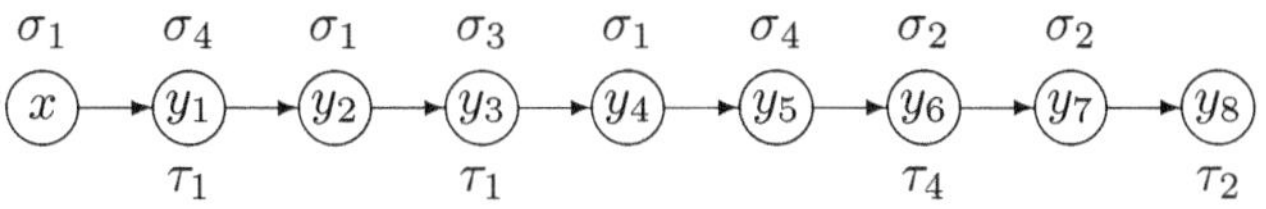

(i) $\vdash \sigma_1 \vee \sigma_2 \vee \cdots \vee \sigma_N$, *and*

(ii) $\vdash \sigma_i \rightarrow \Box\tau_i$, *for* $i = 1, 2, \ldots, N$,

where $N \geq 2$. *Then we have* $\vdash \mathbf{SP} \Rightarrow (\sigma_1 \rightarrow \Box^+\omega)$.

Before the proof, we give a semantical explanation of this theorem; that is, we show the formula $\Box^+\omega$ is true at a world x in a model $M = \langle W, R, V\rangle$ on the assumption that (I) $\sigma_1 \vee \sigma_2 \vee \cdots \vee \sigma_N$ is valid, (II) $\sigma_i \rightarrow \Box\tau_i$ ($i = 1, 2, \ldots, N$) are all valid, (III) all the special formulas are true at x, and (IV) σ_1 is true at x. For example, let us assume that Figure 1 describes some worlds around x, where $xRy_1Ry_2R\cdots Ry_8$ and the displayed σ_i is true there ($\because$ (I),(IV)). We can verify that ω is true at all y_i ($i = 1, 2, \ldots, 8$). For example, ω is true at y_3 because $x \models \sigma_1 \rightarrow \Box^+(\tau_1 \rightarrow \omega)$ ($\because$ (III)), $x \models \sigma_1$ ($\because$ (IV)), and $y_3 \models \tau_1$ ($\because y_2 \models \sigma_1 \rightarrow \Box\tau_1$ by (II)); and ω is true at y_8 because $x \models \sigma_1 \rightarrow \Box^+(\tau_1 \rightarrow \sigma_4 \rightarrow \Box^+(\tau_4 \rightarrow \sigma_2 \rightarrow \Box^+(\tau_2 \rightarrow \omega)))$ ($\because$ (III)), $x \models \sigma_1$ ($\because$ (IV)), $y_1 \models \tau_1 \wedge \sigma_4$ ($\because x \models \sigma_1 \rightarrow \Box\tau_1$ by (II)), $y_6 \models \tau_4 \wedge \sigma_2$ ($\because y_5 \models \sigma_4 \rightarrow \Box\tau_4$ by (II)), and $y_8 \models \tau_2$ ($\because y_7 \models \sigma_2 \rightarrow \Box\tau_2$ by (II)).

Now we start proving Theorem 3.3. If $N = 2$, this can be done by simple application of Lemma 3.2 (by $\sigma = \sigma_1$, $\sigma' = \sigma_2$, $\tau = \tau_1$, $\tau' = \tau_2$). However, we need a more complicated proof when $N > 2$. For this, we introduce an extra notion of *key formulas*.

A formula is called a *key formula of type I* if and only if it is of the form

$$\sigma_{f(1)} \rightarrow \Box^+\Big(\tau_{g(1)} \rightarrow \sigma_{f(2)} \rightarrow \Box^+\Big(\tau_{g(2)} \rightarrow \cdots \rightarrow \sigma_{f(m)} \rightarrow \Box^+(\underline{\tau_{g(m)} \rightarrow} \,\omega)\cdots\Big)\Big) \tag{5}$$

(the underline will be used later) for some natural number m and some functions f and g that satisfy the following conditions.

- $1 \leq m \leq N$.
- f is an injection from $\{1, 2, \ldots, m\}$ to $\{1, 2, \ldots, N\}$.
- g is a function (not limited to injection) from $\{1, 2, \ldots, m\}$ to $\{1, 2, \ldots, N\}$.
- $f(1) = g(1) = 1$.

♡ $(\forall i \in \{1, \ldots, m\})(\exists j \leq i)(f(j) = g(i))$.

The set of key formulas of type I is called **KeyI**, which is a finite superset of **SP**. For example, if $N = 3$, then **KeyI** is the union of **SP** (see (4)) and

$$\begin{aligned}\{\ &\sigma_1 \to \Box^+(\tau_1 \to \sigma_2 \to \Box^+(\tau_1 \to \omega)),\\ &\sigma_1 \to \Box^+(\tau_1 \to \sigma_3 \to \Box^+(\tau_1 \to \omega)),\\ &\sigma_1 \to \Box^+(\tau_1 \to \sigma_2 \to \Box^+(\tau_1 \to \sigma_3 \to \Box^+(\tau_i \to \omega)))\ (i = 1, 2, 3),\\ &\sigma_1 \to \Box^+(\tau_1 \to \sigma_2 \to \Box^+(\tau_2 \to \sigma_3 \to \Box^+(\tau_j \to \omega)))\ (j = 1, 2)^\dagger,\\ &\sigma_1 \to \Box^+(\tau_1 \to \sigma_3 \to \Box^+(\tau_1 \to \sigma_2 \to \Box^+(\tau_i \to \omega)))\ (i = 1, 2, 3),\\ &\sigma_1 \to \Box^+(\tau_1 \to \sigma_3 \to \Box^+(\tau_3 \to \sigma_2 \to \Box^+(\tau_k \to \omega)))\ (k = 1, 3)^\dagger\}.\end{aligned}$$

(† This is a special formula if $j = 3$ or $k = 2$.)

A formula φ is called a *key formula of type II* if and only if there is a formula ψ that satisfies the following conditions.

- ψ is a key formula of type I as (5) where $m \leq (N - 1)$.
- φ is obtained from ψ by deleting the underlined '$\tau_{g(m)} \to$' in (5).

The natural number m is called the *depth of* φ. For example, if $N = 3$, then there are just three key formulas of type II:

$$\begin{aligned}&\sigma_1 \to \Box^+\omega. && \text{(depth} = 1)\\ &\sigma_1 \to \Box^+(\tau_1 \to \sigma_2 \to \Box^+\omega). && \text{(depth} = 2)\\ &\sigma_1 \to \Box^+(\tau_1 \to \sigma_3 \to \Box^+\omega). && \text{(depth} = 2)\end{aligned}$$

The set of key formulas of type II is called **KeyII**.

The target formula $\sigma_1 \to \Box^+\omega$ of Theorem 3.3 is the shortest element of **KeyII**, and the other elements will be used in the inductive proof of Lemma 3.5 below.

Lemma 3.4 $\vdash$ **SP** $\Rightarrow \varphi$, *for any* $\varphi \in$ **KeyI**.

Lemma 3.5 *Suppose that*

(i) $\vdash \sigma_1 \vee \sigma_2 \vee \cdots \vee \sigma_N$, *and*

(ii) $\vdash \sigma_i \to \Box\tau_i$, *for* $i = 1, 2, \ldots, N$,

where $N \geq 2$. *Then* $\vdash$ **KeyI** $\Rightarrow \varphi$, *for any* $\varphi \in$ **KeyII**.

These two lemmas straightforwardly imply the Main Theorem 3.3. So the rest of this section is devoted to proving these lemmas.

Proof of Lemma 3.4. For any key formula φ of type I, there is a special formula $\varphi*$ embedded in φ such that $\vdash \{\varphi*\} \Rightarrow \varphi$. For example, if φ is

$$\begin{aligned}&\sigma_1 \to \Box^+(\tau_1 \to \sigma_2 \to \Box^+(\tau_1 \to \sigma_3 \to \Box^+(\tau_3 \to\\ &\qquad\qquad \sigma_4 \to \Box^+(\tau_3 \to \sigma_5 \to \Box^+(\tau_1 \to \sigma_6 \to \Box^+(\tau_5 \to \omega)))))),\end{aligned}$$

then $\varphi*$ is

$$\sigma_1 \to \Box^+(\tau_1 \to \sigma_3 \to \Box^+(\tau_3 \to \sigma_5 \to \Box^+(\tau_5 \to \omega)))),$$

which is embedded in φ as

$$\underline{\sigma_1 \to} \Box^+(\tau_1 \to \sigma_2 \to \underline{\Box^+(\tau_1 \to \sigma_3 \to} \Box^+(\tau_3 \to \sigma_4 \to \underline{\Box^+(\tau_3 \to \sigma_5 \to} \Box^+(\tau_1 \to \sigma_6 \to \underline{\Box^+(\tau_5 \to \omega)}))))).$$

In general, $\varphi*$ is defined as follows. Let φ be the formula as (5). Without loss of generality, we suppose $f(i) = i$ for all i. Then, by the property $\heartsuit$, we have

$$g(i) \leq i. \tag{$\heartsuit'$}$$

Now we define a sequence $a_1, a_2, \ldots$ of natural numbers by

$$a_1 = g(m), \quad a_{x+1} = g(a_x - 1) \text{ for } x = 1, 2, \ldots$$

By ($\heartsuit'$), this sequence is strictly decreasing, and $\varphi*$ is

$$\sigma_{a_z} \to \Box^+\Big(\tau_{a_z} \to \sigma_{a_{z-1}} \to \Box^+\Big(\tau_{a_{z-1}} \to \cdots \to \sigma_{a_1} \to \Box^+(\tau_{a_1} \to \omega) \cdots\Big)\Big)$$

where $a_z = 1$. The fact $\vdash \{\varphi*\} \Rightarrow \varphi$ is obtained from $\vdash \{\Box^+(\tau_{g(m)} \to \omega)\} \Rightarrow \Box^+(\tau_{g(m)} \to \omega)$ by appropriate applications of Lemma 3.1(2), 3.1(4) and the fact "$\vdash \{\Box^+\alpha\} \Rightarrow \Box^+\beta$ implies $\vdash \{\Box^+\alpha\} \Rightarrow \Box^+(\tau \to \sigma \to \Box^+\beta)$". □

Proof of Lemma 3.5. The key formula φ of type II is of the form

$$\sigma_{f(1)} \to \Box^+\Big(\tau_{g(1)} \to \cdots \to \sigma_{f(m-1)} \to \Box^+\Big(\tau_{g(m-1)} \to \sigma_{f(m)} \to \Box^+\omega\Big) \cdots\Big).$$

We will abbreviate this to

$$\bullet \to \sigma_{f(m)} \to \Box^+\omega.$$

That is, "$\bullet$" denotes the context " $\sigma_{f(1)} \to \Box^+(\tau_{g(1)} \to \cdots \to \sigma_{f(m-1)} \to \Box^+(\tau_{g(m-1)} \to$ ". Therefore, for example, $\bullet \to \sigma_{f(m)} \to \Box^+(\tau_{g(m)} \to \omega)$ is the formula (5), and $\bullet \to \sigma_1 \to \Box^+\omega$ is just $\sigma_1 \to \Box^+\omega$ when $m = 1$.

We define a set U of natural numbers by

$$U = \{1, 2, \ldots, N\} - \{f(1), f(2), \ldots, f(m)\}.$$

U is not empty because of the definition of key formula of type II. We prove Lemma 3.5 by induction on $|U|$; in other words, we prove this lemma for any φ of depth $(N-1)$, any φ of depth $(N-2)$, ..., any φ of depth 1, successively.

(Case 1: $|U| = 1$; depth of φ is $N - 1$.) For any $i \in \{1, \ldots, m\}$, the formula

$$\bullet \to \sigma_{f(m)} \to \Box^+(\tau_{f(i)} \to \omega)$$

is a key formula of type I. Therefore we have

$$\vdash \textbf{KeyI} \Rightarrow \ \bullet \to \sigma_{f(m)} \to \Box^+((\tau_{f(1)} \vee \tau_{f(2)} \vee \cdots \vee \tau_{f(m)}) \to \omega) \tag{6}$$

because of the fact

$$\vdash \{\tau_{f(1)} \to \omega,\ \tau_{f(2)} \to \omega,\ \ldots,\ \tau_{f(m)} \to \omega\} \Rightarrow (\tau_{f(1)} \vee \tau_{f(2)} \vee \cdots \vee \tau_{f(m)}) \to \omega$$

and Lemma 3.1(4) and 3.1(2). Let u be the only element of U. Similarly to (6), we have

$$\vdash \mathbf{KeyI} \Rightarrow \bullet \to \sigma_{f(m)} \to \Box^+((\tau_{f(1)} \vee \tau_{f(2)} \vee \cdots \vee \tau_{f(m)}) \to \sigma_u \to \Box^+(\tau_u \to \omega)) \quad (7)$$

because the formula

$$\bullet \to \sigma_{f(m)} \to \Box^+(\tau_{f(i)} \to \sigma_u \to \Box^+(\tau_u \to \omega))$$

is a key formula of type I for any $i \in \{1, \ldots, m\}$. On the other hand, by Lemma 3.2 ($\sigma = \sigma_{f(m)}$, $\sigma' = \sigma_u$, $\tau = (\tau_{f(1)} \vee \tau_{f(2)} \vee \cdots \vee \tau_{f(m)})$, $\tau' = \tau_u$), we get

$$\begin{aligned} \vdash \{\ & \sigma_{f(m)} \to \Box^+((\tau_{f(1)} \vee \cdots \vee \tau_{f(m)}) \to \omega), \\ & \sigma_{f(m)} \to \Box^+((\tau_{f(1)} \vee \cdots \vee \tau_{f(m)}) \to \sigma_u \to \Box^+(\tau_u \to \omega))\ \} \Rightarrow \sigma_{f(m)} \to \Box^+\omega. \end{aligned} \quad (8)$$

Note that the hypotheses (a), (b), and (c) of Lemma 3.2 are shown by the hypotheses (i) and (ii) of this Lemma 3.5. Then (6), (7), (8) and Lemma 3.1 imply

$$\vdash \mathbf{KeyI} \Rightarrow \bullet \to \sigma_{f(m)} \to \Box^+\omega, \quad (9)$$

which is the required formula.

(Case 2: $|U| > 1$; depth of φ is less then $N - 1$.) By the same argument as (6), we obtain

$$\vdash \mathbf{KeyI} \Rightarrow \bullet \to \sigma_{f(m)} \to \Box^+((\tau_{f(1)} \vee \tau_{f(2)} \vee \cdots \vee \tau_{f(m)}) \to \omega). \quad (10)$$

On the other hand, for any $i \in \{1, \ldots, m\}$ and any $u \in U$, the formula

$$\bullet \to \sigma_{f(m)} \to \Box^+(\tau_{f(i)} \to \sigma_u \to \Box^+\omega))$$

is a key formula of type II with greater depth. Therefore by the induction hypothesis,

$$\vdash \mathbf{KeyI} \Rightarrow \bullet \to \sigma_{f(m)} \to \Box^+(\tau_{f(i)} \to \sigma_u \to \Box^+\omega)),$$

and then

$$\vdash \mathbf{KeyI} \Rightarrow \bullet \to \sigma_{f(m)} \to \Box^+((\tau_{f(1)} \vee \cdots \vee \tau_{f(m)}) \to (\sigma_{u_1} \vee \cdots \vee \sigma_{u_k}) \to \Box^+(\top \to \omega)) \quad (11)$$

where $U = \{u_1, \ldots, u_k\}$. Now (10), (11), and Lemma 3.2 ($\sigma = \sigma_{f(m)}$, $\sigma' = (\sigma_{u_1} \vee \cdots \vee \sigma_{u_k})$, $\tau = (\tau_{f(1)} \vee \cdots \vee \tau_{f(m)})$, $\tau' = \top$) imply

$$\vdash \mathbf{KeyI} \Rightarrow \bullet \to \sigma_{f(m)} \to \Box^+\omega$$

similarly to (9). □

Fig. 2. A semantic diagram.

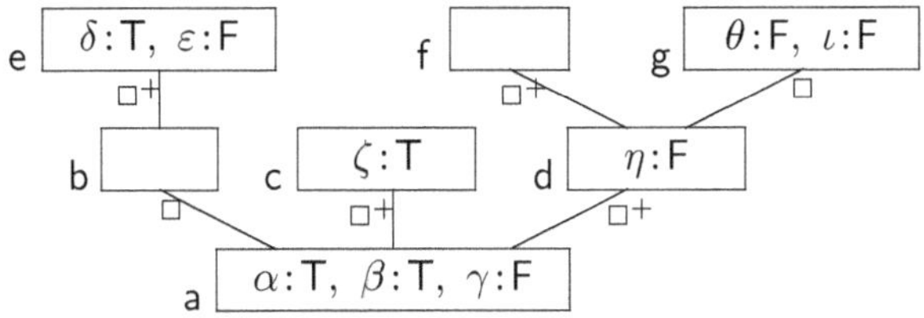

4 Making a countermodel

If φ is a formula, then the expressions $\varphi : \mathsf{T}$ and $\varphi : \mathsf{F}$ are called *signed formulas.* A *semantic diagram* is a finite tree whose nodes are associated with finite sets of signed formulas and whose edges are labeled by $\Box$ or $\Box^+$. Set(a) denotes the set of signed formulas that is associated with the node a. If a node b is a $\Box$-successor (or $\Box^+$-successor) of a node a, then we write $\mathsf{a}<^{\Box}\mathsf{b}$ (or $\mathsf{a}<^{\Box^+}\mathsf{b}$, respectively). Moreover we write $\mathsf{a} < \mathsf{b}$ if and only if $\mathsf{a}<^{\Box}\mathsf{b}$ or $\mathsf{a}<^{\Box^+}\mathsf{b}$. The transitive closure of $<$ is written by $\ll$. Figure 2 is an example of a semantic diagram, in which $\mathrm{Set}(\mathsf{a}) = \{\alpha:\mathsf{T}, \beta:\mathsf{T}, \gamma:\mathsf{F}\}$, $\mathrm{Set}(\mathsf{b}) = \emptyset$, $\mathsf{a}<^{\Box}\mathsf{b}$, $\mathsf{b}<^{\Box^+}\mathsf{e}$, $\mathsf{a} < \mathsf{b}$, $\mathsf{b} < \mathsf{e}$, $\mathsf{a} \not< \mathsf{e}$, $\mathsf{a} \ll \mathsf{b}$, $\mathsf{a} \ll \mathsf{e}$, and $\mathsf{a} \not\ll \mathsf{a}$ hold. In the following, $\Gamma, \Delta, \ldots$ will denote sets of signed formulas, $\mathcal{S}, \mathcal{T}, \ldots$ will denote semantic diagrams, and $\mathsf{a}, \mathsf{b}, \ldots$ will denote nodes of diagrams. By "$\varphi \in_{\mathsf{T}} \mathsf{x}$" (or "$\varphi \in_{\mathsf{F}} \mathsf{x}$"), we mean "$(\varphi:\mathsf{T}) \in \mathrm{Set}(\mathsf{x})$" (or "$(\varphi:\mathsf{F}) \in \mathrm{Set}(\mathsf{x})$", respectively).

For each diagram $\mathcal{S}$, we define a formula $\mathrm{Neg}(\mathcal{S})$ (called the *negation of* $\mathcal{S}$) inductively as follows. If a set $\{\varphi_1:\mathsf{T}, \varphi_2:\mathsf{T}, \ldots, \varphi_m:\mathsf{T},\ \psi_1:\mathsf{F},\ \psi_2:\mathsf{F}, \ldots, \psi_n:\mathsf{F}\}$ is associated with the root of $\mathcal{S}$, and subdiagrams $\mathcal{S}_1, \mathcal{S}_2, \ldots, \mathcal{S}_k$ are connected with the root by $\Box$-edges and $\mathcal{T}_1, \mathcal{T}_2, \ldots, \mathcal{T}_l$ are connected with the root by $\Box^+$-edges, then $\mathrm{Neg}(\mathcal{S})$ is the formula

$$\begin{aligned}\bot \vee \neg\varphi_1 \vee \neg\varphi_2 \vee \cdots \vee \neg\varphi_m \vee \psi_1 \vee \psi_2 \vee \cdots \vee \psi_n \vee \\ \Box(\mathrm{Neg}(\mathcal{S}_1)) \vee \Box(\mathrm{Neg}(\mathcal{S}_2)) \vee \cdots \vee \Box(\mathrm{Neg}(\mathcal{S}_k)) \vee \\ \Box^+(\mathrm{Neg}(\mathcal{T}_1)) \vee \Box^+(\mathrm{Neg}(\mathcal{T}_2)) \vee \cdots \vee \Box^+(\mathrm{Neg}(\mathcal{T}_l)).\end{aligned}$$

For example, the negation of the diagram of Figure 2 is provably equivalent to the formula $\neg\alpha \vee \neg\beta \vee \gamma \vee \Box\Box^+(\neg\delta \vee \varepsilon) \vee \Box^+\neg\zeta \vee \Box^+(\eta \vee \Box^+\bot \vee \Box(\theta \vee \iota))$. A diagram $\mathcal{S}$ is said to be *HK*$^+$*-consistent* if and only if $\not\vdash \mathrm{Neg}(\mathcal{S})$.

Let $\mathcal{S}$ and $\mathcal{T}$ be semantic diagrams and a be a node of $\mathcal{S}$. By $\mathcal{S} \overset{\mathsf{a}}{+} \mathcal{T}$, we mean the diagram obtained by joining $\mathcal{S}$ and $\mathcal{T}$, in which a and the root of $\mathcal{T}$ are merged into one node. Figure 3 describes an example.

Let $\mathbb{L}$ be a finite set $\{\lambda_1, \lambda_2, \ldots, \lambda_k\}$ of formulas. We say that a set Λ of signed formulas is a *valuation of* $\mathbb{L}$ if Λ is $\{\lambda_1 : \bullet_1, \lambda_2 : \bullet_2, \ldots, \lambda_k : \bullet_k\}$ ($\bullet_i$ is T or F). There are 2^k distinct valuations of $\mathbb{L}$.

For a set Γ of signed formulas, we define a formula $\langle\Gamma\rangle$ and a set $\Gamma^{\mathsf{T}}_{\Box}$ of signed formulas as follows.

$$\langle\Gamma\rangle = \bigwedge\{\varphi \mid (\varphi:\mathsf{T}) \in \Gamma\} \wedge \bigwedge\{\neg\varphi \mid (\varphi:\mathsf{F}) \in \Gamma\}.$$

$$\Gamma^{\mathsf{T}}_{\Box} = \{\varphi:\mathsf{T} \mid (\Box\varphi:\mathsf{T}) \in \Gamma\}.$$

Fig. 3. Semantic diagrams $\mathcal{S}$, $\mathcal{T}$ and $\mathcal{S} \overset{\mathsf{a}}{+} \mathcal{T}$.

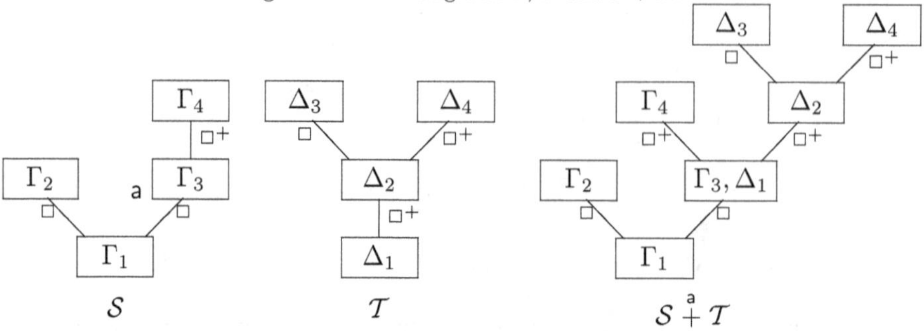

For example, if $\Gamma = \{\Box\varphi_1 : \mathsf{T},\ \Box\Box\varphi_2 : \mathsf{T},\ \neg\neg\Box\varphi_3 : \mathsf{T},\ \Box^+\varphi_4 : \mathsf{T},\ \Box\varphi_5 : \mathsf{F}\}$, then $\langle\Gamma\rangle$ is $\Box\varphi_1 \wedge \Box\Box\varphi_2 \wedge \neg\neg\Box\varphi_3 \wedge \Box^+\varphi_4 \wedge \neg\Box\varphi_5$, and $\Gamma^{\mathsf{T}}_{\Box}$ is $\{\varphi_1 : \mathsf{T},\ \Box\varphi_2 : \mathsf{T}\}$. Note that

$$\vdash \langle\Lambda\rangle \rightarrow \Box\langle\Lambda^{\mathsf{T}}_{\Box}\rangle \tag{12}$$

holds for any set Λ; for example, $\vdash (\Box\varphi_1 \wedge \Box\Box\varphi_2 \wedge \neg\neg\Box\varphi_3 \wedge \Box^+\varphi_4 \wedge \neg\Box\varphi_5) \rightarrow \Box(\varphi_1 \wedge \Box\varphi_2)$ if Λ is the above Γ.

We give some basic lemmas on diagrams.

Lemma 4.1 *Let $\mathcal{S}, \mathcal{T}, \mathcal{T}_1, \mathcal{T}_2, \ldots, \mathcal{T}_n$ be semantic diagrams ($n \geq 0$) and* a *be a node of $\mathcal{S}$. If*

$$\vdash \{\mathrm{Neg}(\mathcal{T}_1),\ \mathrm{Neg}(\mathcal{T}_2),\ \ldots,\ \mathrm{Neg}(\mathcal{T}_n)\} \Rightarrow \mathrm{Neg}(\mathcal{T}),$$

then

$$\vdash \{\mathrm{Neg}(\mathcal{S} \overset{\mathsf{a}}{+} \mathcal{T}_1),\ \mathrm{Neg}(\mathcal{S} \overset{\mathsf{a}}{+} \mathcal{T}_2),\ \ldots,\ \mathrm{Neg}(\mathcal{S} \overset{\mathsf{a}}{+} \mathcal{T}_n)\} \Rightarrow \mathrm{Neg}(\mathcal{S} \overset{\mathsf{a}}{+} \mathcal{T}).$$

Proof. By Lemma 3.1 and the definition of Neg(). □

Lemma 4.2 (Maximalization) *Let $\mathbb{L} = \{\lambda_1, \lambda_2, \ldots, \lambda_k\}$ ($k \geq 1$) be a finite set of formulas. If a semantic diagram $\mathcal{S}$ is HK$^+$-consistent and* a *is a node of it, then there exists a valuation Λ of $\mathbb{L}$ such that the diagram $\mathcal{S} \overset{\mathsf{a}}{+} \Lambda$ (i.e., the diagram obtained from $\mathcal{S}$ by adding Λ to the node* a*) is HK$^+$-consistent. The process of making $\mathcal{S} \overset{\mathsf{a}}{+} \Lambda$ from $\mathcal{S}$ will be called "maximalization for* a *with respect to $\mathbb{L}$".*

Proof. Since $\vdash \{\neg\lambda_i, \lambda_i\} \Rightarrow \bot$, one of the diagrams $\mathcal{S} \overset{\mathsf{a}}{+} \{\lambda_i : \mathsf{T}\}$ and $\mathcal{S} \overset{\mathsf{a}}{+} \{\lambda_i : \mathsf{F}\}$ is HK$^+$-consistent (otherwise $\vdash \mathrm{Neg}(\mathcal{S})$ by Lemma 4.1). By iterating this argument, we can chose $\bullet_1, \bullet_2, \ldots, \bullet_k$ ($\bullet_i \in \{\mathsf{T}, \mathsf{F}\}$) such that $\mathcal{S} \overset{\mathsf{a}}{+} \{\lambda_1 : \bullet_1, \lambda_2 : \bullet_2, \ldots, \lambda_k : \bullet_k\}$ is HK$^+$-consistent. □

Lemma 4.3 (Fulfillment of □) *If a diagram $\mathcal{S}$ of Figure 4 is HK$^+$-consistent, then also the diagram $\mathcal{T}$ of Figure 4 is HK$^+$-consistent. (In the Figure, $\mathcal{U}$, $\mathcal{V}_1$, $\mathcal{V}_2$, …, $\mathcal{V}_n$ are subdiagrams, where $\mathcal{U}$ may be null and $n \geq 0$ — this means that the node* a *may be the root or a leaf.) The process of making $\mathcal{T}$ from $\mathcal{S}$ will be called "fulfillment of $\Box\varphi : \mathsf{F}$ for* a*", and the added node* b *will be called the "witness node".*

Fig. 4. Diagrams $\mathcal{S}$ and $\mathcal{T}$ of Lemma 4.3.

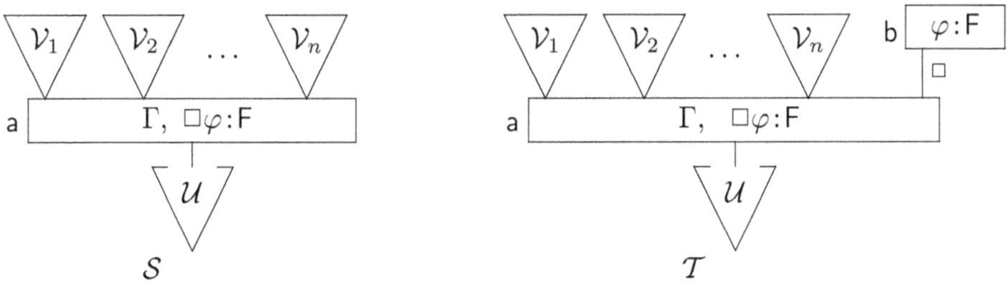

Fig. 5. Diagram $\mathcal{S}$ of Proposition 4.4.

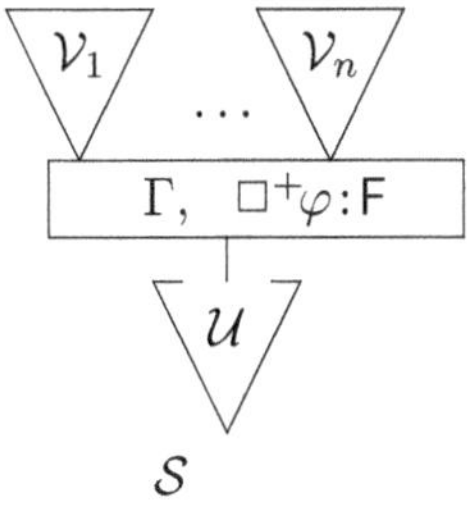

Proof. Neg($\mathcal{S}$) and Neg($\mathcal{T}$) are provably equivalent. $\Box$

Let us explain the outline and the point of our completeness proof.

The goal is to construct a finite countermodel for a given unprovable formula α_0. When α_0 does not contain the operator $\Box^+$, the argument is equivalent to the well-known completeness proof for the smallest normal modal logic K, and it is arranged as follows. If $\nvdash \alpha_0$, then the one-node diagram $\{\alpha_0 : \mathsf{F}\}$ is HK$^+$-consistent. We extend it by iterated applications of maximalization (Lemma 4.2) and fulfillment of $\Box$ (Lemma 4.3), and we eventually get a "saturated diagram" $\mathcal{T}$. Then a model $M = \langle W, R, V\rangle$ is defined by: W is the set of nodes in $\mathcal{T}$; $R = <^{\Box}$; and $V(\mathsf{a}, p) = \mathsf{True} \iff p \in_{\mathsf{T}} \mathsf{a}$. This is the required countermodel, because "$\varphi \in_{\mathsf{T}} \mathsf{a} \Rightarrow M, \mathsf{a} \models \varphi$" and "$\varphi \in_{\mathsf{F}} \mathsf{a} \Rightarrow M, \mathsf{a} \not\models \varphi$" hold for any φ, and the root contains $\alpha_0 : \mathsf{F}$.

When α_0 contains both the operators $\Box$ and $\Box^+$, we need additional constructions to fulfill $\Box^+\varphi : \mathsf{F}$. There are two naive and unsuccessful ways for this. After showing these *bad* ways, we will present our *good* way, which enables us to make a witness of $\Box^+\varphi : \mathsf{F}$ in an HK$^+$-consistent diagram.

The first way uses the following proposition corresponding to Lemma 4.3.

Proposition 4.4 *If a diagram $\mathcal{S}$ of Figure 5 is HK$^+$-consistent, then at least one of the diagram $\mathcal{T}_i$ of Figure 6 is HK$^+$-consistent. Note that Figure 6 contains infinitely many diagrams.*

In this way, we are faced with a difficulty in proving Proposition 4.4. Of course we can prove this proposition using the soundness and completeness of HK$^+$; however, we

Fig. 6. Diagrams $\mathcal{T}_1, \mathcal{T}_2, \ldots$ of Proposition 4.4.

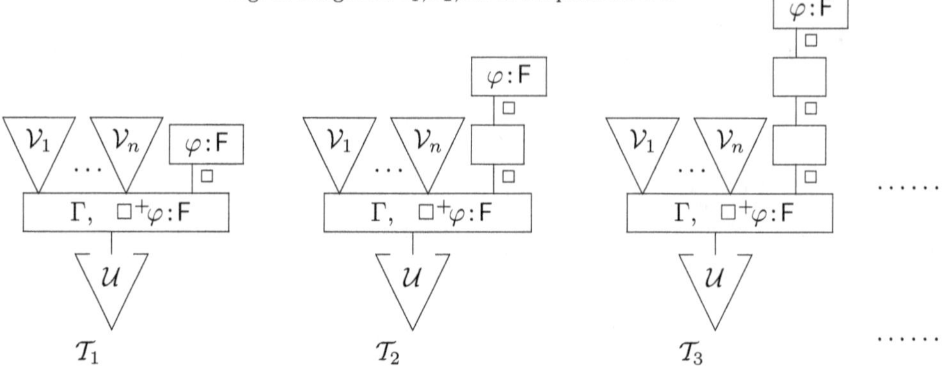

Fig. 7. Diagrams $\mathcal{S}$ and $\mathcal{T}$ of Proposition 4.5.

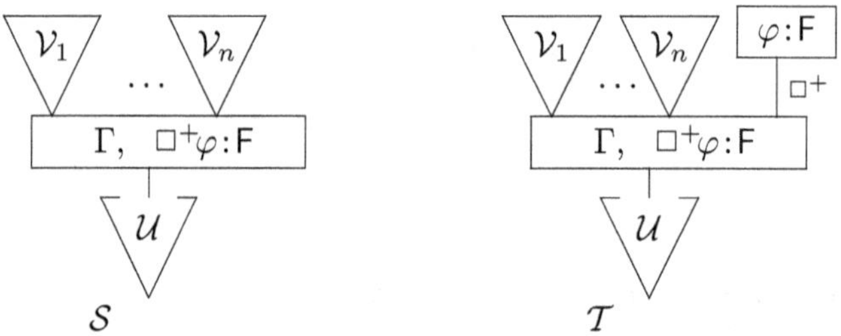

are now in course of proving completeness theorem.

The second way uses the following proposition.

Proposition 4.5 *If a diagram $\mathcal{S}$ of Figure 7 is HK$^+$-consistent, then also the diagram $\mathcal{T}$ of Figure 7 is HK$^+$-consistent.*

This proposition is easily proved in contrast to Proposition 4.4; however, we are faced with another difficulty in making a countermodel — we cannot define a well-behaved accessibility relation on the saturated diagram based on Proposition 4.5.

Then the following lemma is the third and successful way, which is the main contribution of this paper. This is done by enumerating all possible candidates of paths to the witness (called "special paths" below), as (2) in Section 1.

Lemma 4.6 (Fulfillment of $\Box^+$) *Let $\mathbb{L} = \{\lambda_1, \lambda_2, \ldots, \lambda_k\}$ $(k \geq 1)$ be a finite set of formulas. If a diagram $\mathcal{S}$ of Figure 8 is HK$^+$-consistent and Γ_1 is a valuation of $\mathbb{L}$, then there exist valuations $\Gamma_2, \Gamma_3, \ldots, \Gamma_m$ of $\mathbb{L}$ for some $m \geq 1$ such that the diagram $\mathcal{T}$ of Figure 8 is HK$^+$-consistent. The process of making $\mathcal{T}$ from $\mathcal{S}$ will be called "fulfillment of $\Box^+\varphi$:F for* a *with respect to $\mathbb{L}$", and the top node* b *will be called the "witness node".*

Proof. We say that a diagram is a *special path from Γ_1 to φ:*F if and only if it is of the form as in Figure 9 for some valuations $\Gamma_2, \ldots, \Gamma_m$ of $\mathbb{L}$ $(m \geq 1)$ such that $\Gamma_1, \Gamma_2, \ldots, \Gamma_m$ are mutually distinct. There are finitely many distinct valuations of $\mathbb{L}$, say $\Lambda_1, \Lambda_2 \ldots, \Lambda_N$

Fig. 8. Diagrams $\mathcal{S}$ and $\mathcal{T}$ of Lemma 4.6

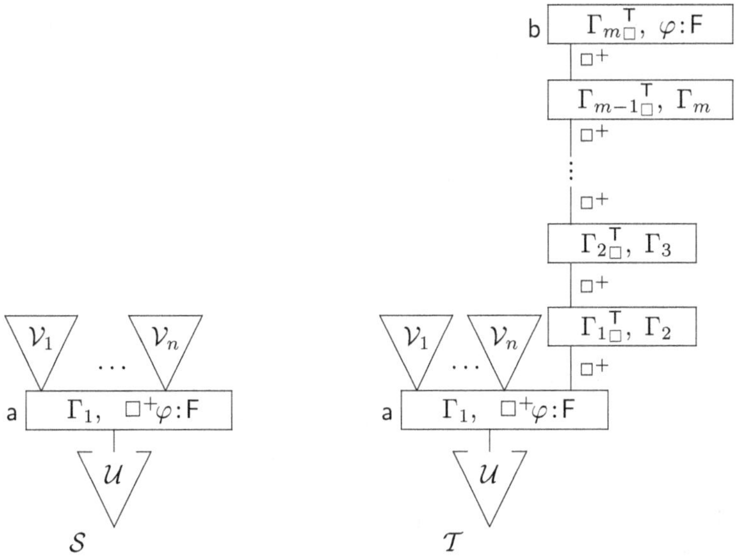

Fig. 9. Special path from Γ_1 to $\varphi{:}\mathsf{F}$

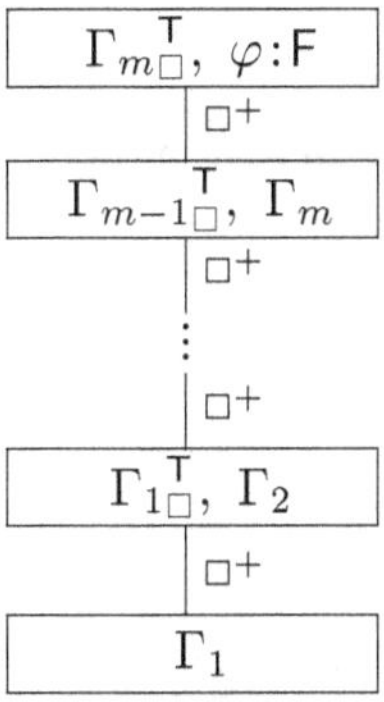

($N - 2^k \geq 2$ because $k \geq 1$); therefore the number of all special paths from Γ_1 to $\varphi{:}\mathsf{F}$ is also finite. Then let $\{\mathcal{W}_1, \mathcal{W}_2, \ldots, \mathcal{W}_P\}$ be the set of special paths. Now we will show

$$\vdash \{\mathrm{Neg}(\mathcal{W}_1), \mathrm{Neg}(\mathcal{W}_2), \ldots, \mathrm{Neg}(\mathcal{W}_P)\} \Rightarrow \mathrm{Neg}(\Gamma_1, \Box^+\varphi{:}\mathsf{F}). \tag{13}$$

The negation of a special path is provably equivalent to the formula

$$\langle\Gamma_1\rangle \to \Box^+\Big(\langle\Gamma_{1\Box}^{\mathsf{T}}\rangle \to \langle\Gamma_2\rangle \to \Box^+\Big(\cdots \to \Box^+\Big(\langle\Gamma_{m-1\Box}^{\mathsf{T}}\rangle \to \langle\Gamma_m\rangle \to \Box^+\Big(\langle\Gamma_{m\Box}^{\mathsf{T}}\rangle \to \varphi\Big)\Big)\Big)\Big),$$

and the formula $\mathrm{Neg}(\Gamma_1, \Box^+\varphi{:}\mathsf{F})$ is provably equivalent to

$$\langle\Gamma_1\rangle \to \Box^+\varphi.$$

Moreover we have

$$\vdash \langle\Lambda_1\rangle \vee \langle\Lambda_2\rangle \vee \cdots \vee \langle\Lambda_N\rangle$$

because this formula is a tautology. Using these facts and (12) (before Lemma 4.1), we can apply Theorem 3.3 ($\{\sigma_1, \sigma_2, \ldots, \sigma_N\} = \{\langle\Lambda_1\rangle, \langle\Lambda_2\rangle, \ldots, \langle\Lambda_N\rangle\}$, $\sigma_1 = \langle\Gamma_1\rangle$, $\sigma_{f(i)} = \langle\Gamma_i\rangle$, $\tau_{f(i)} = \langle\Gamma_{i\Box}^{\mathsf{T}}\rangle$, $\omega = \varphi$), and we get (13). Now the HK$^+$-consistent diagram $\mathcal{S}$ of Figure 8 is equivalent to $\mathcal{S} \overset{\mathsf{a}}{+} \{\Gamma_1, \Box^+\varphi : \mathsf{F}\}$. Then (13) and Lemma 4.1 imply that there is a special path $\mathcal{W}$ such that $\mathcal{S} \overset{\mathsf{a}}{+} \mathcal{W}$ is HK$^+$-consistent. □

Remarks on Lemma 4.6.
(1) If the set $\mathbb{L}$ is closed under subformulas and $\mathcal{T}$ is HK$^+$-consistent, then it must be the case that $\Gamma_{i\Box}^{\mathsf{T}} \subseteq \Gamma_{i+1}$ in the node $\{\Gamma_{i\Box}^{\mathsf{T}}, \Gamma_{i+1}\}$.
(2) Special paths consist of not □-edges, but □$^+$-edges, since the origin of a □$^+$-edge is the R^+-edge between $\{\Sigma_{a_i}\}$ and $\{\Sigma_{a_{(i+1)}}\}$ in the long path (3) from Section 1. On the other hand, the □$^+$-edges will become not R^+-edges but R-edges in the countermodel below. This one-step reachability is justified by the connection between Γ_i and $\Gamma_{i\Box}^{\mathsf{T}}$.

Now let us fix a formula α_0, for which we are going to construct a countermodel. The set of subformulas of α_0 is called $\mathrm{Sub}(\alpha_0)$. We define some conditions on a node a of semantic diagrams as follows.

[$\mathrm{Sub}(\alpha_0)$-**maximality**] $\varphi \in \mathrm{Sub}(\alpha_0) \iff (\varphi \in_{\mathsf{T}} \mathsf{a}$ or $\varphi \in_{\mathsf{F}} \mathsf{a})$.

[□-**correctness**] If $\Box\varphi \in_{\mathsf{T}} \mathsf{a}$ and $\mathsf{a} < \mathsf{b}$, then $\varphi \in_{\mathsf{T}} \mathsf{b}$.

[□-**witness property**] If $\Box\varphi \in_{\mathsf{F}} \mathsf{a}$, then the following condition holds.

$$\exists \mathsf{b}\big(\mathsf{a} <^{\Box} \mathsf{b} \text{ and } \varphi \in_{\mathsf{F}} \mathsf{b}\big). \qquad (\spadesuit)$$

[□$^+$-**witness property**] If $\Box^+\varphi \in_{\mathsf{F}} \mathsf{a}$, then the following condition holds.

$$\exists m \geq 1, \exists \mathsf{b}_1, \exists \mathsf{b}_2, \ldots, \exists \mathsf{b}_m \big(\mathsf{a} <^{\Box^+} \mathsf{b}_1 <^{\Box^+} \mathsf{b}_2 <^{\Box^+} \cdots <^{\Box^+} \mathsf{b}_m \text{ and } \varphi \in_{\mathsf{F}} \mathsf{b}_m\big). \qquad (\clubsuit)$$

We say that a node x is *set-fresh* if and only if the condition ($\mathsf{y} \ll \mathsf{x} \Rightarrow \mathrm{Set}(\mathsf{y}) \neq \mathrm{Set}(\mathsf{x})$) holds for any node y. The following is called the *diagram-model condition for* $\mathcal{T}$ *with respect to* $\mathrm{Sub}(\alpha_0)$, which is the key notion of our completeness proof.

- $\mathcal{T}$ is HK$^+$-consistent;
- all nodes of $\mathcal{T}$ are $\mathrm{Sub}(\alpha_0)$-maximal and □-correct; and
- all set-fresh nodes of $\mathcal{T}$ satisfy □-witness and □$^+$-witness properties.

Lemma 4.7 *If* $\not\vdash \alpha_0$, *then there exists a semantic diagram* $\mathcal{T}$ *such that the diagram-model condition holds with respect to* $\mathrm{Sub}(\alpha_0)$ *and the root contains the signed formula* $\alpha_0 : \mathsf{F}$.

Proof. We define a procedure to construct semantic diagrams $\mathcal{T}_0, \mathcal{T}_1, \mathcal{T}_2 \ldots$, such that $\mathcal{T}_i$ is HK$^+$-consistent and all the nodes of $\mathcal{T}_i$ are $\mathrm{Sub}(\alpha_0)$-maximal and □-correct.

[Construction of $\mathcal{T}_0$]
The one-node diagram $\{\alpha_0 : \mathsf{F}\}$ is HK$^+$-consistent because $\nvdash \alpha_0$. We apply the maximalization with respect to $\mathrm{Sub}(\alpha_0)$ (Lemma 4.2). Then we obtain a diagram whose only node is $\mathrm{Sub}(\alpha_0)$-maximal and contains $\alpha_0 : \mathsf{F}$. This is the diagram $\mathcal{T}_0$.

[Construction of $\mathcal{T}_{i+1}$ from $\mathcal{T}_i$]
If $\mathcal{T}_i$ satisfies the diagram-model condition with respect to $\mathrm{Sub}(\alpha_0)$, then we stop the procedure and we get the required diagram. Otherwise there is a node, say a, which is set-fresh, but the $\Box$-witness (or $\Box^+$-witness) property fails; that is, there is a formula $\Box\varphi$ (or $\Box^+\varphi$) $\in_{\mathsf{F}} \mathsf{a}$ such that the condition ♠ (or ♣) does not hold. Then we apply the fulfillment (Lemma 4.3 or 4.6) of $\Box\varphi : \mathsf{F}$ (or $\Box^+\varphi : \mathsf{F}$) for a and maximalization (Lemma 4.2) with respect to $\mathrm{Sub}(\alpha_0)$ for the witness node (the other nodes are already maximal), and the resulting diagram is $\mathcal{T}_{i+1}$. The node a will be called a *growing point.* Note that all the nodes of $\mathcal{T}_{i+1}$ satisfy $\Box$-correctness; here we show some cases: (Case 1) If $\Box\psi : \mathsf{T}$ is in the growing point of fulfillment of $\Box\varphi : \mathsf{F}$, then $\psi : \mathsf{T}$ must be in the maximalized witness node, say b; otherwise $(\psi : \mathsf{F}) \in \mathsf{b}$ and the diagram would be HK$^+$-inconsistent because $\vdash \neg\Box\psi \vee \Box(\psi \vee \cdots)$. (Case 2) If $\Box\psi : \mathsf{T}$ is in a node $\{\Gamma_{j\Box}^{\mathsf{T}}, \Gamma_{j+1}\}$ in the special path of fulfillment of $\Box^+\varphi : \mathsf{F}$, then $(\Box\psi : \mathsf{T}) \in \Gamma_{j+1}$ (otherwise $(\Box\psi : \mathsf{F}) \in \Gamma_{j+1}$ and the diagram would be HK$^+$-inconsistent), and then $\psi : \mathsf{T}$ is in the next node $\{\Gamma_{j+1\Box}^{\mathsf{T}}, \Gamma_{j+2}\}$.

We show that the above procedure must terminate, and hence we eventually get the required diagram. In fact, otherwise an infinite sequence $\mathcal{T}_0, \mathcal{T}_1, \mathcal{T}_2 \ldots$ is produced. Then consider the infinite diagram $\bigcup_{i=0}^{\infty} \mathcal{T}_i$. This infinite tree is finite branching because we can apply at most p times fulfillment for each growing point where p is the number of $\Box$- or $\Box^+$- formulas in $\mathrm{Sub}(\alpha_0)$. Therefore there is an infinite path which contains infinitely many growing points; however this is impossible because each growing point must be set-fresh and the number of set-fresh nodes in one path cannot be greater than $2^{|\mathrm{Sub}(\alpha_0)|}$. □

Lemma 4.8 *If a semantic diagram $\mathcal{T}$ satisfies the diagram-model condition with respect to* $\mathrm{Sub}(\alpha_0)$, *then the following hold for any node* a *of* $\mathcal{T}$. *(1) If $\varphi \in_{\mathsf{F}} \mathsf{a}$, then $\varphi \notin_{\mathsf{T}} \mathsf{a}$. (2) If $\varphi \wedge \psi \in_{\mathsf{T}} \mathsf{a}$, then $\varphi \in_{\mathsf{T}} \mathsf{a}$ and $\psi \in_{\mathsf{T}} \mathsf{a}$. (3) If $\varphi \wedge \psi \in_{\mathsf{F}} \mathsf{a}$, then $\varphi \in_{\mathsf{F}} \mathsf{a}$ or $\psi \in_{\mathsf{F}} \mathsf{a}$. (4) If $\neg\varphi \in_{\mathsf{T}} \mathsf{a}$, then $\varphi \in_{\mathsf{F}} \mathsf{a}$. (5) If $\neg\varphi \in_{\mathsf{F}} \mathsf{a}$, then $\varphi \in_{\mathsf{T}} \mathsf{a}$. (6) If $\Box^+\varphi \in_{\mathsf{T}} \mathsf{a}$ and $\mathsf{a} < \mathsf{b}$, then $\Box^+\varphi \in_{\mathsf{T}} \mathsf{b}$ and $\varphi \in_{\mathsf{T}} \mathsf{b}$.*

Proof. We check only the clause (6), which is divided into the following four: (6-1) If $\Box^+\varphi \in_{\mathsf{T}} \mathsf{a}$ and $\mathsf{a} <^{\Box} \mathsf{b}$, then $\Box^+\varphi \in_{\mathsf{T}} \mathsf{b}$. (6-2) If $\Box^+\varphi \in_{\mathsf{T}} \mathsf{a}$ and $\mathsf{a} <^{\Box} \mathsf{b}$, then $\varphi \in_{\mathsf{T}} \mathsf{b}$. (6-3) If $\Box^+\varphi \in_{\mathsf{T}} \mathsf{a}$ and $\mathsf{a} <^{\Box^+} \mathsf{b}$, then $\Box^+\varphi \in_{\mathsf{T}} \mathsf{b}$. (6-4) If $\Box^+\varphi \in_{\mathsf{T}} \mathsf{a}$ and $\mathsf{a} <^{\Box^+} \mathsf{b}$, then $\varphi \in_{\mathsf{T}} \mathsf{b}$. The clause (6-1) is verified as follows. If $\Box^+\varphi \in_{\mathsf{T}} \mathsf{a}$, $\mathsf{a} <^{\Box} \mathsf{b}$, and $\Box^+\varphi \notin_{\mathsf{T}} \mathsf{b}$, then $\Box^+\varphi \in_{\mathsf{F}} \mathsf{b}$ by $\mathrm{Sub}(\varphi_0)$-maximality, and then $\mathcal{T}$ would be HK$^+$-inconsistent because $\vdash \neg\Box^+\varphi \vee \Box(\Box^+\varphi \vee \cdots)$ $(\because \vdash \Box^+\varphi \to \Box\Box^+\varphi)$. The clauses (6-2), (6-3) and (6-4) are considered similarly using the facts $\vdash \neg\Box^+\varphi \vee \Box(\varphi \vee \cdots)$ $(\because \vdash \Box^+\varphi \to \Box\varphi)$, $\vdash \neg\Box^+\varphi \vee \Box^+(\Box^+\varphi \vee \cdots)$ $(\because \vdash \Box^+\varphi \to \Box^+\Box^+\varphi)$, and $\vdash \neg\Box^+\varphi \vee \Box^+(\varphi \vee \cdots)$ $(\because \vdash \Box^+\varphi \to \Box^+\varphi)$. □

Theorem 4.9 (Completeness of HK$^+$ with respect to finite models) *If* $\nvdash \alpha_0$, *then there exists a finite model M such that $M, x \not\models \alpha_0$ for some world x.*

Proof. Let $\mathcal{T}$ be the diagram obtained by Lemma 4.7. We define $M = \langle W, R, V\rangle$ as follows.

- W is the set of nodes in $\mathcal{T}$.
- $\mathsf{a}R\mathsf{b} \iff \mathsf{a} < \mathsf{b}$ or $\exists \mathsf{a}_0(\mathsf{a}_0 \ll \mathsf{a},\ \mathrm{Set}(\mathsf{a}_0) = \mathrm{Set}(\mathsf{a}),$ and $\mathsf{a}_0 < \mathsf{b})$.
- $V(\mathsf{a}, p) = \mathsf{True} \iff p \in_{\mathsf{T}} \mathsf{a}$.

Using the diagram-model condition of $\mathcal{T}$ and (6) of Lemma 4.8, we can show the following:

(i) If $\Box\varphi \in_{\mathsf{T}} \mathsf{a}$ and $\mathsf{a}R\mathsf{b}$, then $\varphi \in_{\mathsf{T}} \mathsf{b}$.

(ii) If $\Box\varphi \in_{\mathsf{F}} \mathsf{a}$, then there is a node b such that $\mathsf{a}R\mathsf{b}$ and $\varphi \in_{\mathsf{F}} \mathsf{b}$.

(iii) If $\Box^+\varphi \in_{\mathsf{T}} \mathsf{a}$ and $\mathsf{a}R^+\mathsf{b}$, then $\varphi \in_{\mathsf{T}} \mathsf{b}$.

(iv) If $\Box^+\varphi \in_{\mathsf{F}} \mathsf{a}$, then there is a node b such that $\mathsf{a}R^+\mathsf{b}$ and $\varphi \in_{\mathsf{F}} \mathsf{b}$.

Then we have "$\varphi \in_{\mathsf{T}} \mathsf{a} \Rightarrow M, \mathsf{a} \models \varphi$" and "$\varphi \in_{\mathsf{F}} \mathsf{a} \Rightarrow M, \mathsf{a} \not\models \varphi$", which are proved by induction on φ using (1)–(5) of Lemma 4.8 and (i)–(iv) above. M is the required model because the root of $\mathcal{T}$ contains $\alpha_0 : \mathsf{F}$. □

5 Concluding remarks

This paper gives a new proof of the completeness theorem for the Hilbert style system of the propositional modal logic with two operators $\Box$ and $\Box^+$. Our method is "semantic diagram", and the point is how to construct the witness of $\neg\Box^+\varphi$. We enumerate all the possible candidates of paths to the witness ("special paths"), and search them using the Hilbert system as an oracle. The two-facedness of $\Box^+$-edges (Remark (2) on Lemma 4.6) is also remarkable.

A feature of our method is that we do not need extra operators other than $\Box$ and $\Box^+$. If the 'until' operator is allowed, there may be another possible way as in Brünnler and Lange [1] and Gaintzarain et al. [3]. Although our method seems to be ineffective for more complex logics like modal μ-calculus, it may be useful for certain logics without the 'until' operator, for example, epistemic logics.

Acknowledgement

I give great thanks to anonymous reviewers for their valuable comments on an earlier version of this paper.

References

[1] Kai Brünnler and Martin Lange. Cut-free sequent systems for temporal logic. *J. Log. Algebr. Program.*, 76(2):216–225, 2008.

[2] Ronald Fagin, Joseph Y. Halpern, Yoram Moses, and Moshe Y. Vardi. *Reasoning about Knowledge.* MIT Press, 1995.

[3] Joxe Gaintzarain, Montserrat Hermo, Paqui Lucio, Marisa Navarro, and Fernando Orejas. A cut-free and invariant-free sequent calculus for PLTL. In Jacques Duparc and Thomas A. Henzinger, editors, *CSL*, volume 4646 of *Lecture Notes in Computer Science*, pages 481–495. Springer, 2007.

[4] Robert Goldblatt. *Logics of Time and Computation.* Number 7 in CSLI Lecture Notes. Center for the Study of Language and Information, Stanford, CA, 2. edition, 1992.

A Proof of Lemma 3.2

Define formulas $\alpha, \beta, \gamma, \delta, \delta', \delta'', \varepsilon, \varepsilon', \varepsilon'', \zeta, \zeta'$:

$\alpha = \Box\tau.$ $\quad \beta = \Box^{+}(\sigma' \to \Box\tau').$ $\quad \gamma = \Box^{+}(\neg\sigma' \to \Box\tau).$
$\delta = \Box^{+}(\tau \to \omega).$ $\quad \delta' = \Box^{+}\Box(\tau \to \omega).$ $\quad \delta'' = \Box(\tau \to \omega).$
$\zeta = \sigma' \to \Box^{+}(\tau' \to \omega).$ $\quad \zeta' = \sigma' \to \Box(\tau' \to \omega).$
$\varepsilon = \Box^{+}(\tau \to \zeta).$ $\quad \varepsilon' = \Box(\tau \to \zeta).$ $\quad \varepsilon'' = \Box^{+}(\Box\tau \to \Box\zeta).$

An outline of the proof is as follows.

(i) $\vdash \Box^{+}(\tau' \to \omega) \to \Box\Box^{+}(\tau' \to \omega).$ $\quad$ ($\because$ A4)

(ii) $\vdash \Box^{+}(\tau' \to \omega) \to \Box\zeta.$ $\quad$ ($\because$ i)

(iii) $\vdash \Box^{+}(\Box^{+}(\tau' \to \omega) \to \Box\zeta).$ $\quad$ ($\because$ ii, R2),

(iv) $\vdash \{\alpha,\ \varepsilon'\} \Rightarrow \Box\zeta.$ $\quad$ ($\because$ A2)

(v) $\vdash \{\gamma,\ \varepsilon''\} \Rightarrow \Box^{+}(\neg\sigma' \to \Box\zeta).$ $\quad$ ($\because$ Lemma 3.1(4))

(vi) $\vdash \{\gamma,\ \varepsilon''\} \Rightarrow \Box^{+}(\neg\sigma' \vee \Box^{+}(\tau' \to \omega) \to \Box\zeta).$ $\quad$ ($\because$ v, iii)

(vii) $\vdash \{\gamma,\ \varepsilon''\} \Rightarrow \Box^{+}(\zeta \to \Box\zeta).$ $\quad$ ($\because$ vi)

(viii) $\vdash \{\alpha,\ \gamma,\ \varepsilon',\ \varepsilon''\} \Rightarrow \Box^{+}\zeta.$ $\quad$ ($\because$ iv, vii, A5)

(ix) $\vdash \{\alpha,\ \delta''\} \Rightarrow \Box\omega.$ $\quad$ ($\because$ A2)

(x) $\vdash \{\Box\tau',\ \Box(\tau' \to \omega)\} \Rightarrow \Box\omega.$ $\quad$ ($\because$ A2)

(xi) $\vdash \{\beta,\ \Box^{+}\zeta'\} \Rightarrow \Box^{+}(\sigma' \to \Box\omega).$ $\quad$ ($\because$ x, Lemma 3.1(2,4))

(xii) $\vdash \{\Box\tau,\ \Box(\tau \to \omega)\} \Rightarrow \Box\omega.$ $\quad$ ($\because$ A2)

(xiii) $\vdash \{\gamma,\ \delta'\} \Rightarrow \Box^{+}(\neg\sigma' \to \Box\omega).$ $\quad$ ($\because$ xii, Lemma 3.1(2,4))

(xiv) $\vdash \{\beta,\ \gamma,\ \delta',\ \Box^{+}\zeta'\} \Rightarrow \Box^{+}\Box\omega.$ $\quad$ ($\because$ xi, xiii)

(xv) $\vdash \{\alpha,\ \beta,\ \gamma,\ \delta',\ \delta'',\ \Box^{+}\zeta'\} \Rightarrow \Box^{+}\omega.$ $\quad$ ($\because$ ix, xiv, A5)

(xvi) $\vdash \delta \to \delta',$ $\quad \vdash \delta \to \delta'',$ $\quad \vdash \varepsilon \to \varepsilon',$ $\quad \vdash \varepsilon \to \varepsilon'',$ $\quad \vdash \zeta \to \zeta'.$

(xvii) $\vdash \{\alpha,\ \beta,\ \gamma,\ \delta,\ \varepsilon\} \Rightarrow \Box^{+}\omega.$ $\quad$ ($\because$ viii, xv, xvi)

(xviii) $\vdash \Box^{+}(\sigma' \to \Box\tau').$ $\quad$ ($\because$ (b), R2)

(xix) $\vdash \Box^{+}(\neg\sigma' \to \Box\tau).$ $\quad$ ($\because$ (c), R2)

(xx) $\vdash \{\alpha,\ \delta,\ \varepsilon\} \Rightarrow \Box^{+}\omega.$ $\quad$ ($\because$ xvii, xviii, xix)

(xxi) $\vdash \{\sigma \to \alpha,\ \sigma \to \delta,\ \sigma \to \varepsilon\} \Rightarrow \sigma \to \Box^{+}\omega.$ $\quad$ ($\because$ xx, Lemma 3.1(2))

(xxii) $\vdash \{\sigma \to \Box^{+}(\tau \to \omega),\ \sigma \to \Box^{+}(\tau \to \sigma' \to \Box^{+}(\tau' \to \omega))\} \Rightarrow \sigma \to \Box^{+}\omega.$
($\because$ (a), xxi)

Semantic Characterization of Kracht Formulas

Stanislav Kikot

Moscow State University
Moscow, Vorobjovy Gory, 1

Abstract

Kracht formulas are first-order correspondents of modal Sahlqvist formulas. In this paper we present a model-theoretic characterization of Kracht formulas similar to Van Benthem's theorem saying that a first-order formula is equivalent to a modal formula iff it is invariant under bisimulation. Our characterization yields a method to prove that a given first-order formula is not equivalent to any Kracht formula. In particular, we prove that the first-order formula, expressing the 'cubic property' of a 3-dimensional modal frame does not have a Kracht equivalent.

Keywords: Kracht Formulas, Semantic Characterization, Kracht-simulation

1 Introduction

Sahlqvist theorem on completeness and correspondence [12] is one of the basic tools in modal logic. Completeness and decidability of many modal calculi can be proved using this theorem together with other methods.

Kracht's theorem [8],[9] gives a syntactic characterization of first-order correspondents of Sahlqvist formulas; they are called Kracht formulas. [1] This theorem also describes an algorithm constructing a Sahlqvist correspondent for a given Kracht formula.

However, we do not know any sufficiently general method to decide whether a given first-order formula is equivalent to a Kracht formula. In general this problem is undecidable [3], and even particular cases of this problem may be hard.

For instance, recall a standard argument showing that not all first-order definable modal formulas are Sahlqvist. It is based on the fact that unlike Sahlqvist formulas, the formula $(\Diamond p \to \Diamond\Diamond p) \wedge (\Box\Diamond p \to \Diamond\Box p)$ is locally undefinable.

---

[1] Correspondence Theory and, in particular, Krachts formulas are applicable to knowledge base query answering, cf. [13].

When generalized Sahlqvist formulas appeared [5], there was a question if they really semantically extend the class of standard Sahlqvist formulas. The question was solved positively by D. Vakarelov and V. Goranko in [6], where they introduced the notion of a-persistence, showed that all standard Sahlqvist formulas are a-persistent and gave an example of a generalized Sahlqvist formula without this property.

In [13] some syntactic extensions of Kracht's fragment were proposed, and there is a question if these new formulas extend Kracht formulas semantically.

This paper proposes a general method for distinguishing Kracht formulas and thus for proofs that certain first-order formulas do not have Kracht equivalents. We point out that unlike [6], we use the methods of classical first-order model theory (elementary chains, ultraproducts and ω-saturated models) and deal with classical first-order formulas.

Our characterization is obtained as a combination of two well-known ideas. The first is Van Benthem's theorem [1] — a first-order formula is equivalent to a modal formula iff it is bisimulation-invariant. The second is preservation of first-order positive formulas under homomorphisms.

2 Kracht formulas.

Let Λ be a set of indices. We use the standard first-order language $\mathcal{L}f_\Lambda$ containing countably many individual variables x_i, binary predicates R_λ for every $\lambda \in \Lambda$, equality, boolean connectives $\wedge, \vee, \rightarrow, \neg$ and quantifiers $\exists x_i$, $\forall x_i$. To avoid subscripts, we often denote individual variables by $x, y, \ldots$.

The definitions from this section are almost the same as in [2] and originate from [8], [9].

Definition 2.1 *We use the following abbreviations*

$$(\forall x_i \rhd_\lambda x_j)A \equiv \forall x_i(x_j R_\lambda x_i \rightarrow A);$$

$$(\exists x_i \rhd_\lambda x_j)A \equiv \exists x_i(x_j R_\lambda x_i \wedge A).$$

$(\forall x_i \rhd_\lambda x_j)$, $(\exists x_i \rhd_\lambda x_j)$ *are called restricted quantifiers.*

To define Kracht formulas, we need a fragment $\mathcal{R}f_\Lambda$ of $\mathcal{L}f_\Lambda$. Informally, to obtain $\mathcal{R}f_\Lambda$ we take the positive fragment of $\mathcal{L}f_\Lambda$, add new relation symbols for compositions of R_λ and use only restricted quantifiers. We consider the symbols $\rhd_\lambda$ as elements of our language.

Formally, the formulas of $\mathcal{R}f_\Lambda$ are defined by recursion [9]:

- $\bot, \top$ are formulas of $\mathcal{R}f_\Lambda$;
- if $\varepsilon = \lambda_1 \ldots \lambda_n$, where $n \geq 0$, $\lambda_i \in \Lambda$, and x_i, x_j are individual variables, then $x_i R^\varepsilon x_j$ is a formula of $\mathcal{R}f_\Lambda$ (if ε is empty, we obtain equality);
- if A, B are formulas of $\mathcal{R}f_\Lambda$, then $(A \wedge B), (A \vee B)$ are formulas of $\mathcal{R}f_\Lambda$;
- if x_i, x_j are individual variables, $\lambda \in \Lambda$ and A is a formula of $\mathcal{R}f_\Lambda$, then $(\forall x_i \rhd_\lambda x_j)A$ and $(\exists x_i \rhd_\lambda x_j)A$ are formulas of $\mathcal{R}f_\Lambda$.

Definition 2.2 *A formula A of $\mathcal{R}f_\Lambda$ is called clean if A does not contain variables occurring both free and bound and every two different occurrences of quantifiers in A bind different variables.*

Henceforth we consider only clean formulas.

$\phi[t/x]$ denotes the substitution of the term t for all free occurrences of the variable x in ϕ. In general $\phi[t/x]$ is not necessary clean (even if ϕ is clean), but in this paper all such substitutions generate clean formulas.

Consider an $\mathcal{L}f_\Lambda$-structure $M = (W, (R^M_\lambda : \lambda \in \Lambda))$. $\mathcal{R}f^M_\Lambda$ denotes the language obtained from $\mathcal{R}f_\Lambda$ by adding the constants c_w for all $w \in W$. A formula ϕ of $\mathcal{R}f^M_\Lambda$ is called an $\mathcal{R}f^M_\Lambda$-sentence if ϕ does not contain free variables. The truth of $\mathcal{R}f^M_\Lambda$-sentences in M is defined in a standard way. In particular, the formula $c_w R^\varepsilon c_v$, where $\varepsilon = \lambda_1 \dots \lambda_n$, is true in M iff there is a sequence of points $w_0, w_1, \dots, w_n$ of M such that $w_0 = w$, $w_n = v$ and for all i from 1 to n we have $x_{i-1} R^M_{\lambda_i} x_i$. In particular, if $n = 0$ (i.e. ε is an empty sequence), then $c_w R^\varepsilon c_v \iff w = v$.

Definition 2.3 *A variable x in a formula ϕ is called inherently universal for ϕ if either x is free, or x is bound by a universal quantifier, which is not within the scope of an existential quantifier.*

A formula ϕ of $\mathcal{R}f_\Lambda$ is called a parametrized Kracht formula if in every its atomic subformula of the form $x_i R^\varepsilon x_j$ at least one of the variables x_i and x_j is inherently universal for ϕ.

A parametrized Kracht formula with a single free variable is called a Kracht formula.

Definition 2.4 *Consider an $\mathcal{L}f_\Lambda$-structure $\hat{T} = (W^T, (R^T_\lambda : \lambda \in \Lambda))$, where for all $\lambda \in \Lambda$ $R^T_\lambda \subseteq W^T \times W^T$. A sequence $x_1, \lambda_1, x_2, \lambda_2, \dots x_n$, where $x_i \in W^T$, $\lambda_i \in \Lambda$ and $(x_i, x_{i+1}) \in R^T_{\lambda_i}$ for $1 \leq i \leq n-1$ is called a path from x_1 to x_n in $\hat{T}$. A tuple $T = (\hat{T}, r^T)$ is called a tree with a root r^T if the following holds:*

- $r^T \in W^T$,
- *W^T is finite;*
- $(R^T_\lambda)^{-1}(r^T) = \emptyset$ *for all* $\lambda \in \Lambda$,
- *for every point $x^T \neq r^T$ there exists a unique path from r^T to x^T.*

Consider a tree $T = (W^T, (R^T_\lambda : \lambda \in \Lambda), r^T)$ and an $\mathcal{L}f_\Lambda$-structure $F = (W^F, (R^F_\lambda : \lambda \in \Lambda))$. A mapping $f : W^T \to W^F$ is called *monotonic* if for all $x, y \in W^T$, $\lambda \in \Lambda$, $x R^T_\lambda y$ implies $f(x) R^F_\lambda f(y)$.

3 Semantic Characterization

Definition 3.1 *Consider two $\mathcal{L}f_\Lambda$-structures $G = (W^G, (R^G_\lambda : \lambda \in \Lambda))$ and $F = (W^F, (R^F_\lambda : \lambda \in \Lambda))$, a tree $T = (W^T, (R^T_\lambda : \lambda \in \Lambda))$, monotonic mappings $g : T \to G$ and $f : T \to F$. A relation $Z \subseteq W^G \times W^F$ is called a Kracht-simulation if Z satisfies the following conditions:*

(KB1) For every $t \in W^T$, $(g(t), f(t)) \in Z$;

(KB2) For any $x^G \in W^G$, $x^F \in W^F$, $t \in W^T$, for arbitrary sequence $\varepsilon \in \Lambda^*$ if

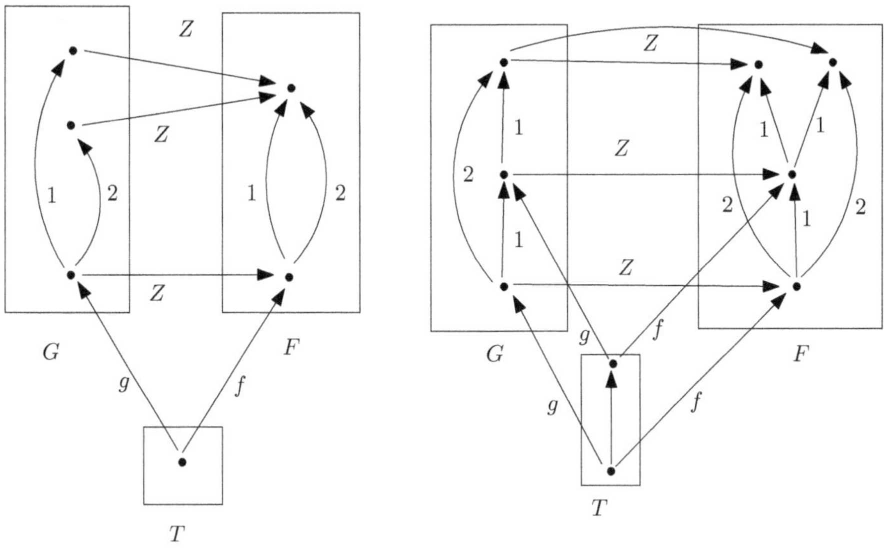

Fig. 1. Some examples of Kracht-simulations.

$(x^G, x^F) \in Z$ and $g(t)(R^G)^\varepsilon x^G$, then $f(t)(R^F)^\varepsilon x^F$.

(KB3) For any points $x^F \in W^F$ and $x^G \in W^G$ such that $(x^G, x^F) \in Z$ for any $(x')^G \in R^G_\lambda(x^G)$ there exists a point $(x')^F \in R^F_\lambda(x^F)$ such that $(x'^G, x'^F) \in Z$.

(KB4) For any points $x^F \in W^F$ and $x^G \in W^G$ such that $(x^G, x^F) \in Z$, for any $(x')^F \in R^F_\lambda(x^F)$ there exists a point $(x')^G \in R^G_\lambda(x^G)$, such that $(x'^G, x'^F) \in Z$.

In this case we say that the triple (G, T, g) is Kracht-reducible to (F, T, f) by Z, in symbols: $(G, T, g) \gg_Z (F, T, f)$.

Note that this notion generalizes modal bisimulations. (KB3) and (KB4) are the classical "forth" and "back" properties of bisimulations. (KB2) replaces the additional property of bisimulation saying that bisimilar worlds satisfy the same proposition letters. Note that unlike classical modal bisimulations, the relation $\gg_Z$ is not symmetric, since (KB2) acts only in one direction.

Examples of Kracht-simulations can be found in Figures 1 and 2.

Definition 3.2 *A tuple $G_\circ = (G, x_0^G)$ is called an $\mathcal{L}f_\Lambda$-structure with a designated point if $G = (W^G, (R^G_\lambda : \lambda \in \Lambda))$ is an $\mathcal{L}f_\Lambda$-structure and $x_0^G \in W^G$.*

Definition 3.3 *Consider two $\mathcal{L}f_\Lambda$-structures with designated points $G_\circ = (G, x_0^G)$ and $F_\circ = (F, x_0^F)$. We say that $G_\circ$ is Kracht-reducible to $F_\circ$ (notation: $G_\circ \ggg F_\circ$) if for any tree $T = (W^T, (R^T_\lambda : \lambda \in \Lambda), x_0^T)$ for all monotonic mappings $f : T \to F$ sending x_0^T to x_0^F, there exists a monotonic mapping $g : T \to G$, sending x_0^T to x_0^G, and a relation $Z \subseteq W^G \times W^F$ such that $(G, T, g) \gg_Z (F, T, f)$.*

The intuition underlying this definition is the following. For $\mathcal{L}f_\Lambda$-structures with designated points $G_\circ$ and $F_\circ$ we can regard a tuple (T, f), where T is a tree, and $f : T \to F$ is a monotonic mapping, sending x_0^T to x_0^F, as a test checking if $F_\circ$ really simulates $G_\circ$. The pair $(G_\circ, F_\circ)$ passes the test if there exist g and Z such that $(G, T, g) \gg_Z (F, T, f)$.

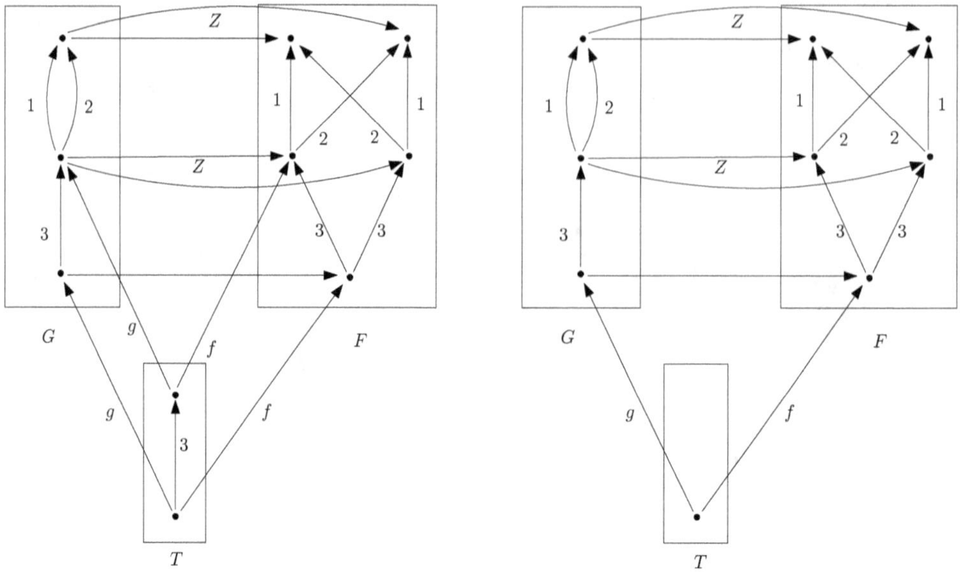

Fig. 2. The left-hand picture is not a Kracht-simulation while the right-hand picture is.

And $G_\circ \ggg F_\circ$ if the pair $(G_\circ, F_\circ)$ passes all possible tests.

Definition 3.4 *We say that a formula $\phi(x_0)$ of $\mathcal{L}f_\Lambda$ is preserved under Kracht-reducibility if for every pair of $\mathcal{L}f_\Lambda$-structures with designated points $G_\circ = (G, x_0^G)$ and $F_\circ = (F, x_0^F)$ if $G_\circ \ggg F_\circ$ and $G \models \phi[x_0^G/x_0]$, then $F \models \phi[x_0^F/x_0]$.*

Theorem 3.5 *A formula $\phi(x_0)$ of $\mathcal{L}f_\Lambda$ is equivalent to a Kracht formula iff $\phi(x_0)$ is preserved under Kracht-reducibility.*

This theorem is proved in Sections 4 – 6.

4 Soundness

The aim of this section is to prove that every Kracht formula is preserved under Kracht-reducibility.

To begin with, we give the definition of a local Kracht formula. The idea of this definition is to characterize a point in any frame 'up to K-reducibility'. To define these formulas we use the following convention. We assume that every local Kracht formula ϕ has a fixed free variable v (however, v may be dummy in ϕ), and v differs from $x_0, x_1, x_2, \ldots$. The notation $\phi(t)$ denotes the substitution of t for v in ϕ.

These formulas are similar to standard translations of modal formulas $ST_x(\chi)$ [2], where x is a fixed variable. To emphasise this similarity, for a point w of an $\mathcal{L}f_\Lambda$-structure $F = (W^F, (R_\lambda^F : \lambda \in \Lambda))$ and a local Kracht formula ϕ it is tempting to write $F, w \models \phi$ instead of $F \models \phi[w/v]$.

But there is yet another difference between local Kracht formulas and the standard translation. In $ST_x(\chi)$ the only free variable is x. Local Kracht formulas may contain some additional free variables $x_0, x_1, \ldots x_n$. Suppose that in addition to an

$\mathcal{L}f_\Lambda$-structure F we have a tree $T = (W^T, (R^T_\Lambda : \lambda \in \Lambda))$ and a monotonic mapping $f : W^T \to W^F$. Suppose also that $W^T = \{x^T_0, x^T_1, \dots, x^T_n\}$. In this case the notation $(F, T, f), w \models \phi(v)$ means $F \models \phi[f(x^T_i)/x_i][w/v]$.

Definition 4.1 *Local Kracht formulas are defined recursively:*

- $\top$, $\bot$ *and* $x_i R^\varepsilon v$ *are local Kracht formulas;*
- *if* $\phi(v)$ *is a local Kracht formula, then* $(\exists v' \rhd_\lambda v)\phi(v')$ *and* $(\forall v' \rhd_\lambda v)\phi(v')$ *are local Kracht formulas. As in the standard translation, here* v' *is a new variable not occurring in* ϕ. *This restriction is crucial, e. g. the formula* $(\forall v_1 \rhd v)(\forall v_2 \rhd v)(v_1 R v_2)$ *is not a local Kracht formula, while* $(\forall v_2 \rhd v)(x_1 R v_2)$ *is.*
- *if* $\phi(v)$ *and* $\psi(v)$ *are local Kracht formulas, then* $\phi(v) \wedge \psi(v)$ *and* $\phi(v) \vee \psi(v)$ *are local Kracht formulas.*

Example 4.2 *Here are some examples of local Kracht formulas:* $x_1 R^\varepsilon v$, $\exists v_1 \rhd_\lambda v(x_1 R^\varepsilon v_1)$, $(\exists v_1 \rhd_\lambda v(x_1 R^{\varepsilon_1} v_1)) \wedge (x_2 R^{\varepsilon_2} v)$.

Lemma 4.3 *If* $(G, T, g) \gg_Z (F, T, f)$ *and* $(x^G, x^F) \in Z$, *then for any local Kracht formula* $\psi(v)$ $(G, T, g), x^G \models \psi(v)$ *implies* $(F, T, f), x^F \models \psi(v)$.

Proof. The proof by induction on the length of ψ.

The base of induction follows from (KB2); the cases $\psi = \psi_1 \wedge \psi_2$ and $\psi = \psi_1 \vee \psi_2$ are trivial.

The case $\psi(v) = \exists v' \rhd_\lambda v\psi'[v'/v]$ follows immediately from (KB3).

Consider the case $\psi(v) = \forall v' \rhd_\lambda v\psi'[v'/v]$. Suppose that $(G, T, g), x^G \models \psi$, but $(F, T, f), x^F \not\models \psi$. The latter means that there exists a point $x'^F \in R^F_\lambda(x^F)$, such that $(F, T, f), x'^F \not\models \psi'$. By $(KB4)$, there is a point $x'^G \in R^G_\lambda(x^G)$, such that $(x'^G, x'^F) \in Z$. By the induction hypothesis, $(G, T, g), x'^G \not\models \psi'$. This contradicts $(G, T, g), x^G \models \psi$. Thus the claim holds. □

Now we reduce arbitrary Kracht formulas to local Kracht formulas.

Definition 4.4 *If a formula* ϕ *is built from formulas of the form* $\psi(x_i)$, *where* ψ *is a local Kracht formula, using* $\wedge$, $\vee$ *and restricted universal quantifiers, we say that* ϕ *is a parametrized decomposed Kracht formula.*

Lemma 4.5 *Let* $\xi(x_1, \dots, x_n, y_1, \dots, y_r)$ *be a formula of* $\mathcal{R}f_\Lambda$. *Suppose that any subformula of* ξ *of the form* $v_1 R^\varepsilon v_2$ *contains at least one* x*-variable. Then there exist local Kracht formulas* $\psi_1(v), \dots, \psi_k(v)$, *not containing* y*-variables, such that* ξ *is equivalent to a formula built from* $\psi_i(x'_i)$ $(x'_i \in \{x_1, \dots, x_n, y_1, \dots, y_r\})$ *using only* $\wedge$ *and* $\vee$.

Proof. At first we replace each subformula of ϕ of the form $zR^\varepsilon x$, where $\varepsilon = \lambda_1 \dots \lambda_n$, and the variable x (but not z) is inherently universal, with an equivalent formula $(\exists z_1 \rhd_{\lambda_1} z)(\exists z_2 \rhd_{\lambda_2} z_1) \dots (\exists z_n \rhd_{\lambda_n} z_{n-1})(x = z_n)$, where all z_i are new variables. This operation gives us a formula with all atoms of the form $x_i R^\varepsilon z$.

Then we argue by induction on the length of ξ.

If $\xi = x_j R^\varepsilon z$, then $\xi = \psi(z)$ for $\psi(v) = x_j R^\varepsilon v$. The cases of booleans are trivial.

Suppose $\xi = (\exists z \rhd_\lambda x_j)\xi'(x_1, \dots, x_n, z)$. By the induction hypothesis and distributivity, ξ' is of the form $K_1 \vee \ldots \vee K_m$, where $K_i = \psi_1^i(x_1^i) \wedge \ldots \wedge \psi_{m_i}^i(x_{m_i}^i)$ ($x_l^i \in \{x_1, \ldots, x_k, z\}$). So, $\xi \equiv (\exists z \rhd_\lambda x_j)K_1 \vee \ldots \vee (\exists z \rhd_\lambda x_j)K_m$. But

$$(\exists z \rhd_\lambda x_j)K_i = \bigwedge_{x_l^i \neq z} \psi_l^i(x_l^i) \wedge \psi(x_j)$$

for the local Kracht formula

$$\psi(v) = (\exists v' \rhd_\lambda v) \bigwedge_{x_l^i = v} \psi_l^i(v').$$

The case of universal quantifier is proved dually. □

Lemma 4.6 *Every parametrized Kracht formula ϕ is equivalent to a parametrized decomposed formula.*

Proof. By induction on the length of ϕ relying upon Lemma 4.5 for atoms and for formulas beginning with an existential quantifier. □

Example 4.7 *Kracht formula*

$$(\forall x_2 \rhd_1 x_1)(x_1 R_2 x_2 \vee x_1 R_3 x_2) \wedge (\forall x_3 \rhd_1 x_1)(\exists x_4 \rhd_1 x_1) x_3 R_1 x_4$$

is equivalent to a decomposed Kracht formula

$$(\forall x_2 \rhd_1 x_1)(\psi_1(x_2) \vee \psi_2(x_2)) \wedge (\forall x_3 \rhd_1 x_1)\psi_3(x_1)$$

for local Kracht formulas $\psi_1(v) = x_1 R_2 v$, $\psi_2(v) = x_1 R_3 v$, $\psi_3(v) = (\exists v' \rhd_1 v) x_3 R_1 v'$.

Definition 4.8 *Consider a model F and a parametrized decomposed Kracht formula ϕ. Suppose that we substitute points $x_i^F \in W^F$ for all free variables x_i in ϕ. The result of such substitution is called a Kracht F-sentence. A local Kracht F-sentence is defined in a similar way.*

Definition 4.9 *We say that a set of F-sentences S witnesses the falsity of an F-sentence ϕ if S satisfies the following conditions:*

(i) *$\phi \in S$;*

(ii) *if $\phi_1 \vee \phi_2 \in S$, then $\phi_1 \in S$ and $\phi_2 \in S$;*

(iii) *if $\phi_1 \wedge \phi_2 \in S$, then $\phi_1 \in S$ or $\phi_2 \in S$;*

(iv) *if $(\forall x_i \rhd_\lambda x_j^F)\phi' \in S$, then there exists a point x_i^F in F such that $x_j^F R_\lambda^F x_i^F$ and $\phi'[x_i^F/x_i] \in S$;*

(v) *if $\psi(v)$ is a local Kracht sentence and $\psi(x_i^F) \in S$, then $F \not\models \psi(x_i^F)$.*

Lemma 4.10 *Let ϕ be a Kracht F-sentence. Then*

$F \not\models \phi \iff$ *there exists a set of Kracht F-sentences S^ϕ witnessing the falsity of ϕ.*

Proof. ($\Longrightarrow$) Suppose that $F \not\models \phi$. We argue by induction on the length of ϕ.

- if ϕ is of the form $\psi(x_k^F)$ for a local Kracht formula $\psi(v)$, put $S^\phi = \{\phi\}$;
- if $\phi = \phi_1 \vee \phi_2$, then both ϕ_1 and ϕ_2 are false, so we put $S^\phi = \{\phi\} \cup S^{\phi_1} \cup S^{\phi_2}$;
- if $\phi = \phi_1 \wedge \phi_2$, then one of ϕ_1 or ϕ_2 is false so we can put either $S^\phi = \{\phi\} \cup S^{\phi_1}$ or $S^\phi = \{\phi\} \cup S^{\phi_2}$ depending on the falsity of ϕ_1 or ϕ_2;
- if $\phi = (\forall x_i \rhd_\lambda x_j^F)\phi' \in S$, then there is a point x_i^F in a model F such that $x_j^F R_\lambda^F x_i^F$ and $F \not\models \phi'[x_i^F/x_i]$, hence we can put $S^\phi = \{\phi\} \cup S^{\phi'[x_i^F/x_i]}$.

($\Longleftarrow$) A trivial induction shows that if S witnesses the falsity of any F-sentence, then for all $\phi' \in S$ $F \not\models \phi'$. But $\phi \in S$. □

Now we are ready to prove that every Kracht formula $\phi(x_0)$ is preserved under Kracht-reducibility, that is the soundness part of Theorem 3.5.

Proof. According to Lemma 4.6, without any loss of generality we can assume that $\phi(x_0)$ is decomposed.

Assume that $\phi(x_0)$ is not preserved under Kracht-reducibility, that is there are $\mathcal{L}f_\Lambda$-structures with designated points $G_\circ$ and $F_\circ$ such that $G_\circ \ggg F_\circ$ and

(1) $$G \models \phi[x_0^G/x_0],$$

but $F \not\models \phi[x_0^F/x_0]$. By Lemma 4.10 there exists a set of F-sentences S_F witnessing the falsity of $\phi[x_0^F/x_0]$ in F.

Let $\Gamma = \{\psi(x_k^F) \mid \psi(v)$ be a local Kracht sentence and $\psi(x_k^F) \in S_F\}$. By W^T we denote the set of all inherently universal variables x_k such that the corresponding constants x_k^F occur in Γ. In other words, W^T consists of those inherently universal variables of ϕ, whose valuations are essential for the falsity of $\phi[x_0^F/x_0]$ in F. Consider the tree $T = (W^T, (R_\lambda^T : \lambda \in \Lambda), x_0)$, where $x_j R_\lambda^T x_i$ iff the variable x_i is bound by a quantifier of the form $(\forall x_i \rhd_\lambda x_j)$.

The construction of S_F gives us a monotonic mapping $f : T \to F$ sending each point $x_i \in W^T$ to a point $x_i^F \in F$. By Definition 3.3, there exists a monotonic mapping $g : T \to G$, sending x_0 to x_0^G such that $(G, T, g) \gg_Z (F, T, f)$ for some Z. Let S_G be the set of G-sentences obtained by replacing all x_i^F with x_i^G in all formulas of S_F.

We claim that S_G witnesses the falsity of the G-sentence $\phi[x_0^G/x_0]$ in G. In fact, the items (i)–(iii) of Definition 4.9 hold by the construction of S_G. The item (v) is true due to Lemma 4.3. Let us show that the item (iv) holds for S_G. Suppose that $(\forall x_i \rhd_\lambda x_j^G)\phi' \in S_G$. This may happen only in the case when $(\forall x_i \rhd_\lambda x_j^F)\phi'[x_k^F/x_k^G] \in S_F$. So there exists $x_i^F \in F$ such that $(F, T, f), x_i^F \not\models \phi'[x_k^F/x_k^G]$. Put $x_i^G = g(x_i)$. By (KB1), $(x_i^G, x_i^F) \in Z$, hence due to (KB2), $(G, T, g), x_i^G \not\models \phi'$. And $x_j^G R_\lambda^G x_i^G$ by the monotonicity of g.

Therefore, by Lemma 4.10, we conclude that $G \not\models \phi[x_0^G/x_0]$. This contradicts our initial assumption (1). □

5 Model-theoretic background

In the next section we assume that the reader is familiar with standard model-theoretic tools such as elementary extensions, ω-saturated models and ultrapowers. However, we recall the latter two notions and their basic properties.

Definition 5.1 *An $\mathcal{L}f_\Lambda$-structure $F = (W, (R_\lambda : \lambda \in \Lambda))$ is called n-saturated, if for any set Γ of first-order formulas with at most n free variables $\gamma(x_1, x_2, \ldots, x_n)$ the following holds:*

IF $x_1^0, \ldots, x_{n-1}^0$ is a sequence of points from W such that for all finite $\Delta \subseteq \Gamma$ there is a point $(x_n^0)_\Delta$ such that

$$F \models \gamma(x_1^0, x_2^0, \ldots, x_{n-1}^0, (x_n^0)_\Delta)$$

for all formulas $\gamma \in \Delta$,

THEN there exists a point x_n^0 such that $F \models \gamma(x_1^0, x_2^0, \ldots, x_{n-1}^0, x_n^0)$ for all formulas $\gamma \in \Gamma$.

A model F is called ω-saturated if it is n-saturated for all n.

Definition 5.2 *Consider a model $F = (W, (R_\lambda : \lambda \in \Lambda))$ and a non-principal ultrafilter u over the set of all natural numbers $\mathbb{N}$.*

We say that two sequences of points from W $\bar{\alpha} = (\alpha_1, \alpha_2, \alpha_3, \ldots)$ and $\bar{\beta} = (\beta_1, \beta_2, \beta_3, \ldots)$ are u-equivalent (denoted by $\bar{\alpha} \sim_u \bar{\beta}$), if $\{i \mid \alpha_i = \beta_i\} \in u$. The equivalence class of a sequence α is denoted by $\lceil \alpha \rceil$.

The $\mathcal{L}f_\Lambda$-structure $F = (W', (R'_\lambda : \lambda \in \Lambda))$, where

$$W' = \{ \text{ all sequences of points from } W\} / \sim_u$$

and

$$\lceil \bar{\alpha} \rceil R'_\lambda \lceil \bar{\beta} \rceil \iff \{i \mid \alpha_i R_\lambda \beta_i\} \in u.$$

is called an ultrapower *of F (with respect to u) and denoted by $\prod_u F$.*

Proposition 5.3 *A natural embedding $i : F \to \prod_u F$ such that*

$$i(w) = \lceil (w, w, w, w, \ldots) \rceil,$$

is elementary.

Proposition 5.4 *For any $\mathcal{L}f_\Lambda$-structure $F = (W, (R_\lambda : \lambda \in \Lambda))$ and any non-principal ultrafilter u over $\mathbb{N}$ the ultrapower $\prod_u F$ is ω-saturated.*

6 Completeness

We follow the plan of the proof of van Benthem's theorem from [2]. The keystone of the proof is Lemma 6.5 (an analogue of Detour lemma from [2]). Its proof is carried out step by step in Lemmas 6.1 – 6.4. After that the theorem is proved in a standard way.

Lemma 6.1 *Consider $\mathcal{L}f_\Lambda$-structures with designated points $G_\circ = (G, x_0^G)$ and $F_\circ = (F, x_0^F)$ such that for any Kracht formula $\phi(x_0)$ $G \models \phi[x_0^G/x_0]$ implies $F \models \phi[x_0^F/x_0]$. Suppose there is a tree T and a monotonic mapping $f : T \to F$ sending r^T to x_0^F.*

Then there exists an elementary extension G' of G and a monotonic mapping $g : T \to G'$ sending r^T to x_0^G, such that for any local Kracht formula $\psi(v)$ for any point $t \in W^T$ $G' \models \psi(g(t))$ implies $F \models \psi(f(t))$.

Proof. Let $W^T = \{x_0, x_1, \ldots, x_n\}$, let $x_{p(i)}$ be the unique predecessor of x_i in T, and let $x_{p(i)} R_{\lambda_i} x_i$. Suppose that for all $0 \leq i \leq n$

$$\Psi_i = \{\psi(v) \mid \psi(v) \text{ is a local Kracht formula and } F \not\models \psi(f(x_i))\}.$$

We enumerate all formulas of Ψ_i:

$$\Psi_i = \{\psi_1^i(v), \psi_2^i(v), \ldots\}.$$

Fix $m \in \mathbb{N}$. Consider the formula

$$\gamma_m = (\exists x_1 \rhd_{\lambda_1} x_0) \ldots (\exists x_n \rhd_{\lambda_n} x_{p(n)}) \bigwedge_{x_i \in W^T} ((\neg\psi_1^i \wedge \ldots \wedge \neg\psi_m^i)(x_i)).$$

It is clear that $F \models \gamma_m[x_0^F/x_0]$ and γ_m is a negation of a Kracht formula.

Hence, due to the assumption of the lemma, $G \models \gamma_m[x_0^G/x_0]$, that is there exist points $g_0^m, \ldots, g_n^m \in W^G$ (here $g_0^m = x_0^G$ for all m) such that the mappings $g^m : T \to G$ $g^m(x_i) = g_i^m$ are monotonic and for all $0 \leq i \leq n$

$$G, g_i^m \models (\neg\psi_1^i \wedge \ldots \neg\psi_m^i).$$

Let u be a non-principal ultrafilter over $\mathbb{N}$. Put $G' = \Pi_u G$.

This guarantees that G' is an elementary extension of G.

Now define $g(x_i)$ as the equivalence class of the sequence $(g_i^1, g_i^2, g_i^3, \ldots)$.

It is clear that if $x_i R_\lambda^T x_j$, then for all m $g_i^m R_\lambda^G g_j^m$, therefore, $g(x_i) R_\lambda^{G'} g(x_j)$.

Now we show that for any local Kracht formula $\psi(v)$ for any point $x_i \in W^T$ if $G' \models \psi(g(x_i))$ then $F \models \psi(f(x_i)$.

In fact, suppose that $F \not\models \psi(f(x_i))$ for some i. Then $\psi \in \Psi_i$, hence there is m_0 such that $\psi = \psi_i^{m_0}$. Then for all $m \geq m_0$ $g_i^m \models \neg\psi$. This contradicts $G' \models \psi(g(x_i))$. □

Lemma 6.2 *Consider $\mathcal{L}f_\Lambda$-structures with designated points $G_\circ$ and $F_\circ$ such that W^F is countable and for any Kracht formula $\phi(x_0)$ $G \models \phi[x_0^G/x_0]$ implies $F \models \phi[x_0^F/x_0]$. Then there is an elementary extension G^* of G, such that*

(2) *for any tree T for any monotonic mapping $f : T \to F$, sending r^T to x_0^F, there exists a monotonic mapping $g : T \to G^*$ sending r^T to x_0^G, such that for any local Kracht formula $\psi(v)$ for any point $x_i \in W^T$ $G^* \models \psi(g(x_i))$ implies $F \models \psi(f(x_i))$*

Proof. Consider all possible pairs (T, f) consisting of a tree T and a monotonic mapping $f : T \to F$ sending r^T to x_0^F. There are countably many such pairs, we enumerate them as $(T_1, f_1), (T_2, f_2), \ldots$.

We construct an elementary chain $G_0 \prec G_1 \prec G_2 \prec \ldots$, where $G_0 = G$, and for $i > 0$ G_i is obtained by applying Lemma 6.1 to frames G_{i-1}, F and the pair (T_i, f_i). The condition of Lemma 6.1 holds due to the elementarity of G_{i-1} over G_0.

Put G^* to be the limit of this chain. By the elementary chain principle, G^* is an elementary extension of G_0, so (2) obviously holds. □

Lemma 6.3 *Consider $\mathcal{L}f_\Lambda$-structures with designated points $G_\circ$ and $F_\circ$ such that W^F is countable and for any Kracht formula $\phi(x_0)$, $G \models \phi[x_0^G/x_0]$ implies $F \models \phi[x_0^F/x_0]$. Then there exist elementary embeddings $G \prec \bar{G}$ and $F \prec \bar{F}$ such that $\bar{G}$ and $\bar{F}$ are ω-saturated and satisfy 6.2 (2).*

Proof. Due to Lemma 6.2, there exists an elementary embedding $G \prec G^*$, such that (2) holds.

Take a non-principal ultrafilter u over $\mathbb{N}$ and put $\bar{G} = \Pi_u G^*$, $\bar{F} = \Pi_u F$. Let us verify that (2) still holds for $\bar{G}$ and $\bar{F}$.

In fact, consider $f : T \to \bar{F}$. Suppose that $f(x_i) = \lceil(f_i^1, f_i^2, f_i^3, \ldots)\rceil$. Then for each monotonic mapping $f^j : T \to f$ we find a corresponding monotone mapping $g^j : T \to G^*$, and finally we put $g(x_i) = \lceil(g^1(x_i), g^2(x_i), \ldots)\rceil$. It is clear that g is well defined. Let us prove that for any local Kracht formula $\psi(v)$ for any point t of T $\bar{G} \models \psi(g(x_i))$ implies $F \models \psi(f(x_i))$. If $\bar{G} \models \psi(g(x_i))$, then $A = \{j \mid G \models \psi(g^j(x_i))\} \in u$, therefore $A \subseteq \{j \mid F \models \psi(f^j(x_i))\}$, and so $\{j \mid F \models \psi(f^j(x_i))\} \in u$, hence $\bar{F} \models \psi(f(x_i))$.

Due to Proposition 5.4, the $\mathcal{L}f_\Lambda$-structures $\bar{G}$ and $\bar{F}$ are ω-saturated. □

Lemma 6.4 *Let $\bar{G}$, $\bar{F}$ be ω-saturated $\mathcal{L}f_\Lambda$-structures. Suppose there is a tree T and monotonic mappings $g : T \to \bar{G}$ and $f : T \to \bar{F}$. Suppose also that $W^T = \{x_1, \ldots, x_n\}$. Let us define a relation $Z \subseteq W^G \times W^F$, by putting for $x \in \bar{G}$ and $y \in \bar{F}$*

$$(x,y) \in Z \iff \begin{array}{l}\textit{for any local Kracht formula } \psi(v) \textit{ such that all its free variables}\\ \textit{except } v \textit{ are in } W^T,\ (\bar{G},T,g) \models \psi(x) \textit{ implies } (\bar{F},T,f) \models \psi(y).\end{array}$$

Then the relation Z satisfies $(KB2)$–$(KB4)$.

Proof. $(KB2)$ follows readily from the definition of Z.

Let us check $(KB3)$. Suppose that $x^F \in W^F$, $x^G \in W^G$, $x^G Z x^F$ and there exists $(x')^G \in R_\lambda^G(x^G)$. Let Ψ be the set of all local Kracht formulas true at the point $(x')^G$ of the frame G under the valuation $[g(x_i)/x_i]$. Consider the set of formulas $\Psi' = \{x^F R_\lambda v\} \cup \{\psi[f(x_i)/x_i] \mid \psi \in \Psi\}$ in the first-order language $\mathcal{L}f_\Lambda$ enriched with the constants naming the points of W^F. Let us show that every set of the form $\Psi'_m = \{x^F R_\lambda v\} \cup \{\psi_1, \ldots, \psi_m \mid \psi_i \in \Psi\}$ is realized in F. Consider the local Kracht formula $\phi_m = (\exists v' \rhd_\lambda v)(\psi_1(v') \wedge \ldots \wedge \psi_m(v'))$. It is clear that $G \models \phi_m(x^G)$, therefore $F \models \phi_m(x^F)$, that is there exists a point x'^F realizing Ψ_m. Hence, due to the saturation of F, we can conclude that the frame F realizes the whole set Ψ', that is there is a point $(x')^F \in R_\lambda^F(x^F)$ such that $(x'^G, x'^F) \in Z$.

Let us check $(KB4)$. Suppose that $x^F \in W^F$, $x^G \in W^G$, and $x^G Z x^F$. Take a point $(x')^F \in R_\lambda^F(x^F)$.

Let $\Psi = \{\neg\psi \mid \psi$ is a local Kracht formula and $F \models \neg\psi[f(x_i)/x_i][(x')^F/v]\}$. Consider the set of formulas $\Psi' = \{x^G R_\lambda v\} \cup \{\psi[g(x_i)/x_i] \mid \psi \in \Psi\}$ in the first-order language $\mathcal{L}f_\Lambda$ enriched with the constants from W^G. Let us show that any set of the form $\Psi'_m = \{x^G R_\lambda v\} \cup \{\neg\psi_1, \ldots, \neg\psi_m \mid \neg\psi_i \in \psi\}$ is realized in G.

Suppose the contrary, i. e. the formula $\psi_1 \vee \ldots \vee \psi_m$ is true at every point $v \in R_\lambda(x^G)$. Then the local Kracht formula $\phi_m(v) = \forall v' \rhd_\lambda v(\psi_1 \vee \ldots \vee \psi_m)$ is true at the point x^G . Hence ϕ_m is true at the point x^F of the frame F. This means that the formula $\psi_1 \vee \ldots \vee \psi_m$ is true at $(x')^F$. This contradicts the definition of Ψ.

Due to the saturation of G, we conclude that the whole set Ψ' is realized in G. This means that there is a point $(x')^G \in R^G_\lambda(x^G)$ such that $(x'^G, x'^F) \in Z$. □

Lemma 6.5 *Consider $\mathcal{L}f_\Lambda$-structures with designated points $G_\circ$ and $F_\circ$ such that W^F is countable and for any Kracht formula $\phi(x_0)$ $G \models \phi[x_0^G/x_0]$, implies $F \models \phi[x_0^F/x_0]$. Then there exist elementary embeddings $G \prec \bar{G}$ and $F \prec \bar{F}$, such that $(\bar{G}, x_0^G) \ggg (\bar{F}, x_0^F)$.*

Proof. Apply Lemma 6.3 and then Lemma 6.4. □

Proof. [The proof of theorem 3.5 (completeness)] Consider an arbitrary first-order formula with a single free variable $\phi(v)$ which is preserved under Kracht-reducibility. Let us show that $\phi(v)$ is equivalent to some Kracht formula.

To this end we following the plan in [2], consider the set of first-order formulas with a single free variable v $KC(\phi) = \{\psi(v) \mid \phi(v) \models \psi(v), \psi(v)$ is a Kracht formula$\}$.

(1) Note that if $KC(\phi) \models \phi$, then ϕ is equivalent to a Kracht formula. In fact, if $KC(\phi) \models \phi$, then there exist $\psi_1, \ldots, \psi_n$ such that $\vdash_{PC} \psi_1 \wedge \ldots \wedge \psi_n \to \phi$. Therefore the formula ϕ is equivalent to $\psi_1 \wedge \ldots \wedge \psi_n$.

(2) Let us show that $KC(\phi) \models \phi$. To this end, take a countable model N, a point $y \in N$ and suppose that $N \models KC[y/v]$. Consider the set of formulas

$$NKT(N, y) = \{\neg\delta \mid \delta \text{ is a Kracht formula and } N \models \neg\delta(y)\}.$$

(3) We claim that the set $NKT(N, y) \cup \{\phi\}$ is consistent. In fact, suppose the contrary. Then there is a finite subset $NKT_0 \subset NKT(N, y)$ such that $\vdash_{PC} \phi \to \neg \bigwedge NKT_0$. This means that there are Kracht formulas $\delta_1, \ldots, \delta_n$ such that $\vdash_{PC} \phi \to \delta_1 \vee \ldots \vee \delta_n$. Then $\delta_1 \vee \ldots \vee \delta_n \in KC(\phi)$, therefore $N, y \models \delta_1 \vee \ldots \vee \delta_n$. This contradicts the fact that $N, y \models \neg\delta_i$ for all i.

(4) Hence, due to the Gödel Completeness Theorem, there is a model M and a point $x \in M$ such that $(M, x) \models NKT(N, y) \cup \{\phi\}$. We claim that for every Kracht formula ψ if $M, x \models \psi$ then $N, y \models \psi$. In fact, if $N, y \models \neg\psi$, then $\neg\psi \in NKT(N, y)$, therefore $M, x \models \neg\psi$. This is the contradiction.

(5) Now we apply Lemma 6.5. It states that there exist elementary extensions $\bar{M}$ and $\bar{N}$ of the models M and N such that $\bar{M} \ggg \bar{N}$. But $M \models \phi$, therefore $\bar{M} \models \phi$, hence $\bar{N} \models \phi$, and so $N \models \phi$. So we have proved that $KC(\phi) \models \phi$. □

7 "Cubic" property

In this section we apply our semantic characterization of Kracht formulas to show that a certain formula fc (well-known in many-dimensional modal logic) does not have a Kracht equivalent.

Consider unimodal Kripke frames $F_1 = (W_1, \hat{R}_1), \ldots, F_n = (W_n, \hat{R}_n)$. Recall that their product $F_1 \times \ldots \times F_n$ is the frame $(W_1 \times \ldots \times W_n, R_1, \ldots, R_n)$, where

$$(x_1, \ldots, x_n) R_i (y_1, \ldots, y_n) \Longleftrightarrow x_j = y_j \text{ for } i \neq j \text{ and } x_i \hat{R}_i y_i.$$

All 3-modal frames of the form $F_1 \times F_2 \times F_3$, satisfy [4] the "cubic" formula $\forall x_0 fc(x_0)$, where

$$fc(x_0) = \forall x_1 \forall x_2 \forall x_3 (x_0 R_1 x_1 \wedge x_0 R_2 x_2 \wedge x_0 R_3 x_3 \rightarrow$$
$$\exists y_{12} \exists y_{13} \exists y_{23} \exists y_{123} (x_1 R_2 y_{12} \wedge x_1 R_3 y_{13} \wedge$$
$$\wedge x_2 R_1 y_{12} \wedge x_2 R_3 y_{23} \wedge x_3 R_1 y_{13} \wedge$$
$$x_3 R_2 y_{23} \wedge y_{23} R_1 y_{123} \wedge y_{13} R_2 y_{123} \wedge y_{12} R_3 y_{123})).$$

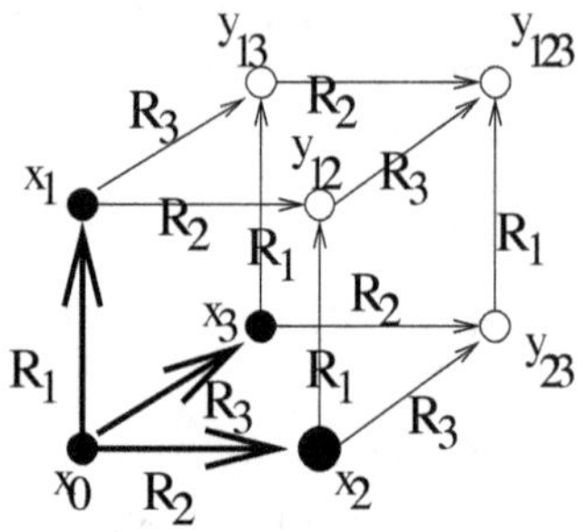

In [10] and [11] modifications of the formula fc are used to obtain negative results on axiomatizing modal logics of dimensions ≥ 3.

Theorem 7.1 *The formula $fc(x_0)$ is not equivalent to any Kracht formula.*

To prove Theorem 7.1, we fix $\Lambda = \{1, 2, 3\}$ and construct two $\mathcal{L}f_\Lambda$-structures with designated points $G_\circ$ and $F_\circ$ such that $G_\circ \ggg F_\circ$, $G \models fc[x_0^G/x_0]$, but $F \not\models fc[x_0^F/x_0]$.

Put $F_\circ = (F, r)$ where $F = (W, R_1, R_2, R_3)$, (see Fig. 3)

$$W = \{r\} \cup \{a_1, a_2, a_3\} \cup \{b_j^i \mid i, j = 1, 2, 3\} \cup \{c_1, c_2, c_3\},$$

and R_l are defined by the following conditions:

$$r R_l a_i \Longleftrightarrow l = i;$$

$$a_i R_l b_k^j \Longleftrightarrow i \neq l, i \neq k, k \neq l, j \neq l$$
$$b_k^j R_l c_i \Longleftrightarrow k = l \text{ and } (j = i = k \text{ or } (j \neq k \text{ and } i \neq k)).$$

Put $G_\circ = (G, r)$, where $G = (W', R_1', R_2', R_3')$ and

$$W' = W \cup \{\bar{b}_j^i \mid i, j = 1, 2, 3\} \cup \{\bar{c}_1, \bar{c}_2, \bar{c}_3, \bar{c}_{123}\},$$

$$R_l' = R_l \cup \{(a_i, \bar{b}_k^j) \mid a_i R_l b_k^j\} \cup \{(\bar{b}_k^j, \bar{c}_i) \mid b_k^j R_l c_i\} \cup \{(\bar{b}_j^j, \bar{c}_{123}) \mid 1 \leq j \leq 3\}.$$

One can see that F is a part of G, that is there exists a natural embedding $\iota : W \to W'$.

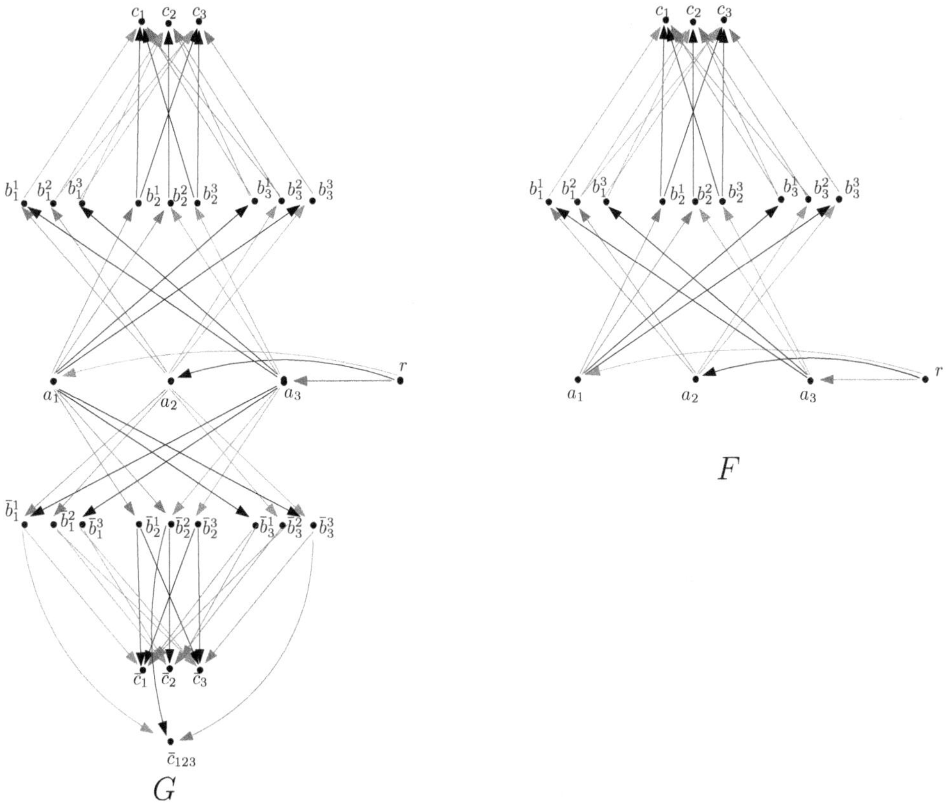

Fig. 3. Frames G and F; here R_1 is shown in red, R_2 in black and R_3 in blue

Lemma 7.2 *In $\mathcal{L}f_\Lambda$-structures F and G for all $i, l \in \{1,2,3\}$ for all $\varepsilon \in \{1,2,3\}^*$*

(3) $$a_i R^\varepsilon c_l \iff \varepsilon = jk, \text{ where } |\{i,j,k\}| = 3,$$

(4) $$r R^\varepsilon c_l \iff \varepsilon = ijk, \text{ where } |\{i,j,k\}| = 3;$$

In G for all $h \in \{1,2,3,123\}$ for all $\varepsilon \in \{1,2,3\}^*$ in G

(5) $$a_i R^\varepsilon \bar{c}_h \iff \varepsilon = jk \text{ and } \{i,j,k\} = \{1,2,3\}.$$

Proof. At first, we prove (3) for F and G simultaneously. Consider the point a_1. One can see that

$$a_1 R_2 b_3^1 R_3 c_1;$$

$$a_1 R_3 b_2^1 R_2 c_1;$$

$$a_1 R_2 b_3^1 R_3 c_2;$$

$$a_1 R_3 b_2^2 R_3 c_2;$$

$$a_1 R_2 b_3^3 R_3 c_3;$$

$$a_1 R_3 b_2^1 R_2 c_3;$$

and $R_1(a_1) = \emptyset$, $R_1((R_2 \cup R_3)(a_1)) = \emptyset$.

For the points a_2, a_3 the statement (3) is true, due to the symmetry of F. The statement (4) follows readily from (3); (5) can be checked similarly. □

Lemma 7.3 *In F*

$$R_3(R_2(a_1) \cap R_1(a_2)) \cap R_2(R_3(a_1) \cap R_1(a_3)) \cap R_1(R_2(a_3) \cap R_3(a_2)) = \emptyset$$

(and hence $F \not\models fc[x_0^F/x_0]$*).*

Proof. In fact, $R_2(a_1) \cap R_1(a_2) = b_3^3$, $R_3(a_1) \cap R_1(a_3) = b_2^2$, $R_2(a_3) \cap R_3(a_2) = b_1^1$, therefore

$$R_3(R_2(a_1) \cap R_1(a_2)) \cap R_2(R_3(a_1) \cap R_1(a_3)) \cap R_1(R_2(a_3) \cap R_3(a_2)) =$$

$$= R_3(b_3^3) \cap R_2(b_2^2) \cap R_1(b_1^1) = \{c_1\} \cap \{c_2\} \cap \{c_3\} = \emptyset.$$

□

Lemma 7.4 *$G_\circ$ is Kracht-reducible to $F_\circ$.*

Proof. Take a tree T with a root x_0^T and a monotonic mapping $f : T \to F$ sending x_0^T to r. Our goal is to construct a monotonic mapping $g : T \to G$ sending x_0^T to r and a relation $Z \subseteq W' \times W$ such that $(G, T, g) \gg (F, T, f)$.

To this end we use an embedding $\iota : W \to W'$ and put $g = \iota \cdot f$. Then we construct Z (uniformly for all (T, f)).

Put $(\alpha, \beta) \in Z$ if one of the following holds:

(i) $\alpha, \beta \in W$ and $\alpha = \beta$

(ii) $\alpha \in W', \beta \in W$ and $\alpha = \bar{\beta}$;

(iii) $\alpha = \bar{c}_{123}$, $\beta = c_i$, where $i \in \{1, 2, 3\}$.

Now our goal is to check the conditions KB1) – KB4).

KB1) Follows immediately from Item 1 of definition of Z.

KB2) Take $t \in W^T$ and $(\alpha, \beta) \in Z$. We have to ensure that for all $\varepsilon \in \{1, 2, 3\}^*$

(6) $$\text{if } g(t) R^\varepsilon \alpha \text{ then } f(t) R^\varepsilon \beta.$$

Consider the following two possibilities.

(i) $f(t) \in \{r, a_1, a_2, a_3\}$. Then $g(t) = f(t)$ and the only non-trivial instance of (6) (namely, for $\alpha = \bar{c}_{123}$) holds by (3), (4), and (5).

(ii) $f(t) \in \{b_i^j \mid i, j \in \{1, 2, 3\}\} \cup \{c_i \mid i \in \{1, 2, 3\}\}$. In this case if $g(t) R^\varepsilon \alpha$, then $\alpha \in W$. Hence $\beta = \alpha$, and (6) is evident, since F is a part of G.

KB3) Let $\alpha, \alpha' \in G$, $\alpha R_\lambda \alpha'$, and $\alpha Z \beta$. The only non-trivial case is when $\alpha' = \bar{c}_{123}$ (otherwise, we can take $\beta' = \sigma(\alpha')$, where $\sigma(w) = w$ for $w \in W$ and $\sigma(\bar{b}_i^j) = b_i^j$, $\sigma(\bar{c}_i) = c_i$). In this case β' also depends on α. But in G there are only three points seeing $\bar{c}_{123}$, namely, $\bar{b}_1^1$, $\bar{b}_2^2$ and $\bar{b}_3^3$. So we act as follows: if $\alpha = \bar{b}_k^k$, put $\beta' = c_k$.

KB4) Suppose that $(\alpha, \beta) \in Z$ and $\beta R_\lambda \beta'$. Consider the following two cases.

- If $\alpha = \beta$, then take $\alpha' = \beta'$;
- If $\alpha = \bar{\beta}, \beta \in \{b^i_j \mid i, j \in \{1, 2, 3\}$, then take $\alpha' = \bar{\beta}'$.

□

So due to Lemma 7.4, $G_\circ \ggg F_\circ$. But $G \models fc(r)$ and $F \not\models fc(r)$. Therefore the formula fc is not equivalent to any Kracht formula.

8 Final Remarks

(i) The proposed characterization looks more transparent in terms of games.

Two players, $\forall$ and $\exists$, play over a pair of $\mathcal{L}f_\Lambda$-structures with designated points (G, x_0^G) and (F, x_0^F). The player $\forall$ (he) wants to show that there exists a Kracht formula $\phi(x_0)$ such that $G \models \phi(x_0^G)$, but $F \not\models \phi(x_0^F)$, and the goal of $\exists$ is to prevent it. A position in the game is a triple (T, g, f), where T is a tree with a root x_0^T and $g : T \to G, f : T \to F$ are monotonic mappings sending x_0^T respectively to x_0^G and x_0^F.

When the game starts, $\forall$ announces the number of rounds n and constructs a tree T_0 and a monotonic mapping $f : T_0 \to F$ sending x_0^T to x_0^F. His opponent replies with a monotonic mapping $g : T_0 \to G$ sending x_0^T to x_0^G. Then n round follow indexed by numbers $1, \ldots, n$. Let T_{i-1} be a tree at the position before round i. The round i consists of the following actions. At first, $\forall$ adds a new leaf to T_{i-1}, and obtains a tree T_i. Then he chooses either F or G and extends respectively f or g to T_i. Then $\exists$ extends the other mapping to T_i (if she cannot, she looses). The game is won by $\exists$ if in the final position (T_n, g_n, f_n) for all $t \in T_0, s \in T_n, \varepsilon \in \Lambda^*$, if $g_n(t)(R^G)^\varepsilon g_n(s)$, then $f_n(t)(R^F)^\varepsilon f_n(s)$. An infinite generalization of such a game is obvious. In fact, these rules are easily extracted from the very definition of Kracht formulas.

In these terms a Kracht-simulation is nothing but a winning strategy for $\exists$ in a game with infinitely many rounds. Some lemmas from Section 6 can also be reformulated in terms of games. For example, Lemma 6.5 states that if $\exists$ has a winning strategy in every finite game between (G, x_0^G) and (F, x_0^F), then she has a winning strategy in an infinite game between suitable elementary extensions $\bar{G}$ and $\bar{F}$.

(ii) In [7] Kracht's theorem was extended to generalized Sahlqvist formulas of D. Vakarelov and V. Goranko [6]. The semantic characterization from the present paper can be easily transferred to the generalized Kracht formulas from [7].

(iii) Besides our characterization, there is another semantic property of Sahlqvist formulas [6]. A general frame $(W, (R_\lambda : \lambda \in \Lambda), \mathcal{A})$ is called *ample* if $\mathcal{A}$ contains the sets $R^\varepsilon(w)$ for all $\varepsilon \in \Lambda^*$, $w \in W$. A modal formula ϕ is called *locally a-persistent* if for every ample general frame $\mathfrak{F} = (F, \mathcal{A})$, for any world w in F, $\mathfrak{F}, w \models \phi \iff F, w \models \phi$. Clearly, all Sahlqvist modal formulas are a-persistent, and this fact is used in [6] to show that the first-order definable modal formula $p \land \Box(\Diamond p \to \Box q) \to \Diamond\Box\Box q$ does not have a Sahlqvist equivalent. It is interesting whether the converse holds:

Question 8.1 *Does a-persistence and first-order definability of a modal formula ϕ imply its frame-equivalence to a Sahlqvist formula?*

Anyway it is interesting to understand how these two properties (i.e., a-persistence of a modal formula and Kracht-reducibility-invariance of its first-order equivalent) are related.

References

[1] Benthem, J., "Modal Logic and Classical Logic," Bibliopolis, Naples, 1983.

[2] Blackburn, P., M. de Rijke and Y. Venema, "Modal Logic," Cambridge University Press, 2002.

[3] Chagrov, A. and L. Chagrova, *The truth about algorithmic problems in correspondence theory*, in: G. Governatori, I. Hodkinson and Y. Venema, editors, *Advances in Modal Logic 6* (2006), pp. 121–138.

[4] Gabbay, D., A. Kurucz, F. Wolter and M. Zakharyaschev, "Many-dimensional modal logics: theory and applications," Studies in Logic and the Foundations of Mathematics, 148, Elsevier, 2003.

[5] Goranko, V. and D. Vakarelov, *Sahlqvist formulas unleashed in polyadic modal languages*, in: *Advances in Modal Logic 3* (2000), pp. 221–240.

[6] Goranko, V. and D. Vakarelov, *Elementary canonical formulae: extending Sahlqvist's theorem*, Annals of Pure and Applied Logic **141** (2006), pp. 180–217.

[7] Kikot, S., *An extension of Kracht's theorem to generalized Sahlqvist formulas*, Journal of Applied Non-Classical Logic **19/2** (2009), pp. 227–251.

[8] Kracht, M., *How completeness and correspondence theory got married*, in: *M. de Rijke (Ed.), Diamonds and Defaults* (1993), pp. 175–214.

[9] Kracht, M., "Tools and Techniques in Modal Logic," Studies in Logic and the Foundations of Mathematics, 142, Elsevier, 1999.

[10] Kurucz, A., *On axiomatising products of Kripke frames*, Journal of Symbolic Logic **65** (2000), pp. 923–945.

[11] Kurucz, A., *On axiomatising products of Kripke frames, part II*, in: C. Areces and R. Goldblatt, editors, *Advances in Modal Logic 7* (2008), pp. 219–230.

[12] Sahlqvist, H., *Completeness and correspondence in the first and second order semantics for modal logic*, in: *Proceedings of the Third Scandinavian Logic Symposium, Amsterdam, North-Holland*, 1975, pp. 110–143.

[13] Zolin, E., *Query answering based on modal correspondence theory*, in: *Proceedings of the 4th "Methods for modalities" Workshop (M4M-4)*, 2005, pp. 21–37.

On Modal Logics of Linear Inequalities

Clemens Kupke and Dirk Pattinson

Department of Computing
Imperial College London

Abstract

We consider probabilistic modal logic, graded modal logic and stochastic modal logic, where linear inequalities may be used to express numerical constraints between quantities. For each of the logics, we construct a cut-free sequent calculus and show soundness with respect to a natural class of models. The completeness of the associated sequent calculi is then established with the help of coalgebraic semantics which gives completeness over a (typically much smaller) class of models. With respect to either semantics, it follows that the satisfiability problem of each of these logics is decidable in polynomial space.

Keywords: Probabilistic modal logic, graded modal logic, linear inequalities

1 Introduction

In this paper, we consider three different, but closely related, modal logics. The first logic that we consider is probabilistic modal logic, where we think of every formula A as denoting an event $[\![A]\!]$ in a probability space. The variant of probabilistic modal logic that we consider here allows explicit comparisons of the likelihoods of individual formulas by means of linear inequalities. If $A_1, \dots, A_m$ are formulas and $p, c_1, \dots, c_m \in \mathbb{Q}$, then the expression $\sum_{j=1}^{m} c_j \cdot \mu([\![A_j]\!]) \geq p$ can be denoted a formula which is satisfied at a point x of a probability space if the local probability measure μ associated with point x satisfies the above inequality. An expression of this form is written as an m-ary modal operator, $L_p(c_1, \dots, c_m)$, applied to $A_1, \dots, A_m$.

The second logic that we consider is graded modal logic, where we may again use linear inequalities to express constraints on successors with certain properties. As before, we use m-ary modal operators of the form $L_p(c_1, \dots, c_m)$ to express that the inequality $\sum_{j=1}^{m} c_j \cdot \sharp A_j \geq p$ holds at a particular point in a Kripke model, where $\sharp A_j$ is the number of successors of that state satisfying property A_j.

Finally, we consider a third logic, stochastic modal logic, that is a hybrid between the two. To get from probabilistic modal logic to stochastic modal logic, one needs to generalise from probability measures to arbitrary measures. To get to stochastic logic from graded modal logic, one gives up the idea of always having an integer number of successors and replaces the transition relation by a family of local real-valued measures that determine the total weight of a successor set.

For each of the logics mentioned above, we give an axiomatisation in terms of a cut-free sequent calculus and prove soundness with respect to a natural class of models: Markov models, Kripke models and what we call measurable models – the natural generalization of Markov models, where one drops the requirement of dealing with probability measures. We then establish completeness of the sequent calculus with respect to coalgebraic models. For each of the logics, we isolate a natural, coalgebraic semantics and show how the general results of coalgebraic modal logics can be used to give a rather simple completeness proof. In a third step, we relate both types of semantics, and show that the coalgebraic semantics embeds into the 'natural' semantics considered initially. Our treatment thus combines the best of both worlds for each of the logics: we establish soundness for a large class of models, whereas the logics are proved complete for a much smaller class. The complexity of each of the logics then follows by analysing the complexity of backwards proof search in the given sequent calculus.

The main contributions of this paper are the cut-free axiomatisation of three different modal logics and the completeness proof of this axiomatisation using coalgebraic methods. The sequent calculi appear to be new in each case. While the soundness proofs are certainly standard, completeness relies on coalgebraic techniques. Rather than exhibiting a fully fledged (canonical) model construction, we can make do with showing that the rules that generate the sequent systems are *one-step complete*: we interpret all logics over T-coalgebras $(X, \gamma : X \to TX)$ for suitably chosen T, where γ is the transition function. One-step completeness now stipulates that all sequents valid over the set of 'successors' TX should be derivable via modal rules whose premises are already valid over X, where X is an arbitrary set. For probabilistic and stochastic modal logic, the question of one-step completeness can be translated into a linear programming problem over the rational domain, which fails for the case of graded modal logic, where we use maximal consistent sets, but only at the level of one-step successors.

The coalgebraisation of all three logics moreover allows us to apply a number of generic (coalgebraic) results: with the help of [2] we obtain completeness and EXPTIME decidability of an extension of each logic with least/greatest fixpoints, [7] allows us to construct generic (tableau) algorithms for the global consequence problem, and [17,8] provides an EXPTIME complexity bound and optimal tableau algorithm, respectively, of hybrid extensions over arbitrary sets of global assumptions. As such, the paper does not present any new results concerning the coalgebraic interpretation of modal logics. Rather, we show how coalgebraic methods can be used to obtain results about existing modal logics.

Related Work. Probabilistic modal logic, as studied in this paper, can be seen as an extension of the probabilistic modal logic presented in [9] with linear inequalities and is a notational variant of the probabilistic logic considered in [5,4], where a complete axiomatisation in a Hilbert-style proof system and a proof of PSPACE-decidability is presented. Our contribution here is a cut-free sequent system that allows for purely syntax driven implementations of satisfiability checking that are amenable to standard optimisations [10,20].

The extension of graded modal logic extended with linear inequalities considered here is a fragment of Presburger modal logic with regularity constraints [3] but subsumes Majority logic [11] and the both (standard form of) graded modal logic [6] and description logics with qualified number restrictions [1]. In absence of linear inequalities, this logic is known to be PSPACE complete [18] and PSPACE-completeness in presence of linear inequalities was shown in [3], but no complete axiomatisation appears to be known so far, which is provided here.

For stochastic modal logic, we are not aware of any results concerning completeness and complexity.

2 Preliminaries and Notation

2.1 Preliminaries on Sequent Calculi

Throughout the paper, we fix a set V of propositional variables. As we will be dealing with three different modal logics, it is convenient to isolate their syntactical differences into a *modal similarity type*, i.e. a set of modal operators with associated arities. Given a modal similarity type Λ, the set $\mathcal{F}(\Lambda)$ of Λ-formulas is given by the grammar

$$\mathcal{F}(\Lambda) \ni A, B ::= p \mid \neg A \mid A \wedge B \mid \heartsuit(A_1, \dots, A_n)$$

where $p \in \mathsf{V}$ and $\heartsuit \in \Lambda$ is n-ary. If $F \subseteq \mathcal{F}(\Lambda)$ is a set of formulas, then we write

$$\Lambda(F) = \{\heartsuit(A_1, \dots, A_n) \mid \heartsuit \in \Lambda \text{ } n\text{-ary}, A_1, \dots, A_n \in F\}$$

for the set of formulas consisting of modalities applied to elements of F. If $\sigma : \mathsf{V} \to \mathcal{F}(\Lambda)$ is a substitution, then $A\sigma$ denotes the result of replacing all occurrences of $p \in \mathsf{V}$ in A by $\sigma(p)$.

A *sequent* is a finite multiset (so that contraction is made explicit) of formulas that we read disjunctively. We identify $A \in \mathcal{F}(\Lambda)$ with the sequent $\{A\}$ and write Γ, Δ for the (multiset) union of Γ and Δ. If $F \subseteq \mathcal{F}(\Lambda)$ is a set of formulas, we write $\mathcal{S}(F)$ for the set of those sequents that only contain elements of F, possibly negated. Substitution applies pointwise to sequents, respecting multiplicity so that $\Gamma\sigma = \{A\sigma \mid A \in \Gamma\}$. The three logics we consider in this paper can be axiomatised by *one-step rules*, that is, rules of the form

$$\frac{\Gamma_1 \quad \dots \quad \Gamma_n}{\Gamma_0}$$

where $\Gamma_1, \dots, \Gamma_n \in \mathcal{S}(\mathsf{V})$ and $\Gamma_0 \in \mathcal{S}(\Lambda(\mathsf{V}))$. If $\mathcal{R}$ is a set of one-step rules, we write $\mathcal{R} \vdash \Gamma$ if Γ is an element of the least set of sequents that is closed under the propositional rules and all substitution instances of one-step rules, that is under the rules

$$\frac{}{p, \neg p, \Delta} \qquad \frac{A, \Gamma}{\neg\neg A, \Gamma} \qquad \frac{\neg A, \neg B, \Delta}{\neg(A \wedge B), \Delta} \qquad \frac{A, \Delta \qquad B, \Delta}{A \wedge B, \Delta} \qquad \frac{\Gamma_1\sigma \quad \dots \quad \Gamma_n\sigma}{\Gamma_0\sigma, \Delta}$$

where $p \in \mathsf{V}$, $\sigma : \mathsf{V} \to \mathcal{F}(\Lambda)$ and $\Delta \in \mathcal{S}(\mathcal{F}(\Lambda))$ is a weakening context. It is easy to see that the propositional part of this calculus can be embedded into the system **GS3p** of [19] which is known to be sound and complete. The concrete syntactical presentation of the modal rules for the logics considered here is most conveniently expressed using the following notation. If $c_1, \dots, c_m, k \in \mathbb{Q}$ are rational numbers, $a_1, \dots, a_m \in \mathsf{V}$ are propositional variables and $\Gamma = \{a_i \mid i \in I\} \cup \{\neg a_j \mid j \notin I\}$ then

$$\Gamma \in \sum_{j=1}^{m} c_j a_j \geq k \iff \sum_{i \in I} c_i \geq k$$

so that we may use $\sum_{i=1}^{m} c_i a_i \geq k$ to denote a set of sequents that we think of the set of premises of a proof rule. We write $\mathsf{sign}(q)$ for the sign of a rational number q and, if A is a formula and $r \neq 0$, we put $\mathsf{sg}(r)A = A$ if $r > 0$ and $\mathsf{sg}(r)A = \neg A$, otherwise.

2.2 *Coalgebraic Preliminaries*

If $T : \mathsf{Set} \to \mathsf{Set}$ is an endofunctor, a *T-coalgebra* is a pair (X, γ) where X is a (carrier) set and $\gamma : X \to TX$ is a (transition) function. We think of T-coalgebras as playing the role of frames, and take a *T-model* to be a T-coalgebra equipped with a valuation, i.e. a triple (X, γ, π) where (X, γ) is a T-coalgebra and $\pi : \mathsf{V} \to \mathcal{P}(X)$ is a valuation of the propositional variables.

Given a similarity type Λ, we can interpret Λ-formulas over T-models provided T extends to a Λ-structure, i.e. T comes equipped with a predicate lifting (a set-indexed family of maps)

$$(\llbracket \heartsuit \rrbracket_X : \mathcal{P}(X)^n \to \mathcal{P}(TX))_{X \in \mathsf{Set}}$$

for every n-ary $\heartsuit \in \Lambda$ that satisfies the *naturality requirement*

$$(Tf)^{-1} \circ \llbracket \heartsuit \rrbracket_Y(S_1, \dots, S_n) = \llbracket \heartsuit \rrbracket_X(f^{-1}(S_1), \dots, f^{-1}(S_n))$$

for all $f : X \to Y$ and all $S_1, \dots S_n \subseteq Y$. If $M = (X, \gamma, \pi)$ is a T-model, the semantics of modal formulas is now defined as expected for propositional connectives

$$\llbracket p \rrbracket_M = \pi(p) \quad \llbracket \neg A \rrbracket_M = X \setminus \llbracket A \rrbracket_M \quad \llbracket A \wedge B \rrbracket_M = \llbracket A \rrbracket_M \cap \llbracket B \rrbracket_M$$

together with the clause

$$\llbracket \heartsuit(A_1, \dots, A_n) \rrbracket_M = \gamma^{-1} \circ \llbracket \heartsuit \rrbracket_X(\llbracket A_1 \rrbracket_M, \dots, \llbracket A_n \rrbracket_M)$$

for the modal operators. We write $M, x \models A$ in case $x \in \llbracket A \rrbracket_M$ and $M \models A$ if $M, x \models A$ for all $x \in X$. Finally, we write $T \models \Gamma$ if $M \models \Gamma$ for all T-models M. The glue between the axiomatisation (in terms of one-step rules) and the modal semantics is provided by the following notions:

Definition 2.1 *Suppose that Λ is a modal similarity type, T a Λ-structure and $\mathcal{R}$ a set of one-step rules over Λ. We introduce the following notions in case X is a set and $\tau : \mathsf{V} \to \mathcal{P}(X)$ is a valuation:*

- *(i) If $\Gamma \in \mathcal{S}(\mathsf{V})$ is a propositional sequent, we write $[\![\Gamma]\!]_{(X,\tau)} = \bigcup\{[\![A]\!]_{(X,\tau)} \mid A \in \Gamma\}$ (where $[\![p]\!]_{(X,\tau)} = \tau(p)$) for the interpretation of a sequent $\Gamma \in \mathcal{S}(\mathsf{V})$ with respect to τ and $(X,\tau) \models \Gamma$ in case Γ is τ-*valid*, i.e. $[\![\Gamma]\!]_{(X,\tau)} = X$.*
- *(ii) Similarly, if $\Gamma \in \mathcal{S}(\Lambda(V))$ we write $[\![\Gamma]\!]_{(TX,\tau)} = \bigcup\{[\![A]\!] \mid A \in \Gamma\}$ (where $[\![\heartsuit(a_1,\ldots,a_n)]\!]_{(TX,\tau)} = [\![\heartsuit]\!]_X(\tau(a_1),\ldots,\tau(a_n)))$ for the interpretation of a modalised sequent with respect to τ, and $(TX,\tau) \models \Gamma$ in case Γ is τ-*valid*, i.e. $[\![\Gamma]\!]_{(TX,\tau)} = TX$.*
- *(iii) Finally, a sequent $\Gamma \in \mathcal{S}(\Lambda(\mathsf{V}))$ is τ-*derivable* (with respect to $\mathcal{R}$) if there exists $\Gamma_1 \ldots \Gamma_n/\Gamma_0 \in \mathcal{R}$ and $\sigma : \mathsf{V} \to \mathsf{V}$ such that all $\Gamma_i\sigma$ are τ-valid for $1 \leq i \leq n$ and $\Gamma_0\sigma \subseteq \Gamma$.*

We can now justify the sum notation introduced earlier:

Lemma 2.2 *Suppose that $\tau : \mathsf{V} \to \mathcal{P}(X)$ is a valuation. Then*

$$\forall x \in X \left(\sum_{j=1}^{m} c_j \mathbb{1}_{\tau(a_j)}(x) \geq k \right) \iff \forall\, \Gamma \in \sum_{j=1}^{m} c_j a_j \geq k \, ((X,\tau) \models \Gamma).$$

Moreover, we can relate one-step rules and coalgebraic semantics as follows:

Definition 2.3 *Suppose Λ is a modal similarity type and $T : \mathsf{Set} \to \mathsf{Set}$ is a Λ-structure. We say that a set $\mathcal{R}$ of one-step rules is* one-step sound *(resp.* one-step cut-free complete*) if, for all valuations $\tau : \mathsf{V} \to \mathcal{P}(X)$ and all $\Gamma \in \mathcal{S}(\Lambda(\mathsf{V}))$: Γ is τ-derivable if (resp. only if) Γ is τ-valid.*

We note that the notions of one-step soundness and one-step (cut-free) completeness do not quantify over models: both conditions can be checked locally. Importantly, these notions give rise to soundness and cut-free completeness in the standard way.

Theorem 2.4 *Suppose $\mathcal{R}$ is a set of one-step rules over a modal similarity type Λ and let T be a Λ-structure. If $\Gamma \in \mathcal{S}(\Gamma)$ and*

- *(i) $\mathcal{R}$ is one-step sound, then $\models \Gamma$ whenever $\mathcal{R} \vdash \Gamma$*
- *(ii) $\mathcal{R}$ is one-step cut-free complete, then $\mathcal{R} \vdash \Gamma$ whenever $T \models \Gamma$.*

The proof of the last theorem can be found in [13] but it should be remarked that this type of coherence condition between syntax and semantics is well studied: [12,15] use similar (weaker) coherence conditions to obtain soundness and completeness of a Hilbert system and [16] uses *strict completeness* to obtain what essentially amounts to a cut-free sequent system.

3 Probabilistic Modal Logic

We start our investigation into modal logics of linear inequalities by considering probabilistic modal logic where we may allow ourselves linear inequalities to specify the relationships between individual formulas. That is, we consider the modal similarity type

$$\Lambda = \{L_p(c_1, \dots, c_m) \mid m \in \mathbb{N}, p, c_1, \dots, c_m \in \mathbb{Q}\}$$

where the arity of $L_p(c_1, \dots, c_m)$ is m. We interpret probabilistic modal logic over state spaces X where every point $x \in X$ induces a probability distribution μ over successor states. Informally, validity of $L_p(c_1, \dots, c_m)(A_1, \dots, A_m)$ at point $x \in X$ means that the linear inequality $\sum_{j=1}^{m} c_j \mu(A_j) \geq p$ holds, where $\mu(A_j)$ is the measure of the truth-set of the formula A_j, seen from point x. In particular this allows us to compare the probabilities of events:

Example 3.1 According to a recent experience of the second author with a well-known budget airline, we may consider a state space comprising all European airports, and we may think of the probability distribution associated with a particular city as giving us the probability of landing at a particular airport when boarding any flight of this carrier. In this logic, which we refrain from calling EasyLogic, we can for instance express that landing in England is 5 times as likely as landing in Scotland as $L_0(1, -5)(\mathsf{England}, \mathsf{Scotland})$ (which is reasonable to assume for carriers that are based in England). The second author however doubts that the business model of said budget airline can be axiomatised in any logic.

We axiomatise probabilistic modal logic with linear inequalities and prove soundness and completeness with respect to two different classes of models. The (complete, cut-free) axiomatisation is induced by the set $\mathcal{R}$ of one-step rules that comprises all instances of

$$(P)\frac{\sum_{i=1}^{n} r_i \left(\sum_{j=1}^{m_i} c_j^i \cdot a_j^i\right) \geq k}{\{\mathsf{sg}(r_i) L_{p_i}(c_1^i, \dots, c_{m_i}^i)(a_1^i, \dots, a_{m_i}^i) \mid i = 1, \dots, n\}}$$

where $r_1, \dots, r_n \in \mathbb{Z} \setminus \{0\}$ and $k \in \mathbb{Z}$ that satisfy the side condition

$$\sum_{i=1}^{n} r_i p_i < k \quad \text{if all } r_i < 0, \text{ and} \quad \sum_{i=1}^{n} r_i p_i \leq k \quad \text{otherwise.}$$

We first treat soundness of probabilistic modal logic, interpreted over Markov chain models before showing completeness over a class of coalgebraic models that corresponds to finitely supported Markov chains.

3.1 Markov Chain Semantics and Soundness

The first semantics of probabilistic modal logic is given with respect to Markov models.

Definition 3.2 *A* Markov model *is a triple* (X, μ, π) *where* X *is a measurable space with* σ*-algebra* Σ_X*,* $\pi : \mathsf{V} \to \Sigma_X$ *is a valuation and* $\mu : X \times \Sigma \to [0,1]$ *is a Markov kernel, that is,* $\mu(x, \cdot) : \Sigma \to [0,1]$ *is a probability measure for all* $x \in X$ *and* $\mu(\cdot, S) : X \to [0,1]$ *is measurable for all* $S \in \Sigma_X$.

If $M = (X, \mu, \pi)$ is a Markov model, then the semantics $[\![A]\!]_M \in \Sigma_X$ is given as expected for the propositional connectives (where atomic propositions are mapped to measurable sets) and the clause for modal operators is

$$[\![L_p(c_1, \dots, c_m)(A_1, \dots, A_m)]\!]_M = \{x \in X \mid \sum_{j=1}^{m} c_j \mu(x, [\![A_j]\!]_M) \geq p]\!]$$

where we write $M, x \models A$ if $x \in [\![A]\!]_M$ and $M \models A$ if $M, x \models A$ for all $x \in X$. Finally $\mathsf{Mark} \models \Gamma$ if $M \models \bigvee \Gamma$ for all Markov models M. Note that the measurability conditions guarantee that the truth-set $[\![A]\!]_M$ of a formula is always measurable. We now show soundness of probabilistic modal logic with respect to Markov models.

Proposition 3.3 $\mathsf{Mark} \models \Gamma$ *whenever* $\mathcal{R} \vdash \Gamma$.

Proof. Suppose that $M = (X, \mu, \pi)$ is a Markov model and $\mathcal{R} \vdash \Gamma$. We show that $M \models \Gamma$ by induction on the proof of $\mathcal{R} \vdash \Gamma$ where the application of an instance of (P) is the only interesting case.

Consider the sequent Γ appearing as the conclusion of the rule

$$\frac{\sum_{i=1}^{n} r_i \sum_{j=1}^{m_i} c_i^j A_i^j \geq k}{\{\mathsf{sg}(r_i) L_{p_i}(c_i^1, \dots, c_i^{m_i})(A_1, \dots, A_m) \mid i = 1, \dots, n\}}$$

the applicability of which is ensured by the side condition

$$\sum_{i=1}^{n} r_i p_i < k \quad \text{if all } r_i < 0, \text{ and} \quad \sum_{i=1}^{n} r_i p_i \leq k \quad \text{otherwise.}$$

By induction hypothesis,

$$\sum_{i=1}^{n} r_i \sum_{j=1}^{m_i} c_i^j \mathbb{1}_{[\![A_i^j]\!]_M}(x) \geq k$$

for all $x \in X$. Now suppose for a contradiction that there exists an $x \in X$ so that $M, x \not\models \Gamma$. If $\mu = \mu(x, \cdot)$ then this implies that

$$\sum_{i=1}^{n} r_i \sum_{j=1}^{m_i} c_i^j \mu([\![A_i^j]\!]_M) \geq k$$

by integrating both sides with respect to μ, and

$$\sum_{j=1}^{m_i} c_i^j \mu[\![A_i^j]\!]_M \geq p_i \quad (\text{if } r_i < 0), \text{ and} \quad \sum_{j=1}^{m_i} c_i^j \mu[\![A_i^j]\!]_M < p_i \quad (\text{if } r_i > 0)$$

for all $i = 1, \dots, n$. In summary, this implies that

$$k \leq \sum_{i=1}^{n} r_i \sum_{j=1}^{m_i} c_i^j \mu(\llbracket A_i^j \rrbracket_M) \leq \sum_{i=1}^{n} r_i p_i \leq k$$

where either the last or the penultimate inequality are strict so that $k < k$ in both cases, contradicting $M, x \not\models \Gamma$ and therefore proving the claim. □

We next establish completeness over a smaller class of models, that is, Markov chains where the transition measures are finitely supported. Crucially, these fit into the framework of coalgebraic semantics:

3.2 Coalgebraic Semantics and Completeness

We write $\mathrm{supp}(f) = \{x \in X \mid f(x) \neq 0\}$ for the support of a function $f : X \to \mathbb{R}$ and consider the functor $\mathcal{D} : \mathsf{Set} \to \mathsf{Set}$ where

$$\mathcal{D}(X) = \{\mu : X \to [0,1] \mid \mathrm{supp}(\mu) \text{ finite}, \sum_{x \in X} \mu(x) = 1\}$$

that extends to a Λ-structure by stipulating that

$$\llbracket L_p(c_1, \dots, c_n) \rrbracket_X (S_1, \dots, S_n) = \{\mu \in \mathcal{D}(X) \mid \sum_{i=1}^{n} c_i \cdot \mu(S_i) \geq p\}$$

for $S_1, \dots, S_n \subseteq X$ where $\mu(S) = \sum_{x \in X} \mu(x)$. As spelled out in Section 2.2 this induces an interpretation $\llbracket A \rrbracket_M \subseteq X$ of Λ-formulas over $\mathcal{D}$-models $M = (X, \gamma, \pi)$. We now show that the set $\mathcal{R}$ of one-step rules consisting of all instances of (P) is indeed one-step complete, which is the content of the next lemma.

Lemma 3.4 *Consider a valuation $\tau : \mathsf{V} \to \mathcal{P}(X)$ and suppose that $\Gamma \in \mathcal{S}(\Lambda(\mathsf{V}))$ is τ-valid. Then Γ is τ-derivable.*

Proof. Suppose that $\Gamma = \{\mathsf{sg}(\epsilon_i) L_{p_i}(c_i^1, \dots, c_i^{m_i})(a_i^1, \dots, a_i^{m_i}) \mid i = 1, \dots, n\}$ where $\epsilon_1, \dots, \epsilon_n \in \{-1, 1\}$, the p_i and $c_i^j \in \mathbb{Q}$ and the $a_i^j \in \mathsf{V}$. Furthermore let $\tau : \mathsf{V} \to \mathcal{P}(X)$ be a valuation such that Γ is τ-valid. To see that Γ is τ-derivable, we show that there exist $k, r_1, \dots, r_n \in \mathbb{Z}$ so that

(i) $\sum_{i=1}^{n} r_i^2 > 0$ (i.e. at least one of the $r_1, \dots, r_n$ is non-zero)

(ii) $\mathsf{sign}(r_i) = \mathsf{sign}(\epsilon_i)$ for all $i = 1, \dots, n$ with $r_i \neq 0$

(iii) $\sum_{i=1}^{n} r_i \left(\sum_{j=1}^{m_i} c_j^i \cdot \mathbb{1}_{\tau(a_j^i)}(x) \right) \geq k$ for all $x \in X$

(iv) $\sum_{i=1}^{n} r_i p_i \leq k$ if at least one ϵ_i is positive, and $\sum_{i=1}^{n} r_i p_i < k$ otherwise.

We define an equivalence relation $\sim$ on X by

$$x \sim y \iff \left(x \in \tau(a_i^j) \iff y \in \tau(a_i^j) \right)$$

for all $i = 1, \ldots, n$ and all $j = 1, \ldots, m_i$. Assume that $x_1, \ldots, x_k \in X$ are the (finitely many) representatives of the equivalence classes of X under $\sim$. Consider the matrices

$$A_0 = \begin{pmatrix} -\epsilon_1 & & 0 & 0 \\ & \ddots & & \vdots \\ 0 & & -\epsilon_n & 0 \\ -f_1(x_1) & \ldots & -f_n(x_1) & 1 \\ \vdots & & \vdots & \\ -f_1(x_k) & \ldots & -f_n(x_k) & 1 \end{pmatrix}$$
$$A_1 = \begin{pmatrix} p_1 & \ldots & p_n & -1 \end{pmatrix}$$

where $f_i = \sum_{j=1}^{m_i} c_i^j \cdot \mathbb{1}_{\tau(a_i^j)}$ and let $A = \begin{pmatrix} A_0 \\ A_1 \end{pmatrix}$. We note the following properties, where $y = (y_1, \ldots, y_n, \hat{y}_1, \ldots, \hat{y}_k, y_0) \in \mathbb{Q}_{\geq 0}^{n+k+1}$:

(i) if $b = (b_1, \ldots, b_n, 0, \ldots, 0)$ with $\sum_{i=1}^n b_i^2 > 0$, $y^T A = 0$ and $y^T b < 0$ then $y_0 > 0$.

(ii) if $y_0 = 1$ and $y^T A = 0$, then the assignment $\mu_y(x_i) = \hat{y}_i$ and $\mu_y(x) = 0$ if $x \notin \{x_1, \ldots, x_k\}$ is a finitely supported probability distribution with

$$\sum_{j=1}^{m_i} c_i^j \cdot \mu_y(\tau(a_i^j)) = p_i - \epsilon_i y_i$$

for all $i = 1, \ldots, n$.

For item (i) we assume (for a contradiction) that $y_0 = 0$ and consider the last column of A to obtain $\sum_{l=1}^k \hat{y}_l - y_0 = 0$ hence $\hat{y}_1 = \cdots = \hat{y}_k = 0$ as $y \in \mathbb{Q}_{\geq 0}^{n+k+1}$. Now, considering the i-th column of A, we have that $0 = -\epsilon_i y_i - \sum_{l=1}^k \hat{y}_l f_l(x_i) + p_i y_0 = -\epsilon_i y_i$ whence $y_i = 0$ for all $i = 1, \ldots, n$ so that, in summary, $y = 0$, contradicting $yb < 0$.

Finally, for item (ii), we first consider the last column of A and deduce from $yA = 0$ that $\sum_{l=1}^k \hat{y}_l - y_0 = 0$ so that $\sum_{l=1}^k \hat{y}_l = 1$ and μ_y is a finitely supported probability distribution as $y \in \mathbb{Q}_{\geq 0}^{n+k+1}$. Moreover, considering the i-th column of A, the equality

$y \cdot A = 0$ gives

$$
\begin{aligned}
0 &= -\epsilon_i y^i - \sum_{l=1}^{k} \hat{y}^l f_i(x_l) + y^0 p_i \\
&= -\epsilon_i y^i - \sum_{l=1}^{k} \hat{y}^l \cdot \sum_{j=1}^{m_i} c_i^j \cdot \mathbb{1}_{\tau(a_i^j)}(x_l) + p_i \\
&= -\epsilon_i y^i - \sum_{j=1}^{m_i} c_i^j \cdot \sum_{l=1}^{k} \hat{y}^l \cdot \mathbb{1}_{\tau(a_i^j)}(x_l) + p_i \\
&= -\epsilon_i y^i - \sum_{j=1}^{m_i} c_i^j \cdot \mu(\tau(a_i^j)) + p_i
\end{aligned}
$$

so that $\sum_{j=1}^{m_i} c_i^j \mu_y(\tau(a_i^j)) = p_i - \epsilon_i y_i$ as required.

According to the statement of the theorem, we distinguish the following cases.

Case 1: At least one ϵ_i is positive. The claim follows (by multiplying with a common denominator) if there exists $b = (b_1, \dots, b_n, 0, \dots, 0) \in \mathbb{Q}^n_{\leq 0}$ with $\sum_{i=1,\dots,n} b_i^2 \neq 0$ so that the system of linear inequalities

$$Ar \leq b^T \tag{1}$$

has a solution $r = (r_1, \dots, r_n, k)^T$.

Now suppose, for a contradiction, that Equation (1) does *not* have a solution for any choice of $b_1, \dots, b_n \in \mathbb{Q}_{\leq 0}$ with $\sum_{i=1}^n b_i^2 > 0$. Then, by Farkas' Lemma in the form of [14, Corollary 7.1 (e)], there exists, for every $b = (b_1, \dots, b_n, 0, \dots, 0) \in \mathbb{Q}^{n+k+1}_{\leq 0}$ with $\sum_{i=1}^n b_i^2 > 0$, a vector $y_b \in \mathbb{Q}^{n+k+1}_{\geq 0}$ such that $y_b^T A = 0$ and $y_b^T \cdot b < 0$. Now consider the (non-empty) set

$$I^+ = \{i \in \{1, \dots, n\} \mid \epsilon_i > 0\}$$

and, for $i \in I^+$, the vector

$$b_i = (0, \dots, 0, -1, 0, \dots, 0)^T \in \mathbb{Q}^{n+k+1}$$

where -1 appears in the i-th coordinate. By Farkas' Lemma, this gives a vector y^i so that $(y^i)^T A = 0$ and $(y^i)^T b < 0$. In particular, $y_0^i \neq 0$ by item (i) so that we may assume that $y_0^i = 1$ by linearity for all $i \in I^+$. Moreover, $y^T b < 0$ implies that $y_i^i > 0$ for all $i \in I^+$. Now consider $y = \frac{1}{\sharp I^+} \sum_{i \in I^+} y_i$ and let $y = (y_1, \dots, y_n, \hat{y}_1, \dots, \hat{y}_k, y_0)$. By linearity, we have $y_0 = 1$, $y^T A = 0$ and $y_i = \sum_{h \in I^+} y_h^i \geq y_i^i > 0$. By item (ii) above the vector y induces a finitely supported probability measure μ so that

- $\sum_{j=1}^{m_i} c_i^j \cdot \mu(\tau(a_i^j)) = p_i - \epsilon_i y^i > p_i$ if $i \in I^+$ as $y^i > 0$ and $\epsilon_i = +1$
- $\sum_{j=1}^{m_i} c_i^j \cdot \mu(\tau(a_i^j)) = p_i - \epsilon_i y^i \leq p_i$ if $i \notin I^+$ since $\epsilon_i = -1$ and $y^i \geq 0$.

As a consequence, we have that $\mu \notin [\![\mathsf{sg}(\epsilon_i) L_{p_i}(c_i^1, \dots, c_i^{m_i})(a_i^1, \dots, a_i^{m_i})]\!]_{\mathcal{D}(X),\tau}$ for all $i = 1, \dots, n$ which contradicts our assumption that $\mathcal{D}(X), \tau \models \Gamma$. This finishes our treatment of the first case.

Case 2: $\epsilon_1 = \cdots = \epsilon_n = -1$. The claim follows if we can show that there exists $b = (b_1, \ldots, b_n, 0, \ldots, 0) \in \mathbb{Q}^{n+k}_{\leq 0}$ with $\sum_{i=1}^n b_i^2 > 0$ so that the system

$$A_0 r \leq^T b \quad A_1 r < 0 \tag{2}$$

has a solution $r = (r_1, \ldots, r_n, k)$.

Suppose for a contradiction, that (2) has no solution for all $b = (b_1, \ldots, b_n, 0, \ldots, 0) \in \mathbb{Q}^{n+k}_{\leq 0}$ with $\sum_{i=1}^n b_i^2 > 0$. In particular, (2) has no solution for $b = (-1, \ldots, -1, 0, \ldots, 0)$. By Motzkin's transposition theorem in the form of [14, Corollary 7.1 (k)], there exists

$$y = (y_1, \ldots, y_n, \hat{y}_1, \ldots, \hat{y}_k, y_0) \in \mathbb{Q}^n_{\geq 0}$$

so that $y^T A = y^T A_0 + y^T A_1 = 0$ and *either* $y_0 = 0$ and $yb < 0$ *or* $y_0 \neq 0$ and $yb \leq 0$.

By (i) the case $y_0 = 0$ and $y^T b < 0$ is impossible, so we may assume that $y_0 \neq 0$ and $y^T b \leq 0$, and, without loss of generality that $y_0 = 1$.

By item (ii) the vector y induces a finitely supported probability measure μ so that

$$\sum_{j=1}^{m_i} c_i^j \cdot \mu(\tau(a_i^j)) = p_i + y_i \geq p_i.$$

Hence $\mu \notin [\![\neg L_{p_i}(c_i^1, \ldots, c_i^{m_i})(\tau(a_i^1), \ldots, \tau(a_i^{m_i})]\!]_{\mathcal{D}(X),\tau}$ for any $i = 1, \ldots, n$ which implies that $\mathcal{D}(X), \tau \not\models \Gamma$, again contradicting our assumption that $\mathcal{D}(X), \tau \models \Gamma$. Having reached a contradiction in both cases finishes the proof. □

We obtain completeness of probabilistic modal logic with respect to $\mathcal{D}$-models as a corollary of Theorem 2.4.

Corollary 3.5 $\mathcal{R} \vdash \Gamma$ *whenever* $\mathcal{D} \models \Gamma$.

We summarise our results about probabilistic modal logic with linear inequalities in the next theorem, that ties the two different semantics together.

Theorem 3.6 *Let* $\Gamma \in \mathcal{S}(\Lambda)$. *Then* $\mathcal{D} \models \Gamma$ *whenever* $\mathsf{Mark} \models \Gamma$. *As a consequence, the following are equivalent:*

(i) $\mathcal{R} \vdash \Gamma$ *(ii)* $\mathsf{Mark} \models \Gamma$ *(iii)* $\mathcal{D} \models \Gamma$.

Proof. We just need to show that $\mathcal{D} \models \Gamma$ whenever $\mathsf{Mark} \models \Gamma$ as the other assertions are covered in Corollary 3.5 and Proposition 3.3. So suppose that $\mathsf{Mark} \models \Gamma$ and take a $\mathcal{D}$-model $M = (X, \gamma, \pi)$ and equip X with the trivial σ-algebra $\Sigma_X = \mathcal{P}(X)$. Then $\mu(x, S) = \gamma(x)(S)$ is a Markov kernel. If $M' = (X, \mu, \pi)$ one shows by induction on the structure of formulas that $[\![A]\!]_M = [\![A]\!]'_M$ whence $M \models \Gamma$. This proves that $\mathcal{D} \models \Gamma$ as M was arbitrary. □

4 Graded Modal Logic

As for probabilistic modal logic, the graded modal logic features linear inequalities comprising the number of successors in a Kripke model. As for probabilistic modal logic, we consider modal operators $L_p(c_1, \ldots, c_n)$ but $p, c_1, \ldots, c_n$ are now required to be integers. In other words, we consider the modal similarity type

$$\Lambda = \{L_p(c_1, \ldots, c_m) \mid m \in \mathbb{N}, p, c_1, \ldots, c_m \in \mathbb{Z}\}$$

that defines the set $\mathcal{F}(\Lambda)$ of formulas of graded modal logic. For $p \in \mathbb{Z}$ we write L_p as a shorthand for the unary modality $L_p(1)$. If $A_1, \ldots, A_n \in \mathcal{F}(\Lambda)$, then $L_p(c_1, \ldots, c_n)(A_1, \ldots, A_n)$ is valid at a point c if the linear inequality $\sum_{j=1}^m c_j \sharp A_j \geq p$ holds, where $\sharp A_j$ is the number of successors of c that satisfy A_j.

Example 4.1 We may use graded modal logic to reason about supporters of different football teams. Consider a Kripke model $M = (X, \gamma : X \to \mathcal{P}(X), \pi)$ where X represents individuals. We think of $x' \in \gamma(x)$ as representing that individual x "knows" x'. If $\mathsf{Arsenal}$ and $\mathsf{Chelsea}$ are propositional variables that hold for those individuals that support the respective football team, then the second author (living in North London) would satisfy the formula $L_0(1, -5)(\mathsf{Arsenal}, \mathsf{Chelsea})$ that stipulates that the individual in question knows at least 5 times as many Arsenal than Chelsea supporters – which is not valid for the first author (who resides in South London).

To obtain a sound and complete axiomatisation of graded modal logic with linear inequalities, we consider the set $\mathcal{R}$ of one-step rules that consists of all instances of

$$(G)\frac{\sum_{i=1}^n r_i \cdot \sum_{j=1}^{m_i} c_i^j a_i^j \geq 0}{\{\mathsf{sg}(r_i) L_{p_i}(c_1^i, \ldots, c_{m_i}^i)(a_1^i, \ldots, a_{m_i}^i) \mid i = 1, \ldots, n\}}$$

where $r_1, \ldots, r_n \in \mathbb{Z} \setminus \{0\}$ under the side condition

$$\sum_{r_i > 0} r_i(p_i - 1) + \sum_{r_i < 0} r_i p_i < 0$$

As before, we first discuss soundness of the ensuing sequent system with respect to Kripke frames, then provide completeness with respect to a coalgebraic semantics in terms of multigraphs, and then compare the two classes of models.

4.1 Kripke Semantics and Soundness

We define the semantics of graded modal logic with respect to image finite Kripke models, essentially following Demri and Lugiez [3] and thus generalising the definitions of Fine [6] and of Pacuit and Salame [11]. By an image finite Kripke model, we mean a triple (X, γ, π) where, as usual, X is the set of worlds, $\gamma : X \to \mathcal{P}_f(X)$ assigns a finite set of successors to every $x \in X$ and $\pi : \mathsf{V} \to \mathcal{P}(X)$ is a valuation. The semantics of $\mathcal{F}(\Lambda)$ with respect to a Kripke model $M = (X, \gamma, \pi)$ is given by the usual propositional rules, together with

$$M, x \models L_p(c_1, \ldots, c_m)(A_1, \ldots, A_m) \iff \sum_{j=1}^m c_j \cdot \sharp(\gamma(x) \cap [\![A_j]\!]_M) \geq p$$

where $[\![A]\!]_M = \{x \in X \mid M, x \models A\}$ is the truth-set of A, and $\sharp$ denotes cardinality. As usual, $M \models A$ if $M, x \models A$ for all $x \in X$. We write $\mathsf{Krip} \models \Gamma$ for $\Gamma \in \mathcal{S}(\mathcal{F}(\Lambda))$ if $M \models \bigvee \Gamma$ for all Kripke models $M = (X, \gamma, \pi)$.

Proposition 4.2 $\mathsf{Krip} \models \Gamma$ *whenever* $\mathcal{R} \vdash \Gamma$.

Proof. Consider a Kripke model $M = (X, \gamma, \pi)$ and suppose that $\mathcal{R} \vdash \Gamma$. We show that $M \models \Gamma$ by induction on the proof of $\mathcal{R} \vdash \Gamma$, where the application of an instance of (G) is the only interesting case. Consider the modal rule

$$\frac{\sum_{i=1}^{n} r_i \sum_{j=1}^{m_i} c_i^j \cdot A_i^j \geq 0}{\{\mathsf{sg}(r_i) L_{p_i}(c_i^1, \ldots, c_i^{m_i})(A_i^1, \ldots, A_i^{m_i}) \mid i = 1, \ldots, n\}}$$

the applicability of which assumes that the side condition

$$\sum_{r_i < 0} r_i p_i + \sum_{r_i > 0} r_i (p_i - 1) > 0 \tag{3}$$

holds. By induction hypothesis, we may assume that

$$\sum_{i=1}^{n} r_i \sum_{j=1}^{m_i} c_i^j \cdot \mathbb{1}_{[\![A_i^j]\!]_M}(x) \geq 0$$

for all $x \in X$. To see that $M, x \models \Gamma$, note that the above inequality implies that

$$\sum_{i=1}^{n} r_i \sum_{j=1}^{m_i} c_i^j \cdot \sharp(\gamma(x) \cap [\![A_i^j]\!]_M) = \sum_{x' \in \gamma(x)} \sum_{i=1}^{n} r_i \sum_{j=1}^{m_i} c_i^j \cdot \mathbb{1}_{[\![A_i^j]\!]_M}(x') \geq 0 \tag{4}$$

Now suppose, for a contradiction, that $M, x \not\models \Gamma$. Then we have

$$\sum_{j=1}^{m_i} c_i^j \sharp(\gamma(x) \cap [\![A_i^j]\!]_M) < p$$

in case $r_i > 0$ and

$$\sum_{j=1}^{m_i} c_i^j \sharp(\gamma(x) \cap [\![A_i^j]\!]_M) \geq p$$

if $r_i < 0$. Combining the side condition (3) with (4) this gives

$$0 \leq \sum_{r_i < 0} r_i p_i + \sum_{r_i > 0} r_i (p_i - 1) < 0$$

i.e. the desired contradiction. □

This shows that graded modal logic is sound with respect to Kripke frames. We next establish completeness of graded modal logic with respect to multigraphs before we relate the different types of semantics.

4.2 Coalgebraic Semantics and Completeness

We consider the functor

$$\mathcal{B}(X) = \{f : X \to \mathbb{N} \mid \mathrm{supp}(f) \text{ is finite}\}$$

that extends to a Λ-structure by stipulating that

$$[\![L_p(c_1, \ldots, c_m)]\!]_X(S_1, \ldots, S_m) = \{f \in \mathcal{B}(X) \mid \sum_{j=1}^{m} c_j \cdot f(S_j) \leq p\}$$

where X is a set and $S_1, \ldots, S_m \subseteq X$ and $f(S) = \sum_{x \in S} f(x)$ for $S \subseteq X$. We may think of a $\mathcal{B}$-coalgebra $(X, \gamma : X \to \mathcal{B}(X))$ as a multigraph where every edge is assigned an integer weight.

For the whole section, we fix the set $\mathcal{R}$ of one-step rules that comprises all instances of (G). Completeness of graded modal logic with linear inequalities is proved using a variant of a canonical model construction, but at the level of one-step formulas. We note two simple properties that correspond to admissibility of contraction of cut, but at the level of one-step derivations.

Lemma 4.3 *Suppose $\tau : \mathsf{V} \to \mathcal{P}(X)$ is a valuation.*

(i) if Γ, A, A is τ-derivable, then so is Γ, A.

(ii) if Γ, A and $\Gamma, \neg A$ are τ-derivable, then so is Γ.

Proof. Both are immediate from the rule format: for the first item, we obtain a new instance of (G) that witnesses derivability of Γ, A by simply adding the coefficients that induce both occurrences of A. For the second item, we are given two rule instances that witness derivability of Γ, A and $\Gamma, \neg A$ that we normalise so that the coefficients associated with A and $\neg A$ have the same magnitude and then add (the coefficients of) both rules. □

The next lemma ensures that every consistent set of formulas can be satisfied, at the one-step level, by an assignment of integer weights that is bounded.

Lemma 4.4 *Suppose that $\tau : \mathsf{V} \to \mathcal{P}(X)$ is a valuation and let $\Gamma \in \mathcal{S}(\Lambda(\mathsf{V}))$ so that Γ is not τ-derivable.*

(i) for all $a \in \mathsf{V}$, there exists $p \geq 0$ so that $\Gamma, L_p a$ is not *τ-derivable.*

(ii) If $L_p a \in \Gamma$, then $\Gamma, \neg L_{p+n} a$ is τ-derivable for all $n \geq 0$.

Proof. The first item is by contraposition: If $\Gamma, L_p A$ were derivable for all $p \geq 0$ we obtain a contradiction in terms of the inequalities in premise and side condition of the rules. For the second item, one shows that $L_p a, \neg L_{p+n} a$ is derivable for all $n \geq 0$. □

One-step completeness is now content of the following lemma.

Lemma 4.5 *Suppose that $\tau : \mathsf{V} \to \mathcal{P}(X)$ is a valuation, X is finite and $\Gamma \in \mathcal{S}(\Lambda(\mathsf{V}))$. If $\mathcal{B}(X), \tau \models \Gamma$, then Γ is τ-derivable.*

Proof. Suppose, for a contradiction, that Γ is *not* τ-derivable. Pick, for all $x \in X$, pairwise distinct propositional variables b_x not occurring in Γ and let $\tau(b_x) = \{x\}$. Repeated application of Lemma 4.4 gives, for all (finitely many) $x \in X$, a number $k_x \in \mathbb{N}$ so that

$$\Gamma' = \Gamma \cup \{L_{k_x} b_x \mid x \in X\}$$

is not τ-derivable. By Lemma 4.3 the same holds for the sequent $\mathrm{supp}(\Gamma)$ that we may extend to a maximal subset $\mathcal{M} \subseteq \Lambda(\mathsf{V}) \cup \neg\Lambda(\mathsf{V})$ with the property that no finite subset $\Delta \subseteq \mathcal{M}$, viewed as a multiset where every element has multiplicity one, is derivable. We now define a measure $\mu : X \to \mathbb{N}$ by

$$\mu(x) = \max\{p \in \mathbb{N} \mid \neg L_p(b_x) \in \mathcal{M}\}$$

for all $x \in X$ and write $\mu(A) = \sum_{x \in A} \mu(x)$ as usual. Note that $\mu(x) \in \mathbb{N}$ by Lemma 4.4. We now claim that

$$\neg L_{\mu(\tau(a))} a \in \mathcal{M} \quad \text{and} \quad L_{\mu(\tau(a))+1} a \in \mathcal{M}$$

for all $a \in \mathsf{V}$. For the first point, note that $\neg L_{\mu(x)} b_x \in \mathcal{M}$ by definition of μ and consider the rule

$$\frac{-\sum_{x \in \tau(a)} b_x + a \geq 0}{\{\neg L_{\mu(x)} b_x \mid x \in \tau(a)\} \cup \{L_{\mu(\tau(a))} a\}}$$

that witnesses $L_{\mu(\tau(a))} a \notin \mathcal{M}$ as $\mathcal{M}$ is not derivable, and hence $\neg L_{\mu(\tau(a))} a \in \mathcal{M}$ by Lemma 4.3. The proof of the second point is entirely dual.

We now establish that

$$\mathsf{sg}(\epsilon) A \in \mathcal{M} \implies \mu \notin [\![\mathsf{sg}(\epsilon) A]\!]_{(\mathcal{B}(X), \tau)}$$

for all $A \in \mathcal{F}(\Lambda)$ and all $\epsilon \in \{-1, +1\}$. Let $A = L_p(c_1, \dots, c_m)(a_1, \dots, a_m)$. For the case $\epsilon = +1$ assume for a contradiction that $\mu \in [\![A]\!]_{(\mathcal{B}(X), \tau)}$ so that

$$\sum_{j=1}^{m} c_j \mu(a_j) \geq p.$$

Consider the rule (the side condition of which is readily established)

$$\frac{c_1 a_1 + \cdots + c_m a_m + \sum_{j=1}^{m} c_j a_j \geq 0}{\{\neg L_{\mu(\tau(a_j))} a_j \mid c_j < 0\} \cup \{L_{\mu(\tau(a_j))+1} a_j \mid c_j > 0\} \cup \{L_p(c_1, \dots, c_j)(a_1, \dots, a_j)\}}$$

that witnesses $A \notin \mathcal{M}$ as $\mathcal{M}$ is not derivable, contradicting $A \in \mathcal{M}$. The case for $\epsilon = -1$ is entirely dual.

In summary our assumption that Γ is not τ-derivable, we obtain that $\mu \notin [\![\Gamma]\!]_{(\mathcal{B}(X), \tau)}$ contradicting that Γ is τ-valid.

□

As a corollary, we obtain that the set $\mathcal{R}$ comprising all instances of (G) is one-step complete.

Proposition 4.6 *$\mathcal{R} \vdash \Gamma$ whenever $\mathcal{B} \models \Gamma$.*

Proof. By Lemma 4.5 we have that $\mathcal{R}$ is one-step cut-free complete over finite sets, which implies that $\mathcal{R}$ is one-step complete. This can either be seen as a consequence of [13, Proposition 4.5] or directly: Given that $\mathcal{B}X, \tau \models \Gamma$ let V_0 denote the propositional variables that occur in Γ and define an equivalence relation $\sim$ on X by $x \sim y \iff \forall p \in \mathsf{V}_0(x \in \tau(p) \iff y \in \tau(p))$. Let $X_0 = X/\sim$ and $\tau_0(p) = \{[x]_\sim \mid x \in \tau(p)\}$. Then $\mathcal{B}(X_0), \tau_0 \models \Gamma$ by naturality of predicate liftings whence Γ is τ_0-derivable by Lemma 4.5 which implies τ-derivability of Γ. Completeness now follows from one-step completeness (Theorem 2.4). □

Theorem 4.7 *Let $\Gamma \in \mathcal{S}(\Lambda)$. Then $\mathcal{B} \models \Gamma$ whenever $\mathsf{Krip} \models \Gamma$. In particular, the following are equivalent:*

(i) $\mathcal{R} \vdash \Gamma$ *(ii)* $\mathsf{Krip} \models \Gamma$ *(iii)* $\mathcal{B} \models \Gamma$

witnessing soundness and completeness of graded modal logic with linear inequalities both over Kripke frames and multigraphs.

Proof. We only need to show that $\mathcal{B} \models \Gamma$ whenever $\mathsf{Krip} \models \Gamma$, as the other claims are consequences of Proposition 4.2 and Proposition 4.6. So suppose that $\Gamma \in \mathcal{S}(\mathcal{F}(\Lambda))$ and $M = (X, \gamma, \pi)$ is a $\mathcal{B}$-model so that $M \not\models \Gamma$, i.e. there exists $x_0 \in X$ so that $M, x_0 \not\models \Gamma$. We construct a Kripke model $M' = (X', \gamma', \pi')$ by unravelling at x_0: we put

- $X' = \{x_0 \stackrel{w_1}{\to} x_1 \stackrel{w_2}{\to} \cdots \stackrel{w_n}{\to} x_n \mid n \geq 0, 0 \leq w_i < \gamma(x_i)(x_{i+1})\}$
- $\gamma'(x_0 \stackrel{w_1}{\to} \cdots \stackrel{w_n}{\to} x_n) = \{x_0 \stackrel{w_1}{\to} \cdots \stackrel{w_n}{\to} x_n \stackrel{w_{n+1}}{\to} x_{n+1} \mid 0 \leq w_{n+1} < \gamma(x_n)(x_{n+1})\}$
- $\pi'(p) = \{x_0 \stackrel{w_1}{\to} \cdots \stackrel{w_n}{\to} x_n \mid x_n \in \pi(p)\}$.

In other words, the worlds of the Kripke model (X', γ', π') are the paths from the initial point $x_0 \in X$ where we make duplicates of states according to the multiplicity of the transition. It now follows by induction on the structure of formulas that

$$M, x \models A \implies M', x' \models A$$

whenever x' is of the form $x_0 \stackrel{m_1}{\to} \cdots \stackrel{w_n}{\to} x_n$ with $x_n = x$. In particular, $M' \not\models \Gamma$ as we had to show. □

5 Stochastic Logic

We may think of stochastic logic as a hybrid between probabilistic modal logic and graded modal logic. As for probabilistic modal logic, every state of a model is equipped with a measure, but we do not insist that this measure be a probability measure. If c is a state in a stochastic model and μ the ensuing measure, we may think of $\mu([\![A]\!])$ as the *total* cost of observing event A in the next transition state. As before, our formulas are linear inequalities in terms of the measures of the (truth sets of) formulas. In other words, we consider the similarity type

$$\Lambda = \{L_p(c_1, \ldots, c_m) \mid m \in \mathbb{N}, p, c_1, \ldots, c_m \in \mathbb{Q}\}$$

defining the formulas $\mathcal{F}(\Lambda)$ of stochastic modal logic. (Note that the syntax of stochastic modal logic is identical to that of probabilistic modal logic.) Informally speaking, the formula $L_p(c_1, \ldots, c_m)(A_1, \ldots, A_m)$ is valid at a point c, if the linear inequality $\sum_{j=1}^{m} c_j \mu(A_j) \geq p$ holds, where $\mu(A_j)$ is the measure of the truth-set of A_j as seen from point c.

Example 5.1 At the time of writing this paper, both authors frequently discussed the outcome of the (then) upcoming general election in their country of residence. To this effect, one may consider a stochastic model based on the set of inhabitants of said country. To every inhabitant c, we associate a measure that – applied to a subset S of the population – yields the overall amount of persuasion (measured as a non-negative real number) that c would have to apply in order to swing the votes of all elements of S into a particular direction. If Tory and Labour are propositional variables that denote the respective political angle, there was a heated debate whether the formula $\neg L_0(1, -1)(\mathsf{Tory}, \mathsf{Labour})$, $L_0(1, -1)(\mathsf{Labour}, \mathsf{Tory})$ or $L_0(1, -1)(\mathsf{Tory}, \mathsf{Labour}) \wedge L_0(1, -1)(\mathsf{Labour}, \mathsf{Tory})$ yields the most realistic model (both authors still hope that this did apply to $L_0(-1)(\top)$).

A sound and complete axiomatisation of stochastic modal logic will be provided by the set $\mathcal{R}$ of one-step rules comprising all instances of

$$(S) \frac{\sum_{i=1}^{n} r_j \sum_{j=1}^{m_i} c_i^j a_i^j \geq 0}{\{\mathsf{sg}(\epsilon_i) L_{p_i}(c_i^1, \ldots, c_i^{m_i})(a_i^1, \ldots, a_i^{m_i}) \mid i = 1, \ldots, n\}}$$

where $r_1, \ldots, r_n \in \mathbb{Z} \setminus \{0\}$ that satisfy the side condition

$$\sum_{i=1}^{n} r_i p_i < 0 \quad \text{if all } r_i < 0, \text{ and} \quad \sum_{i=1}^{n} r_i p_i \leq 0 \quad \text{otherwise.}$$

We now establish soundness of stochastic modal logic with respect to the class of finite measures, prove completeness of stochastic modal logic with respect to finitely based measures, and then align both views.

5.1 Measurable Semantics and Soundness

Definition 5.2 *A* measurable model *is a triple* (X, μ, π) *where* X *is a measurable space with* σ*-algebra* Σ_X, $\pi : \mathsf{V} \to \Sigma_X$ *is a valuation and* $\mu : X \times \Sigma_X \to [0, \infty)$ *is a measurable kernel, i.e. a function so that* $\mu(\cdot, B) : X \to [0, \infty)$ *is measurable for all* $B \in \Sigma_X$ *and* $\mu(x, \cdot) : \Sigma_X \to [0, \infty)$ *is a measure on* X.

Note that we require the measure that is induced my a measurable kernel is always finite. The semantics of $\mathcal{F}(\Lambda)$ with respect to measurable models is as expected for the propositional rules (note that atomic propositions are mapped to measurable sets) and the clause for a modal operator is

$$[\![L_p(c_1, \ldots, c_m)(A_1, \ldots, A_m)]\!]_M = \{x \in X \mid \sum_{j=1}^{m} c_j \cdot \mu(x, [\![A_j]\!]_M) \geq p\}$$

so that $[\![A]\!]_M$ is a measurable set for all $A \in \mathcal{F}(\Lambda)$. We write $M, x \models A$ if $x \in [\![A]\!]_M$ and $M \models A$ if $M, x \models A$ for all $x \in X$. Finally, $\mathsf{Meas} \models \Gamma$ if $M \models \bigvee \Gamma$ for all measurable models M. Soundness of stochastic modal logic over measurable models is similar to soundness of probabilistic modal logic and takes the following form.

Proposition 5.3 $\mathsf{Meas} \models \Gamma$ *whenever* $\mathcal{R} \vdash \Gamma$.

Proof. By induction on the proof of $\mathcal{R} \vdash \Gamma$ analogous to the proof of Proposition 3.3.□

This shows that stochastic modal logic is sound with respect to measurable models. We now look upon stochastic modal logic coalgebraically and establish completeness.

5.2 Coalgebraic Semantics and Completeness

To interpret stochastic modal logic over coalgebraic models, we consider the functor

$$\mathcal{M}(X) = \{\mu : X \to [0, \infty) \mid \mathrm{supp}(\mu) \text{ finite}\}$$

and we write $\mu(S) = \sum_{x \in S} \mu(x)$ whenever $\mu \in \mathcal{M}(X)$ and $S \subseteq X$. The functor $\mathcal{M}$ extends to a Λ-structure by virtue of

$$[\![L_p(c_1, \dots, c_m)]\!]_X(S_1, \dots, S_m) = \{\mu \in \mathcal{M}(X) \mid \sum_{j=1}^{m} c_j \cdot \mu(S_j) \geq p\}$$

where $S_1, \dots, S_m \subseteq X$. We may think (modulo currying) of $\mathcal{M}$-coalgebras as measurable kernels with finite support. Completeness of stochastic modal logic over $\mathcal{M}$-coalgebras is established by means of the following lemma that again uses results from linear programming:

Lemma 5.4 *Consider a valuation* $\tau : \mathsf{V} \to \mathcal{P}(X)$ *and suppose that* $\Gamma \in \mathcal{S}(\Lambda(\mathsf{V}))$ *is* τ*-valid. Then* Γ *is* τ*-derivable.*

Proof. We proceed as in the proof of Lemma 3.4 but instead consider the matrix

$$A_0 = \begin{pmatrix} -\epsilon_1 & & 0 \\ & \ddots & \vdots \\ 0 & & -\epsilon_n \\ -f_1(x_1) & \dots & -f_n(x_1) \\ \vdots & \vdots & \\ -f_1(x_k) & \dots & -f_n(x_k) \end{pmatrix}$$

$$A_1 = \begin{pmatrix} p_1 & \dots & p_n \end{pmatrix}$$

where $f_i = \sum_{j=1}^{m_i} c_i^j \cdot \mathbb{1}_{\tau(a_i^j)}$ and let $A = \begin{pmatrix} A_0 \\ A_1 \end{pmatrix}$ and proceed as in the proof of Lemma 3.4 (where the absence of the last column means that we do not define a probability distribution). □

As a corollary, we obtain that stochastic modal logic is complete over $\mathcal{M}$-coalgebras.

Corollary 5.5 $\mathcal{R} \vdash \Gamma$ *whenever* $\mathcal{M} \models \Gamma$.

In summary, we obtain the following theorem for stochastic modal logic:

Theorem 5.6 *Let* $\Gamma \in \mathcal{S}(\Lambda)$. *Then* $\mathcal{M} \models \Gamma$ *whenever* $\mathsf{Meas} \models \Gamma$. *In particular, the following are equivalent:*

(i) $\mathcal{R} \vdash \Gamma$ *(ii)* $\mathsf{Meas} \models \Gamma$ *(iii)* $\mathcal{M} \models \Gamma$.

Proof. We only need to show that $\mathcal{M} \models \Gamma$ whenever $\mathsf{Meas} \models \Gamma$, as the remaining assertions are the content corollary 5.5 and Proposition 5.3. So suppose that (X, γ, π) is a $\mathcal{M}$-model and $\mathsf{Meas} \models \Gamma$. We equip X with the trivial σ-algebra $\mathcal{P}(X)$ and consider the measurable kernel $\mu(x, S) = \sum_{x' \in S} \gamma(x)(x')$. Note that μ is well-defined as $\gamma(x)$ has finite support, and so defines a measurable kernel. Let $M' = (X, \mu, \pi)$. One now shows by induction on the structure of formulas that $M, x \models A \iff M', x \models A$ for all $x \in X$ which finishes the proof. □

6 Complexity

Given that we have coalgebraised all three logics by equipping them with a sound and complete coalgebraic semantics, we are now in a position to use generic (coalgebraic) methods to establish complexity bounds. As we have a characterisation of universal validity in terms of a cut-free sequent calculus where the size of the formulas strictly decreases when we move from conclusion to premise, we can map the decidability problem onto backwards proof search, which we can be seen as the problem of searching a tree the length of whose branches is polynomially bounded. To see that this problem is in polynomial space, we have to agree on representations for modal operators. Here, we represent numbers in binary, that is, we put $\mathsf{size}(n) = \lceil \log_2 n \rceil$ and $\mathsf{size}(p/q) = \mathsf{size}(p) + \mathsf{size}(q)$ which allows us to define the size of a modal operator as $\mathsf{size}(L_p(c_1, \ldots, c_m))$ as $\mathsf{size}(p) + \sum_{j=1}^{m} \mathsf{size}(c_i)$. To show decidability in PSPACE, we have to show that we can encode rules into strings of polynomial length so that all premises can be decided in NP. The formal definition is as follows.

Definition 6.1 *A set* $\mathcal{R}$ *of one-step rules is* PSPACE-tractable *if there exists a polynomial p such that all substitution instances of rules with conclusion* $\Gamma \in \mathcal{S}(\Lambda(\mathcal{F}(\Lambda))$ *can be encoded into a string of length at most* $p(|\Gamma|)$ *and it can be decided in* NP *whether*

- *a code represents a (substitution instance) of a rule with a given conclusion* Γ
- *a sequent belongs to the set of premises of a rule given as a code.*

We can use the methods presented in in [16] to show that the rule sets comprising of (P), (G) and (S) are indeed PSPACE-tractable.

Lemma 6.2 *If $\mathcal{R}$ comprises all instances of (P), (G) or (S), then $\mathcal{R}$ is* PSPACE-*tractable.*

Proof. It has been argued in [16, Lemma 6.16] that the coefficients r_i that occur in the rule sets (P), (G) and (S) can be polynomially bounded in the size of the linear inequalities (and hence in the size of the rule conclusions), and our argument is essentially identical to Example 6.17 of *op.cit.*. □

As a consequence, we obtain a PSPACE upper bound for all three logics considered in this paper.

Theorem 6.3 *The satisfiability problem of probabilistic modal logic, graded modal logic and stochastic modal logic (each considered with linear inequalities) is decidable in* PSPACE.

Proof. One can either invoke Theorem 6.13 of [16] or directly argue in terms of proof search where the branches of every putative proof tree are polynomially bounded in length, their nodes can be represented by strings of polynomial length, and membership in nodes can be decided in NP, all of which are consequences of tractability. This gives decidability of satisfiability in connection with the completeness (Theorem 3.6, Theorem 4.7 and Theorem 5.6). □

7 Conclusions

In this paper, we have given complete, cut-free axiomatisations of three modal logics that use linear inequalities to express constraints between probabilities of events, the number of successors in a Kripke model or, more generally, the measure of a successor set. In each case, completeness was established with the help of coalgebraic semantics, where we just had to show that a given set of (one-step) rules is one-step complete: the actual statement of completeness then follows from the general (coalgebraic) theory. As such, this paper tries to demonstrate the usefulness of the coalgebraic approach per se – we did not develop the general theory of coalgebraic logics, but just used off-the-shelf results to obtain completeness and complexity bounds. The semantics of graded and stochastic modal logic was given here in terms of image finite Kripke frames and bounded measures. It is an open problem whether the semantics can be extended to the general case in a sound fashion.

References

[1] Baader, F., D. Calvanese, D. McGuinness, D. Nardi and P. Patel-Schneider, editors, "The Description Logic Handbook," CUP, 2003.

[2] Cîrstea, C., C. Kupke and D. Pattinson, *EXPTIME tableaux for the coalgebraic μ-calculus*, in: E. Grädel and R. Kahle, editors, *Proc. CSL 2009*, number 5771 in Lect. Notes in Comp. Sci., 2009, pp. 179–193.

[3] Demri, S. and D. Lugiez, *Presburger modal logic is only PSPACE-complete*, in: U. Furbach and N. Shankar, editors, *Proc. IJCAR 2006*, Lect. Notes in Artif. Intell. **4130** (2006), pp. 541–556.

[4] Fagin, R. and J. Halpern, *Reasoning about knowledge and probability*, J. ACM **41** (1994), pp. 340–367.

[5] Fagin, R., J. Halpern and N. Megiddo, *A Logic for Reasoning about Probabilities*, Information and Computation **87** (1990), pp. 78–128.

[6] Fine, K., *In so many possible worlds*, Notre Dame J. Formal Logic **13** (1972), pp. 516–520.

[7] Goré, R., C. Kupke and D. Pattinson, *Optimal tableau algorithms for coalgebraic logics*, in: J. Esparza and R. Majumdar, editors, *Proc. TACAS 2010*, Lect. Notes in Comp. Sci., 2010.

[8] Goré, R., C. Kupke, D. Pattinson and L. Schröder, *Global caching for coalgebraic description logics*, in: J. Giesl and R. Haehnle, editors, *Proc. IJCAR 2010*, 2010, to appear.

[9] Heifetz, A. and P. Mongin, *Probabilistic logic for type spaces*, Games and Economic Behavior **35** (2001), pp. 31–53.

[10] Horrocks, I. and P. Patel-Schneider, *Optimising description logic subsumption*, J. Logic Comput. **9** (1999), pp. 267–293.

[11] Pacuit, E. and S. Salame, *Majority logic*, in: D. Dubois, C. Welty and M.-A. Williams, editors, *Proc. KR 2004* (2004), pp. 598–605.

[12] Pattinson, D., *Coalgebraic modal logic: Soundness, completeness and decidability of local consequence*, Theoret. Comput. Sci. **309** (2003), pp. 177–193.

[13] Pattinson, D. and L. Schröder, *Cut elimination in coalgebraic logics*, Information and Computation (2010), accepted for publication.

[14] Schrijver, A., "Theory of linear and integer programming," Wiley Interscience, 1986.

[15] Schröder, L., *A finite model construction for coalgebraic modal logic*, J. Log. Algebr. Program. **73** (2007), pp. 97–110.

[16] Schröder, L. and D. Pattinson, *PSPACE bounds for rank-1 modal logics*, ACM Transactions on Computational Logics **10** (2009).

[17] Schröder, L., D. Pattinson and C. Kupke, *Nominals for everyone*, in: C. Boutilier, editor, *Proc. IJCAI 2009*, 2009, pp. 917–922, online proceedings.

[18] Tobies, S., *PSPACE reasoning for graded modal logics*, J. Logic Comput. **11** (2001), pp. 85–106.

[19] Troelstra, A. and H. Schwichtenberg, "Basic Proof Theory," Number 43 in Cambridge Tracts in Theoretical Computer Science, Cambridge University Press, 1996.

[20] Tsarkov, D., I. Horrocks and P. Patel-Schneider, *Optimizing terminological reasoning for expressive description logics*, J. Autom. Reasoning **39** (2007), pp. 277–316.

On the Complexity of Modal Axiomatisations over Many-dimensional Structures

Agi Kurucz

Department of Computer Science
King's College London, U.K.
agi.kurucz@kcl.ac.uk

Abstract

We show that all the complexities of a possible axiomatisation of $\mathbf{S5}^n$, the n-modal logic of products of n equivalence frames, are already present in any axiomatisation of $\mathbf{K}^n$. Then we show that if $3 \leq n < \omega$ then, for any set L of n-modal formulas between $\mathbf{K}^n$ and $\mathbf{S5}^n$, the class of all frames for L is not closed under ultraproducts and is therefore not elementary. So any modal axiomatisation for a Kripke complete logic in the interval between $\mathbf{K}^n$ and $\mathbf{S5}^n$ must contain modal formulas with no first-order correspondents. The proof is based on a construction of Hirsch and Hodkinson [15] showing that the class of strongly representable n-dimensional cylindric algebra atom structures is not closed under ultraproducts. We show that this construction can be carried through in a diagonal-free setting.

Keywords: many-dimensional modal logic, products of Kripke frames, ultraproducts

1 Introduction

As usual in any area of logic, when one considers the "logic" or "theory" of a class $\mathcal{C}$ of structures (the "intended models"), then there are always "non-intended", "non-standard" models of this "logic". These non-standard structures are often hard to describe. In this paper we discuss this problem in the setting of *n-modal logics*: propositional multi-modal logics having finitely many unary modal operators $\Diamond_0, \dots, \Diamond_{n-1}$ (and their duals $\Box_0, \dots, \Box_{n-1}$), where n is a non-zero natural number. Formulas of this language, using propositional variables from some fixed countably infinite set, are called *n-modal formulas.* Frames for n-modal logics — *n-frames* — are structures of the form $\mathfrak{F} = (W, T_i)_{i<n}$ where W is a non-empty set and each T_i is a binary relation on W, for $i < n$. *Validity* of a set Σ of n-modal formulas in an n-frame $\mathfrak{F}$ (in symbols: $\mathfrak{F} \models \Sigma$) is defined as usual. If $\mathfrak{F} \models \Sigma$ then we also say that $\mathfrak{F}$ is a *frame for* Σ. Given a class $\mathcal{C}$ of

n-frames, we denote by $\mathsf{Log}(\mathcal{C})$ the set of all n-modal formulas that are valid in every n-frame in $\mathcal{C}$.

Our "intended" structures are the following special n-frames. Given 1-frames $\mathfrak{F}_i = (W_i, R_i)$, $i < n$, their *product* is the n-frame

$$\mathfrak{F}_0 \times \cdots \times \mathfrak{F}_{n-1} = (W_0 \times \cdots \times W_{n-1}, \bar{R}_i)_{i<n},$$

where $W_0 \times \cdots \times W_{n-1}$ is the Cartesian product of the W_i and for all $\mathbf{u}, \mathbf{v} \in W_0 \times \cdots \times W_{n-1}$ and $i < n$,

$$\mathbf{u}\bar{R}_i\mathbf{v} \quad \text{iff} \quad u_i R_i v_i \text{ and } u_j = v_j \text{ for } j \neq i,\ j < n.$$

Such n-frames we call *n-dimensional product frames.* They have been introduced in [9,24] and have been extensively studied both in pure modal logic and in applications, see [8,21] and the references therein.

Two examples of classes of n-dimensional product frames are:

$$\begin{aligned} \mathcal{C}^n_{all} &= \text{the class of all } n\text{-dimensional product frames,} \\ \mathcal{C}^n_{equiv} &= \text{the class of all } n\text{-dimensional products of equivalence frames.} \end{aligned}$$

Let us also introduce notations for the n-modal logics they determine:

$$\begin{aligned} \mathbf{K}^n &= \mathsf{Log}(\mathcal{C}^n_{all}), \\ \mathbf{S5}^n &= \mathsf{Log}(\mathcal{C}^n_{equiv}). \end{aligned}$$

It can be hard to describe an arbitrary n-frame for $\mathbf{K}^n$ or $\mathbf{S5}^n$. As is shown in [16], if $n \geq 3$ and L is any set of n-modal formulas such that $\mathbf{K}^n \subseteq L \subseteq \mathbf{S5}^n$, then it is undecidable whether a finite n-frame is a frame for L or not. (So no such logic L can be finitely axiomatisable.) Here we show that these non-standard n-frames are hard to "catch" in an other sense: They cannot be described in the first-order "frame language", that is, in the language having n binary predicate symbols and equality.

Theorem 1.1 *Let $3 \leq n < \omega$ and let L be any set of n-modal formulas such that $\mathbf{K}^n \subseteq L \subseteq \mathbf{S5}^n$. Then the class of all frames for L is not closed under ultraproducts, and so is not elementary.*

Note that both $\mathbf{K}^2$ and $\mathbf{S5}^2$ are (finitely) axiomatisable by Sahlqvist-formulas (see [9,14]), so the respective classes of all their frames *are* elementary. Also note that Theorem 1.1 only says that the class of *all* frames for certain modal logics is not closed under ultraproducts. Such a logic can still be determined by some *smaller*, ultraproduct-closed class of n-frames. This is indeed the case for many, see Prop. 2.9 below. As is shown in [20], $\mathbf{K}^n$ is even determined by a class of n-frames that can be *finitely* axiomatised in the first-order frame language.

However, as a consequence of Theorem 1.1 we obtain the following quite discouraging result, as far as finding an explicit axiomatisation for the logics in question is concerned:

Corollary 1.2 *Let $3 \leq n < \omega$ and let L be any Kripke-complete n-modal logic such that $\mathbf{K}^n \subseteq L \subseteq \mathbf{S5}^n$. Then any axiomatisation for L must contain n-modal formulas with no first-order correspondents.*

We conjecture that, for canonical logics L in the interval between $\mathbf{K}^n$ and $\mathbf{S5}^n$, a combination of the techniques of the present paper with those of Hodkinson and Venema [17] might result in an even stronger statement: Any axiomatisation for such an L must contain infinitely many non-canonical n-modal formulas.

The structure of the paper is as follows. In Section 2 we give a general characterisation of arbitrary frames of multi-modal logics determined by frame-classes satisfying some closure conditions. Using this we show that if we could "deal" with non-standard n-frames for $\mathbf{K}^n$, then we could do that with arbitrary n-frames for $\mathbf{S5}^n$ as well. In particular, we show that $\mathbf{S5}^n$ is finitely axiomatisable over $\mathbf{K}^n$. Then in Sections 3 and 4 we prove Theorem 1.1. The proof is based on a construction of Hirsch and Hodkinson [15] showing that the class of strongly representable n-dimensional cylindric algebra atom structures is not closed under ultraproducts. We show that this construction can be carried through in a diagonal-free setting, and then apply the results of Section 2.

2 Non-standard frames for logics determined by classes of n-dimensional product frames

We begin with proving some general results on modal logics determined by special classes of relational structures of *any* signature. In what follows we use the words *frame* and *relational structure* as synonyms. (So the n-frames introduced in Section 1 are special frames.) We use without explicit reference standard notions and results from basic modal logic and universal algebra; such as *p-morphisms*, *generated subframes*, *Sahlqvist formulas* and *canonicity*, duality between relational structures and *Boolean algebras with operators* (*BAOs*), *homomorphisms*, *subalgebras*, *direct products*, *ultraproducts*, *varieties*, *subdirect embeddings* and *subdirectly irreducible* algebras. For notions and statements not defined or proved here, see [3,4,10,13].

If x is a point in a relational structure $\mathfrak{F}$ then we denote by $\mathfrak{F}^x$ the smallest generated subframe of $\mathfrak{F}$ containing x. We call $\mathfrak{F}^x$ a *point-generated subframe* of $\mathfrak{F}$. If $\mathfrak{F} = \mathfrak{F}^x$ for some x, then $\mathfrak{F}$ is called *rooted*. Apart from the usual operators $\mathbf{H}$, $\mathbf{S}$ and $\mathbf{P}$ on classes of algebras (denoting homomorphic images, subalgebras, and isomorphic copies of direct products, respectively), we use the following operators on classes of frames of the same signature:

$$\mathbb{G}\mathsf{sf}\,\mathcal{C} = \text{isomorphic copies of generated subframes of frames in } \mathcal{C},$$

$$\mathbb{G}\mathsf{sf}_{\mathsf{p}}\,\mathcal{C} = \text{isomorphic copies of point-generated subframes of frames in } \mathcal{C}.$$

The *(full) complex algebra* of a frame $\mathfrak{F} = (W, R_i)_{i \in I}$ is denoted by $\mathfrak{Cm}\,\mathfrak{F}$. That is, $\mathfrak{Cm}\,\mathfrak{F} = (\mathcal{P}(W), \cap, -^W, f_i)_{i \in I}$, where $(\mathcal{P}(W), \cap, -^W)$ is the Boolean algebra of all subsets of W, and for each $k+1$-ary relation R_i, f_i is a k-ary function defined by taking, for every $X_1, \dots, X_k \subseteq W$,

$$f_i(X_1, \dots, X_k) = \{w \in W : R_i(w, x_1, \dots, x_k) \text{ for some } x_1 \in X_1, \dots, x_k \in X_k\}.$$

Given a class $\mathcal{C}$ of frames of the same signature, we denote by $\mathsf{Cm}\,\mathcal{C}$ the class of complex algebras of frames in $\mathcal{C}$. The starting point of the duality between Kripke complete modal logics and BAOs is the following well-known property. For any class $\mathcal{C}$ of frames, and for any frame $\mathfrak{F}$ of the same signature,

$$\mathfrak{F} \models \mathsf{Log}(\mathcal{C}) \quad \iff \quad \mathfrak{Cm}\,\mathfrak{F} \in \mathbf{H}\,\mathbf{S}\,\mathbf{P}\,\mathsf{Cm}\,\mathcal{C}. \tag{1}$$

The following general result shows that if $\mathcal{C}$ satisfies some closure conditions, then $\mathbf{H}$ is not needed in generating the variety corresponding to $\mathsf{Log}(\mathcal{C})$:

Theorem 2.1 (Goldblatt [11]) *If $\mathcal{C}$ is a class of frames that is closed under ultraproducts, then $\mathbf{S}\,\mathbf{P}\,\mathsf{Cm}\,\mathbb{G}\mathsf{sf}\,\mathcal{C}$ is a canonical variety.*

Let us have a closer look at the subdirectly irreducible algebras of these varieties.

Lemma 2.2 *For any class $\mathcal{C}$ of frames, the subdirectly irreducible members of $\mathbf{S}\,\mathbf{P}\,\mathsf{Cm}\,\mathbb{G}\mathsf{sf}\,\mathcal{C}$ belong to $\mathbf{S}\,\mathsf{Cm}\,\mathbb{G}\mathsf{sf}_{\mathsf{p}}\,\mathcal{C}$.*

Proof. Let $\mathfrak{A} \in \mathbf{S}\,\mathbf{P}\,\mathsf{Cm}\,\mathbb{G}\mathsf{sf}\,\mathcal{C}$ and let $\mathfrak{A} \rightarrowtail \prod_{i\in I} \mathfrak{A}_i$ be a subdirect embedding, for some $\mathfrak{A}_i \in \mathbf{S}\,\mathsf{Cm}\,\mathbb{G}\mathsf{sf}\,\mathcal{C}$, $i \in I$. If $\mathfrak{A}$ is subdirectly irreducible then there is an $i \in I$ such that $\mathfrak{A}$ is isomorphic to $\mathfrak{A}_i$, and so $\mathfrak{A}$ is isomorphic to a subalgebra of $\mathfrak{Cm}\,\mathfrak{F}$ for some $\mathfrak{F} \in \mathbb{G}\mathsf{sf}\,\mathcal{C}$. Then for each point x in $\mathfrak{F}$, $\mathfrak{F}^x \in \mathbb{G}\mathsf{sf}_{\mathsf{p}}\,\mathbb{G}\mathsf{sf}\,\mathcal{C} \subseteq \mathbb{G}\mathsf{sf}_{\mathsf{p}}\,\mathcal{C}$. It is not hard to show (see e.g. [10, 3.3]) that $\mathfrak{Cm}\,\mathfrak{F} \rightarrowtail \prod_{x\in\mathfrak{F}} \mathfrak{Cm}\,\mathfrak{F}^x$ is a (subdirect) embedding. So there exist subalgebras $\mathfrak{B}_x$ of $\mathfrak{Cm}\,\mathfrak{F}^x$ such that $\mathfrak{A} \rightarrowtail \prod_{x\in\mathfrak{F}} \mathfrak{B}_x$ is a subdirect embedding as well. As $\mathfrak{A}$ is subdirectly irreducible, there is some x in $\mathfrak{F}$ such that $\mathfrak{A}$ is isomorphic to $\mathfrak{B}_x$, and so $\mathfrak{A} \in \mathbf{S}\,\mathsf{Cm}\,\mathbb{G}\mathsf{sf}_{\mathsf{p}}\,\mathcal{C}$. □

Now Theorem 2.1 and Lemma 2.2 imply the following characterisation of varieties generated by certain classes of complex algebras.

Theorem 2.3 *If $\mathcal{C}$ is a class of frames that is closed under ultraproducts and point-generated subframes, then $\mathbf{S}\,\mathbf{P}\,\mathsf{Cm}\,\mathcal{C} = \mathbf{H}\,\mathbf{S}\,\mathbf{P}\,\mathsf{Cm}\,\mathcal{C}$ is a canonical variety.*

We can also have a 'dual' structural characterisation of subdirectly irreducible algebras of these varieties. Recall that an *ultrafilter* of a BAO $\mathfrak{A} = (A, \wedge, -, f_i)_{i\in I}$ is any subset μ of A such that, for all $a, b \in A$,

- if $a \in \mu$ and $a \wedge b = a$ then $b \in \mu$;
- if $a, b \in \mu$ then $a \wedge b \in \mu$;
- $a \in \mu$ iff $-a \notin \mu$.

Let $Uf(A)$ denote the set of all such ultrafilters. Given a BAO $\mathfrak{A} = (A, \wedge, -, f_i)_{i\in I}$, we denote by $\mathfrak{Uf}\,\mathfrak{A} = (Uf(A), R_i)_{i\in I}$ its *ultrafilter frame*, where for each k-ary function f_i,

R_i is the following $k+1$-ary relation: for any $\mu, \nu_1, \ldots, \nu_k \in Uf(A)$,

$$R_i(\mu, \nu_1, \ldots, \nu_k) \quad \text{iff} \quad \forall a_1 \in \nu_1, \ldots, a_k \in \nu_k \; f_i(a_1, \ldots, a_k) \in \mu.$$

The *ultrafilter extension* of a frame $\mathfrak{F}$ is $\mathfrak{Ue}\,\mathfrak{F} = \mathfrak{Uf}\,\mathfrak{Cm}\,\mathfrak{F}$.

Theorem 2.4 *Let $\mathcal{C}$ be a class of frames that is closed under ultraproducts and point-generated subframes. Then for every subdirectly irreducible algebra $\mathfrak{A}$,*

$$\mathfrak{A} \in \mathbf{SP}\,\mathsf{Cm}\,\mathcal{C} \iff \mathfrak{A} \in \mathbf{S}\,\mathsf{Cm}\,\mathcal{C} \iff \mathfrak{Uf}\,\mathfrak{A} \textit{ is a p-morphic image of some } \mathfrak{G} \in \mathcal{C}.$$

Proof. $\Leftarrow$: By Jónsson and Tarski's [19] theorem, $\mathfrak{A}$ is embeddable into $\mathfrak{Cm}\,\mathfrak{Uf}\,\mathfrak{A}$. And by duality, $\mathfrak{Cm}\,\mathfrak{Uf}\,\mathfrak{A}$ is embeddable into $\mathfrak{Cm}\,\mathfrak{G} \in \mathsf{Cm}\,\mathcal{C}$.

$\Rightarrow$: If $\mathfrak{A} \in \mathbf{SP}\,\mathsf{Cm}\,\mathcal{C}$ then there is a subdirect embedding $\mathfrak{A} \rightarrowtail \prod_{i\in I} \mathfrak{A}_i$, for some $\mathfrak{A}_i \in \mathbf{S}\,\mathsf{Cm}\,\mathcal{C}$, $i \in I$. As $\mathfrak{A}$ is subdirectly irreducible, there is an $i \in I$ such that $\mathfrak{A}$ is isomorphic to $\mathfrak{A}_i$, that is, $\mathfrak{A}$ is isomorphic to a subalgebra of $\mathfrak{Cm}\,\mathfrak{F}$ for some $\mathfrak{F} \in \mathcal{C}$. By duality, $\mathfrak{Uf}\,\mathfrak{A}$ is a p-morphic image of $\mathfrak{Ue}\,\mathfrak{F}$. As $\mathfrak{Ue}\,\mathfrak{F}$ is a p-morphic image of an ultrapower of $\mathfrak{F}$ (see [7,1,2]) and $\mathcal{C}$ is closed under taking ultraproducts, the proof is completed. □

As a consequence, we obtain a characterisation of "non-standard" frames for certain logics of the form $\mathsf{Log}(\mathcal{C})$:

Corollary 2.5 *Let $\mathcal{C}$ be a class of frames that is closed under ultraproducts and point-generated subframes. Then for every rooted frame $\mathfrak{F}$,*

$$\mathfrak{F} \models \mathsf{Log}(\mathcal{C}) \iff \mathfrak{Ue}\,\mathfrak{F} \textit{ is a p-morphic image of some } \mathfrak{G} \in \mathcal{C}.$$

Proof. By (1) and Theorem 2.3,

$$\mathfrak{F} \models \mathsf{Log}(\mathcal{C}) \iff \mathfrak{Cm}\,\mathfrak{F} \in \mathbf{SP}\,\mathsf{Cm}\,\mathcal{C}.$$

As the complex algebra of a rooted frame is subdirectly irreducible [10], the statement follows from Theorem 2.4. □

As the ultrafilter extension of a finite frame is isomorphic to the frame itself, we obtain:

Corollary 2.6 *Let $\mathcal{C}$ be a class of frames that is closed under ultraproducts and point-generated subframes. Then for every finite rooted frame $\mathfrak{F}$,*

$$\mathfrak{F} \models \mathsf{Log}(\mathcal{C}) \iff \mathfrak{F} \textit{ is a p-morphic image of some } \mathfrak{G} \in \mathcal{C}.$$

Now we would like to apply these general results to various classes of n-dimensional product frames, whenever $0 < n < \omega$. To this end, observe that the product operation commutes with ultraproducts and point-generated subframes:

Claim 2.7 *Let U be an ultrafilter over some index set I, and let $\mathfrak{F}^i_k$ be a 1-frame, for $i \in I$, $k < n$. Then:*

$$\prod_{i\in I}(\mathfrak{F}^i_0 \times \cdots \times \mathfrak{F}^i_{n-1})/U \quad \textit{is isomorphic to} \quad (\prod_{i\in I}\mathfrak{F}^i_0/U) \times \cdots \times (\prod_{i\in I}\mathfrak{F}^i_{n-1}/U).$$

Claim 2.8 *Let $\mathfrak{F} = \mathfrak{F}_0 \times \cdots \times \mathfrak{F}_{n-1}$ and $\mathbf{x}$ be a point in $\mathfrak{F}$. Then:*

$$\mathfrak{F}^{\mathbf{x}} = \mathfrak{F}_0^{x_0} \times \cdots \times \mathfrak{F}_{n-1}^{x_{n-1}}.$$

Given classes $\mathcal{C}_i$ of 1-frames, for $i < n$, let us define

$$\mathcal{C}_0 \times \cdots \times \mathcal{C}_{n-1} = \{\mathfrak{F}_0 \times \cdots \times \mathfrak{F}_{n-1} : \mathfrak{F}_i \in \mathcal{C}_i,\ i < n\}.$$

As a consequence of Claims 2.7 and 2.8, we obtain:

Proposition 2.9 *If, for $i < n$, $\mathcal{C}_i$ is a class of 1-frames that is closed under ultraproducts and point-generated subframes, then the class $\mathcal{C}_0 \times \cdots \times \mathcal{C}_{n-1}$ of n-dimensional product frames is closed under ultraproducts and point-generated subframes.*

Now, by Theorem 2.3, (1) and Corollary 2.5, we have:

Theorem 2.10 *If, for $i < n$, $\mathcal{C}_i$ is a class of 1-frames that is closed under ultraproducts and point-generated subframes, then:*
(i) $\mathbf{SP}\,\mathsf{Cm}\,(\mathcal{C}_0 \times \cdots \times \mathcal{C}_{n-1}) = \mathbf{HSP}\,\mathsf{Cm}\,(\mathcal{C}_0 \times \cdots \times \mathcal{C}_{n-1})$ *is a canonical variety.*
(ii) $\mathsf{Log}(\mathcal{C}_0 \times \cdots \times \mathcal{C}_{n-1})$ *is a canonical n-modal logic.*
(iii) *For every rooted n-frame $\mathfrak{F}$,*

$$\mathfrak{F} \models \mathsf{Log}(\mathcal{C}_0 \times \cdots \times \mathcal{C}_{n-1}) \iff \mathfrak{Ue}\,\mathfrak{F} \textit{ is a p-morphic image of some } \mathfrak{G} \in \mathcal{C}_0 \times \cdots \times \mathcal{C}_{n-1}.$$

Remark 2.11 The condition of Theorem 2.10 clearly holds if each $\mathcal{C}_i$ is defined by a set of 1-modal formulas having first-order correspondents, such as the classes of all frames of well-known modal logics like **K**, **K4**, **K4.3**, **S4.3**, **S5**, $\mathsf{Log}\{(\mathbb{Q}, <)\}$.

In particular, the classes $\mathcal{C}^n_{all}$ and $\mathcal{C}^n_{equiv}$ introduced in Section 1 are examples of classes of the form $\mathcal{C}_0 \times \cdots \times \mathcal{C}_{n-1}$ within the scope of Theorem 2.10. So, for every rooted n-frame $\mathfrak{F}$,

$$\mathfrak{F} \models \mathbf{K}^n \iff \mathfrak{Ue}\,\mathfrak{F} \text{ is a p-morphic image of some } \mathfrak{G} \in \mathcal{C}^n_{all}, \tag{2}$$

$$\mathfrak{F} \models \mathbf{S5}^n \iff \mathfrak{Ue}\,\mathfrak{F} \text{ is a p-morphic image of some } \mathfrak{G} \in \mathcal{C}^n_{equiv}. \tag{3}$$

Also, $\mathbf{SP}\,\mathsf{Cm}\,\mathcal{C}^n_{all}$ and $\mathbf{SP}\,\mathsf{Cm}\,\mathcal{C}^n_{equiv}$ are canonical varieties. The latter is a variety well-known in algebraic logic: the variety of *n-dimensional representable diagonal-free cylindric algebras* [14].

The following lemma shows that any n-frame having n equivalence relations and being a p-morphic image of an arbitrary n-dimensional product frame is also a p-morphic image of a product of n equivalence frames.

Lemma 2.12 *Let $n > 0$ be an arbitrary natural number, and let $\mathfrak{F} = (W, T_i)_{i<n}$ be an n-frame such that every T_i is an equivalence relation, for $i < n$. Suppose that $f : \mathfrak{G}_0 \times \cdots \times \mathfrak{G}_{n-1} \to \mathfrak{F}$ is a surjective p-morphism, for some 1-frames $\mathfrak{G}_i = (U_i, R_i)$, $i < n$. Then there exist 1-frames $\mathfrak{G}_i^* = (U_i, R_i^*)$, $i < n$, such that*

- *each R_i^* is an equivalence relation extending R_i, and*
- *$f : \mathfrak{G}_0^* \times \cdots \times \mathfrak{G}_{n-1}^* \to \mathfrak{F}$ is still a surjective p-morphism.*

Proof. In order to obtain the 'equivalence-closure' R_i^* of each R_i, one can add the missing pairs step by step, like it is done for the $n = 2$ case in the proof of [8, Lemma 5.8]. The fact that now n is an arbitrary natural number does not make any difference. □

Remark 2.13 Note that a similar proof would prove a stronger statement. The property of each T_i being an equivalence relation can be replaced with any property of T_i that can be defined by a set of *universal Horn* formulas in the first-order language having a binary predicate symbol and equality (and there can be different such properties for different i).

As a consequence of Theorem 2.10 and Lemma 2.12 we obtain:

Theorem 2.14 *Let L be any canonical n-modal logic with $\mathbf{K}^n \subseteq L \subseteq \mathbf{S5}^n$. Then $\mathbf{S5}^n$ is finitely axiomatisable over L: $\mathbf{S5}^n$ is the smallest n-modal logic containing L and the $\mathbf{S5}$-axioms for $\Diamond_i$, $i < n$.*

Proof. One inclusion is clear, let us prove the other. The **S5**-axioms are well-known examples of Sahlqvist formulas, and their first-order correspondent is the property of being an equivalence relation. So, by Sahlqvist's completeness theorem, the smallest n-modal logic containing L and the **S5**-axioms for $\Diamond_i$, $i < n$ is canonical, and so Kripke complete. So it is enough to show that every rooted n-frame $\mathfrak{F}$ for this logic is a frame for $\mathbf{S5}^n$.

Take such an n-frame $\mathfrak{F}$. As $\mathfrak{F}$ is a frame for $\mathbf{K}^n = \mathsf{Log}(\mathcal{C}^n_{all})$, by (2), $\mathfrak{Ue}\,\mathfrak{F}$ is a p-morphic image of some n-dimensional product frame $\mathfrak{G}$. As $\mathfrak{F}$ validates the canonical **S5**-axioms, they also hold in $\mathfrak{Ue}\,\mathfrak{F}$, and so all the relations in $\mathfrak{Ue}\,\mathfrak{F}$ are equivalence relations. Now by Lemma 2.12, $\mathfrak{Ue}\,\mathfrak{F}$ is a p-morphic image of some $\mathfrak{G}^* \in \mathcal{C}^n_{equiv}$, and so by (3), $\mathfrak{F}$ is a frame for $\mathbf{S5}^n = \mathsf{Log}(\mathcal{C}^n_{equiv})$. □

Remark 2.15 By Remarks 2.11 and 2.13 we can have similar statements for any $\mathsf{Log}(\mathcal{K})$ in place of $\mathbf{S5}^n$, whenever $\mathcal{K} = \mathcal{C}_0 \times \cdots \times \mathcal{C}_{n-1}$ for some classes $\mathcal{C}_i$ of 1-frames, each of which is definable by Sahlqvist formulas having universal Horn first-order correspondents.

Theorem 2.14 shows that any negative result on the equational axiomatisation of the variety on n-dimensional representable diagonal-free cylindric algebras (such as its non-finiteness [18], for $n \geq 3$) transfers not only to its logic counterpart $\mathbf{S5}^n$, but also to other many-dimensional modal logics like $\mathbf{K}^n$. In other words, this theorem also means that all the complexities of a possible axiomatisation of $\mathbf{S5}^n$ come from the many-dimensional nature of the product frames and are already present in an axiomatisation of $\mathbf{K}^n$. Though, by a general result of [9], $\mathbf{K}^n$ is known to be recursively enumerable, an axiomatisation of $\mathbf{K}^n$ should be quite complex, whenever $n \geq 3$: any such axiomatisation should contain modal formulas of arbitrary modal depth for each modality [20], and infinitely many propositional variables [22]. (At the moment we cannot use Theorem 2.14 to infer the latter, as it is not known whether $\mathbf{S5}^n$ can be axiomatised using finitely many variables, whenever $n \geq 3$.) As Theorem 1.1 above shows, it will be quite hard to find an explicit axiomatisation for $\mathbf{K}^n$, as any such must contain n-modal formulas having no first-order correspondents.

3 Frames constructed from graphs

This and the next section are devoted to the proof of Theorem 1.1. Throughout, we fix a natural number $n \geq 3$. We will use n as a notation for both this number and for the set $\{0, \ldots, n-1\}$. In order to show Theorem 1.1, we will give n-frames $\mathfrak{G}_k$, for $k < \omega$, such that each $\mathfrak{G}_k$ is a frame for $\mathbf{S5}^n$, but any non-principal ultraproduct of the $\mathfrak{G}_k$s is not a frame for $\mathbf{K}^n$.

We will use a construction of Hirsch and Hodkinson [15], so let us introduce the necessary notions. To begin with, let us enrich n-frames by adding some unary relations. An *$n\delta$-frame* is a relational structure of the form $\mathfrak{F} = (W, T_i, E_{ij})_{i,j<n}$ where $(W, T_i)_{i<n}$ is an n-frame and $E_{ij} \subseteq W$ for all $i, j < n$. For any n-dimensional product frame $\mathfrak{F} = (W_0 \times \cdots \times W_{n-1}, \bar{R}_i)_{i<n}$, we define an $n\delta$-frame $\mathfrak{F}^\delta$ by taking

$$\mathfrak{F}^\delta = (W_0 \times \cdots \times W_{n-1}, \bar{R}_i, \delta_{ij})_{i,j<n},$$

where $\delta_{ij} = \{\mathbf{w} \in W_0 \times \cdots \times W_{n-1} : w_i = w_j\}$, for $i, j < n$. These δ_{ij}s are called *diagonal elements*. Now let

$$\mathcal{C}^{n\delta}_{cube} = \{(\underbrace{\mathfrak{F} \times \cdots \times \mathfrak{F}}_{n})^\delta : \mathfrak{F} = (U, U \times U) \text{ for some non-empty set } U\}.$$

Note that if $\mathfrak{F}^\delta \in \mathcal{C}^{n\delta}_{cube}$ then $\mathfrak{F} \in \mathcal{C}^n_{equiv}$. Using Claims 2.7 and 2.8, it is not hard to see that $\mathcal{C}^{n\delta}_{cube}$ is closed under ultraproducts and point-generated subframes. So, by Theorem 2.3, $\mathbf{S}\,\mathbf{P}\,\mathsf{Cm}\,\mathcal{C}^{n\delta}_{cube}$ is a canonical variety, well-known in algebraic logic: the variety of *n-dimensional representable cylindric algebras* [14].

Next, we define special $n\delta$-frames with the help of graphs. By a *graph* we mean a pair (Γ, E), where Γ is non-empty set and E is an irreflexive and symmetric binary relation on Γ (the *edges*). We identify a graph with its underlying set Γ of *nodes*. Given a graph $\Gamma = (\Gamma, E)$, a set $X \subseteq \Gamma$ is called *independent*, if $(x, y) \notin E$ whenever $x, y \in X$. The *chromatic number* $\chi(\Gamma)$ of Γ is the smallest $k < \omega$ such that Γ can be partitioned into k independent sets, and ∞ is there is no such k. An *ultrafilter on* Γ is an ultrafilter of the Boolean algebra of all subsets of Γ. For any graph Γ and $n < \omega$, we define the graph $\Gamma \times n$ as n disjoint copies of Γ, with all possible edges between distinct copies being added. For notions not defined here and general information on graphs, see [5].

Given a graph Γ, Hirsch and Hodkinson [15] define an $n\delta$-frame

$$\mathfrak{F}_\Gamma = (H_\Gamma, \equiv_i, D_{ij})_{i,j<n}$$

as follows.

- H_Γ is the set of all pairs $(K, \sim)$, where $K : n \to \Gamma \times n$ is a partial map, and $\sim$ is an equivalence relation on n, satisfying one of the following properties:
  · Either: all distinct $i, j < n$ are not $\sim$-equivalent, $K(i)$ is defined for all $i < n$, and $\{K(0), \ldots, K(n-1)\}$ is not an independent set in $\Gamma \times n$.
  · Or: $\{i, j\}$ is a 2-element $\sim$-class, all other $\sim$-classes are singletons, $K(i)$ and $K(j)$ are both defined and $K(i) = K(j)$, and $K(k)$ is not defined for $k \neq i, j$.
  · Or: the number of $\sim$-classes is $\leq n-2$ and $K = \emptyset$.
- For every $i < n$, $\equiv_i$ is a binary relation on H_Γ defined by

$$(K, \sim) \equiv_i (K', \sim') \quad \text{iff} \quad \begin{array}{l} \sim|_{n-\{i\}} = \sim'|_{n-\{i\}}, \text{ and} \\ \text{either both } K(i) \text{ and } K'(i) \text{ are undefined,} \\ \text{or both } K(i) \text{ and } K'(i) \text{ are defined and } K(i) = K'(i). \end{array}$$

- For all $i, j < n$, D_{ij} is the following subset of H_Γ:

$$D_{ij} = \{(K, \sim) \, : \, i \sim j\}.$$

The following two propositions are proved in [15]:

Proposition 3.1 [15, Prop.5.2]
If $\chi(\Gamma) = \infty$ then $\mathfrak{Cm}\,\mathfrak{F}_\Gamma$ is an n-dimensional representable cylindric algebra.

Proposition 3.2 [15, Prop.5.4]
If Γ is infinite and $\chi(\Gamma) < \omega$, then $\mathfrak{Cm}\,\mathfrak{F}_\Gamma$ is not an n-dimensional representable cylindric algebra.

Observe that $\mathfrak{Cm}\,\mathfrak{F}_\Gamma$ is a BAO of the form $(A, \wedge, -, c_i, d_{ij})_{i,j<n}$, where each c_i is a unary function on A and each d_{ij} is an element of A. If we forget about the d_{ij}s, we obtain what is called the *diagonal-free reduct* of $\mathfrak{Cm}\,\mathfrak{F}_\Gamma$. It should be clear that this diagonal-free reduct is in fact $\mathfrak{Cm}\,\mathfrak{F}_\Gamma^-$, where $\mathfrak{F}_\Gamma^-$ is the n-frame $(H_\Gamma, \equiv_i)_{i<n}$.

We would like to have the diagonal-free "analogues" of Propositions 3.1 and 3.2. On the one hand, it is straightforward to see that if $\mathfrak{Cm}\,\mathfrak{F}_\Gamma$ is an n-dimensional representable cylindric algebra, that is, it belongs to $\mathbf{S}\,\mathbf{P}\,\mathsf{Cm}\,\mathcal{C}^{n\delta}_{cube}$, then its diagonal-free reduct $\mathfrak{Cm}\,\mathfrak{F}_\Gamma^-$ belongs to $\mathbf{S}\,\mathbf{P}\,\mathsf{Cm}\,\mathcal{C}^{n}_{equiv}$. So by (1) and Prop. 3.1 we obtain:

Proposition 3.3 *If $\chi(\Gamma) = \infty$ then $\mathfrak{F}_\Gamma^-$ is a frame for $\mathbf{S5}^n$.*

On the other hand, having the analogue of Prop. 3.2 is not so easy. As is well-known in algebraic logic, there are $n\delta$-frames $\mathfrak{G}$ such that though $\mathfrak{Cm}\,\mathfrak{G}$ is *not* an n-dimensional representable cylindric algebra, yet its diagonal-free reduct $\mathfrak{Cm}\,\mathfrak{G}^-$ *is* an n-dimensional representable diagonal-free cylindric algebra [14]. We will show that if Γ is infinite and $\chi(\Gamma) < \infty$ then for $\mathfrak{G} = \mathfrak{F}_\Gamma$ this is not the case: $\mathfrak{Cm}\,\mathfrak{F}_\Gamma^-$ is not an n-dimensional representable diagonal-free cylindric algebra, and so $\mathfrak{F}_\Gamma^-$ is not a frame for $\mathbf{S5}^n$.

Let us begin with showing some further properties of $\mathfrak{F}_\Gamma$:

Claim 3.4 (i) *For every $i < n$, $\equiv_i$ is an equivalence relation, and $D_{ii} = H_\Gamma$.*
(ii) *For all $i, j < n$, $\equiv_i$ and $\equiv_j$ commute.*
(iii) *For all $i, j, k < n$, $i \neq j$, $k \neq i, j$ and for all $(K \sim) \in H_\Gamma$,*

$$(K, \sim) \in D_{ij} \quad \textit{iff} \quad \textit{there is } (K', \sim') \in D_{ik} \cap D_{kj} \textit{ such that } (K, \sim) \equiv_k (K', \sim').$$

(iv) *For all $i, j < n$, $i \neq j$, if $(K, \sim), (K', \sim') \in D_{ij}$ and $(K, \sim) \equiv_i (K', \sim')$, then $(K, \sim) = (K', \sim')$.*
(v) *$\mathfrak{F}_\Gamma$ is rooted.*

Proof. The proofs of items (i) and (ii) are tiresome at places, but straightforward.

(iii): Fix some $k \neq i, j$. First, let $(K, \sim) \in D_{ij}$. Then $i \sim j$ and $K(k)$ is not defined for $k \neq i, j$. Let $K' = \emptyset$ and $\sim'$ such that $\sim'|_{n-\{k\}} = \sim|_{n-\{k\}}$ and $k \sim' i \sim' j$. Then $(K', \sim') \in H_\Gamma$ as required. For the other direction, let $(K', \sim') \in D_{ik} \cap D_{kj}$ and $(K, \sim) \equiv_k (K', \sim')$. Then $i \sim' k \sim' j$ and $\sim'|_{n-\{k\}} = \sim|_{n-\{k\}}$, so $i \sim j$, thus $(K, \sim) \in D_{ij}$.

(iv): If $(K, \sim), (K', \sim') \in D_{ij}$ and $(K, \sim) \equiv_i (K', \sim')$, then $i \sim j$, $i \sim' j$ and $\sim|_{n-\{i\}} = \sim'|_{n-\{i\}}$. Therefor $\sim = \sim'$ follows. Then there are two cases: either all of $K(i)$, $K(j)$, $K'(i)$, $K'(j)$ are defined and equal, or none of them is defined. In either case, $K = K'$ follows.

(v): (cf. [15, proof of Lemma 5.1]) We show that $(\emptyset, n \times n) \in H_\Gamma$ is suitable as root. To this end, take any $(K, \sim) \in H_\Gamma$. For any $i < n$, define a partial function $K_i : n \to \Gamma \times n$ by taking

$$K_i(j) = \begin{cases} K(i), & \text{if } j = 0 \text{ or } j = i, \text{ and } K(i) \text{ is defined,} \\ \text{undefined,} & \text{else.} \end{cases}$$

Let $\sim_i$ be the unique equivalence relation such that $\sim_i|_{n-\{i\}} = \sim|_{n-\{i\}}$ and $i \sim_i 0$. Then $(K_i, \sim_i) \in H_\Gamma$ and $(K, \sim) \equiv_i (K_i, \sim_i)$. So we have

$$(K, \sim) \equiv_1 (K_1, \sim_1) \equiv_2 (K_{12}, \sim_{12}) \cdots \equiv_{n-1} (K_{12\ldots n-1}, \sim_{12\ldots n-1}).$$

As $n \geq 3$, we have $0 \sim_{12} 1 \sim_{12} 2$, so $K_{12} = \cdots = K_{12\ldots n-1} = \emptyset$. Also, $\sim_{12\ldots n-1} = n \times n$. Therefore, by item (i), $(\emptyset, n \times n)$ is a root of $\mathfrak{F}_\Gamma$. □

Properties (i)–(iv) above form the definition of what is called in algebraic logic an *n-dimensional cylindric atom structure* (see [13, 2.7.40]). Complex algebras of these special $n\delta$-frames belong to the variety of *n-dimensional cylindric algebras*. The interested reader can find the definition of this class in e.g. [13]. Here we only use that, being a variety, the class of n-dimensional cylindric algebras is closed under subalgebras. So, in particular, by Claim 3.4 we have that

$$\text{any subalgebra of } \mathfrak{Cm}\,\mathfrak{F}_\Gamma \text{ is an } n\text{-dimensional cylindric algebra.} \tag{4}$$

An element a in an algebra $\mathfrak{A} = (A, \wedge, -, c_i, d_{ij})_{i,j<n}$ is called *$< n$-dimensional*, if there is some $i < n$ such that $c_i(a) = a$. We will use the following result:

Theorem 3.5 (Johnson [18], see also [12,14])
Let $\mathfrak{A}$ be an n-dimensional cylindric algebra that is generated by its $<n$-dimensional elements. If the diagonal-free reduct $\mathfrak{A}^-$ of $\mathfrak{A}$ is an n-dimensional representable diagonal-free cylindric algebra, then $\mathfrak{A}$ is an n-dimensional representable cylindric algebra.

In Section 4 below we will define a subalgebra $\mathfrak{A}_\Gamma$ of $\mathfrak{Cm}\,\mathfrak{F}_\Gamma$ and show the following two statements:

Proposition 3.6 *$\mathfrak{A}_\Gamma$ is an n-dimensional cylindric algebra generated by its $<n$-dimensional elements.*

Proposition 3.7 (cf. [15, Prop.5.4])
If Γ is infinite and $\chi(\Gamma) < \omega$, then $\mathfrak{A}_\Gamma$ is not an n-dimensional representable cylindric algebra.

Now if Γ is infinite and $\chi(\Gamma) < \omega$ then, by Theorem 3.5, the diagonal-free reduct $\mathfrak{A}_\Gamma^-$ of $\mathfrak{A}_\Gamma$ is not an n-dimensional representable diagonal-free cylindric algebra, that is, it does not belong to $\mathbf{S}\,\mathbf{P}\,\mathsf{Cm}\,\mathcal{C}^n_{equiv}$. As $\mathfrak{A}_\Gamma^-$ is a subalgebra of $\mathfrak{Cm}\,\mathfrak{F}_\Gamma^-$, it follows that $\mathfrak{Cm}\,\mathfrak{F}_\Gamma^-$ does not belong to $\mathbf{S}\,\mathbf{P}\,\mathsf{Cm}\,\mathcal{C}^n_{equiv}$ either. So, by Claim 3.4(v) and (3), $\mathfrak{Ue}\,\mathfrak{F}_\Gamma^-$ is not a p-morphic image of a product of n equivalence frames. On the other hand, by Claim 3.4(i), all the relations $\equiv_i$ in $\mathfrak{F}_\Gamma^-$ are equivalence relations, for $i < n$. Therefore, the n-frame $\mathfrak{F}_\Gamma^-$ validates the canonical **S5**-axioms, for all $i < n$, so they also hold in $\mathfrak{Ue}\,\mathfrak{F}_\Gamma^-$, meaning that all its relations are equivalence relations as well. So, by Lemma 2.12, $\mathfrak{Ue}\,\mathfrak{F}_\Gamma^-$ is not a p-morphic image of *any* product frame. So, by (2), we have the required analogue of Prop. 3.2:

Proposition 3.8 *If Γ is infinite and $\chi(\Gamma) < \omega$, then $\mathfrak{F}_\Gamma^-$ is not a frame for $\mathbf{K}^n$.*

Now we can complete the proof of Theorem 1.1 precisely as it is done in the proof of [15, Thm.6.1]: It is not hard to see that if U is a non-principal ultrafilter over some index set I, then

$$\prod_{i\in I} \mathfrak{F}_{\Gamma_i}^-/U \quad \text{is isomorphic to} \quad \mathfrak{F}^-_{\prod_{i\in I}\Gamma_i/U}\,. \tag{5}$$

So what is left is to have a sequence $(\Gamma_k)_{k<\omega}$ of graphs such that

- $\chi(\Gamma_k) = \infty$ for all $k < \omega$.
- If Γ is any non-principal ultraproduct of the Γ_k, then Γ is infinite and $\chi(\Gamma) < \omega$.

As is shown in [15], one can have such a sequence of graphs by using Erdős's famous theorem [6]. Now let L be any set of n-modal formulas such that $\mathbf{K}^n \subseteq L \subseteq \mathbf{S5}^n$. Then, by Prop. 3.3, each $\mathfrak{F}_{\Gamma_k}^-$ is a frame for L. On the other hand, by (5) and Prop. 3.8, any non-principal ultraproduct of the $\mathfrak{F}_{\Gamma_k}^-$ is not a frame for $\mathbf{K}^n$, and so not a frame for L.

4 The algebra $\mathfrak{A}_\Gamma$

What is left is to define a subalgebra $\mathfrak{A}_\Gamma$ of $\mathfrak{Cm}\,\mathfrak{F}_\Gamma$, and prove Propositions 3.6 and 3.7 about it. We define $\mathfrak{A}_\Gamma$ using notions introduced in [15, Defs. 4.1, 4.4]. To this end, for

$i < n$, let

$$F_i = \bigcap_{j,k \neq i,\ j \neq k} (H_\Gamma - D_{jk}) = \{(K, \sim) \in H_\Gamma : K(i) \text{ is defined}\}.$$

Now, for any $X \subseteq \Gamma \times n$, put

$$X^{(i)} = \{(K, \sim) \in F_i : K(i) \in X\},$$

and let $\mathfrak{A}_\Gamma$ be the subalgebra of $\mathfrak{Cm}\,\mathfrak{F}_\Gamma$ generated by the set

$$\{X^{(i)} : i < n,\ X \subseteq \Gamma \times n\}.$$

Proof of Prop. 3.6. By (4), $\mathfrak{A}_\Gamma$ is an n-dimensional cylindric algebra. Now take any $i < n$ and $X \subseteq \Gamma \times n$. Let $(K, \sim) \in X^{(i)}$ and $(K', \sim') \in H_\Gamma$ such that $(K, \sim) \equiv_i (K', \sim')$. Then both $K(i)$ and $K'(i)$ are defined, $(K', \sim') \in F_i$ and $K(i) = K'(i)$, so $(K', \sim') \in X^{(i)}$ as well. This shows that $c_i(X^{(i)}) = X^{(i)}$, so $X^{(i)}$ is $<n$-dimensional.

Proof of Prop. 3.7. We establish a connection between ultrafilters of $\mathfrak{A}_\Gamma$ and ultrafilters over $\Gamma \times n$, just like it is done in [15] between ultrafilters of $\mathfrak{Cm}\,\mathfrak{F}_\Gamma$ and ultrafilters over $\Gamma \times n$.

For any $i < n$, let E_i denote the binary relation corresponding to c_i in the ultrafilter frame of $\mathfrak{A}_\Gamma$. For any $S \subseteq F_i$, put $S(i) = \{K(i) : (K, \sim) \in S\}$. For any $i < n$, and any ultrafilter μ of $\mathfrak{A}_\Gamma$, let

$$\mu(i) = \{S(i) : S \in \mu,\ S \subseteq F_i\}.$$

Claim 4.1 (analogue of [15, Lemma 4.6])
Let μ be an ultrafilter of $\mathfrak{A}_\Gamma$ such that $F_i \in \mu$ for some $i < n$. Then:
(i) *$\mu(i)$ is an ultrafilter on $\Gamma \times n$.*
(ii) *If $j < n$ and $D_{ij} \in \mu$, then $F_j \in \mu$ and $\mu(j) = \mu(i)$.*
(iii) *For any ultrafilter ν of $\mathfrak{A}_\Gamma$, we have $\mu E_i \nu$ iff $F_i \in \nu$ and $\mu(i) = \nu(i)$.*

Proof. (i): An arbitrary element of $\mu(i)$ is of the form $S(i)$ for some $S \in \mu$, $S \subseteq F_i$. Suppose that $S(i) \subseteq X \subseteq \Gamma \times n$. Then it is not hard to see that $S \subseteq S(i)^{(i)} \subseteq X^{(i)}$. As $X^{(i)}$ is an element of $\mathfrak{A}_\Gamma$ and μ is an ultrafilter of $\mathfrak{A}_\Gamma$, $X^{(i)} \in \mu$ follows. We also have $X^{(i)} \subseteq F_i$. So $X = X^{(i)}(i) \in \mu(i)$.

The proofs of the other two ultrafilter-properties, and of (ii) and (iii) are the same as those of the corresponding items in [15, Lemma 4.6]. □

Now we can complete the proof of Prop. 3.7 by following precisely the same steps as in the proof of [15, Prop.5.4]), using ultrafilters of $\mathfrak{A}_\Gamma$ in place of ultrafilters of $\mathfrak{Cm}\,\mathfrak{F}_\Gamma$. If $\chi(\Gamma) < \omega$, then also $\chi(\Gamma \times n) < \omega$. So $\Gamma \times n = I_0 \cup \dots \cup I_{k-1}$ for some natural number k and independent sets I_j, for $j < k$. So, for every ultrafilter μ on $\Gamma \times n$, there is a unique $j < k$ such that $I_j \in \mu$. As Γ is infinite, so is H_Γ, and so is $\mathfrak{A}_\Gamma$.

Now suppose that $\mathfrak{A}_\Gamma$ is an n-dimensional representable cylindric algebra. As is shown in [15, Lemma 5.1], every subalgebra of $\mathfrak{Cm}\,\mathfrak{F}_\Gamma$ is subdirectly irreducible, therefore so is $\mathfrak{A}_\Gamma$. Thus, by Theorem 2.4, $\mathfrak{Uf}\,\mathfrak{A}_\Gamma$ is a p-morphic image of some frame from $C^{n\delta}_{cube}$, that is, there exist an infinite set U and a surjective function $h : U^n \to \{\text{ultrafilters of } \mathfrak{A}_\Gamma\}$ such that

(h1) for all $i < n$, $\mathbf{a}, \mathbf{b} \in U^n$, if $a_j = b_j$ for all $j < n$, $j \neq i$, then $h(\mathbf{a})E_i h(\mathbf{b})$,

(h2) for all $i, j < n$, $\mathbf{a} \in U^n$, $a_i = a_j$ iff $D_{ij} \in h(\mathbf{a})$.

(We will not use the 'backward' condition w.r.t. E_i.) So if $\mathbf{a} \in U^n$ is such that all the a_i are different for $i < n$ then, by (h2) and Claim 4.1(i),

$$(h(\mathbf{a})(0), \ldots, h(\mathbf{a})(n-1))$$

is an n-tuple of $(\Gamma \times n)$-ultrafilters. We show that for each $i < n$, $h(\mathbf{a})(i)$ depends only on the set $\{a_0, \ldots, a_{n-1}\} - \{a_i\}$. That is, such a function h determines of what is called in [15] a *patch system.*

Claim 4.2 *Let $i, j < n$ and $\mathbf{a}, \mathbf{b} \in U^n$ be such that*

- *$a_k \neq a_\ell$ whenever $k, \ell \neq i$, $k, \ell < n$,*
- *$b_k \neq b_\ell$ whenever $k, \ell \neq j$, $k, \ell < n$, and*
- *$\{a_k : k < n, k \neq i\} = \{b_k : k < n, k \neq j\}$.*

Then $h(\mathbf{a})(i) = h(\mathbf{b})(j)$.

Proof. This claim is claimed and proved in the proof of [15, Lemma 4.12(2)]. Using ultrafilters of $\mathfrak{A}_\Gamma$ instead of ultrafilters of $\mathfrak{Cm}\,\mathfrak{F}_\Gamma$ does not make any difference. □

As a consequence we obtain:

Claim 4.3 (cf. [15, Def. 4.11, Lemma 4.12(2)])
Given h as above, define a function

$$\partial h : \{n-1\text{-element subsets of } U\} \to \{\text{ultrafilters on } \Gamma \times n\}$$

by taking, for every n-element subset A of U an n-tuple $\mathbf{a} \in U^n$ such that $A = \{a_0, \ldots, a_{n-1}\} - \{a_i\}$ for some $i < n$ and putting

$$\partial h(A) = h(\mathbf{a})(i).$$

Then ∂h is well-defined.

Take the functions h and ∂h as defined above. As $\mathfrak{A}_\Gamma$ is infinite, the domain U^n of h should also be infinite. Choose an infinite sequence $a_0, a_1, \ldots$ of distinct elements from U, and define a function

$$f : \{n-1\text{-element subsets of } \omega\} \to k$$

by taking

$$f(\{i_1, \ldots, i_{n-1}\}) = j \quad \text{iff} \quad I_j \in h(\{a_{i_1}, \ldots, a_{i_{n-1}}\}).$$

By Ramsey's theorem [23], we may assume that the value of f is constant, say, c. Let $A = \{a_0, \ldots, a_{n-1}\}$ and $\mathbf{a} = (a_0, \ldots, a_{n-1})$. Then $I_c \in \partial h(A - \{a_i\}) = h(\mathbf{a})(i)$, for each

$i < n$. So for every $i < n$ there exists some $S_i \in h(\mathbf{a})$ such that $S_i \subseteq F_i$ and $S_i(i) = I_c$. As $h(\mathbf{a})$ is an ultrafilter of $\mathfrak{A}_\Gamma$ and $\bigcap_{i<n} S_i \in h(\mathbf{a})$, we have that $\bigcap_{i<n} S_i \neq \emptyset$. Take any $(K, \sim) \in \bigcap_{i<n} S_i$. Then on the one hand, $K(i)$ is defined for all $i < n$, so the set $\{K(0), \ldots, K(n-1)\}$ is not independent. (This argument is written in the proof of [15, Lemma 4.10].) On the other hand, as $S_i(i) = I_c$, we have $\{K(0), \ldots, K(n-1)\} \subseteq I_c$, so it is independent, a contradiction, completing the proof of Prop. 3.7.

Acknowledgements. I am grateful to Ian Hodkinson for discussions and for his many comments on draft versions. Thanks are also due to Rob Goldblatt, András Simon and Misha Zakharyaschev for discussions, and to the anonymous referees for their suggestions.

References

[1] Benthem, J., *Canonical modal logics and ultrafilter extensions*, Journal of Symbolic Logic **44** (1979), pp. 1–8.

[2] Benthem, J., *Some kinds of modal completeness*, Studia Logica **39** (1980), pp. 125–141.

[3] Blackburn, P., M. de Rijke and Y. Venema, "Modal Logic," Cambridge University Press, 2001.

[4] Chagrov, A. and M. Zakharyaschev, "Modal Logic," Oxford Logic Guides **35**, Clarendon Press, Oxford, 1997.

[5] Diestel, R., "Graph Theory," Graduate Texts in Mathematics **173**, Springer-Verlag, 1997.

[6] Erdős, P., *Graph theory and probability*, Canadian Journal of Mathematics **11** (1959), pp. 34–38.

[7] Fine, K., *Some connections between elementary and modal logic*, in: S. Kanger, editor, *Proceedings of the Third Scandinavian Logic Symposium*, North-Holland, Amsterdam, 1975 pp. 15–31.

[8] Gabbay, D., A. Kurucz, F. Wolter and M. Zakharyaschev, "Many-Dimensional Modal Logics: Theory and Applications," Studies in Logic and the Foundations of Mathematics **148**, Elsevier, 2003.

[9] Gabbay, D. and V. Shehtman, *Products of modal logics. Part I*, Journal of the IGPL **6** (1998), pp. 73–146.

[10] Goldblatt, R., *Varieties of complex algebras*, Annals of Pure and Applied Logic **38** (1989), pp. 173–241.

[11] Goldblatt, R., *Elementary generation and canonicity for varieties of Boolean algebras with operators*, Algebra Universalis **34** (1995), pp. 551–607.

[12] Halmos, P., *Algebraic logic, IV*, Transactions of the AMS **86** (1957), pp. 1–27.

[13] Henkin, L., J. Monk and A. Tarski, "Cylindric Algebras, Part I," Studies in Logic and the Foundations of Mathematics **64**, North-Holland, 1971.

[14] Henkin, L., J. Monk and A. Tarski, "Cylindric Algebras, Part II," Studies in Logic and the Foundations of Mathematics **115**, North-Holland, 1985.

[15] Hirsch, R. and I. Hodkinson, *Strongly representable atom structures of cylindric algebras*, Journal of Symbolic Logic **74** (2009), pp. 811–828.

[16] Hirsch, R., I. Hodkinson and A. Kurucz, *On modal logics between* $\mathbf{K} \times \mathbf{K} \times \mathbf{K}$ *and* $\mathbf{S5} \times \mathbf{S5} \times \mathbf{S5}$, Journal of Symbolic Logic **67** (2002), pp. 221–234.

[17] Hodkinson, I. and Y. Venema, *Canonical varieties with no canonical axiomatisation*, Trans. Amer. Math. Soc. **357** (2005), pp. 4579–4605.

[18] Johnson, J., *Nonfinitizability of classes of representable polyadic algebras*, Journal of Symbolic Logic **34** (1969), pp. 344–352.

[19] Jónsson, B. and A. Tarski, *Boolean algebras with operators. I*, American Journal of Mathematics **73** (1951), pp. 891–939.

[20] Kurucz, A., *On axiomatising products of Kripke frames*, Journal of Symbolic Logic **65** (2000), pp. 923–945.

[21] Kurucz, A., *Combining modal logics*, in: P. Blackburn, J. van Benthem and F. Wolter, editors, *Handbook of Modal Logic*, Studies in Logic and Practical Reasoning **3**, Elsevier, 2007 pp. 869–924.

[22] Kurucz, A., *On axiomatising products of Kripke frames, part II*, in: C. Areces and R. Goldblatt, editors, *Advances in Modal Logic, Volume 7*, College Publications, 2008 pp. 219–230.

[23] Ramsey, F., *On a problem of formal logic*, Proceedings of the London Mathematical Society **30** (1930), pp. 264–286.

[24] Segerberg, K., *Two-dimensional modal logic*, Journal of Philosophical Logic **2** (1973), pp. 77–96.

Islands of Tractability for Relational Constraints: Towards Dichotomy Results for the Description Logic $\mathcal{EL}$

Agi Kurucz

Department of Computer Science
King's College London, UK
agi.kurucz@kcl.ac.uk

Frank Wolter

Department of Computer Science
University of Liverpool, UK
frank@csc.liv.ac.uk

Michael Zakharyaschev

Department of Computer Science and Information Systems
Birkbeck College London, UK
michael@dcs.bbk.ac.uk

Abstract

$\mathcal{EL}$ is a tractable description logic serving as the logical underpinning of large-scale ontologies. We launch a systematic investigation of the boundary between tractable and intractable reasoning in $\mathcal{EL}$ under relational constraints. For example, we show that there are (modulo equivalence) exactly 3 universal constraints on a transitive and reflexive relation under which reasoning is tractable: being a singleton set, an equivalence relation, or the empty constraint. We prove a number of results of this type and discuss a spectrum of open problems including generalisations to the algebraic semantics for $\mathcal{EL}$ (semi-lattices with monotone operators).

Keywords: Description logic, tractability, frame condition.

1 Introduction

Standard modal logics are usually based on propositional logic and therefore cannot be tractable: unless P = NP, no algorithm is capable of checking validity (or satisfiability) for such a logic in polynomial time. In most cases, the computational complexity is even higher: with the notable exception of **S5**, basic modal logics like **K**, **K4**, **S4**, the Gödel–Löb logic **GL** and the Grzegorczyk logic **Grz**, as well as their polymodal variants, are all PSPACE-complete as far as the 'local' reasoning problem 'if φ is true in a world, then ψ is true in that world' is concerned. The 'global' reasoning problem 'if φ is true in all worlds, then ψ is true all worlds' is EXPTIME-complete for all polymodal fusions of these logics and even unimodal **K** [7].

Very few attempts have been made to understand the complexity of *sub-Boolean* modal logics, which do not have all propositional connectives or use them in a restricted way. For example, Hemaspaandra [10] considered satisfiability of the 'poor man's formulas,' built from literals, $\wedge$, $\Box$ and $\Diamond$, over various classes of frames. A complete classification of the complexity of modal satisfiability for finite sets of propositional connectives (without any constraints on frames) was obtained in [4]. More recently, the computational complexity of sub-Boolean hybrid logics has been considered in [13].

In description logic (DL), the situation is quite different.[1] Until the mid-1990s, sub-Boolean DLs were the rule rather than exception, and mapping out the border between DLs with tractable and non-tractable reasoning problems was one of the main research goals [5]. This changed drastically in the second half of the 1990s when the focus was shifted to DLs with all Booleans (the so-called *expressive DLs*) due to the development of highly optimised tableau decision procedures and reasoning systems exhibiting satisfactory performance on real-world ontologies given in expressive DLs [11]. As a consequence, the DL-based web ontology language OWL,[2] which became a W3C standard in 2003, was based solely on expressive DLs with (at least) EXPTIME-hard TBox reasoning. Since then, however, two developments have led to a massive resurgence of interest in sub-Boolean and tractable DLs.

First, very large ontologies like SNOMED CT[3] (with $\geq 300,000$ axioms) have been designed and used in every day practice. These ontologies represent application domains at such a high level of abstraction that the full power of propositional connectives is not required. On the other hand, the enormous size of the ontologies makes tractability of reasoning a crucial factor. Second, realising the idea of employing ontologies for data access requires query answering to be tractable, at least in the size of the typically very large data sets. The two main families of tractable DLs currently evolving are $\mathcal{EL}$ and *DL-Lite*. $\mathcal{EL}$ is tailored towards representing large ontologies; it is the logical underpinning of the OWL 2 profile OWL 2 EL. *DL-Lite* is designed for ontology-based data access; it is the basis of OWL 2 QL.[4]

In this paper, we focus on the DL $\mathcal{EL}$, where concepts are constructed using intersection $\sqcap$ and existential restriction $\exists r.C$ ($\wedge$ and $\Diamond_r\varphi$, in the modal logic parlance) interpreted over relational (or Kripke) models. The fundamental *subsumption problem for general TBoxes* in $\mathcal{EL}$—whether every model of an $\mathcal{EL}$ TBox (a set of concept in-

[1] We refer to differences between research communities and their activities rather than differences between modal and description logics. The view taken in this paper is that DLs form a class of modal logics [3].

[2] `http://www.w3.org/TR/owl-overview/`

[3] `http://www.nlm.nih.gov/research/umls/Snomed/snomed_main.html`

[4] `http://www.w3.org/TR/owl2-profiles/`

clusions $C \sqsubseteq D$) satisfies a given concept inclusion $C' \sqsubseteq D'$—is decidable in polynomial time. In modal logic, this inference corresponds to the *global consequence* relation 'if a set of implications $\varphi \to \psi$ between $\mathcal{EL}$-formulas is true in every world of a Kripke model, then an implication $\varphi' \to \psi'$ is true in every world of the model.' In algebraic terms, this problem is equivalent to the validity problem for *quasi-identities* in the variety of semi-lattices with monotone operators [15].

In DL applications, the intended models are rarely arbitrary; more often they have to satisfy certain constraints. Of particular importance are constraints imposed on the interpretation of relations. For example, the Gene Ontology GO[5] is an $\mathcal{EL}$ ontology with one transitive relation. SNOMED CT is an $\mathcal{EL}$-ontology interpreted over models where certain relations are included in each other (e.g., causative_agent is a subrelation of associated_with). Other standard OWL constraints (also familiar from modal logic) include (ir)reflexivity, (a)symmetry and functionality. The complexity of reasoning in $\mathcal{EL}$ under some of such concrete relational constraints is well understood [1,2,15]. For example, the subsumption problem for general TBoxes in $\mathcal{EL}$ is tractable for any finite set of constraints of the form

$$r_1(x_1, x_2) \land \cdots \land r_n(x_n, x_{n+1}) \to r_{n+1}(x_1, x_{n+1}) \tag{1}$$

(the order of the variables is essential). On the other hand, subsumption becomes ExpTime-complete in the presence of symmetry or functionality constraints [2].

Nevertheless, from a theoretical point of view, the selection of constraints on $\mathcal{EL}$ models investigated so far is rather *ad hoc* and narrow. In fact, no attempt has been made to *classify* constraints according to tractability of $\mathcal{EL}$-reasoning. The aim of this paper is to start filling in this gap by mapping out the border between tractability and intractability of TBox reasoning in $\mathcal{EL}$ under *arbitrary relational constraints.*

Our initial findings indicate that informative dichotomy results can indeed be obtained. We establish transparent P/coNP dichotomies for finite classes of finite relational structures, classes of quasi-orders with universal first-order definitions, and classes of Noetherian partial orders closed under substructures. Not every relational constraint is 'visible' to $\mathcal{EL}$: for example, as in modal logic, TBox reasoning over irreflexive relations coincides with TBox reasoning over arbitrary relations. To obtain basic insights into relational constraints 'visible' to $\mathcal{EL}$, we show that, for universal classes of relational constraints, there is no difference between modal definability and definability in $\mathcal{EL}$. On the other hand, a typical condition definable in modal logic but not in $\mathcal{EL}$ is the Church-Rosser property.

2 Description logic $\mathcal{EL}$

Fix two disjoint countably infinite sets NC of *concept names* and NR of *role names.* We use arbitrary concept names in NC for constructing complex concepts, but often restrict the set of available role names to some subset R of NR. Thus, for $R \subseteq$ NR, the

[5] `http://www.geneontology.org/`

$\mathcal{EL}$*-concepts* C *over* R are defined inductively as follows:

$$C \quad ::= \quad \top \quad | \quad \bot \quad | \quad A \quad | \quad C_1 \sqcap C_2 \quad | \quad \exists r.C,$$

where $A \in \mathsf{NC}$, $r \in R$ and C, C_1, C_2 range over $\mathcal{EL}$-concepts over R. An R-*TBox* is a finite set of *concept inclusions* (CIs) $C \sqsubseteq D$, where C and D are $\mathcal{EL}$-concepts over R. An R-*interpretation* is a structure of the form $\mathcal{I} = (\Delta^{\mathcal{I}}, \cdot^{\mathcal{I}})$, where $\Delta^{\mathcal{I}} \neq \emptyset$ is the *domain* of interpretation and $\cdot^{\mathcal{I}}$ is an *interpretation function* assigning to each concept name $A \in \mathsf{NC}$ a set $A^{\mathcal{I}} \subseteq \Delta^{\mathcal{I}}$ and to each role name $r \in R$ a binary relation $r^{\mathcal{I}} \subseteq \Delta^{\mathcal{I}} \times \Delta^{\mathcal{I}}$. Complex concepts over R are interpreted in $\mathcal{I}$ as follows:

$$\top^{\mathcal{I}} = \Delta^{\mathcal{I}}, \qquad \bot^{\mathcal{I}} = \emptyset,$$
$$(C_1 \sqcap C_2)^{\mathcal{I}} = C_1^{\mathcal{I}} \cap C_2^{\mathcal{I}}, \qquad (\exists r.C)^{\mathcal{I}} = \{x \in \Delta^{\mathcal{I}} \mid \exists y \in C^{\mathcal{I}}\ (x,y) \in r^{\mathcal{I}}\}.$$

If $C^{\mathcal{I}} \subseteq D^{\mathcal{I}}$, we say that $\mathcal{I}$ *satisfies* $C \sqsubseteq D$ and write $\mathcal{I} \models C \sqsubseteq D$. $\mathcal{I}$ is a *model* of a R-TBox $\mathcal{T}$, $\mathcal{I} \models \mathcal{T}$ in symbols, if it satisfies all the CIs in $\mathcal{T}$.

We now formally define what we understand by constraints on interpretations. An R-*frame* is a structure $\mathfrak{F} = (\Delta^{\mathfrak{F}}, \cdot^{\mathfrak{F}})$ where $\Delta^{\mathfrak{F}} \neq \emptyset$ and $\cdot^{\mathfrak{F}}$ is a map associating with each $r \in R$ a relation $r^{\mathfrak{F}} \subseteq \Delta^{\mathfrak{F}} \times \Delta^{\mathfrak{F}}$. We say that an R-interpretation $\mathcal{I}$ is *based on* an R-frame $\mathfrak{F}$ if $\Delta^{\mathcal{I}} = \Delta^{\mathfrak{F}}$ and $r^{\mathcal{I}} = r^{\mathfrak{F}}$ for all $r \in R$. A class $\mathcal{K}$ of R-frames closed under isomorphic copies is called an R-*constraint*, or an R-*frame condition*. For example, a constraint for $R = \{r_1, r_2, r_3\}$ can consist of all R-frames $\mathfrak{F} = (\Delta^{\mathfrak{F}}, \cdot^{\mathfrak{F}})$ with arbitrary $r_1^{\mathfrak{F}}$, transitive $r_2^{\mathfrak{F}}$ and functional $r_3^{\mathfrak{F}}$. We say that an interpretation $\mathcal{I}$ *satisfies* an R-constraint $\mathcal{K}$ if $\mathcal{I}$ is based on some $\mathfrak{F} \in \mathcal{K}$.

A pair $(\mathcal{T}, C \sqsubseteq D)$ with an R-TBox $\mathcal{T}$ and an R-CI $C \sqsubseteq D$ will be called an R-*entailment query* in $\mathcal{EL}$. Given an R-constraint $\mathcal{K}$, we say that $C \sqsubseteq D$ *follows from* $\mathcal{T}$ *with respect to* $\mathcal{K}$ and write

$$\mathcal{T} \models_{\mathcal{K}} C \sqsubseteq D$$

if $\mathcal{I} \models C \sqsubseteq D$ for every model $\mathcal{I}$ of $\mathcal{T}$ based on an R-frame in $\mathcal{K}$. For singleton $\mathcal{K} = \{\mathfrak{F}\}$, we sometimes write $\mathcal{T} \models_{\mathfrak{F}} C \sqsubseteq D$. The *TBox theory* $\mathrm{Th}_{\mathcal{T}}\mathcal{K}$ *of* $\mathcal{K}$ is the set of all R-entailment queries $(\mathcal{T}, C \sqsubseteq D)$ for which $\mathcal{T} \models_{\mathcal{K}} C \sqsubseteq D$. The reasoning problem we consider in this paper, known in description logic as the *subsumption problem for* $\mathcal{K}$, is the decision problem for $\mathrm{Th}_{\mathcal{T}}\mathcal{K}$: given an R-entailment query $(\mathcal{T}, C \sqsubseteq D)$, decide whether $\mathcal{T} \models_{\mathcal{K}} C \sqsubseteq D$.

Example 2.1 In the extension $\mathcal{EL}^+$ of $\mathcal{EL}$ [1], along with a TBox one can also define an *RBox* containing inclusions of the form $r_1 \circ \cdots \circ r_n \sqsubseteq r_{n+1}$, where $r_1, \ldots, r_{n+1}$ are role names. In this case we write $(\mathcal{T}, \mathcal{R}) \models C \sqsubseteq D$ if $\mathcal{I} \models C \sqsubseteq D$ holds whenever $\mathcal{I} \models \mathcal{T}$ and $\mathcal{I}$ satisfies constraint (1) for every $r_1 \circ \cdots \circ r_n \sqsubseteq r_{n+1} \in \mathcal{R}$. Reasoning with RBoxes $\mathcal{R}$ as defined above is clearly captured by the frame condition $\mathcal{K}_{\mathcal{R}}$ containing all NR-frames $\mathfrak{F}$ in which constraint (1) is valid for all $r_1 \circ \cdots \circ r_n \sqsubseteq r_{n+1}$ in $\mathcal{R}$. According to [1,15], the subsumption problem for any such $\mathcal{K}_{\mathcal{R}}$ is decidable in polynomial time.

Example 2.2 It follows from Example 2.1 that the subsumption problem for the class of transitive frames is in P. Similarly, it is straightforward to extend existing proofs to show that the subsumption problem for the classes of reflexive or reflexive and transitive frames is also in P. On the other hand, the subsumption problem for the class of symmetric frames is ExpTime-complete [2].

3 TBox definability

To better understand the frame conditions in the context of $\mathcal{EL}$, let us take a look at frame classes that can be defined using TBoxes and compare them with modally definable frame classes. Thus, we take a brief detour into what is known in modal logic as *correspondence theory* [17].

Call R-frame conditions $\mathcal{K}_1$ and $\mathcal{K}_2$ *TBox-equivalent* if $\mathrm{Th}_T\mathcal{K}_1 = \mathrm{Th}_T\mathcal{K}_2$. For example, the standard unravelling argument from modal logic shows that the TBox theory of the class of all frames coincides with the TBox theory of the class of all irreflexive frames. Similarly, the finite model property of the TBox theory of all frames [1] means that it coincides with the TBox theory of all finite frames.

Given a set Γ of R-entailment queries, denote by $\mathrm{Fr}\Gamma$ the class of R-frames $\mathfrak{F}$ such that $\mathcal{T} \models_{\mathfrak{F}} C \sqsubseteq D$ for all $(\mathcal{T}, C \sqsubseteq D) \in \Gamma$. An R-frame condition $\mathcal{K}$ is *TBox definable* if $\mathcal{K} = \mathrm{Fr}\Gamma$ for a suitable set Γ of R-entailment queries. For example, the class of transitive $\{r\}$-frames is defined by $\Gamma = \{(\emptyset, \exists r.\exists r.A \sqsubseteq \exists r.A)\}$. Observe that in this definition the TBox is empty. Such R-frame conditions are called *concept definable.* Density is another example of a concept definable frame condition: it is defined by $\Gamma = \{(\emptyset, \exists r.A \sqsubseteq \exists r.\exists r.A)\}$.

The class of R-frames defined by $(\emptyset, C \sqsubseteq D)$ is clearly the class of R-frames validating the modal formula $C^\sharp \to D^\sharp$, where $\cdot^\sharp$ replaces each $A \in \mathsf{NC}$ with a propositional variable and each $\exists r$ with $\Diamond_r$. As all formulas of the form $C^\sharp \to D^\sharp$ are Sahlqvist, every concept definable class is first-order definable, and its first-order definition can be computed effectively [14]. More generally, a class $\mathcal{K}$ of R-frames is *modally definable* if there is a set Γ of modal formulas such that $\mathfrak{F} \in \mathcal{K}$ iff $\mathfrak{F} \models \Gamma$. $\mathcal{K}$ is called *globally definable* if there is a set Γ of pairs (φ, ψ) of modal formulas such that $\mathfrak{F} \in \mathcal{K}$ iff $\mathfrak{F} \models \Box_u\varphi \to \Box_u\psi$, where $\Box_u$ is the universal modality [9]. One can easily show that every TBox definable class is globally definable.

Recall from modal logic that a *p-morphism* from an R-frame $\mathfrak{F}_1$ to an R-frame $\mathfrak{F}_2$ is a function $f\colon \Delta^{\mathfrak{F}_1} \to \Delta^{\mathfrak{F}_2}$ such that, for every $r \in R$, (i) $(v_1, v_2) \in r^{\mathfrak{F}_1}$ implies $(f(v_1), f(v_2)) \in r^{\mathfrak{F}_2}$ and (ii) if $(f(v_1), w) \in r^{\mathfrak{F}_2}$, then there is v_2 with $(v_1, v_2) \in r^{\mathfrak{F}_1}$ and $f(v_2) = w$. If there is a p-morphism from $\mathfrak{F}_1$ onto $\mathfrak{F}_2$, then $\mathfrak{F}_2$ is called a *p-morphic* image of $\mathfrak{F}_1$. An R-frame $\mathfrak{F}_1$ is called a *subframe* of an R-frame $\mathfrak{F}_2$ if $\Delta^{\mathfrak{F}_1} \subseteq \Delta^{\mathfrak{F}_2}$ and $r^{\mathfrak{F}_1}$ is the restriction of $r^{\mathfrak{F}_2}$ to $\Delta^{\mathfrak{F}_1}$, for every $r \in R$. A subframe $\mathfrak{F}_1$ of $\mathfrak{F}_2$ is said to be *generated* if whenever $u \in \Delta^{\mathfrak{F}_1}$ and $(u, v) \in r^{\mathfrak{F}_2}$, for some $r \in R$, then $v \in \Delta^{\mathfrak{F}_1}$. Finally, $u \in \Delta^{\mathfrak{F}}$ is a *root* of a frame $\mathfrak{F}$ if the subframe of $\mathfrak{F}$ generated by u coincides with $\mathfrak{F}$.

The following result is straightforward and left to the reader:

Lemma 3.1 *TBox definable frame conditions are closed under p-morphic images and disjoint unions.*

However, unlike modally definable frame classes, TBox definable classes are not necessarily closed under generated subframes.

Example 3.2 Let $\Gamma = (\{\top \sqsubseteq \exists r.\top\}, \top \sqsubseteq \bot)$. Then the $\{r\}$-frame condition $\mathrm{Fr}\Gamma$ contains the $\{r\}$-frame $\mathfrak{F}$, which is the disjoint union of an r-reflexive point and an r-irreflexive point, as no interpretation based on $\mathfrak{F}$ is a model of $\top \sqsubseteq \exists r.\top$. However, the subframe of $\mathfrak{F}$ generated by the r-reflexive point does not belong to $\mathrm{Fr}\Gamma$.

A *universal R-frame condition* is a class of R-frames definable by universal first-order sentences in the signature R. Equivalently, by [16], a universal frame condition is a first-order definable class of frames closed under taking (not necessarily generated) subframes. The vast majority of frame conditions considered in modal and description logics are universal: transitivity, reflexivity, symmetry, weak linearity, just to mention a few. Typical examples of non-universal (first-order) conditions are the Church-Rosser property and density.

To characterise TBox definable universal frame conditions, with every R-frame $\mathfrak{F}$ we associate the 'TBox' $\mathcal{T}_S(\mathfrak{F})$ (here we slightly abuse notation as $\mathcal{T}_S(\mathfrak{F})$ is infinite whenever $\mathfrak{F}$ or R is infinite) containing the following CIs, where the A_u, for $u \in \Delta^{\mathfrak{F}}$, are distinct concept names:

- $A_u \sqsubseteq \exists r.A_v$, for $(u, v) \in r^{\mathfrak{F}}$, $r \in R$;
- $A_u \sqcap A_v \sqsubseteq \bot$, for $u \neq v$;
- $A_u \sqcap \exists r.A_v \sqsubseteq \bot$, for $(u, v) \notin r^{\mathfrak{F}}$, $r \in R$.

The meaning of $\mathcal{T}_S(\mathfrak{F})$ is explained by the following lemma (the standard proof of which is left to the reader):

Lemma 3.3 *Let $\mathfrak{F}$ be an R-frame with root w. Then, for every R-frame $\mathfrak{G}$, we have $\mathcal{T}_S(\mathfrak{F}) \not\models_{\mathfrak{G}} A_w \sqsubseteq \bot$ iff $\mathfrak{F}$ is a p-morphic image of a subframe of $\mathfrak{G}$.*

Using this lemma we obtain a characterisation of TBox definable universal frame conditions:

Theorem 3.4 *Let $\mathcal{K}$ be a universal class of R-frames, for some $R \subseteq \mathsf{NR}$. Then the following conditions are equivalent:*

(1) *$\mathcal{K}$ is TBox definable;*

(2) *$\mathcal{K}$ is closed under p-morphic images and disjoint unions;*

(3) *$\mathcal{K}$ is modally definable;*

(4) *$\mathcal{K}$ is globally definable.*

Proof. By Lemma 3.1, (1) $\Rightarrow$ (2) and, as shown in [18], (2) $\Leftrightarrow$ (3) $\Leftrightarrow$ (4). To prove that (2) $\Rightarrow$ (1) it suffices to show that $\mathrm{FrTh}_T\mathcal{K} \subseteq \mathcal{K}$. So suppose that $\mathfrak{F} \in \mathrm{FrTh}_T\mathcal{K}$. We will have $\mathfrak{F} \in \mathcal{K}$ if we can show that all rooted generated subframes of $\mathfrak{F}$ are in $\mathcal{K}$ (because $\mathfrak{F}$ is a p-morphic image of the disjoint union of these frames). So let $\mathfrak{F}_w$ be the rooted subframe of $\mathfrak{F}$ with root w. If $\mathfrak{F}_w \notin \mathcal{K}$ then, by Lemma 3.3, $\mathcal{T}_S(\mathfrak{F}_w) \models_{\mathcal{K}} A_w \sqsubseteq \bot$. By compactness—as $\mathcal{K}$ is first-order definable—there exists a finite subset $\mathcal{T}$ of $\mathcal{T}_S(\mathfrak{F}_w)$ with $\mathcal{T} \models_{\mathcal{K}} A_w \sqsubseteq \bot$. But then $(\mathcal{T}, A_w \sqsubseteq \bot) \in \mathrm{Th}_T\mathcal{K}$ and $\mathcal{T} \not\models_{\mathfrak{F}_w} A_w \sqsubseteq \bot$, which is a contradiction. □

We conjecture that the equivalence of (1) and (4) in Theorem 3.4 can be generalised to arbitrary (not necessarily first-order definable) classes of R-frames closed under subframes. Note that without the subframe condition there are modally but not TBox definable classes of frames. One example is the *Church-Rosser property*

$$\forall x, y_1, y_2 \left(r(x, y_1) \wedge r(x, y_2) \rightarrow \exists z(r(y_1, z) \wedge r(y_2, z))\right),$$

which is modally definable by $\Diamond\Box p \rightarrow \Box\Diamond p$, but not TBox definable; see Section A for details.

It is beyond the scope of this paper to develop correspondence theory any further. The main conclusion, however, is clear: as far as TBox definability is concerned, $\mathcal{EL}$ is still a very powerful language, and one has to go beyond subframe conditions to find natural classes of frames definable in modal logic but not in $\mathcal{EL}$.

4 P/coNP dichotomy for tabular frame conditions

An R-frame condition $\mathcal{K}$ is called *tabular* if there is a number $n > 0$ such that $|\Delta^{\mathfrak{F}}| \leq n$ for all $\mathfrak{F} \in \mathcal{K}$. The aim of this section is to characterise the tabular R-frame conditions $\mathcal{K}$ for which the subsumption problem is tractable, that is, there is an algorithm which, given an R-entailment query $(\mathcal{T}, C \sqsubseteq D)$, can decide whether $\mathcal{T} \models_{\mathcal{K}} C \sqsubseteq D$ in time polynomial in the size $|(\mathcal{T}, C \sqsubseteq D)|$ of $(\mathcal{T}, C \sqsubseteq D)$. Note that, for any tabular $\mathcal{K}$, $\mathrm{Th}_{\mathcal{T}}\mathcal{K}$ belongs to coNP. Our proofs of coNP-hardness in this and subsequent sections are by reduction of the following *set splitting problem*, which is known to be NP-complete [8]:

- given a family I of subsets of a finite set S, decide whether there exists a *splitting* of (S, I), that is, a partition S_1, S_2 of S such that each set $G \in I$ is split by S_1 and S_2 in the sense that it is not the case that $G \subseteq S_i$ for $i \in \{1, 2\}$.

The characterisation of tabular frame conditions we are about to prove dichotomises them into functional and non-functional. An R-frame condition $\mathcal{K}$ is called *R-functional* if, for every $\mathfrak{F} \in \mathcal{K}$, every $r \in R$ and every $w \in \Delta^{\mathfrak{F}}$, we have $|\{v \in \Delta^{\mathfrak{F}} \mid (w, v) \in r^{\mathfrak{F}}\}| \leq 1$. For R-interpretations $\mathcal{I}_1$ and $\mathcal{I}_2$ based on a functional frame $\mathfrak{F}$, we say that $\mathcal{I}_1$ is *smaller* than $\mathcal{I}_2$ and write $\mathcal{I}_1 \leq \mathcal{I}_2$ if $A^{\mathcal{I}_1} \subseteq A^{\mathcal{I}_2}$ for all $A \in \mathsf{NC}$. Clearly, $\leq$ is a partial order on the set of interpretations based on $\mathfrak{F}$. A simple proof of the following lemma is given in Section B.

Lemma 4.1 *Suppose that $\mathcal{I}$ is an interpretation based on a finite R-functional frame $\mathfrak{F}$ and $w \in \Delta^{\mathcal{I}}$. Given any R-concept C, one can decide in polynomial time in $|C|$ whether there exists an R-interpretation $\mathcal{J}$ such that $\mathcal{I} \leq \mathcal{J}$ and $w \in C^{\mathcal{J}}$. If such an interpretation exists, then there is a unique minimal (with respect to $\leq$) R-interpretation $\mathcal{I}(w, C) \geq \mathcal{I}$ with $w \in C^{\mathcal{I}(w,C)}$; moreover, this minimal interpretation can be constructed in polynomial time in $|C|$.*

We are now in a position to formulate the main result of this section.

Theorem 4.2 *Let $\mathcal{K}$ be a tabular R-frame condition for a finite $R \subseteq \mathsf{NR}$. Then either $\mathcal{K}$ is functional, in which case $\mathrm{Th}_{\mathcal{T}}\mathcal{K}$ is in* P, *or $\mathrm{Th}_{\mathcal{T}}\mathcal{K}$ is* coNP*-complete.*

Proof. Assume first that $\mathcal{K}$ is functional and that we are given an R-TBox $\mathcal{T}$ and and R-CI $C' \sqsubseteq D'$. Our polynomial time algorithm checking whether $\mathcal{T} \models_{\mathcal{K}} C' \sqsubseteq D'$ runs as follows. Let $\mathfrak{F}_1, \ldots, \mathfrak{F}_n$ be a list of all frames in $\mathcal{K}$ (up to isomorphism). For each $\mathfrak{F}_i$ and each $w \in \mathfrak{F}_i$, we do the following:

1. Let $\mathcal{I}$ be the R-interpretation based on $\mathfrak{F}_i$ with $A^{\mathcal{I}} = \emptyset$ for all $A \in \mathsf{NC}$.
2. Compute $\mathcal{I} := \mathcal{I}(w, C')$ if it exists (cf. Lemma 4.1). If it does not exist, return 'yes' and stop.
3. Apply the following rule exhaustively: for $C \sqsubseteq D \in \mathcal{T}$ and $v \in \Delta^{\mathcal{I}}$, if $v \in C^{\mathcal{I}}$ and $\mathcal{I}(v, D)$ does not exist, return 'yes' and stop; otherwise, if $\mathcal{I}(v, D) \neq \mathcal{I}$, set $\mathcal{I} = \mathcal{I}(v, D)$.
4. If $w \in (D')^{\mathcal{I}}$, return 'yes.' Otherwise, return 'no.'

It is easy to see that $\mathcal{T} \models_{\mathcal{K}} C' \sqsubseteq D'$ iff the output is 'yes' for all $\mathfrak{F}_i$ and all $w \in \Delta^{\mathfrak{F}_i}$.

Suppose now that $\mathcal{K}$ is not R-functional. Then there exists $\mathfrak{F} \in \mathcal{K}$ with $w \in \Delta^{\mathfrak{F}}$ such that $|\{v \mid (w, v) \in r^{\mathfrak{F}}\}| \geq 2$. Let m be the maximal number for which there exist $r \in R$, $\mathfrak{F} \in \mathcal{K}$ and $w \in \Delta^{\mathfrak{F}}$ with $|\{v \mid (w, v) \in r^{\mathfrak{F}}\}| = m$. Fix such r, $\mathfrak{F}$ and w.

It should be clear that the complement of $\mathrm{Th}_{\mathcal{T}}\mathcal{K}$ is decidable in nondeterministic polynomial time. We show now that $\mathrm{Th}_{\mathcal{T}}\mathcal{K}$ is coNP-hard by reduction of the set splitting problem. Suppose we are given an instance (S, I) of this problem. It will be convenient for us to assume that the members of S are concept names. Consider the $\{r\}$-TBox $\mathcal{T}$ containing the following CIs:

(a) $B_i \sqcap B_j \sqsubseteq \bot$, for $1 \leq i < j \leq m$;

(b) $A \sqcap B_i \sqsubseteq \bot$, for $3 \leq i \leq m$ and $A \in S$;

(c) $\exists r.(B_i \sqcap \bigsqcap_{A \in G} A) \sqsubseteq \bot$, for $i = 1, 2$ and $G \in I$.

The meaning of these CIs will become clear from the following:

Claim *There exists a splitting of (S, I) iff*

$$\mathcal{T} \not\models_{\mathcal{K}} \bigsqcap_{A \in S} \exists r.A \sqcap \bigsqcap_{1 \leq i \leq m} \exists r.B_i \sqsubseteq \bot.$$

Proof of claim. Suppose S_1, S_2 is a splitting of (S, I). Let $w_1, \ldots, w_m$ be the r-successors of w in $\mathfrak{F}$. Define an interpretation $\mathcal{I}$ based on $\mathfrak{F}$ by setting $B_i^{\mathcal{I}} = \{w_i\}$ and

$$A^{\mathcal{I}} = \begin{cases} \{w_1\}, & \text{if } A \in S_1; \\ \{w_2\}, & \text{if } A \in S_2. \end{cases}$$

The reader can check that $w \in (\bigsqcap_{A \in S} \exists r.A \sqcap \bigsqcap_{1 \leq i \leq m} \exists r.B_i)^{\mathcal{I}}$ and $\mathcal{I} \models \mathcal{T}$.

Conversely, suppose that there is a model $\mathcal{I}$ of $\mathcal{T}$ based on a frame $\mathfrak{F} \in \mathcal{K}$ and such that $v \in (\bigsqcap_{A \in S} \exists r.A \sqcap \bigsqcap_{1 \leq i \leq m} \exists r.B_i)^{\mathcal{I}}$. By the choice of m and (a), v has exactly m r-successors, say $w_1, \ldots, w_m$, such that $w_i \in B_i^{\mathcal{I}}$. Now let

$$S_1 = \{A \in S \mid w_1 \in A^{\mathcal{I}}\}, \qquad S_2 = \{A \in S \setminus S_1 \mid w_2 \in A^{\mathcal{I}}\}.$$

By (b) and $v \in (\exists r.A)^{\mathcal{I}}$, $A^{\mathcal{I}} \cap \{w_1, w_2\} \neq \emptyset$ for any $A \in S$, and so S_1, S_2 is a partition of S. We show that S_1, S_2 is a splitting of (S, I). Indeed, let $G \in I$. By (c), there are $A_1, A_2 \in G$ such that $w_1 \notin A_1^{\mathcal{I}}$, $w_2 \in A_1^{\mathcal{I}}$ and $w_2 \notin A_2^{\mathcal{I}}$, $w_1 \in A_2^{\mathcal{I}}$, i.e., $A_1 \in S_2$ and $A_2 \in S_1$.

As the set splitting problem is NP-complete, $\mathrm{Th}_T\mathcal{K}$ is coNP-hard. □

Note that this proof of coNP-hardness goes through for many other constraints:

Theorem 4.3 *Let $\mathcal{K}$ be an R-frame condition such that there are $r \in R$ and $n \geq 2$ for which (i) no point in frames from $\mathcal{K}$ has $> n$ r-successors, and (ii) at least one point in a frame from $\mathcal{K}$ has ≥ 2 r-successors. Then $\mathrm{Th}_T\mathcal{K}$ is coNP-hard.*

5 P/coNP-hardness dichotomy for quasi-order constraints

In this section we start analysing the border between tractability and intractability of subsumption for important classes of *quasi-orders*, i.e., reflexive and transitive frames. Throughout, we assume that $R = \{r\}$ and omit R from our terminology. A *cluster* in a quasi-order $\mathfrak{F}$ is a set of the form $\{v \mid (u, v), (v, u) \in r^{\mathfrak{F}}\}$, for some $u \in \Delta^{\mathfrak{F}}$. Single-point clusters are called *simple*. A *partial order* is a quasi-order in which all clusters are simple. A quasi-order is called *Noetherian* if it is a partial-order without infinite ascending chains.

The main result to be proved in this section is the following:

Theorem 5.1 *Let $\mathcal{K} \neq \emptyset$ be a class of quasi-orders closed under isomorphic copies.*

(a) *If $\mathcal{K}$ is universal, then $\mathrm{Th}_T\mathcal{K}$ is in P if one of the following holds:*

(a.1) *$\mathcal{K}$ is TBox-equivalent to the class of all quasi-orders;*

(a.2) *$\mathcal{K}$ is TBox-equivalent to the class of all equivalence relations;*

(a.3) *$\mathcal{K}$ is TBox-equivalent to the singleton class consisting of a single-point frame.*

If none of (a.1)–(a.3) *holds then $\mathrm{Th}_T\mathcal{K}$ is coNP-hard.*

(b) *If $\mathcal{K}$ is a class of Noetherian partial orders (e.g., a class of finite partial orders) closed under subframes, then $\mathrm{Th}_T\mathcal{K}$ is in P if one of the following holds:*

(b.1) *$\mathcal{K}$ is TBox-equivalent to the class of all Noetherian partial orders;*

(b.2) *$\mathcal{K}$ is TBox-equivalent to the singleton class consisting of a single-point frame.*

If neither (b.1) *nor* (b.2) *holds then $\mathrm{Th}_T\mathcal{K}$ is coNP-hard.*

Remark 5.2 Observe that there are uncountably many distinct $\mathrm{Th}_T\mathcal{K}$, where $\mathcal{K}$ is a universal class of quasi-orders, and exactly three of them are in P. This follows from Theorem 3.4 and the fact that there are uncountably many distinct universal modally definable classes of quasi-orders [19]. The same applies to classes of Noetherian partial orders. To show this, one can again observe that there are uncountably many modally definable classes of Noetherian quasi-orders closed under subframes [19] and prove that they are non-TBox equivalent by using their finite model property [6] and the finite TBoxes $\mathcal{T}_S(\mathfrak{F})$ for finite rooted $\mathfrak{F}$.

The remainder of this section contains the proof of Theorem 5.1. First we concentrate on statement **(b)**. Call a finite rooted partial order a *finite transitive tree* if every point except the root has exactly one immediate predecessor. The proof of **(b)** consists of proving the following three claims:

Claim B1 *If neither* (b.1) *nor* (b.2) *holds for a non-empty class $\mathcal{K}$ of Noetherian partial orders, then there exists a finite transitive tree $\mathfrak{F} \notin \mathrm{FrTh}_T\mathcal{K}$ such that $|\Delta^{\mathfrak{F}}| \geq 3$ and every proper subframe of $\mathfrak{F}$ is in $\mathrm{FrTh}_T\mathcal{K}$.*

Claim B2 *If there is a finite transitive tree $\mathfrak{F} \notin \mathrm{FrTh}_T\mathcal{K}$ such that $|\Delta^{\mathfrak{F}}| \geq 3$ and every proper subframe of $\mathfrak{F}$ is in $\mathrm{FrTh}_T\mathcal{K}$, then $\mathrm{Th}_T\mathcal{K}$ is* coNP*-hard.*

Claim B3 *If either of* (b.1) *or* (b.2) *holds, then $\mathrm{Th}_T\mathcal{K}$ is in P.*

Proof of **B1**. Let $\mathcal{K}$ be a non-empty class of Noetherian partial orders such that neither (b.1) nor (b.2) holds. Since (b.1) does not hold, we have $\mathcal{T} \models_{\mathcal{K}} C \sqsubseteq D$, for some $\mathcal{T}$, C and D such that $\mathcal{T} \not\models_{\mathcal{K}'} C \sqsubseteq D$, where $\mathcal{K}'$ is the class of all Noetherian partial orders. The proof of Theorem 5.3 below shows that we can find a finite interpretation $\mathcal{I}$ based on a Noetherian partial order such that $\mathcal{I} \not\models C \sqsubseteq D$ and $\mathcal{I} \models \mathcal{T}$. (This can also be proved using the finite model property of **Grz**.) Further, by applying the standard unravelling argument to $\mathcal{I}$, we can find a finite transitive tree $\mathfrak{F}$ such that $\mathfrak{F} \notin \mathrm{FrTh}_T\mathcal{K}$ but $\mathfrak{F}' \in \mathrm{FrTh}_T\mathcal{K}$ for all proper subtrees $\mathfrak{F}'$ of $\mathfrak{F}$.

If $\mathfrak{F}$ is a single-point frame then $\mathfrak{F}$ is a p-morphic image of any quasi-order, and so we must have $\mathcal{K} = \emptyset$, which is a contradiction. Suppose next that $\mathfrak{F}$ is a two-point chain. Then $\mathfrak{F}$ is a subframe of any rooted Noetherian frame with at least two points, and so $\mathcal{K}$ is TBox-equivalent to a single-point frame, contrary to our assumption that (b.2) does not hold. It follows that $|\Delta^{\mathfrak{F}}| \geq 3$.

Proof of **B2**. We actually prove a slightly stronger claim covering all classes of quasi-orders closed under subframes. This claim will also be used in the proof of Theorem 5.1 **(a)**. The precise formulation is as follows:

Claim B2* *Let $\mathcal{K}$ be a non-empty class of quasi-orders closed under subframes. If there is a finite transitive tree $\mathfrak{F} \notin \mathrm{FrTh}_T\mathcal{K}$ such that $|\Delta^{\mathfrak{F}}| \geq 3$ and every proper subframe of $\mathfrak{F}$ is in $\mathrm{FrTh}_T\mathcal{K}$, then $\mathrm{Th}_T\mathcal{K}$ is* coNP*-hard.*

The proof of this claim is by reduction of the set splitting problem. Suppose that we are given a family I of subsets of a finite set S. As before, we assume that the elements of S are concept names. Two cases are possible.

Case 1: *$\mathfrak{F}$ contains a point w_1 with exactly one successor w_2, which is a leaf.* Denote by $\mathfrak{F}'$ the tree obtained from $\mathfrak{F}$ by removing the leaf w_2. Then $\mathfrak{F}' \in \mathrm{FrTh}_T\mathcal{K}$. Denote by w the immediate predecessor of w_1 in $\mathfrak{F}'$; it must exist because $|\Delta^{\mathfrak{F}}| \geq 3$. Denote by w_0 the root of $\mathfrak{F}'$ and consider the TBox $\mathcal{T}$ containing the following CIs:

- $\mathcal{T}_S(\mathfrak{F}')$ defined in Section 3;
- $A \sqcap \exists r.A_{w'} \sqsubseteq \exists r.A_w$, for $(w, w') \in r^{\mathfrak{F}'}$, $w' \neq w_1$, $A \in S$;

- $A_w \sqsubseteq \exists r.(A \sqcap \exists r.A_{w_1})$ for $A \in S$;
- $\exists r.(A \sqcap \exists r.A_w) \sqcap \exists r.(A_{w_1} \sqcap \exists r.A) \sqsubseteq \bot$, for $A \in S$;
- $\displaystyle\bigsqcap_{A \in G} \exists r.(A \sqcap \exists r.A_w) \sqsubseteq \bot$, for $G \in I$;
- $\displaystyle\bigsqcap_{A \in G} \exists r.(A_{w_1} \sqcap \exists r.A) \sqsubseteq \bot$, for $G \in I$.

Intuitively, we distribute the $A \in S$ over w and w_1, which represent S_1 and S_2: if $\exists r.(A \sqcap \exists r.A_w) \neq \emptyset$ we put A in S_1, and if $\exists r.(A_{w_1} \sqcap \exists r.A) \neq \emptyset$ we put A in S_2.

Claim *There exists a splitting of* (S, I) *iff* $\mathcal{T} \not\models_{\mathcal{K}} A_{w_0} \sqsubseteq \bot$.

Proof of claim. Let S_1, S_2 be a splitting of (S, I). Define an interpretation $\mathcal{I}$ based on $\mathfrak{F}'$ by taking $A_v^{\mathcal{I}} = \{v\}$ for $v \in \Delta^{\mathfrak{F}'}$, $w \in A^{\mathcal{I}}$ for $A \in S_1$, and $w_1 \in A^{\mathcal{I}}$ for $A \in S_2$. One can check that $\mathcal{I} \models \mathcal{T}$ and $\mathcal{I} \not\models A_{w_0} \sqsubseteq \bot$, from which $\mathcal{T} \not\models_{\mathcal{K}} A_{w_0} \sqsubseteq \bot$ as $\mathfrak{F}' \in \mathrm{FrTh}_T\mathcal{K}$.

Conversely, let $\mathcal{I}$ be a model of $\mathcal{T}$ based on a frame $\mathfrak{G} \in \mathcal{K}$ and let $d_0 \in A_{w_0}^{\mathcal{I}}$. Since $\mathfrak{F}'$ is a finite transitive tree, one can use Lemma 3.3 to show that there is an embedding f of $\mathfrak{F}'$ into $\mathfrak{G}$ such that $f(w_0) = d_0$, $(v, v') \in r^{\mathfrak{F}'}$ iff $(f(v), f(v')) \in r^{\mathfrak{G}}$, and $f(v) \in A_v^{\mathcal{I}}$, for all $v, v' \in \Delta^{\mathfrak{F}'}$. We claim that, for every $A \in S$, we have either $d_0 \in (\exists r.(A \sqcap \exists r.A_w))^{\mathcal{I}}$ or $d_0 \in (\exists r.(A_{w_1} \sqcap \exists r.A))^{\mathcal{I}}$. Indeed, suppose that this is not the case for some $A \in S$. Take the point $d = f(w) \in A_w$ with $(d_0, d) \in r^{\mathfrak{G}}$. By the definition of $\mathcal{T}$, we have $d \in (\exists r.(A \sqcap \exists r.A_{w_1}))^{\mathcal{I}}$, and so, in view of reflexivity of $r^{\mathfrak{G}}$ and our assumption, there must exist points d' and d'' such that $(d, d'), (d', d'') \in r^{\mathfrak{G}}$, $(d', d), (d'', d') \notin r^{\mathfrak{G}}$; $d' \in A^{\mathcal{I}}$, $d'' \in A_{w_1}$; and $d' \notin (\exists r.A_w)^{\mathcal{I}}$. As $d' \notin (\exists r.A_w)^{\mathcal{I}}$, by the definition of $\mathcal{T}$, we must have $d' \notin (\exists r.A_{w'})^{\mathcal{I}}$, for all w' with $w' \neq w_1$. Consider now the map $f' : \Delta^{\mathfrak{F}} \to \Delta^{\mathfrak{G}}$ defined by taking

$$f'(u) = \begin{cases} f(u), & \text{if } u \notin \{w_1, w_2\}; \\ d', & \text{if } u = w_1; \\ d'', & \text{if } u = w_2. \end{cases}$$

Clearly, f' is an embedding of $\mathfrak{F}$ into $\mathfrak{G}$, contrary to $\mathfrak{F} \notin \mathrm{FrTh}_T\mathcal{K}$ and $\mathcal{K}$ being closed under subframes.

Thus, we have shown that, for every $A \in S$, either (i) $d_0 \in (\exists r.(A \sqcap \exists r.A_w))^{\mathcal{I}}$ or (ii) $d_0 \in (\exists r.(A_{w_1} \sqcap \exists r.A))^{\mathcal{I}}$, but not both, as stated in the definition of $\mathcal{T}$. Define S_1 and S_2 by putting A in the former if (i) holds and in the latter if (ii) holds. The last two items in the definition of $\mathcal{T}$ guarantee that S_1, S_2 is a splitting of (S, I).

This completes the proof for Case 1. The complement of Case 1 is the following:

Case 2: *$\mathfrak{F}$ contains a point w with at least two successors, and all successors of w are leaves.* Take a proper successor w_3 of w and denote by $\mathfrak{F}'$ the frame obtained from $\mathfrak{F}$ by removing w_3. Let w_1 be one of the remaining successors of w in $\mathfrak{F}'$. Denote by $\mathfrak{F}''$ the frame obtained from $\mathfrak{F}'$ by adding a fresh successor w_2 to w_1. Clearly, both $\mathfrak{F}'$ and $\mathfrak{F}''$ are finite transitive trees; as before, we denote by w_0 the root of $\mathfrak{F}''$. Two cases are possible now.

Case 2.1: $\mathfrak{F}'' \in \mathrm{FrTh}_T\mathcal{K}$. To encode set splitting for (S, I), we need additional concept names $\bar{A}$, for $A \in S$. This time the intuition behind the encoding is as follows:

$A \in S_1$ will be encoded by $\exists r(A' \sqcap \exists r.\bar{A}')$ and $A \in S_2$ by $\exists r.(\bar{A}' \sqcap \exists r.A')$, where $A' = A_{w_1} \sqcap A$ and $\bar{A}' = A_{w_1} \sqcap \bar{A}$. Let $\mathcal{T}$ be the TBox with the following CIs:

- $\mathcal{T}_S(\mathfrak{F}'')$;
- $A_w \sqsubseteq \exists r.A'$, for $A \in S$;
- $A_w \sqsubseteq \exists r.\bar{A}'$, for $A \in S$;
- $\exists r.(A' \sqcap \exists r.\bar{A}') \sqcap \exists r.(\bar{A}' \sqcap \exists r.A') \sqsubseteq \bot$. for $A \in S$;
- $\displaystyle\bigsqcap_{A \in G} \exists r.(A' \sqcap \exists r.\bar{A}') \sqsubseteq \bot$, for $G \in I$;
- $\displaystyle\bigsqcap_{A \in G} \exists r.(\bar{A}' \sqcap \exists r.A') \sqsubseteq \bot$, for $G \in I$.

Claim *There exists a splitting of* (S, I) *iff* $\mathcal{T} \not\models_{\mathcal{K}} A_{w_0} \sqsubseteq \bot$.

Proof of claim. Suppose S_1, S_2 is a splitting of (S, I). Define an interpretation $\mathcal{I}$ based on $\mathfrak{F}''$ by taking $A_v^{\mathcal{I}} = \{v\}$ for $v \in \Delta^{\mathfrak{F}''} \setminus \{w_1, w_2\}$, $A_{w_1}^{\mathcal{I}} = \{w_1, w_2\}$, $w_1 \in A^{\mathcal{I}}$ and $w_2 \in \bar{A}^{\mathcal{I}}$ for $A \in S_1$, $w_2 \in A^{\mathcal{I}}$ and $w_1 \in \bar{A}^{\mathcal{I}}$ for $A \in S_2$. It is readily checked that $\mathcal{I} \models \mathcal{T}$ and $\mathcal{I} \not\models_{\mathcal{K}} A_{w_0} \sqsubseteq \bot$. Thus, $\mathcal{T} \not\models_{\mathcal{K}} A_{w_0} \sqsubseteq \bot$.

Conversely, let $\mathcal{I}$ be a model of $\mathcal{T}$ based on a frame $\mathfrak{G} \in \mathcal{K}$ and $d_0 \in A_{w_0}^{\mathcal{I}}$. Since $\mathfrak{F}'$ is a finite transitive tree, there is an embedding f of $\mathfrak{F}'$ into $\mathfrak{G}$ such that $f(w_0) = d_0$, $(v, v') \in r^{\mathfrak{F}'}$ iff $(f(v), f(v')) \in r^{\mathfrak{G}}$ and $f(v) \in A_v^{\mathcal{I}}$ for all $v, v' \in \Delta^{\mathfrak{F}'}$. We claim that, for every $A \in S$, either $d_0 \in (\exists r.(A' \sqcap \exists r.\bar{A}'))^{\mathcal{I}}$ or $d_0 \in (\exists r.(\bar{A}' \sqcap \exists r.A'))^{\mathcal{I}}$. Indeed, assume that this is not the case for $A \in S$. Let $d = f(w) \in A_w$ with $(d_0, d) \in r^{\mathfrak{G}}$. Then there are $r^{\mathfrak{G}}$-incomparable $d_1, d_2 \in A_{w_1}^{\mathcal{I}}$ such that $(d, d_1), (d, d_2) \in r^{\mathfrak{G}}$. Now we modify f to a map f' from $\mathfrak{F}$ into $\mathfrak{G}$ by taking $f'(w_1) = d_1$ and $f'(w_3) = d_2$, where w_3 is the point removed from $\mathfrak{F}$ in the definition of $\mathfrak{F}'$. Clearly, f' is an embedding of $\mathfrak{F}$ into $\mathfrak{G}$, contrary to $\mathfrak{F} \notin \mathrm{FrTh}_{\mathcal{T}}\mathcal{K}$ and $\mathcal{K}$ being closed under subframes.

Case 2.2: $\mathfrak{F}'' \notin \mathrm{FrTh}_{\mathcal{T}}\mathcal{K}$. As $\mathfrak{F}' \in \mathrm{FrTh}_{\mathcal{T}}\mathcal{K}$, we can deal with $\mathfrak{F}''$ in precisely the same way as in Case 1.

This completes the proof of **B2***.

Proof of **B3**. If (b.2) holds, then $\mathrm{Th}_{\mathcal{T}}\mathcal{K}$ is in P, by Theorem 4.2. The case (b.1) is proved in Theorem 5.3 below.

The proof of Theorem 5.1 **(a)** proceed via the following four claims:

Claim A1 *Let* $\mathcal{K} \neq \emptyset$ *be a universal class of quasi-orders. If none of* (a.1)–(a.3) *holds, then either*

(eq) *$\mathcal{K}$ is a class of equivalence relations such that the size of equivalence classes is bounded by some $n > 1$ and at least one equivalence relation in $\mathcal{K}$ is different from identity, or*

(tr) *there is a finite transitive tree* $\mathfrak{F} \notin \mathrm{FrTh}_{\mathcal{T}}\mathcal{K}$ *such that* $|\Delta^{\mathfrak{F}}| \geq 3$ *and every proper subframe of* $\mathfrak{F}$ *is in* $\mathrm{FrTh}_{\mathcal{T}}\mathcal{K}$.

Claim A2 *If* **(tr)** *holds, then* $\mathrm{Th}_{\mathcal{T}}\mathcal{K}$ *is* coNP*-hard by Claim* **B2***.

Claim A3 *If* **(eq)**, *then* $\mathrm{Th}_{\mathcal{T}}\mathcal{K}$ *is* coNP*-hard.*

Claim A4 *If one of* (a.1), (a.2) *or* (a.3) *holds, then* $\mathrm{Th}_{\mathcal{T}}\mathcal{K}$ *is in* P.

The proof of **A1** is similar to the proof of **B1** and is given in Section C. **A3** is an immediate consequence of Theorem 4.3. For **A4**, the case (a.3) follows from Theorem 4.3 and the case (a.1) is a straightforward modification of the polynomial time algorithm for transitive frames [1]. It thus remains to consider the case (a.2) in which $\mathcal{K}$ is TBox-equivalent to the class of all equivalence relations. This is proved in Theorem 5.3 below.

Theorem 5.3 *Let* $\mathcal{K}$ *be the class of Noetherian partial orders or the class of equivalence relations. Then* $\mathrm{Th}_{\mathcal{T}}\mathcal{K}$ *is in* P.

The proof of this theorem uses the notion of canonical interpretation, which was introduced and investigated in [1,12].

Canonical interpretation for the class of all frames. For the class $\mathcal{K}$ of *all* NR-frames, every satisfiable TBox $\mathcal{T}$ and every concept name A_0, the canonical interpretation $\mathcal{I}_{\mathcal{T},A_0}$ is an interpretation with a designated $d_{A_0} \in \Delta^{\mathcal{I}_{\mathcal{T},A_0}}$, which can be constructed in polynomial time in such a way that for all concepts D,

$$d_{A_0} \in D^{\mathcal{I}_{\mathcal{T},A_0}} \quad \text{iff} \quad \mathcal{T} \models_{\mathcal{K}} A_0 \sqsubseteq D.$$

Thus, one can check in polynomial time whether $\mathcal{T} \models_{\mathcal{K}} A_0 \sqsubseteq D$ by inspecting $\mathcal{I}_{\mathcal{T},A_0}$.

We now describe the construction of $\mathcal{I}_{\mathcal{T},A_0}$ and its properties in more detail. Without loss of generality, we assume that all TBoxes $\mathcal{T}$ in this section are *normalised* in the sense that in every $C \sqsubseteq D \in \mathcal{T}$, the concept D is either a concept name or of the form $\exists r.A$, for a concept name A, and in every subconcept $\exists r.E$ of C, E is a concept name. Moreover, when deciding whether $\mathcal{T} \models_{\mathcal{K}} C \sqsubseteq D$ we can assume that C is a concept name. An easy polynomial reduction of the general subsumption problem to this case by adding 'abbreviations' $A \equiv C$ (i.e., $A \sqsubseteq C$ and $C \sqsubseteq A$) to TBoxes can be found in [1].

Assume now that we are given a normalised TBox $\mathcal{T}$ and a concept name A_0. We consider first the case when $\bot$ does not occur in $\mathcal{T}$. Denote by $sub(\mathcal{T})$ the set of subconcepts of concepts in $\mathcal{T}$. First, define an interpretation $\mathcal{I}_0$ by taking

$$\Delta^{\mathcal{I}_0} = \{d_{A_0}\} \cup \{d_A \mid \exists r.A \in sub(\mathcal{T})\},$$

where the d_A and d_{A_0} are fresh objects. Set $d \in A^{\mathcal{I}_0}$ iff $d = d_A$, for all $d_A \in \Delta^{\mathcal{I}_0}$, and $r^{\mathcal{I}_0} = \emptyset$. Next, we apply exhaustively the following two rules to $\mathcal{I} := \mathcal{I}_0$:

- for $C \sqsubseteq A \in \mathcal{T}$ and $d \in \Delta^{\mathcal{I}_0}$, if $d \in C^{\mathcal{I}}$ and $d \notin A^{\mathcal{I}}$, then update $\mathcal{I}$ by setting $A^{\mathcal{I}} := A^{\mathcal{I}} \cup \{d\}$ and leaving the interpretation of all remaining symbols unchanged;
- for $C \sqsubseteq \exists r.A \in \mathcal{T}$ and $d \in \Delta^{\mathcal{I}_0}$, if $d \in C^{\mathcal{I}}$ and $d \notin (\exists r.A)^{\mathcal{I}}$, then update $\mathcal{I}$ by setting $r^{\mathcal{I}} := r^{\mathcal{I}} \cup \{(d, d_A)\}$ and leaving it unchanged for the remaining symbols.

The resulting interpretation is denoted by $\mathcal{I}_{\mathcal{T},A_0}$ and called the *canonical interpretation* of $\mathcal{T}$ and A_0. Clearly, it can be constructed in polynomial time. It will be convenient to employ a characterisation of $\mathcal{I}_{\mathcal{T},A_0}$ in terms of simulations. Recall that a relation

$S \subseteq \Delta^{\mathcal{I}_1} \times \Delta^{\mathcal{I}_2}$ is a *simulation* between interpretations $\mathcal{I}_1$ and $\mathcal{I}_2$ if the following conditions hold:

(i) for all concept names A and all $(e_1, e_2) \in S$, if $e_1 \in A^{\mathcal{I}_1}$ then $e_2 \in A^{\mathcal{I}_2}$;

(ii) for all role names r, all $(e_1, e_2) \in S$ and all $e_1' \in \Delta^{\mathcal{I}_1}$ with $(e_1, e_1') \in r^{\mathcal{I}_1}$, there exists $e_2' \in \Delta^{\mathcal{I}_2}$ such that $(e_2, e_2') \in r^{\mathcal{I}_2}$ and $(e_1', e_2') \in S$.

For interpretations $\mathcal{I}_1, \mathcal{I}_2$ with $d_1 \in \Delta^{\mathcal{I}_1}$, $d_2 \in \Delta^{\mathcal{I}_2}$, we write $(\mathcal{I}_1, d_1) \leq (\mathcal{I}_2, d_2)$ and say that $(\mathcal{I}_1, d_1)$ is *simulated* by $(\mathcal{I}_2, d_2)$ if there is a simulation S between $\mathcal{I}_1$ and $\mathcal{I}_2$ such that $(d_1, d_2) \in S$.

The role of simulations in $\mathcal{EL}$ is explained by the following two lemmas the proofs of which can be found in [12].

Lemma 5.4 *If $(\mathcal{I}_1, d_1) \leq (\mathcal{I}_2, d_2)$ and $d_1 \in C^{\mathcal{I}_1}$ then $d_2 \in C^{\mathcal{I}_2}$, for any C.*

Now, the canonical interpretation $\mathcal{I}_{\mathcal{T},A_0}$ can be characterised as an interpretation simulated by any other interpretation satisfying the TBox $\mathcal{T}$ and the appropriate concept names:

Lemma 5.5 *$\mathcal{I}_{\mathcal{T},A_0} \models \mathcal{T}$ and, for all interpretations $\mathcal{I}$ with $\mathcal{I} \models \mathcal{T}$, all $d_A \in \Delta^{\mathcal{I}_{\mathcal{T},A_0}}$ and $d \in A^{\mathcal{I}}$, we have $(\mathcal{I}_{\mathcal{T},A_0}, d_A) \leq (\mathcal{I}, d)$.*

It follows immediately that, as claimed above, $\mathcal{T} \models A_0 \sqsubseteq D$ iff $d_{A_0} \in D^{\mathcal{I}_{\mathcal{T},A_0}}$.

Canonical interpretation for equivalence relations. We introduce a canonical interpretation, denoted by $\mathcal{I}^{e}_{\mathcal{T},A_0}$, which characterises TBox reasoning over equivalence relations in the same way as $\mathcal{I}_{\mathcal{T},A_0}$ characterises TBox reasoning over arbitrary frames. Set

$$\mathfrak{E}_n = (\{1, \ldots, n\}, r^{\mathfrak{E}_n} = \{1, \ldots, n\} \times \{1, \ldots, n\}), \quad \mathfrak{E}_\omega = (\omega, r^{\mathfrak{E}_\omega} = \omega \times \omega).$$

Clearly, for the class $\mathcal{E}$ of all equivalence relations, we have

$$\mathcal{T} \models_{\mathcal{E}} C \sqsubseteq D \quad \text{iff} \quad \mathcal{T} \models_{\mathfrak{E}_\omega} C \sqsubseteq D \quad \text{iff} \quad \mathcal{T} \models_{\{\mathfrak{E}_i \mid i < \omega\}} C \sqsubseteq D.$$

Lemma 5.6 *Given A_0 and a normalised $\mathcal{T}$ not containing $\bot$, one can construct in polynomial time, starting from $\mathcal{I}_{\mathcal{T},A_0}$, an interpretation $\mathcal{I}^{e}_{\mathcal{T},A_0}$ based on some $\mathfrak{E}_n$ such that*

(i) *$\mathcal{I}^{e}_{\mathcal{T},A_0} \models \mathcal{T}$ and $d_{A_0} \in A_0^{\mathcal{I}^{e}_{\mathcal{T},A_0}}$, and*

(ii) *if $\mathcal{J}$ is an interpretation based on $\mathfrak{E}_\omega$ with $d \in A_0^{\mathcal{J}}$, then $(\mathcal{I}^{e}_{\mathcal{T},A_0}, d_{A_0}) \leq (\mathcal{J}, d)$.*

Proof. Given an interpretation $\mathcal{I}$ and $d \in \Delta^{\mathcal{I}}$, we define a new interpretation $\mathcal{I}^{d}_{\sim}$ which coincides with $\mathcal{I}$ except that $(e_1, e_2) \in r^{\mathcal{I}^{d}_{\sim}}$, for all e_1, e_2 reachable from d via an $r^{\mathcal{I}}$-path $d_1, \ldots, d_n$ with $d = d_1$ and $(d_i, d_{i+1}) \in r^{\mathcal{I}}$ for $i < n$. We now apply exhaustively the following rules to $\mathcal{I} = \mathcal{I}_{\mathcal{T},A_0}$:

(s1) if $\mathcal{I} \neq \mathcal{I}^{d_{A_0}}_{\sim}$ then set $\mathcal{I} := \mathcal{I}^{d_{A_0}}_{\sim}$;

(s2) for $C \sqsubseteq A \in \mathcal{T}$ and $d \in \Delta^{\mathcal{I}}$, if $d \in C^{\mathcal{I}}$ and $d \notin A^{\mathcal{I}}$ then update $\mathcal{I}$ by setting $A^{\mathcal{I}} := A^{\mathcal{I}} \cup \{d\}$ and leaving the interpretation of all remaining symbols unchanged;

(s3) for $C \sqsubseteq \exists r.A \in \mathcal{T}$ and $d \in \Delta^{\mathcal{I}}$, if $d \in C^{\mathcal{I}}$ and $d \notin (\exists r.A)^{\mathcal{I}}$ then update $\mathcal{I}$ by setting $r^{\mathcal{I}} := r^{\mathcal{I}} \cup \{(d, d_A)\}$ and leaving it unchanged for the remaining symbols.

Denote by $\mathcal{I}^e_{\mathcal{T},A_0}$ the restriction of the resulting interpretation to the subframe generated by d_{A_0}. Clearly, it can be constructed in polynomial time. One can show that $\mathcal{I}^e_{\mathcal{T},A_0}$ is as required (for details see Section D). □

Using Lemma 5.6, we can decide whether $\mathcal{T} \models_{\mathfrak{C}_\omega} A_0 \sqsubseteq D$ by checking, in polynomial time, whether $d_{A_0} \in D^{\mathcal{I}^e_{\mathcal{T},A_0}}$. If $\mathcal{T}$ contains $\bot$, we replace every occurrence of $\bot$ in $\mathcal{T}$ by the concept name $A_\bot$ and denote the resulting TBox by $\mathcal{T}^\bot$. By Lemma 5.6, the following conditions are equivalent:

- $\mathcal{T}^\bot \models_{\mathfrak{C}_\omega} A_0 \sqsubseteq \exists r^n.A_\bot$ for some n;
- $A_\bot^{\mathcal{I}^e_{\mathcal{T}^\bot,A_0}} \neq \emptyset$;
- $\mathcal{T} \models A_0 \sqsubseteq \bot$.

Thus, $\mathcal{T} \models A_0 \sqsubseteq D$ iff $\mathcal{T}^\bot \models A_0 \sqsubseteq D$ or $A_\bot^{\mathcal{I}^e_{\mathcal{T}^\bot,A_0}} \neq \emptyset$, and both conditions can be checked in polynomial time.

Canonical interpretation for Noetherian partial orders. Finally, we define a canonical interpretation $\mathcal{I}^N_{\mathcal{T},A_0}$ which characterises TBox reasoning over Noetherian partial orders.

Lemma 5.7 *Given A_0 and $\mathcal{T}$ not containing $\bot$, one can construct in polynomial time, starting from $\mathcal{I}_{\mathcal{T},A_0}$, an interpretation $\mathcal{I}^N_{\mathcal{T},A_0}$ based on a finite partial order with root $d^*_{A_0}$ and such that*

(i) $\mathcal{I}^N_{\mathcal{T},A_0} \models \mathcal{T}$ *and* $d^*_{A_0} \in A_0^{\mathcal{I}^N_{\mathcal{T},A_0}}$, *and*

(ii) *if $\mathcal{J}$ is based on a partial order and $d \in A_0^{\mathcal{J}}$, then $(\mathcal{I}^N_{\mathcal{T},A_0}, d^*_{A_0}) \leq (\mathcal{J}, d)$.*

Proof. Let $\mathcal{I}^+_{\mathcal{T},A_0}$ be the interpretation obtained from $\mathcal{I}_{\mathcal{T},A_0}$ by adding a copy $d^*_{A_0}$ of d_{A_0} to its domain. More precisely, we set $(d^*_{A_0}, d) \in r^{\mathcal{I}^+_{\mathcal{T},A_0}}$ whenever $(d_{A_0}, d) \in r^{\mathcal{I}_{\mathcal{T},A_0}}$ or $d = d^*_{A_0}$ (note that $d^*_{A_0}$ has no proper predecessors). We set $d_{\{A\}} = d_A$, for all $d_A \in \Delta^{\mathcal{I}_{\mathcal{T},A_0}}$, and define two operators on interpretations $\mathcal{I}$ whose domains consist of points d_X, where X is a nonempty set of concept names, and the point $d^*_{A_0}$.

First, define $\mathcal{I}^*$ by replacing $r^{\mathcal{I}}$ with its transitive and reflexive closure $r^{\mathcal{I}^*}$. Second, if $r^{\mathcal{I}}$ is transitive and reflexive and $d \in \Delta^{\mathcal{I}}$, then define $\mathcal{I}_d$ by removing the cluster $[d] = \{d' \in \Delta^{\mathcal{I}} \mid (d, d'), (d', d) \in r^{\mathcal{I}}\}$ generated by d from $\mathcal{I}$, replacing it with a single point d_X, where $X = \bigcup_{d_Y \in [d]} Y$, and setting $d_X \in A^{\mathcal{I}_d}$ iff $d' \in A^{\mathcal{I}}$, for some $d' \in [d]$. (This operation has no effect for singleton $[d]$.) Now, we apply exhaustively the following rules to $\mathcal{I} = \mathcal{I}_{\mathcal{T},A_0}$:

(r1) if $r^{\mathcal{I}}$ is transitive and reflexive and $\mathcal{I} \neq \mathcal{I}_d$, for some $d \in \Delta^{\mathcal{I}}$, then set $\mathcal{I} := \mathcal{I}_d$;

(r2) if $\mathcal{I} \neq \mathcal{I}^*$ then set $\mathcal{I} := \mathcal{I}^*$;

(r3) for $C \sqsubseteq A \in \mathcal{T}$ and $d \in \Delta^{\mathcal{I}}$, if $d \in C^{\mathcal{I}}$ and $d \notin A^{\mathcal{I}}$ then update $\mathcal{I}$ by setting $A^{\mathcal{I}} := A^{\mathcal{I}} \cup \{d\}$ and leaving the interpretation of all remaining symbols unchanged;

(r4) for $C \sqsubseteq \exists r.A \in \mathcal{T}$ and $d \in \Delta^{\mathcal{I}}$, if $d \in C^{\mathcal{I}}$ and $d \notin (\exists r.A)^{\mathcal{I}}$ then update $\mathcal{I}$ by setting

$r^{\mathcal{I}} := r^{\mathcal{I}} \cup \{(d, d_X)\}$ for the (unique) X with $A \in X$ and leaving the interpretation of all remaining symbols unchanged.

Denote by $\mathcal{I}^N_{\mathcal{T},A_0}$ the restriction of the resulting interpretation to the subframe generated by $d^*_{A_0}$. One can show that $\mathcal{I}^N_{\mathcal{T},A_0}$ is as required; see Section D. □

We can now apply Lemma 5.7—in the same way as Lemma 5.6—to obtain a polynomial time decision procedure for $\mathrm{Th}_{\mathcal{T}}\mathcal{K}$, $\mathcal{K}$ the class of Noetherian partial orders, and TBoxes with and without $\bot$.

6 Future directions

Our primary aim in this paper was to start investigating—from a purely theoretical standpoint—the difference between tractable and intractable relational constraints in the context of the sub-Boolean DL $\mathcal{EL}$ (a finer classification of the intractable constraints could also be very interesting). As a next step, one can consider classes of transitive frames or general frame conditions closed under subframes. We note, however, that even for classes of *irreflexive* transitive frames without infinite ascending chains (aka Noetherian transitive frames) closed under subframes, the dichotomy appears to be much more involved than for Noetherian partial orders. For example, using the technique developed above one can show that $\mathrm{Th}_{\mathcal{T}}\mathcal{K}$ is in P not only for the class $\mathcal{K}$ of all such frames (the $\mathcal{EL}$ analogue of **GL**) but also for the class of irreflexive transitive frames of depth $\leq n$, for any $n < \omega$. We conjecture that there are other 'polynomial classes' of Noetherian transitive frames.

Although DLs come equipped with the intended semantics, generalisations to the algebraic setting would also be of interest. In Section 3, we gave first 'correspondence' results for $\mathcal{EL}$, aiming to demonstrate the type of relational constraints 'visible' to $\mathcal{EL}$. It turned out that essentially all 'standard' modal conditions were TBox definable. Here are two more illustrative examples:

- $\Gamma_0 = \{(\emptyset, \exists r.\exists r.A \sqsubseteq \exists r.A), (\mathcal{T}_S(\circ), A_w \sqsubseteq \bot)\}$ defines the class of Noetherian transitive frames ($\circ$ is a single reflexive point w);
- $\Gamma_1 = \{(\emptyset, \exists r.\exists r.A \sqsubseteq \exists r.A), (\emptyset, A \sqsubseteq \exists r.A), (\mathcal{T}_S((\circ\circ)), A_w \sqsubseteq \bot)\}$ defines the class of Noetherian partial orders ($(\circ\circ)$ is a two-point cluster containing w).

Despite the insights provided by such results, their applicability is somewhat limited. The main problem is that correspondence alone does not build a bridge between the algebraic/syntactic and the first-order views of modal logic. Ideally, correspondence results should come together with *completeness* results, like in Sahlqvist's theorem [14]. For instance, we would like to know whether the Γ_i above actually *axiomatise* (in some equational or Hilbert-style calculus) the classes of frames they define. Unfortunately, but not surprisingly, the gap between correspondence and completeness in $\mathcal{EL}$ is even wider than in classical modal logic. To be a bit more precise, we can regard $\mathcal{EL}$-concepts to be *terms* in the language of bounded semi-lattices with monotone operators (see, e.g., [15]). Then every CI $C \sqsubseteq D$ can be identified with the *identity* $C \sqcap D = C$, and every

R-entailment query $(\mathcal{T}, C' \sqsubseteq D')$ with the *quasi-identity*

$$\bigsqcap_{C \sqsubseteq D \in \mathcal{T}} C \sqcap D = C \quad \Rightarrow \quad C' \sqcap D' = C'.$$

Now, we call a set Γ of R-entailment queries *complete* (for relational models) if, for every R-entailment query $q = (\mathcal{T}, C \sqsubseteq D)$, we have $\mathcal{T} \models_{\mathrm{Fr}\Gamma} C \sqsubseteq D$ iff q is valid in all bounded semi-lattices with monotone operators validating Γ. $\Gamma = \emptyset$ was shown to be complete in [15] by reduction of TBox reasoning in $\mathcal{EL}$ to validity of quasi-identities in semi-lattices with distributive operators. It is also shown in [15] that $(\emptyset, \{\exists r_1. \cdots \exists r_n.A \sqsubseteq \exists r_{n+1}.A\})$ is complete for the R-frames defined in (1). However, numerous completeness questions (e.g., for Γ_0 and Γ_1 above) remain open.

The P/NP dichotomy problem can be extended to the algebraic setting. It is to be noted, however, that there are 'many more' quasi-varieties of semilattices with monotone operators than TBox non-equivalent relational constraints. In contrast to modal logic, this is already the case for tabular logics. Indeed, consider the 3-element set-semilattice $\{\emptyset, \{a\}, \{a,b\}\}$ with $\Diamond_r(\emptyset) = \emptyset$, $\Diamond_r(\{a\}) = \emptyset$ and $\Diamond_r(\{a,b\}) = \{a\}$ induced by the 2-element irreflexive $\{r\}$-frame $\mathfrak{F} = (\{a,b\}, r^{\mathfrak{F}} = \{(a,b)\})$. This semilattice $\mathfrak{A}$ validates $(\emptyset, \exists r.A \sqsubseteq A)$. One can readily show that (i) the TBox theory corresponding to the quasi-variety generated by $\mathfrak{A}$ is 'incomplete' for relational models as it is not TBox equivalent to any TBox theory of any class of $\{r\}$-frames, and (ii) that $\{(\emptyset, \exists r.A \sqsubseteq A)\}$ is not complete either. Thus, even simple Sahlqvist inequalities such as $\Diamond x \leq x$ become incomplete when added as axioms to the theory of bounded semilattices with monotone operators. It follows that we cannot obtain dichotomy results for (even tabular) quasi-varieties of semi-lattices with monotone operators as immediate consequences of the results presented in this paper.

References

[1] Baader, F., S. Brandt and C. Lutz, *Pushing the $\mathcal{EL}$ envelope*, in: L. Kaelbling and A. Saffiotti, editors, *IJCAI*, 2005, pp. 364–369.

[2] Baader, F., S. Brandt and C. Lutz, *Pushing the $\mathcal{EL}$ envelope further*, in: K. Clark and P. Patel-Schneider, editors, *Proc. of OWLED*, 2008.

[3] Baader, F. and C. Lutz, *Description logic*, in: P. Blackburn, J. van Benthem and F. Wolter, editors, *The Handbook of Modal Logic*, Elsevier, 2007 pp. 757–820.

[4] Bauland, M., E. Hemaspaandra, H. Schnoor and I. Schnoor, *Generalized modal satisfiability*, in: *STACS*, 2006, pp. 500–511.

[5] Donini, F., M. Lenzerini, D. Nardi and W. Nutt, *The complexity of concept languages*, in: *KR*, 1991, pp. 151–162.

[6] Fine, K., *Logics containing* **K4***, part II*, Journal of Symbolic Logic **50** (1985), pp. 619–651.

[7] Gabbay, D., A. Kurucz, F. Wolter and M. Zakharyaschev, "Many-Dimensional Modal Logics: Theory and Applications," Studies in Logic and the Foundations of Mathematics **148**, Elsevier, 2003.

[8] Garey, M. and D. Johnson, "Computers and Intractability: A Guide to the Theory of NP-Completeness," W.H. Freeman & Co., 1979.

[9] Goranko, V. and S. Passy, *Using the universal modality: gains and questions*, Journal of Logic and Computation **2** (1992), pp. 5–30.

[10] Hemaspaandra, E., *The complexity of poor man's logic*, Journal of Logic and Computation **11** (2001), pp. 609–622.

[11] Horrocks, I., *Using an expressive description logic: FaCT or fiction?*, in: *KR*, 1998, pp. 636–649.

[12] Lutz, C. and F. Wolter, *Deciding inseparability and conservative extensions in the description logic* $\mathcal{EL}$, Journal of Symbolic Computation **45** (2010), pp. 194–228.

[13] Meier, A., M. Mundhenk, T. Schneider, M. Thomas, V. Weber and H. Vollmer, *The complexity of satisfiability for fragments of hybrid logic, part I*, in: *Proc. of MFCS*, 2009, pp. 587–599.

[14] Sahlqvist, H., *Completeness and correspondence in the first and second order semantics for modal logic*, in: S. Kanger, editor, *Proc. of the 3rd Scandinavian Logic Symp.*, North-Holland, 1975 pp. 110–143.

[15] Sofronie-Stokkermans, V., *Locality and subsumption testing in* $\mathcal{EL}$ *and some of its extensions*, in: C. Areces and R. Goldblatt, editors, *Advances in Modal Logic, Vol.7*, 2008, pp. 315–339.

[16] Tarski, A., *Contributions to the theory of models I, II*, Indag. Mathematicae **16** (1954), pp. 572–588.

[17] van Benthem, J., *Correspondence theory*, in: D. Gabbay and F. Guenthner, editors, *Handbook of Philosophical Logic, Vol.2*, Reidel, Dordrecht, 1984 pp. 167–247.

[18] Wolter, F., *The structure of lattices of subframe logics*, Annals of Pure and Applied Logic **86** (1997), pp. 47–100.

[19] Zakharyaschev, M., *Canonical formulas for* **K4**. *Part II: Cofinal subframe logics*, Journal of Symbolic Logic **61** (1996), pp. 421–449.

A Church-Rosser property is not TBox definable

To show that the Church-Rosser property is not TBox definable, we prove a more general closure property. Call a subframe $\mathfrak{F}'$ of $\mathfrak{F}$ *downward closed* if whenever $v \in \Delta^{\mathfrak{F}'}$ and $(v', v) \in r^{\mathfrak{F}}$ then $v' \in \Delta^{\mathfrak{F}'}$.

Lemma A.1 *TBox definable $\{r\}$-frame conditions are closed under downward closed subframes of Noetherian partial orders.*

Proof. Suppose that $\mathfrak{F} \in \mathrm{Fr}\Gamma$ is a Noetherian partial order and $\mathfrak{F}'$ is a downward closed subframe of $\mathfrak{F}$. Assume also that $\mathcal{I}'$ is based on $\mathfrak{F}'$, $\mathcal{I}' \models \mathcal{T}$ and $\mathcal{I}' \not\models C' \sqsubseteq D'$. We have to show that there exists a model $\mathcal{I}$ based on $\mathfrak{F}$ such that $\mathcal{I} \models \mathcal{T}$ and $\mathcal{I} \not\models C' \sqsubseteq D'$. We construct $\mathcal{I}$ by extending $\mathcal{I}'$ to $\mathfrak{F}$ in the following way:

$$A^{\mathcal{I}} = A^{\mathcal{I}'} \cup (\Delta^{\mathfrak{F}} \setminus \Delta^{\mathfrak{F}'}), \quad \text{for all } A \text{ with } \mathcal{I}' \models \top \sqsubseteq \exists r.A.$$

For the remaining concept names A, we set $A^{\mathcal{I}} = A^{\mathcal{I}'}$. Using the condition that $\mathfrak{F}$ is Noetherian, one can prove by induction that, for all concepts C and all $v \in \Delta^{\mathfrak{F}} \setminus \Delta^{\mathfrak{F}'}$,

$$v \in C^{\mathcal{I}} \quad \text{iff} \quad \mathcal{I}' \models \top \sqsubseteq \exists r.C.$$

It follows that $v \in C^{\mathcal{I}'}$ iff $v \in C^{\mathcal{I}}$, for all $v \in \Delta^{\mathfrak{F}'}$. Moreover, suppose that there exists $v \in \Delta^{\mathfrak{F}} \setminus \Delta^{\mathfrak{F}'}$ such that $v \in C^{\mathcal{I}} \setminus D^{\mathcal{I}}$, for some C, D. Then $w \in C^{\mathcal{I}'}$, for all $w \in \Delta^{\mathfrak{F}'}$

without proper r-successors in $\Delta^{\mathfrak{F}'}$, and there exists such a w_0 with $w_0 \in D^{\mathcal{I}'}$. It follows that $\mathcal{I} \models \mathcal{T}$ and $\mathcal{I} \not\models C' \sqsubseteq D'$. □

The Church-Rosser property is not TBox definable because it is not closed under downward closed subframes of Noetherian partial orders.

B Proof of Lemma 4.1

Lemma B.1 *Given any R-concept C, one can decide in polynomial time in $|C|$ whether there exists an R-interpretation $\mathcal{J}$ such that $\mathcal{I} \leq \mathcal{J}$ and $w \in C^{\mathcal{J}}$. If such an interpretation does exist, then one can construct, again in polynomial time in $|C|$, the smallest (with respect to $\leq$) R-interpretation $\mathcal{I}(w, C) \geq \mathcal{I}$ such that $w \in C^{\mathcal{I}(w,C)}$.*

Proof. If $w \notin C^{\mathcal{I}}$, we 'saturate' $\mathcal{I}$ in the following way. Let $e(w)$ be the set of all conjuncts of C and $e(u) = \emptyset$ for $u \neq w$. If $\exists r.D \in e(u)$ and $(u, v) \in r^{\mathcal{I}}$, for some v, we remove $\exists r.D$ from $e(u)$ and add all the conjuncts of D to $e(v)$. If there is no such v, then the required interpretation does not exist. Otherwise, we repeat the construction. After at most $|C|$ steps, every $e(u)$ will either be empty or contain only atomic concepts. Then we define $\mathcal{I}(w, C)$ by taking $A^{\mathcal{I}(w,C)} = A^{\mathcal{I}} \cup \{u \mid A \in e(u)\}$, for every concept name A. □

C Proof of Claim A1

Claim A1 *Let $\mathcal{K} \neq \emptyset$ be a universal class of quasi-orders. If none of* (a.1)–(a.3) *holds, then either*

- **(eq)** *$\mathcal{K}$ is a class of equivalence relations such that the size of equivalence classes is bounded by some $n > 1$ and at least one equivalence relation in $\mathcal{K}$ is different from identity, or*
- **(tr)** *there is a finite transitive tree $\mathfrak{F} \notin \mathrm{FrTh}_{\mathcal{T}}\mathcal{K}$ such that $|\Delta^{\mathfrak{F}}| \geq 3$ and every proper subframe of $\mathfrak{F}$ is in $\mathrm{FrTh}_{\mathcal{T}}\mathcal{K}$.*

Proof. As (a.1) does not hold, there are $\mathcal{T}$, C and D such that $\mathcal{T} \models_{\mathcal{K}} C \sqsubseteq D$ and $\mathcal{T} \not\models_{\mathcal{K}'} C \sqsubseteq D$ for the class $\mathcal{K}'$ of all quasi-orders. Using the finite model property of **S4**, one can readily show that there exists a finite interpretation $\mathcal{I}$ based on a quasi-order such that $\mathcal{I} \models \mathcal{T}$ but $\mathcal{I} \not\models C \sqsubseteq D$. Applying the unravelling argument to $\mathcal{I}$ provides us with a finite transitive tree of clusters $\mathfrak{G}$ with $\mathfrak{G} \notin \mathrm{FrTh}_{\mathcal{T}}\mathcal{K}$. By replacing every cluster in $\mathfrak{G}$ with an infinite ascending chain, we obtain an infinite $\mathfrak{G}' \notin \mathrm{FrTh}_{\mathcal{T}}\mathcal{K}$ all rooted finite subframes of which are transitive trees. But then, using the fact that $\mathcal{K}$ is universal and employing Tarski's finite embedding property [16] (see also [6,19]), we can show that there is a finite transitive tree $\mathfrak{F}$ with $\mathfrak{F} \notin \mathrm{FrTh}_{\mathcal{T}}\mathcal{K}$. Take a minimal $\mathfrak{F}$ of this kind. Now, if $\mathfrak{F}$ contains only one point then $\mathfrak{F}$ is a p-morphic image of any quasi-order, and therefore $\mathcal{K} = \emptyset$, which is a contradiction. If $\mathfrak{F}$ is a rooted frame with two points then $\mathfrak{F}$ is a subframe of every rooted quasi-order with at least two clusters. Thus, $\mathcal{K}$ can only be a class of equivalence relations. As (a.3) does not hold, $\mathcal{K}$ cannot consist only frames with the identity relation. It follows that either $\mathcal{K}$ is a class of equivalence relations with equivalence classes of size bounded by some $n > 1$ and containing at least one equivalence relation not identical to the identity relation or $\mathrm{Th}_T\mathcal{K}$ is the TBox theory of all equivalence relations, contrary to our assumption that (a.2) does not hold. The only remaining case is $|\Delta^{\mathfrak{F}}| \geq 3$. □

D Proofs of Lemmas 5.6 and 5.7

Lemma D.1 *Given A_0 and $\mathcal{T}$ not containing $\perp$, one can construct in polynomial time, starting from $\mathcal{I}_{\mathcal{T},A_0}$, an interpretation $\mathcal{I}^e_{\mathcal{T},A_0}$ based on some $\mathfrak{E}_n$ such that*

(i) $\mathcal{I}^e_{\mathcal{T},A_0} \models \mathcal{T}$ *and* $d_{A_0} \in A_0^{\mathcal{I}^e_{\mathcal{T},A_0}}$, *and*

(ii) *if $\mathcal{J}$ is an interpretation based on $\mathfrak{E}_\omega$ with $d \in A_0^{\mathcal{J}}$, then $(\mathcal{I}^e_{\mathcal{T},A_0}, d_{A_0}) \leq (\mathcal{J}, d)$.*

Proof. Let $\mathcal{I}_{\mathcal{T},A_0} = \mathcal{I}_0, \mathcal{I}_1, \ldots$ be a sequence obtained from $\mathcal{I}_{\mathcal{T},A_0}$ by applying the rules (s1), (s2), (s3). We show by induction on $n \geq 0$ that if $\mathcal{J}$ is based on $\mathfrak{E}_\omega$, $\mathcal{J} \models \mathcal{T}$ and $A_0^{\mathcal{J}} \neq \emptyset$, then the relation

$$S = \bigcup_{d_A \in \Delta^{\mathcal{I}_n}} \{(d_A, d) \mid d \in A^{\mathcal{J}}\}$$

is a simulation between $\mathcal{I}_n$ and $\mathcal{J}$. For $\mathcal{I}_0$ this follows from Lemma 5.4 (can). Now suppose that the claim holds for $\mathcal{I}_n$. Observe that $\Delta^{\mathcal{I}_n} = \Delta^{\mathcal{I}_{n+1}}$, and so the relation S does not depend on n.

Case 1: $\mathcal{I}_{n+1} = \mathcal{I}^{d_{A_0}}_{\sim}$ for $\mathcal{I} = \mathcal{I}_n$. By IH, S is a simulation between $\mathcal{I}_n$ and $\mathcal{J}$. As the interpretation of concept names coincides for $\mathcal{I}_n$ and $\mathcal{I}_{n+1}$, it is sufficient to show that, for $(d_A, d_B) \in r^{\mathcal{I}_{n+1}}$ and $(d_A, d') \in S$, there exists $d'' \in \Delta^{\mathcal{J}}$ such that $(d_B, d'') \in S$. This follows from IH if $(d_A, d_B) \in r^{\mathcal{I}_n}$. Otherwise, d_A, d_B are both reachable from d_{A_0} in $\mathcal{I}_n$. In view of $A_0^{\mathcal{J}} \neq \emptyset$ and IH, there exists d such that $(d_{A_0}, d) \in S$. Since S is a simulation between $\mathcal{I}_n$ and $\mathcal{J}$ and d_B is reachable from d_{A_0}, there exists d'' with $(d_B, d'') \in S$, as required.

Case 2: $\mathcal{I}_{n+1}$ is obtained from $\mathcal{I}_n$ using (s2). This case follows from $\mathcal{J} \models \mathcal{T}$.

Case 3: $\mathcal{I}_{n+1}$ is obtained from $\mathcal{I}_n$ using (s3). Let $C \sqsubseteq \exists r.B \in \mathcal{T}$, $d_0 \in C^{\mathcal{I}_n}$ and $r^{\mathcal{I}_{n+1}} = r^{\mathcal{I}_n} \cup \{(d_0, d_B)\}$. By IH, it is sufficient to show that if $(d_0, d) \in S$, then there exists d' with $(d_B, d') \in S$. Suppose $(d_0, d) \in S$. Since $d_0 \in C^{\mathcal{I}_n}$ and S is a simulation between $\mathcal{I}_n$ and $\mathcal{J}$, we obtain $d \in C^{\mathcal{J}}$ (Lemma 5.4). Since $\mathcal{J} \models \mathcal{T}$, there exists $d' \in \Delta^{\mathcal{J}}$ such that $d' \in B^{\mathcal{J}}$. But then $(d_B, d') \in S$, as required. □

Lemma D.2 *Given A_0 and $\mathcal{T}$ not containing $\perp$, one can construct in polynomial time, starting from $\mathcal{I}_{\mathcal{T},A_0}$, an interpretation $\mathcal{I}^N_{\mathcal{T},A_0}$ based on a finite partial order with root $d^*_{A_0}$ such that*

(i) $\mathcal{I}^N_{\mathcal{T},A_0} \models \mathcal{T}$ *and* $d^*_{A_0} \in A_0^{\mathcal{I}^N_{\mathcal{T},A_0}}$, *and*

(ii) *if $\mathcal{J}$ is based on a partial order and $d \in A_0^{\mathcal{J}}$, then $(\mathcal{I}^N_{\mathcal{T},A_0}, d^*_{A_0}) \leq (\mathcal{J}, d)$.*

Proof. Let $\mathcal{I}_{\mathcal{T},A_0} = \mathcal{I}_0, \mathcal{I}_1, \dots$ be a sequence obtained from $\mathcal{I}^+_{\mathcal{T},A_0}$ by applying the rules (r1), (r2), (r3), (r4). For a Noetherian partial order $\mathcal{J}$ and a concept name A, we set

$$m(A)^{\mathcal{J}} = \{d \in A^{\mathcal{J}} \mid \forall d' \; [(d' \in A^{\mathcal{J}} \wedge (d, d') \in r^{\mathcal{J}}) \Rightarrow d = d']\}$$

and call the elements of $m(A)^{\mathcal{J}}$ *maximal* in $A^{\mathcal{J}}$. We show by induction on $n \geq 0$ that, for every interpretation $\mathcal{J}$ based on a Noetherian partial order and such that $\mathcal{J} \models \mathcal{T}$,

$$S_n = \{(d^*_{A_0}, d) \mid d \in A_0^{\mathcal{J}}\} \cup \bigcup_{d_X \in \Delta^{\mathcal{I}_n}} \{(d_X, d) \mid \exists A \in X \; d \in m(A)^{\mathcal{J}}\}$$

is a simulation between $\mathcal{I}_n$ and $\mathcal{J}$, and for every $d_X \in \Delta^{\mathcal{I}_n}$, $m(A)^{\mathcal{J}} = m(B)^{\mathcal{J}}$ for all $A, B \in X$. For $\mathcal{I}_0$ this is readily shown using Lemma 5.4 (can) and the fact that $\mathcal{J}$ is a Noetherian partial order.

Case 1: $\mathcal{I}_{n+1} = \mathcal{I}_d$ for $\mathcal{I} = \mathcal{I}_n$. Let $X = \bigcup_{d_Y \in [d]} Y$. We first show that $m(A)^{\mathcal{J}} = m(B)^{\mathcal{J}}$ for all $A, B \in X$. Suppose that $d \in m(A)^{\mathcal{J}}$. Let $A \in X_1$, $B \in X_2$ be such that $d_{X_1}, d_{X_2} \in [d]$. Then $(d_{X_1}, d) \in S_n$. Since S_n is a simulation and $(d_{X_1}, d_{X_2}), (d_{X_2}, d_{X_1}) \in r^{\mathcal{I}_n}$, there exist d', d'' with $(d, d'), (d', d'') \in r^{\mathcal{J}}$ and $(d_{X_2}, d'), (d_{X_1}, d'') \in S_n$. By IH, $d' \in m(B)^{\mathcal{J}}$ and $d'' \in m(A)^{\mathcal{J}}$. Then $d = d''$ and, therefore, $d = d'$ and $d \in m(B)^{\mathcal{J}}$, as required. It is now straightforward to show that S_{n+1} is a simulation between $\mathcal{I}_{n+1}$ and $\mathcal{J}$.

Case 2: $\mathcal{I}_{n+1} = \mathcal{I}^*_n$. This case is straightforward in view of transitivity of $\mathcal{J}$.

Case 3: $\mathcal{I}_{n+1}$ is obtained from $\mathcal{I}_n$ using (r3). This case follows from $\mathcal{J} \models \mathcal{T}$.

Case 4. $\mathcal{I}_{n+1}$ is obtained from $\mathcal{I}_n$ using (r4). Let $C \sqsubseteq \exists r.B \in \mathcal{T}$, $d_0 \in C^{\mathcal{I}_n}$ and $r^{\mathcal{I}_{n+1}} = r^{\mathcal{I}_n} \cup \{(d_0, d_X)\}$, where $B \in X$. By IH, it is sufficient to show that if $(d_0, d) \in S_{n+1}$, then there exists d' with $(d, d') \in r^{\mathcal{J}}$ and $(d_X, d') \in S_{n+1}$. Suppose that $(d_0, d) \in S_{n+1}$. Then $(d_0, d) \in S_n$. Since $d_0 \in C^{\mathcal{I}_n}$ and S_n is a simulation between $\mathcal{I}_n$ and $\mathcal{J}$, we obtain $d \in C^{\mathcal{J}}$ by Lemma 5.4. Since $\mathcal{J} \models \mathcal{T}$, there exists $d' \in \Delta^{\mathcal{J}}$ such that $d' \in B^{\mathcal{J}}$ and $(d, d') \in r^{\mathcal{J}}$. Since $\mathcal{J}$ is Noetherian, we may assume that $d' \in m(B)^{\mathcal{J}}$. But then $(d_X, d') \in S_{n+1}$, as required. □

Coalgebraic Lindström Theorems

Alexander Kurz

University of Leicester, UK

Yde Venema

ILLC, Amsterdam, The Netherlands

Abstract

We study modal Lindström theorems from a coalgebraic perspective. We provide three different Lindström theorems for coalgebraic logic, one of which is a direct generalisation of de Rijke's result for Kripke models. Both the other two results are based on the properties of bisimulation invariance, compactness, and a third property: ω-bisimilarity, and expressive closure at level ω, respectively. These also provide new results in the case of Kripke models. Discussing the relation between our work and a recent result by van Benthem, we give an example showing that only requiring bisimulation invariance together with compactness does not suffice to characterise basic modal logic.

Keywords: coalgebra, modal logic, Lindström theorem

1 Introduction

Lindström's theorem [19,12] states that every 'abstract logic' extending first-order logic and satisfying Löwenheim-Skolem and compactness is equivalent to first-order logic. The notion of abstract logic is a technical one capturing some fundamental properties one expects from the set $\mathcal{L}$ of sentences of any legitimate logic. Thus, roughly speaking, Lindström's theorem says that first-order logic is the strongest logic satisfying Löwenheim-Skolem and compactness.

In modal logic, de Rijke's theorem [23,7] states that every 'abstract modal logic' extending basic modal logic and having finite depth (or finite degree) is equivalent to basic modal logic. Later, van Benthem [5] showed that the finite depth condition can be replaced by compactness and relativisation. In collaboration with ten Cate and Väänänen, van Benthem expanded on this result in [6], where Lindström theorems are discussed for various fragments of first-order logic, including for instance graded modal logic. Inspired by this work, Otto and Piro [21] establish a Lindström type characterisation of the extension of basic modal logic by a global modality, and of the guarded fragment of first-order logic, with the corresponding bisimulation invariance, and the Tarski Union Property replacing the closure under relativisation.

Our approach to Lindström-type theorems for modal logic is coalgebraic. The notion of a coalgebra $X \to TX$ for a functor T encompasses, for particular instantiations of T, Kripke frames and models, but also many other structures of importance in computer science. Accordingly, we generalise de Rijke's theorem to a wide range of functors T. But our analysis also gives an elegant explanation of de Rijke's theorem and leads to new variations in the case of Kripke frames and models.

There are two main differences between our approach and that of van Benthem et al [5,6]. In the mentioned papers, abstract modal logics are *not* supposed to be bisimulation invariant, and, second, they *are* supposed to be closed under relativisation. Neither of these assumptions is very natural in the coalgebraic setting, and therefore, our starting point will be rather different from that of the cited authors. Similarly, we do not see how to translate the Tarski Union Property of Otto and Piro [21] to a coalgebraic setting.

The technique used to obtain our results is based on the final coalgebra sequence. It can be explained as follows.

$$1 \xleftarrow{p_0^1} T1 \xleftarrow{p_1^2} T^2 1 \xleftarrow{\;p_2^\alpha\;} \cdots \; T^\alpha 1 \xleftarrow{\;p_\alpha\;} \cdots \; Z \qquad (1)$$

An element of Z (the final coalgebra) is an equivalence class of pointed models up to bisimilarity. Similarly, for each ordinal α, there is a notion of α-bisimilarity and the 'approximants' $T^\alpha 1$ classify pointed models up to α-bisimilarity. The 'projections' p identify elements that cannot be distinguished at a coarser level.

¿From our perspective, an abstract coalgebraic logic $\mathcal{L}$ then will just be a collection of subsets of Z, see eg [11,16] for more on this perspective. $\mathcal{L}$ will have finite depth if all these subsets are determined at a finite stage $\alpha < \omega$. Since it is known that basic modal logic (over finitely many variables) allows us to define all subsets of all finitary approximants, de Rijke's theorem is an immediate consequence: An abstract coalgebraic logic extending basic modal logic and having a notion of finite depth must be equivalent to basic modal logic (because the formulas of such a logic, up to logical equivalence, correspond precisely to the subsets of the finitary approximants $T^n 1, n < \omega$).

In a further analysis, we investigate ways to replace the notion of finite depth by compactness. Whereas the argument in the previous paragraph was based on all subsets of all finitary approximants being definable in $\mathcal{L}$, we now look at higher approximants, such as $T^\omega 1$ (or Z itself) and consider the topologies generated by (extensions of) formulas. Then topological compactness and logical compactness coincide and we can combine topological results with properties of the final coalgebra sequence.

Summary and Structure of the paper. Section 7 presents our three Lindström theorems, stating that a logic invariant under bisimilarity, closed under Boolean operations and at least as expressive as the 'basic modal logic' $\mathcal{L}_T^{\mathrm{F}}$ is actually equivalent to $\mathcal{L}_T^{\mathrm{F}}$ if additionally one of three conditions hold:

- $\mathcal{L}$ has finite depth,
- $\mathcal{L}$ is compact and invariant under ω-bisimilarity,
- $\mathcal{L}$ is compact and expressively closed at ω.

We also give an example showing that in the second item ω-bisimilarity cannot be replaced by bisimilarity.

Sections 3-5 are devoted to these three conditions, respectively. Section 6 presents two lemmas on the final coalgebra sequence, which we believe are of independent interest.

2 Preliminaries on Coalgebras and Modal Logic

2.1 Coalgebras

Coalgebras generalise Kripke frames and models.

Definition 2.1 The category $\mathsf{Coalg}(T)$ of coalgebras for a functor T on a category $\mathcal{X}$ has as objects arrows $\xi : X \to TX$ in $\mathcal{X}$ and morphisms $f : (X, \xi) \to (X', \xi')$ are arrows $f : X \to X'$ such that $Tf \circ \xi = \xi' \circ f$.

The main examples of functors of interest to us in this paper are

Definition 2.2 A Kripke polynomial functor (KPF) $T : \mathsf{Set} \to \mathsf{Set}$ is built according to

$$T ::= Id \mid C \mid T + T \mid T \times T \mid T \circ T \mid \mathcal{P} \tag{2}$$

where *Id* is the identity functor, C is the constant functor that maps all sets to a finite set C, $+$ is disjoint union, $\times$ is cartesian product, $\circ$ is composition, and $\mathcal{P}$ maps a set to the collection of its subsets.

Remark 2.3
- In the above definition, we insist on constants C being finite and coproducts being finite as well, hence a KPF in our sense maps finite sets to finite sets.
- A coalgebra $X \to \mathcal{P}X$ is a Kripke frame and a coalgebra $X \to 2^P \times \mathcal{P}X$ for a finite set P (of atomic propositions) is a Kripke model.
- An element x_0 of a coalgebra $\xi : X \to C \times X$ specifies a stream (infinite list) $(c_1, c_2, \ldots)$ via $(c_{n+1}, x_{n+1}) = \xi(x_n)$, $n < \omega$.
- Similarly, an element x_0 of a coalgebra $X \to C_1 + C_2 \times X \times X$ specifies a possibly infinite binary tree with leaves labelled from C_1 and the other nodes labelled from C_2.
- Consider a coalgebra $X \to 2 \times X^C$ where $2 = \{0, 1\}$. It can be understood as a deterministic automaton over the alphabet C, where the elements in 2 are used to label states as accepting or non-accepting.
- The reader can easily extend this list, for example, non-deterministic automata are coalgebras $X \to 2 \times (\mathcal{P}X)^C$

Definition 2.4 We say that T has the properties (wp) and (fs) if T is a functor $\mathsf{Set} \to \mathsf{Set}$ and, respectively,

(wp) preserves weak pullbacks,

(fs) restricts to finite sets.

Example 2.5 A KPF T satisfies (wp) and (fs). Further examples of functors are obtained by extending (2) with

$$\ldots \mid \mathcal{P}_\omega \mid \mathsf{List} \mid \mathsf{Mult} \mid \mathcal{D} \mid \mathcal{H} \tag{3}$$

where $\mathcal{P}_\omega X$ is the set of finite subsets of X, $\mathsf{List}X$ the set of finite lists over X, $\mathsf{Mult}X$ the set of finite multisets over X, $\mathcal{D}X$ the set of discrete probability distributions over X, and $\mathcal{H}X = 2^{2^X}$. They all satisfy (wp) with the exception of $\mathcal{H}$ and they do not satisfy (fs) with the exception of $\mathcal{P}_\omega$ and $\mathcal{H}$. $\mathcal{H}$-coalgebras coincide with neighbourhood frames in modal logic and are investigated, from a coalgebraic point of view, in Hansen and Kupke [13].

Final Coalgebra and Bisimilarity. Given $T : \mathsf{Set} \to \mathsf{Set}$ the **final coalgebra** (Z, ζ) is determined by the property that for any coalgebra $\xi : X \to TX$ there is a unique coalgebra morphism $f : X \to Z$. Instantiating T as $\mathcal{P}$, we find that f identifies two elements of X iff they are bisimilar in the usual sense. So we may take this as a definition: Two elements of a coalgebra are **bisimilar** if they are identified by the unique morphism into the final coalgebra. This definition extends to elements in two different coalgebras (X, ξ), (X', ξ') by considering their disjoint union (or coproduct) $X + X' \xrightarrow{\xi+\xi'} TX + TX' \to T(X + X')$. **Bisimilarity** is the smallest relation between elements of (possibly different) coalgebras containing all pairs of bisimilar elements. Clearly, bisimilarity is an equivalence relation.

Remark 2.6 (Existence of the final coalgebra) The final coalgebra does always exist if we allow its carrier to be a proper class as in [2] or if we assume the existence of an appropriate inaccessible cardinal as in [4]. Both approaches are equivalent, but we find the latter point of view more convenient as we then know that the final sequence, discussed below, always converges against the final coalgebra.

Example 2.7 (i) The final coalgebra $Z \to \mathcal{P}Z$ is Aczel's universe of non-well-founded sets [1]. The elements of Z can be understood as precisely the non-well-founded sets, that is, the equivalence classes of pointed Kripke frames under bisimilarity.

(ii) The final coalgebra for $TX = C \times X$ is $\zeta : C^\omega \to C \times C^\omega$ where ζ maps a stream $(c_i)_{i<\omega}$ to the pair $(c_0, (c_i)_{1 \le i < \omega})$ consisting of the 'head' and the 'tail' of the stream.

(iii) Similarly, the final coalgebra for $TX = C_1 + C_2 \times X \times X$ consists of the set of all, possibly infinite, binary trees with leaves labelled from C_1 and the other nodes labelled from C_2.

(iv) In case of $TX = 2 \times X^C$, the final coalgebra Z is the set of all languages (ie subsets of C^*) and the unique coalgebra morphism $X \to Z$ maps a state of the automaton

X to its accepted language [24].

Final Sequence and α-Bisimilarity. The final coalgebra is approximated by the final coalgebra sequence

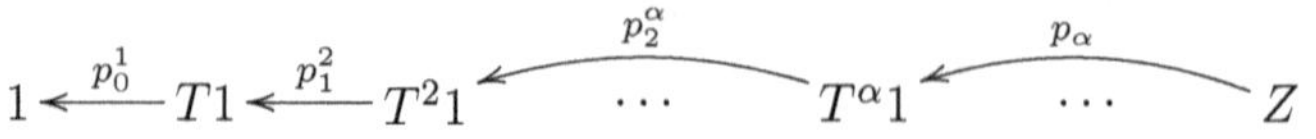

$$1 \xleftarrow{p_0^1} T1 \xleftarrow{p_1^2} T^2 1 \xleftarrow{p_2^\alpha} \cdots T^\alpha 1 \xleftarrow{p_\alpha} \cdots Z \qquad (4)$$

where 1 denotes a one-element set and

- $T^{\alpha+1}1 = T(T^\alpha 1)$ and $p_{\beta+1}^{\alpha+1} = T(p_\beta^\alpha)$ for all $\beta < \alpha$,
- $p_\gamma^\alpha = p_\gamma^\beta \circ p_\beta^\alpha$ for $\gamma < \beta < \alpha$,
- if α is a limit ordinal then $(T^\alpha 1, (p_\beta^\alpha))_{\beta < \alpha}$ is a limiting cone.

Since the final coalgebra (Z, ζ) does exist [4] we know from [3, Thm 2] that the final sequence converges, that is, there is an isomorphism $p_\beta^{\beta+1} = \zeta^{-1}$. As in (4) we write Z for $T^\beta 1$ and p_α for p_α^β.

For any coalgebra $\xi : X \to TX$ there are maps $f_\alpha : X \to T^\alpha 1$

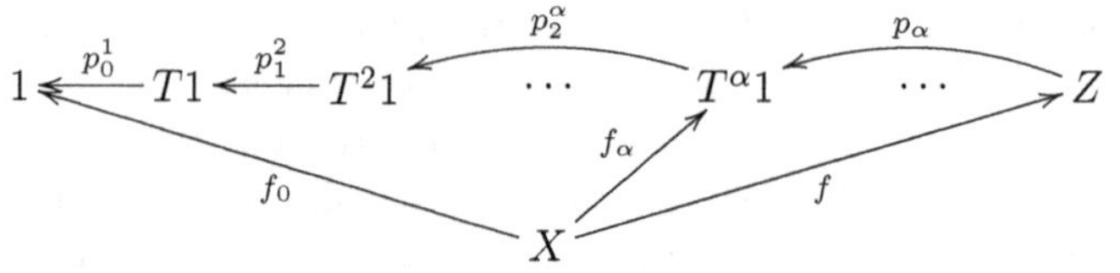

(5)

with f_α given by $T^\alpha 1$ being a limit if α is a limit ordinal and

$$f_{\alpha+1} = T f_\alpha \circ \xi \qquad (6)$$

otherwise. We say that $x, y \in X$ are **α-bisimilar** iff $f_\alpha(x) = f_\alpha(y)$. In the case of $T = \mathcal{P}$ this notion coincides with the notion of bounded bisimulation of Gerbrandy [9]. It was investigated from the point of view of coalgebraic logic in [18].

Remark 2.8 • The limit $T^\omega 1$ can be calculated explicitly as the set of sequences $\{t \in \prod_{n < \omega} T^n 1 \mid \forall m, n < \omega . p_n^m(t_m) = t_n\}$. For example, in the case of streams ($TX = C \times X$), we have that $T^n 1$ is C^n and $T^\omega 1$ is (isomorphic to) C^ω. The $p_n : C^\omega \to C^n$ are then the projections. In this case $T^\omega 1$ is also the final coalgebra and its elements classify bisimilarity.

- In general, the elements of $T^\omega 1$ classify behaviour up to ω-bisimilarity. In the case of Kripke models ($TX = 2^P \times \mathcal{P}X$), we have that $T^\omega 1$ is the canonical model of the modal logic **K**. Here, $T^\omega 1$ is not the final coalgebra (and this cannot be mended by replacing $\mathcal{P}$ by the finite powerset).
- The final sequence for $TX = \mathcal{P}X$ has been studied eg in Ghilardi [10] and Worrell [26]. Elements can be seen as α-bisimilar classes of pointed Kripke frames.

- In case of $TX = 2 \times X^C$, the inductive definition (6) of the maps $f_n : X \to T^n 1$ provides the standard partition-refinement algorithm of minimising deterministic automata: Starting with a finite X, we are sure to find some $n < \omega$ such that the image of f_n is isomorphic to the image of f_{n+1}, which then is the minimisation of X. The coalgebraic view on this algorithm allows to generalise this, for example, to π-calculus processes [8].

2.2 *Coalgebraic Logic*

The aim of this section is just to give a snapshot of how modal logics for coalgebras can be set up parametric in the functor T. We cannot give full details here and it mainly serves to substantiate the notion of abstract coalgebraic logic below.

Example 2.9

(i) We denote by $\mathcal{L}_T^{\mathrm{P}}$ the logic given by all n-ary predicate liftings, $n < \omega$, see [22,25]. An n-ary predicate lifting is a natural transformation $\lambda_X : 2^X \to 2^{TX}$. Each predicate lifting gives rise to a modal operator $[\lambda]$. It is interpreted (unary case) on a coalgebra (X, ξ) via

$$x \Vdash [\lambda]\varphi \;\Leftrightarrow\; \xi(x) \in \lambda_X(\{x \in X \mid x \Vdash \varphi\}).$$

Conversely, modal operators in the usual sense can be described by predicate liftings. For example, if $T = \mathcal{P}$, the usual modal $\Box$ arises from the predicate lifting $(\lambda_\Box)_X(S) = \{S' \in \mathcal{P}X \mid S' \subseteq S\}$ and, similarly, $(\lambda_\Diamond)_X(S) = \{S' \in \mathcal{P}X \mid S' \cap S \neq \emptyset\}$.

(ii) If T has (wp), we denote by $\mathcal{L}_T^{\mathrm{M}}$ the smallest logic closed under Boolean operations and the rule that $\nabla\psi \in \mathcal{L}_T^{\mathrm{M}}$ whenever $\psi \in T(\Phi)$ for some finite $\Phi \subseteq \mathcal{L}_T^{\mathrm{M}}$. We put $x \Vdash \nabla\psi \;\Leftrightarrow\; \exists w \in T(\Vdash) \,.\, T(\pi_1)(w) = \xi(x) \,\&\, T(\pi_2)(w) = \psi$ where $\Vdash \subseteq X \times \mathcal{L}_T^{\mathrm{M}}$ and π_1, π_2 are the two projections. This logic was introduced by Moss in [20], but see also [15] for the finitary version.

If T has (fs) and (wp) both logics are equally expressive, see [17] for explicit translations in both directions. For $T = \mathcal{P}$ both are equally expressive with basic modal logic (in the sense of $\equiv$ of Definition 2.10).

2.3 *Abstract coalgebraic logic*

We take the point of view that any coalgebraic logic should be invariant under bisimilarity. That is, we can identify (the meaning of) a logical formula with its extension on the final coalgebra.

Definition 2.10

- An abstract coalgebraic logic for a functor $T : \mathsf{Set} \to \mathsf{Set}$ is given by a class of formulas $\mathcal{L}$ and a function $[\![-]\!] : \mathcal{L} \to 2^Z$ to subsets of the carrier Z of the final coalgebra. We usually identify $\varphi \in \mathcal{L}$ with its extension $[\![\varphi]\!]$.

- We write $[\![\varphi]\!]_\alpha$ for the direct image $p_\alpha[\varphi]$ where p_α is as in (5).
- A subset $S \subseteq T^\alpha 1$ is definable in $\mathcal{L}$ iff there is $\varphi \in \mathcal{L}$ such that $[\![\varphi]\!]_\alpha = S$ and $\varphi = p_\alpha^{-1}([\![\varphi]\!]_\alpha)$.
- $\mathcal{L}_1 \leq \mathcal{L}_2$ iff for all $\varphi_1 \in \mathcal{L}_1$ there is $\varphi_2 \in \mathcal{L}_2$ such that $[\![\varphi_1]\!] = [\![\varphi_2]\!]$. $\mathcal{L}_1 \equiv \mathcal{L}_2$ means $\mathcal{L}_1 \leq \mathcal{L}_2$ and $\mathcal{L}_2 \leq \mathcal{L}_1$.
- Given a coalgebra $\xi : X \to TX$ and $\varphi \in \mathcal{L}$, we define

$$x \Vdash \varphi \;\Leftrightarrow\; f(x) \in [\![\varphi]\!] \tag{7}$$

where f is as in (5).

Remark 2.11 (Invariance under bisimilarity) It is immediate from (7) that any abstract coalgebraic logic is invariant under bisimilarity (for the notion of bisimilarity defined on page 295). Conversely, if a logic for coalgebras is invariant under bisimilarity then (7) holds. To summarise, invariance under bisimulation means precisely that formulas can be identified with their extensions on the final coalgebra (Z, ζ).

In particular, we see that the logics of Example 2.9 become abstract coalgebraic logics simply by putting $[\![\varphi]\!] = \{z \in Z \mid z \Vdash \varphi\}$. The definitions of $\Vdash$ from Example 2.9 then agree with (7).

Example 2.12 We denote by $\mathcal{L}_T^{\mathrm{F}}$ the abstract coalgebraic logic in which precisely all subsets of $T^n 1, n < \omega$, are definable, that is, $\mathcal{L}_T^{\mathrm{F}} = \bigcup_{n<\omega}\{p_n^{-1}(S) \mid S \subseteq T^n 1\}$.

For a detailed comparison of $\mathcal{L}_T^{\mathrm{M}}, \mathcal{L}_T^{\mathrm{P}}, \mathcal{L}_T^{\mathrm{F}}$ we refer to [17], here we only record the following.

Proposition 2.13 *We have*

$$\mathcal{L}_T^{\mathrm{M}} \leq \mathcal{L}_T^{\mathrm{P}} \leq \mathcal{L}_T^{\mathrm{F}},$$

and $\mathcal{L}_T^{\mathrm{P}} \equiv \mathcal{L}_T^{\mathrm{F}}$ *if* T *has (fs) and* $\mathcal{L}_T^{\mathrm{M}} \equiv \mathcal{L}_T^{\mathrm{P}}$ *if* T *has (fs) and (wp).*

Definition 2.14 $\mathcal{L}$ has the properties (Sep), (fSep), (Full), (fFull), (Bool) if $\mathcal{L}$ is an abstract coalgebraic logic and, respectively,

(Sep) any two distinct behaviours in the final coalgebra are separated by two formulas, that is, for all $z \neq z' \in Z$ there are $\varphi, \varphi' \in \mathcal{L}$ such that $\varphi \cap \varphi' = \emptyset$ and $z \in \varphi$ and $z' \in \varphi'$,

(fSep) for all $n < \omega$, any two behaviours in $T^n 1$ are separated by formulas, that is, for all $t \neq t' \in T^\omega 1$ there is $n < \omega$ and there are definable (see Definition 2.10) $S, S' \subseteq T^n 1$ such that $S \cap S' = \emptyset$ and $p_n^\omega(t) \in S$ and $p_n^\omega(t') \in S'$,

(Full) for ordinals α, any subset of $T^\alpha 1$ is definable,

(fFull) for all $n < \omega$, any subset of $T^n 1$ is definable,

(Bool) $\mathcal{L}$ is closed under Boolean operations.

(fSep) holds iff any two behaviours in $T^\omega 1$ can be separated by a formula. (fFull) does not imply that all subsets of $T^\omega 1$ are definable: For example, in the basic modal logic, having an infinite path is a property corresponding to a subset of $\mathcal{P}^\omega 1$ which cannot be expressed by the finitary approximants ($\bigwedge_{n<\omega} \Diamond^n \top$ only implies that the lengths of paths cannot be bounded).

(Full) implies (Sep) and (fFull) implies (fSep).

All of $\mathcal{L}_T^{\mathrm{M}}, \mathcal{L}_T^{\mathrm{P}}, \mathcal{L}_T^{\mathrm{F}}$ enjoy (fSep) and (Bool), $\mathcal{L}_T^{\mathrm{F}}$ also (fFull).

3 Finite depth

Definition 3.1 We say that an abstract coalgebraic logic $\mathcal{L}$ has finite depth if all formulas are determined by some subset of $T^n 1$ for some $n < \omega$, that is, for all $\varphi \in \mathcal{L}$ there is $n < \omega$ such that $[\![\varphi]\!] = p_n^{-1}([\![\varphi]\!]_n) = p_n^{-1}(p_n[\varphi])$.

Remark 3.2 (i) $\mathcal{L}_T^{\mathrm{M}}, \mathcal{L}_T^{\mathrm{P}}, \mathcal{L}_T^{\mathrm{F}}$ have finite depth, the μ-calculus does not have finite depth.

(ii) Every logic with finite depth is invariant under ω-bismilarity. The converse is not true.

4 Compactness

This section is needed for Theorems 7.3 and 7.8.

Notation We already overloaded the symbol $\mathcal{L}$ to denote the logic as well as the collection of definable subsets of the final coalgebra. For example, if we say that the intersection of a set of formulas is non-empty, we refer to the extensions of the formulas on the final coalgebra.

Similarly we now write

$$\mathcal{L}_\alpha = \{p_\alpha[\varphi] \subseteq T^\alpha 1 \mid \varphi \in \mathcal{L}\}.$$

Note that $\mathcal{L}_\alpha$ contains (possibly strictly) the set of definable subsets $S \subseteq T^\alpha 1$ in the sense of Definition 2.10. Also note that even if $\mathcal{L}$ is closed under Boolean operations, $\mathcal{L}_\alpha$ need not be (because the direct image only preserves unions but neither intersections nor complements).

Furthermore, when we treat $T^\alpha 1$ (or the carrier Z of the final coalgebra) as a topological space in the following, then we do this with respect to the topology generated by $\mathcal{L}_\alpha$ (or $\mathcal{L}$). For example, in the presence of (Bool), the following definition of compactness coincides with the topological definition of Z being compact (that is, every cover of opens has a finite subcover). Recall that a collection $\mathcal{C}$ of subsets of the final coalgebra has the finite intersection property (f.i.p.) if all finite subcollections have non-empty intersection.

Definition 4.1 $\mathcal{L}$ is compact if each collection of formulas with the finite intersection property has a non-empty intersection.

Later we will use results from topology to show that there is only one possible logic on $T^\omega 1$ satisfying certain restrictions. But let us first take a look at the logics we considered so far.

Proposition 4.2 (i) *If $\mathcal{L}$ satisfies (fSep), then $T^\omega 1$ is Hausdorff.*

(ii) *Let $\mathcal{L}$ be one of $\mathcal{L}_T^{\mathrm{M}}, \mathcal{L}_T^{\mathrm{P}}, \mathcal{L}_T^{\mathrm{F}}$. Then $T^\omega 1$ is Hausdorff and has a basis of clopens.*

(iii) *Moreover, $(\mathcal{L}_T^{\mathrm{M}})_\omega \equiv (\mathcal{L}_T^{\mathrm{P}})_\omega \equiv (\mathcal{L}_T^{\mathrm{F}})_\omega$ are compact if T has (fs).*

Proof. (i) is immediate from the definitions. (ii) follows from (Bool). (iii) holds since under (fs) the approximants $T^n 1$, $n < \omega$, are finite (hence compact Hausdorff) and a limit of compact Hausdorff spaces is compact Hausdorff. □

To reason about compactness we use the following standard topological facts. The first is an auxiliary statement. The second shows that a compact Hausdorff topology cannot be made smaller (without loosing Hausdorff) nor bigger (without loosing compactness). The third says that if compact Hausdorff spaces have a basis closed under Boolean operations, then this basis is uniquely determined.

Proposition 4.3 (i) *In a compact space, every closed set is compact. In a Hausdorff space, every compact set is closed.*

(ii) *Let $\tau \subseteq \tau'$ be two topologies on a set X and assume that τ is Hausdorff and τ' is compact (and hence both are compact Hausdorff). Then $\tau = \tau'$.*

(iii) *Let (X, τ) be a Stone space, ie, τ is compact Hausdorff and has a basis of clopens. Then τ has one and only one basis closed under the Boolean operations.*

Proof. For the second statement, assume there is an open $o \in \tau'$, $o \notin \tau$. The complement o' of o is not closed in τ, hence not compact. Hence there is an open cover o'_i of o' from τ that has no finite subcover. This cover is also a cover from τ', hence o' is not compact in τ', hence not closed in τ', contradicting $o \in \tau$. □

5 Expressively closed logics

This section is needed for Theorem 7.8.

The following definition makes precise the requirement that if some property P is definable in the logic, then the weaker property P_α of 'P up to α-bisimilarity' is also definable.

Definition 5.1 $\mathcal{L}$ is expressively closed at α if for all $\varphi \in \mathcal{L}$ there is $\psi \in \mathcal{L}$ such that $p_\alpha^{-1}([\![\varphi]\!]_\alpha) = [\![\psi]\!]$. $\mathcal{L}$ is expressively closed if it is expressively closed at α for all ordinals α.

In other words, $\mathcal{L}$ is expressively closed at α iff $\mathcal{L}_\alpha$ coincides with the definable subsets of $T^\alpha 1$ (see Definition 2.10).

Example 5.2 (i) $\mathcal{L}_T^{\mathrm{M}}, \mathcal{L}_T^{\mathrm{P}}, \mathcal{L}_T^{\mathrm{F}}$ are expressively closed. Every logic extending $\mathcal{L}_T^{\mathrm{F}}$ is expressively closed at n for $n < \omega$.

(ii) Expressively closed at ω does not imply expressively closed at $n < \omega$. For example, in the case of $T = \mathcal{P}$, take a logic which has exactly 4 pairwise different formulas (false, true, φ, $\neg\varphi$) and where φ is definable at stage ω but not at any $n < \omega$.

(iii) The μ-calculus is not expressively closed, since (in the case of $T = \mathcal{P}$) the formula $\mu x.\Box x$ expresses termination of each execution sequence (well-foundedness) and this property cannot be defined at any ordinal level.

Remark 5.3 If $\mathcal{L}$ is expressively closed at α then $p_\alpha : Z \to T^\alpha 1$ is continuous (with respect to the topology generated by $\mathcal{L}_\alpha$).

Recall that if X is compact and $f : X \to Y$ is continuous and onto, then Y is compact. The previous remark then implies the first part of the proposition below.

Proposition 5.4 (i) *If $\mathcal{L}$ is compact, expressively closed at α, and $p_\alpha : Z \to T^\alpha 1$ is onto, then $\mathcal{L}_\alpha$ is compact.*

(ii) *If $\mathcal{L}$ is expressively closed and satisfies (Bool) and $p_\alpha : Z \to T^\alpha 1$ is onto, then $\mathcal{L}_\alpha$ is closed under Boolean operations.*

Proof. For the second item, because direct image preserves unions, $\mathcal{L}_\alpha$ is closed under unions. To see closure under complements let S be a definable subset and $\varphi = p_\alpha^{-1}(S)$. Then $[\![\neg\varphi]\!]_\alpha$ is the complement of S. □

Proposition 5.5 *If $\mathcal{L}$ is compact and extends $\mathcal{L}_T^{\mathrm{F}}$, then*

(i) $\bigcap\{(p_n^\omega)^{-1}([\![\varphi]\!]_n) \mid n < \omega\} = [\![\varphi]\!]_\omega$

(ii) $\bigcap\{p_n^{-1}([\![\varphi]\!]_n) \mid n < \omega\} = p_\omega^{-1}([\![\varphi]\!]_\omega)$

(iii) $p_\omega : Z \to T^\omega 1$ *is onto.*

Proof. (ii) is immediate from (i). Putting $\varphi = \mathit{true}$ in (i) yields $T^\omega 1 = [\![\mathit{true}]\!]_\omega$, which implies (iii). To show (i), "$\supseteq$" is immediate. For "$\subseteq$", assume that $t \in \bigcap\{(p_n^\omega)^{-1}([\![\varphi]\!]_n) \mid n < \omega\}$, that is, for each $n < \omega$ there is $z_n \in \varphi$ with $p_n^\omega(t) = p_n(z_n)$. Since $\mathcal{L}$ extends $\mathcal{L}_T^{\mathrm{F}}$, we have $p_n^{-1}(\{p_n(z_n)\}) \in \mathcal{L}$. By construction, $\{p_n^{-1}(\{p_n(z_n)\}) \mid n < \omega\} \cup \{\varphi\}$ has the f.i.p. and so, by compactness, there is $z \in \varphi$ such that $p_n^\omega(t) = p_n(z)$ for all $n < \omega$. Hence $t = p_\omega(z) \in [\![\varphi]\!]_\omega$. □

Remark 5.6 The proof also works for compact $\mathcal{L}$ that have for each $t \in T^\omega 1$ a collection of definable $S_n \subseteq T^n 1$, $n < \omega$, such that $\{t\}$ is the intersection of the $(p_n^\omega)^{-1}(S_n)$. For example, the proposition holds if the singleton subsets of $T^n 1$, $n < \omega$ are definable, or if $\mathcal{L}$ satisfies (fSep) and is expressively closed at n for all $n < \omega$.

The topological analysis above gives us a Lindström theorem 'at ordinal level ω', roughly saying: an abstract coalgebraic logic that is compact and expressively closed at each finite n agrees with $\mathcal{L}_T^{\mathrm{F}}$ at ω. The Lindström theorems of Section 7 can be seen as variations where some further work is put into dropping the restriction 'at ordinal level ω'.

Theorem 5.7

(i) *Let $\mathcal{L}$ be an abstract coalgebraic logic that satisfies (fSep), (Bool) and is compact and expressively closed at n, $n < \omega$. Then $T^\omega 1$ is a Stone space.*

(ii) *If, moreover, T satisfies (fs), then $\mathcal{L}_\omega = (\mathcal{L}_T^{\mathrm{M}})_\omega = (\mathcal{L}_T^{\mathrm{P}})_\omega = (\mathcal{L}_T^{\mathrm{F}})_\omega$.*

Proof. (i). By (fSep) and Proposition 4.2, $\mathcal{L}_\omega$ is Hausdorff. By compactness of $\mathcal{L}$ and (Bool), and Propositions 5.5 and 5.4, $\mathcal{L}_\omega$ is compact and has a basis of clopens. (ii). If T preserves finite sets, then (fSep) and (Bool), together with expressively closed, imply (fFull), that is, $\mathcal{L}$ extends $\mathcal{L}_T^{\mathrm{F}}$. Hence (the topology generated by) $\mathcal{L}_\omega$ is a Stone topology which extends the Stone topology (Proposition 4.2) (generated by) $(\mathcal{L}_T^{\mathrm{F}})_\omega$. By Proposition 4.3(ii) the two topologies agree and by Proposition 4.3(iii) the same holds for the bases, that is, $\mathcal{L}_\omega = (\mathcal{L}_T^{\mathrm{F}})_\omega$. □

Concerning the assumptions of the theorem, we recall that a logic that extends $\mathcal{L}_T^{\mathrm{F}}$ has (fSep) and is expressively closed at n. Also $\mathcal{L}_T^{\mathrm{M}}$, $\mathcal{L}_T^{\mathrm{P}}$, $\mathcal{L}_T^{\mathrm{F}}$ satisfy these two properties, as well as (Bool). But, depending on T, neither need to be compact.

6 Two lemmas on the final coalgebra sequence

For the proof of Theorem 7.8 we will need two lemmas about the final coalgebra sequence. We study the diagram

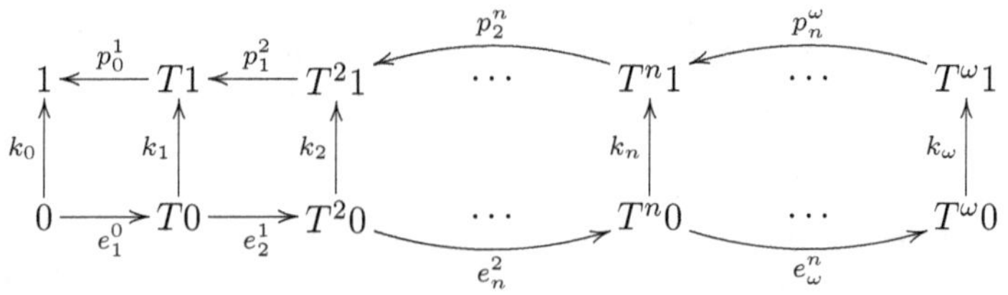

(8)

where the upper row is the final coalgebra sequence (4) and the lower row is its dual, the initial algebra sequence. The arrows k_n are the unique ones induced by initiality/finality: k_0 is the empty map and $k_{n+1} = Tk_n$; for each $n < \omega$ this induces over the terminal sequence $(T^m1)_{m<\omega}$ a cone[1] $k_{nm} : T^n0 \to T^m1$ given by $k_{nm} = k_m \circ e_m^n$ for $m > n$ and by $k_{nm} = p_m^n \circ k_n$ for $m \leq n$; since $T^\omega 1$ is a limit, the cones $(k_{nm})_{m<\omega}$ induce maps $k_{n\omega} : T^n0 \to T^\omega 1$, which in turn form a co-cone[2] over the initial sequence $(T^n0)_{n<\omega}$; since $T^\omega 0$ is a colimit, the co-cone $(k_{n\omega})_{n<\omega}$ induces k_ω.

In the situation above, the arrows p_n^ω induce a metric d on $T^\omega 1$ via $d(x, y) = 2^{-n}$ where n is the smallest number such that $p_n^\omega(x) \neq p_n^\omega(y)$. Moreover, $T^\omega 0$ inherits this metric via the injective k_ω.

Proposition 6.1 (Barr [4, Proposition 3.1]) *Let* $T : \mathsf{Set} \to \mathsf{Set}$ *with* $T0 \neq 0$. *Then* k_ω *in diagram (4) is a Cauchy completion.*

The topology on $T^\omega 1$ induced by the metric coincides with the limit topology over the discrete spaces $T^n1, n < \omega$. The topology is compact, if the $T^n1, n < \omega$, are finite. Then the collection of clopens of this topology is $(\mathcal{L}_T^{\mathrm{F}})_\omega$. Since $T^\omega 0$ is dense in $T^\omega 1$ we have

---

[1] That is, $p_l^m \circ k_{nm} = k_{nl}$ for all $l \leq m < \omega$.

[2] That is, $k_{m\omega} \circ e_m^n = k_{n\omega}$ for all $n \leq m < \omega$.

Lemma 6.2 *Let $T : \mathsf{Set} \to \mathsf{Set}$ with $T0 \neq 0$. Then for all $\varphi \in \mathcal{L}_T^{\mathrm{F}}$, either $[\![\varphi]\!]_\omega = \emptyset$ or $[\![\varphi]\!]_\omega \cap T^\omega 0 \neq \emptyset$.*

For the next lemma, recall the notation $p_\omega : Z \to T^\omega 1$ from Diagram (4).

Lemma 6.3 *Let T be a weak pullback preserving functor $\mathsf{Set} \to \mathsf{Set}$ with $T0 \neq 0$. If for two elements z, z' of the final coalgebra satisfying $p_\omega(z) = p_\omega(z')$ we have that $p_\omega(z) = p_\omega(z')$ is in the image of k_ω in (8), then $z = z'$.*

Proof. Without loss of generality, assume that k_ω is an inclusion. Consider

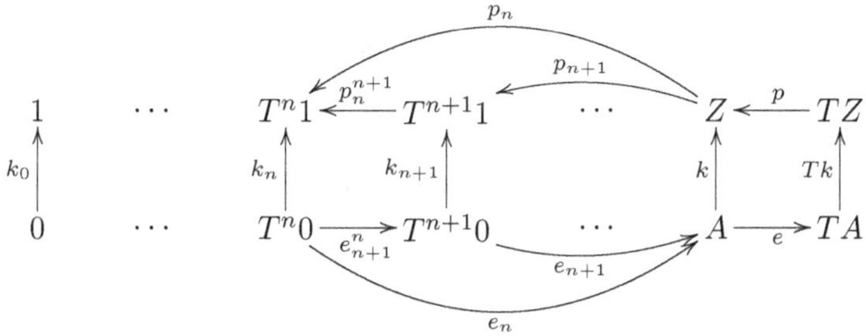

which is as (8) but extended transfinitely through the ordinals until reaching the initial algebra $(A, \alpha) = (A, e^{-1})$ and the final coalgebra $(Z, \zeta) = (Z, p^{-1})$ [3]. We write ";" for relational composition and converse of R as R^o, eg, if f and g are functions, $(g \circ f)^o = g^o; f^o$ and also $p^{-1} = p^o, e^{-1} = e^0$ (p and e are isos). The claim follows once we have shown $e_n; k = k_n; p_n^o$ for all $n < \omega$. Indeed, $p_\omega(z) = p_\omega(z') \in T^\omega 0$ means there is $n < \omega$ and $x \in T^n 0$ such that $x(k_n; p_n^o)z$ and $x(k_n; p_n^o)z'$. Since $e_n; k$ is a functional relation, we must have $z = z'$.

To show $e_n; k = k_n; p_n^o$ first observe that it holds for $n = 0$ since then both relations are the empty relation. Further, $e_{n+1}; k =$ (by definition of $e = \alpha^o$ and (A, α) being initial) $Te_n; e^o; k =$ (by definition of k) $Te_n; Tk; p =$ (by ind.hyp. and T preserving weak pullbacks) $Tk_n; (Tp_n)^o; p =$ (by definition of $p = \zeta^o$ and (Z, ζ) being final) $Tk_n; (p_{n+1} \circ p)^o; p =$ (by $p^o = \zeta$ being iso and definition of k_{n+1}) $k_{n+1}; p_{n+1}^o$. □

7 Lindström theorems

The theorems below state that a logic $\mathcal{L}$ invariant under bisimilarity, closed under Boolean operations and at least as expressive as $\mathcal{L}_T^{\mathrm{F}}$ is actually equivalent to $\mathcal{L}_T^{\mathrm{F}}$ if additionally one of the following conditions is satisfied

- $\mathcal{L}$ has finite depth,
- $\mathcal{L}$ is compact and invariant under ω-bisimilarity,
- $\mathcal{L}$ is compact and expressively closed at ω.

7.1 Finite depth

The essence of de Rijke's Lindström theorem [23] is coalgebraic and, in view of the final sequence, its proof is almost obvious: That $\mathcal{L}$ has finite depth means that formulas are determined by their extensions on the approximants $T^n 1$, $n < \omega$, and that $\mathcal{L}$ extends $\mathcal{L}_T^{\mathrm{F}}$ means, conversely, that all subsets of $T^n 1$, $n < \omega$, are definable.

Theorem 7.1 *If an abstract coalgebraic logic $\mathcal{L}$ extends $\mathcal{L}_T^{\mathrm{F}}$ and has finite depth, then $\mathcal{L} \equiv \mathcal{L}_T^{\mathrm{F}}$.*

Proof. Because $\mathcal{L}$ is invariant under bisimilarity and has finite depth, each formula φ is determined by some $T^n 1, n < \omega$, that is, $\varphi = p_n^{-1}([\![\varphi]\!]_n)$ for some $n < \omega$. Since all subsets of $T^n 1$ are definable in $\mathcal{L}_T^{\mathrm{F}}$, it follows $\mathcal{L} \leq \mathcal{L}_T^{\mathrm{F}}$. □

Corollary 7.2 (i) *If $T = \mathcal{P}$, or $T = \mathcal{P} \times 2^P$ for a finite set P of atomic propositions, we recover de Rijke's theorem [23], modulo the simplification that we do not consider infinitely many atomic propositions.*[3]

(ii) *If $T = \mathcal{H}$, or $T = \mathcal{H} \times 2^P$, we obtain a Lindström theorem for non-normal modal logics and neighbourhood frames/models.*

(iii) *More generally, if T satisfies (fs) [and (wp)], the conclusion of the theorem can be strengthened to $\mathcal{L} \equiv \mathcal{L}_T^{\mathrm{P}}$ [and $\mathcal{L} \equiv \mathcal{L}_T^{\mathrm{M}}$].*

7.2 Compactness and ω-bisimilarity

Since Lindström's theorem characterising first-order logic [19] makes crucial use of compactness, it is natural to follow van Benthem [5] and search for characterisations of modal logic involving compactness.

The proof of the following theorem is similar to Theorem 7.1, replacing the role of the sequence of approximants $(T^n 1)_{n<\omega}$ by $T^\omega 1$: That $\mathcal{L}$ is invariant under ω-bisimilarity means that formulas are determined by their extensions on $T^\omega 1$ and that $\mathcal{L}$ is compact and extends $\mathcal{L}_T^{\mathrm{F}}$ means that all clopen (closed and open) subsets of $T^\omega 1$ are extensions of some $\mathcal{L}$-formula. Finally, we use a result from topology to establish that the topologies on $T^\omega 1$ induced by the definable subsets of $\mathcal{L}$ and $\mathcal{L}_T^{\mathrm{F}}$ coincide.

Theorem 7.3 *If T preserves finite sets and an abstract coalgebraic logic $\mathcal{L}$ has the following properties: (i) invariant under ω-bisimilarity, (ii) closed under Boolean operations, (iii) $\mathcal{L}$ extends $\mathcal{L}_T^{\mathrm{F}}$, (iv) $\mathcal{L}$ is compact, then $\mathcal{L} \equiv \mathcal{L}_T^{\mathrm{F}}$.*

Proof. By (i), we can consider $\mathcal{L}$ as a collection of subsets of $T^\omega 1$. By (ii), $\mathcal{L}$ is a Boolean algebra. The topology τ generated by the basis $\mathcal{L}$ is Hausdorff by (iii). By (iv), $(T^\omega 1, \tau)$ is a Stone space, by Proposition 4.3(ii) we have that τ is the topology generated by $\mathcal{L}_T^{\mathrm{F}}$. By Proposition 4.3(iii), the basis of clopens of this space is uniquely determined, hence $\mathcal{L} \equiv \mathcal{L}_T^{\mathrm{F}}$. □

---

[3] For infinite P the logic $\mathcal{L}_{\mathcal{P}\times 2^P}^{\mathrm{F}}$ is more expressive than basic modal logic (it has formulas with infinitely many atomic propositions). This problem can be overcome in several ways, but the issues arising are orthogonal to the interests of this paper.

7.3 *Compactness and invariance under bisimilarity is not enough*

We investigate what can be said if, in the assumptions of Theorem 7.3, invariance under ω-bisimilarity is weakened to invariance under bisimilarity. The following example shows that we need to look for some additional condition.

Example 7.4 The logic $\mathcal{L}^{cb}$ is obtained from basic modal logic (without propositional variables) by adding one constant θ, which may not appear under a $\Box$. Explicitly, formulas are constructed according to

$$\mathsf{phi} ::= \theta \mid \mathsf{psi} \mid \neg\mathsf{phi} \mid \mathsf{phi} \wedge \mathsf{phi} \tag{9}$$

$$\mathsf{psi} ::= \bot \mid \neg\mathsf{psi} \mid \mathsf{psi} \wedge \mathsf{psi} \mid \Box\mathsf{psi} \tag{10}$$

Given any von Neumann ordinal α, let M_α be the pointed Kripke frame which has root (distinguished point) α, carrier $\alpha + 1 = \alpha \cup \{\alpha\}$ and the accessibility relation given by the converse of $\in$. The semantics of θ is then given by

$$(W,R), w \Vdash \theta \;\Leftrightarrow\; (W,R), w \text{ bisimilar to } M_{\omega+2}$$

Remark 7.5 With respect to Theorem 7.3, note that $M_{\omega+2}$ and $M_{\omega+1}$ are ω-bisimilar but not bisimilar (an observation that can be found in [9]) and hence $\mathcal{L}^{cb}$ is not invariant under ω-bisimilarity.

In our approach we identify formulas with their extension on the final coalgebra Z. In this view, θ is the singleton subset of the final coalgebra containing the equivalence class of $M_{\omega+2}$ up to bisimilarity. (If we take Z to be the non-well founded sets of Aczel [1], then $\theta = \{\omega + 2\}$.) This observation allows us to replace (9) by

$$\mathsf{phi} ::= \mathsf{psi} \mid \mathsf{psi} \vee \theta \mid \mathsf{psi} \wedge \neg\theta \tag{11}$$

To show that (9) and (11) are equivalent, we need to show that the language described by (11) is closed under Boolean operations. Closure under negation (complement) is immediate. For closure under conjunction (intersection) we have to check 6 cases. For instance, using that θ is a singleton, we calculate on the final coalgebra

$$(\psi_1 \vee \theta) \wedge (\psi_2 \vee \theta) = (\psi_1 \wedge \psi_2) \vee \theta \tag{12}$$

since for any φ we have that $\varphi \wedge \theta$ is either $\emptyset$ or θ.

Proposition 7.6 *$\mathcal{L}^{cb}$ is compact.*

Proof. To check compactness we (identify formulas with their extension on the final coalgebra and) assume that we are given a set $\Psi \cup \Psi_\vee \cup \Psi_\wedge$ of formulas enjoying the f.i.p. and where $\Psi, \Psi_\vee, \Psi_\wedge$ contain formulas of type $\mathsf{psi}, \mathsf{psi} \vee \theta, \mathsf{psi} \wedge \neg\theta$, respectively.

Case 1: $\Psi_\wedge$ is non-empty. Then from the f.i.p. we can assume without loss of generality that $\Psi_\vee$ is empty and that $\Psi_\wedge = \{\neg\theta\}$. For a contradiction assume that $\bigcap \Psi \cup \Psi_\vee \cup \Psi_\wedge$ is empty, ie, $\bigcap \Psi \subseteq \theta$. Then, since θ is a singleton and Ψ has the f.i.p.

(and $\mathcal{L}^{\mathrm{F}}_{\mathcal{P}}$ is compact), it follows $\bigcap \Psi = \theta$. This contradicts the fact that θ is not definable by an infinite conjunction of formulas in $\mathcal{L}^{\mathrm{F}}_{\mathcal{P}}$.

Case 2: $\Psi_\wedge$ is empty. Then, reasoning as in (12), we see that $\bigcap \Psi_\vee$ is semantically equivalent to $C \vee \theta$ for some $C \subseteq Z$ where C is the intersection of some formulas in $\mathcal{L}^{\mathrm{F}}_{\mathcal{P}}$. Note that $\Psi \cup \{C \vee \theta\}$ has the f.i.p. iff $\Psi \cup \{C\}$ or $\Psi \cup \{\theta\}$ have the f.i.p. If $\Psi \cup \{C\}$ has the f.i.p., we have $\bigcap \Psi \cap C \neq \emptyset$ since $\mathcal{L}^{\mathrm{F}}_{\mathcal{P}}$ is compact; hence also $\bigcap \Psi \cap \{C \vee \theta\} \neq \emptyset$. If $\Psi \cup \{\theta\}$ has the f.i.p., we have $\theta \subseteq \psi$ for all $\psi \in \Psi$ since θ is a singleton; hence $\bigcap \Psi \cap \{C \vee \theta\} \neq \emptyset$. □

To summarise:

Theorem 7.7 *$\mathcal{L}^{cb}$ is closed under Boolean operations, extends the basic modal logic $\mathcal{L}^{\mathrm{F}}_{\mathcal{P}}$, is invariant under bisimilarity and is compact, but is not equivalent to $\mathcal{L}^{\mathrm{F}}_{\mathcal{P}}$.*

7.4 Compact and expressively closed at ω

The previous section showed that compactness needs to be complemented by some further condition. In Theorem 7.3 it was invariance under ω-bisimilarity, which we replace now by the weaker condition of being expressively closed at ω (Definition 5.1), meaning that if some property P is definable in the logic, then the weaker property P_ω of 'P up to ω-bisimilarity' is also definable.

Theorem 7.8 *Let T : $\mathsf{Set} \to \mathsf{Set}$ preserve finite sets and weak pullbacks. If an abstract coalgebraic logic $\mathcal{L}$ (i) extends $\mathcal{L}^{\mathrm{F}}_{T}$, (ii) is closed under Boolean operations, (iii) is compact and (iv) is expressively closed at ω, then $\mathcal{L} \equiv \mathcal{L}^{\mathrm{F}}_{T}$.*

Proof. Suppose $\mathcal{L}$ extends $\mathcal{L}^{\mathrm{F}}_{T}$, that is, there is a formula $\theta \in \mathcal{L}$ such that $\theta \subsetneq p_n^{-1}([\![\theta]\!]_n)$ for all $n < \omega$. If $\theta = \bigcap\{p_n^{-1}([\![\theta]\!]_n) \mid n < \omega\}$ we obtain a contradiction to compactness, so we assume that $\theta \subsetneq \bigcap\{p_n^{-1}([\![\theta]\!]_n) \mid n < \omega\} = p_\omega^{-1}([\![\theta]\!]_\omega)$, where the latter equality is due to Proposition 5.5. By (iv), there is $\psi \in \mathcal{L}$ such that (the extension of) ψ is $p_\omega^{-1}([\![\theta]\!]_\omega)$. By (ii), we have $\psi \wedge \neg\theta \in \mathcal{L}$ and due to $\theta \subsetneq p_\omega^{-1}([\![\theta]\!]_\omega)$ we know that $[\![\psi \wedge \neg\theta]\!]_\omega$ is non-empty. Using that $\mathcal{L}$ has (wp), it now follows from Lemma 6.3 that $[\![\psi \wedge \neg\theta]\!]_\omega \cap T^\omega 0 = \emptyset$. Indeed, $t \in [\![\psi \wedge \neg\theta]\!]_\omega$ implies, on the one hand, $t \in [\![\psi]\!]_\omega$ and hence the existence of a $z \in \theta$ with $p_\omega(z) = t$ and, on the other hand, $t \in [\![\neg\theta]\!]_\omega$ and hence the existence of a $z' \notin \theta$ with $p_\omega(z') = t$; we thus have $p_\omega(z) = p_\omega(z')$ and, since $z \neq z'$, can apply Lemma 6.3 to conclude that $t \notin T^\omega 0$. But $[\![\psi \wedge \neg\theta]\!]_\omega \cap T^\omega 0 = \emptyset$ now contradicts $\mathcal{L}_\omega = (\mathcal{L}^{\mathrm{F}}_{T})_\omega$ (using that $\mathcal{L}$ has (fs) and Theorem 5.7) and the fact that all non-empty sets in $(\mathcal{L}^{\mathrm{F}}_{T})_\omega$ intersect $T^\omega 0$ (Lemma 6.2). □

Remark 7.9 One way to analyse Example 7.4 in the light of Theorems 7.3 and 7.8 is as follows. In Example 7.4, the extension $[\![\theta]\!]_\omega$ of θ on $T^\omega 1$ is closed and not open in the topology generated by $\mathcal{L}^{cb}_\omega$. From the uniqueness of compact Hausdorff topologies (Proposition 4.3), we know that the complement of $[\![\theta]\!]_\omega$ cannot be in $\mathcal{L}^{cb}_\omega$, which suggests a contradiction to $\mathcal{L}^{cb}$ being closed under Boolean operations. Unfortunately, the complement of $[\![\theta]\!]_\omega$ is not definable in $\mathcal{L}^{cb}$ (note that $[\![\neg\theta]\!]_\omega = T^\omega 1$). As we have seen in the proof of Theorem 7.8, the notion of expressive closure at ω, together with the two lemmas of Section 6, gives us a way to exploit $\neg\theta$.

The following specialises Theorem 7.8 to Kripke models.

Corollary 7.10 *Let $\mathcal{L}$ be a logic which extends basic modal logic, is invariant under bisimilarity, is closed under Boolean operations, is compact and is expressively closed at ω, then $\mathcal{L}$ is basic modal logic.*

Similarly the result applies to all Kripke polynomial functors.

7.5 Discussion

In [5], van Benthem presents a modal Lindström theorem stating that an *abstract modal logic* $\mathcal{L}$ extending basic modal logic is equivalent to basic modal logic if $\mathcal{L}$ is invariant under bisimilarity and $\mathcal{L}$ is compact. To compare this result to ours, we recall that the definition of abstract modal logic includes the property

(rl) closed under relativisation,

which means that for every formula φ and every proposition letter p, the logic contains a formula $rel(\varphi, p)$ which is true at a state x in a Kripke model M iff φ is true at x in the Kripke model we obtain from M by throwing away all the states where p is false. This condition is also part of Lindström's definition of an abstract logic, but from our coalgebraic perspective, it is not so natural. In particular, given a coalgebra $X \to TX$ and a subset $X' \subseteq X$, it is not a priori clear what the induced coalgebra with carrier X' would be. (In the case where $T = P$ one can simply take for $\xi'(x)$ the intersection of $\xi(x)$ with X').

In addition, there are natural logics that do not satisfy (rl). Consider for instance the diamond $\langle\star\rangle$, which is to be interpreted over the reflexive/transitive closure of the accessibility relation of the diamond $\Diamond$. If we consider the language with $\langle\star\rangle$ but *without* $\Diamond$, we obtain a natural logic which is bisimulation invariant and compact, but does not have the relativisation property, as can easily be verified.

Finally, there is the work of Otto and Piro [21] on Lindström theorems for the extension of basic modal logic by a global modality, and of the guarded fragment of first-order logic. This work revolves around the

(tup) Tarski Union Property,

which requires the logic L to be closed under unions of L-elementary chains. Without going into the details, we just mention that the definition of this notion also involves *substructures*, which, as opposed to *generated substructures*, do not provide a coalgebraic notion. For this reason, the property (tup) does not seem to be a natural candidate for coalgebraic generalisations.

8 Conclusion

We showed that de Rijke's modal Lindström theorem is coalgebraic in nature and generalises from Kripke models to T-coalgebras for a large class of functors T. De Rijke's theorem is based on the notion of finite depth, whereas Lindström's original theorem makes crucial use of compactness. We therefore presented two coalgebraic Lindström theorems, replacing finite depth by compactness plus an additional condition. We showed that some additional condition is needed, but there may be other conditions still to be discovered (ideally such a condition would be enjoyed by all important non-compact logics extending basic modal logic). Further open questions include

- a Lindström theorem that covers basic modal logic, implies van Benthem's result, and can be generalised to coalgebra of arbitrary type,
- a Lindström theorem for modal logics extended with fixpoint operators, in particular, modal μ-calculus,
- Lindström theorems that do not mention compactness and work for modal languages smaller than $\mathcal{L}_T^F$ such as probabilistic modal logic [14].

References

[1] Aczel, P., "Non-Well-Founded Sets," CSLI, Stanford, 1988.

[2] Aczel, P. and N. P. Mendler, *A final coalgebra theorem*, in: *Category Theory and Computer Science*, LNCS **389**, 1989.

[3] Adámek, J. and V. Koubek, *On the greatest fixed point of a set functor*, Theoret. Comput. Sci. **150** (1995).

[4] Barr, M., *Terminal coalgebras in well-founded set theory*, Theoret. Comput. Sci. **114** (1993).

[5] van Benthem, J., *A new modal Lindström theorem*, Logica universalis **1** (2007).

[6] van Benthem, J., B. ten Cate and J. A. Väänänen, *Lindström theorems for fragments of first-order logic*, Logical Methods in Computer Science **5** (2009).

[7] Blackburn, P., M. de Rijke and Y. Venema, "Modal Logic," CUP, 2001.

[8] Ferrari, G., U. Montanari and M. Pistore, *Minimizing transition systems for name passing calculi: A co-algebraic formulation*, in: *FoSSaCS'02*.

[9] Gerbrandy, J., "Bisimulations on Planet Kripke," Ph.D. thesis, University of Amsterdam (1999).

[10] Ghilardi, S., *An algebraic theory of normal forms*, Ann. Pure Appl. Logic **71** (1995).

[11] Goldblatt, R., *Final coalgebras and the Hennessy-Milner property*, Ann. Pure Appl. Logic **183** (2006).

[12] H.-D. Ebbinghaus, J. F. and W. Thomas, "Mathematical Logic," Springer, 1994.

[13] Hansen, H. and C. Kupke, *A coalgebraic perspective on monotone modal logic*, in: *CMCS'04*, ENTCS 106, 2004.

[14] Heifetz, A. and P. Mongin, *Probabilistic logic for type spaces*, Games and Economic Behavior **35** (2001).

[15] Kupke, C., A. Kurz and Y. Venema, *Completeness of the finitary Moss logic*, in: *Advances in Modal Logic*, 2008.

[16] Kupke, C. and R. A. Leal, *Characterising behavioural equivalence: Three sides of one coin*, in: *CALCO'09*.

[17] Kurz, A. and R. Leal, *Equational coalgebraic logic*, in: *MFPS'09*.

[18] Kurz, A. and D. Pattinson, *Coalgebraic modal logic of finite rank*, Math. Structures Comput. Sci. **15** (2005).

[19] Lindström, P., *On extensions of elementary logic*, Theoria **35** (1969).

[20] Moss, L., *Coalgebraic logic*, Ann. Pure Appl. Logic **96** (1999).

[21] Otto, M. and R. Piro, *A Lindström characterisation of the guarded fragment and of modal logic with a global modality*, in: *Advances in Modal Logic*, 2008.

[22] Pattinson, D., *Coalgebraic modal logic: Soundness, completeness and decidability of local consequence*, Theoret. Comput. Sci. **309** (2003).

[23] de Rijke, M., *A Lindström theorem for modal logic*, in: *Modal Logic and Process Algebra* (1995).

[24] Rutten, J., *Automata and coinduction - an exercise in coalgebra*, in: *CONCUR'98.*

[25] Schröder, L., *Expressivity of Coalgebraic Modal Logic: The Limits and Beyond*, in: *FoSSaCS'05.*

[26] Worrell, J., *On the final sequence of a finitary set functor*, Theoret. Comput. Sci. **338** (2005).

Complexity of the Lambek Calculus and Its Fragments

Mati Pentus

Department of Mathematical Logic and Theory of Algorithms
Faculty of Mechanics and Mathematics
Moscow State University
119991, Moscow, Russia

Abstract

We consider the original Lambek calculus and its natural modification called *the Lambek calculus allowing empty premises*. Both calculi have three binary connectives: an associative product operator and its two residuals, the left and right division. This paper contains a short survey of complexity results concerning fragments of these calculi obtained by restricting the set of connectives and/or the number of variables.

Keywords: Lambek calculus, complexity

Introduction

The Lambek syntactic calculus L (introduced in [8]) is one of the logical calculi used in the paradigm of categorial grammar for deriving reduction laws of syntactic types (also called "categories") in natural and formal languages. In categorial grammars based on the Lambek calculus (or its variants), an expression is assigned to category B / A (resp. $A \setminus B$) if and only if the expression produces an expression of category B whenever it is followed (resp. preceded) by an expression of category A. An expression is assigned to category $A \cdot B$ if and only if the expression can be obtained by concatenation of an expression of category A and an expression of category B. The reduction laws derivable in this calculus are of the form $A \to B$ (meaning "every expression of category A is also assigned to category B"). An overview of logical frameworks used in categorial grammars can be found, e.g., in [2] and [11].

There is a natural modification of the original Lambek calculus, which we call *the Lambek calculus allowing empty premises* and denote by L^* (see [21, p. 44]). Intuitively, the modified calculus allows the empty expression to be assigned to some categories.

The calculus L^* is in fact a fragment of noncommutative linear logic (introduced by V. M. Abrusci in [1]). Essentially the same logic was called BL2 by J. Lambek [9] (it was also studied by several other authors). Also the cyclic linear logic proposed by J.-Y. Girard and expounded by D. N. Yetter (see [5,22]) is conservative over L^*. In the propositional multiplicative fragments of all these logics, the cut rule can be eliminated and all cut-free proofs are of polynomial size. Thus, the derivability problem for these fragments is in NP.

In 2003, it was proved that for L (and L^*) the derivability problem is NP-hard and thus NP-complete. This was done by establishing a polynomial-time reduction from the Boolean satisfiability problem to the derivability problem for L (this reduction works also for L^*, and thus also for the multiplicative fragment of noncommutative linear logic and for the multiplicative fragment of cyclic linear logic). The derivability problem for all fragments obtained from the original Lambek calculus L (or L^*) by restricting the set of connectives (except for the trivial fragment with only the multiplication) remained open. All these complexity problems were solved in 2006–2009 by Yury Savateev who established polynomial-time reductions from the Boolean satisfiability problem to the derivability problems for those fragments that contain two of the three connectives of the Lambek calculus. It turned out that for the unidirectional Lambek calculus (the fragment that has only one division operator and no multiplication) the derivability problem is decidable in deterministic polynomial time. To establish this theorem, Y. Savateev invented an efficient method for testing derivability in the unidirectional Lambek calculus. At first, one decomposes the given types into atomic building blocks, labels them with natural numbers (which indicate the "Horn depth" in the original type), and puts them in a certain order (which reflects the group-theoretic interpretation of the division operator). Next, one evaluates an auxiliary predicate of 'acceptability' for all substrings of the string of labelled atomic building blocks. Doing this in the manner of dynamic programming we obtain a straightforward cubic algorithm for deciding derivability in the unidirectional Lambek calculus.

For each of the above fragments, the derivability problem remains in the same complexity class if we allow for only one variable.

The proofs of all these results are sketched in this survey. We also propose a modification of Y. Savateev's construction for the product-free fragments. The modified construction uses less variables and yields slightly shorter sequents.

This paper is organized as follows. The first section contains definitions of the calculi L and L^*. In Section 2, we define the calulus CMLL and show that it is conservative over L^*. In Section 3, we formulate a criterion for derivability in CMLL. In Section 4, we sketch the proof of NP-completeness for L and L^*. Complexity results for fragments of L and L^* are presented in the last two sections.

1 Lambek Calculus

First we define *the Lambek calculus allowing empty premises* (denoted by L^*).

Assume that an enumerable set of *variables* Var is given. The *types* of L^* are built of variables (also called *primitive types* in the context of the Lambek calculus) and three binary connectives $\cdot$, $/$, and $\backslash$. The set of all types is denoted by Tp. The letters p, $q, \ldots$ range over the set Var, capital letters A, $B, \ldots$ range over types, and capital Greek letters range over finite (possibly empty) sequences of types. For notational convenience, we assume that $\cdot$ has higher priority than $\backslash$ and $/$.

The *sequents* of L^* are of the form $\Gamma \to A$ (note that Γ can be the empty sequence). The calculus L^* has the following axioms and rules of inference:

$$A \to A, \qquad \frac{\Phi \to B \quad \Gamma B \Delta \to A}{\Gamma \Phi \Delta \to A}\ (\text{cut}),$$

$$\frac{\Pi A \to B}{\Pi \to B \,/\, A}\ (\to/), \qquad \frac{\Phi \to A \quad \Gamma B \Delta \to C}{\Gamma (B \,/\, A) \Phi \Delta \to C}\ (/\to),$$

$$\frac{A \Pi \to B}{\Pi \to A \backslash B}\ (\to\backslash), \qquad \frac{\Phi \to A \quad \Gamma B \Delta \to C}{\Gamma \Phi (A \backslash B) \Delta \to C}\ (\backslash\to),$$

$$\frac{\Gamma \to A \quad \Delta \to B}{\Gamma \Delta \to A \cdot B}\ (\to\cdot), \qquad \frac{\Gamma A B \Delta \to C}{\Gamma (A \cdot B) \Delta \to C}\ (\cdot\to).$$

As usual, we write $\mathrm{L}^* \vdash \Gamma \to A$ to indicate that the sequent $\Gamma \to A$ is derivable in L^*.

Example 1.1 The sequent $(p \backslash p) \backslash p \to p$ can be derived in L^* as follows:

$$\frac{\dfrac{p \to p}{\to p \backslash p}\ (\to\backslash) \qquad p \to p}{(p \backslash p) \backslash p \to p}\ (\backslash\to).$$

The calculus L has the same axioms and rules with the only exception that in the rules $(\to\backslash)$ and $(\to/)$ we require Π to be nonempty. Thus, if $\mathrm{L} \vdash \Gamma \to A$, then Γ is nonempty. In fact, L is the original syntactic calculus introduced in [8]. Evidently, if $\mathrm{L} \vdash \Gamma \to A$, then $\mathrm{L}^* \vdash \Gamma \to A$.

It is known that the cut-elimination theorem holds for both L and L^*.

Example 1.2 The sequent $(p \backslash p) \backslash p \to p$ cannot be derived in L.

Example 1.3 The sequent $r \cdot (q \backslash s) \to (q \,/\, r) \backslash s$ can be derived in L as follows:

$$\frac{\dfrac{\dfrac{\dfrac{r \to r \quad q \to q}{(q \,/\, r)\, r \to q}\ (/\to) \qquad s \to s}{(q \,/\, r)\, r\, (q \backslash s) \to s}\ (\backslash\to)}{r\, (q \backslash s) \to (q \,/\, r) \backslash s}\ (\to\backslash)}{r \cdot (q \backslash s) \to (q \,/\, r) \backslash s}\ (\cdot\to).$$

The intended linguistic use of the Lambek calculus is demonstrated by the following example.

Example 1.4 Let s and np be two primitive types (s is intended to be the type of sentences and np is intended to be the type of noun phrases). In an example from [8], the English words *Jane* and *likes* are assigned the types np, and $(np \backslash s) / np$, respectively. The type of *likes* shows that *likes* yields a sentence if two noun phrases are added to it, one on the right-hand side, another one on the left-hand side. Further, in this toy grammar the pronouns *he* and *she* are assigned the type $s \,/\, (np \setminus s)$, since any of them yields a sentence when combined with a sentence missing a noun phrase on the left-hand side. To check the grammaticality of a string of words, say *he likes Jane*, we attempt to derive a sequent whose left-hand side consists of types corresponding to these words. Since the sequent

$$(s \,/\, (np \setminus s))\,((np \setminus s) \,/\, np)\; np \to s$$

is derivable in L, we see that *he likes Jane* is accepted as a grammatically correct sentence by this grammar.

Similarly, *him* and *her* are assigned the type $(s \,/\, np) \setminus s$. In view of

$$\mathrm{L} \vdash np\,((np \setminus s) \,/\, np)\,((s \,/\, np) \setminus s) \to s,$$

we see that *Jane likes him* is a grammatically correct sentence. Moreover, *she likes him* is also accepted, since

$$\mathrm{L} \vdash (s \,/\, (np \setminus s))\,((np \setminus s) \,/\, np)\,((s \,/\, np) \setminus s) \to s.$$

The rules $(\to\setminus)$, $(\to/)$, and $(\cdot\to)$ are reversible in both L and L* (the converse rules are easy to derive with the help of the cut rule).

If $\mathrm{L} \vdash A \to B$ and $\mathrm{L} \vdash B \to A$, then we write $A \underset{\mathrm{L}}{\leftrightarrow} B$ and say that A and B are equivalent. Replacing a type by an equivalent type in a sequent does not affect the derivability of the sequent. It is easy to verify that $(A \cdot B) \cdot C \underset{\mathrm{L}}{\leftrightarrow} A \cdot (B \cdot C)$ and $(A \setminus B) \,/\, C \underset{\mathrm{L}}{\leftrightarrow} A \setminus (B \,/\, C)$, which allows us to omit parentheses in certain types.

2 Cyclic Linear Logic

The cyclic linear logic was proposed by J.-Y. Girard and expounded by D. N. Yetter [5,22]. It is conservative over L*. We consider its multiplicative fragment without the constants $\bot$ and $\mathbf{1}$. There are several equivalent sequent calculi for this fragment; here we consider only one of them and denote it by CMLL. The calculus CMLL may also be considered as a fragment of Lambek's bilinear logic BL3 from [9]; however, we use $\parr$ instead of $\oplus$.

In the definition of formulas of CMLL we shall employ the same enumerable set Var that was used in the definition of Lambek calculus types. It is well known that in CMLL every formula has an equivalent normal form, where the negation operator is only applied to variables. For simplicity, we shall only consider formulas in normal form. The negation of a variable p will be denoted by $\overline{p}$.

The set of formulas Fm of the calculus CMLL is defined as the smallest set satisfying the following conditions:

- if $p \in \mathrm{Var}$, then $p \in \mathrm{Fm}$ and $\overline{p} \in \mathrm{Fm}$;
- if $A \in \mathrm{Fm}$ and $B \in \mathrm{Fm}$, then $(A \otimes B) \in \mathrm{Fm}$ and $(A \parr B) \in \mathrm{Fm}$.

The binary connective $\otimes$ is called 'tensor', and $\wp$ is called 'par'. Variables and their negations (i.e., the formulas shown in the first item of the definition of formulas) are called *atoms*.

In the linear logic context, capital letters A, B, ... range over the set Fm and capital Greek letters range over the set of finite (possibly empty) sequences of formulas from Fm.

The sequents of the calculus CMLL are of the form $\to \Gamma$.

We need an operation $(\,\cdot\,)^{\perp}\colon \mathrm{Fm} \to \mathrm{Fm}$ defined on the set Fm. It maps each formula to its *negation* as follows:

$$\begin{aligned}(p)^{\perp} &= \overline{p},\\ (\overline{p})^{\perp} &= p,\\ (A \otimes B)^{\perp} &= (B)^{\perp} \wp (A)^{\perp},\\ (A \wp B)^{\perp} &= (B)^{\perp} \otimes (A)^{\perp}\end{aligned}$$

(as usual, p ranges over Var). Evidently, $A^{\perp\perp} = A$.

Example 2.1 Let $s \in \mathrm{Var}$ and $np \in \mathrm{Var}$. According to the definition,

$$(s \wp (\overline{s} \otimes np))^{\perp} = (\overline{np} \wp s) \otimes \overline{s} \quad \text{and} \quad ((\overline{np} \wp s) \wp \overline{np})^{\perp} = np \otimes (\overline{s} \otimes np).$$

The axioms of the calculus CMLL are $\to \overline{p}\,p$ and $\to p\,\overline{p}$, where $p \in \mathrm{Var}$.

The calculus CMLL has the following rules of inference:

$$\frac{\to \Gamma A B \Delta}{\to \Gamma (A \wp B) \Delta}\,(\wp),$$

$$\frac{\to \Gamma A \Gamma' \quad \to B \Delta}{\to \Gamma (A \otimes B) \Delta \Gamma'}\,(\otimes_1), \qquad \frac{\to \Gamma A \quad \to \Delta B \Delta'}{\to \Delta \Gamma (A \otimes B) \Delta'}\,(\otimes_2),$$

$$\frac{\to \Gamma A \Gamma' \quad \to (A)^{\perp} \Delta}{\to \Gamma \Delta \Gamma'}\,(\mathrm{cut}_1), \qquad \frac{\to \Gamma A \quad \to \Delta (A)^{\perp} \Delta'}{\to \Delta \Gamma \Delta'}\,(\mathrm{cut}_2).$$

We shall write $\mathrm{CMLL} \vdash \to \Gamma$ if the sequent $\to \Gamma$ is derivable in CMLL.

The cut-elimination theorem holds for CMLL, i.e., the set of derivable sequents does not change if we drop the rules (cut_1) and (cut_2).

It can be proved by induction on derivation length that if $\to \Phi\Psi$ is derivable, then $\to \Psi\Phi$ is derivable too. Thus, only cyclic permutations of formulas in a sequent are allowed. This justifies the word 'cyclic' in the name of the logic.

To embed L^* into CMLL, we shall map each type $A \in \mathrm{Tp}$ to a formula $\widehat{A} \in \mathrm{Fm}$:

$$\begin{aligned}\widehat{p} &= p,\\ \widehat{A \,/\, B} &= \widehat{A} \wp (\widehat{B})^{\perp},\\ \widehat{A \setminus B} &= (\widehat{A})^{\perp} \wp \widehat{B},\\ \widehat{A \cdot B} &= \widehat{A} \otimes \widehat{B}.\end{aligned}$$

In [7], this mapping is denoted by $(\,\cdot\,)^\flat$.

Lemma 2.2 *Let $C_1, \ldots, C_n, D \in \mathrm{Tp}$. The sequent $C_1 \ldots C_n \to D$ is derivable in L^* if and only if the sequent $\to \widehat{C_n}^\perp \ldots \widehat{C_1}^\perp \widehat{D}$ is derivable in* CMLL.

Proof. This lemma was proved in [12, Sec. 7]. □

Example 2.3 Let $s \in \mathrm{Var}$ and $np \in \mathrm{Var}$. According to the definition, $\widehat{s/(np\backslash s)} = s \parr (\overline{s} \otimes np)$, $\widehat{(np\backslash s)/np} = (\overline{np} \parr s) \parr \overline{np}$, and $\widehat{(s/np)\backslash s} = (np \otimes \overline{s}) \parr s$. Thus, the sequent $(s\,/\,(np \setminus s))\,((np \setminus s)\,/\,np)\,((s\,/\,np) \setminus s) \to s$ from Example 1.4 corresponds to the CMLL-sequent

$$\to (\overline{s} \otimes (s \parr \overline{np}))\,(np \otimes (\overline{s} \otimes np))\,((\overline{np} \parr s) \otimes \overline{s})\; s,$$

which is indeed derivable in CMLL.

In view of Lemma 2.2, any algorithm for deciding derivability in CMLL also provides an algorithm for deciding derivability in L^*.

3 Proof Nets

To characterize derivability in CMLL, we shall use a proof-net-based criterion introduced in [13], where it was exposed for the multiplicative fragment of Abrusci's noncommutative linear logic (with the multiplicative constants). Essentially the same criterion is given in [3], where it is formulated for the constant-free multiplicative fragment of the cyclic linear logic. This criterion is quite similar to the one presented in Section 4 of the survey [7] and credited to Chapter 5 of [4]. In all these criteria, a planar graph consisting of a parse tree and axiom links divides the plane into regions, a binary relation is defined on the set of all regions (this is where the definitions from [13] and [7] differ), and the acyclicity condition is imposed on the graph corresponding to this binary relation.

In this section, we give an informal exposition of the criterion from [13] for CMLL.

Given a sequent $\to A_1 A_2 \ldots A_{k-1} A_k$, we replace it by the sequent

$$\to A_1 \parr (A_2 \parr \ldots \parr (A_{k-1} \parr A_k) \ldots)$$

(it is well known that they are either both derivable or both nonderivable in CMLL). Thus, we may assume that we have a sequent $\to A$, containing n occurrences of atoms. Evidently, it contains $n-1$ occurrences of binary connectives.

On the set of all occurrences of connectives and atoms in A, we define the binary relation $\prec$ so that $\alpha \prec \beta$ if and only if α is in the scope of β.

Now we draw the parse tree of the formula A so that (1) moving from left to right we visit all nodes in the infix order, (2) if $\alpha \prec \beta$, then α is placed higher than β (thus, the root of the parse tree is at the bottom), and (3) the edges of the parse tree are drawn as straight line segments, and they do not intersect. We add one dummy occurrence of $\parr$ at the bottom left corner and connect it with the root by a dummy edge (again a straight line segment). This dummy occurrence becomes the new root of the tree. The *extended parse tree* (i.e., the parse tree together with the dummy node and the dummy edge) is the first component of a proof net. It is uniquely determined by the given sequent.

Example 3.1 Let p, q, and s be different elements of Var. Let us consider the sequent

$$\to (\overline{p} \otimes (((q \parr (\overline{s} \otimes s)) \otimes \overline{s}) \parr s)) \parr ((((\overline{s} \otimes s) \parr (\overline{s} \parr s)) \otimes \overline{q}) \parr p).$$

Its extended parse tree is shown in Figure 1.

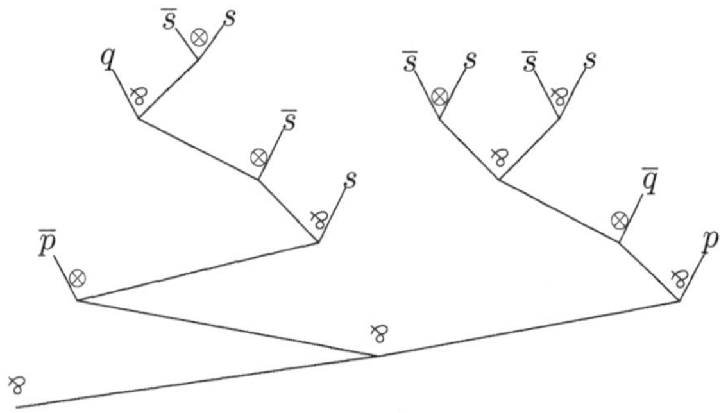

Fig. 1. An extended parse tree

The second component of a proof net is a set of axiom links. In general, axiom links are not uniquely determined by the given sequent. Each axiom link is an edge between two occurrences of atoms that are negations of each other. Each atom occurrence must be incident to exactly one axiom link. The axiom links must be drawn above the parse tree without intersecting each other or edges from the extended parse tree. Thus, the first two components together form a planar graph, which divides the plane into regions (in [7,3] they are called *faces*). It is easy to see that we obtain $\frac{n}{2}+1$ regions.

Example 3.2 One possible set of axiom links for the sequent from Example 3.1 is shown in Figure 2 using dotted lines.

For each occurrence of a binary connective, we specify to which adjacent region it "belongs". The dummy occurrence belongs to its only adjacent region, the outer region of the graph (in [7] this region is called the *infinite face*). Each nondummy occurrence of $\parr$ or $\otimes$ belongs to the region adjacent to both edges leading from this occurrence to its children (for an occurrence of $\parr$, in [7] this region is called the *Inner face* of the occurrence). Thus, every occurrence of a binary connective belongs to the region immediately above it.

Now we come to the third component of a proof net. We require that in each region there must be exactly one occurrence of $\parr$ (i.e., there must be exactly $\frac{n}{2}+1$ occurrences of $\parr$ and they must all belong to different regions). The third component of a proof tree consists of arcs leading from each tensor occurrence to the par occurrence in the same region. These arcs (we call them *region arcs*) together with the arcs specified by the

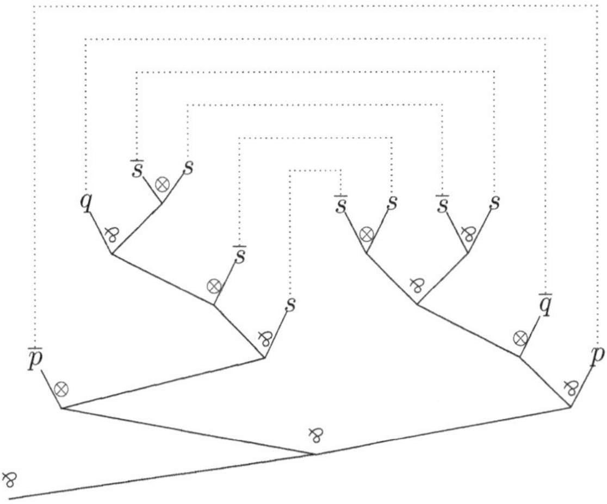

Fig. 2. Axiom links

partial order $\prec$ (directed towards the root) form a directed graph. This graph should not contain cycles (of course, we consider only directed cycles). In other words, if we denote the set of all region arcs by $\mathcal{A}$, then the binary relation $\prec \cup \mathcal{A}$ must be acyclic (we call a binary relation *acyclic* if its transitive closure is irreflexive). Note that if the first two components of a proof net are given, then the third component either does not exist or is uniquely determined.

If all the above conditions are satisfied, then we have a proof net for $\to A$. We shall call such proof nets *region proof nets*.

Example 3.3 Let us consider the extended proof tree and axiom links from Example 3.2. Indeed, in each of the seven regions there is exactly one occurrence of $\invamp$. The region arcs are shown in Figure 3 (we indicate the left-to-right order of occurrences of connectives by subscripts). However, there is a cycle in the directed graph consisting of region arcs and arcs specified by the partial order $\prec$ (the cycle consists of the vertices denoted by $\invamp_2$, $\otimes_4$, $\invamp_8$, and $\otimes_{10}$). Thus, the structure shown in Figure 3 is not a region proof net.

Theorem 3.4 *A sequent $\to A$ is derivable in* CMLL *if and only if there exists a region proof net for $\to A$.*

Proof. The proof is similar to that of Theorem 7.12 from [13]. □

4 The Complexity of L* and L

We give a sketch of the proof of the NP-completeness for L* and L that can be found in [14,15].

In this and the next section, we construct mappings that take Boolean formulas in conjunctive normal form to sequents of the Lambek calculus. In both sections, we assume that we are given a Boolean formula in the conjunctive normal form $c_1 \wedge \ldots \wedge c_m$,

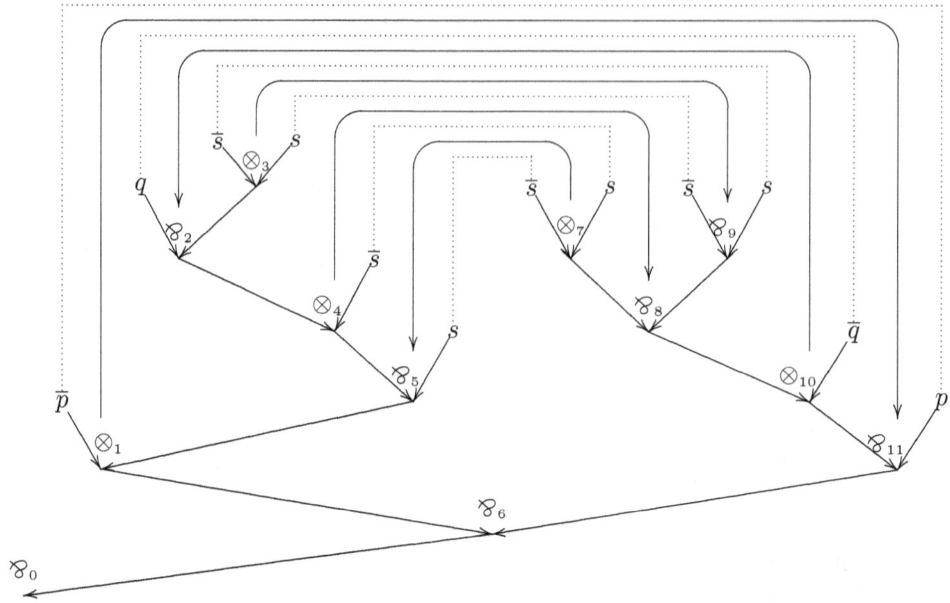

Fig. 3. Region arcs

with clauses $c_1, \ldots, c_m$ and variables $x_1, \ldots, x_n$. The reduction we are going to present in this section maps the formula to a sequent that is derivable in L* (and in L) if and only if the formula $c_1 \wedge \ldots \wedge c_m$ is satisfiable.

For any Boolean variable x_i let $\neg_0 x_i$ stand for the literal $\neg x_i$ and $\neg_1 x_i$ stand for the literal x_i. Note that $\langle t_1, \ldots, t_n \rangle \in \{0,1\}^n$ is a satisfying assignment for the Boolean formula $c_1 \wedge \ldots \wedge c_m$ if and only if for every positive integer $j \leq m$ there exists a positive integer $i \leq n$ such that the literal $\neg_{t_i} x_i$ occurs in the clause c_j (as usual, 1 stands for "true" and 0 stands for "false").

Let p_i^j (where $0 \leq i \leq n$ and $0 \leq j \leq m$) be distinct primitive types from Var.

We define three families of types:

$$
\begin{aligned}
G_i^0 &\rightleftharpoons p_0^0 \setminus p_i^0 \quad \text{if } 1 \leq i \leq n, \\
G_i^j &\rightleftharpoons (p_0^j \setminus G_i^{j-1}) \cdot p_i^j \quad \text{if } 1 \leq i \leq n \text{ and } 1 \leq j \leq m, \\
H_i^0 &\rightleftharpoons p_{i-1}^0 \setminus p_i^0 \quad \text{if } 1 \leq i \leq n, \\
H_i^j &\rightleftharpoons p_{i-1}^j \setminus (H_i^{j-1} \cdot p_i^j) \quad \text{if } 1 \leq i \leq n \text{ and } 1 \leq j \leq m, \\
E_i^0(t) &\rightleftharpoons p_{i-1}^0 \setminus p_i^0 \quad \text{if } 1 \leq i \leq n \text{ and } t \in \{0,1\}, \\
E_i^j(t) &\rightleftharpoons \begin{cases} (p_{i-1}^j \setminus E_i^{j-1}(t)) \cdot p_i^j & \text{if the literal } \neg_t x_i \text{ occurs in the clause } c_j, \\ p_{i-1}^j \setminus (E_i^{j-1}(t) \cdot p_i^j) & \text{otherwise} \end{cases} \\
&\qquad \text{if } 1 \leq i \leq n,\ 1 \leq j \leq m, \text{ and } t \in \{0,1\}.
\end{aligned}
$$

For convenience we introduce the following abbreviations:

$$
\begin{aligned}
G &\leftrightharpoons G_n^m, \\
H_i &\leftrightharpoons H_i^m \quad \text{if } 1 \leq i \leq n, \\
E_i(t) &\leftrightharpoons E_i^m(t) \quad \text{if } 1 \leq i \leq n \text{ and } t \in \{0,1\}, \\
F_i &\leftrightharpoons (E_i(1) \,/\, H_i) \cdot H_i \cdot (H_i \setminus E_i(0)) \quad \text{if } 1 \leq i \leq n.
\end{aligned}
$$

Example 4.1 Let $n = 1$, $m = 2$, $c_1 = x_1$, and $c_2 = \neg x_1$. Then

$$
\begin{aligned}
G &= (p_0^2 \setminus ((p_0^1 \setminus (p_0^0 \setminus p_1^0)) \cdot p_1^1)) \cdot p_1^2, \\
H_1 &= p_0^2 \setminus ((p_0^1 \setminus ((p_0^0 \setminus p_1^0) \cdot p_1^1)) \cdot p_1^2), \\
E_1(0) &= (p_0^2 \setminus (p_0^1 \setminus ((p_0^0 \setminus p_1^0) \cdot p_1^1))) \cdot p_1^2, \\
E_1(1) &= p_0^2 \setminus (((p_0^1 \setminus (p_0^0 \setminus p_1^0)) \cdot p_1^1) \cdot p_1^2), \\
F_1 &= (E_1(1) \,/\, H_1) \cdot H_1 \cdot (H_1 \setminus E_1(0)).
\end{aligned}
$$

The proofs of the following lemmas can be found in [14] and [15].

Lemma 4.2 *Let $1 \leq i \leq n$ and $t \in \{0,1\}$. Then $\mathrm{L} \vdash F_i \to E_i(t)$.*

Lemma 4.3 *Suppose $\langle t_1, \ldots, t_n \rangle$ is a satisfying assignment for the Boolean formula $c_1 \wedge \ldots \wedge c_m$. Then $\mathrm{L} \vdash E_1(t_1) \ldots E_n(t_n) \to G$.*

Example 4.4 Let $n = 2$, $m = 1$, $c_1 = x_1 \vee x_2$, $t_1 = 0$, and $t_2 = 1$. In view of Lemma 4.3, $\mathrm{L} \vdash E_1(0)E_2(1) \to G$, where

$$
\begin{aligned}
G &= (p_0^1 \setminus (p_0^0 \setminus p_2^0)) \cdot p_2^1, \\
E_1(0) &= p_0^1 \setminus ((p_0^0 \setminus p_1^0) \cdot p_1^1), \\
E_2(1) &= (p_1^1 \setminus (p_1^0 \setminus p_2^0)) \cdot p_2^1.
\end{aligned}
$$

Lemma 4.5 *If the formula $c_1 \wedge \ldots \wedge c_m$ is satisfiable, then $\mathrm{L} \vdash F_1 \ldots F_n \to G$.*

Proof. Suppose $\langle t_1, \ldots, t_n \rangle$ is a satisfying assignment for the formula $c_1 \wedge \ldots \wedge c_m$. According to Lemma 4.3 $\mathrm{L} \vdash E_1(t_1) \ldots E_n(t_n) \to G$. It remains to apply Lemma 4.2 and the cut rule n times. □

Example 4.6 Let $n = 1$, $m = 1$, and $c_1 = x_1$. In view of Lemma 4.5, we have $\mathrm{L} \vdash (E_1(1) \,/\, H_1) \cdot H_1 \cdot (H_1 \setminus E_1(0)) \to G$, where

$$
\begin{aligned}
G &= (p_0^1 \setminus (p_0^0 \setminus p_1^0)) \cdot p_1^1, \\
H_1 &= p_0^1 \setminus ((p_0^0 \setminus p_1^0) \cdot p_1^1), \\
E_1(0) &= p_0^1 \setminus ((p_0^0 \setminus p_1^0) \cdot p_1^1), \\
E_1(1) &= (p_0^1 \setminus (p_0^0 \setminus p_1^0)) \cdot p_1^1.
\end{aligned}
$$

Lemma 4.7 *If $\mathrm{L}^* \vdash F_1 \ldots F_n \to G$, then $\mathrm{L}^* \vdash E_1(t_1) \ldots E_n(t_n) \to G$ for some $\langle t_1, \ldots, t_n \rangle \in \{0,1\}^n$.*

Lemma 4.8 *Let $\langle t_1, \ldots, t_n \rangle \in \{0,1\}^n$. If $\mathrm{L}^* \vdash E_1(t_1) \ldots E_n(t_n) \to G$, then the assignment $\langle t_1, \ldots, t_n \rangle$ is a satisfying assignment for the Boolean formula $c_1 \wedge \ldots \wedge c_m$.*

Example 4.9 Let $n = 2$, $m = 1$, and $c_1 = x_1 \vee x_2$. We consider the assignment $\langle 0, 0 \rangle$, which is not a satisfying assignment for the Boolean formula $x_1 \vee x_2$. In view of Lemma 4.8, $\mathrm{L}^* \nvdash E_1(0)E_2(0) \to G$, where

$$\begin{aligned} G &= (p_0^1 \setminus (p_0^0 \setminus p_2^0)) \cdot p_2^1, \\ E_1(0) &= p_0^1 \setminus ((p_0^0 \setminus p_1^0) \cdot p_1^1), \\ E_2(0) &= p_1^1 \setminus ((p_1^0 \setminus p_2^0) \cdot p_2^1). \end{aligned}$$

Lemma 4.10 *If $\mathrm{L}^* \vdash F_1 \ldots F_n \to G$, then the formula $c_1 \wedge \ldots \wedge c_m$ is satisfiable.*

Proof. Immediate from Lemma 4.7 and Lemma 4.8. □

Theorem 4.11 *The* L*-derivability problem is* NP*-complete.*

Proof. The number of variable occurrences in a cut-free derivation in L can not exceed the square of the number of variable occurrences in the final sequent. Thus, the L-derivability problem is in NP.

According to Lemma 4.5 and Lemma 4.10, the mapping that takes $c_1 \wedge \ldots \wedge c_m$ to $F_1 \ldots F_n \to G$ yields a reduction from the classical satisfiability problem SAT to the L-derivability problem. The problem SAT is known to be NP-hard. Thus the L-derivability problem is NP-hard as well. □

Theorem 4.12 *The* L^**-derivability problem is* NP*-complete.*

Proof. The theorem follows immediately from Lemma 4.5 and Lemma 4.10. The proof is similar to that of the previous theorem. □

All the lemmas in Sections 4 and 5 can be proved using the region proof nets defined in Section 3. To reformulate any result concerning L^* in terms of proof nets, we use Lemma 2.2 and Theorem 3.4. When we deal with L, to reformulate $\mathrm{L} \vdash C_1 \ldots C_n \to D$, we need a variant of region proof nets with the additional stipulation that the connective occurrence preceding a (not necessarily proper) subformula of $\widehat{D}$ or $\widehat{C_i}^{\perp}$ is not in the same region as the connective occurrence succeeding this subformula (in [6] such proof nets are called *strong*).

Example 4.13 To illustrate Lemma 4.8, we consider the sequent $E_1(0)\, E_2(0) \to G$ from Example 4.9. This sequent corresponds to the Boolean formula $x_1 \vee x_2$ and the assignment where both x_1 and x_2 are false. Our aim is to show that $\mathrm{L}^* \nvdash E_1(0)\, E_2(0) \to G$. In view of Lemma 2.2, this is equivalent to

$$\mathrm{CMLL} \nvdash \; \to \widehat{E_2(0)}^{\perp}\, \widehat{E_1(0)}^{\perp}\, \widehat{G}.$$

Note that

$$\begin{aligned}
\widehat{G} &= (\overline{p_0^1} \parr (\overline{p_0^0} \parr p_2^0)) \otimes p_2^1, \\
\widehat{E_1(0)} &= \overline{p_0^1} \parr ((\overline{p_0^0} \parr p_1^0) \otimes p_1^1), \\
\widehat{E_2(0)} &= \overline{p_1^1} \parr ((\overline{p_1^0} \parr p_2^0) \otimes p_2^1), \\
\widehat{E_1(0)}^{\perp} &= (\overline{p_1^1} \parr (\overline{p_1^0} \otimes p_0^0)) \otimes p_0^1, \\
\widehat{E_2(0)}^{\perp} &= (\overline{p_2^1} \parr (\overline{p_2^0} \otimes p_1^0)) \otimes p_1^1.
\end{aligned}$$

In view of Theorem 3.4, it suffices to show that there is no proof net for the sequent $\rightarrow \widehat{E_2(0)}^{\perp}\, \widehat{E_1(0)}^{\perp}\, \widehat{G}$.

The extended parse tree for $\widehat{E_2(0)}^{\perp} \parr (\widehat{E_1(0)}^{\perp} \parr \widehat{G})$ is shown in Figure 4.

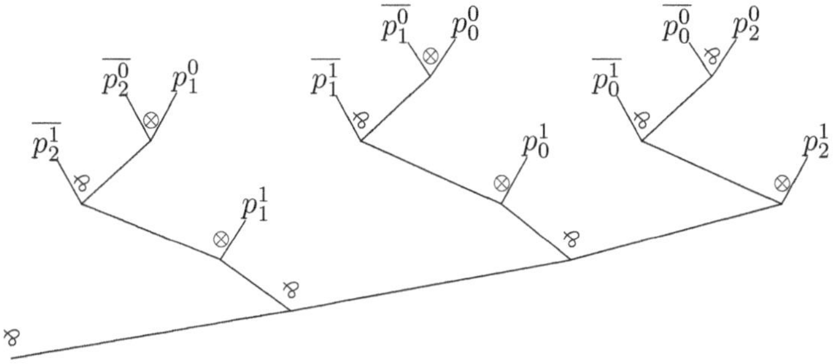

Fig. 4. The extended parse tree corresponding to $x_1 \vee x_2$ and $\langle 0, 0 \rangle$

There is a unique way to establish axiom links between occurrences of atoms that are negations of each other. The axiom links are shown in Figure 5. We indicate the left-to-right order of occurrences of connectives by subscripts.

Finally, we add the region arcs. In order to demonstrate some regularities of extended parse trees and axiom links corresponding to sequents of the form $E_1(t_1) \ldots E_n(t_n) \rightarrow G$, we distort the graph layout in a continuous way, as shown in Figure 6. The occurrences denoted by $\parr_1$, $\otimes_3$, $\parr_5$, $\otimes_7$, $\parr_9$, and $\otimes_{11}$ form a cycle. Hence the sequent $\rightarrow \widehat{E_2(0)}^{\perp}\, \widehat{E_1(0)}^{\perp}\, \widehat{G}$ is not derivable in CMLL.

Example 4.14 To illustrate Lemma 4.3, we consider the sequent $E_1(0)\, E_2(1) \rightarrow G$ from Example 4.4. This sequent corresponds to the Boolean formula $x_1 \vee x_2$ and the assignment where x_1 is false and x_2 is true. Our aim is to show that $\mathrm{L} \vdash E_1(0)\, E_2(1) \rightarrow G$. It suffices to find a strong proof net for the sequent

$$\rightarrow \widehat{E_2(1)}^{\perp}\, \widehat{E_1(0)}^{\perp}\, \widehat{G}.$$

The types $\widehat{E_1(0)}^{\perp}$ and $\widehat{G}$ are the same as in Example 4.13, and

$$\widehat{E_2(1)}^{\perp} = \overline{p_2^1} \parr ((\overline{p_2^0} \otimes p_1^0) \otimes p_1^1).$$

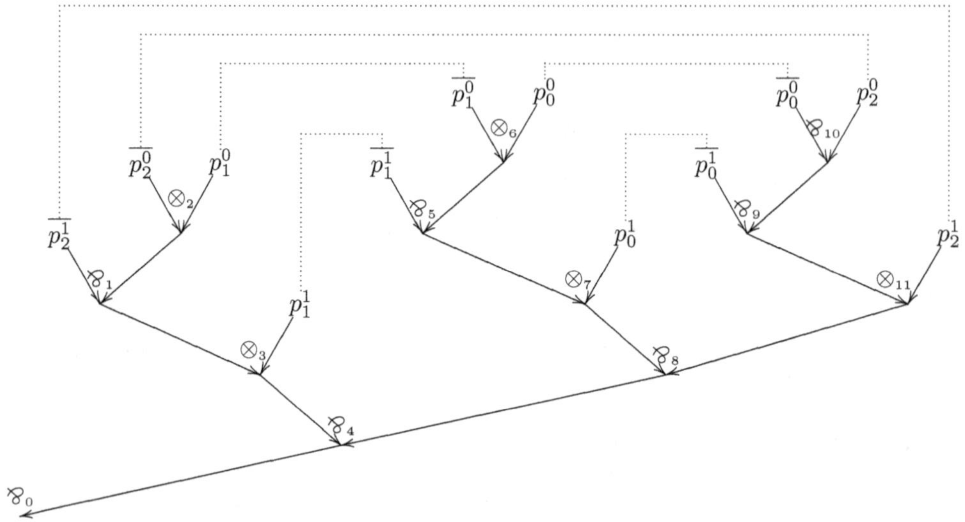

Fig. 5. The extended parse tree and axiom links corresponding to $x_1 \vee x_2$ and $\langle 0, 0 \rangle$

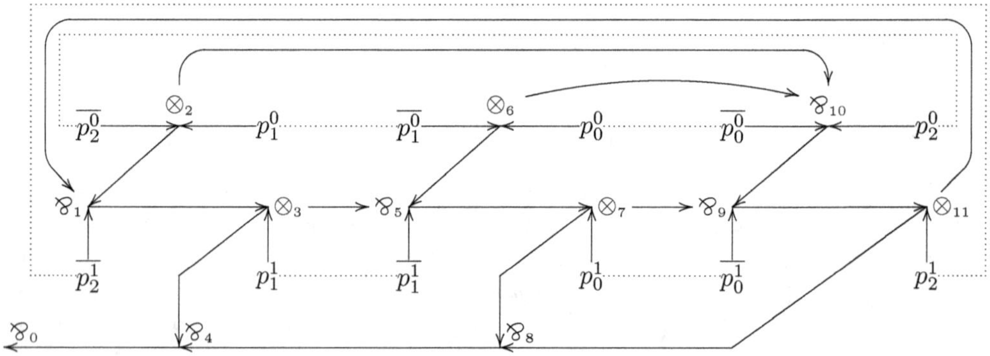

Fig. 6. The extended parse tree, axiom links, and region arcs corresponding to $x_1 \vee x_2$ and $\langle 0, 0 \rangle$

The proof net is shown in Figure 7. Evidently, it is strong.

Example 4.15 The Boolean formula $x_1 \wedge \neg x_1$ is not satisfiable. This means that $\mathrm{L} \nvdash E_1(0) \to G$ and $\mathrm{L} \nvdash E_1(1) \to G$ (the types G, $E_1(0)$, and $E_1(1)$ are shown in Example 4.1). For both assignments, we can draw the axiom links and region arcs, but there is a cycle consisting of region arcs and arcs specified by the partial order $\prec$. For the sequent $E_1(0) \to G$, this can be seen in Figure 8, where the occurrences denoted by $\parr_2$, $\otimes_4$, $\parr_8$, and $\otimes_{10}$ form a cycle.

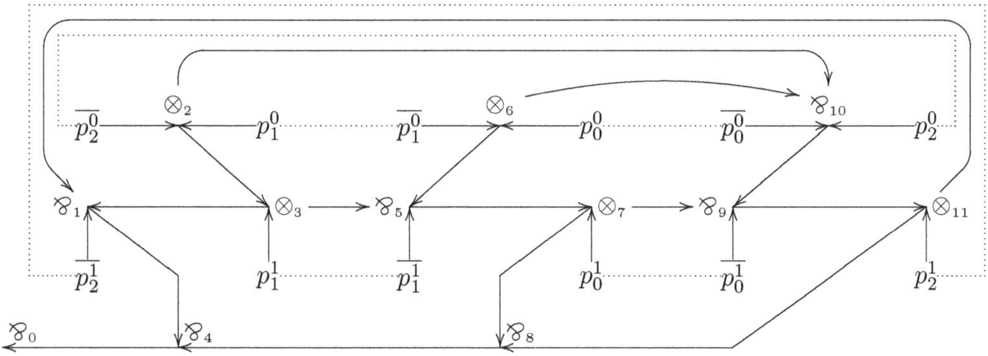

Fig. 7. The proof net corresponding to $x_1 \vee x_2$ and $\langle 0, 1 \rangle$

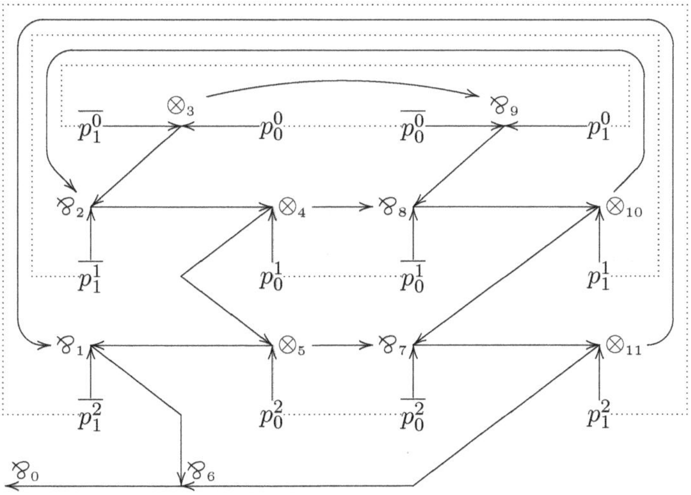

Fig. 8. The extended parse tree, axiom links, and region arcs corresponding to $x_1 \wedge \neg x_1$ and $\langle 0 \rangle$

5 The Complexity of Fragments with Restricted Sets of Connectives

5.1 The Product-Free Fragments of L* *and* L

A proof of the NP-completeness for $L^*(\backslash, /)$ and $L(\backslash, /)$ was discovered by Y. Savateev and published in [18]. Here we follow the exposition from [20], but slightly modify the notation in order to avoid name collision.

Let p_i^j, q_i^j, a_i^j, and b_i^j (where $0 \leq i \leq n$ and $0 \leq j \leq m$) be distinct primitive types from Var.

We define the following families of types:

$$\begin{aligned}
\breve{G}^0 &\leftrightharpoons (p_0^0 \setminus p_n^0),\\
\breve{G}^j &\leftrightharpoons (q_n^j \,/\, ((q_0^j \setminus p_0^j) \setminus \breve{G}^{j-1})) \setminus p_n^j,\\
\breve{G} &\leftrightharpoons \breve{G}^m\\
\breve{A}_i^0 &\leftrightharpoons (a_i^0 \setminus p_i^0),\\
\breve{A}_i^j &\leftrightharpoons (q_i^j \,/\, ((b_i^j \setminus a_i^j) \setminus \breve{A}_i^{j-1})) \setminus p_i^j,\\
\breve{A}_i &\leftrightharpoons \breve{A}_i^m,\\
\breve{E}_i^0(t) &\leftrightharpoons p_{i-1}^0,\\
\breve{E}_i^j(t) &\leftrightharpoons \begin{cases} q_i^j \,/\, (((q_{i-1}^j \,/\, \breve{E}_i^{j-1}(t)) \setminus p_{i-1}^j) \setminus p_i^{j-1}), & \text{if } \neg_t x_i \text{ occurs in } c_j,\\ (q_{i-1}^j \,/\, (q_i^j \,/\, (\breve{E}_i^{j-1}(t) \setminus p_i^{j-1}))) \setminus p_{i-1}^j, & \text{otherwise,}\end{cases}\\
\breve{F}_i(t) &\leftrightharpoons (\breve{E}_i^m(t) \setminus p_i^m),\\
\breve{B}_i^0 &\leftrightharpoons a_i^0,\\
\breve{B}_i^j &\leftrightharpoons q_{i-1}^j \,/\, (((b_i^j \,/\, \breve{B}_i^{j-1}) \setminus a_i^j) \setminus p_{i-1}^{j-1}),\\
\breve{B}_i &\leftrightharpoons \breve{B}_i^m \setminus p_{i-1}^m.
\end{aligned}$$

Let Π_i denote the sequence of types $(\breve{F}_i(0) \,/\, (\breve{B}_i \setminus \breve{A}_i))\, \breve{F}_i(0)\, (\breve{F}_i(0) \setminus \breve{F}_i(1))$.

Lemma 5.1 *If the formula* $c_1 \wedge \ldots \wedge c_m$ *is satisfiable, then* $\mathrm{L}(\setminus, /) \vdash \Pi_1 \ldots \Pi_n \to \breve{G}$.

Lemma 5.2 *If* $\mathrm{L}^*(\setminus, /) \vdash \Pi_1 \ldots \Pi_n \to \breve{G}$, *then the formula* $c_1 \wedge \ldots \wedge c_m$ *is satisfiable.*

Theorem 5.3 *The derivability problems for* $\mathrm{L}^*(\setminus, /)$ *and* $\mathrm{L}(\setminus, /)$ *are* NP*-complete.*

The length of the sequent $\Pi_1 \ldots \Pi_n \to \breve{G}$ (i.e., the total number of variable occurrences) equals $(6n+1)(4m+2)$.

We propose a modification of Savateev's construction. In this modification, a Boolean formula in conjunctive normal form with n clauses and m variables is mapped to a sequent of length $(5n+1)(4m+2)$.

We use the types $\breve{G}$ and $\breve{F}_i(t)$ defined above and introduce the following family of types:

$$\begin{aligned}
\breve{C}_i^0 &\leftrightharpoons p_i^0,\\
\breve{C}_i^j &\leftrightharpoons (q_i^j \,/\, \breve{C}_i^{j-1}) \setminus p_i^j.
\end{aligned}$$

We denote $\breve{C}_i \leftrightharpoons \breve{C}_i^m$.

Let Γ_i denote the sequence of types $(\breve{F}_i(0) \,/\, (\breve{C}_{i-1} \setminus \breve{C}_i))\, \breve{F}_i(0)\, (\breve{F}_i(0) \setminus \breve{F}_i(1))$.

Lemma 5.4 *If the formula* $c_1 \wedge \ldots \wedge c_m$ *is satisfiable, then* $\mathrm{L}(\setminus, /) \vdash \Gamma_1 \ldots \Gamma_n \to \breve{G}$.

Lemma 5.5 *If* $\mathrm{L}^*(\setminus, /) \vdash \Gamma_1 \ldots \Gamma_n \to \breve{G}$, *then the formula* $c_1 \wedge \ldots \wedge c_m$ *is satisfiable.*

The above lemmas show that our modification of Savateev's construction provides another polynomial time reduction of the classical satisfiability problem SAT to the $\mathrm{L}^*(\setminus, /)$-derivability problem and the $\mathrm{L}(\setminus, /)$-derivability problem.

5.2 The Fragments with Multiplication and One Division

A proof of the NP-completeness for $\mathrm{L}^*(\cdot,\backslash)$, $\mathrm{L}^*(\cdot,/)$, $\mathrm{L}(\cdot,\backslash)$, and $\mathrm{L}(\cdot,/)$ can be found in Y. Savateev's thesis [19].

Using the types introduced in Section 4 and additional primitive types r_i (where $1 \leq i \leq n$), we define the following family of types:

$$\begin{aligned} \dot{F}_1 &\leftrightharpoons E_1(0) \cdot ((E_1(0) \setminus E_1(1)) \cdot (H_1 \setminus r_1)), \\ \dot{F}_i &\leftrightharpoons ((E_{i-1}(0) \setminus r_{i-1}) \setminus E_i(0)) \cdot (E_i(0) \setminus E_i(1)) \cdot (H_i \setminus r_i) \quad \text{if } 1 < i \leq n, \\ \dot{F}_{n+1} &\leftrightharpoons (E_n(0) \setminus r_n) \setminus H_{n+1}. \end{aligned}$$

Lemma 5.6 *If the formula $c_1 \wedge \ldots \wedge c_m$ is satisfiable, then $\mathrm{L}(\cdot,\backslash) \vdash \dot{F}_1 \ldots \dot{F}_{n+1} \to G$.*

Lemma 5.7 *If $\mathrm{L}^*(\cdot,\backslash) \vdash \dot{F}_1 \ldots \dot{F}_{n+1} \to G$, then the formula $c_1 \wedge \ldots \wedge c_m$ is satisfiable.*

Theorem 5.8 *The derivability problems for $\mathrm{L}^*(\cdot,\backslash)$, $\mathrm{L}^*(\cdot,/)$, $\mathrm{L}(\cdot,\backslash)$, and $\mathrm{L}(\cdot,/)$ are* NP*-complete.*

5.3 The Fragments with One Connective

The fragments $\mathrm{L}^*(\cdot)$ and $\mathrm{L}(\cdot)$ are obviously decidable in polynomial time: a sequent is derivable if its antecedent and succedent yield the same sequence of primitive types after we remove all parentheses and all occurrences of the connective $\cdot$.

A deterministic polynomial-time algorithm for $\mathrm{L}(\backslash)$ and $\mathrm{L}(/)$ was discovered by Y. Savateev and published in [16] and [17]. It relies on a graph-based presentation of derivations. Here we shall call these graphs *strong unidirectional proof nets.* We give an informal exposition of the criterion from [16] in a slightly modified form.

With each type of $\mathrm{L}(\backslash)$ we associate a directed tree together with a linear order on its vertices; we draw the tree so that this linear order corresponds to the left-to-right direction. The vertices are the occurrences of primitive types in the given type. To obtain the tree corresponding to $A \setminus B$, we take the tree corresponding to B, add the left-right converse of the tree for A on the left, and draw an arc from the root of B to the root of A.

In a directed tree, each vertex has *depth,* the number of edges in the unique path from the root to this vertex. In the following diagrams, we shall indicate the depth of a vertex by the number of short horizontal lines over the primitive type.

Example 5.9 The following three trees correspond to $q \setminus r$, $s \setminus (q \setminus r)$, and $(q \setminus r) \setminus t$.

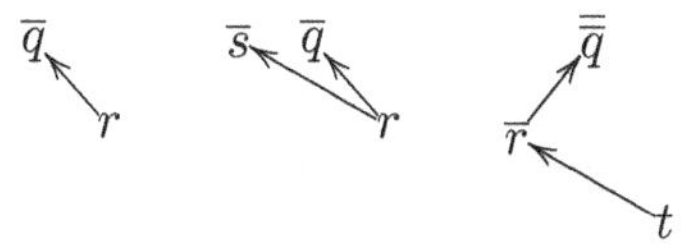

Now we can give the definition of a strong unidirectional proof net for a sequent $C_1 \ldots C_n \to D$. The first component of such a proof net is the tree that corresponds to $C_n \setminus (C_{n-1} \setminus \ldots \setminus (C_1 \setminus D) \ldots)$. The second component of the proof net is a set of axiom links. Each axiom link is an edge between two occurrences of the same primitive type. Each vertex must be incident to exactly one axiom link. The axiom links must be drawn above the tree without intersecting each other or edges from the tree.

We require that for each axiom link the depth of its left end be one greater than the depth of its right end. Moreover, if the right end of an axiom link has non-zero even depth, then between the two ends of the axiom link there must be a vertex of smaller depth.

If all the above conditions are satisfied, then we have a strong unidirectional proof net for $C_1 \dots C_n \to D$.

Example 5.10 Consider the sequent $(r \setminus p)\,((s \setminus p) \setminus t) \to (s \setminus r) \setminus t$, which is derivable in L. A strong unidirectional proof net for this sequent is shown in Figure 9.

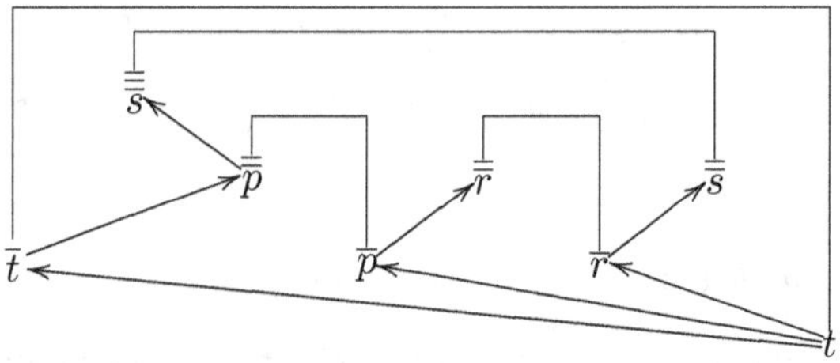

Fig. 9. A strong unidirectional proof net

Theorem 5.11 *A sequent $C_1 \dots C_n \to D$ is derivable in* L($\setminus$) *if and only if there exists a strong unidirectional proof net for $C_1 \dots C_n \to D$.*

The proof of the theorem can be found in [16].

Now a polynomial-time algorithm for the derivability problem in L($\setminus$) is easy to design. We construct the tree corresponding to a given sequent, and for each interval of vertices (in the sense of the left-to-right order), we find out whether there exists a set of axiom links that involves all the vertices of the interval and does not involve other vertices (each axiom link must satisfy the conditions from the definition of a strong unidirectional proof net). The maximal interval covers all the tree and gives an answer whether a strong unidirectional proof net exists.

A deterministic polynomial-time algorithm for L*($\setminus$) and L*($/$) was also discovered by Y. Savateev. To obtain a derivability criterion for L*($\setminus$) it suffices to modify the definition of a proof net. The only difference is in the last condition (about axiom links whose right end has non-zero even depth). For L*($\setminus$) the corresponding condition is the following. If the right end of an axiom link has depth $2k$, where $k > 0$, and the depth of the left predecessor of the left end is greater than $2k$, then there must be a vertex of depth less than $2k$ between the two ends of the axiom link. We shall call proof nets of this kind *unidirectional*. It is easy to see that each strong unidirectional proof net is a unidirectional proof net.

Example 5.12 Consider the sequent $(q \setminus q) \setminus s \to \setminus s$, which is derivable in L*, but not in L. A unidirectional proof net for this sequent is shown in Figure 10.

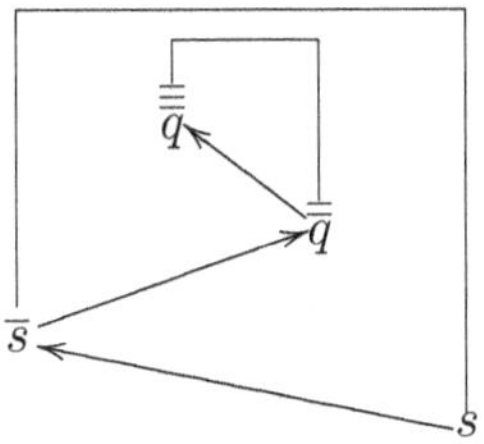

Fig. 10. A unidirectional proof net

Theorem 5.13 *A sequent* $C_1 \dots C_n \to D$ *is derivable in* $\mathrm{L}^*(\backslash)$ *if and only if there exists a unidirectional proof net for* $C_1 \dots C_n \to D$.

Theorem 5.13 provides a deterministic polynomial-time algorithm for the derivability problem for $\mathrm{L}^*(\backslash)$. We use the method of dynamic programming, similarly to the case of $\mathrm{L}(\backslash)$.

6 The Complexity of Fragments with Restricted Number of Variables

For each of the above fragments, the derivability problem remains in the same complexity class if we allow for only one variable (or restrict the number of variables by any positive integer).

We introduce the abbreviation

$$A^k \leftrightharpoons \underbrace{A \cdot \ldots \cdot A}_{k \text{ times}},$$

where A is a type and k is a positive integer.

If a fragment contains both $\backslash$ and $/$ and a sequent contains only the variables q_i, where $0 < i < N$, then we can replace each variable q_i by the type $p^i \backslash p \,/\, p^{N-i}$, where $p \in \mathrm{Var}$, and this does not affect derivability (this substitution is based on a construction from [10]). Since $(A \cdot B) \backslash C \underset{\mathrm{L}}{\leftrightarrow} B \backslash (A \backslash C)$ and $C \,/\, (A \cdot B) \underset{\mathrm{L}}{\leftrightarrow} (C \,/\, B) \,/\, A$, the above types can be replaced by equivalent types that do not contain $\cdot$.

Example 6.1 It is easy to see that $\mathrm{L} \nvdash q_1 \backslash q_2 \to q_1$. In the one-variable fragment, this corresponds to

$$\mathrm{L} \nvdash ((p \backslash p \,/\, p) \,/\, p) \backslash (p \backslash (p \backslash p \,/\, p)) \to (p \backslash p \,/\, p) \,/\, p.$$

Here $N = 3$.

To prove the NP-completeness of the one-variable fragments of $\mathrm{L}(\cdot, \backslash)$ and $\mathrm{L}^*(\cdot, \backslash)$, we use a substitution discovered by S. Kuznetsov (see [6]) involving only the left division (of course, there is a dual substitution, which fits for $\mathrm{L}(\cdot, /)$ and $\mathrm{L}^*(\cdot, /)$). Let in L or L^* a sequent contain only the variables q_i, where $0 < i < N$. Then we can replace each variable q_i by the type

$$(p^{i+1} \cdot (((p \cdot p) \backslash p) \backslash p) \cdot p^{N-i}) \backslash p,$$

and this does not affect derivability. Obviously, these types can be replaced by equivalent types that do not contain $\cdot$ or $/$.

Example 6.2 If $N = 3$, then Kuznetsov's substitution maps q_1 to

$$p \backslash (p \backslash (((p \backslash (p \backslash p)) \backslash p) \backslash (p \backslash (p \backslash p))))$$

and q_2 to

$$p \backslash (((p \backslash (p \backslash p)) \backslash p) \backslash (p \backslash (p \backslash (p \backslash p)))).$$

Conclusion

The complexity results for the derivability problem for fragments of L and L* are summarized in Table 1, where NP denotes NP-completeness. One-variable fragments are denoted by $L(p_1)$ and $L^*(p_1)$.

<table>
<tr><th></th><th>L</th><th>L*</th><th>$L(p_1)$</th><th>$L^*(p_1)$</th></tr>
<tr><td>$\cdot, \backslash, /$</td><td>NP</td><td>NP</td><td>NP</td><td>NP</td></tr>
<tr><td>$\cdot, \backslash$</td><td rowspan="2">NP</td><td rowspan="2">NP</td><td rowspan="2">NP</td><td rowspan="2">NP</td></tr>
<tr><td>$\cdot, /$</td></tr>
<tr><td>$\cdot$</td><td>P</td><td>P</td><td>P</td><td>P</td></tr>
<tr><td>$\backslash, /$</td><td>NP</td><td>NP</td><td>NP</td><td>NP</td></tr>
<tr><td>$\backslash$</td><td rowspan="2">P</td><td rowspan="2">P</td><td rowspan="2">P</td><td rowspan="2">P</td></tr>
<tr><td>$/$</td></tr>
</table>

Table 1
The complexity of fragments of L and L*

References

[1] Abrusci, V. M., *Phase semantics and sequent calculus for pure noncommutative classical linear propositional logic*, Journal of Symbolic Logic **56** (1991), pp. 1403–1451.

[2] Buszkowski, W., *Mathematical linguistics and proof theory*, in: J. van Benthem and A. ter Meulen, editors, *Handbook of Logic and Language*, Elsevier/MIT Press, Amsterdam/Cambridge, MA, 1997 pp. 683–736.

[3] de Groote, P., *A dynamic programming approach to categorial deduction*, in: H. Ganzinger, editor, *Proc. Automated deduction—CADE-16*, Springer, Berlin, 1999 pp. 1–15.

[4] Fleury, A., "La règle d'échange: logique linéaire multiplicative tréssée," Ph.D. thesis, Université Paris 7 (1996), thèse de Doctorat, spécialité Mathématiques.

[5] Girard, J.-Y., *Towards a geometry of interaction*, in: J. W. Gray and A. Scedrov, editors, *Categories in Computer Science and Logic*, American Mathematical Society, Providence, RI, 1989 pp. 69–108.

[6] Kuznetsov, S. L., *Lambek calculus with one division and one primitive type permitting empty antecedents*, Moscow University Mathematics Bulletin **64** (2009), pp. 76–79.

[7] Lamarche, F. and C. Retoré, *Proof nets for the Lambek calculus—an overview*, in: V. M. Abrusci and C. Casadio, editors, *Proofs and Linguistic Categories, Proc. 1996 Roma Workshop*, CLUEB, Bologna, 1996 pp. 241–262.

[8] Lambek, J., *The mathematics of sentence structure*, American Mathematical Monthly **65** (1958), pp. 154–170.

[9] Lambek, J., *From categorial grammar to bilinear logic*, in: K. Došen and P. Schroeder-Heister, editors, *Substructural Logics*, Oxford University Press, Oxford, 1993 pp. 207–237.

[10] Métayer, F., *Polynomial equivalence among systems* LLNC, LLNC_a *and* LLNC_0, Theoretical Computer Science **227** (1999), pp. 221–229.

[11] Moortgat, M., *Categorial type logics*, in: J. van Benthem and A. ter Meulen, editors, *Handbook of Logic and Language*, Elsevier/MIT Press, Amsterdam/Cambridge, MA, 1997 pp. 93–177.

[12] Pentus, M., *Equivalent types in Lambek calculus and linear logic*, MIAN Prepublication Series for Logic and Computer Science LCS-92-02, Steklov Mathematical Institute, Moscow (1992).

[13] Pentus, M., *Free monoid completeness of the Lambek calculus allowing empty premises*, in: J. M. Larrazabal, D. Lascar and G. Mints, editors, *Proc. Logic Colloquium '96*, Springer, Berlin, 1998 pp. 171–209.

[14] Pentus, M., *Lambek calculus is* NP*-complete*, CUNY Ph.D. Program in Computer Science Technical Report TR-2003005, CUNY Graduate Center, New York (2003).

[15] Pentus, M., *Lambek calculus is* NP*-complete*, Theoretical Computer Science **357** (2006), pp. 186–201.

[16] Savateev, Y., *The derivability problem for Lambek calculus with one division*, Artificial Intelligence Preprint Series 56, Utrecht University (2006).

[17] Savateev, Y., *Lambek grammars with one division are decidable in polynomial time*, in: E. A. Hirsch, A. A. Razborov, A. L. Semenov and A. Slissenko, editors, *Computer Science—Theory and Applications, Third International Computer Science Symposium in Russia, CSR 2008, Moscow, Russia, June 7–12, 2008, Proceedings*, Springer, Berlin, 2008 pp. 273–282.

[18] Savateev, Y., *Product-free Lambek calculus is* NP*-complete*, CUNY Ph.D. Program in Computer Science Technical Report TR-2008012, CUNY Graduate Center, New York (2008).

[19] Savateev, Y., "Algorithmic Complexity of Fragments of the Lambek Calculus," Ph.D. thesis, Moscow State University (2009), (in Russian).

[20] Savateev, Y., *Product-free Lambek calculus is* NP*-complete*, in: S. N. Artemov and A. Nerode, editors, *Proc. Logical Foundations of Computer Science 2009*, Springer, Berlin, 2009 pp. 380–394.

[21] van Benthem, J., "Language in Action: Categories, Lambdas and Dynamic Logic," North-Holland, Amsterdam, 1991.

[22] Yetter, D. N., *Quantales and (noncommutative) linear logic*, Journal of Symbolic Logic **55** (1990), pp. 41–64.

Goldblatt-Thomason-style Theorems for Graded Modal Language

Katsuhiko Sano

JSPS Research Fellow
Department of Humanistic Informatics, Kyoto University, Japan
ILLC, Universiteit van Amsterdam

Minghui Ma

Department of Philosophy, Tsinghua University, Beijing, China
ILLC, Universiteit van Amsterdam

Abstract

We prove two main Goldblatt-Thomason-style Theorems for graded modal language in Kripke semantics: full Goldblatt-Thomason Theorem for elementary classes and relative Goldblatt-Thomason Theorem within the class of finite transitive frames. Two different semantic views on GML allow us to prove these results: neighborhood semantics and graph semantics. By neighborhood semantic view, we can define a natural generalization of Jankov-Fine formula for GML and establish relative Goldblatt-Thomason Theorem. By extracting graph semantics from Fine's completeness proof of GML (1972), we introduce a new notion of graded ultrafilter images and establish full Goldblatt-Thomason Theorem. Therefore we revive Fine's old idea in the new context of Goldblatt-Thomason-style characterization.

Keywords: graded modal logic, Goldblatt-Thomason theorem, graded Jankov-Fine formula, graph semantics, graded ultrafilter images.

1 Introduction

Graded modal logic (GML) is one of extended modal logics. It was originally proposed by Kit Fine [10] to express modal analogues to counting quantifiers $\exists_k xP(x)$ in first-order logic explored by A. Tarski [19]. The modal analogue to $\exists_k xP(x)$ is written as $\Diamond_k p$; it is true at a state w in a Kripke model iff the number of accessible p-worlds is at least k. In GML, we add the family $\{ \Diamond_k : k \in \omega \}$ of modal operators to the basic modal logic. In particular, $\Diamond_k$ is non-normal: $\Diamond_k \bot \leftrightarrow \bot$ $(k > 1)$ and $(\Diamond_k p \vee \Diamond_k q) \to \Diamond_k (p \vee q)$ are valid but $\Diamond_k (p \vee q) \to (\Diamond_k p \vee \Diamond_k q)$ is not valid $(k > 1)$. Such modalities are used when it comes to counting successors. For example, GML was applied to epistemic logic [25] and description logic [1].

The model theory for GML has been explored since the 1970s. Some normal graded modal logics were shown to be strongly complete by Kit Fine [10], and later canonical models were constructed by M. Fattorosi-Barnaba and C. Cerrato [8,9], F. de Caro [6], and C. Cerrato [3]. M. de Rijke [7] defined the notion of graded bisimulation and proved Van Benthem-style Characterization Theorem. De Rijke also noted that graded modalities $\Diamond_{n+2}$ are not definable in basic modal logic since they are not invariant under ordinary notion bisimulations. Hence GML is a proper extension of basic modal logic with more expressive power at the level of Kripke models. Recently, Ten Cate et.al. [21] observed that GML can describe finite tree models up to isomorphism. If we turn to Kripke frames, we can define the frame property "there exist at least two successors", by $\Diamond_2\top$, which is undefinable in basic modal logic. So, we can also state that GML is more expressive than basic modal logic at the level of Kripke frames.

Goldblatt-Thomason Theorem [12] allows us to characterize the modal definability of elementary classes by four frame constructions: generated subframes, disjoint unions, bounded morphic images, and ultrafilter extensions. By this theorem, we can state that the semantic essence for frame-definability of basic modal language consists of these four frame constructions. Later, Van Benthem [24] gave a model-theoretical proof of Goldblatt-Thomason Theorem. Since [12,24], Goldblatt-Thomason-style Theorem has been investigated also for extended modal logics: difference logic [11], hybrid logic [20], etc. The first author and Sato [18] (see also [17]) provided a uniform Goldblatt-Thomason-style characterization of frame definability for *any* modal language extended with a set of *normal* modal operators, whose accessibility relations are defined by Boolean combinations of a (binary) accessibility relation R and the equality, that is, by quantifier-free formulas.

As for Goldblatt-Thomason-style characterization of definability in GML, de Rijke asked the following question.

> Obvious questions to be answered next include the following: Can g-bisimulations be used to prove a Goldblatt-Thomason style results about the classes of frames definable in $\mathcal{L}_{\text{GML}}$? [7, p.282]

As far as the authors know, we still lack Goldblatt-Thomason theorem for GML in Kripke semantics. In this paper, we provide two Goldblatt-Thomason-style Theorems, thus answering de Rijke's question positively (though we will not use the notion of g-bisimulation). One is Goldblatt-Thomason theorem for elementary classes of frames (Theorem 6.3), and the other is relative Goldblatt-Thomason theorem within the class of finite transitive frames (Theorem 4.3).

Our results for GML correspond to the results from [2, Theorem 3.21] and [2, Theorem 3.19] for basic modal logic. This generalization, however, is not straightforward. One of the main reasons is the non-normal character of $\Diamond_k$ mentioned above. In order to deal with modalities of this kind and obtain our main results, we need to apply two different semantic approaches to GML: *neighborhood semantics* and *graph semantics*. The neighborhood semantic view leads us to the notion of *g-bounded morphism* and to a natural generalization of Jankov-Fine formulas [2, pp.144-5] for GML. By applying these formulas, we establish relative Goldblatt-Thomason Theorem for GML in terms of generated subframes, (finite) disjoint unions, and g-bounded morphisms.

A difficulty for establishing full Goldblatt-Thomason Theorem for GML is also in finding an appropriate notion of ultrafilter extension for GML. The method used in [17] does not seem to work here, because of the non-normal behavior of $\Diamond_k$. However, following the observation of [18] about the completeness proof and Goldblatt-Thomason-style characterization, we can extract an appropriate frame construction from Kit Fine's original completeness proof for GML [10]. By analyzing his proof carefully, we demonstrate that Fine's construction allows us to define a new graph semantics for GML. Moreover, we rewrite Fine's notion of *canonical mapping* [10, p.518] as the new frame construction *graded ultrafilter image* via a graph frame defined on the set of all ultrafilters on (W, R). In other words, a key semantic idea for our full Goldblatt-Thomason-style Theorem already appeared implicitly in the first study of GML by Fine. So we *revive* Fine's old idea in the *new* context of Goldblatt-Thomason-style characterization for GML.

2 Kripke Semantics for Graded Modal Language

Graded modal language (*GML*, for short) consists of (i) a countable set Prop of proposition letters, (ii) $\wedge$, $\neg$, $\bot$, and (iii) a set $\{\,\Diamond_k : k \in \omega\,\}$ of modal operators called *graded modalities.* The set of formulas in GML is defined as follows:

$$\varphi ::= p \,|\, \bot \,|\, \neg\varphi \,|\, \varphi \wedge \psi \,|\, \Diamond_k \varphi,$$

where $p \in \mathsf{Prop}$ and $k \in \omega$. The intuitive meaning of $\Diamond_k\varphi$ is 'the number of accessible φ-worlds is at least k'. In addition to the usual abbreviations like $\rightarrow$, $\vee$, etc., we define $\Diamond\varphi := \Diamond_1\varphi$ and $\Box\varphi := \neg\Diamond\neg\varphi$. We also define $\Box^n p$ inductively as: $\Box^0 p := p$ and $\Box^{n+1} p := \Box\Box^n p$. By the *basic modal language*, we mean the sublanguage $\{\,\wedge, \neg, \bot, \Diamond_1\,\} \cup \mathsf{Prop}$ of GML.

GML is interpreted in Kripke structures. A *Kripke frame* $\mathfrak{F}$ (or, just a *frame*) is a pair (W, R) of a non-empty set W and a binary relation $R \subseteq W^2$. A *Kripke model* $\mathfrak{M}$ (or, just a *model*) consists of a frame $\mathfrak{F} = (W, R)$ and a valuation $V : \mathsf{Prop} \rightarrow \mathcal{P}(W)$. The domain of a Kripke frame $\mathfrak{F}$ (or a Kripke model $\mathfrak{M}$) is denoted by $|\mathfrak{F}|$ (or $|\mathfrak{M}|$, respectively).

Given any model $\mathfrak{M} = (W, R, V)$, any $w \in W$ and any formula φ of GML, we define the *satisfaction relation* $\Vdash$ as standard except the clause for graded modalities:

$$\mathfrak{M}, w \Vdash \Diamond_k\varphi \ \text{ iff } \ \#(R(w) \cap [\![\varphi]\!]) \geq k,$$

where $R(w) := \{\, v \in W : wRv\,\}$, $[\![\varphi]\!]_{\mathfrak{M}} := \{\, v \in W : \mathfrak{M}, v \Vdash \varphi\,\}$ (when the context is clear, we usually drop the subscript), and $\#X$ means the cardinality of X. When $k = 0$, it is easy to see that $\Diamond_0\varphi$ is true at any state w of any model $\mathfrak{M}$. Remark that the satisfaction for $\Diamond\varphi := \Diamond_1\varphi$ is equivalent to $R(w) \cap [\![\varphi]\!] \neq \emptyset$. Based on the satisfaction relation, we can define the notion of *frame validity, frame definability, satisfiability,* etc.

as usual, cf. [2].

Fact 2.1 (Fine [10]) *The following formulas are valid in all frames:*

(i) $\Box(p \to q) \to (\Box p \to \Box q)$,

(ii) $\Diamond_k p \to \Diamond_l p$ $(l < k)$,

(iii) $\Diamond_k p \leftrightarrow \bigvee_{i=0}^{k}(\Diamond_i(p \wedge q) \wedge \Diamond_{k-i}(p \wedge \neg q))$,

(iv) $\Box(p \to q) \to (\Diamond_k p \to \Diamond_k q)$.

Here is another form of the truth condition for $\Diamond_k \varphi$ at $w \in W$ in (W, R, V):

$$\#(R(w) \cap [\![\varphi]\!]) \geq k \text{ iff } \exists X \subseteq R(w). (\#X = k \text{ and } X \subseteq [\![\varphi]\!]).$$

This form allows us to define *neighborhood maps* $\tau_k : W \to \mathcal{PP}(W)$ $(k \in \omega)$ for any given Kripke frame (W, R) as:

$$\tau_k(w) := \{ Y \subseteq W : \exists X \subseteq R(w). (\#X = k \text{ and } X \subseteq Y) \}.$$

τ_k is closed under unions, i.e., $Y_1, Y_2 \in \tau_k(w)$ implies $Y_1 \cup Y_2 \in \tau_k(w)$, for any $w \in W$ and $Y_1, Y_2 \subseteq W$. Thus τ_k is *monotonic*, i.e., closed under set-inclusion $\subseteq$. However, $\tau_k(w)$ does not satisfy the following property in general (if $k > 1$): $Y_1 \cup Y_2 \in \tau_k(w)$ implies $Y_1, Y_2 \in \tau_k(w)$, for any $w \in W$ and $Y_1, Y_2 \subseteq W$. For any valuation V, it is clear that $\#(R(w) \cap [\![\varphi]\!]) \geq k$ iff $[\![\varphi]\!] \in \tau_k(w)$. This observation enables us to use the notion of *bounded morphism* between neighborhood structures for GML in the next section.

3 Preservation under Frame Constructions

Definition 3.1 (g-bounded morphism) *Given any $\mathfrak{F} = (W, R)$ and $\mathfrak{F}' = (W', R')$, we say that $f : W \to W'$ is a* g-bounded morphism *if: for any $k \in \omega$ and any $Y \subseteq W'$,*

$$\#(R(w) \cap f^{-1}[Y]) \geq k \text{ iff } \#(R'(f(w)) \cap Y) \geq k.$$

If there is a surjective g-bounded morphism from $\mathfrak{F}$ and $\mathfrak{F}'$, then we say that $\mathfrak{F}'$ is a g-bounded morphic image *of $\mathfrak{F}$ (notation: $\mathfrak{F} \twoheadrightarrow_g \mathfrak{F}'$).*

In terms of derived neighborhood structures, this definition can be rewritten as: $f^{-1}[Y] \in \tau_k(w)$ iff $Y \in \tau_k(f(w))$ for any $Y \subseteq W'$, where τ_k and τ'_k are the neighborhood structures derived from $\mathfrak{F}$ and $\mathfrak{F}'$, respectively. If we restrict our attention to $k = 1$ in the definition of g-bounded morphism, then we can easily obtain the notion of *bounded morphism* [2, p.59]:

Definition 3.2 *Given any $\mathfrak{F} = (W, R)$ and $\mathfrak{F}' = (W', R')$, we say that $f : W \to W'$ is a* bounded morphism *if $f[R(w)] = R'(f(w))$ for any $w \in W$.*

Proposition 3.3 *Assume that $\mathfrak{F} \twoheadrightarrow_g \mathfrak{F}'$. If $\mathfrak{F} \Vdash \varphi$, then $\mathfrak{F}' \Vdash \varphi$, for any φ of GML.*

Proof. We show the contrapositive implication. Assume $\mathfrak{F}' \nVdash \varphi$. That is, $(\mathfrak{F}', V'), w' \nVdash \varphi$ for some $w' \in |\mathfrak{F}'|$ and some valuation V'. Define a valuation V on $\mathfrak{F}$ by: $V(p) :=$

$f^{-1}[V'(p)]$ for any $p \in \mathsf{Prop}$. Put $\mathfrak{M} := (\mathfrak{F}, V)$ and $\mathfrak{M}' := (\mathfrak{F}', V')$. Then we can establish that $[\![\varphi]\!]_{\mathfrak{M}} = f^{-1}[[\![\varphi]\!]_{\mathfrak{M}'}]$ by induction on φ. Since f is surjective, $f(w) = w'$ for some $w \in |\mathfrak{F}|$. Therefore, we obtain $w \notin [\![\varphi]\!]_{\mathfrak{M}}$ hence $\mathfrak{F} \not\Vdash \varphi$. □

When we try to check that a given mapping is a g-bounded morphism, the equivalent notion of *locally injective bounded morphism* is quite helpful, since it does not involve any quantification over $Y \subseteq W'$.

Definition 3.4 *Given any $\mathfrak{F} = (W, R)$ and $\mathfrak{F}' = (W', R')$, we say that $f : W \to W'$ is* locally injective *if $f \upharpoonright R(w)$ is injective for any $w \in W$.*

Proposition 3.5 *Given any $\mathfrak{F} = (W, R)$, $\mathfrak{F}' = (W', R')$, and $f : W \to W'$, f is a g-bounded morphism* iff *f is a locally injective bounded morphism.*

Proof. We only establish the left-to-right direction, since the converse direction is obvious. Suppose that f is a g-bounded morphism. It is easy to show that f is a bounded morphism (it suffices to use the clause of $k = 1$ in the definition of g-bounded morphism). We show that $f \upharpoonright R(w)$ is injective. Let us fix any $w_1,\ w_2 \in R(w)$ with $w_1 \neq w_2$. Our goal is to establish that $f(w_1) \neq f(w_2)$. Since $\{w_1, w_2\} = R(w) \cap \{w_1, w_2\} \subseteq R(w) \cap f^{-1}[f[\{w_1, w_2\}]]$, we obtain $\#(R(w) \cap f^{-1}[f[\{w_1, w_2\}]]) \geq 2$. From the clause of $k = 2$ in the definition of g-bounded morphism it follows that $\#(R'(f(w)) \cap f[\{w_1, w_2\}]) \geq 2$. Since $R'(f(w)) = f[R(w)]$ (here we use the fact that f is a bounded morphism), we can establish that $\#(f[R(w)] \cap f[\{w_1, w_2\}]) \geq 2$. Equivalently, $\#f[\{w_1, w_2\}] \geq 2$. This means that $f(w_1) \neq f(w_2)$, as required. □

Surprisingly, this proof tells us that the clauses of $k = 1$ and 2 are enough to define the notion of g-bounded morphism.

Given any cardinal κ, let us denote by $\mathrm{NS}(\kappa)$ ('NS' means the number of successors) the following property of Kripke frames: $\#R(w) \geq \kappa$ for any $w \in W$. The next proposition shows the expressive strength of GML over the basic modal language.

Proposition 3.6 *Let $k \in \omega$ and $k \geq 2$. Then, $\mathrm{NS}(k)$ is undefinable in the basic modal language. However, $\mathrm{NS}(k)$ is definable in GML by $\Diamond_k \top$.*

Proof. The second part is easy to check. So we only show the first part. Define $\mathfrak{F}_k = (W, R)$ by: $W = \{1, ..., k\}^{<\omega}$ and $\langle l_1, ..., l_m \rangle R \langle r_1, ..., r_n \rangle$ iff $n = m + 1$ and $l_i = r_i$ for any i with $1 \leq i \leq m$. So $\mathfrak{F}_k$ is a tree with k-branches and ω-height. Define $\mathfrak{G}$ as a one-point reflexive frame $(\{*\}, \{(*, *)\})$. Let f be the unique mapping from W to $\{*\}$. It is easy to see that f is a surjective bounded morphism. Then by the validity-preservation under bounded morphic images [2, Theorem 3.14 (iii)], we can establish the undefinablity of $\mathrm{NS}(k)$. □

Proposition 3.7 *$\mathrm{NS}(\omega)$ is definable in GML by $\{\Diamond_k \top : k \in \omega\}$.*

However, irreflexivity (wRw fails for any $w \in W$) is still undefinable in GML.

Proposition 3.8 *Irreflexivity is undefinable in GML.*

Proof. Let us use the frame $\mathfrak{F}_1 = (W, R)$ and $\mathfrak{G}$ from the proof of Proposition 3.6 (we fix $k = 1$ here). Note that $\mathfrak{F}_1$ is irreflexive, but $\mathfrak{G}$ is not irreflexive. Then we can show that the unique surjective $f : W \to \{*\}$ is a g-bounded morphism from $\mathfrak{F}_1$ to $\mathfrak{G}$. By Proposition 3.3, we can establish the undefinability of irreflexivity in GML. □

Definition 3.9 (Generated subframes) *Given any frames $\mathfrak{F} = (W, R)$ and $\mathfrak{F}' = (W', R')$, we say that $\mathfrak{F}'$ is a* generated subframe *of $\mathfrak{F}$ if* (i) $W' \subseteq W$, (ii) $R' = R \cap (W')^2$, (iii) $R(w') \subseteq W'$ *for any $w' \in W'$. We say that $\mathfrak{F}'$ is a* point-generated subframe *of $\mathfrak{F}$ by a* root *w in $\mathfrak{F}$ (notation: $\mathfrak{F}_w$) if $\mathfrak{F}'$ is the smallest generated subframe of $\mathfrak{F}$ whose domain contains w.*

Proposition 3.10 *If $\mathfrak{F}'$ is a generated subframe of $\mathfrak{F}$, then $\mathfrak{F} \Vdash \varphi$ implies $\mathfrak{F}' \Vdash \varphi$, for any φ of GML.*

Definition 3.11 (Disjoint unions) *Let $\{\mathfrak{F}_i : i \in I\}$ be a pairwise disjoint family of frames, where $\mathfrak{F}_i = (W_i, R_i)$. We define the disjoint union $\biguplus_{i\in I} \mathfrak{F}_i = (W, R)$ of $\{\mathfrak{F}_i : i \in I\}$ as: $W = \bigcup_{i\in I} W_i$ and $R = \bigcup_{i\in I} R_i$.*

Proposition 3.12 *For any pairwise disjoint family $\{\mathfrak{F}_i : i \in I\}$ of frames and any φ of GML, if $\mathfrak{F}_i \Vdash \varphi$ $(i \in I)$ then $\biguplus_{i\in I} \mathfrak{F}_i \Vdash \varphi$.*

Note that we may assume that, up to isomorphism, any family of frames is pairwise disjoint.

Proposition 3.13 *Any frame $\mathfrak{F}$ is a g-bounded morphic image of the disjoint union of some family of generated subframes of $\mathfrak{F}$.*

Proof. It suffices to note that $\biguplus_{w\in|\mathfrak{F}|} \mathfrak{F}_w \twoheadrightarrow_g \mathfrak{F}$. □

4 Graded Modal Classes of Finite Transitive Frames

First, we define the graded Jankov-Fine formulas as follows.

Definition 4.1 *Let $\mathfrak{F}_w = (W, R)$ be a finite transitive frame with the root w. Put $W = \{w_0, \dots, w_n\}$ and $w = w_0$. Associate each $w_i \in W$ with a new proposition letter p_i. Define $p_X := \bigvee\{p_i : w_i \in X\}$ for each finite $X \subseteq W$. Let $\Box^+\varphi := \varphi \wedge \Box\varphi$. The* graded Jankov-Fine formula *$\varphi_{\mathfrak{F},w}$ is defined as the conjunction of all the following formulas:*

(i) p_0

(ii) $\Box(p_0 \vee \cdots \vee p_n)$

(iii) $\bigwedge\{\Box^+(p_i \to \neg p_j) : i \neq j\}$

(iv) $\bigwedge\{\Box^+(p_i \to \Diamond_k p_X) : X \in \tau_k(w_i)\}$

(v) $\bigwedge\{\Box^+(p_i \to \neg\Diamond_k p_X) : X \notin \tau_k(w_i)\}$

Clearly, $\varphi_{\mathfrak{F},w}$ is true at w of (W, R) under the natural valuation $V(p_i) = \{w_i\}$ (remark that $V(p_X) = X$ for finite $X \subseteq W$).

Lemma 4.2 *Let $\mathfrak{F} = (W, R)$ be a finite transitive point-generated frame with the root w. Then for any transitive $\mathfrak{G}$, the following are equivalent:*

(A) *$\varphi_{\mathfrak{F},w}$ is satisfiable in $\mathfrak{G}$,*

(**B**) *there exists* $v \in |\mathfrak{G}|$ *such that* $\mathfrak{F}$ *is a g-bounded morphic image of* $\mathfrak{G}_v$.

Proof. We can easily show the direction from (B) to (A) by Proposition 3.10 and 3.13, since $\varphi_{\mathfrak{F},w}$ is satisfiable in $\mathfrak{F}$ at w under the natural valuation V such that $V(p_i) = \{ w_i \}$. Conversely, let us assume (A). It follows that $(\mathfrak{G}_v, U), v \Vdash \varphi_{\mathfrak{F},w}$ for some $v \in |\mathfrak{G}|$ and some valuation U on $|\mathfrak{G}_v|$. Now put $\mathfrak{G}_v = (G, S)$. By the conjuncts (ii) and (iii) of $\varphi_{\mathfrak{F},w}$, we have $G = \bigcup_{0 \leq 1 \leq n} U(p_i)$ and $U(p_i) \cap U(p_j) = \emptyset$ for any i, j with $i \neq j$, respectively. By the rootedness of $\mathfrak{F}$ and (iv), we can also establish that $U(p_i) \neq \emptyset$ for any i with $0 \leq i \leq n$. This allows us to define a *surjective* mapping $f : G \to W$.

Now we show that f is a g-bounded morphism. Consider $k \in \omega$ and $X \subseteq W$ and $x \in G$ (note that X is finite). Let us put $f(x) := w_i$. We show the equivalence: $\#(S(x) \cap f^{-1}[X]) \geq k$ iff $\#(R(w_i) \cap X) \geq k$. First, we establish the left-to-right direction. We show the contrapositive implication. So, suppose $\#(R(w_i) \cap X) < k$ hence $X \notin \tau_k(w_i)$. Then, by the conjunct (v) of $\varphi_{\mathfrak{F},w}$ and our assumption: $(\mathfrak{G}_v, U), v \Vdash \varphi_{\mathfrak{F},w}$, we deduce that $(\mathfrak{G}_v, U), v \Vdash \Box^+(p_i \to \neg\Diamond_k p_X)$. Since $x \in |\mathfrak{G}_v|$ and $\mathfrak{G}_v$ is still transitive, from $x \in U(p_i)$ we deduce that $(\mathfrak{G}_v, U), x \nVdash \Diamond_k p_X$, hence $\#(S(x) \cap U(p_X)) < k$. It is easy to show that $f^{-1}[X] = U(p_X)$ (recall $p_X := \bigvee \{ p_i : w_i \in X \}$ and X is finite). Therefore, $\#(S(x) \cap f^{-1}[X]) < k$, as desired. Second, we can also establish the right-to-left direction similarly to the argument above. It suffices to use the conjunct (iv) instead of (v). □

Theorem 4.3 *Let* C *be the class of all finite transitive frames and* $\mathsf{F} \subseteq \mathsf{C}$. *Then* F *is definable by a set of formulas in GML within* C iff F *is closed under taking* (i) *generated subframes,* (ii) *(finite) disjoint unions,* (iii) *g-bounded morphic images.*

Proof. We can easily establish the left-to-right direction by Propositions 3.3, 3.10 and 3.12. Conversely, suppose that F satisfies the closure conditions. Define $\mathrm{Log}(\mathsf{F}) := \{ \varphi : \mathsf{F} \Vdash \varphi \}$. Let us show that $\mathrm{Log}(\mathsf{F})$ defines F within C. Consider $\mathfrak{F} \in \mathsf{C}$, i.e., $\mathfrak{F}$ is finite and transitive. We show the equivalence: $\mathfrak{F} \in \mathsf{F}$ iff $\mathfrak{F} \Vdash \mathrm{Log}(\mathsf{F})$. The left-to-right direction is easy to show. Let us establish the right-to-left direction. Assume $\mathfrak{F} \Vdash \mathrm{Log}(\mathsf{F})$. We subdivide our argument into the following two cases: (a) $\mathfrak{F}$ is point-generated and (b) $\mathfrak{F}$ is not point-generated.

First, let us consider the case (a). Let w be the root of $\mathfrak{F}$. Consider the Jankov-Fine formula $\varphi_{\mathfrak{F},w}$ of $\mathfrak{F}$ and w (note that $\mathfrak{F}$ is finite, transitive, and point-generated). Since $\varphi_{\mathfrak{F},w}$ is satisfiable in $\mathfrak{F}$ (i.e., $\mathfrak{F} \nVdash \neg\varphi_{\mathfrak{F},w}$), we have $\neg\varphi_{\mathfrak{F},w} \notin \mathrm{Log}(\mathsf{F})$. Then there exists $\mathfrak{G} \in \mathsf{F}$ such that $\varphi_{\mathfrak{F},w}$ is satisfiable in $\mathfrak{G}$. From Lemma 4.2 it follows that $\mathfrak{F}$ is a g-bounded morphic image of $\mathfrak{G}_v$ for some v in $\mathfrak{G}$. Since $\mathfrak{G} \in \mathsf{F}$, $\mathfrak{G}_v \in \mathsf{F}$ by the closure property (i) of F. Therefore, from the closure property (iii) of F and $\mathfrak{G}_v \twoheadrightarrow_g \mathfrak{F}$ we deduce that $\mathfrak{F} \in \mathsf{F}$, as required.

Second, let us consider the case (b). By Proposition 3.13 and our assumption, it is enough to show that each point generated subframe of $\mathfrak{F}$ is in F. But each of these frames validates $\mathrm{Log}(\mathsf{F})$ by Proposition 3.10, and hence belongs to F by the same argument as in the case (a). □

Let us say that $\mathfrak{F} = (W, R)$ is m-*transitive* if $R^{\leq m} := \bigcup_{1 \leq k \leq m} R^k$ is transitive, where R^n is defined inductively by: $R^1 = R$ and $R^{n+1} := R^n \circ R$. We can define m-transitivity by $\Box^{\leq m}p \to \Box^{\leq m}\Box^{\leq m}p$, where $\Box^{\leq m}p := \Box p \wedge \cdots \wedge \Box^m p$. We can easily generalize Theorem 4.3 to cover the class C of all finite *m-transitive* frames, since it suffices to modify our Jankov-Fine formula as follows: replace each occurrence of $\Box$ with $\Box^{\leq m}$.

Van Benthem [23, p.29] showed that the assumption 'C is transitive' is crucial in the corresponding theorem [2, Theorem 3.21] for basic modal logic. We can use the same example to show that the transitivity assumption for C is also crucial in GML. Intuitively, this is because his proof uses frames, where every world has at most one successor [23, p.29], and so $\Diamond_k$ $(k > 1)$ does not matter. Let F be the class of finite linear orders with immediate succession as in [23, p.29] and F' be the finite disjoint union closure of F. It is easy to see that F' is closed under taking finite disjoint unions, generated subframes and also *g-bounded morphic images*. By the same argument, however, we can show that a one-point reflexive frame validates $\mathrm{Log}(\mathsf{F}') = \{\varphi : \mathsf{F}' \Vdash \varphi\}$. This shows the indispensability of transitivity in Theorem 4.3.

5 Graph Semantics and Graded Ultrafilter Images

5.1 Graph Semantics and Fine Mapping

Definition 5.1 *Define $X \subseteq_\omega Y$ if $X \subseteq Y$ and $\#X < \omega$.*

Definition 5.2 (Graph semantics for GML) *A* (directed) graph frame *is a pair of a non-empty set W and a family $(R_k)_{k\in\omega}$ of binary relations on W such that if $k > l$ then $R_k \subseteq R_l$. A graph model is a pair of a graph frame and a valuation on it. Given any graph model $(W, (R_k)_{k\in\omega}, V)$, we define the satisfaction relation $w \models_V \varphi$ as follows:*

$$
\begin{array}{lll}
w \models_V p & \text{iff} & w \in V(p),\\
w \models_V \bot & & \text{Never},\\
w \models_V \neg\varphi & \text{iff} & w \not\models_V \varphi,\\
w \models_V \varphi \wedge \psi & \text{iff} & w \models_V \varphi, \textit{ and } w \models_V \psi\\
w \models_V \Diamond_k\varphi & \text{iff} & \exists X \subseteq_\omega W.\, \exists l : X \to \omega.\\
 & & \left(\sum_{v\in X} l(v) \geq k \text{ and } \forall v \in X.\, (wR_{l(v)}v \text{ and } v \in |\varphi|)\right),
\end{array}
$$

where $|\varphi| := \{w \in W : w \models_V \varphi\}$ (if $X = \emptyset$, we put $\sum_{v\in X} l(v) = 0$). If $w \models_V \varphi$ for all $w \in W$ and all $V : \mathsf{Prop} \to \mathcal{P}(W)$, we say that φ is valid *on $(W, (R_k)_{k\in\omega})$ and denote it by $(W, (R_k)_{k\in\omega}) \models \varphi$.*

For any $k \in \omega$, we can easily check that $w \models_V \Diamond_k\varphi$ is also equivalent to:

$$\exists X \subseteq_\omega W.\, \exists l : X \to \omega \setminus \{0\}.\, \left(\sum_{v\in X} l(v) \geq k \text{ and } \forall v \in X.\, (wR_{l(v)}v \text{ and } v \in |\varphi|)\right).$$

In this sense, R_0 does not play any role in the truth condition of $\Diamond_0\varphi$. As in Kripke semantics for GML, $(W, (R_k)_{k\in\omega}) \models \Diamond_0\varphi$ for any graph frame $(W, (R_k)_{k\in\omega})$, since it suffices to take $\emptyset \subseteq_\omega W$ and the empty function from $\emptyset$ to ω as our witness.

The following notion is our renewal of Fine's *canonical mapping* [10, pp.518-9].

Definition 5.3 (Fine mapping) *Let $(W, (R_k)_{k\in\omega})$ be a graph frame and $\mathfrak{G} = (G, S)$ a Kripke frame. We say that $f : G \to W$ is a* Fine mapping *if, for any $n \in \omega$, any $x \in G$ and any $w \in W$,*

$$\#\{\, y \in G : f(y) = w \text{ and } xSy \,\} \geq n \text{ iff } f(x)R_n w.$$

We call the left-to-right direction (**Forth**) *and the right-to-left direction* (**Back**).

A surjective Fine mapping allows us to associate a graph frame with a Kripke frame in a validity-preserving way (Proposition 5.7). Before showing this, let us see some examples of Fine mapping.

Example 5.4 (Fine [10]) Let $(W, (R_k)_{k\in\omega})$ be a graph frame. Let us define a Kripke frame $\mathfrak{G} = (G, S_1)$ by (i) $G := W \times \omega$ and (ii) $(w, l)S_1(w', k)$ iff $k > l$ and $wR_{k-l}w'$. Then the projection $\pi_1 : W \times \omega \to W$ is a Fine mapping. This is verified as follows: Consider any $(w, l) \in G$ and $w' \in W$. Take any $n \in \omega$. Then:

$$\begin{aligned} &\#\{\, (w', k) \in G : \pi_1((w', k)) = w' \text{ and } (w, l)S_1(w', k) \,\} \geq n \\ \text{iff } &\#\{\, (w', k) \in G : k > l \text{ and } wR_{k-l}w' \,\} \geq n \text{ iff } wR_n w' \end{aligned}$$

Let us check the latter equivalence. First, let us show the left-to-right direction. Assume that $\#\{\, (w', k) \in G : k > l \text{ and } wR_{k-l}w' \,\} \geq n$. Fix $k_1, \ldots, k_n$ such that $(w', k_i) \in G$ and $k_i > l$ and $wR_{k_i-l}w'$ $(1 \leq i \leq n)$. Since $\max\{\, k_i - l : 1 \leq i \leq n \,\} \geq n$, we have $wR_n w'$ by $R_{k_1} \subseteq R_{k_2}$ $(k_1 > k_2)$. In order to establish the right-to-left direction, assume $wR_n w'$. From $R_{k_1} \subseteq R_{k_2}$ $(k_1 > k_2)$ we deduce that $wR_{n-1}w'$, $\cdots$, $wR_1 w'$. Then it is easy to see that $(w', l+1)$, $\ldots$, $(w', l+n)$ belong to $\{\, (w', k) \in G : k > l \text{ and } wR_{k-l}w' \,\}$.

Example 5.5 (Fine [10]) In Example 5.4, we can replace S_1 with the following S_2: $(w, l)S_2(w', k)$ iff $wR_k w'$. Then the projection $\pi_1 : W \times \omega \to W$ is still a Fine mapping. This is verified as follows: Consider any $(w, l) \in G$ and $w' \in W$. Take any $n \in \omega$. Then:

$$\begin{aligned} &\#\{\, (w', k) \in G : \pi_1((w', k)) = w' \text{ and } (w, l)S_2(w', k) \,\} \geq n \\ \text{iff } &\#\{\, (w', k) \in G : wR_k w' \,\} \geq n \text{ iff } wR_n w' \quad (\text{by } R_{k_1} \subseteq R_{k_2}\ (k_1 > k_2)) \end{aligned}$$

We will see some specific application of these constructions later on, in Examples 5.22 and 5.24.

Lemma 5.6 *Let $(W, (R_k)_{k\in\omega})$ be a graph frame and $\mathfrak{G} = (G, S)$ a Kripke frame, and V a valuation on W. Assume that $f : G \to W$ is a Fine mapping. Define a valuation V' on $\mathfrak{G}$ by $V'(p) = f^{-1}[V(p)]$. Then for any formula φ and $x \in G$,*

$$(\mathfrak{G}, V'), x \Vdash \varphi \text{ iff } f(x) \models_V \varphi.$$

Proof. By induction on φ. It suffices to check the case where $\varphi \equiv \Diamond_k\psi$. Consider any $x \in G$ and put $w := f(x)$. First, assume that $w \models_V \Diamond_k\psi$. That is,

$$\exists\, X \subseteq_\omega W.\, \exists\, l : X \to \omega.\, \left(\sum\nolimits_{v\in X} l(v) \geq k \text{ and } \forall\, v \in X.\, (wR_{l(v)}v \text{ and } v \in |\psi|)\right).$$

By definition of f and I.H. ($[\![\psi]\!]_{(\mathfrak{G},V')} = f^{-1}[|\psi|]$, the inverse image of $|\psi|$ by f), it follows that:

$$\exists\, X \subseteq_\omega W.\, \exists\, l : X \to \omega. \tag{†}$$
$$\left(\sum_{v\in X} l(v) \geq k \text{ and } \forall\, v \in X.\, (\#(S(x) \cap f^{-1}[\{\, v \,\}]) \geq l(v) \text{ and } f^{-1}[\{\, v \,\}] \subseteq [\![\psi]\!]_{(\mathfrak{G},V')})\right).$$

Then it follows that $\#(S(x) \cap [\![\psi]\!]_{(\mathfrak{G},V')}) \geq k$ hence $(\mathfrak{G}, V'), x \Vdash \Diamond_k\psi$, as required.

Conversely, assume that $(\mathfrak{G}, V'), x \Vdash \Diamond_k\psi$, i.e., $\#(S(x) \cap [\![\psi]\!]_{(\mathfrak{G},V')}) \geq k$ (we drop the subscript $(\mathfrak{G}, V')$ from $[\![\psi]\!]_{(\mathfrak{G},V')}$ below and write $[\![\psi]\!]'$). It suffices to derive (†). Note that $S(x) \cap [\![\psi]\!]'$ consists of the partitions $\{\, f^{-1}[\{\, v \,\}] : v \in f[S(x) \cap [\![\psi]\!]'] \,\}$. The number of these partitions is $\alpha := \#f[S(x) \cap [\![\psi]\!]']$. If $\alpha \geq k$, we can easily derive (†): it suffices to choose some $X \subseteq_\omega f[S(x) \cap [\![\psi]\!]']$ such that $\#X = k$ and define $l : X \to \omega$ by $l(v) := \#(S(x) \cap f^{-1}[\{\, v \,\}])$. So, assume that $\alpha < k$. Then $\#(S(x) \cap [\![\psi]\!]') \geq k$ allows us to derive (†): let us put $X := f[S(x) \cap [\![\psi]\!]']$ and define $l(v) := \#(S(x) \cap f^{-1}[\{\, v \,\}])$. □

Proposition 5.7 *Let $(W, (R_k)_{k\in\omega})$ be a graph frame and $\mathfrak{G} = (G, S)$ a frame. If $f : G \to W$ is a surjective Fine mapping and $\mathfrak{G} \Vdash \varphi$, then $(W, (R_k)_{k\in\omega}) \models \varphi$.*

Proof. Assume $w \not\models_V \varphi$ for some V on W and some $w \in W$. Since f is surjective, $f(x) = w$ for some $x \in G$. Define V' on $\mathfrak{G}$ by $V'(p) := f^{-1}[V(p)]$ for all $p \in \mathsf{Prop}$. By Proposition 5.6, $(\mathfrak{G}, V'), x \nVdash \varphi$ hence $\mathfrak{G} \nVdash \varphi$. □

This proposition shows soundness of graph semantics for GML.

Corollary 5.8 *All the formulas from Fact 2.1 are valid in any graph frame $(W, (R_k)_{k\in\omega})$.*

Proof. By Proposition 5.7 and Example 5.4. □

5.2 Ultrafilter Graph Model and Graded Ultrafilter Images

Given any $\mathfrak{F} = (W, R)$ and any $k \in \omega$, define $m_k : \mathcal{P}(W) \to \mathcal{P}(W)$ by $m_k(X) := \{\, w \in W : \#(R(w) \cap X) \geq k \,\}$. Let us write $m_R(X) := m_1(X)$ and define $l_R : \mathcal{P}(W) \to \mathcal{P}(W)$ by $l_R(X) := W \setminus m_R(W \setminus X)$.

Definition 5.9 *Let $X, Y \subseteq W$. We define $X \Rightarrow Y := (W \setminus X) \cup Y$.*

Proposition 5.10 *For any $X \subseteq W$,*

(i) $l_R(X \Rightarrow Y) \cap l_R(X) \subseteq l_R(Y)$.

(ii) $m_k(X) \subseteq m_l(X)$ $(l < k)$.

(iii) $m_k(X) = \bigcup_{i=0}^{k}(m_i(X \cap Y) \cap m_{k-i}(X \cap (W \setminus Y)))$ *for any $Y \subseteq W$.*

(iv) $l_R(X \Rightarrow Y) \cap m_k(X) \subseteq m_k(Y)$.

Proof. It is easy to see that $\mathfrak{M}, w \Vdash \Diamond_k \varphi$ iff $w \in m_k([\![\varphi]\!])$, i.e., $[\![\Diamond_k \varphi]\!] = m_k([\![\varphi]\!])$. Then all four items are clear from Fact 2.1. □

Definition 5.11 *We define the binary relation R_k^{ue} on the set* Uf(W) *of all ultrafilters on W as: $\mathcal{U} R_k^{\mathrm{ue}} \mathcal{U}'$* iff *$X \in \mathcal{U}'$ implies $m_k(X) \in \mathcal{U}$, for any $X \subseteq W$.*

Proposition 5.12 $(\mathrm{Uf}(W), (R_k^{\mathrm{ue}})_{k \in \omega})$ *is a graph frame.*

Proof. It suffices to check that $R_k \subseteq R_l$ $(l < k)$. This follows trivially from Proposition 5.10 (ii). □

Then we can define the final frame construction to characterize the definability of GML for elementary classes as follows.

Definition 5.13 (Graded ultrafilter images) *Let $\mathfrak{F} = (W, R)$ and $\mathfrak{G} = (G, S)$ be frames. We say that $f : G \to \mathrm{Uf}(W)$ is a* graded ultrafilter mapping *if f is a Fine mapping from $\mathfrak{G}$ to $(\mathrm{Uf}(W), (R_k^{\mathrm{ue}})_{k \in \omega})$. $\mathfrak{F}$ is a* graded ultrafilter image *of $\mathfrak{G}$ if there exists a graded ultrafilter mapping $f : G \to \mathrm{Uf}(W)$ such that f is surjective.*

The rest of this subsection is devoted to establishing the preservation result for graded ultrafilter images. Our strategy is as follows. First, we show the implication: if $(\mathrm{Uf}(W), (R_k^{\mathrm{ue}})_{k \in \omega}) \models \varphi$ then $(W, R) \Vdash \varphi$ (Proposition 5.20). Only for this purpose, we use the notion called *truth-set function* originating from Fine [10] [1]. Second, by combining Proposition 5.7 with the implication above, we will establish our desired preservation result (Theorem 5.21).

Definition 5.14 (Fine [10]) *Given any frame (W, R) and an ultrafilter $\mathcal{U}$ on W, we define the* truth-set function $\mathrm{T}_{\mathcal{U}}(-) : \mathcal{P}(W) \to \mathcal{P}(\mathrm{Uf}(W) \times \omega)$ *by:*

$$\mathrm{T}_{\mathcal{U}}(X) := \{ (\mathcal{U}', l) : l > 0 \text{ and } X \in \mathcal{U}' \text{ and } \mathcal{U} R_l^{\mathrm{ue}} \mathcal{U}' \}.$$

Proposition 5.15 *For any $X, Y \subseteq W$,*

(i) $\mathrm{T}_{\mathcal{U}}(X \cap Y) \cap \mathrm{T}_{\mathcal{U}}(X \cap (W \setminus Y)) = \emptyset$.

(ii) $\mathrm{T}_{\mathcal{U}}(X) = \mathrm{T}_{\mathcal{U}}(X \cap Y) \cup \mathrm{T}_{\mathcal{U}}(X \cap (W \setminus Y))$.

Lemma 5.16 *For any $(k, \mathcal{U}) \in \omega \times \mathrm{Uf}(W)$, $m_k(X) \in \mathcal{U}$ implies $\#\mathrm{T}_{\mathcal{U}}(X) \geq k$.*

Proof. See Appendix A. □

Lemma 5.17 *For any $(k, \mathcal{U}) \in \omega \times \mathrm{Uf}(W)$, $\#\mathrm{T}_{\mathcal{U}}(X) \geq k$ implies $m_k(X) \in \mathcal{U}$.*

Proof. See Appendix A. □

Lemma 5.18 *For any $k \in \omega$ and any $\mathcal{U} \in \mathrm{Uf}(W)$, $\#\mathrm{T}_{\mathcal{U}}(X) \geq k$* iff *:*

$$\exists \mathbb{X} \subseteq_\omega \mathrm{Uf}(W).\ \exists l : \mathbb{X} \to \omega.\ \left(\sum\nolimits_{\mathcal{V} \in \mathbb{X}} l(\mathcal{V}) \geq k \text{ and } \forall \mathcal{V} \in \mathbb{X}.\, (\mathcal{V}, l(\mathcal{V})) \in \mathrm{T}_{\mathcal{U}}(X) \right).$$

[1] If the reader checks Fine's paper [10], he might first feel that the truth-set function $\mathrm{T}_{\mathcal{U}}(-)$ plays the main role in Fine's completeness proof. However, from our viewpoint the graph semantics is the most essential in his proof.

Proof. The right-to-left direction is easy to show by Proposition 5.10 (ii). So let us establish the left-to-right direction. Assume $\#\mathrm{T}_{\mathcal{U}}(X) \geq k$. Remark that $\mathrm{T}_{\mathcal{U}}(X)$ consists of the partitions $\{\pi_1^{-1}[\{\mathcal{V}\}] \cap \mathrm{T}_{\mathcal{U}}(X) : \mathcal{V} \in \pi_1[\mathrm{T}_{\mathcal{U}}(X)]\}$, where $\pi_1 : \mathrm{Uf}(W) \times \omega \to \mathrm{Uf}(W)$ is the projection [2]. We can regard $\#\pi_1[\mathrm{T}_{\mathcal{U}}(X)]$ as the number of all these partitions. If $\#\pi_1[\mathrm{T}_{\mathcal{U}}(X)] \geq k$, we are done: it suffices to choose some $\mathbb{X} \subseteq_\omega \pi_1[\mathrm{T}_{\mathcal{U}}(X)]$ such that $\#\mathbb{X} = k$ and define $l : \mathbb{X} \to \omega$ such that $(\mathcal{V}, l(\mathcal{V})) \in \pi_1^{-1}[\{\mathcal{V}\}] \cap \mathrm{T}_{\mathcal{U}}(X)$. So let us consider the case where $\#\pi_1[\mathrm{T}_{\mathcal{U}}(X)] < k$. Define $\mathbb{X} := \pi_1[\mathrm{T}_{\mathcal{U}}(X)]$. We divide our argument into the following two cases: (a) there exists $\mathcal{V} \in \mathbb{X}$ such that $\#\pi_1^{-1}[\{\mathcal{V}\}] \cap \mathrm{T}_{\mathcal{U}}(X) \geq \omega$, (b) for all $\mathcal{V} \in \mathbb{X}$, $\#\pi_1^{-1}[\{\mathcal{V}\}] \cap \mathrm{T}_{\mathcal{U}}(X) < \omega$. First, let us consider the case (a). Fix such $\mathcal{V} \in \mathbb{X}$. Then for our purpose, it suffices to choose some $l : \mathbb{X} \to \omega$ such that $l(\mathcal{V}) \geq k$. Finally, let us turn to the case (b). For any $\mathcal{V} \in \mathbb{X}$, we can define $l(\mathcal{V}) := \max\{n : (\mathcal{V}, n) \in \mathrm{T}_{\mathcal{U}}(X)\}$. Then from $\#\mathrm{T}_{\mathcal{U}}(X) \geq k$ it is clear that $\sum_{\mathcal{V} \in \mathbb{X}} l(\mathcal{V}) \geq k$. □

Proposition 5.19 *Given any Kripke model $\mathfrak{M} := (W, R, V)$, define the graph model $(\mathrm{Uf}(W), (R_k^{\mathrm{ue}})_{k \in \omega}, V^{\mathrm{ue}})$, where $V^{\mathrm{ue}}(p) := \{\mathcal{U} \in \mathrm{Uf}(W) : V(p) \in \mathcal{U}\}$. Then for any φ and any $\mathcal{U} \in \mathrm{Uf}(W)$,*

$$[\![\varphi]\!] \in \mathcal{U} \text{ iff } \mathcal{U} \models_{V^{\mathrm{ue}}} \varphi.$$

Proof. By induction on φ. It suffices to consider the case where $\varphi \equiv \Diamond_k \psi$. Then $[\![\Diamond_k \psi]\!] \in \mathcal{U}$ iff $m_k([\![\psi]\!]) \in \mathcal{U}$ iff $\#\mathrm{T}_{\mathcal{U}}([\![\psi]\!]) \geq k$ by Lemma 5.16 and Lemma 5.17. Equivalently, by Lemma 5.18 and the definition of $\mathrm{T}_{\mathcal{U}}$, we obtain:

$$\exists \mathbb{X} \subseteq_\omega \mathrm{Uf}(W).\ \exists l : \mathbb{X} \to \omega.\ \left(\sum_{\mathcal{V} \in \mathbb{X}} l(\mathcal{V}) \geq k \text{ and } \forall \mathcal{V} \in \mathbb{X}.\ (\mathcal{U} R_{l(\mathcal{V})}^{\mathrm{ue}} \mathcal{V} \text{ and } [\![\psi]\!] \in \mathcal{V}) \right).$$

By I.H. ($[\![\psi]\!] \in \mathcal{V}$ iff $\mathcal{V} \in |\psi|$), we establish $\mathcal{U} \models_{V^{\mathrm{ue}}} \Diamond_k \varphi$, as required. □

Proposition 5.20 *For any $\mathfrak{F} = (W, R)$, if $(\mathrm{Uf}(W), (R_k^{\mathrm{ue}})_{k \in \omega}) \models \varphi$, then $\mathfrak{F} \Vdash \varphi$.*

Proof. Assume $\mathfrak{F} \nVdash \varphi$, i.e., $(\mathfrak{F}, V), w \nVdash \varphi$ for some V and some $w \in W$. Take the principal ultrafilter $\mathcal{U}_w := \{D \subseteq W : w \in D\}$. It follows that $[\![\varphi]\!] \notin \mathcal{U}_w$. It follows from Proposition 5.19 that $\mathcal{U}_w \not\models_{V^{\mathrm{ue}}} \varphi$; hence $(\mathrm{Uf}(W), (R_k^{\mathrm{ue}})_{k \in \omega}) \not\models \varphi$. □

Theorem 5.21 *If $\mathfrak{F}$ is a graded ultrafilter image of $\mathfrak{G}$, then $\mathfrak{G} \Vdash \varphi$ implies $\mathfrak{F} \Vdash \varphi$.*

Proof. By Proposition 5.7 and Proposition 5.20. □

Example 5.22 By definition of S_1 in Example 5.4, $\mathfrak{G}$ is irreflexive. Let us consider an example of graded ultrafilter images. Take a one-point reflexive frame $\mathfrak{F} = (\{*\}, \{(*, *)\})$. Then $\mathrm{Uf}(|\mathfrak{F}|)$ consists only of the principal ultrafilter $\mathcal{U}_*$ generated by $*$. We have $\mathcal{U}_* R_0^{\mathrm{ue}} \mathcal{U}_*$ and $\mathcal{U}_* R_1^{\mathrm{ue}} \mathcal{U}_*$ by definition. However, if $k \geq 2$, $\mathcal{U}_* R_k^{\mathrm{ue}} \mathcal{U}_*$ fails. By the construction of Example 5.4, we can construct Kripke frame $\mathfrak{G} := (\{\mathcal{U}_*\} \times \omega, S_1)$.

---

[2] Remark that $\mathrm{T}_{\mathcal{U}}(X)$ is contained in (the union of) $\{\pi_1^{-1}[\{\mathcal{V}\}] : \mathcal{V} \in \pi_1[\mathrm{T}_{\mathcal{U}}(X)]\}$. However, $\{\pi_1^{-1}[\{\mathcal{V}\}] \cap \mathrm{T}_{\mathcal{U}}(X) : \mathcal{V} \in \pi_1[\mathrm{T}_{\mathcal{U}}(X)]\}$ gives us a partition of $\mathrm{T}_{\mathcal{U}}(X)$.

It is easy to see that $\mathfrak{G}$ is isomorphic to (ω, Suc^{-1}), where Suc^{-1} is the inverse of the successor relation Suc on ω.

This example also shows the undefinability of irreflexivity in GML by Theorem 5.21. Also by Example 5.22 and Theorem 5.21, we can establish:

Proposition 5.23 *The existence of a distinct predecessor (for any w, there exists w' such that $w'Rw$ and $w \neq w'$) is undefinable in GML.*

Example 5.24 In Example 5.22, let us start from the one point *irreflexive* frame $\mathfrak{F}' := (\{ * \}, \emptyset)$. While $\mathrm{Uf}(|\mathfrak{F}|)$ consists only of the principal ultrafilter $\mathcal{U}_*$ generated by $*$ as before, $\mathcal{U}_* R_0^{\mathfrak{ue}} \mathcal{U}_*$ holds but $\mathcal{U}_* R_k^{\mathfrak{ue}} \mathcal{U}_*$ fails $(k > 0)$. Here let us use the definition S_2 of Example 5.5. Then $\mathfrak{G} := (\{ \mathcal{U}_* \} \times \omega, S_2)$ is isomorphic to (ω, R'), where $nR'm$ iff $m = 0$. Therefore a reflexive state is accessible from all the states in $\mathfrak{G}$. However, this is not the case in $\mathfrak{F}'$. This example uses the relation $R_0^{\mathfrak{ue}}$ crucially.

By Example 5.24 and Theorem 5.21, we can establish:

Proposition 5.25 $\forall w.\, \exists w'.\, (wRw' \text{ and } w'Rw')$ *is undefinable in GML.*

6 Goldblatt-Thomason-style Characterization of Elementary Graded Modal Classes

In this section we use some notions from first-order model theory, e.g., elementary embedding, ω-saturation, etc. The reader unfamiliar with them can refer to [4]. The original Goldblatt-Thomason Theorem for basic modal logic was proved via duality between algebras and frames [12]. The proof of our Goldblatt-Thomason Theorem for GML modifies the model-theoretic proof given by Van Benthem [24] for basic modal logic.

Definition 6.1 *Let $\mathfrak{F} = (W, R)$ be a generated subframe with a root w. We expand our language GML with the (possibly uncountable) set $\{ p_X : X \subseteq W \}$ of new proposition letters. Let Δ be the set consisting of:*

$$p_{X \cap Y} \leftrightarrow p_X \wedge p_Y,$$
$$p_{W \setminus X} \leftrightarrow \neg p_X,$$
$$p_{m_k(X)} \leftrightarrow \Diamond_k p_X,$$
$$p_W,$$

where $X, Y \subseteq W$ and $k \in \omega$. Then we define $\Delta_{\mathfrak{F}}$ as follows:

$$\Delta_{\mathfrak{F}} := \{ p_{\{ w \}} \} \cup \{ \Box^n \varphi : \varphi \in \Delta \text{ and } n \in \omega \}.$$

Note that $\Delta_{\mathfrak{F}}$ is satisfiable in $\mathfrak{F}$ under the natural valuation V such that $V(p_X) = X$. Let F be an elementary class of frames. Similarly to our graded Jankov-Fine formula in Definition 4.2, by this 'complete description' of $\mathfrak{F}$, for a given $\mathfrak{G} \in \mathsf{F}$ such that $\Delta_{\mathfrak{F}}$ is satisfiable in $\mathfrak{G}$, we can extract the following semantic information.

Lemma 6.2 *Let* F *be an elementary class of frames and* $\mathfrak{F} = (W, R)$ *a generated subframe with a root* w. *Also let* $\mathfrak{G} \in \mathsf{F}$. *If* $\Delta_{\mathfrak{F}}$ *is satisfiable in* $\mathfrak{G}$, *then there exists some* $v \in |\mathfrak{G}|$ *and some elementary extension* $(\mathfrak{G}_v)^*$ *of* $\mathfrak{G}_v$ *such that* $\mathfrak{F}$ *is a graded ultrafilter image of* $(\mathfrak{G}_v)^*$.

Proof. Let us assume that $\Delta_{\mathfrak{F}}$ is satisfiable in $\mathfrak{G}$. Thus we have $(\mathfrak{G}, V), v \Vdash \Delta_{\mathfrak{F}}$ for some valuation V and $v \in |\mathfrak{G}|$. It follows that $(\mathfrak{G}_v, V_1), v \Vdash \Delta_{\mathfrak{F}}$, where $V_1(p) := V(p) \cap |\mathfrak{G}_v|$. By construction of $\Delta_{\mathfrak{F}}$, we can show that $(\mathfrak{G}_v, V_1) \Vdash \Delta$ and $(\mathfrak{G}_v, V_1), v \Vdash p_X$ for any X with $w \in X \subseteq W$. Let us take some ω-saturated elementary extension $((\mathfrak{G}_v)^*, V_1^*)$ of $(\mathfrak{G}_v, V_1)$. Let v^* be the element in $(\mathfrak{G}_v)^*$ corresponding to v. Then, we obtain $((\mathfrak{G}_v)^*, V_1^*), v^* \Vdash \Delta_{\mathfrak{F}}$. We can also establish that $((\mathfrak{G}_v)^*, V_1^*) \Vdash \Delta$ and $((\mathfrak{G}_v)^*, V_1^*), v^* \Vdash p_X$ for any X with $w \in X \subseteq W$.

Let us define a mapping $f : |(\mathfrak{G}_v)^*| \to \mathrm{Uf}(W)$ by

$$f(s) := \{ X \subseteq W : ((\mathfrak{G}_v)^*, V_1^*), s \Vdash p_X \}.$$

Now we are going to show the following: (a) $f(s)$ is an ultrafilter; (b) f is surjective; (c) f is a Fine mapping. Of course, the most important step for our purpose is to establish (c). So, we concentrate on showing (c). For the proof of (a) and (b), the reader can refer to [24]. Let us denote the accessibility relation of $(\mathfrak{G}_v)^*$ by S. We establish (**Forth**)- and (**Back**)-conditions for a Fine mapping (recall Definition 5.3). Consider any $s \in |(\mathfrak{G}_v)^*|$ and any $\mathcal{V} \in \mathrm{Uf}(W)$.

First, let us show (**Forth**). Assume that $\#\{ s' : f(s') = \mathcal{V} \text{ and } sSs' \} \geq k$. We want to show $f(s) R_k^{\mathfrak{ue}} \mathcal{V}$. So, consider any $X \in \mathcal{V}$. We show $m_k(X) \in f(s)$, or equivalently, $((\mathfrak{G}_v)^*, V_1^*), s \Vdash p_{m_k(X)}$. Since $f(s') = \mathcal{V}$ and $X \in \mathcal{V}$ implies $X \in f(s')$, we obtain: $\{ s' : f(s') = \mathcal{V} \text{ and } sSs' \} \subseteq \{ s' : X \in f(s) \text{ and } sSs' \}$. By our assumption, $\#\{ s' : X \in f(s) \text{ and } sSs' \} \geq k$. This gives us

$$\#\{ s' : ((\mathfrak{G}_v)^*, V_1^*), s \Vdash p_X \text{ and } sSs' \} \geq k.$$

Thus $((\mathfrak{G}_v)^*, V_1^*), s \Vdash \Diamond_k p_X$. Since Δ is valid on $((\mathfrak{G}_v)^*, V_1^*)$, we can deduce that $((\mathfrak{G}_v)^*, V_1^*), s \Vdash p_{m_k(X)}$, as desired. We have shown (**Forth**).

Next, let us establish (**Back**). In what follows, we assume for simplicity that $k = 3$. Our assumption is $f(s) R_3^{\mathfrak{ue}} \mathcal{V}$. We show that $\#\{ s' : f(s') = \mathcal{V} \text{ and } sSs' \} \geq 3$. Define

$$\begin{aligned}\Gamma :=& \{ \mathbf{R}(\underline{s}, z_1) \wedge \mathbf{R}(\underline{s}, z_2) \wedge \mathbf{R}(\underline{s}, z_3) \wedge z_1 \neq z_2 \wedge z_1 \neq z_3 \wedge z_2 \neq z_3 \} \\ &\cup \{ \mathbf{P}_Y(z_1) \wedge \mathbf{P}_Y(z_2) \wedge \mathbf{P}_Y(z_3) : Y \in \mathcal{V} \},\end{aligned}$$

where $\mathbf{R}(x, y)$ is the binary symbol corresponding to S and each $\mathbf{P}_Y(x)$ is the unary predicate symbol corresponding to p_Y. By ω-saturation of $((\mathfrak{G}_v)^*, V_1^*)$, it suffices to show that Γ is finitely satisfiable. Consider $Y_1, \ldots, Y_n \in \mathcal{V}$. We show:

$$\begin{aligned}\Gamma' :=& \{ \mathbf{R}(\underline{s}, z_1) \wedge \mathbf{R}(\underline{s}, z_2) \wedge \mathbf{R}(\underline{s}, z_3) \wedge z_1 \neq z_2 \wedge z_1 \neq z_3 \wedge z_2 \neq z_3 \} \\ &\cup \{ \mathbf{P}_{Y_i}(z_1) \wedge \mathbf{P}_{Y_i}(z_2) \wedge \mathbf{P}_{Y_i}(z_3) : 1 \leq i \leq n \}\end{aligned}$$

is satisfiable in $((\mathfrak{G}_v)^*, V_1^*)$. Clearly, $Y_1 \cap \cdots \cap Y_n \in \mathcal{V}$. From $f(s) R_k^{\mathfrak{ue}} \mathcal{V}$ we have: $m_3(Y_1 \cap \cdots \cap Y_n) \in f(s)$, i.e., $((\mathfrak{G}_v)^*, V_1^*), s \Vdash p_{m_3(Y_1 \cap \cdots \cap Y_n)}$. Since Δ is valid on $((\mathfrak{G}_v)^*, V_1^*)$, we can deduce that $((\mathfrak{G}_v)^*, V_1^*), s \Vdash \Diamond_3(p_{Y_1} \wedge \cdots \wedge p_{Y_n})$, i.e., $\#(S(s) \cap [\![p_{Y_1} \wedge \cdots \wedge p_{Y_n}]\!]) \geq 3$. This means that Γ' is satisfiable. So we can conclude that Γ is finitely satisfiable in $((\mathfrak{G}_v)^*, V_1^*)$. We can easily generalize the above argument to other cases, where $k \neq 3$.

Therefore we have established the desired statement. □

Theorem 6.3 *An elementary class* F *of frames is definable by a set of formulas in GML* iff F *is closed under taking* (i) *generated subframes,* (ii) *disjoint unions,* (iii) *g-bounded morphic images, and* (iv) *graded ultrafilter images.*

Proof. The left-to-right direction is easy to show by Propositions 3.3, 3.10, 3.12 and Theorem 5.21. Conversely, assume that F satisfies the closure properties. Define $\mathrm{Log}(\mathsf{F}) := \{\varphi : \mathsf{F} \Vdash \varphi\}$. We show that, for any $\mathfrak{F} \in \mathsf{F}$, $\mathfrak{F} \in \mathsf{F}$ iff $\mathfrak{F} \Vdash \mathrm{Log}(\mathsf{F})$. Consider $\mathfrak{F} \in \mathsf{F}$. It is trivial to show the Only-If-direction. Let us show the If-direction. Assume that $\mathfrak{F} \Vdash \mathrm{Log}(\mathsf{F})$. Similarly to the proof of Theorem 4.3, by our assumptions (i), (ii), (iii) and Proposition 3.13, we can assume without any loss of generality that $\mathfrak{F}$ is point-generated by a root $w \in |\mathfrak{F}|$.

Construct the 'complete description' $\Delta_{\mathfrak{F}}$ of $\mathfrak{F}$ (see Definition 6.1) [3]. Then we show that $\Delta_{\mathfrak{F}}$ is satisfiable in F as follows. It suffices to show that $\Delta_{\mathfrak{F}}$ is finitely satisfiable in F, because F is elementary. Let Γ be a finite subset of $\Delta_{\mathfrak{F}}$. To get a contradiction, suppose $\mathsf{F} \Vdash \neg \bigwedge \Gamma$. Then $\neg \bigwedge \Gamma \in \mathrm{Log}(\mathsf{F})$. Since $\mathfrak{F} \Vdash \mathrm{Log}(\mathsf{F})$, $\mathfrak{F} \Vdash \neg \bigwedge \Gamma$. However, $\bigwedge \Gamma$ is clearly satisfiable in $\mathfrak{F}$ under the natural valuation, which implies a contradiction. Therefore, $\Delta_{\mathfrak{F}}$ is satisfiable in some $\mathfrak{G} \in \mathsf{F}$.

By Lemma 6.2, there exists some $v \in |\mathfrak{G}|$ and some elementary extension $(\mathfrak{G}_v)^*$of $\mathfrak{G}_v$ such that $\mathfrak{F}$ is a graded ultrafilter image of $(\mathfrak{G}_v)^*$. By our closure properties and $\mathfrak{G} \in \mathsf{F}$, we can conclude that $\mathfrak{F} \in \mathsf{F}$ as required. □

In our proof of Theorem 6.3 (esp. Lemma 6.2), we require that a class F of frames is elementary. We essentially use a compactness argument that requires F to be elementary. So we can replace the assumption that F is elementary with closure under ultraproducts. Moreover, we can also reduce this condition to closure under ultrapowers as in the case of basic modal language as follows (cf. [13, Proposition 85]): an ultraproduct of frames is isomorphic to a generated subframe of the ultrapower, with respect to the same ultrafilter, of the disjoint union of the original frames. Since any graded modally definable class is closed under taking disjoint unions, we can apply the same argument to GML.

In basic modal language, we can also get even weaker closure condition: closure under ultrafilter extensions. Closure under ultrafilter extensions implies closure under ultraproducts, provided the intended class F is also closed under disjoint unions, generated subframes and isomorphisms. Thus, we can regard closure under ultrafilter extensions as the essential assumption of F in the proof of Goldblatt-Thomason Theorem in basic

[3] Remark that we can still assume that $\mathfrak{F} \Vdash \mathrm{Log}(\mathsf{F})$ in spite of our expansion of the original language in Definition 6.1. This is intuitively clear, because the choice of Prop is irrelevant, whenever we consider *frame validity*.

modal language. Then, the next question is: can we get a similar kind of closure condition even in GML? As we explained in the introduction, however, it seems difficult to find an appropriate notion of ultrafilter extension for GML in Kripke semantics. One possible way to avoid this difficulty is to change the semantics into the coalgebraic one [4].

7 Further Directions

7.1 The Scope of Our Guiding Idea

The first author and Sato [18,17] observed that there is a strong connection between the (strong) completeness proof of an extended modal logic and Goldblatt-Thomason characterization for it and they extracted an essential part of this connection, the notion of *realizer* [18]. The result of this paper gives us a further evidence of this observation. We can regard the notion of *Fine mapping* as the corresponding notion of realizer. Then our next question is: what is the scope of this observation? For example, we have not obtained a well-established model-theoretical study of conditional logic for preference frames [22]. Is it possible to apply the idea of [18] to obtain Goldblatt-Thomason-style results for conditional logic over preference frames?

7.2 Extensions of GML

We may consider extensions of GML with more expressive power. One general problem is definability of frame classes in extended GMLs. Lethinen [16] studied the extension of GML obtained by adding the path quantifier, which allows us to talk about the truth on all paths starting from a certain state, the modal μ-extension, and an infinite GML. Some excellent results on relative definability of frame classes are proved there. Another nice extension is graded hybrid logic (GHL). Kaminski, et. al. [14] studied terminating tableaux for graded hybrid logic. However we still lack a Goldblatt-Thomason-style characterization for GHL. Is it possible to merge our idea from this paper with Ten Cate's Goldblatt-Thomason-style characterization for hybrid logic [20], in particular, his idea of *ultrafilter morphic images* [20, Definition 4.2.5]? This would be a promising further direction.

7.3 Coalgebraic GML

We have shown that there is an alternative semantics which interprets GML on directed graphs. Another excellent (but related) alternative semantics is the coalgebraic one. This was shown by D'Agostino and Visser [5] and it was claimed that graded modal logic is subsumed under coalgeraic modal logics. We may define a functor Ω on the category of sets such that $\Omega(X) = (\omega + 1)^X$, the set of all functions from X to $\omega + 1$. A Ω-coalgebra is a pair (X, σ) where $\sigma : X \to \Omega(X)$ is a transition map. Recently the second author has found out that there is a natural Goldblatt-Thomason theorem for

[4] Recently, the second author has found the notion of ultrafilter extension in the coalgebraic semantics and proved a natural Goldbatt-Thomason Theorem for GML (see also Section 7.3)

coalgebraic GML, by applying duality between algebras and Ω-coalgebras. Remark that this is not a consequence of the general Goldblatt-Thomason theorem given by Kurz and Rosický [15], because they restrict functors to those preserving finiteness, i.e., mapping finite sets to finite sets, while the functor Ω above does not preserve finiteness.

References

[1] Baader, F. and W. Nutt, *Basic description logics*, in: F. B. et.al, editor, *The description logic handbook: theory, implementation, and applications*, Cambridge University Press, 2003 pp. 43–95.

[2] Blackburn, P., M. de Rijke and Y. Venema, "Modal Logic," Cambridge Tracts in Theoretical Computer Science, Cambridge University Press, Cambridge, 2001.

[3] Cerrato, C., *General canonical models for graded normal logics*, Studia Logica **49** (1990), pp. 241–252.

[4] Chang, C. C. and H. J. Keisler, "Model Theory," North-Holland Publishing Company, Amsterdam, 1990, 3 edition.

[5] D'Agostino, G. and A. Visser, *Finality regained: a coalgebraic study of Scott-sets and multisets*, Archive for Mathematical Logic **41** (2002), pp. 267–298.

[6] de Caro, F., *Graded modalities II (canonical models)*, Studia Logica **47** (1988), pp. 1–10.

[7] De Rijke, M., *A note on graded modal logic*, Studia Logica **64** (2000), pp. 271–283.

[8] Fattorosi-Banarba, M. and C. Cerrato, *Graded modalities I*, Studia Logica **44** (1985), pp. 197–221.

[9] Fattorosi-Banarba, M. and C. Cerrato, *Graded modalities III (the completeness and compactness of S40)*, Studia Logica **47** (1988), pp. 99–110.

[10] Fine, K., *In so many possible worlds*, Notre Dame Journal of Formal Logic **13** (1972), pp. 516–520.

[11] Gargov, G. and V. Goranko, *Modal logic with names*, Journal of Philosophical Logic **22** (1993), pp. 607–36.

[12] Goldblatt, R. I. and S. K. Thomason, *Axiomatic classes in propositional modal logic*, in: J. N. Crossley, editor, *Algebra and Logic*, Springer-Verlag, 1975 pp. 163–73.

[13] Goranko, V. and M. Otto, *Model theory of modal logic*, in: F. W. P. Blackburn, J. van Benthem, editor, *Handbook of Modal Logic*, Kluwer, 2007 pp. 249–329.

[14] Kaminski, M., S. Schneider and G. Smolka, *Terminating tableaux for graded hybrid logic with global modalities and role hierarchies*, in: *Automated Reasoning with Analytic Tableaux and Related Methods*, Lecture Notes in Computer Science **5607**, 2009.

[15] Kurz, A. and J. Rosický, *The Goldblatt-Thomason theorem for coalgebras*, in: T. Mossakowski, editor, *CALCO 2007*, number 4624 in LNCS (2007).

[16] Lethinen, S., "Generalizing the Goldblatt-Thomason Theorem and Modal Definability," Ph.D. thesis, University of Tampere (2008).

[17] Sano, K., "Semantical Investigation into Extended Modal Languages," Ph.D. thesis, Graduate School of Letters, Kyoto University (2010).

[18] Sano, K. and K. Sato, *Semantical characterization for irreflexive and generalized modal languages*, Notre Dame Journal of Formal Logic **48** (2007), pp. 205–228.

[19] Tarski, A., "Introduction to Logic and to the Methodology of Deductive Sciences," Oxford University Press, New York, 1941.

[20] Ten Cate, B., "Model theory for extended modal languages," Ph.D. thesis, University of Amsterdam, Institute for Logic, Language and Computation (2005).

[21] Ten Cate, B., J. Van Benthem and J. Vänäanen, *Lindström theorems for fragments of first-order logic*, in: *22nd Annual IEEE Symposium on Logic in Computer Science, LICS*, 2007, pp. 280–292.

[22] Van Benthem, J., *Correspondence theory*, in: D. Gabbay and F. Guenthner, editors, *Handbook of Philosophical Logic: Volume II: Extensions of Classical Logic*, Reidel, Dordrecht, 1984 pp. 167–247.

[23] Van Benthem, J., *Notes on modal definability*, Notre Dame Journal of Formal Logic **30** (1988), pp. 20–35.

[24] Van Benthem, J., *Modal frame classes revisited*, Fundamenta Informaticae **18** (1993), pp. 307–17.

[25] Van der Hoek, W. and J. J. C. Meyer, *Graded modalities in epistemic logic*, in: *Logical Foundations of Computer Science -Tver'92*, Lecture Notes in Computer Science (1992), pp. 503–514.

Acknowledgements

The work presented here was carried out during both authors' stay at ILLC, Universiteit van Amsterdam. We would like to thank Yde Venema for discussions and also for maintaining a nice research environment. We would like to thank the three reviewers, whose comments and corrections were all very helpful for revising an earlier version of this paper. In particular, we owe Definition 3.4, Proposition 3.5 and the remark just after Theorem 6.3 to them.

A Proofs of Lemmas

The proof of Lemmas 5.16 and 5.17 adapts and extends the proof of Theorem 1 from [10].

A.1 Proof of Lemma 5.16

By induction on k.

(Base) Since $\#\mathrm{T}_{\mathcal{U}}(X) \geq 0$, there is nothing to prove in the case where $k = 0$. Consider the case where $k = 1$. Assume that $m_1(X) \in \mathcal{U}$. Since $m_1(X) := \{ w \in W : R(w) \cap X \neq \emptyset \}$, we can identify $(\mathrm{Uf}(W), R_1^{\mathfrak{ue}})$ with the ultrafilter extension of $\mathfrak{F}$ for the basic modal language. Then from $m_1(X) \in \mathcal{U}$ we easily deduce that there exists an ultrafilter $\mathcal{U}'$ such that $\mathcal{U}' \subseteq \{ Y \subseteq W : m_1(Y) \in \mathcal{U} \}$ and $X \in \mathcal{U}'$. So we have $\mathcal{U} R_1^{\mathfrak{ue}} \mathcal{U}'$. This yields us $(\mathcal{U}', 1) \in \mathrm{T}_{\mathcal{U}}(X)$, hence $\#\mathrm{T}_{\mathcal{U}}(X) \geq 1$.

(Induction Step) Consider the case where $k > 1$. First, note that our induction hypothesis is: $\forall l < k.\ \forall X \subseteq W.\ m_l(X) \in \mathcal{U}$ implies $\#\mathrm{T}_{\mathcal{U}}(X) \geq l$. Assume that $m_k(X) \in \mathcal{U}$. By Proposition 5.10 (iii), we have:

$$\forall Y \subseteq W.\ \exists i \leq k.\ m_i(X \cap Y) \cap m_{k-i}(X \cap (W \setminus Y)) \in \mathcal{U}. \qquad (*)$$

For simplicity, let us abbreviate $(*)$ as $\forall Y \subseteq W.\ \exists i \leq k.\ \Psi_{\mathcal{U}}(Y, i)$. We subdivide our argument into the following two cases:
(a) $\exists Y \subseteq W.\ \exists i \leq k.\ (\Psi_{\mathcal{U}}(Y, i)$ and $i \neq 0$ and $i \neq k)$;
(b) $\forall Y \subseteq W.\ \forall i \leq k.\ (\Psi_{\mathcal{U}}(Y, i)$ implies $(i = 0$ or $i = k))$.
First, let us consider the easier case (a). Fix Y and i with $\Psi_{\mathcal{U}}(Y, i)$. Then we have $1 < i < k$. So by our I.H. and $\Psi_{\mathcal{U}}(Y, i)$, we can state that: $\#\mathrm{T}_{\mathcal{U}}(X \cap Y) \geq i$ and

$\#\mathrm{T}_{\mathcal{U}}(X \cap (W \setminus Y)) \geq k - i$. By Proposition 5.15, we can calculate as follows:

$$\#\mathrm{T}_{\mathcal{U}}(X) = \#\mathrm{T}_{\mathcal{U}}(X \cap Y) + \#\mathrm{T}_{\mathcal{U}}(X \cap (W \setminus Y)) \geq i + (k - i) = k,$$

as desired. This finishes the case (a).

Let us consider the case (b). First, we show the following claim:

Claim A.1 *For any* $Y \subseteq W$, $l_R(X \Rightarrow (W \setminus Y)) \cup l_R(X \Rightarrow Y) \in \mathcal{U}$.

(Proof of Claim) Let us consider any $Y \subseteq W$. By our assumption $(*)$, we can find $j \leq k$ such that $\Psi_{\mathcal{U}}(Y, j)$, i.e., $m_j(X \cap Y) \in \mathcal{U}$ and $m_{k-j}(X \cap (W \setminus Y)) \in \mathcal{U}$. Then, (b) teaches us that we have $j = 0$ or $j = k$. Here we only check the desired conclusion in the case $j = 0$, because we can similarly show it in the case $j = k$. Let us assume that $j = 0$. Then we have $m_k(X \cap (W \setminus Y)) \in \mathcal{U}$ (note: $m_0(X \cap Y) \in \mathcal{U}$ always holds). We are going to establish $l_R(X \Rightarrow (W \setminus Y)) \in \mathcal{U}$, i.e., $W \setminus m_R(X \cap Y) \in \mathcal{U}$. So suppose that on the contrary, $m_R(X \cap Y) \in \mathcal{U}$, hence $m_1(X \cap Y) \in \mathcal{U}$. But by Proposition 5.10 (ii) and $m_k(X \cap (W \setminus Y)) \in \mathcal{U}$, we obtain $m_{k-1}(X \cap (W \setminus Y)) \in \mathcal{U}$ (note that $k > 1$). Together with $m_1(X \cap Y) \in \mathcal{U}$, this means that we have shown $\Psi_{\mathcal{U}}(Y, 1)$. Then from (b) it follows that $1 = 0$ or $1 = k$. Both of the disjuncts, however, are impossible. So we get the desired contradiction. Therefore, $l_R(X \Rightarrow (W \setminus Y)) \in \mathcal{U}$ hence $l_R(X \Rightarrow (W \setminus Y)) \cup l_R(X \Rightarrow Y) \in \mathcal{U}$. ⊣

By this claim, we can also establish the following.

Claim A.2 $\{X\} \cup \{Y \subseteq W : W \setminus m_k(W \setminus Y) \in \mathcal{U}\}$ *satisfies the finite intersection property.*

(Proof of Claim) Suppose the contrary — that this set does not have the finite intersection property. It follows that there exists finite $Y_1, \ldots Y_n$ such that $W \setminus m_k(W \setminus Y_i) \in \mathcal{U}$ $(1 \leq i \leq n)$ and $X \cap Y_1 \cap \cdots \cap Y_n = \emptyset$. We subdivide our argument into the following two cases:

(i) $\forall i. (1 \leq i \leq n$ implies $l_R(X \Rightarrow Y_i) \in \mathcal{U})$;

(ii) $\exists i. (1 \leq i \leq n$ and $l_R(X \Rightarrow Y_i) \notin \mathcal{U})$.

First, let us consider the case (i). Then from (i) we derive that $l_R(X \Rightarrow (Y_1 \cap \cdots \cap Y_n)) \in \mathcal{U}$. By our assumption $X \cap Y_1 \cap \cdots \cap Y_n = \emptyset$ (i.e., $Y_1 \cap \cdots \cap Y_n \subseteq (W \setminus X)$), however, we have: $l_R((Y_1 \cap \cdots \cap Y_n) \Rightarrow (W \setminus X)) \in \mathcal{U}$. Thus we obtain $l_R(W \setminus X) \in \mathcal{U}$ hence $W \setminus m_R(X) \in \mathcal{U}$. Recall that our assumption for the induction step is $m_k(X) \in \mathcal{U}$, which implies a contradiction by Proposition 5.10 (ii). This finishes the case (i).

Finally, consider the case (ii). Fix i with $l_R(X \Rightarrow Y_i) \notin \mathcal{U}$. By Claim A.1, $l_R(X \Rightarrow (W \setminus Y_i)) \in \mathcal{U}$. Recall that Y_i satisfies $W \setminus m_k(W \setminus Y_i) \in \mathcal{U}$. Then by Proposition 5.10 (iv), we can establish that $W \setminus m_k(X) \in \mathcal{U}$, which gives us a contradiction to our assumption: $m_k(X) \in \mathcal{U}$. This finishes the case (ii). ⊣

By Claim A.2, we can find an ultrafilter $\mathcal{U}'$ such that $X \in \mathcal{U}'$ and $Y \in \mathcal{U}'$ for any Y with $W \setminus m_k(W \setminus Y) \in \mathcal{U}$. Thus, we have $\mathcal{U} R_k^{\mathtt{ue}} \mathcal{U}'$ and $X \in \mathcal{U}'$. From Proposition 5.10 (ii) it follows that $\mathcal{U} R_l^{\mathtt{ue}} \mathcal{U}'$ for any l with $1 \leq l \leq k$, i.e., $(\mathcal{U}', l) \in \mathrm{T}_{\mathcal{U}}(X)$ for any l with $1 \leq l \leq k$. Therefore we have shown $\#\mathrm{T}_{\mathcal{U}}(X) \geq k$, as required. This completes our proof for the induction step. □

A.2 Proof of Lemma 5.17

By induction on k.

(Base) Since $m_0(X) \in \mathcal{U}$ always holds, there is nothing to prove in the case where $k = 0$. Consider the case $k = 1$ and assume that $l > 1$ and $\#\mathrm{T}_{\mathcal{U}}(X) \geq 1$. So we can find $(\mathcal{U}', l) \in \mathrm{T}_{\mathcal{U}}(X)$. Equivalently, $\mathcal{U}R^{\mathrm{ue}}_{l}\mathcal{U}'$ and $X \in \mathcal{U}'$. From Proposition 5.10 (ii) we can derive that $\mathcal{U}R^{\mathrm{ue}}_{1}\mathcal{U}'$ and $X \in \mathcal{U}'$ hence $m_1(X) \in \mathcal{U}$, as required.

(Induction Step) Consider the case $k > 1$. Assume that $\#\mathrm{T}_{\mathcal{U}}(X) \geq k$. We split our argument into the following cases:

(a) $\exists (\mathcal{U}_1, l_1) \in \mathrm{T}_{\mathcal{U}}(X). \exists (\mathcal{U}_2, l_2) \in \mathrm{T}_{\mathcal{U}}(X). (\mathcal{U}_1 \neq \mathcal{U}_2)$.

(b) $\forall (\mathcal{U}_1, l_1) \in \mathrm{T}_{\mathcal{U}}(X). \forall (\mathcal{U}_2, l_2) \in \mathrm{T}_{\mathcal{U}}(X). (\mathcal{U}_1 = \mathcal{U}_2)$.

First, we consider the case (b). Our assumption $\#\mathrm{T}_{\mathcal{U}}(X) \geq k$ and (b) allow us to establish $(\mathcal{U}_1, l_1) \in \mathrm{T}_{\mathcal{U}}(X)$ for some ultrafilter $\mathcal{U}_1$ and some $l_1 \geq k$. Fix such $\mathcal{U}_1$ and $l_1 \geq k$. Since $\mathcal{U}R^{\mathrm{ue}}_{l_1}\mathcal{U}_1$ and $X \in \mathcal{U}_1$, from Proposition 5.10 (ii) we deduce that $\mathcal{U}R^{\mathrm{ue}}_{k}\mathcal{U}_1$ and $X \in \mathcal{U}_1$ hence $m_k(X) \in \mathcal{U}$, as desired.

Second, let us consider the case (a). Take some $(\mathcal{U}_1, l_1)$, $(\mathcal{U}_2, l_2) \in \mathrm{T}_{\mathcal{U}}(X)$ with $\mathcal{U}_1 \neq \mathcal{U}_2$. From $\mathcal{U}_1 \neq \mathcal{U}_2$ it follows that $Y \in \mathcal{U}_1$, but $W \setminus Y \in \mathcal{U}_2$ for some $Y \subseteq W$. Since $(\mathcal{U}_1, l_1)$, $(\mathcal{U}_2, l_2) \in \mathrm{T}_{\mathcal{U}}(X)$, we can state that: (i) $\mathcal{U}R^{\mathrm{ue}}_{l_1}\mathcal{U}_1$ and $X \cap Y \in \mathcal{U}_1$; (ii) $\mathcal{U}R^{\mathrm{ue}}_{l_2}\mathcal{U}_2$ and $X \cap (W \setminus Y) \in \mathcal{U}_2$. So $\mathrm{T}_{\mathcal{U}}(X \cap Y) \neq \emptyset$ and $\mathrm{T}_{\mathcal{U}}(X \cap (W \setminus Y)) \neq \emptyset$. Hence by Proposition 5.15 (ii), we can establish the following:

$$\exists i. (0 < i < k \text{ and } \#\mathrm{T}_{\mathcal{U}}(X \cap Y) \geq i \text{ and } \#\mathrm{T}_{\mathcal{U}}(X \cap (W \setminus Y)) \geq k - i).$$

Then from I.H. it follows that $m_i(X \cap Y) \in \mathcal{U}$ and $m_{k-i}(X \cap (W \setminus Y)) \in \mathcal{U}$. Therefore, by Proposition 5.10 (iii), $m_k(X) \in \mathcal{U}$, as required. This finishes our proof of the induction step. □

Uniform Interpolation for Monotone Modal Logic

Luigi Santocanale

Laboratoire d'Informatique Fondamentale de Marseille, Université de Provence, 39 rue F. Joliot Curie, 13453 Marseille Cedex 13, France. Email: luigi.santocanale@lif.univ-mrs.fr

Yde Venema

Institute for Logic, Language and Computation, Universiteit van Amsterdam, Science Park 904 1098 XH Amsterdam, Netherlands. Email: Y.Venema@uva.nl

Abstract

We reconstruct the syntax and semantics of monotone modal logic, in the style of Moss' coalgebraic logic. To that aim, we replace the box and diamond with a modality ∇ which takes a finite collection of finite sets of formulas as its argument. The semantics of this modality in monotone neighborhood models is defined in terms of a version of relation lifting that is appropriate for this setting.
We prove that the standard modal language and our ∇-based one are effectively equi-expressive, meaning that there are effective translations in both directions. We prove and discuss some algebraic laws that govern the interaction of ∇ with the Boolean operations. These laws enable us to rewrite each formula into a special kind of disjunctive normal form that we call transparent. For such transparent formulas it is relatively easy to define the bisimulation quantifiers that one may associate with our notion of relation lifting. This allows us to prove the main result of the paper, viz., that monotone modal logic enjoys the property of uniform interpolation.

Keywords: monotone modal logic, uniform interpolation, neighborhood semantics, coalgebra

1 Introduction

Monotone modal logic is a generalization of normal modal logic in which the distribution of $\Box$ over conjunctions has been weakened to a monotonicity condition, which can either be expressed as an axiom $((\Box(p \wedge q) \to \Box p)$, or as a rule (from $p \to q$ derive $\Box p \to \Box q$). The standard semantics for such logics is provided by so-called monotone neighborhood models. Here the binary relation over a state space S is generalized to a so-called *monotone neighborhood function*, that is, a map $\sigma : S \to \mathcal{PP}S$ which is closed under taking supersets (if $X \in \sigma(s)$ and $X \subseteq Y$ then $Y \in \sigma(s)$). The interpretation of the modality in a monotone neighborhood model $\mathbb{S} = \langle S, \sigma, V\rangle$ (with V a valuation) is then given by

$$\mathbb{S}, s \Vdash \Box a \iff \exists U \in \sigma(s)\, \forall u \in U.\, \mathbb{S}, u \Vdash a\,.$$

Together with its polymodal versions and its fixed-point extensions, this logic has found applications in settings where the use of normal modalities would have undesirable consequences, such as in deontic [7], epistemic [35], or game logic [27]. The latter application, in which the monotone neighborhood function encodes the power of a player to achieve a certain outcome in an interactive system or game, has received some attention in computer science lately [1,28]. Monotone modal logic also appears to be a crucial link between lattice theory and modal logic, through the new covering semantics of lattices and substructural logics [15,31].

Although the monotone variant has never taken a central place in modal logic, various technical results are known. Two sources of information are the textbook [7], and the more recent survey [16], which also contains many original results. We mention here some facts concerning **M**, the monotone variant of the basic modal logic **K**. **M** is sound and complete with respect to the neighborhood semantics and satisfies the finite model property [7]; its satisfiability problem is NP-complete [35]. Finally, **M** has the Craig interpolation property [16]: Given two modal formulas a, b, if $\models a \to b$ (meaning that $a \to b$ holds in every state of every monotone neighborhood model), then there is a formula c, which may only use propositional variables that appear both in a and in b, such that $\models a \to c$ and $\models c \to b$.

The main contribution of this paper will be to add *uniform* interpolation to the list of properties of monotone modal logic. Uniform interpolation is a very strong version of interpolation, where the interpolant c does not really depend on b itself, but only on the *language* it shares with a (that is, the set of variables occurring both in a and in b). More precisely, we shall prove the following result. Here P_a denotes the set of proposition letters occurring in a given formula a, and $\mathcal{L}_\Diamond(\mathsf{Q})$ denotes the set of modal formulas a with $\mathsf{P}_a \subseteq \mathsf{Q}$.

Theorem 1 (Uniform interpolation for M) *For any modal formula a and any set $\mathsf{Q} \subseteq \mathsf{P}_a$ of proposition letters, there is a formula $a_\mathsf{Q} \in \mathcal{L}_\Diamond(\mathsf{Q})$, effectively constructible from a, such that for any formula b with $\mathsf{P}_a \cap \mathsf{P}_b \subseteq \mathsf{Q}$ we have*

$$\models a \to b \ \textit{ iff } \ \models a_\mathsf{Q} \to b. \tag{1}$$

Observe that by (1) it follows from $\models a_\mathsf{Q} \to a_\mathsf{Q}$ and $a_\mathsf{Q} \in \mathcal{L}_\Diamond(\mathsf{Q})$ that $\models a \to a_\mathsf{Q}$, and so a_Q is indeed an interpolant for every b with $\mathsf{P}_a \cap \mathsf{P}_b \subseteq \mathsf{Q}$: if $\models a \to b$ then $\models a \to a_\mathsf{Q}$ and $\models a_\mathsf{Q} \to b$.

A survey on (uniform) interpolation and the tools used to prove this property appears in [9, Section 4]. While it is easily argued that classical propositional logic has uniform interpolation, not many logics have this property, for example, first-order logic lacks it [19]. Recent interest in the property was initiated by the seminal work by A. Pitts [29] who proved that intuitionistic logic has the uniform interpolation property. In modal logic, Shavrukov [33] independently proved that the provability logic (also known as Gödel-Löb logic) **GL** has uniform interpolation. Subsequently, the property was established for modal logic **K**, independently by Ghilardi [13] and Visser [36], while [14] contains negative results on modal logics like **S4**. Finally, in the theory of modal fixed-point logics, it was realized in [10] that the logical property of uniform interpolation corresponds to the automata-theoretic property of *closure under projection*. In the same paper it was proved that the full modal μ-calculus [21] has uniform interpolation, in contrast to the fact that for instance PDL lacks the property [23].

Proofs of uniform interpolation property either follow a proof-theoretic or a semantic road. Notable examples of the proof-theoretic approach are Pitts' work [29] and, for modal logics, [5]. The semantic approach towards proving uniform interpolation is based on proving that a certain nonstandard second-order quantifier is definable in the language [29,36]. This *bisimulation quantifier* is interpreted as follows:

$$\mathbb{S}, s \Vdash \exists p.a \ \text{ iff } \ \mathbb{S}', s' \Vdash a, \ \text{for some } \mathbb{S}', s' \text{ with } \mathbb{S}, s \simeq_p \mathbb{S}', s', \tag{2}$$

where $\simeq_p$ denotes the relation of bisimilarity *up to proposition letter* p (see Definition 2.5 for a precise definition). Intuitively, (2) says that we can make the formula a true by, indeed, changing the interpretation of p, although not necessarily here, but in an up-to-p bisimilar state. For an detailed study of bisimulation quantifiers in modal logic, see [12].

In the case of the normal modal logic **K**, the proof simplifies considerably if we reconstruct the modal language on the basis of the so-called *cover modality*, here written as $\nabla_{\mathcal{P}}$. This modality, which takes a finite set of formulas as its argument, was introduced as a primitive operator, independently by Barwise & Moss [4] and by Janin & Walukiewicz [20]. Moss [25] observed that the semantics of this modality, which takes a *set* of formulas as its argument, can be defined in terms of *relation lifting*. More precisely, given a relation $R \subseteq S \times S'$, define

$$\begin{aligned}
\overrightarrow{\mathcal{P}}(R) &:= \{ (A, A') \in \mathcal{P}(S) \times \mathcal{P}(S') \mid \forall a \in A\, \exists a' \in A'.(a, a') \in R \}, \\
\overleftarrow{\mathcal{P}}(R) &:= \{ (A, A') \in \mathcal{P}(S) \times \mathcal{P}(S') \mid \forall a' \in A'\, \exists a \in A.(a, a') \in R \}, \\
\overline{\mathcal{P}}(R) &:= \overrightarrow{\mathcal{P}}(R) \cap \overleftarrow{\mathcal{P}}(R).
\end{aligned}$$

The relation $\overline{\mathcal{P}}(R)$ is called the Egli-Milner lifting of R — note that this relation underlies the back-and-forth clauses in the definition of a bisimulation between Kripke models. Returning to the semantics of $\nabla_{\mathcal{P}}$, given a Kripke model $\mathbb{S}$, we may consider the Egli-Milner lifting $\overline{\mathcal{P}}(\Vdash) \subseteq \mathcal{P}(S) \times \mathcal{P}(\mathcal{L})$ of the satisfaction relation $\Vdash \subseteq S \times \mathcal{L}$ between states and formulas, and define:

$$\mathbb{S}, s \Vdash \nabla_{\mathcal{P}} A \ \text{ iff } \ (\rho(s), A) \in \overline{\mathcal{P}}(\Vdash), \tag{3}$$

where $\rho(s)$ is the set of successors of s. Recently, axiomatic bases and proof systems have been found for languages based on $\nabla_{\mathcal{P}}$ and its generalization [6,22].

Two properties make the cover modality $\nabla_{\mathcal{P}}$ very useful: First, the connectives $\nabla_{\mathcal{P}}$ and $\vee$ have in some sense the same expressive power as the set $\{ \vee, \wedge, \Diamond, \Box \}$. And second, the bisimulation quantifier distributes both over $\vee$ and over $\nabla_{\mathcal{P}}$:

$$\exists p.\nabla_{\mathcal{P}} A \equiv \nabla_{\mathcal{P}} \{ \exists p.a \mid a \in A \}.$$

As a consequence, once we have 'reconstructed' the modal language on the basis of $\vee$ and $\nabla_{\mathcal{P}}$, the bisimulation quantifiers can be defined by a trivial inductive definition. This approach to uniform interpolation goes back to [10,11]. In [34] an algorithm is given computing uniform interpolants of size (singly) exponential in the size of the original formula.

We shall mimick this approach in our proof of Theorem 1. There is a natural notion of bisimilarity associated with monotone neighborhood models, and so we may naturally interpret the bisimulation quantifiers along this relation. Monotone bisimilarity can be expressed in terms of a relation lifting as well [18]. This lifting $\widetilde{\mathcal{M}}$ is given by defining, for a relation $R \subseteq S \times S'$, the relation $\widetilde{\mathcal{M}}(R) \subseteq \mathcal{PP}(S) \times \mathcal{PP}(S')$ as follows:

$$\widetilde{\mathcal{M}}(R) := \overrightarrow{\mathcal{P}}\overleftarrow{\mathcal{P}}(R) \cap \overleftarrow{\mathcal{P}}\overrightarrow{\mathcal{P}}(R).$$

Our idea is now to introduce a modality ∇ for monotone modal logic, which, analogous to the cover modality for relational models, is interpreted in neighborhood models by means of the lifting $\widetilde{\mathcal{M}}(\Vdash)$ of the satisfaction relation $\Vdash$.

Taking this idea as our guideline, we arrive at the following 'reconstruction' of monotone modal logic. Our language $\mathcal{L}_\nabla$ is based on a nonstandard modality ∇ which takes finite collections of finite sets of formulas as its argument:

- $\nabla\alpha$ is a formula of $\mathcal{L}_\nabla$, for each $\alpha \in \mathcal{P}_\omega\mathcal{P}_\omega\mathcal{L}_\nabla$.

The semantics of this operator is expressed in terms of the relation lifting $\widetilde{\mathcal{M}}$. That is, in every monotone neighborhood model $\mathbb{S}$, we have

$$\mathbb{S}, s \Vdash \nabla\alpha \ \text{ iff } \ (\sigma(s), \alpha) \in \widetilde{\mathcal{M}}(\Vdash). \tag{4}$$

The main aim of this paper is to show that this alternative way to set up monotone modal logic makes sense: With some work we can prove results analogous to the relational case. To start with, Theorem 3.5 below states that the standard modal language and our ∇-based one are effectively equi-expressive, meaning that there are effective translations in both directions. We prove and discuss some algebraic laws that guide the interaction of ∇ with the Boolean operations. These laws may not be as straightforward as in the relational case, but still they enable us to rewrite each formula into a special kind of disjunctive normal form that we call transparent (Proposition 5.3). For such transparent formulas it is relatively easy to define the bisimulation quantifiers associated with our notion $\widetilde{\mathcal{M}}$ of relation lifting (Definition 5.4). This allows us to prove the main result of the paper, viz., Theorem 1 above.

Finally, our approach has been very much influenced by the *coalgebraic* perspective on modal logic. The theory of (Universal) Coalgebra [30] provides a general mathematical framework for studying behavior of state-based evolving systems. Key examples of such systems are Kripke frames and models, together with many other structures from the theory of modal logic, such as (weighted/probabilistic) transition systems, general frames, and neighborhood structures. The link between modal logic and coalgebra is in fact very tight: Modal logic, suitably generalized and modified, provides natural languages and derivation systems for specifying and reasoning about behavior at a coalgebraic level of generality [8]. For a coalgebraic perspective on monotone modal logic the reader is referred to [17].

The link between modal logic and coalgebra goes back to the work of Moss [25], who observed that modal logic, once formulated in terms of the cover modality $\nabla_{\mathcal{P}}$, can be generalized to coalgebras of arbitrary type. Each coalgebraic type T, formally given as a functor $T : \mathsf{Set} \to \mathsf{Set}$, canonically induces an operation $\overline{T}$, which lifts a relation $R \subseteq S \times S'$ to a relation $\overline{T}(R) \subseteq TS \times TS'$. On this basis, Moss develops a modal formalism, based on a modality ∇_T, of which the semantics is defined in terms of the lifting $\overline{T}(\Vdash)$. Our modality ∇ is inspired by Moss' approach, instantiated by the type $\mathcal{M}$ of neighborhood frames. However, our notion of relation lifting, $\widetilde{\mathcal{M}}$, *differs* from the canonically defined relation lifting, $\overline{\mathcal{M}}$. We return to this issue in the final section of the paper.

Overview In the next section we recall some definitions, introduce some basic notions, and discuss some new concepts, including some properties of our relation lifting. In section 3 we introduce the monotone nabla modality, and we prove the equi-expressiveness result. In section 4 we discuss some algebraic laws that govern the interaction of ∇ with the Boolean connectives. Section 5 is the main part of the paper: here we show how to define bisimulation quantifiers in the language $\mathcal{L}_\nabla$, and we show how to derive uniform interpolation from that. We finish with drawing some conclusions and listing some future work.

2 Preliminaries

2.1 Neighborhood models and monotone modal logic

We first recall some basic facts on monotone modal logic. It will be convenient for us to base our language on formulas in negation normal form, in which the use of negations is restricted to atomic formulas. As a consequence, all our primitive connectives will come in pairs of Boolean duals. In particular, next to the modality $\Box$ we also have to take its Boolean dual, $\Diamond$, as a primitive connective.

Definition 2.1 Given a set Prop of proposition letters, the set $\mathcal{L}_\Diamond(\mathsf{Prop})$ of *modal formulas* over Prop, is given by the following grammar:

$$a \ := \ p \mid \neg p \mid \bot \mid \top \mid a \wedge a \mid a \vee a \mid \Diamond a \mid \Box a$$

where $p \in \mathsf{Prop}$.

Definition 2.2 A *neighborhood frame* is a pair $\langle S, \sigma \rangle$ such that $\sigma : S \to \mathcal{PP}(S)$ is a map assigning to a state $s \in S$ a collection $\sigma(s)$ of *neighborhoods* of s. In case each $\sigma(s)$ is closed under taking supersets (that is, if $Y \supseteq X \in \sigma(s)$ implies $Y \in \sigma(s)$), we say that the neighborhood frame is *monotone*. A *neighborhood model* is a triple $\mathbb{S} = \langle S, \sigma, V \rangle$ such that $\langle S, \sigma \rangle$ is a neighborhood frame, and $V : S \to \mathcal{P}(\mathsf{Prop})$ is a *coloring*. Neighborhood models based on monotone frames will simply be called *models*. A *pointed model* is just a pair $(\mathbb{S}, s)$ consisting of a model $\mathbb{S}$ and a point s in $\mathbb{S}$.

Definition 2.3 Given a model $\mathbb{S} = \langle S, \sigma, V \rangle$, we define the satisfaction relation $\Vdash \subseteq S \times \mathcal{L}_{\diamond}(\mathsf{Prop})$ by induction. For the classical connectives the definition is as usual:

$$\begin{array}{ll} \mathbb{S}, s \Vdash p \text{ iff } p \in V(s) & \mathbb{S}, s \Vdash \neg p \text{ iff } p \notin V(s) \\ \mathbb{S}, s \Vdash \top \text{ iff } \mathtt{true} & \mathbb{S}, s \Vdash \bot \text{ iff } \mathtt{false} \\ \mathbb{S}, s \Vdash \phi \wedge \psi \text{ iff } \mathbb{S}, s \Vdash \phi \text{ and } \mathbb{S}, s \Vdash \psi & \mathbb{S}, s \Vdash \phi \vee \psi \text{ iff } \mathbb{S}, s \Vdash \phi \text{ or } \mathbb{S}, s \Vdash \psi . \end{array}$$

For the modal connectives we define:

$$\begin{array}{l} \mathbb{S}, s \Vdash \Box\phi \text{ iff } \exists U \in \sigma(s) \text{ such that } \forall u \in U\ \mathbb{S}, u \Vdash \phi \\ \mathbb{S}, s \Vdash \Diamond\phi \text{ iff } \forall U \in \sigma(s)\ \exists u \in U \text{ such that } \mathbb{S}, u \Vdash \phi . \end{array}$$

Convention 2.4 *In order to recall the logical structure of the satisfaction relation of the two modal connectors, we shall write* $\langle \exists\forall \rangle$ *in place of* $\Box$*, and* $\langle \forall\exists \rangle$ *in place of* $\Diamond$.

Definition 2.5 Let $\mathbb{S} = \langle S, \sigma, V \rangle$ and $\mathbb{S}' = \langle S', \sigma', V' \rangle$ be two models, and let $\mathsf{P} \subseteq \mathsf{Prop}$ be a set of proposition letters. A relation $Z \subseteq S \times S'$ is a P-*bisimulation*, if for all $(s, s') \in Z$:

(prop) $V(s) \cap \mathsf{P} = V'(s') \cap \mathsf{P}$;

(forth) $\forall U \in \sigma(s) \exists U' \in \sigma'(s') \forall u' \in U' \exists u \in U. uZu'$;

(back) $\forall U' \in \sigma'(s') \exists U \in \sigma(s) \forall u \in U \exists u' \in U'. uZu'$.

Prop-bisimulations are simply called *bisimulations*, and $(\mathsf{Prop} \setminus \{ p \})$-bisimulations will be called *bisimulations up to p*. If Z is a P-bisimulation between $\mathbb{S}$ and $\mathbb{S}'$ linking s to s', we write $Z : \mathbb{S}, s \simeq_{\mathsf{P}} \mathbb{S}', s'$.

2.2 Functors and coalgebras

While we shall generally suppress the use of category theory in this paper, we need the set functors, $\mathcal{P}$, $\mathcal{Q}$ and $\mathcal{M}$, and their finitary versions, $\mathcal{P}_\omega$, $\mathcal{Q}_\omega$ and $\mathcal{M}_\omega$. As mentioned, neighborhood frames are coalgebras for the functor $\mathcal{M}$.

Definition 2.6 On the category Set (with sets as objects and functions as arrows) we let $\mathcal{P}$ denote the covariant *power set functor*; $\mathcal{Q} := \mathcal{P} \circ \mathcal{P}$ is the *double power set functor.* We write $\mathcal{P}_\omega(S)$ for the collection of all finite subsets of S. As the direct image of a finite subset is finite, $\mathcal{P}_\omega$ is itself a functor. We define $\mathcal{Q}_\omega := \mathcal{P}_\omega \circ \mathcal{P}_\omega$.

Given an element $\alpha \in \mathcal{Q}S$, we define

$$\alpha^{\uparrow} := \{ X \in \mathcal{P}(S) \mid X \supseteq Y \text{ for some } Y \in \alpha \},$$

and we say that α is upward closed if $\alpha = \alpha^{\uparrow}$. The functor $\mathcal{M}$ is given by $\mathcal{M}(S) := \{ \alpha \in \mathcal{Q}S \mid \alpha \text{ is upward closed } \}$, while for $f : S \to S'$, we define $\mathcal{M}f : \mathcal{M}S \to \mathcal{M}S'$ by $(\mathcal{M}f)(\alpha) := ((\mathcal{Q}f)(\alpha))^{\uparrow}$. For $\mathcal{M}_\omega$ we define $\mathcal{M}_\omega(S) := \{\alpha \in \mathcal{M}(S) \mid \alpha = \beta^{\uparrow} \text{ for some } \beta \in \mathcal{Q}_\omega(S)\}$. It is easily verified that by putting $\mathcal{M}_\omega f(\alpha) := \mathcal{M}f(\alpha)$, we turn $\mathcal{M}_\omega$ itself into a functor.

Instead of working with an element $\alpha \in \mathcal{M}_\omega S$ it will often be convenient to work with a finite generating set, that is, a set $\beta \in \mathcal{Q}_\omega S$ such that $\alpha = \beta^\uparrow$. Fortunately there is always a canonical choice for such a β: we leave it for the reader to verify that the following is well-defined.

Definition 2.7 An element $\beta \in \mathcal{Q}S$ is an *anti-chain* if $X \subseteq Y$ for no $X, Y \in \beta$. Given a set $\alpha \in \mathcal{M}_\omega S$, we let $\alpha_\downarrow$ denote the unique anti-chain $\beta \in \mathcal{Q}_\omega S$ with $\alpha = \beta^\uparrow$.

2.3 The exchange operator

One key argument in our proofs shall be a principle of quantifier exchange, that we state in Proposition 2.12. This principle can be understood algebraically as providing a direct characterization of the closure operator arising from a Galois connection. Typically, such a closure operator is defined in terms of universal quantifiers, while the characterization we provide of the same operator is an existential statement.

Definition 2.8 Given some set S, for $A, B \subseteq \mathcal{P}(S)$, we write

$$A \perp_S B \text{ iff } A \cap B \neq \emptyset,$$

and for $\alpha \in \mathcal{Q}S$ we define

$$\alpha^{\perp_S} := \{ B \in \mathcal{P}(S) \mid B \perp_S A \text{ for all } A \in \alpha \}.$$

In lattice theory, the operation $(\cdot)^{\perp_S}$ is known as the *polarity* associated with the relation $\perp_S$, and as such it is well-known to have some nice properties. To ease the reading we shall omit the subscript S from $(\cdot)^{\perp_S}$, if no confusion is likely to arise. The following observation is straightforward.

Proposition 2.9 *Given a set S, the operation $(\cdot)^{\perp\perp}$ is a closure operation on $\mathcal{Q}S$, with $\mathcal{M}S \subseteq \mathcal{Q}S$ forming the set of closed elements. For $\alpha \in \mathcal{Q}S$, we have*
(1) $(\alpha^\perp)^\perp = \alpha^\uparrow$;
(2) $\alpha^\perp = (\alpha^\uparrow)^\perp$;
(3) $\alpha = \varnothing$ iff $\varnothing \in \alpha^\perp$;
(4) $\varnothing \in \alpha$ iff $\alpha^\perp = \varnothing$.

Given $\alpha \in \mathcal{Q}_\omega S$, the set $\alpha^{\perp_S} \in \mathcal{Q}S$ need not belong to $\mathcal{Q}_\omega S$ (unless S is finite), but fortunately we can make the following observations.

Proposition 2.10 *Given some set S, assume that $\alpha \in \mathcal{M}_\omega S \cup \mathcal{Q}_\omega S$. Then*
(1) $\alpha^\perp \in \mathcal{M}_\omega S$;
(2) $(\alpha^\perp)_\downarrow \in \mathcal{Q}_\omega(\bigcup \alpha)$;
(3) $\alpha = ((\alpha^\perp)_\downarrow)^\perp$.

This justifies the following definition of the antichain representation of $\alpha^{\perp_S}$ in $\mathcal{Q}_\omega S$:

Definition 2.11 Given a set S and an element $\alpha \in \mathcal{Q}_\omega S \cup \mathcal{M}_\omega S$, we define

$$\alpha^\bullet := (\alpha^\perp)_\downarrow.$$

Observe that the above definition does not depend on S: namely, if $S \subseteq S'$ and $\alpha \in \mathcal{Q}_\omega S \subseteq \mathcal{Q}_\omega S'$, then we can index the operation $\perp$ either by S or by S'. However, a straightforward calculation shows that $(\alpha^{\perp_S})_\downarrow = (\alpha^{\perp_{S'}})_\downarrow$. Consequently, the computation of $\alpha^\bullet$ can be executed relative to the least S such that $\alpha \subseteq \mathcal{Q}_\omega S$, which clearly is $\bigcup \alpha$. Together with Proposition 2.10, this also implies that $(\alpha^\bullet)^\bullet = (((\alpha^\perp)_\downarrow)^\perp)_\downarrow = \alpha_\downarrow$.

This paper will see a few key applications for the following *exchange principle*. For its proof, notice that the principle is almost a restatement of Proposition 2.9(1).

Proposition 2.12 (Exchange Principle) *Given a set S, the following are equivalent for any $\alpha \in \mathcal{Q}S$ and $P \in \mathcal{P}(S)$:*
(a) $\exists A \in \alpha \forall a \in A.\, a \in P$;
(b) $\forall B \in \alpha^\perp \exists b \in B.\, b \in P$.

Similarly, for any $\alpha \in \mathcal{Q}_\omega S$ we have the following equivalence:
(a′) $\exists A \in \alpha \forall a \in A.\, a \in P$;
(b′) $\forall B \in \alpha^\bullet \exists b \in B.\, b \in P$.

Remark 2.13 It is well known that $\mathcal{M}$ corresponds to the functor of taking free completely distributive lattice, see for example [24]. In a similar manner, $\mathcal{M}_\omega$ is the free distributive lattice functor [26]. To see this, it is enough to put the set $\alpha \in \mathcal{P}_\omega S$ in correspondence with the term $t_\alpha := \bigwedge_{A \in \alpha_\downarrow} \bigvee A$. Behind the operations $(\cdot)^\bullet$ and $(\cdot)^\perp$ we may recognize the action of *dualizing* the term t_α (that is, exchanging meets and joins), followed by rewriting the result, which is now in conjunctive normal form, back into disjunctive normal form.

2.4 Relation Lifting

In the introduction we already defined the operations $\overrightarrow{\mathcal{P}}$, $\overleftarrow{\mathcal{P}}$, $\overline{\mathcal{P}}$, and $\widetilde{\mathcal{M}}$. As mentioned, the notion of a bisimulation between neighborhood models can be nicely expressed using these definitions:

Proposition 2.14 *Let $\mathbb{S}$ and $\mathbb{S}'$ be two models, and let $Z \subseteq S \times S'$. Then Z is a P-bisimulation iff $V(s) \cap \mathsf{P} = V(s') \cap \mathsf{P}$ and $(\sigma(s), \sigma'(s')) \in \widetilde{\mathcal{M}}(Z)$, for all $(s, s') \in Z$.*

The following preliminary observations, which can be proved via a straightforward verification, will be used throughout the paper. We shall use ';' and '$\smile$' to denote relational composition and converse, respectively; Δ denotes the identity/diagonal relation, and we write $Gr(f)$ to denote the graph of a function f.

Proposition 2.15 *The operation $\widetilde{\mathcal{M}}$ has the following properties:*

(1) $\widetilde{\mathcal{M}}$ is monotone: if $R \subseteq R'$ then $\widetilde{\mathcal{M}}(R) \subseteq \widetilde{\mathcal{M}}(R')$,
(2) $\widetilde{\mathcal{M}}$ commutes with converse: $\widetilde{\mathcal{M}}(R^\smile) = (\widetilde{\mathcal{M}}(R))^\smile$,
(3) $\widetilde{\mathcal{M}}$ is lax functorial: $\Delta_{\mathcal{Q}S} \subseteq \widetilde{\mathcal{M}}(\Delta_S)$ and $(\widetilde{\mathcal{M}}R_0);(\widetilde{\mathcal{M}}R_1) \subseteq \widetilde{\mathcal{M}}(R_0;R_1)$,
(4) $\widetilde{\mathcal{M}}$ is well defined for $\mathcal{M}$: $(\alpha, \alpha') \in \widetilde{\mathcal{M}}R$ iff $(\alpha^\uparrow, \alpha') \in \widetilde{\mathcal{M}}R$,
(5) $\widetilde{\mathcal{M}}$ commutes with restrictions: $(\widetilde{\mathcal{M}}R) \cap (\mathcal{Q}Y \times \mathcal{Q}Y') = \widetilde{\mathcal{M}}(R \cap (Y \times Y'))$,
(6) $\widetilde{\mathcal{M}}$ is a lax extension of $\mathcal{M}$: $Gr(\mathcal{Q}f) \subseteq \widetilde{\mathcal{M}}(Gr(f))$,

(7) $\widetilde{\mathcal{M}}$ distributes over composition to the left with function graphs:
$\widetilde{\mathcal{M}}(Gr(f);R) = Gr(\mathcal{Q}f);\widetilde{\mathcal{M}}(R)$.

Remark 2.16 It is shown in [32] that, given the properties of the functors $\mathcal{P}$ and $\mathcal{M}$ as *monads* on the category Set, the associated pairs of *directed* relation liftings, $\overrightarrow{\mathcal{P}}/\overleftarrow{\mathcal{P}}$ and $\overrightarrow{\mathcal{P}}\overleftarrow{\mathcal{P}}/\overleftarrow{\mathcal{P}}\overrightarrow{\mathcal{P}}$ respectively, arise in a canonical way. It is interesting to note that, from the logical point of view, the *intersections* $\overline{\mathcal{P}}$ and $\widetilde{\mathcal{M}}$ of these canonically obtained relation liftings turn out to be relevant. The interested reader is referred to [3], where coalgebraic logics are developed on the basis of such directed notions of relation lifting.

3 A Monotone ∇

In this section we introduce the syntax and semantics of $\mathcal{L}_\nabla$, the ∇-based version of monotone modal logic. We prove that this language has the same expressive power as $\mathcal{L}_\diamond$ by showing the interdefinability of ∇ with the pair of modalities $\{\diamond, \Box\}$.

Definition 3.1 Given a set Prop of proposition letters, the set $\mathcal{L}_\nabla(\mathsf{Prop})$ of ∇*-formulas* over Prop, is given by the following (pseudo-)grammar:

$$a ::= p \mid \neg p \mid \bot \mid \top \mid a \wedge a \mid a \vee a \mid \nabla\alpha,$$

where $p \in \mathsf{Prop}$ and $\alpha \in \mathcal{Q}_\omega\mathcal{L}_\nabla(\mathsf{Prop})$. Given a formula $a \in \mathcal{L}(\mathsf{Prop})$, we let P_a denote the set of variables occurring in a and $d_\nabla(a)$, the ∇-*depth* of a. The latter notion is defined via a straightforward formula induction, with $d_\nabla(\nabla\alpha) := 1 + \max\{ d_\nabla(a) \mid a \in \bigcup\alpha \}$.

Definition 3.2 Let $\mathbb{S} = \langle S, \sigma, V\rangle$ be a monotone neighborhood model. We define the *truth* or *satisfaction relation* $\Vdash \subseteq S \times \mathcal{L}_\nabla$ by induction on the complexity of formulas, the only nontrivial clause (4) already been given in the introduction.

As an immediate consequence of (4) we see that

$$\mathbb{S}, s \Vdash \nabla\alpha \text{ iff } (\sigma(s), \alpha) \in \overrightarrow{\mathcal{P}}\overleftarrow{\mathcal{P}}(\Vdash) \cap \overleftarrow{\mathcal{P}}\overrightarrow{\mathcal{P}}(\Vdash), \tag{5}$$

or in words: $\nabla\alpha$ holds at s if every neighborhood U of s supports some set $A \in \alpha$, and every set $A \in \alpha$ holds throughout some neighborhood $U \in \sigma(s)$. Here we say that U *supports* a set A, notation $U \rhd A$, if every formula in A is true at some point in U, and, conversely, that A *holds throughout* U, notation: $U \Vdash \bigvee A$, if every point in U makes some formula in A true:

$$U \Vdash \bigvee A \text{ iff } (U, A) \in \overrightarrow{\mathcal{P}}(\Vdash) \qquad\qquad U \rhd A \text{ iff } (U, A) \in \overleftarrow{\mathcal{P}}(\Vdash).$$

Definition 3.3 Let a and b be ($\mathcal{L}_\diamond$- or $\mathcal{L}_\nabla$-)formulas. We write $a \models b$ if $\mathbb{S}, s \Vdash a$ implies $\mathbb{S}, s \Vdash b$, for all pointed models $(\mathbb{S}, s)$. We say that a and b are *equivalent*, notation: $a \equiv b$, if $a \models b$ and $b \models a$, and that a is *valid*, notation: $\models a$, if $\top \models a$.

The following elementary properties of ∇ will turn out to be handy.

Proposition 3.4 *The following hold for any pointed model* $(\mathbb{S}, s)$*:*
(1) $\mathbb{S}, s \Vdash \nabla\varnothing$ *iff* $\sigma(s) = \varnothing$,
(2) $\mathbb{S}, s \Vdash \nabla\{\varnothing\}$ *iff* $\varnothing \in \sigma(s)$.

The main result of this section states that the two languages, $\mathcal{L}_\diamond$ and $\mathcal{L}_\nabla$, are effectively equi-expressive.

Theorem 3.5 *There are effectively defined translations* $(\cdot)^\diamond : \mathcal{L}_\nabla(\mathsf{Prop}) \to \mathcal{L}_\diamond(\mathsf{Prop})$ *and* $(\cdot)^\nabla : \mathcal{L}_\diamond(\mathsf{Prop}) \to \mathcal{L}_\nabla(\mathsf{Prop})$ *such that* $a \equiv a^\diamond$ *for each formula* $a \in \mathcal{L}_\nabla$*, and* $b \equiv b^\nabla$ *for each formula* $b \in \mathcal{L}_\diamond$.

In order to prove this theorem, a direct verification will reveal that the modalities $\langle\exists\forall\rangle$ and $\langle\forall\exists\rangle$ are definable in the language $\mathcal{L}_\nabla$.

Proposition 3.6 *The following equivalences hold, for any formula a:*

$$\langle\exists\forall\rangle a \equiv \nabla\{\{a\},\{\top\}\} \vee \nabla\{\varnothing\}, \tag{6}$$
$$\langle\forall\exists\rangle a \equiv \nabla\{\{a,\top\}\} \vee \nabla\varnothing. \tag{7}$$

Conversely, the nabla modality can be expressed using the box and diamond of monotone modal logic.

Proposition 3.7 *The following equivalence holds for any collection* α *of formula sets:*

$$\nabla\alpha \equiv \bigwedge_{A\in\alpha} \langle\exists\forall\rangle\bigvee A \wedge \bigwedge_{B\in\alpha^\bullet} \langle\forall\exists\rangle\bigvee B. \tag{8}$$

Proof. Fix a pointed model $(\mathbb{S}, s)$. It is immediate by the definitions that

$$(\sigma(s), \alpha) \in \overleftarrow{\mathcal{P}}\overrightarrow{\mathcal{P}}(\Vdash) \text{ iff } \mathbb{S}, s \Vdash \bigwedge_{A\in\alpha} \langle\exists\forall\rangle \bigvee A. \tag{9}$$

Also observe that

$$\begin{array}{lll}
(\sigma(s),\alpha) \in \overrightarrow{\mathcal{P}}\overleftarrow{\mathcal{P}}(\Vdash) & & \\
& \text{iff } \forall U \in \sigma(s)\exists A \in \alpha\forall a \in A\exists u \in U \text{ with } u \Vdash a & \text{(definition)} \\
& \text{iff } \forall U \in \sigma(s)\forall B \in \alpha^\bullet\exists b \in B\exists u \in U \text{ with } u \Vdash b & \text{(exchange principle)} \\
& \text{iff } \forall B \in \alpha^\bullet\forall U \in \sigma(s)\exists u \in U\exists b \in B \text{ with } u \Vdash b & \text{(quantifier swaps)} \\
& \text{iff } \mathbb{S}, s \Vdash \bigwedge_{B\in\alpha^\bullet} \langle\forall\exists\rangle\bigvee B. & \text{(definitions)}
\end{array}$$

By (5), the combination of these observations yields the desired equivalence (8). □

Finally, the proof of Theorem 3.5 follows an obvious induction on formulas, based on the Propositions 3.6 and 3.7.

4 Some algebraic laws

In this section we will see how ∇ interacts with, respectively, the consequence relation $\models$, and the Boolean connectives: $\wedge, \vee$, and $\neg$. (The proofs of this section are deferred to the appendix.)

First we show that the monotone nabla modality is monotone indeed.

Proposition 4.1 *For any $\alpha, \alpha' \in \mathcal{Q}\mathcal{L}_\nabla$ we have*

$$(\alpha, \alpha') \in \widetilde{\mathcal{M}}(\models) \text{ implies } \nabla\alpha \models \nabla\alpha'. \tag{10}$$

Proof. Assume that $(\alpha, \alpha') \in \widetilde{\mathcal{M}}(\models)$ and let $\mathbb{S}, s$ be a pointed model such that $\mathbb{S}, s \Vdash \nabla\alpha$. Then by definition of $\Vdash$, we have $(\sigma(s), \alpha) \in \widetilde{\mathcal{M}}(\Vdash)$, and so by Proposition 2.15(3) we find that $(\sigma(s), \alpha') \in \widetilde{\mathcal{M}}(\Vdash); \widetilde{\mathcal{M}}(\models) \subseteq \widetilde{\mathcal{M}}(\Vdash;\models)$. Then by $\Vdash;\models \subseteq \Vdash$ and Proposition 2.15(1) we obtain $(\sigma(s), \alpha') \in \widetilde{\mathcal{M}}(\Vdash)$ which shows that $\mathbb{S}, s \Vdash \nabla\alpha'$, as required. □

Next we prove the following distributive law for conjunction.

Definition 4.2 Given a set $F \subseteq \mathcal{L}_\nabla$ of formulas, we let $\mathsf{Conj}(F)$ denote the set of (finite) conjunctions of formulas in F.

Proposition 4.3 *For any $\alpha, \alpha' \in \mathcal{Q}_\omega\mathcal{L}_\nabla$, we have*

$$\nabla\alpha \wedge \nabla\alpha' \equiv \bigvee\{\, \nabla\beta \mid \beta \in \mathcal{Q}(\mathsf{Conj}(\bigcup\alpha \cup \bigcup\alpha')) \text{ with } (\beta, \alpha), (\beta, \alpha') \in \widetilde{\mathcal{M}}(\models) \,\}. \tag{11}$$

Proof. It follows by Proposition 4.1 that the left hand side of (11) is a semantic consequence of the right hand side. For the converse, assume that $\mathbb{S}, s \Vdash \nabla\alpha \wedge \nabla\alpha'$. It suffices to come up with a $\beta \in \mathcal{Q}(\mathsf{Conj}(\bigcup\alpha \cup \bigcup\alpha'))$ such that $\mathbb{S}, s \Vdash \nabla\beta$ and $(\beta, \alpha), (\beta, \alpha') \in \widetilde{\mathcal{M}}(\models)$. To this aim, define, for $t \in S$, and $U \in \sigma(s)$,

$$\begin{aligned} A_t &:= \{\, a \in \textstyle\bigcup\alpha \mid \mathbb{S}, t \Vdash a \,\}, & A'_t &:= \{\, a' \in \textstyle\bigcup\alpha' \mid \mathbb{S}, t \Vdash a' \,\}, \\ b_t &:= \textstyle\bigwedge A_t \wedge \bigwedge A'_t, \\ B_U &:= \{\, b_t \mid t \in U \,\}, \\ \beta &:= \{\, B_U \mid U \in \sigma(s) \,\}. \end{aligned}$$

We verify that β has the desired properties. First we check that

$$\mathbb{S}, s \Vdash \nabla\beta. \tag{12}$$

To prove this, think of b as a map $b : S \to \mathcal{L}_\nabla$. It is easy to see that $Gr(b) \subseteq \Vdash$, and that $\beta = (\mathcal{Q}b)(\sigma(s))$. But then it follows from the properties of relation lifting, Proposition 2.15(6), that $(\sigma(s), \beta) \in Gr(\mathcal{Q}b) \subseteq \widetilde{\mathcal{M}}(\Vdash)$. This proves (12).

Second, we prove that

$$(\beta, \alpha) \in \widetilde{\mathcal{M}}(\models) \text{ and } (\beta, \alpha') \in \widetilde{\mathcal{M}}(\models). \tag{13}$$

We show that $(\beta,\alpha) \in \widetilde{\mathcal{M}}(\models)$. If $B \in \beta$, then there is a $U \in \sigma(s)$ such that $B = B_U$. Since $\mathbb{S}, s \Vdash \nabla\alpha$, take an $A \in \alpha$ such that $U \rhd A$. For such an A, if $a \in A$, then $a \in A_u$ for some $u \in U$, and hence the formula $b_u \in B_U = B$ satisfies $b_u \models a$. Conversely, for $A \in \alpha$, let $U \in \sigma(s)$ be such that $U \Vdash \bigvee A$, and consider the set $B_U \in \beta$. For each $b \in B_U$ there is a $u \in U$ with $b = b_u$. From $U \Vdash \bigvee A$, we see that there is an $a \in A$ with $u \Vdash a$. Then $b = b_u \models a$. This proves (13), and hence finishes the proof of the Proposition. □

As a corollary of Proposition 4.3, we may almost eliminate conjunctions from the language $\mathcal{L}_\nabla$, restricting their occurrence to special ones of the form $\bigwedge \Pi \wedge \nabla\alpha$, where Π is a set of literals. This issue will play an important role in the next section.

We now study the behavior of disjunctions occurring directly under the ∇-modality. Observe that an arbitrary such formula can be represented as $\nabla(\alpha \cup \{C \cup \{\bigvee B\}\})$. First we consider nonempty disjunctions, that is, the case where $B \neq \varnothing$. The following propositions shows how such disjunctions can be *eliminated.*

Proposition 4.4 *For any $\alpha \in \mathcal{Q}_\omega\mathcal{L}_\nabla$ and $B, C \in \mathcal{P}_\omega\mathcal{L}_\nabla$ such that $B \neq \varnothing$, we have*

$$\nabla(\alpha \cup \{C \cup \{\bigvee B\}\}) \equiv \nabla(\alpha \cup \{C \cup B\} \cup \{C \cup \{b, \top\} \mid b \in B\}). \tag{14}$$

Proof. Fix a pointed model $\mathbb{S}, s$ and abbreviate

$$\begin{aligned}\gamma &= \alpha \cup \{C \cup \{\bigvee B\}\},\\ \delta &= \alpha \cup \{C \cup B\} \cup \{C \cup \{b, \top\} \mid b \in B\}.\end{aligned}$$

Recall that

$$\begin{aligned}\mathbb{S}, s \Vdash \nabla\gamma \text{ iff } \quad & \forall U \in \sigma(s).\Big(\exists A \in \alpha.\, U \rhd A \text{ or } U \rhd C \cup \{\bigvee B\}\Big)\\ & \text{and } \forall A \in \alpha\, \exists U \in \sigma(s).\, U \Vdash \bigvee A\\ & \text{and } \exists U \in \sigma(s).\, U \Vdash \bigvee C \vee \bigvee B,\end{aligned}$$

while

$$\begin{aligned}\mathbb{S}, s \Vdash \nabla\delta \text{ iff } \quad & \forall U \in \sigma(s).\Big(\exists A \in \alpha.\, U \rhd A\\ & \qquad \text{or } U \rhd C \cup B \text{ or } \exists b \in B.\, U \rhd C \cup \{b, \top\}\Big)\\ & \text{and } \forall A \in \alpha\, \exists U \in \sigma(s).\, U \Vdash \bigvee A\\ & \text{and } \exists U \in \sigma(s).\, U \Vdash \bigvee(C \cup B)\\ & \text{and } \forall b \in B\, \exists U \in \sigma(s).\, U \Vdash \bigvee C \vee b \vee \top.\end{aligned}$$

In order to prove that $\nabla\gamma \equiv \nabla\delta$ we first argue that

$$\begin{aligned}U \rhd C \cup \{\bigvee B\} \text{ iff } & \exists b \in B \text{ such that } U \rhd C \cup \{b\}\\ \text{iff } & U \rhd C \cup B \text{ or } \exists b \in B \text{ such that } U \rhd C \cup \{b, \top\},\end{aligned}$$

where the second equivalence holds because B is nonempty. Then we note that $\bigvee C \vee \bigvee B \equiv \bigvee(C \cup B)$, and that for a formula b and a neighborhood U of s we *always* have $U \Vdash \bigvee C \vee b \vee \top$. From these obervations, the desired equivalence (14) is immediate. □

The next proposition states how we may eliminate the empty disjunction $\bigvee \varnothing = \bot$, if it occurs directly under a ∇.

Proposition 4.5 *Let $\alpha \in \mathcal{Q}_\omega \mathcal{L}_\nabla$ and $\beta \in \mathcal{P}_\omega \mathcal{L}_\nabla$.*
(1) If $\alpha = \varnothing$ then we have

$$\nabla(\alpha \cup \{B \cup \{\bot\}\}) \equiv \bot. \tag{15}$$

(2) If $\alpha \neq \varnothing$, then for any $A \in \alpha$ and $Z \subseteq A \times B$ with $(A, B) \in \overrightarrow{\mathcal{P}}(Z)$, we have

$$\nabla(\alpha \cup \{B \cup \{\bot\}\}) \equiv \nabla(\alpha \cup \{B \cup \{a \wedge b \mid (a,b) \in Z\}\}). \tag{16}$$

(3) If the left hand side of (16) is satisfiable, then there are at least one $A \in \alpha$ and $Z \subseteq A \times B$ with $(A, B) \in \overrightarrow{\mathcal{P}}(Z)$, and the formula $a \wedge b$ satisfiable for each $(a, b) \in Z$.

Proof. Abbreviate

$$\gamma = \alpha \cup \{B \cup \{\bot\}\},$$
$$\delta = \alpha \cup \{B \cup \{a \wedge b \mid (a,b) \in Z\}\}.$$

(1) If $\alpha = \varnothing$ then $\mathbb{S}, s \Vdash \nabla\gamma$ implies the existence of some $U \in \sigma(s)$ such that $U \Vdash \bigvee B \vee \bot$; in particular, we find that $\sigma(s) \neq \varnothing$. However, from $(\sigma(s), \gamma) \in \overrightarrow{\mathcal{P}}\overleftarrow{\mathcal{P}}(\Vdash)$ we obtain that $U \rhd B \cup \{\bot\}$ for each $U \in \sigma(s)$. Clearly this is impossible, which shows that $\nabla\gamma$ is not satisfiable, and hence, equivalent to $\bot$.

(2) We argue first that $(\gamma, \delta) \in \widetilde{\mathcal{M}}(\models)$. Take an arbitrary $C \in \gamma$, then we need to find a $D \in \delta$ such that $(C, D) \in \overleftarrow{\mathcal{P}}(\models)$. This is easy: if $C \in \alpha$, then we take $D := C$, and if $C = B \cup \{\bot\}$, then we choose $D := B \cup \{a \wedge b \mid (a,b) \in Z\}$. Conversely, given $D \in \delta$, we need to come up with a $C \in \gamma$ such that $(C, D) \in \overrightarrow{\mathcal{P}}(\models)$. Again, if $D \in \alpha$ we simply take $C := D$. Consider next the case that $D = B \cup \{a \wedge b \mid (a,b) \in Z\}$, and distinguish cases: if $B \neq \varnothing$, then we may take $C := B \cup \{\bot\}$, and if $B = \varnothing$ then by totality of Z we also have $A = \varnothing$; in this case we take $C := A$.

We argue next that $(\delta, \gamma) \in \widetilde{\mathcal{M}}(\models)$. First take some $D \in \delta$. If $D \in \alpha$ then define $C := D$, and if $D = B \cup \{a \wedge b \mid (a,b) \in Z\}$ then put $C := A$. In both cases we see that $(D, C) \in \overleftarrow{\mathcal{P}}(\models)$. Conversely, consider an arbitrary $C \in \gamma$. Again, if $C \in \alpha$ define $D := C$, and if $C = B \cup \{\bot\}$ take $D := B \cup \{a \wedge b \mid (a,b) \in Z\}$. In either case it is easily verified that $(D, C) \in \overrightarrow{\mathcal{P}}(\models)$.

(3) Finally, suppose that $\mathbb{S}, s \Vdash \nabla\gamma$. Then there is a $U \in \sigma(s)$ such that $U \Vdash \bigvee B$, and for this U there is an $A \in \gamma$ such that $U \rhd A$. Clearly then $\bot \notin A$, which means that $A \neq B \cup \{\bot\}$ and so A belongs to α. Define

$$Z := \{(a, b) \in A \times B \mid U \rhd a \wedge b\}.$$

It is straightforward to verify that this A and this Z have the desired properties: It is obvious that for every $(a, b) \in Z$ the formula $a \wedge b$ is satisfiable. To see that $(A, B) \in \overrightarrow{\mathcal{P}} Z$, take an arbitrary formula $a \in A$. Since $U \rhd A$, there is a $u \in U$ with $\mathbb{S}, u \Vdash a$. Also, from $U \Vdash \bigvee B$, there is a $b \in B$ with $\mathbb{S}, u \Vdash b$. Clearly then $(a, b) \in Z$. □

Remark 4.6 To see why this covers all cases of formulas $\nabla\gamma$ with $\bot \in \bigcup \gamma$, first observe that any such set γ is of the form $\alpha \cup \{ B \cup \{ \bot \} \}$. If the formula $\nabla(\alpha \cup \{ B \cup \{ \bot \} \})$ is *not* satisfiable, then it is equivalent to $\bot$. On the other hand, if it *is* satisfiable, then by Proposition 4.5(3) we may rewrite it to the RHS of (16) for some $A \in \alpha$ and $Z \subseteq A \times B$, with none of the new conjunctions being equivalent to $\bot$. Observe that in the latter case, if B is empty then A and Z must be empty as well. The equation (16) then becomes

$$\nabla(\alpha \cup \{ \{ \bot \} \}) \equiv \nabla(\alpha \cup \{ \emptyset \}) .$$

Finally, because of space limitations, we postpone a discussion of the interaction of ∇ with the negation operator, to an extended version of the paper.

5 Bisimulation Quantifiers and Uniform Interpolation

In this section we will show that, as announced in the Introduction, Monotone Modal Logic enjoys *uniform* interpolation, and we will prove this result, Theorem 1, by showing that the so-called *bisimulation quantifiers* $\exists p$ are effectively expressible in $\mathcal{L}_\nabla$ (and hence in $\mathcal{L}_\Diamond$).

Definition 5.1 Given a modal language $\mathcal{L}(\mathsf{Prop})$ and a notion of bisimulation between pointed models of the language, we obtain the extension $\mathcal{L}^\exists(\mathsf{Prop})$ by adding, for each proposition letter $p \in \mathsf{Prop}$, the *bisimulation quantifier* $\exists p$ to the language. The semantics of this quantifier is given by (2).

When we say that the bisimulation quantifiers are effectively expressible in $\mathcal{L}$, we mean that there is an effective translation mapping formulas in $\mathcal{L}^\exists$ to equivalent formulas in $\mathcal{L}$. In our case, we will define such a translation in two steps. First we use the results of the previous section to rewrite an $\mathcal{L}_\nabla$-formula a into its so-called *transparent* normal form a^n. Then we show that transparent formulas admit a simple inductive definition of the translation.

Definition 5.2 The fragment $\mathcal{L}_\nabla^-$ of *disjunctive* formulas in $\mathcal{L}_\nabla$ is given by the following grammar:

$$a \ ::= \ \top \mid \bot \mid \bigwedge\Pi \mid \bigwedge\Pi \wedge \nabla\alpha \mid a \vee a,$$

where Π is a set of *literals*, and $\alpha \in \mathcal{Q}_\omega \mathcal{L}_\nabla^-$. A formula $a \in \mathcal{L}_\nabla^-$ is called *transparent* if in every subformula $\nabla\alpha$ of a, every formula in $\bigcup \alpha$ is satisfiable.

Proposition 5.3 *There is an effective algorithm rewriting any formula $a \in \mathcal{L}_\nabla$ into an equivalent transparent formula a^n.*

The *proof* of this proposition proceeds via a straightforward induction on the ∇-depth of a, on the basis of the Propositions 4.3 and 4.5. We skip the details, and move on to our inductive definition of the bisimulation quantifiers.

Definition 5.4 Given $p \in \mathsf{Prop}$, we inductively define the map $\tau_p : \mathcal{L}^{-}_{\nabla}(\mathsf{Prop}) \to \mathcal{L}^{-}_{\nabla}(\mathsf{Prop})$ by:

$$
\begin{array}{lll}
\tau_p(\top) & := & \top \\
\tau_p(\bot) & := & \bot \\
\tau_p\left(\bigwedge\Pi\right) & := & \begin{cases} \bot & \text{if } \{p, \neg p\} \subseteq \Pi \\ \bigwedge(\Pi \setminus \{p, \neg p\}) & \text{otherwise} \end{cases} \\
\tau_p\left(\bigwedge\Pi \wedge \nabla\alpha\right) & := & \tau_p\left(\bigwedge\Pi\right) \wedge \nabla(\mathcal{Q}\tau_p)(\alpha) \\
\tau_p(a \vee b) & := & \tau_p(a) \vee \tau_p(b)\,.
\end{array}
$$

The following proposition shows that for transparent formulas, the above definition satisfies the required properties.

Proposition 5.5 *Let $p \in \mathsf{Prop}$ be some proposition letter. For any disjunctive formula $a \in \mathcal{L}^{-}_{\nabla}(\mathsf{Prop})$ we have $\tau_p(a) \in \mathcal{L}^{-}_{\nabla}(\mathsf{P}_a \setminus \{p\})$. Moreover, if a is transparent, then $\tau_p(a) \equiv \exists p.a$.*

Proof. The first statement of the Proposition is a straightforward consequence of the definitions. The second statement is proved by induction on the definition of transparent disjunctive formulas; the inductive clauses are immediate consequences of Proposition 5.6 below. $\square$

The following Proposition is the key technical lemma of this paper. In particular, we show that the bisimulation quantifiers distributes over ∇, provided that all formulas under the nabla are *satisfiable*. This proviso explains why in Proposition 5.5 we can only prove that $\tau_p(a) \equiv \exists p.a$ for transparent a.

Proposition 5.6 *The bisimulation quantifier $\exists p$ satisfies the following properties:*

(B1) $\exists p.(a \vee b) \equiv \exists p.a \vee \exists p.b$;

(B2) $\exists p.\nabla\alpha \equiv \nabla(\mathcal{Q}\exists p)(\alpha)$, *provided every $a \in \bigcup\alpha$ is satisfiable;*

(B3) $\exists p.\left(\bigwedge\Pi \wedge \nabla\alpha\right) \equiv \begin{cases} \bot & \textit{if } \{p, \neg p\} \subseteq \Pi, \\ \bigwedge(\Pi \setminus \{p, \neg p\}) \wedge \exists p.\nabla\alpha & \textit{otherwise.} \end{cases}$

In the formulation of condition (B2) above, it is convenient to see the quantifier $\exists p$ as a function on the language $\mathcal{L}$ (mapping a formula a to the formula $\exists p.a$), so that we may apply the functor $\mathcal{Q}$ to it and obtain a map $\mathcal{Q}\exists p : \mathcal{QL} \to \mathcal{QL}$.

Proof. Since the items (B1) and (B3) follow by a routine argument, we focus on the proof of (B2). Fix a pointed model $\mathbb{S}, s_0$. We will show that

$$\mathbb{S}, s_0 \Vdash \exists p.\nabla\alpha \text{ iff } \mathbb{S}, s_0 \Vdash \nabla(\mathcal{Q}\exists p)(\alpha). \tag{17}$$

For the direction from left to right of (17), assume that $\mathbb{S}, s_0 \Vdash \exists p.\nabla\alpha$. Then there is a pointed model $\mathbb{S}', s'_0$ and an up-to-p bisimulation Z such that $Z : \mathbb{S}, s_0 \simeq_p \mathbb{S}', s'_0$ and $\mathbb{S}', s'_0 \Vdash \nabla\alpha$. It follows that $(\sigma(s_0), \sigma'(s'_0)) \in \widetilde{\mathcal{M}}(Z)$ and $(\sigma'(s'_0), \alpha) \in \widetilde{\mathcal{M}}(\Vdash)$, and so by Proposition 2.15(3,1) we find $(\sigma(s_0), \alpha) \in \widetilde{\mathcal{M}}(Z;\Vdash)$. However, since Z is an up-to-p bisimulation, it follows that $Z;\Vdash \subseteq \Vdash; Gr(\exists p)^{\smile}$. From this we may infer that $(\alpha, \sigma(s_0)) \in \left(\widetilde{\mathcal{M}}(Z;\Vdash)\right)^{\smile} = \left(\widetilde{\mathcal{M}}(\Vdash; Gr(\exists p)^{\smile})\right)^{\smile} = \widetilde{\mathcal{M}}(Gr(\exists p); \Vdash^{\smile}) = Gr(\mathcal{Q}\exists p); \widetilde{\mathcal{M}}(\Vdash^{\smile}) = Gr(\mathcal{Q}\exists p); \left(\widetilde{\mathcal{M}}(\Vdash)\right)^{\smile}$. But $(\alpha, \sigma(s_0)) \in (Gr(\mathcal{Q}\exists p)); \left(\widetilde{\mathcal{M}}(\Vdash)\right)^{\smile}$ is another way of saying that $((\mathcal{Q}\exists p)(\alpha), \sigma(s_0)) \in \left(\widetilde{\mathcal{M}}(\Vdash)\right)^{\smile}$, or equivalently, $(\sigma(s_0), (\mathcal{Q}\exists p)(\alpha)) \in \widetilde{\mathcal{M}}(\Vdash)$. This means that $\mathbb{S}, s_0 \Vdash \nabla(\mathcal{Q}\exists p)(\alpha)$, as required.

For the converse direction of (17), assume that $\mathbb{S}, s_0 \Vdash \nabla(\mathcal{Q}\exists p)(\alpha)$. In order to prove that $\mathbb{S}, s_0 \Vdash \exists p.\nabla\alpha$, we need to construct some pointed model $(\mathbb{S}', s'_0)$ such that $\mathbb{S}, s_0 \simeq_p \mathbb{S}', s'_0$ and $\mathbb{S}', s'_0 \Vdash \nabla\alpha$. For this purpose, consider the set

$$P := \{ (s,a) \in \bigcup\sigma(s_0) \times (\bigcup\alpha \cup \{\top\}) \mid \mathbb{S}, s \Vdash \exists p.a \}.$$

For $(s,a) \in P$, pick a pointed model $(\mathbb{T}_{s,a}, t_{s,a})$ with $\mathbb{S}, s \simeq_p \mathbb{T}_{s,a}, t_{s,a}$ and $\mathbb{T}_{s,a}, t_{s,a} \Vdash a$. (In the case that $a = \top \notin \bigcup\alpha$, since $\exists p.\top$ is equivalent to $\top$, we may chose $\mathbb{T}_{s,a} := \mathbb{S}$ and $t_{s,a} := s$.) Also, for $a \in \bigcup\alpha$, let (T_a, t_a) be an arbitrary pointed model of a — thus $T_a, t_a \Vdash a$. Note that here we use the fact that all formulas in $\bigcup\alpha$ are satisfiable.

We define the model $\mathbb{S}'$ as follows. Its domain is given as the disjoint union

$$S' := \{ s'_0 \} \uplus \biguplus\{ T_{s,a} \mid (s,a) \in P \} \uplus \biguplus\{ T_a \mid a \in \bigcup\alpha \}.$$

The neighborhood map $\sigma' : S' \to \mathcal{M}S'$ can be described as follows. For $s' \neq s'_0$, we put

$$\sigma'(s') := (\sigma_x(s'))^{\uparrow},$$

where $x \in P \cup \bigcup\alpha$ is the unique x such that $s' \in T_x$, σ_x is the neighborhood map of $\mathbb{T}_x$, and $(\cdot)^{\uparrow}$ denotes the operation of closing under supersets of S', cf. Definition 2.6. For the definition of $\sigma'(s'_0)$, we first define, for $U \in \sigma(s_0)$, the set

$$U' := \{ t_{s,a} \mid (s,a) \in P, s \in U \}.$$

Second, by $\mathbb{S}, s \Vdash \nabla(\mathcal{Q}\exists p)(\alpha)$, we may pick, for each $A \in \alpha$, some $U_A \in \sigma(s_0)$ such that $U_A \Vdash \bigvee\{ \exists p.a \mid a \in A \}$. Define

$$U^{\bullet}_A := \{ t_{s,a} \mid (s,a) \in P,\ s \in U_A, a \in A \} \cup \{ t_a \mid a \in A \}$$

and

$$\sigma'(s'_0) := (\{ U' \mid U \in \sigma(s_0) \} \cup \{ U^{\bullet}_A \mid A \in \alpha \})^{\uparrow}.$$

With these definitions, each $\sigma(s')$ is indeed an upward closed collection of subsets of S'. We complete the construction of the model $\mathbb{S}'$ by defining the following coloring $V' : S' \to \mathcal{P}(\mathsf{Prop})$:

$$V'(s') := \begin{cases} V_x(s') & \text{if } s' \in T_x, \\ V(s_0) & \text{if } s' = s'_0. \end{cases}$$

In the sequel, we will use without proof the fact that for all points $s' \in S' \setminus \{ s'_0 \}$, there is a unique index x such that $s \in T_x$, and that $\mathbb{S}', s' \simeq \mathbb{T}_x, s'$. From this it follows that $\mathbb{S}', t_{s,a} \Vdash a$ for each $(s, a) \in P$, and that $\mathbb{S}', t_a \Vdash a$ for each $a \in \bigcup \alpha$.

In order to show that $\mathbb{S}, s_0 \Vdash \exists p.\nabla\alpha$, it suffices to prove (18) and (20) below. First we claim that

$$\mathbb{S}, s_0 \simeq_p \mathbb{S}', s'_0. \tag{18}$$

To prove this, take some up-to-p bisimulation $Z_{s,a} : \mathbb{S}, s \simeq_p \mathbb{T}_{s,a}, t_{s,a}$ for each $(s, a) \in P$, and let

$$Z := \{ (s_0, s'_0) \} \cup \bigcup \{ Z_{s,a} \mid (s, a) \in P \}.$$

We claim that $Z \subseteq S \times S'$ is an up-to-p bisimulation between $\mathbb{S}$ and $\mathbb{S}'$.

By Proposition 2.14, it suffices to verify, for each pair $(s, s') \in Z$, that the colorings of s and s' agree on all proposition letters other than p, and that $(\sigma(s), \sigma'(s')) \in \widetilde{\mathcal{M}}(Z)$. These facts require only a routine check for the pairs $(s, s') \neq (s_0, s'_0)$, and so we focus on the pair $(s_0, s'_0) \in Z$. Since $V'(s'_0) = V(s_0)$ by definition, it remains to show that

$$(\sigma(s_0), \sigma'(s'_0)) \in \widetilde{\mathcal{M}}(Z). \tag{19}$$

For a proof of (19), first take an arbitrary neighborhood U of s_0. To see that $(U, U') \in \overleftarrow{\mathcal{P}}(Z)$, consider some $u' \in U'$. By definition of U' we must have $u' = t_{s,a}$ for some $(s, a) \in P$ with $s \in U$. From this it is immediate that $(s, u') \in Z_{s,a} \subseteq Z$, as required. This proves that $(\sigma(s_0), \sigma'(s'_0)) \in \overrightarrow{\mathcal{P}}\overleftarrow{\mathcal{P}}(Z)$. Conversely, consider a set $W \in \sigma'(s'_0)$ and distinguish cases. If $W = U'$ for some $U \in \sigma(s_0)$, then for any $u \in U$ we have $t_{u,\top} \in U'$, and $(u, t_{u,\top}) \in Z_{u,\top} \subseteq Z$, whence $(U, W) \in \overrightarrow{\mathcal{P}}(Z)$. Otherwise, we have $W = U_A^\bullet$ for some $A \in \alpha$, and we claim that $(U_A, W) \in \overrightarrow{\mathcal{P}}(Z)$. To see this, recall that by definition, for any $u \in U_A$ there is some $a \in A$ such that $\mathbb{S}, u \Vdash a$. From this it follows that $t_{u,a} \in U_A^\bullet$ and since $(u, t_{u,a}) \in Z_{u,a} \subseteq Z$, we find that $(U_A, W) \in \overrightarrow{\mathcal{P}}(Z)$ indeed. This means that $(\sigma(s_0), \sigma'(s'_0)) \in \overleftarrow{\mathcal{P}}\overrightarrow{\mathcal{P}}(Z)$, which finishes the proof of (19). Thus we have proved that $Z : \mathbb{S}, s_0 \simeq_p \mathbb{S}', s'_0$, which establishes (18).

Our second claim is that

$$\mathbb{S}', s'_0 \Vdash \nabla\alpha. \tag{20}$$

To see this, first consider an arbitrary set $A \in \alpha$. Any point $s' \in U_A^\bullet$ is either of the form $t_{s,a}$ for a pair $(s, a) \in P$ with $s \in U_A$ and $a \in A$, or of the form t_a for some $a \in A$. In both cases, we see that there exists some $a \in A$ such that $\mathbb{S}', s' \Vdash a$. This suffices to show that $(U^\bullet, A) \in \overrightarrow{\mathcal{P}}(\Vdash)$, and since $U^\bullet \in \sigma'(s'_0)$, it follows that $(\sigma'(s_0), \alpha) \in \overleftarrow{\mathcal{P}}\overrightarrow{\mathcal{P}}(\Vdash)$.

Conversely, take an arbitrary $W \in \sigma(s'_0)$, so that W is either of the form U' for some $U \in \sigma(s_0)$, or of the form $U_A^\bullet$ for some $A \in \alpha$. In the first case, there is some $A' \in \alpha$ such that $U \rhd_{\mathbb{S}} \{ \exists p.a \mid a \in A' \}$. It follows that, for each $a \in A'$, there is a $u \in U$ such that $\mathbb{S}, u \Vdash a$, and so the point $t_{u,a}$ belongs to U' and $\mathbb{S}', t_{u,a} \Vdash a$. That is, $W \rhd_{\mathbb{S}'} A'$. In the second case, for any $a \in A$ we may take the point $t_a \in U^{\bullet}_{A}$, which satisfies $\mathbb{S}', t_a \Vdash a$; thus we see that $W \rhd_{\mathbb{S}'} A$. This means that $(\sigma'(s'_0), \alpha) \in \overrightarrow{\mathcal{P}}\overleftarrow{\mathcal{P}}(\Vdash)$, and so we have proved that $(\sigma'(s'_0, \alpha) \in \widetilde{\mathcal{M}}(\Vdash)$, which suffices for proving (20).

This finishes the proof of the direction from right to left of (17). □

Finally, the Uniform Interpolation Theorem 1 is an immediate consequence of Proposition 5.5.

Proof of Theorem 1 Fix a formula $a \in \mathcal{L}_{\diamond}(\mathsf{Prop})$. Given a set $\mathsf{Q} \subseteq \mathsf{P}_a$, write $\mathsf{P}_a \setminus \mathsf{Q} = \{p_0, \ldots, p_{n-1}\}$. We define

$$a_{\mathsf{Q}} := \left(\tau_{p_0} \cdots \tau_{p_{n-1}} \left(\left(a^{\nabla}\right)^n\right)\right)^{\diamond}.$$

It follows that $a_{\mathsf{Q}} \in \mathcal{L}_{\diamond}(\mathsf{Q})$, and that

$$a_{\mathsf{Q}} \equiv \exists p_0. \cdots \exists p_{n-1}.a.$$

So in order to verify that a_{Q} is the required uniform interpolant of a with respect to Q, it suffices to check (1) for an arbitrary b with $\mathsf{P}_a \cap \mathsf{P}_b \subseteq \mathsf{Q}$. First assume $a \models b$. In order to prove that $a_{\mathsf{Q}} \models b$, take an arbitrary pointed model $\mathbb{S}, s$ such that $\mathbb{S}, s \Vdash a_{\mathsf{Q}}$. It follows that there are pointed models $(\mathbb{S}_i, s_i)_{0 \le i \le n}$ such that $(\mathbb{S}, s) = (\mathbb{S}_0, s_0)$, $\mathbb{S}_i, s_i \simeq_{p_i} \mathbb{S}_{i+1}, s_{i+1}$ for all i, and $\mathbb{S}_n, s_n \Vdash a$. Then by assumption we have $\mathbb{S}_n, s_n \Vdash b$, and since none of the p_i occurs in b, it follows by bisimulation invariance that $\mathbb{S}_i, s_i \Vdash b$, for all i. In particular, we find that $\mathbb{S}, s \Vdash b$, as required. Conversely, if $a_{\mathsf{Q}} \models b$ then by $a \models a_{\mathsf{Q}}$ we immmediately obtain that $a \models b$. □

6 Conclusions & Questions

In this paper we have introduced, for monotone modal logic, a modality ∇ that intuitively simulates in this context the cover modality for modal logic **K**. We have then defined a modal language based on ∇ and proved that this language is equi-expressive with the standard one. Using some algebraic laws satisfied by ∇ we have shown that each formula is equivalent to a formula which is transparent. Transparent formulas should be thought of as formulas in a rather special, disjunctive normal form. For such formulas it is relatively easy to compute uniform interpolants. Consequently, we arrived at our main result stating that all formulas of monotone modal logic have uniform interpolants in **M**.

On the basis of our results we see various way to continue. First of all, we might improve on the results presented here. For instance, we are curious after the optimal *size* of the uniform interpolant is, and after the computational *complexity* of computing it. Note that the present construction is based on rewriting $\mathcal{L}_{\nabla}$-formulas into transparent normal form, and this process, involving satisfiability checks of very complex formulas (Proposition 4.5), and exponential blow-ups each time a conjunction is pushed down

(Proposition 4.3), is probably not optimal in terms of efficiency. We hope to come back to this issue in an extended version of this paper.

Another natural direction for future research would be to look for variations and extensions of our uniform interpolation result. To start with, we are very much interested whether, analogous to the results of d'Agostino and Hollenberg on the modal μ-calculus [10], the extension of monotone modal logic with fixpoint operators has uniform interpolation as well. A related direction would be to investigate the existence of uniform interpolants in specific monotone logics, whether they are defined as axiomatic extensions of **M**, or semantically by means of classes of monotone neighborhood frames. Conversely, we would like to know whether our results generalize to classical modal logic **C**, the logic of arbitrary (that is, not necessarily monotone) neighborhood models. However, taking in account that in normal modal logic, uniform interpolation does not transpose easily from one variety to another, one should expect the same sort of phenomenon in the generalized setting.

On a more abstract level, the approach we have followed might be considered naive, as we mimicked, within monotone modal logic the approach towards coalgebraic logic taken by Moss [25], but based on a different notion of relation lifting than the canonical one. However, the fact that we obtained such a powerful result, may indicate there are some general categorical principles underlying our naive approach. This would be in accordance with the fact that for a functor T that does not preserve weak pullbacks, the notion of bisimilarity based on the standard relation lifting $\overline{T}$ is not the appropriate one. Therefore, our work suggests new directions of research in the area of coalgebras and category theory. Ideas on relation lifting from [18,32] might be useful here.

Finally, we believe that ∇-based monotone modal logic is of interest in its own right, and we plan to study of it in more detail. In particular, we conjecture that the ∇-laws of Section 4, augmented with appropriate axioms expressing the interaction between ∇ and the Boolean negation, provide a sound and complete *axiomatization* for the set of valid $\mathcal{L}_\nabla$-formulas, and we hope to report on this in an extended version of this paper.

References

[1] Alur, R., T. Henzinger and O. Kupferman, *Alternating-time temporal logics*, in: *Compositionality: the significant difference*, LNCS **1536**, 1998, pp. 23–60.

[2] Areces, C. and R. Goldblatt, editors, "Advances in Modal Logic 7," College Publications, 2008.

[3] Baltag, A., *A logic for coalgebraic simulation*, in: H. Reichel, editor, *Coalgebraic Methods in Computer Science (CMCS'00)*, Electronic Notes in Theoretical Computer Science **33**, 2000, pp. 41–60.

[4] Barwise, J. and L. Moss, "Vicious Circles," CSLI Lecture Notes **60**, CSLI Publications, 1996.

[5] Bílková, M., *Uniform interpolation and propositional quantifiers in modal logics*, Studia Logica **85** (2007), pp. 1–31.

[6] Bílková, M., A. Palmigiano and Y. Venema, *Proof systems for the coalgebraic cover modality*, in: Areces and Goldblatt [2], pp. 1–21.

[7] Chellas, B., "Modal Logic, an Introduction," Cambridge University Press, 1980.

[8] Cîrstea, C., A. Kurz, D. Pattinson, L. Schröder and Y. Venema, *Modal logics are coalgebraic*, The Computer Journal (2009).

[9] D'Agostino, G., *Interpolation in non-classical logics*, Synthese **164** (2008), pp. 421–435.

[10] D'Agostino, G. and M. Hollenberg, *Logical questions concerning the μ-calculus: interpolation, Lyndon and Łoś-Tarski*, J. Symbolic Logic **65** (2000), pp. 310–332.

[11] d'Agostino, G. and G. Lenzi, *On modal μ-calculus with explicit interpolation*, Journal of Applied Logic **4** (2006), pp. 256–278.

[12] French, T., "Bisimulation quantifiers for modal logic," Ph.D. thesis, School of Computer Science and Software Engineering, University of Western Australia (2006).

[13] Ghilardi, S., *An algebraic theory of normal forms*, Annals of Pure and Applied Logic **71** (1995), pp. 189–245.

[14] Ghilardi, S. and M. Zawadowski, *Undefinability of propositional quantifiers in the modal system* S4, Studia Logica **55** (1995), pp. 259–271.

[15] Goldblatt, R., *A Kripke-Joyal semantics for noncommutative logic in quantales*, in: *Advances in modal logic. Vol. 6*, Coll. Publ., London, 2006 pp. 209–225.

[16] Hansen, H., "Monotonic Modal Logics," Master's thesis, Institute for Logic, Language and Computation, University of Amsterdam (2003), ILLC Prepublication Series PP-2003-24.

[17] Hansen, H. H. and C. Kupke, *A coalgebraic perspective on monotone modal logic*, in: *Proceedings of the Workshop on Coalgebraic Methods in Computer Science*, Electron. Notes Theor. Comput. Sci. **106** (2004), pp. 121–143 (electronic).

[18] Hansen, H. H., C. Kupke and E. Pacuit, *Neighbourhood structures: bisimilarity and basic model theory*, Logical Methods in Computer Science **5** (2009), pp. 1–38.

[19] Henkin, L., *An extension of the Craig-Lyndon interpolation theorem*, J. Symbolic Logic **28** (1963), pp. 201–216.

[20] Janin, D. and I. Walukiewicz, *Automata for the modal μ-calculus and related results*, in: *Mathematical foundations of computer science 1995 (Prague)*, Lecture Notes in Comput. Sci. **969**, Springer, Berlin, 1995 pp. 552–562.

[21] Kozen, D., *Results on the propositional μ-calculus*, Theoret. Comput. Sci. **27** (1983), pp. 333–354.

[22] Kupke, C., A. Kurz and Y. Venema, *Completeness of the finitary Moss logic*, in: Areces and Goldblatt [2], pp. 193–217.

[23] Maksimova, L. L., *The absence of the interpolation and Beth properties in temporal logics with "the next" operator*, Sibirsk. Mat. Zh. **32** (1991), pp. 109–113, 205.

[24] Markowsky, G., *Free completely distributive lattices*, Proc. Amer. Math. Soc. **74** (1979), pp. 227–228.

[25] Moss, L. S., *Coalgebraic logic*, Ann. Pure Appl. Logic **96** (1999), pp. 277–317.

[26] Nerode, A., *Composita, equations, and freely generated algebras*, Trans. Amer. Math. Soc. **91** (1959), pp. 139–151.

[27] Parikh, R., *The logic of games and its applications*, in: M. Karpinski and J. Leeuwen, editors, *Topics in the Theory of Computation*, Annals of Discrete Mathematics **24**, 1985, pp. 111–140.

[28] Pauly, M. and R. Parikh, *Game logic—an overview*, Studia Logica **75** (2003), pp. 165–182, game logic and game algebra (Helsinki, 2001).

[29] Pitts, A. M., *On an interpretation of second-order quantification in first-order intuitionistic propositional logic*, J. Symbolic Logic **57** (1992), pp. 33–52.

[30] Rutten, J. J. M. M., *Universal coalgebra: a theory of systems*, Theoret. Comput. Sci. **249** (2000), pp. 3–80.

[31] Santocanale, L., *A duality for finite lattices* (2009), available at `http://hal.archives-ouvertes.fr/hal-00432113/`.

[32] Schubert, C. and G. J. Seal, *Extensions in the theory of lax algebras*, Theory Appl. Categ. **21** (2008), pp. No. 7, 118–151.

[33] Shavrukov, V., "Adventures in Diagonizable Algebras," Ph.D. thesis, Institute for Logic, Language and Computation, Universiteit van Amsterdam (1994), (Part I of this thesis appeared as Dissertationes Mathematicae 323 (1993)).

[34] ten Cate, B., W. Conradie, M. Marx and Y. Venema, *Definitorially complete description logics*, in: P. Doherty, J. Mylopoulos and C. Welty, editors, *Proceedings of KR 2006* (2006), pp. 79–89.

[35] Vardi, M. Y., *On the complexity of epistemic reasoning*, in: *Proceedings of the Fourth Annual IEEE Symposium on Logic in Computer Science (LICS 1989)* (1989), pp. 243–252.

[36] Visser, A., *Uniform interpolation and layered bisimulation*, in: *Gödel '96 (Brno, 1996)*, Lecture Notes Logic **6**, Springer, Berlin, 1996 pp. 139–164.

Simulation of Two Dimensions in Unimodal Logics

Ilya Shapirovsky

Institute of Information Transmission Problems
Russian Academy of Sciences
B.Karetny 19, Moscow, Russia, 127994

Abstract

In this paper, we prove undecidability and the lack of finite model property for a certain class of unimodal logics. To do this, we adapt the technique from [7], where products of transitive modal logics were investigated, for the unimodal case. As a particular corollary, we present an undecidable unimodal fragment of Halpern and Shoham's Interval Temporal Logic.

Keywords: products of modal logics, undecidable modal logics, logics without the finite model property, locally one-component frames, Halpern and Shoham's Interval Temporal Logic.

1 Introduction

In the recent paper [7], it was shown that products of transitive modal logics are usually undecidable and lack the finite model property. In the present paper we adapt the technique from [7] for the unimodal case.

We consider logics of $\curlyvee$-*products*: if (W, R_1, R_2) is the product of frames F_1 and F_2, we put $\mathsf{F}_1 \curlyvee \mathsf{F}_2 = (W, R_1 \cup R_2)$. We show that for a certain class of frames this operation allows us to 'maintain' relations R_1 and R_2. Namely, we consider $\curlyvee$-products of transitive *locally one-component frames*: a frame (W, R) is locally one-component at a point w, if the set of all points R-accessible from w cannot be split into the disjoint union of two R-incomparable non-empty sets. In particular, if a transitive frame is linear or directed, then it is locally one-component.

We show that products of unimodal locally one-component frames can be simulated in $\curlyvee$-products. Also, by presenting a set of unimodal axioms, we define a class of unimodal frames, which allows us to 'encode' the modalities of the commutator $[\mathrm{K4}, \mathrm{K4}]$. For various unimodal logics (defined syntactically or semantically), it leads to undecidability and the lack of finite model property.

It is known that modal logics of products are related to modal logics of intervals (see e.g. [10]), namely – to fragments of Halpern and Shoham's Interval Temporal Logic HS [8]. This allows us to prove similar results for a unimodal fragment of HS. It is known that HS and many of its fragments are undecidable over various classes of intervals, for the latest results see [3,2]; these results were obtained for fragments with two or more modalities. Also, in the very recent paper [4], undecidability for a fragment of HS with a single modality was obtained: it was shown that the logic of the *'overlap'* relation is undecidable over discrete linear orders. We obtain another result of this kind: we show the undecidability and the lack of finite model property for the $\langle\overline{B}\vee\overline{E}\rangle$-fragment of HS interpreted over various classes of intervals (including intervals on real and rational numbers), where the modality $\langle\overline{B}\vee\overline{E}\rangle$ corresponds to the inverse of the union of Allen's relations *'begins'* and *'ends'* . As far as we know, this is the first example of an undecidable unimodal fragment of HS over dense linear orders.

The paper is organized as follows. In Section 2 we introduce some basic notions and notations. Section 3 contains some auxiliary observations on modal-to-modal translations that allow us to adapt the technique from [7] to the unimodal case, and also to find some modal axioms for $\curlyvee$-products. In Section 4 we study basic properties of $\curlyvee$-products of locally one-component frames. In Section 5 we formulate and prove results on undecidability and the lack of finite model property. In Section 6 we consider the $\langle\overline{B}\vee\overline{E}\rangle$-fragment of HS.

2 Preliminaries

We consider *propositional normal modal logics* with finitely many modalities. PV denotes the countable set of all propositional variables. The set of all *n-formulas* ML_n is constructed from PV, the classical connectives $\wedge$, $\neg$, and the unary connectives $\Diamond_1,\dots,\Diamond_n$. Other connectives are regarded as abbreviations, in particular, $\Box_i\varphi = \neg\Diamond_i\neg\varphi$. $\Diamond$ and $\Box$ abbreviate $\Diamond_1$ and $\Box_1$, respectively.

Variables are typically denoted by $\mathsf{p},\mathsf{q},\mathsf{r}$, formulas – by φ,ψ, possibly subscripted. For a formula φ, $PV(\varphi)$ denotes the set of all variables of φ. For a set of formulas Γ, $PV(\Gamma)=\bigcup_{\varphi\in\Gamma} PV(\varphi)$. For formulas φ, ψ and a variable p, $[\varphi/\mathsf{p}]\psi$ denotes the result of substitution of φ for p in ψ. Also we use the abbreviations

$$\Diamond_\psi\varphi = \Diamond(\psi\wedge\varphi),\quad \Box_\psi\varphi = \Box(\psi\to\varphi),\quad \varphi^0 = \neg\varphi,\quad \varphi^1=\varphi.$$

An *(n-)frame* is a tuple $\mathsf{F} = (W, R_1,\dots,R_n)$, where $W\neq\varnothing$, $R_i\subseteq W\times W$; an *(n-)model* M *based on* F is a pair (F,θ) or a tuple $(W,R_1,\dots,R_n,\theta)$, where $\theta: PV\to\mathcal{P}(W)$, $\mathcal{P}(W)$ is the powerset of W; θ is called *a valuation on* W. The *truth of a formula at a point in a model*, and also the *validity of a formula in a frame* (or *in a class of frames*) are defined in the standard way, see e.g. [1]. In symbols, $\mathsf{M},w\vDash\varphi$ means that φ is true at w in M, $|\varphi|_{\mathsf{M}} = \{w \mid \mathsf{M},w\vDash\varphi\}$. $\mathsf{F}\vDash\varphi$ means that φ is valid in F. $\mathsf{F},w\vDash\varphi$ means that $(\mathsf{F},\theta),w\vDash\varphi$ for any valuation θ. For a set of formulas Ψ, $\mathsf{F}\vDash\Psi$ means $\mathsf{F}\vDash\varphi$ for all $\varphi \in \Psi$.

φ is *satisfiable in a frame* F *at a point* w, if $(\mathsf{F}, \theta), w \models \varphi$ for some valuation θ. For a class of frames $\mathcal{F}$, φ is *satisfiable in* $\mathcal{F}$ (or $\mathcal{F}$*-satisfiable*), if φ is satisfiable in F for some $\mathsf{F} \in \mathcal{F}$. For a logic L, if $\mathsf{F} \models \mathrm{L}$, we say that F is *an* L*-frame*; φ is L*-satisfiable*, if φ is satisfiable in an L-frame.

For a binary relation R on a set W, $R^{=}$ denotes the reflexive closure of R, i.e., $R^{=} = R \cup \{(w, w) \mid w \in W\}$. wRv means $(w, v) \in R$. For $w \in W$, $V \subseteq W$, put

$$R(w) = \{v \mid wRv\}, \quad R(V) = \bigcup_{w \in V} R(w).$$

For a frame $\mathsf{F} = (W, R_1, \ldots, R_n)$, R_{F}^{cone} denotes the transitive reflexive closure of $R_1 \cup \cdots \cup R_n$. A point w in F is called *a root of* F, if $W = R_{\mathsf{F}}^{cone}(w)$; in this case F is called *rooted*. F^w (M^w) denotes the subframe of F (submodel of M) generated by w, see e.g. [1].

For relations R,S, $R \circ S$ denotes their composition, $R^2 = R \circ R$, $R^{m+1} = R \circ R^m$.

$\mathsf{F} \times \mathsf{G}$ denotes *the product of frames* F, G; for logics $\mathrm{L}_1, \mathrm{L}_2$, $[\mathrm{L}_1, \mathrm{L}_2]$ denotes their *commutator*, see e.g. [5].

The following construction was used in [7] to prove negative results on products of transitive modal logics, and will also play an important role in this paper.

Fix variables h, v and put

$$\Diamond_h \varphi = \bigwedge_{\varepsilon=0,1} (\mathsf{h}^\varepsilon \to \Diamond_1(\neg \mathsf{h}^\varepsilon \wedge (\varphi \vee \Diamond_1 \varphi)),$$

$$\Diamond_v \varphi = \bigwedge_{\varepsilon=0,1} (\mathsf{v}^\varepsilon \to \Diamond_2(\neg \mathsf{v}^\varepsilon \wedge (\varphi \vee \Diamond_2 \varphi))$$

(recall that for a formula ψ, $\psi^0 = \neg\psi$, $\psi^1 = \psi$).

For a 2-model M, put

$$\bar{R}_{h,0}^{\mathsf{M}} = \{(u, w) \mid uR_1w \ \& \ (\mathsf{M}, u \models \mathsf{h} \Leftrightarrow \mathsf{M}, w \models \neg \mathsf{h})\},$$

$$\bar{R}_{v,0}^{\mathsf{M}} = \{(u, w) \mid uR_2w \ \& \ (\mathsf{M}, u \models \mathsf{v} \Leftrightarrow \mathsf{M}, w \models \neg \mathsf{v})\},$$

$$\bar{R}_{h}^{\mathsf{M}} = \bar{R}_{h,0}^{\mathsf{M}} \cup \left(\bar{R}_{h,0}^{\mathsf{M}}\right)^2, \quad \bar{R}_{v}^{\mathsf{M}} = \bar{R}_{v,0}^{\mathsf{M}} \cup \left(\bar{R}_{v,0}^{\mathsf{M}}\right)^2,$$

where R_1, R_2 are the accessability relations of M. Clearly,

$$\bar{R}_{h}^{\mathsf{M}} = \{(u, w) \mid \exists u' \in R_1(u) \, (w \in R_1^{=}(u') \& (\mathsf{M}, u \models \mathsf{h} \Leftrightarrow \mathsf{M}, u' \models \neg \mathsf{h}))\},$$

$$\bar{R}_{v}^{\mathsf{M}} = \{(u, w) \mid \exists u' \in R_2(u) \, (w \in R_2^{=}(u') \& (\mathsf{M}, u \models \mathsf{v} \Leftrightarrow \mathsf{M}, u' \models \neg \mathsf{v}))\}.$$

For any w in M, $\varphi \in ML_2$, we have

$$\mathsf{M}, w \models \Diamond_h \varphi \Leftrightarrow \exists u \in \bar{R}_{h}^{\mathsf{M}}(w)(\mathsf{M}, u \models \varphi), \quad \mathsf{M}, w \models \Diamond_v \varphi \Leftrightarrow \exists u \in \bar{R}_{v}^{\mathsf{M}}(w)(\mathsf{M}, u \models \varphi)$$

(see [7] for more details).

Put

$$\psi_{hv} = \bigwedge_{\varepsilon=0,1} \Box_1\Box_2\left((\mathsf{h}^\varepsilon \vee \Diamond_2\mathsf{h}^\varepsilon \to \Box_2\mathsf{h}^\varepsilon) \wedge (\mathsf{v}^\varepsilon \vee \Diamond_1\mathsf{v}^\varepsilon \to \Box_1\mathsf{v}^\varepsilon)\right).$$

Proposition 2.1 ([7]) *If* M *is a model based on a* [K4, K4]*-frame with a root* w*, and* $\mathsf{M}, w \models \psi_{hv}$*, then* $(W, \bar{R}_h^{\mathsf{M}}, \bar{R}_v^{\mathsf{M}})$ *is a* [K4, K4]*-frame.*

Proof. Straightforward. See [7] for more details. □

3 Modally definable relations in pretransitive frames

In [7] various undecidable problems and infiniteness of a model are encoded by formulas of the form $\psi_{hv} \wedge \psi$, where ψ is built using propositional variables, boolean connectives and derived modal operators $\Diamond_v, \Diamond_h$. Our goal is to describe a unimodal analogue of this fragment.

In this section we consider some syntactic constructions which will be used later to transfer negative results about products to the unimodal case.

3.1 'Diamond-like' formulas

Fix a variable $\mathsf{s} \in PV$. For any formulas ψ, φ, put $\psi(\varphi) = [\varphi/\mathsf{s}]\psi$. Given an n-model $\mathsf{M} = (\mathsf{F}, \theta)$, with every formula $\psi \in ML_n$ we associate a function $\psi^{\mathsf{M}} : 2^W \to 2^W$ defined in the following way:

$$\mathsf{s}^{\mathsf{M}}(V) = V, \qquad \mathsf{p}^{\mathsf{M}}(V) = \theta(\mathsf{p}), \text{ if } \mathsf{p} \in PV \text{ and } \mathsf{p} \neq \mathsf{s},$$
$$(\neg\psi)^{\mathsf{M}}(V) = W - \psi^{\mathsf{M}}(V), \qquad (\psi_1 \wedge \psi_2)^{\mathsf{M}}(V) = \psi_1^{\mathsf{M}}(V) \cap \psi_2^{\mathsf{M}}(V),$$
$$(\Diamond_i\psi)^{\mathsf{M}}(V) = R_i^{-1}(\psi^{\mathsf{M}}(V)).$$

Clearly, for any $\varphi \in ML_n$,

$$\psi^{\mathsf{M}}(|\varphi|_{\mathsf{M}}) = |\psi(\varphi)|_{\mathsf{M}}.$$

Definition 3.1 Consider an n-model M and a relation $\widetilde{R} \subseteq W \times W$. We say that a formula $\psi \in ML_n$ *expresses* $\widetilde{R}$ *in* M, in symbols $\psi \xrightarrow{\mathsf{M}} \widetilde{R}$, if

$$\psi^{\mathsf{M}}(V) = \widetilde{R}^{-1}(V) \text{ for any } V \subseteq W. \tag{1}$$

We say that ψ *expresses* $\widetilde{R}$ *in* F, in symbols $\psi \xrightarrow{\mathsf{F}} \widetilde{R}$, if (1) holds for any M based on F.

Proposition 3.2 *For an* n*-model* $\mathsf{M} = (\mathsf{F}, \theta)$*, the following conditions are equivalent:*

(1) $\psi \xrightarrow{\mathsf{M}} \widetilde{R}$*;*

(2) if $\theta' : PV \to \mathcal{P}(W)$, $\theta'(p) = \theta(p)$ *for any* $p \in (PV(\psi) - \{\mathsf{s}\})$, *then for any* $w \in W$, $\varphi \in ML_n$, *we have*

$$(F, \theta'), w \vDash \psi(\varphi) \;\Leftrightarrow\; \exists u \in \widetilde{R}(w)((F, \theta'), u \vDash \varphi).$$

Proof. Let N denote (F, θ').

(1) $\Rightarrow$ (2). If $\theta'(p) = \theta(p)$ for any $p \in (PV(\psi) - \{\mathsf{s}\})$, then φ^{M} and φ^{N} are the same functions, so $|\psi(\varphi)|_{\mathsf{N}} = \psi^{\mathsf{N}}(|\varphi|_{\mathsf{N}}) = \psi^{\mathsf{M}}(|\varphi|_{\mathsf{N}}) = \widetilde{R}^{-1}(|\varphi|_{\mathsf{N}})$.

(2) $\Rightarrow$ (1). For $V \subseteq W$, put $\theta'(\mathsf{s}) = V$, $\theta'(\mathsf{p}) = \theta(\mathsf{p})$ for $\mathsf{p} \neq \mathsf{s}$. We have: $\psi^{\mathsf{M}}(V) = \psi^{\mathsf{N}}(V) = |\psi|_{\mathsf{N}} = \widetilde{R}^{-1}(|\mathsf{s}|_{\mathsf{N}}) = \widetilde{R}^{-1}(V)$. □

Example 3.3 Consider a model M based on a 2-frame F. Recall that the operators $\Diamond_h$, $\Diamond_v$ and the relations $\bar{R}_h^{\mathsf{M}}, \bar{R}_v^{\mathsf{M}}$ are associated in the following way:

$$\mathsf{M}, w \vDash \Diamond_h\varphi \Leftrightarrow \exists u \in \bar{R}_h^{\mathsf{M}}(w)(\mathsf{M}, u \vDash \varphi), \quad \mathsf{M}, w \vDash \Diamond_v\varphi \Leftrightarrow \exists u \in \bar{R}_v^{\mathsf{M}}(w)(\mathsf{M}, u \vDash \varphi).$$

Moreover, the above equivalences hold for any model based on F, if its valuation on h, v is the same as in M. In other words, for the formulas

$$\psi_h = \bigwedge_{\varepsilon=0,1} (\mathsf{h}^\varepsilon \to \Diamond_1(\neg\mathsf{h}^\varepsilon \wedge (\mathsf{s} \vee \Diamond_1\mathsf{s}))), \; \psi_v = \bigwedge_{\varepsilon=0,1} (\mathsf{v}^\varepsilon \to \Diamond_2(\neg\mathsf{v}^\varepsilon \wedge (\mathsf{s} \vee \Diamond_2\mathsf{s}))),$$

we have

$$\psi_h \overset{\mathsf{M}}{\to} \bar{R}_h^{\mathsf{M}}, \quad \psi_v \overset{\mathsf{M}}{\to} \bar{R}_v^{\mathsf{M}}. \tag{2}$$

This example is important for us: the proofs of our negative results are based on the fact that the relations $\bar{R}_h^{\mathsf{M}}$, $\bar{R}_v^{\mathsf{M}}$ can be expressed in the unimodal language.

To describe fragments of modal logics in different languages it is convenient to use the following construction.

Definition 3.4 Given formulas $\psi_1, \dots \psi_k \in ML_n$, let $[\;\;]_{(\psi_1,\dots\psi_k)}$ denote the following translation from ML_k to ML_n:

$$[\mathsf{p}]_{(\psi_1,\dots\psi_k)} = \mathsf{p} \text{ for } \mathsf{p} \in PV;$$

$$[\phi \wedge \psi]_{(\psi_1,\dots\psi_k)} = [\phi]_{(\psi_1,\dots\psi_k)} \wedge [\psi]_{(\psi_1,\dots\psi_k)};$$

$$[\neg\phi]_{(\psi_1,\dots\psi_k)} = \neg([\phi]_{(\psi_1,\dots\psi_k)});$$

$$[\Diamond_i\phi]_{(\psi_1,\dots\psi_k)} = \psi_i([\phi]_{(\psi_1,\dots\psi_k)}).$$

This definition is explained by the following simple lemmas.

Lemma 3.5 *Consider a model* $\mathsf{M} = (W, R_1, \dots, R_n, \theta)$ *and relations* $\widetilde{R}_1, \dots, \widetilde{R}_k \subseteq W \times W$. *Let* $\varphi \in ML_k$, $\psi_1^\diamond, \dots \psi_k^\diamond \in ML_n$,

$$\psi_1^\diamond \overset{\mathsf{M}}{\to} \widetilde{R}_1, \dots, \psi_k^\diamond \overset{\mathsf{M}}{\to} \widetilde{R}_k.$$

Let θ' be a valuation such that $\theta'(\mathsf{p}) = \theta(\mathsf{p})$ for any $\mathsf{p} \in PV(\varphi, \psi_1^\diamond, \dots \psi_k^\diamond)$. Then for any $w \in W$, we have

$$\mathsf{M}, w \vDash [\varphi]_{(\psi_1^\diamond, \dots \psi_k^\diamond)} \quad \Leftrightarrow \quad (W, \widetilde{R}_1, \dots, \widetilde{R}_k, \theta'), w \vDash \varphi.$$

Proof. By induction on the construction of φ. The basis is trivial.

Suppose $\varphi = \Diamond_i \chi$. Then $[\varphi]_{(\psi_1^\diamond, \dots \psi_k^\diamond)} = \psi_i^\diamond([\chi]_{(\psi_1^\diamond, \dots \psi_k^\diamond)})$. We have:

$$\begin{aligned} \mathsf{M}, w \vDash \psi_i^\diamond([\chi]_{(\psi_1^\diamond, \dots \psi_k^\diamond)}) &\Leftrightarrow \mathsf{M}, v \vDash [\chi]_{(\psi_1^\diamond, \dots \psi_k^\diamond)} \text{ for some } v \in \widetilde{R}_i(w) \text{ (by Proposition 3.2)} \\ &\Leftrightarrow (W, \widetilde{R}_1, \dots, \widetilde{R}_k, \theta'), w \vDash \Diamond_i \chi \text{ (by the induction hypothesis).} \end{aligned}$$

Other cases are trivial. □

Lemma 3.6 *Consider models $\mathsf{M}' = (W, R_1', \dots, R_n', \theta)$, $\mathsf{M}'' = (W, R_1'', \dots, R_m'', \theta)$, and relations $R_1, \dots, R_k \subseteq W \times W$. Let $\psi_1^\diamond, \dots \psi_k^\diamond \in ML_n$, $\phi_1^\diamond, \dots \phi_k^\diamond \in ML_m$,*

$$\psi_i^\diamond \overset{\mathsf{M}'}{\rightarrow} R_i, \quad \phi_i^\diamond \overset{\mathsf{M}''}{\rightarrow} R_i.$$

Then for any $\varphi \in ML_k$, $w \in W$, we have

$$\mathsf{M}', w \vDash [\varphi]_{(\psi_1^\diamond, \dots \psi_k^\diamond)} \Leftrightarrow \mathsf{M}'', w \vDash [\varphi]_{(\phi_1^\diamond, \dots \phi_k^\diamond)}.$$

Proof. Put $\mathsf{M} = (W, R_1, \dots, R_k, \theta)$. By Lemma 3.5,

$$\mathsf{M}', w \vDash [\varphi]_{(\psi_1^\diamond, \dots \psi_k^\diamond)} \Leftrightarrow \mathsf{M}, w \vDash \varphi, \quad \mathsf{M}'', w \vDash [\varphi]_{(\phi_1^\diamond, \dots \phi_k^\diamond)} \Leftrightarrow \mathsf{M}, w \vDash \varphi.$$

□

3.2 Pretransitive frames and cone formulas

Definition 3.7 [6] A frame $\mathsf{F} = (W, R_1, \dots, R_n)$ is called *pretransitive*, if there exists a formula $\psi \in ML_n$ such that $\psi \overset{\mathsf{F}}{\rightarrow} R_{\mathsf{F}}^{cone}$; ψ is called *a cone formula for* F.

For a formula ψ, let

$$\psi^{(0)} = \mathsf{s}, \ \psi^{(1)} = \psi, \ \psi^{(i+1)} = \psi(\psi^{(i)}), \ \psi^{\leq n} = \psi^{(n)} \vee \dots \vee \psi^1 \vee \psi^0.$$

Example 3.8 Clearly, if $\mathsf{F} = (W, R)$ is a transitive frame, then $(\Diamond \mathsf{s})^{\leq 1}$ $(= \Diamond \mathsf{s} \vee \mathsf{s})$ is a cone-formula for F. If F is a product of two transitive 1-frames, then $(\Diamond_1 \mathsf{s} \vee \Diamond_2 \mathsf{s})^{\leq 2}$ is a cone formula for F. These observations are a particular case of the following proposition.

Proposition 3.9 ([6]) *An n-frame F is pretransitive iff there exists l such that $(\Diamond_1 \mathsf{s} \vee \dots \vee \Diamond_n \mathsf{s})^{\leq l}$ is a cone formula for F.*

For a pretransitive n-frame F, put $\psi_{\mathsf{F}}^{cone} = (\Diamond_1 \mathsf{s} \vee \dots \vee \Diamond_n \mathsf{s})^{\leq l_0}$, $\Diamond_{\mathsf{F}}^{cone} \varphi = \psi_{\mathsf{F}}^{cone}(\varphi)$, $\Box_{\mathsf{F}}^{cone} \varphi = \neg \psi_{\mathsf{F}}^{cone}(\neg \varphi)$, where

$$l_0 = \min\{l \mid (\Diamond_1 \mathsf{s} \vee \ldots \vee \Diamond_n \mathsf{s})^{\leq l} \text{ is a cone formula for } \mathsf{F}\}.$$

The following lemma shows how modally definable properties transfer between expressible relations in pretransitive frames.

Lemma 3.10 *Let* $\mathsf{F} = (W, R_1, \ldots, R_n)$ *be a pretransitive frame with a root* w, $\chi, \psi_1^\Diamond, \ldots, \psi_k^\Diamond \in ML_n$, $\varphi \in ML_k$, $PV(\varphi) \cap PV(\chi, \psi_1^\Diamond, \ldots, \psi_k^\Diamond) = \varnothing$. *Suppose* $R_1^\theta, \ldots, R_k^\theta \subseteq W \times W$ *for any valuation* θ *on* W, *and if* $(\mathsf{F}, \theta), w \vDash \chi$, *then*

$$\psi_1^\Diamond \overset{(\mathsf{F},\theta)}{\rightarrow} R_1^\theta, \ldots, \psi_k^\Diamond \overset{(\mathsf{F},\theta)}{\rightarrow} R_k^\theta.$$

Then the following conditions are equivalent:

(1) $\mathsf{F}, w \vDash \chi \to \Box_{\mathsf{F}}^{cone}[\varphi]_{(\psi_1^\Diamond, \ldots, \psi_k^\Diamond)}$;

(2) for any θ, *if* $(\mathsf{F}, \theta), w \vDash \chi$, *then* $(W, R_1^\theta, \ldots, R_k^\theta) \vDash \varphi$.

Proof. Let $\tilde{\mathsf{F}}^\theta$ denote $(W, R_1^\theta, \ldots, R_k^\theta)$. Put $PV_0 = PV(\chi, \psi_1^\Diamond, \ldots, \psi_k^\Diamond)$.

(1) $\Rightarrow$ (2). Let $(\mathsf{F}, \theta), w \vDash \chi$. Suppose that $(\tilde{\mathsf{F}}^\theta, \theta'), u \vDash \neg\varphi$ for some θ', u. Let η coincide with θ' on $PV(\varphi)$, and with θ on all other variables. Then $(\tilde{\mathsf{F}}^\theta, \eta), u \vDash \neg\varphi$ and $(\mathsf{F}, \eta), w \vDash \chi$. Thus $(\mathsf{F}, \eta), w \vDash \Box_{\mathsf{F}}^{cone}[\varphi]_{(\psi_1^\Diamond, \ldots, \psi_k^\Diamond)}$ and $(\mathsf{F}, \eta), u \vDash [\varphi]_{(\psi_1^\Diamond, \ldots, \psi_k^\Diamond)}$. Since $PV_0 \cap \varphi = \varnothing$, then

$$R_1^\theta = R_1^\eta, \ldots, R_k^\theta = R_k^\eta, \text{ and } \psi_1^\Diamond \overset{(\mathsf{F},\eta)}{\rightarrow} R_1^\theta, \ldots, \psi_k^\Diamond \overset{(\mathsf{F},\eta)}{\rightarrow} R_k^\theta.$$

By Lemma 3.5, $(\tilde{\mathsf{F}}^\theta, \eta), u \vDash \varphi$, which is a contradiction.

(1) $\Leftarrow$ (2). Suppose $(\mathsf{F}, \theta), w \vDash \chi$, $u \in W$. φ is valid in $\tilde{\mathsf{F}}^\theta$, so $(\tilde{\mathsf{F}}^\theta, \theta), u \vDash \varphi$, and by Lemma 3.5 $(\mathsf{F}, \theta), u \vDash [\varphi]_{(\psi_1^\Diamond, \ldots, \psi_k^\Diamond)}$. Therefore $(\mathsf{F}, \theta), w \vDash \Box_{\mathsf{F}}^{cone}[\varphi]_{(\psi_1^\Diamond, \ldots, \psi_k^\Diamond)}$. □

4 Υ-products

4.1 Definition and basic properties

Recall that the product of 1-frames (W', R') and (W'', R'') is the frame $(W' \times W'', R_h, R_v)$, where

$$(u_1, w_1) R_h (u_2, w_2) \Leftrightarrow (u_1 R' u_2 \,\&\, w_1 = w_2),$$
$$(u_1, w_1) R_v (u_2, w_2) \Leftrightarrow (u_1 = u_2 \,\&\, w_1 R'' w_2).$$

We consider a monomodal analogue of this operation, the Υ*-product.* [1]

Definition 4.1 The Υ-product of frames $\mathsf{F}' = (W', R')$ and $\mathsf{F}'' = (W'', R'')$ is the frame $\mathsf{F}' \Upsilon \mathsf{F}'' = (W' \times W'', R)$, where

$$(u_1, w_1) R (u_2, w_2) \Leftrightarrow (u_1 R' u_2 \,\&\, w_1 = w_2) \text{ or } (u_1 = u_2 \,\&\, w_1 R'' w_2).$$

---

[1] If frames are considered as transition systems, this operation is called the *asynchronous product*, see e.g. [11].

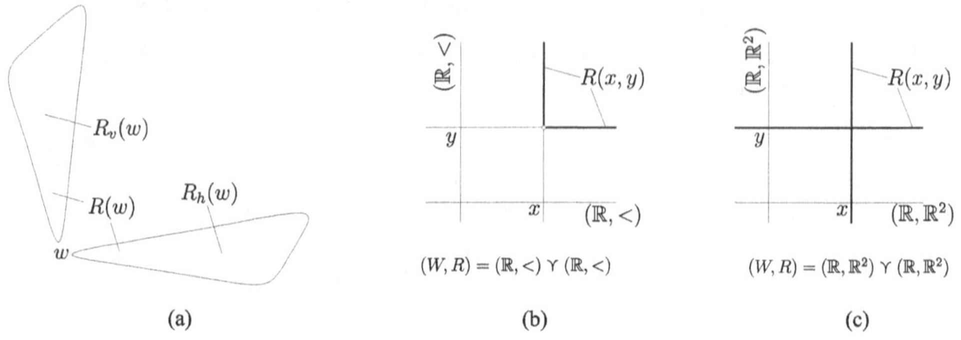

Fig. 1.

Equivalently, if (W, R_h, R_v) is the product of F' and F'', then $\mathsf{F}' \curlyvee \mathsf{F}'' = (W, R_h \cup R_v)$, Fig. 1a.

For classes $\mathcal{F}_1, \mathcal{F}_2$ of n-frames, put $\mathcal{F}_1 \curlyvee \mathcal{F}_2 = \{\mathsf{F}_1 \curlyvee \mathsf{F}_2 \mid \mathsf{F}_1 \in \mathcal{F}_1, \mathsf{F}_2 \in \mathcal{F}_2\}$.

For logics $\mathrm{L}_1, \mathrm{L}_2$, put $\mathrm{L}_1 \curlyvee \mathrm{L}_2 = \mathbf{L}(\{\mathsf{F} \mid \mathsf{F} \vDash \mathrm{L}_1\} \curlyvee \{\mathsf{F} \mid \mathsf{F} \vDash \mathrm{L}_2\})$.

Example 4.2 If $(W, R) = (\mathbb{R}, <) \curlyvee (\mathbb{R}, <)$, then $R(x, y)$ is the union of two open rays $\{(x, t) \mid t > y\} \cup \{(t, y) \mid t > x\}$, Fig. 1b. If $(W, R) = (\mathbb{R}, \mathbb{R}^2) \curlyvee (\mathbb{R}, \mathbb{R}^2)$, then $R(x, y) = \{(t_1, t_2) \mid t_1 = x \text{ or } t_2 = y\}$, Fig. 1c.

These simple examples have a natural geometric interpretation. Recall the notion of the *lightlike* relation λ in Minkowski space $\mathbb{R}^n$, $n \geq 2$: $\overline{x}\lambda\overline{y} \Leftrightarrow \Sigma_{i=1}^{n-1}(y_i - x_i)^2 = (y_n - x_n)^2$, where $\overline{x} = (x_1, \ldots, x_n)$, $\overline{y} = (y_1, \ldots, y_n)$. It is easy to see that $(\mathbb{R}^2, \lambda)$ is isomorphic to the frame $(\mathbb{R}, \mathbb{R}^2) \curlyvee (\mathbb{R}, \mathbb{R}^2)$ (a detailed discussion of the connection between relativistic modalities and modal logics of various geometric structures can be found in [12]). Similarly, $(\mathbb{R}, <) \curlyvee (\mathbb{R}, <)$ is isomorphic to the frame $(\mathbb{R}^2, \lambda^\uparrow)$, where $\overline{x}\lambda^\uparrow\overline{y} \Leftrightarrow \overline{x}\lambda\overline{y} \& y_n > x_n$ (*future directed lightlike* relation).

Since $\mathrm{S5} \times \mathrm{S5}$ is decidable and has the product finite model property (see e.g. [5]), using the above observation, it is easy to show that the logic $\mathrm{L}(\mathbb{R}^2, \lambda)$ is also decidable and has the finite model property. At the same time, it follows from Theorems 5.9 and 5.14 (proved in the next sections) that the logic $\mathrm{L}(\mathbb{R}^2, \lambda^\uparrow)$ is undecidable and does not have the finite model property.

Consider some basic properties of $\curlyvee$-products.

Trivially, $(\mathsf{F}_1 \curlyvee \mathsf{F}_2, \theta), w \vDash \Diamond \mathsf{p} \Leftrightarrow (\mathsf{F}_1 \times \mathsf{F}_2, \theta), w \vDash \Diamond_1 \mathsf{p} \vee \Diamond_2 \mathsf{p}$, so $\mathbf{L}(\mathsf{F}_1) \curlyvee \mathbf{L}(\mathsf{F}_2)$ can be regarded as a fragment of $\mathbf{L}(\mathsf{F}_1 \times \mathsf{F}_2)$. In the next sections we show that these fragments can be very expressive if factors are transitive, so $\curlyvee$-products of many extensions of K4 are quite complex. But first let us consider $\curlyvee$-products of weak logics, like K, $\mathrm{T} = \mathrm{K} + \Diamond p \to p$, $\mathrm{D} = \mathrm{K} + \Diamond \top$.

Recall that w is *serial* in a frame (W, R), if $R(w) \neq \varnothing$; F is *serial*, if all its points is serial.

Due to the Definition 4.1, we have

Proposition 4.3 *For* 1*-frames* F *and* G*, we have:*

- *if one of these frames is serial (reflexive), then* $\mathsf{F} \curlyvee \mathsf{G}$ *is serial (reflexive);*
- *if* G *is an irreflexive singleton, then* $\mathsf{F} \curlyvee \mathsf{G}$ *is isomorphic to* F*;*
- *if* G *is a reflexive singleton, then* $\mathsf{F} \curlyvee \mathsf{G}$ *is isomorphic to the reflexive closure of* F.

These observations yield the following fact for $\curlyvee$-products of unimodal non-transitive logics.

Theorem 4.4

(i) *If an irreflexive singleton validates a logic* L, *then* $\mathrm{K} \curlyvee \mathrm{L} = \mathrm{K}$, $\mathrm{D} \curlyvee \mathrm{L} = \mathrm{D}$.

(ii) *If* $\mathrm{L}_1 \subseteq \mathrm{T}$, L_2 *is consistent, and* $\mathrm{L}_1 \cup \mathrm{L}_2 \supseteq \mathrm{T}$, *then* $\mathrm{L}_1 \curlyvee \mathrm{L}_2 = \mathrm{T}$*; in particular,* $\mathrm{T} \curlyvee \mathrm{L} = \mathrm{T}$ *for any consistent* L.

Proof. (i) Clearly, $\mathrm{K} \curlyvee \mathrm{L} \supseteq \mathrm{K}$, and $\mathrm{D} \curlyvee \mathrm{L} \supseteq \mathrm{D}$ by Proposition 4.3.

To prove the other inclusions, suppose that a formula φ is satisfiable in some (serial) frame F. Consider an irreflexive one-point frame F_0. By Proposition 4.3, $\mathsf{F} \curlyvee \mathsf{F}_0$ is isomorphic to F, so φ is $\mathrm{K}\curlyvee\mathrm{L}$-satisfiable ($\mathrm{D}\curlyvee\mathrm{L}$-satisfiable). Thus $\mathrm{K}\curlyvee\mathrm{L} \subseteq \mathrm{K}$, $\mathrm{D}\curlyvee\mathrm{L} \subseteq \mathrm{D}$.

(ii). Since $\mathrm{L}_1 \cup \mathrm{L}_2 \supseteq \mathrm{T}$, it follows that $\mathrm{L}_1 \curlyvee \mathrm{L}_2 \supseteq \mathrm{T}$ by Proposition 4.3.

To show that $\mathrm{L}_1 \curlyvee \mathrm{L}_2 \subseteq \mathrm{T}$, suppose that a formula φ is satisfiable in some reflexive frame F. Since L_2 is consistent, a one-point frame F_0 validates L_2 (Makinson's Theorem, see e.g. [1]). Since F is reflexive, by Proposition 4.3 we obtain that $\mathsf{F} \curlyvee \mathsf{F}_0$ is isomorphic to F. It follows that $\mathrm{L}_1 \curlyvee \mathrm{L}_2 \supseteq \mathrm{T}$. □

Further on, we will focus on $\curlyvee$-products of unimodal transitive frames and logics.

Consider a 1-frame (W, R). Recall that R is transitive, if $R^2 \subseteq R$. We say that (W, R) is *m-transitive*, if $R^{m+1} \subseteq R^m$. Trivially,

$$\mathsf{F} \text{ is } m\text{-transitive} \Leftrightarrow \mathsf{F} \models \Diamond^{m+1}\mathsf{p} \to \Diamond^m \mathsf{p}.$$

Proposition 4.5 *If* F_1 *and* F_2 *are transitive 1-frames, then* $\mathsf{F}_1 \curlyvee \mathsf{F}_2$ *is 2-transitive.*

Proof. Let $\mathsf{F}_1 \times \mathsf{F}_2 = (W, R_1, R_2)$. Due to commutativity and transitivity,

$$(R_1 \cup R_2)^3 = R_1^3 \cup (R_1^2 \circ R_2) \cup (R_1 \circ R_2^2) \cup R_2^3 \subseteq R_1^2 \cup (R_1 \circ R_2) \cup R_2^2 = (R_1 \cup R_2)^2.$$

□

4.2 Locally n-component frames

Locally n-component frames were studied in [13], and later in [9], in the context of topological modal logics. We use this notion to express the relations $\bar{R}_h^{\mathsf{M}}$, $\bar{R}_v^{\mathsf{M}}$ by unimodal formulas.

Definition 4.6 ([13]) Consider a 1-frame $\mathsf{F} = (W, R)$. For $w \in W$, let $R_w^{\triangle}$ be the following equivalence relation on $R(w)$: $uR_w^{\triangle}v$ iff there exist points $w_0, \dots, w_{k+1} \in R(w)$ such that $u = w_0$, $w_{k+1} = v$ and for every $i = 0, \dots, k$ we have $w_i R w_{i+1}$ or $w_{i+1} R w_i$ or $w_i = w_{i+1}$, see Fig. 2a.

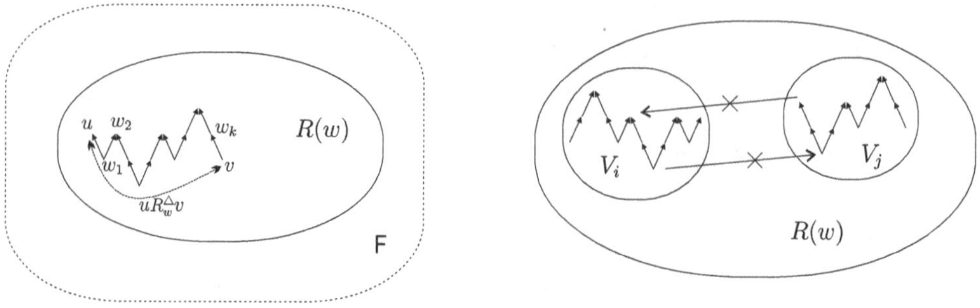

(a) (b)

Fig. 2.

For serial w, let $comp_{\mathsf{F}}(w)$ denote the quotient set of $R(w)$ by $R^{\triangle}_{w}$; if $R(w) = \varnothing$, put $comp_{\mathsf{F}}(w) = \varnothing$. $\#_{\mathsf{F}}(w)$ denotes the cardinality of $comp_{\mathsf{F}}(w)$. F is *locally n-component*, if $\#_{\mathsf{F}}(w) \leq n$ for all $w \in W$.

If $\#_{\mathsf{F}}(w)$ is finite, then $\#_{\mathsf{F}}(w)$ is the maximal k such that for some non-empty $V_1, \ldots, V_k$ we have $R(w) = V_1 \cup \cdots \cup V_k$ and

$$\bigwedge_{1 \leq i \neq j \leq k} (V_i \cap V_j = R(V_i) \cap V_j = R(V_j) \cap V_i = \varnothing) \text{ (Fig. 2b).}$$

In this case $comp_{\mathsf{F}}(w) = \{V_1, \ldots, V_k\}$.

The above described properties are modally definable. For $n \geq 1$, put:

$$\text{COMP}(\mathsf{p}_1, \ldots, \mathsf{p}_n) = \bigwedge_{1 \leq i \leq n} \Diamond \mathsf{p}_i \wedge \Box \bigvee_{1 \leq i \leq n} \mathsf{p}_i \wedge \Box \bigwedge_{1 \leq i \neq j \leq n} (\mathsf{p}_i \to \neg(\mathsf{p}_j \vee \Diamond \mathsf{p}_j));$$

$$\text{AxComp}_n = \neg\text{COMP}(\mathsf{p}_1, \ldots, \mathsf{p}_{n+1}).$$

Proposition 4.7 ([13]) *Consider a frame $\mathsf{F} = (W, R)$. For any $w \in W$, $n > 0$, we have:*

(i) *$\#_{\mathsf{F}}(w) \leq n$ iff $\mathsf{F}, w \models \text{AxComp}_n$; in particular, F is locally n-component iff $\mathsf{F} \models \text{AxComp}_n$;*

(ii) *$\#_{\mathsf{F}}(w) = n$ iff $\mathsf{F}, w \models \text{AxComp}_n$ and $\text{COMP}_n(\mathsf{p}_1, \ldots, \mathsf{p}_n)$ is satisfiable at w in F.*

Locally 1-component frames are especially important for us. Note that frames with properties like reflexivity, linearity or Church–Rosser property (the latter in the transitive case only) are locally 1-component (see Fig. 3).

Now we formulate a number of straightforward propositions that will be used in the next sections.

Proposition 4.8 *Consider 1-frames F, G and points u in F and v in G. If u is reflexive, then $\#_{\mathsf{F}}(u) = \#_{\mathsf{F} \curlyvee \mathsf{G}}(u, v) = 1$; if u and v are irreflexive, then $\#_{\mathsf{F} \curlyvee \mathsf{G}}(u, v) = \#_{\mathsf{F}}(u) + \#_{\mathsf{G}}(v)$.*

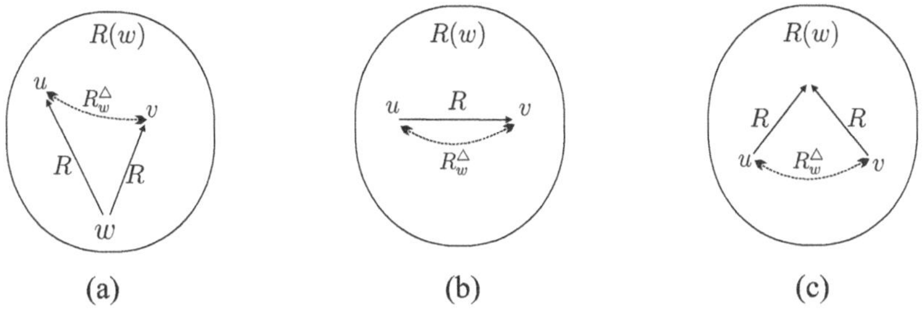

Fig. 3.

Proof. If u is reflexive, then (u, v) is reflexive in $\mathsf{F} \curlyvee \mathsf{G}$.

If u and v are irreflexive, then

$$comp_{\mathsf{F} \curlyvee \mathsf{G}}(u, v) = \{\{u\} \times V \mid V \in comp_{\mathsf{G}}(v)\} \cup \{\{v\} \times U \mid U \in comp_{\mathsf{F}}(u)\}.$$

□

Proposition 4.9 *If* L_1, L_2 *are unimodal logics,* $\mathrm{AxComp}_n \in \mathrm{L}_1$, $\mathrm{AxComp}_m \in \mathrm{L}_2$, *then* $\mathrm{AxComp}_{m+n} \in \mathrm{L}_1 \curlyvee \mathrm{L}_2$.

Proof. Follows from Proposition 4.8. □

Put

$$\mathrm{AxCov} \;=\; comp(\mathsf{p}, \mathsf{q}) \to (\Diamond\Diamond \mathsf{t} \wedge \neg\Diamond \mathsf{t} \to \Diamond(\mathsf{p} \wedge \Diamond \mathsf{t})).$$

By a straightforward argument, we have

Proposition 4.10 *Let* $\mathsf{F} = (W, R)$ *be a locally 2-component frame. Then for any* $w \in W$ *the following conditions are equivalent:*

(1) $\mathsf{F}, w \models \mathrm{AxCov}$;

(2) if $\#_{\mathsf{F}}(w) = 2$ *and* $V \in comp_{\mathsf{F}}(w)$, *then* $R(V) \supseteq R^2(w) - R(w)$.

Proposition 4.11 *If frames* F, G *are locally one-component, then* $\mathsf{F} \curlyvee \mathsf{G} \models \mathrm{AxCov}$.

Proof. Let $\mathsf{F} \times \mathsf{G} = (W, R_1, R_2)$. If $\#_{\mathsf{F} \curlyvee \mathsf{G}}(w) = 2$, then $comp_{\mathsf{F}} = \{R_1(w), R_2(w)\}$. By Proposition 4.10, $\mathsf{F} \curlyvee \mathsf{G} \models \mathrm{AxCov}$. □

Proposition 4.12 *Let* F *be a 2-transitive frame with a root* w *such that* $\#_{\mathsf{F}}(w) = 2$ *and* $\mathsf{F}, w \models \mathrm{AxCov}$. *Then for any valuation* θ *we have:* $(\mathsf{F}, \theta), w \models comp(\mathsf{p}, \mathsf{q})$ *iff* $comp_{\mathsf{F}}(w) = \{\theta(\mathsf{p}), \theta(\mathsf{q})\}$.

Proof. Follows from Proposition 4.10. □

5 Simulation of two dimensions

In this section we define relations $\hat{R}_h^{\mathsf{M}}$, $\hat{R}_v^{\mathsf{M}}$ and operators $\Diamond_v^{\curlyvee}$, $\Diamond_h^{\curlyvee}$ that play the same role as $\bar{R}_h^{\mathsf{M}}$, $\bar{R}_v^{\mathsf{M}}$, $\Diamond_v$, $\Diamond_h$, but 'work' in the unimodal case.

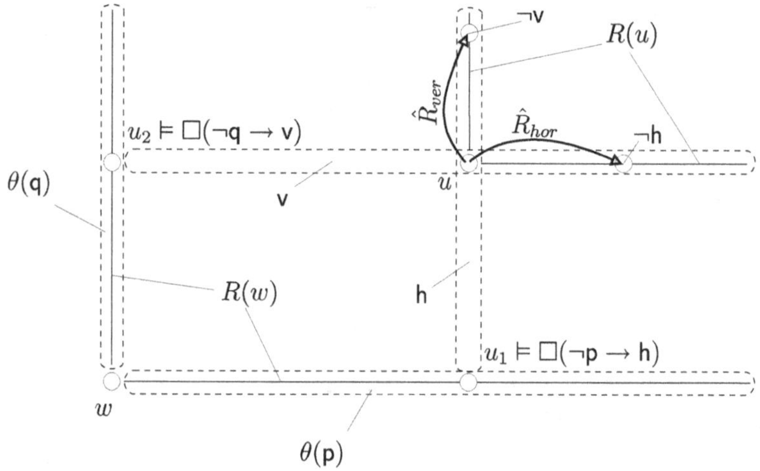

Fig. 4.

The main idea is that if $comp_{\mathsf{F}}(w)$=2, then we can split $R(w)$ into two parts (via the formula COMP(p, q)), and then express two 'directions' by unimodal operators. We show that under some additional restrictions these 'directions' are [K4,K4]-relations.

The formal definition of $\bar{R}^{\mathsf{M}}_h$, $\bar{R}^{\mathsf{M}}_v$ and $\Diamond^{\curlyvee}_v$, $\Diamond^{\curlyvee}_h$ is rather tedious, so first we illustrate the idea with the following example.

Example 5.1 (Fig. 4) Let F_1 and F_2 be rooted strict linear orders, $\mathsf{F} = \mathsf{F}_1 \curlyvee \mathsf{F}_2 = (W, R)$, $\mathsf{F}_1 \times \mathsf{F}_2 = (W, R_1, R_2)$. Let w be the root of F, θ be a valuation on W such that $\theta(\mathsf{p}) = R_1(w)$, $\theta(\mathsf{q}) = R_2(w)$. Suppose

$$(\mathsf{F}, \theta), w \models \Box_{\mathsf{p}}(\Box_{\neg\mathsf{p}}\mathsf{h} \vee \Box_{\neg\mathsf{p}}\neg\mathsf{h}) \wedge \Box_{\mathsf{q}}(\Box_{\neg\mathsf{q}}\mathsf{v} \vee \Box_{\neg\mathsf{q}}\neg\mathsf{v}),$$

or, equivalently, for any $u_1 \in R_1(w), u_2 \in R_2(w)$ we have

$$R_2(u_1) \cap \theta(\mathsf{h}) = \varnothing \text{ or } R_2(u_1) \subseteq \theta(\mathsf{h}), \quad R_1(u_2) \cap \theta(\mathsf{v}) = \varnothing \text{ or } R_1(u_2) \subseteq \theta(\mathsf{v}).$$

In this case we can define 'horizontal' and 'vertical' relations in terms of R and θ in the following way: for any $u \notin R^{=}(w)$, put

$$u\hat{R}_{hor}u' \text{ iff } uRu'\&(u \in \theta(\mathsf{h}) \Leftrightarrow u' \notin \theta(\mathsf{h})), \quad u\hat{R}_{ver}u' \text{ iff } uRu'\&(u \in \theta(\mathsf{v}) \Leftrightarrow u' \notin \theta(\mathsf{v})).$$

Now consider the 2-model $\mathsf{M} = (\mathsf{F}_1 \times \mathsf{F}_2, \theta)$ and observe that

$$(u, u') \in \hat{R}_{hor} \Leftrightarrow (u, u') \in \bar{R}^{\mathsf{M}}_{h,0} \text{ and } (u, u') \in \hat{R}_{ver} \Leftrightarrow (u, u') \in \bar{R}^{\mathsf{M}}_{v,0}$$

for any $u \notin R^{=}(w)$. Due to this observation, it is possible to define unimodal formulas which express $\bar{R}^{\mathsf{M}}_v$, $\bar{R}^{\mathsf{M}}_h$, thus to express [K4, K4]-relations in (F, θ).

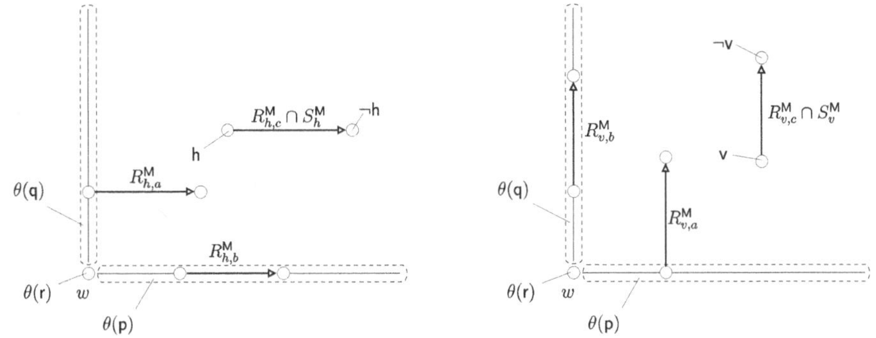

Fig. 5.

Definition 5.2 For a model $\mathsf{M} = (W, R, \theta)$, put

$$S^{\mathsf{M}}_h = \{(u, w) \mid uRw \& (u \in \theta(\mathsf{h}) \Leftrightarrow w \notin \theta(\mathsf{h}))\},$$
$$S^{\mathsf{M}}_v = \{(u, w) \mid uRw \& (u \in \theta(\mathsf{v}) \Leftrightarrow w \notin \theta(\mathsf{v}))\}.$$

Proposition 5.3 *For a unimodal* M, *if* $\psi \overset{\mathsf{M}}{\rightharpoondown} \widetilde{R}$, *then*

$$\bigwedge_{\varepsilon=0,1} (\mathsf{h}^{\varepsilon} \to \psi(\mathsf{s} \wedge \neg \mathsf{h}^{\varepsilon})) \overset{\mathsf{M}}{\rightharpoondown} \widetilde{R} \cap S^{\mathsf{M}}_h, \quad \bigwedge_{\varepsilon=0,1} (\mathsf{v}^{\varepsilon} \to \psi(\mathsf{s} \wedge \neg \mathsf{v}^{\varepsilon})) \overset{\mathsf{M}}{\rightharpoondown} \widetilde{R} \cap S^{\mathsf{M}}_v.$$

Proof. By a straightforward argument using Proposition 3.2. □

Definition 5.4 For a model $\mathsf{M} = (W, R, \theta)$, put

$$uR^{\mathsf{M}}_{h,a}v \Leftrightarrow uRv \& u \in \theta(\mathsf{q}) \cup \theta(\mathsf{r}) \& v \notin \theta(\mathsf{q}), \quad uR^{\mathsf{M}}_{v,a}v \Leftrightarrow uRv \& u \in \theta(\mathsf{p}) \cup \theta(\mathsf{r}) \& v \notin \theta(\mathsf{p}),$$
$$uR^{\mathsf{M}}_{h,b}v \Leftrightarrow uRv \& u, v \in \theta(\mathsf{p}), \quad uR^{\mathsf{M}}_{v,b}v \Leftrightarrow uRv \& u, v \in \theta(\mathsf{q}),$$
$$uR^{\mathsf{M}}_{h,c}v \Leftrightarrow uRv \& u \notin \theta(\mathsf{p}) \cup \theta(\mathsf{q}) \cup \theta(\mathsf{r}), \quad R^{\mathsf{M}}_{v,c} = R^{\mathsf{M}}_{h,c},$$
$$\hat{R}^{\mathsf{M}}_{h,0} = (R^{\mathsf{M}}_{h,a} \cup R^{\mathsf{M}}_{h,b} \cup R^{\mathsf{M}}_{h,c}) \cap S^{\mathsf{M}}_h, \quad \hat{R}^{\mathsf{M}}_{v,0} = (R^{\mathsf{M}}_{v,a} \cup R^{\mathsf{M}}_{v,b} \cup R^{\mathsf{M}}_{v,c}) \cap S^{\mathsf{M}}_v,$$
$$\hat{R}^{\mathsf{M}}_h = \hat{R}^{\mathsf{M}}_{h,0} \cup \left(\hat{R}^{\mathsf{M}}_{h,0}\right)^2, \quad \hat{R}^{\mathsf{M}}_v = \hat{R}^{\mathsf{M}}_{v,0} \cup \left(\hat{R}^{\mathsf{M}}_{v,0}\right)^2.$$

In Fig. 5 these relations are shown for the model described in Example 5.1 where also $\theta(\mathsf{r}) = \{w\}$ is assumed. Note that $R^{\mathsf{M}}_{h,c} \cap S^{\mathsf{M}}_h = \hat{R}_{hor}$, $R^{\mathsf{M}}_{v,c} \cap S^{\mathsf{M}}_v = \hat{R}_{ver}$, so $\hat{R}^{\mathsf{M}}_{h,0} \subseteq R_1$, and $\hat{R}^{\mathsf{M}}_{v,0} \subseteq R_2$; moreover, as we will show later, $\hat{R}^{\mathsf{M}}_h = \bar{R}^{\mathsf{M}}_h$ and $\hat{R}^{\mathsf{M}}_v = \bar{R}^{\mathsf{M}}_v$. It is not hard to see that these relations can be expressed in M by unimodal formulas. For this, we need several simple formulas, namely:

$$\psi^{\curlyvee}_{h,a} = \mathsf{q} \vee \mathsf{r} \to \Diamond(\neg\mathsf{q} \wedge \mathsf{s}), \qquad \psi^{\curlyvee}_{v,a} = \mathsf{p} \vee \mathsf{r} \to \Diamond(\neg\mathsf{p} \wedge \mathsf{s}),$$
$$\psi^{\curlyvee}_{h,b} = \mathsf{p} \to \Diamond(\mathsf{p} \wedge \mathsf{s}), \qquad \psi^{\curlyvee}_{v,b} = \mathsf{q} \to \Diamond(\mathsf{q} \wedge \mathsf{s}),$$
$$\psi^{\curlyvee}_{h,c} = \neg(\mathsf{p} \vee \mathsf{q} \vee \mathsf{r}) \to \Diamond\mathsf{s}, \qquad \psi^{\curlyvee}_{v,c} = \neg(\mathsf{p} \vee \mathsf{q} \vee \mathsf{r}) \to \Diamond\mathsf{s},$$
$$\psi^{\curlyvee}_{h,0} = \bigwedge_{\varepsilon=0,1} \left(\mathsf{h}^{\varepsilon} \to [(\mathsf{s} \wedge \neg\mathsf{h}^{\varepsilon})/\mathsf{s}](\psi^{\curlyvee}_{h,a} \wedge \psi^{\curlyvee}_{h,b} \wedge \psi^{\curlyvee}_{h,c})\right),$$
$$\psi^{\curlyvee}_{v,0} = \bigwedge_{\varepsilon=0,1} \left(\mathsf{v}^{\varepsilon} \to [(\mathsf{s} \wedge \neg\mathsf{v}^{\varepsilon})/\mathsf{s}](\psi^{\curlyvee}_{v,a} \wedge \psi^{\curlyvee}_{v,b} \wedge \psi^{\curlyvee}_{v,c})\right),$$
$$\psi^{\curlyvee}_{h} = \psi^{\curlyvee}_{h,0} \vee \psi^{\curlyvee}_{h,0}(\psi^{\curlyvee}_{h,0}), \qquad \psi^{\curlyvee}_{v} = \psi^{\curlyvee}_{v,0} \vee \psi^{\curlyvee}_{v,0}(\psi^{\curlyvee}_{v,0}).$$

The only subtle case is for the operators $\psi^{\curlyvee}_{h,0}$, $\psi^{\curlyvee}_{v,0}$: here we use Proposition 5.3.

Lemma 5.5 *Consider a model* $\mathsf{M} = (W, R, \theta)$ *such that the sets* $\theta(\mathsf{p})$, $\theta(\mathsf{q})$, $\theta(\mathsf{r})$ *are pairwise disjoint. Then*

$$\psi^{\curlyvee}_{h} \overset{\mathsf{M}}{\to} \hat{R}^{\mathsf{M}}_{h}, \quad \psi^{\curlyvee}_{v} \overset{\mathsf{M}}{\to} \hat{R}^{\mathsf{M}}_{v}.$$

Proof. By a straightforward argument,

$$\psi^{\curlyvee}_{h,a} \wedge \psi^{\curlyvee}_{h,b} \wedge \psi^{\curlyvee}_{h,c} \overset{\mathsf{M}}{\to} R^{\mathsf{M}}_{h,a} \cup R^{\mathsf{M}}_{h,b} \cup R^{\mathsf{M}}_{h,c}, \quad \psi^{\curlyvee}_{v,a} \wedge \psi^{\curlyvee}_{v,b} \wedge \psi^{\curlyvee}_{v,c} \overset{\mathsf{M}}{\to} R^{\mathsf{M}}_{v,a} \cup R^{\mathsf{M}}_{v,b} \cup R^{\mathsf{M}}_{v,c}.$$

By Proposition 5.3,

$$\psi^{\curlyvee}_{h,0} \overset{\mathsf{M}}{\to} \hat{R}^{\mathsf{M}}_{h,0}, \quad \psi^{\curlyvee}_{v,0} \overset{\mathsf{M}}{\to} \hat{R}^{\mathsf{M}}_{v,0},$$

thus $\psi^{\curlyvee}_{h} \overset{\mathsf{M}}{\to} \hat{R}^{\mathsf{M}}_{h}$ and $\psi^{\curlyvee}_{v} \overset{\mathsf{M}}{\to} \hat{R}^{\mathsf{M}}_{v}$. □

Put $\Diamond^{\curlyvee}_{h}\varphi = \psi^{\curlyvee}_{h}(\varphi)$, $\Diamond^{\curlyvee}_{v}\varphi = \psi^{\curlyvee}_{v}(\varphi)$, $\Box^{\curlyvee}_{h}\varphi = \neg\psi^{\curlyvee}_{h}(\neg\varphi)$, $\Box^{\curlyvee}_{v}\varphi = \neg\psi^{\curlyvee}_{v}(\neg\varphi)$.

5.1 Undecidability

The following formula is a unimodal analogue of the formula ψ_{hv}:

$$\psi^{\curlyvee}_{hv} = \mathsf{r} \wedge \neg\Diamond\mathsf{r} \wedge \neg\Diamond\Diamond\mathsf{r} \wedge (\Box_{\mathsf{p}}(\Box_{\neg\mathsf{p}}\mathsf{h} \vee \Box_{\neg\mathsf{p}}\neg\mathsf{h}) \wedge \Box_{\mathsf{q}}(\Box_{\neg\mathsf{q}}\mathsf{v} \vee \Box_{\neg\mathsf{q}}\neg\mathsf{v}))\,.$$

Using it we express the relations $\bar{R}^{\mathsf{M}}_{h}$, $\bar{R}^{\mathsf{M}}_{v}$ in 1-models based on $\curlyvee$-products of transitive locally one-component frames.

Lemma 5.6 *Let* $\mathsf{F}_1, \mathsf{F}_2$ *be transitive locally one-component frames,* $\mathsf{F} = \mathsf{F}_1 \curlyvee \mathsf{F}_2$, $\mathsf{G} = \mathsf{F}_1 \times \mathsf{F}_2 = (W, R_1, R_2)$, θ *be a valuation on* W *such that*

$$\theta(\mathsf{p}) = R_1(w), \ \theta(\mathsf{q}) = R_2(w), \ \theta(\mathsf{r}) = \{w\}. \tag{3}$$

If F *has the irreflexive root* w*, then the following holds.*

(i) $(\mathsf{F}, \theta), w \models \psi^{\curlyvee}_{hv}$ *iff* $(\mathsf{G}, \theta), w \models \psi_{hv}$.

(ii) If $(\mathsf{F}, \theta), w \models \psi^{\curlyvee}_{hv}$*, then* $\hat{R}^{(\mathsf{F},\theta)}_{h} = \bar{R}^{(\mathsf{G},\theta)}_{h}$, $\hat{R}^{(\mathsf{F},\theta)}_{v} = \bar{R}^{(\mathsf{G},\theta)}_{v}$.

(iii) If $(\mathsf{F},\theta), w \vDash \psi^{\curlyvee}_{hv}$, $\varphi \in ML_2$, $u \in W$, *then*

$$(\mathsf{F},\theta), u \vDash [\varphi]_{(\psi^{\curlyvee}_h,\psi^{\curlyvee}_v)} \;\Leftrightarrow\; (\mathsf{G},\theta), u \vDash [\varphi]_{(\psi_h,\psi_v)}.$$

Proof. Put $R = R_1 \cup R_2$, $\mathsf{M} = (\mathsf{F},\theta)$, $\mathsf{N} = (\mathsf{G},\theta)$, $w = (x_0, y_0)$. Note that

$$R_1(R_2(w)) \cap R_1^{=}(w) = R_1(R_2(w)) \cap R_2^{=}(w) = \varnothing. \tag{4}$$

Indeed, for $u = (x,y) \in R_1(R_2(w))$, if $u \in R_1^{=}(w)$, then $y = y_0$ and y_0 is reflexive in F_2, and if $u \in R_2^{=}(w)$, then $x = x_0$ and x_0 is reflexive if F_1; since $\#_{\mathsf{F}}(w) = 2$, then by Proposition 4.8 x_0 is irreflexive in F_1 and y_0 is irreflexive in F_2, that proves (4).

(i) Suppose $\mathsf{M}, w \vDash \psi^{\curlyvee}_{hv}$.

Consider $u = (x,y) \in R_1(R_2(w))$.

Let $\mathsf{M}, u \vDash \mathsf{h}^{\varepsilon} \vee \Diamond_2 \mathsf{h}^{\varepsilon}$, $v \in R_2(u)$ for some $v \in R_2(w)$, $\varepsilon \in \{0,1\}$. Since $wR_1(x,y_0)R_2u$, then $(x,y_0) \in \theta(\mathsf{p})$, and $\mathsf{M}, (x,y_0) \vDash \Box_{\neg\mathsf{p}}\mathsf{h} \vee \Box_{\neg\mathsf{p}}\neg\mathsf{h}$. Due to (4), $u, v \notin \theta(\mathsf{p})$, so $\mathsf{M}, u \vDash \Box_{\neg\mathsf{p}}\mathsf{h}^{\varepsilon}$ and $\mathsf{N}, v \vDash \mathsf{h}^{\varepsilon}$. Thus $\mathsf{M}, u \vDash \mathsf{h}^{\varepsilon} \vee \Diamond_2\mathsf{h}^{\varepsilon} \to \Box_2\mathsf{h}^{\varepsilon}$.

Similarly, $\mathsf{M}, u \vDash \mathsf{v}^{\varepsilon} \vee \Diamond_1\mathsf{v}^{\varepsilon} \to \Box_1\mathsf{v}^{\varepsilon}$, so $\mathsf{N}, w \vDash \psi_{hv}$.

Suppose $\mathsf{N}, w \vDash \psi_{hv}$.

Let us show that $\mathsf{M}, w \vDash \bigwedge_{\varepsilon=0,1} (\Box_{\mathsf{p}}(\Diamond_{\neg\mathsf{p}}\mathsf{h}^{\varepsilon} \to \Box_{\neg\mathsf{p}}\mathsf{h}^{\varepsilon}))$.

Suppose $u \in R(x)$, $u \in \theta(\mathsf{p})$. Then wR_1u, $u = (x, y_0)$ for some x. Put

$$Y_0 = R_2(u) - \theta(\mathsf{h}),\; Y_1 = R_2(u) \cap \theta(\mathsf{h}).$$

Since $\mathsf{N}, (x,y) \vDash \mathsf{h} \vee \Diamond_2\mathsf{h} \to \Box_2\mathsf{h}$ for any $(x,y) \in Y_1$, then $R_2(Y_1) \cap Y_0 = \varnothing$. Similarly, $R_2(Y_0) \cap Y_1 = \varnothing$. Since F_2 is locally one-component, $Y_0 = \varnothing$ or $Y_1 = \varnothing$. Suppose $\mathsf{M}, u \vDash \Diamond_{\neg\mathsf{p}}\mathsf{h}^{\varepsilon}$ for some $\varepsilon \in \{0,1\}$, and let $v \in R(u) - \theta(\mathsf{p})$. Then $\mathsf{M}, u' \vDash \mathsf{h}^{\varepsilon}$ for some $u' \in R(u) - \theta(\mathsf{p})$. It follows that $u', v \notin R_1(u)$, so $u', v \in R_2(u)$. Thus $Y_{\varepsilon} \neq \varnothing$ and $v \in Y_{\varepsilon}$. It follows that $\mathsf{M}, v \vDash \mathsf{h}^{\varepsilon}$, so $\mathsf{M}, u \vDash \Diamond_{\neg\mathsf{p}}\mathsf{h}^{\varepsilon} \to \Box_{\neg\mathsf{p}}\mathsf{h}^{\varepsilon}$.

Similarly, $\mathsf{M}, w \vDash \Box_{\mathsf{q}}(\Box_{\neg\mathsf{q}}\mathsf{v} \vee \Box_{\neg\mathsf{q}}\neg\mathsf{v})$.

Since w is irreflexive in F, $\mathsf{M}, w \vDash \neg\Diamond\mathsf{r}$, and due to (4), $\mathsf{M}, w \vDash \neg\Diamond\Diamond\mathsf{r}$. It follows that $\mathsf{M}, w \vDash \psi^{\curlyvee}_{hv}$.

(ii) Let us show that $\hat{R}^{\mathsf{M}}_{h,0} = \bar{R}^{\mathsf{N}}_{h,0}$. Note that if $u\hat{R}^{\mathsf{M}}_{h,0}v$ or $u\bar{R}^{\mathsf{N}}_{h,0}v$, then

$$uRv \text{ and } (u \in \theta(\mathsf{h}) \Leftrightarrow v \notin \theta(\mathsf{h})). \tag{5}$$

Suppose (5) holds and consider the following cases.

a). $u \in R_2^{=}(w)$ (equivalently, $u \in \theta(\mathsf{q}) \cup \theta(\mathsf{r})$).

In this case, uR_1v iff $v \notin \theta(\mathsf{q})$, so $u\bar{R}^{\mathsf{N}}_{h,0}v$ iff $uR^{\mathsf{M}}_{h,a}v$.

b). $u \in R_1(w)$ (equivalently, $u \in \theta(\mathsf{p})$).

In this case, uR_1v iff $v \in \theta(\mathsf{p})$, so $u\bar{R}^{\mathsf{M}}_{h,0}v$ iff $uR^{\mathsf{M}}_{h,b}v$.

c). $u \in R^2(w) - R^{=}(w)$ (equivalently, $u \notin \theta(\mathsf{p}) \cup \theta(\mathsf{q}) \cup \theta(\mathsf{r})$).

In this case we have $uR^{\mathsf{M}}_{h,c}v$. Let us show that $u\bar{R}^{\mathsf{M}}_{h,0}v$. Due to (5), $u\bar{R}^{\mathsf{M}}_{h,0}v$ iff uR_1v. Since $\mathsf{N}, w \vDash \psi_{hv}$, $\mathsf{N}, u \vDash \bigwedge_{\varepsilon=0,1} (\mathsf{h}^{\varepsilon} \vee \Diamond_2\mathsf{h}^{\varepsilon} \to \Box_2\mathsf{h}^{\varepsilon})$. It follows that if uR_2v then $(u \in \theta(\mathsf{h}) \Leftrightarrow v \in \theta(\mathsf{h}))$, which contradicts (5). Thus uR_1v, and so $u\bar{R}^{\mathsf{M}}_{h,0}v$.

It follows that $\hat{R}^{\mathsf{M}}_{h,0} = \bar{R}^{\mathsf{N}}_{h,0}$. Similarly, $\hat{R}^{\mathsf{M}}_{v,0} = \bar{R}^{\mathsf{N}}_{v,0}$. Thus $\hat{R}^{\mathsf{M}}_{h} = \bar{R}^{\mathsf{N}}_{h}$ and $\hat{R}^{\mathsf{M}}_{v} = \bar{R}^{\mathsf{N}}_{v}$.

(iii) Follows from (2), Lemma 5.5, and Lemma 3.6. □

The above lemma allows to express [K4, K4]-relations in models: to apply the lemma, we need the condition (3). The following key lemma shows how to obtain this result for frames.

Lemma 5.7 *Let $\mathsf{F}_1, \mathsf{F}_2$ be transitive locally one-component frames, (x,y) be an irreflexive point in $\mathsf{F}_1 \curlyvee \mathsf{F}_2$. Let $\varphi \in ML_2$, $PV(\varphi) \cap \{\mathsf{p},\mathsf{q},\mathsf{r}\} = \varnothing$. Then* $\textsc{comp}(\mathsf{p},\mathsf{q}) \wedge \psi^{\curlyvee}_{hv} \wedge [\varphi]_{(\psi^{\curlyvee}_h,\psi^{\curlyvee}_v)}$ *is satisfiable at (x,y) in $\mathsf{F}_1 \curlyvee \mathsf{F}_2$ iff $\psi_{hv} \wedge [\varphi]_{(\psi_h,\psi_v)}$ is satisfiable at (x,y) in $\mathsf{F}_1 \times \mathsf{F}_2$ or at (y,x) in $\mathsf{F}_2 \times \mathsf{F}_1$.*

Proof. Let F denote $(\mathsf{F}_1 \curlyvee \mathsf{F}_2)^{(x,y)}$, $(W, R_1, R_2) = (\mathsf{F}_1 \times \mathsf{F}_2)^{(x,y)}$.

Since (x,y) is irreflexive, $\#_{\mathsf{F}}(x,y) = 2$ (Proposition 4.8).

For $V \subseteq W$, put $V^* = \{(y',x') \mid (x',y') \in V\}$. For a valuation θ on W, let $\theta^*(\mathsf{t}) = (\theta(\mathsf{t}))^*$ for all $\mathsf{t} \in PV$. Trivially, $(\mathsf{F}_1 \curlyvee \mathsf{F}_2, \theta), (x',y') \vDash \psi \Leftrightarrow (\mathsf{F}_2 \curlyvee \mathsf{F}_1, \theta^*), (y',x') \vDash \psi$ for any $\psi \in ML_1$, $(x',y') \in W$.

Suppose $(\mathsf{F},\theta),(x,y) \vDash \textsc{comp}(\mathsf{p},\mathsf{q}) \wedge \psi^{\curlyvee}_{hv} \wedge [\varphi]_{(\psi^{\curlyvee}_h,\psi^{\curlyvee}_v)}$. Since $\#_{\mathsf{F}}(x,y) = 2$, $comp_{\mathsf{F}}(x,y) = \{\theta(\mathsf{p}), \theta(\mathsf{q})\}$. It follows that

$$R_1(x,y) = \theta(\mathsf{p}) \ \&\ R_2(x,y) = \theta(\mathsf{q}) \ \&\ \{(x,y)\} = \theta(\mathsf{r}) \quad \text{or}$$
$$R_2(x,y) = \theta(\mathsf{p}) \ \&\ R_1(x,y) = \theta(\mathsf{q}) \ \&\ \{(x,y)\} = \theta(\mathsf{r}).$$

By Lemma 5.6, in the former case $(\mathsf{F}_1 \times \mathsf{F}_2, \theta), (x,y) \vDash \psi_{hv} \wedge [\varphi]_{(\psi_h,\psi_v)}$, and in the latter case $(\mathsf{F}_2 \times \mathsf{F}_1, \theta^*), (y,x) \vDash \psi_{hv} \wedge [\varphi]_{(\psi_h,\psi_v)}$.

If $(\mathsf{F}_1 \times \mathsf{F}_2, \theta), (x,y) \vDash \psi_{hv} \wedge [\varphi]_{(\psi_h,\psi_v)}$, put

$$\eta(\mathsf{p}) = R_1(x,y),\ \eta(\mathsf{q}) = R_2(x,y),\ \eta(\mathsf{r}) = \{(x,y)\},\ \eta(\mathsf{t}) = \theta(\mathsf{t}) \text{ for } \mathsf{t} \notin \{\mathsf{p},\mathsf{q},\mathsf{r}\};$$

if $(\mathsf{F}_2 \times \mathsf{F}_1, \theta), (y,x) \vDash \psi_{hv} \wedge [\varphi]_{(\psi_h,\psi_v)}$, let η be a valuation on W^* such that

$$\eta(\mathsf{p}) = (R_2(x,y))^*,\ \eta(\mathsf{q}) = (R_1(x,y))^*,\ \eta(\mathsf{r}) = \{(y,x)\},\ \eta(\mathsf{t}) = \theta(\mathsf{t}) \text{ for } \mathsf{t} \notin \{\mathsf{p},\mathsf{q},\mathsf{r}\}.$$

Since $PV([\varphi]_{(\psi_h,\psi_v)}, \psi_{hv}) \cap \{\mathsf{p},\mathsf{q},\mathsf{r}\} = \varnothing$, then $(\mathsf{F}_1 \times \mathsf{F}_2, \eta), (x,y) \vDash \psi_{hv} \wedge [\varphi]_{(\psi_h,\psi_v)}$ or $(\mathsf{F}_2 \times \mathsf{F}_1, \eta), (y,x) \vDash \psi_{hv} \wedge [\varphi]_{(\psi_h,\psi_v)}$. Moreover, if the former case $(\mathsf{F}_1 \times \mathsf{F}_2, \eta), (x,y) \vDash \textsc{comp}(\mathsf{p},\mathsf{q})$, and in the latter case $(\mathsf{F}_2 \times \mathsf{F}_1, \eta), (y,x) \vDash \textsc{comp}(\mathsf{p},\mathsf{q})$. By Lemma 5.6, $\textsc{comp}(\mathsf{p},\mathsf{q}) \wedge \psi^{\curlyvee}_{hv} \wedge [\varphi]_{(\psi^{\curlyvee}_h,\psi^{\curlyvee}_v)}$ is satisfiable at (x,y) in $\mathsf{F}_1 \curlyvee \mathsf{F}_2$ or at (y,x) in $\mathsf{F}_2 \curlyvee \mathsf{F}_1$. To finish the proof, note that $\mathsf{F}_1 \curlyvee \mathsf{F}_2$ and $\mathsf{F}_2 \curlyvee \mathsf{F}_1$ are isomorphic. □

The above lemma allows us to formulate undecidability results for unimodal frames.

Definition 5.8 A transitive locally one-component frame $\mathsf{F} = (W, R)$ is called *an axis-frame*, if there exists an irreflexive point x in F such that F^x contains an infinite descending chain of distinct points, i.e., there exists a sequence $\{x_i\}_{i>0}$ such that $x_{i+1} R x_i$, $x_i \neq x_{i+1}$ and $x R x_i$ for all $i > 0$; x is called *an origin of* F.

Theorem 5.9 *If a class $\mathcal{F}$ of transitive locally one-component frames contains an axis-frame, then $\mathbf{L}(\mathcal{F} \curlyvee \mathcal{F})$ is undecidable.*

Proof. In [7], it was shown that the $\omega \times \omega$-tiling problem is reducible to the [K4, K4]-satisfiability problem. More precisely, there was described a procedure which for a given tile Θ provides φ^Θ with the following properties:

(i) if Θ tiles $\omega \times \omega$ and F is a transitive 1-frame with a root x containing an infinite descending chain, then $\psi_{hv} \wedge [\varphi^\Theta]_{(\psi_h,\psi_v)}$ is satisfiable at (x, x) in $\mathsf{F} \times \mathsf{F}$;

(ii) if $\psi_{hv} \wedge [\varphi^\Theta]_{(\psi_h,\psi_v)}$ is [K4, K4]-satisfiable, then Θ tiles $\omega \times \omega$.[2]

Without any loss of generality we may assume that $PV(\varphi^\Theta) \cap \{\mathsf{p}, \mathsf{q}, \mathsf{r}\} = \varnothing$.

The statement of the theorem follows from the following fact:

$$\Theta \text{ tiles } \omega \times \omega \text{ iff } \textsc{comp}(\mathsf{p}, \mathsf{q}) \wedge \psi^{\curlyvee}_{hv} \wedge [\varphi^\Theta]_{(\psi^{\curlyvee}_h,\psi^{\curlyvee}_v)} \text{ is } \mathcal{F} \curlyvee \mathcal{F}\text{-satisfiable.}$$

To prove it, consider an axis-frame $\mathsf{F} \in \mathcal{F}$ with an origin point x. If Θ tiles $\omega \times \omega$, then $\psi_{hv} \wedge [\varphi^\Theta]_{(\psi_h,\psi_v)}$ is satisfiable at (x, x) in $(\mathsf{F} \times \mathsf{F})^{(x,x)}$, and by Lemma 5.7, $\textsc{comp}(\mathsf{p}, \mathsf{q}) \wedge \psi^{\curlyvee}_{hv} \wedge [\varphi^\Theta]_{(\psi^{\curlyvee}_h,\psi^{\curlyvee}_v)}$ is satisfiable at (x, x) in $\mathsf{F} \curlyvee \mathsf{F}$.

Conversely, suppose $\textsc{comp}(\mathsf{p}, \mathsf{q}) \wedge \psi^{\curlyvee}_{hv} \wedge [\varphi^\Theta]_{(\psi^{\curlyvee}_h,\psi^{\curlyvee}_v)}$ is satisfiable at a point (x, y) in a frame $\mathsf{F}_1 \curlyvee \mathsf{F}_2$ for some $\mathsf{F}_1, \mathsf{F}_2 \in \mathcal{F}$. Then $\#_{\mathsf{F}_1 \curlyvee \mathsf{F}_2}(x, y) = 2$ due to Proposition 4.7. By Lemma 5.7, $\psi_{hv} \wedge [\varphi^\Theta]_{(\psi_h,\psi_v)}$ is satisfiable at (x, y) in $\mathsf{F}_1 \times \mathsf{F}_2$ or at (y, x) in $\mathsf{F}_2 \times \mathsf{F}_1$. Thus Θ tiles $\omega \times \omega$. □

Example 5.10 Clearly, if $\mathsf{F} = (W, R)$ is a strict linear order containing a point x and an infinite descending chain $y_1 R^{-1} y_2 R^{-1} \dots R^{-1} x$, then F is an axis frame. Thus the satisfiability problem for $\mathsf{F} \curlyvee \mathsf{F}$ is undecidable. In particular, the satisfiability problems for $(\mathbb{R}, <) \curlyvee (\mathbb{R}, <)$ and $(\mathbb{Q}, <) \curlyvee (\mathbb{Q}, <)$ are undecidable.

5.2 Lack of the finite model property

In the previous subsection we 'encoded' two dimensions in semantically defined frames – namely, in $\curlyvee$-products. To prove the lack of finite model property, we have to define such frames axiomatically.

For a formula φ, let $\Box^{\leq 2}\varphi$ abbreviate $\Box\Box\varphi \wedge \Box\varphi \wedge \varphi$.

Let $\mathrm{L}^{\curlyvee}_{min}$ be the minimal normal unimodal logic containing the formulas $\Diamond^3\mathsf{p} \to \Diamond^2\mathsf{p}$, $\textsc{AxComp}_2$, $\textsc{AxCov}$, and the following formulas:

$$\begin{aligned}
\textsc{AxTr}^{\curlyvee}_1 &= \textsc{comp}(\mathsf{p}, \mathsf{q}) \wedge \psi^{\curlyvee}_{hv} \to \Box^{\leq 2}(\Diamond^{\curlyvee}_h \Diamond^{\curlyvee}_h \mathsf{t} \to \Diamond^{\curlyvee}_h \mathsf{t}),\\
\textsc{AxTr}^{\curlyvee}_2 &= \textsc{comp}(\mathsf{p}, \mathsf{q}) \wedge \psi^{\curlyvee}_{hv} \to \Box^{\leq 2}(\Diamond^{\curlyvee}_v \Diamond^{\curlyvee}_v \mathsf{t} \to \Diamond^{\curlyvee}_v \mathsf{t}),\\
\textsc{AxCR}^{\curlyvee} &= \textsc{comp}(\mathsf{p}, \mathsf{q}) \wedge \psi^{\curlyvee}_{hv} \to \Box^{\leq 2}(\Diamond^{\curlyvee}_h \Box^{\curlyvee}_v \mathsf{t} \to \Box^{\curlyvee}_h \Diamond^{\curlyvee}_v \mathsf{t}),\\
\textsc{AxComm}^{\curlyvee} &= \textsc{comp}(\mathsf{p}, \mathsf{q}) \wedge \psi^{\curlyvee}_{hv} \to \Box^{\leq 2}(\Diamond^{\curlyvee}_h \Diamond^{\curlyvee}_v \mathsf{t} \leftrightarrow \Diamond^{\curlyvee}_v \Diamond^{\curlyvee}_h \mathsf{t}).
\end{aligned}$$

[2] See the proof of Theorem 2 in [7], where $\psi_{hv} \wedge [\varphi^\Theta]_{(\psi_h,\psi_v)}$ is the conjunction of the formulas denoted by φ_∞, φ_{grid}, and φ_Θ.

Lemma 5.11 *If* $\mathsf{F}_1, \mathsf{F}_2$ *are unimodal transitive locally one-component frames, then* $\mathsf{F}_1 \curlyvee \mathsf{F}_2 \models \mathrm{L}^{\curlyvee}_{min}$.

Proof. Due to Propositions 4.5, 4.9, and 4.11,

$$\mathsf{F}_1 \curlyvee \mathsf{F}_2 \models \{\Diamond^3 \mathsf{p} \to \Diamond^2 \mathsf{p}, \mathrm{AxComp}_2, \mathrm{AxCov}\}.$$

Due to Proposition 2.1 and Lemmas 5.6, 3.10,

$$\mathsf{F}_1 \curlyvee \mathsf{F}_2 \models \{\mathrm{AxTr}_1^{\curlyvee}, \mathrm{AxTr}_2^{\curlyvee}, \mathrm{AxCR}^{\curlyvee}, \mathrm{AxComm}^{\curlyvee}\}.$$

□

Proposition 5.12 *Let* $\mathsf{F} = (W, R_1, R_2)$ *be a* [K4, K4]*-frame,* $\mathsf{M} = (\mathsf{F}, \theta)$, $\mathsf{G} = (W, \bar{R}^{\mathsf{M}}_h, \bar{R}^{\mathsf{M}}_v)$. *Then* $\bar{R}^{(\mathsf{G},\theta)}_h = \bar{R}^{\mathsf{M}}_h$, $\bar{R}^{(\mathsf{G},\theta)}_v = \bar{R}^{\mathsf{M}}_v$.

Proof. Put $\mathsf{N} = (\mathsf{G}, \theta)$.

$u\bar{R}^{\mathsf{M}}_{h,0}v$ iff $uR_1v \& (\mathsf{M}, u \models \mathsf{h} \Leftrightarrow \mathsf{M}, v \models \neg\mathsf{h})$ iff $u\bar{R}^{\mathsf{M}}_{h,0}v \& (\mathsf{N}, u \models \mathsf{h} \Leftrightarrow \mathsf{N}, v \models \neg\mathsf{h})$. It follows that $\bar{R}^{\mathsf{M}}_{h,0} = \bar{R}^{\mathsf{N}}_{h,0}$. Similarly, $\bar{R}^{\mathsf{M}}_{v,0} = \bar{R}^{\mathsf{N}}_{v,0}$. □

Lemma 5.13 *Let* $\mathsf{F} = (W, R) \models \mathrm{L}^{\curlyvee}_{min}$, w *be a root of* F, $\theta : PV \to \mathcal{P}(W)$, $\mathsf{M} = (\mathsf{F}, \theta)$, $\mathsf{G} = (W, \hat{R}^{\mathsf{M}}_h, \hat{R}^{\mathsf{M}}_v)$, $\mathsf{N} = (\mathsf{G}, \theta)$. *Suppose that* $\mathsf{M}, w \models \psi^{\curlyvee}_{hv} \wedge \mathrm{COMP}(\mathsf{p}, \mathsf{q})$. *Then we have:*

(i) $\psi^{\curlyvee}_h \overset{\mathsf{M}}{\to} \hat{R}^{\mathsf{M}}_h$, $\psi^{\curlyvee}_v \overset{\mathsf{M}}{\to} \hat{R}^{\mathsf{M}}_v$;

(ii) G *is a* [K4, K4]*-frame;*

(iii) $\mathsf{N}, w \models \psi_{hv}$;

(iv) *if* $\varphi \in ML_2$, $PV(\varphi) \cap \{\mathsf{p}, \mathsf{q}, \mathsf{r}\} = \varnothing$, *then for any* $u \in W$

$$\mathsf{M}, u \models [\varphi]_{(\psi^{\curlyvee}_h, \psi^{\curlyvee}_v)} \;\Leftrightarrow\; \mathsf{N}, u \models [\varphi]_{(\psi_h, \psi_v)}.$$

Proof. Put $V_1 = \theta(\mathsf{p})$, $V_2 = \theta(\mathsf{q})$. By Proposition 4.12, $comp(w) = \{V_1, V_2\}$. It follows that $\theta(\mathsf{p})$, $\theta(\mathsf{q})$, $\theta(\mathsf{r})$ are pairwise disjoint, so (i) follows from Lemma 5.5. (ii) follows from (i), Lemma 5.11, and Lemma 3.10.

Let us check (iii). Put $R_1 = \bar{R}^{\mathsf{N}}_h$, $R_2 = \bar{R}^{\mathsf{N}}_v$. It follows that if $u \in R_2(R_1(w))$, then $u \notin V_2$, and if $u \in R_1(R_2(w))$, then $u \notin V_1$. Due to the commutativity, we have

$$R_1(R_2(w)) \subseteq R^2(w) - R(w). \tag{6}$$

Let $u \in R_2(R_1(w))$, $\mathsf{N}, u \models \mathsf{h}^{\varepsilon} \vee \Diamond \mathsf{h}^{\varepsilon}$ for some $\varepsilon \in \{0, 1\}$. Then $u_0 R_2 u$ and $\mathsf{N}, u' \models \mathsf{h}^{\varepsilon}$ for some $u_0 \in V_1$, $u_1 \in R_2^{=}(u)$. Thus $\mathsf{M}, u_0 \models \Diamond_{\neg\mathsf{p}} \mathsf{h}^{\varepsilon} \to \Box_{\neg\mathsf{p}} \mathsf{h}^{\varepsilon}$. Suppose $v \in R_2(u)$. Since R_2 is transitive, $u_1, v \in R_2(R_1(w))$. Due to (6), $u_1, v \notin \theta(\mathsf{p})$. Recall that $R \supseteq R_2$, thus $\mathsf{M}, u_0 \models \Diamond_{\neg\mathsf{p}} \mathsf{h}^{\varepsilon}$, so $\mathsf{M}, u_0 \models \Box_{\neg\mathsf{p}} \mathsf{h}^{\varepsilon}$ and $\mathsf{N}, v \models \mathsf{h}^{\varepsilon}$. It follows that $\mathsf{N}, u \models \bigwedge\limits_{\varepsilon=0,1} (\mathsf{h}^{\varepsilon} \vee \Diamond_2 \mathsf{h}^{\varepsilon} \to \Box_2 \mathsf{h}^{\varepsilon})$.

Analogously, $\mathsf{N}, u \models \bigwedge\limits_{\varepsilon=0,1} (\mathsf{v}^{\varepsilon} \vee \Diamond_1 \mathsf{v}^{\varepsilon} \to \Box_1 \mathsf{v}^{\varepsilon})$ for any $u \in R_2(R_1(w))$, which implies (iii).

By Proposition 5.12, $\hat{R}^{\mathsf{M}}_h = \bar{R}^{\mathsf{N}}_h$, $\hat{R}^{\mathsf{M}}_v = \bar{R}^{\mathsf{N}}_v$. Due to (i) we have $\psi^{\curlyvee}_h \overset{\mathsf{M}}{\to} \bar{R}^{\mathsf{N}}_h$, $\psi^{\curlyvee}_v \overset{\mathsf{M}}{\to} \bar{R}^{\mathsf{N}}_v$. Recall that $\psi_h \overset{\mathsf{N}}{\to} \bar{R}^{\mathsf{N}}_h$, $\psi_v \overset{\mathsf{N}}{\to} \bar{R}^{\mathsf{N}}_v$. By Lemma 3.6 we obtain (iv). □

Theorem 5.14 *If a unimodal logic* L *contains* $L^{\curlyvee}_{min}$ *and there exists an axis-frame* F *such that* $\mathsf{F} \curlyvee \mathsf{F} \models \mathrm{L}$, *then* L *has no finite model property.*

Proof. In [7], it was shown that there exists a formula $\varphi^{diag} \in ML_2$ such that

(i) if F is a transitive 1-frame with a root x containing an infinite descending chain, then $\psi_{hv} \wedge [\varphi^{diag}]_{(\psi_h,\psi_v)}$ is satisfiable at (x, x) in $\mathsf{F} \times \mathsf{F}$,

(ii) if G is a [K4, K4]-frame and $\psi_{hv} \wedge [\varphi^{diag}]_{(\psi_h,\psi_v)}$ is satisfiable in G, then G is infinite.[3]

Due to Lemma 5.7, $\textsc{comp}(\mathsf{p},\mathsf{q}) \wedge \psi^{\curlyvee}_{hv} \wedge [\varphi^{diag}]_{(\psi^{\curlyvee}_h,\psi^{\curlyvee}_v)}$ is satisfiable in $\mathsf{F} \curlyvee \mathsf{F}$.

On the other hand, if $\textsc{comp}(\mathsf{p},\mathsf{q}) \wedge \psi^{\curlyvee}_{hv} \wedge [\varphi^{diag}]_{(\psi^{\curlyvee}_h,\psi^{\curlyvee}_v)}$ is satisfiable in a finite L-frame G, then, by Lemma 5.13, $\psi_{hv} \wedge [\varphi^{diag}]_{(\psi_h,\psi_v)}$ is satisfiable in a finite [K4, K4]-frame, which is a contradiction. □

Corollary 5.15 *If a class* $\mathcal{F}$ *of transitive locally one-component frames contains an axis-frame, then* $\mathbf{L}(\mathcal{F} \curlyvee \mathcal{F})$ *does not have the finite model property.*

Example 5.16 The logics $\mathbf{L}((\mathbb{R}, <) \curlyvee (\mathbb{R}, <))$ and $\mathbf{L}((\mathbb{Q}, <) \curlyvee (\mathbb{Q}, <))$ does not have the finite model property.

6 $\langle \overline{B} \vee \overline{E} \rangle$-fragment of HS

In this section we consider modal logics, where modal operators are interpreted by relations between intervals. For known results on these logics see [3,2,4].

We show how Theorems 5.9 and 5.14 can be used to obtain negative results for logics of intervals.

For a (strict or non-strict) partial order $\mathsf{F} = (W, R)$, let $Ints(W)$ denote *the set of all (non-strict) intervals over* F: $Ints(W) = \{(a, b) \mid aR^{=}b\}$. For intervals (a, b), (c, d),

$$(a, b)R_{\langle B\rangle}(c, d) \text{ iff } a = c \ \wedge \ dRb;$$
$$(a, b)R_{\langle E\rangle}(c, d) \text{ iff } aRc \ \wedge \ b = d;$$
$$R_{\langle B \vee E\rangle} = R_{\langle B\rangle} \cup R_{\langle E\rangle}, \quad R_{\langle \overline{B} \vee \overline{E}\rangle} = R^{-1}_{\langle B \vee E\rangle}.$$

For a partial order F, let $\mathbf{L}_{\langle \overline{B} \vee \overline{E}\rangle}(\mathsf{F})$ denote $\mathbf{L}(Ints(W), R_{\langle \overline{B} \vee \overline{E}\rangle})$. For a class F of partial orders, put $\mathbf{L}_{\langle \overline{B} \vee \overline{E}\rangle}(\mathsf{F}) = \bigcap_{\mathsf{F} \in \mathcal{F}} \mathbf{L}_{\langle \overline{B} \vee \overline{E}\rangle}(\mathsf{F})$.

Lemma 6.1 *Let* $\mathsf{F} = (W, R)$ *be a partial order,* $\mathsf{G} = (W, R^{-1})$. *Then* $\mathbf{L}_{\langle \overline{B} \vee \overline{E}\rangle}(\mathsf{F}) = \bigcap_{aR^{=}b} \mathbf{L}(\mathsf{G}^a \curlyvee \mathsf{F}^b)$.

Proof. The statement of the lemma is based on the following observation (see e.g. [10]): $(Ints(W), R_{\langle \overline{E}\rangle}, R_{\langle \overline{B}\rangle})^i = \mathsf{G}^a \times \mathsf{F}^b$ for any $i = (a, b) \in Ints(W)$; therefore,

$$(Ints(W), R_{\langle \overline{B} \vee \overline{E}\rangle})^i = \mathsf{G}^a \curlyvee \mathsf{F}^b. \tag{7}$$

---

[3] In [7], $\psi_{hv} \wedge [\varphi^{diag}]_{(\psi_h,\psi_v)}$ is denoted by ψ_∞.

We have:

$$\mathbf{L}(Ints(W), R_{\langle \overline{B} \vee \overline{E} \rangle}) = \bigcap_{aR^=b} (Ints(W), R_{\langle \overline{B} \vee \overline{E} \rangle})^{(a,b)} = \bigcap_{aR^=b} \mathbf{L}(\mathsf{G}^a \curlyvee \mathsf{F}^b).$$

□

Theorem 6.2 *Let $\mathcal{F}$ be a class of strict linear orders such that (W, R) is isomorphic to (W, R^{-1}) and $(W, R)^a$ is isomorphic to $(W, R)^b$ for any $(W, R) \in \mathcal{F}$, $a, b \in W$. If $\mathcal{F}$ contains an axis-frame, then the following holds:*

(i) $\mathbf{L}_{\langle \overline{B} \vee \overline{E} \rangle}(\mathcal{F}) = \mathbf{L}(\mathcal{F}) \curlyvee \mathbf{L}(\mathcal{F})$;

(ii) $\mathbf{L}_{\langle \overline{B} \vee \overline{E} \rangle}(\mathcal{F})$ *is undecidable;*

(iii) $\mathbf{L}_{\langle \overline{B} \vee \overline{E} \rangle}(\mathcal{F})$ *lacks the finite model property.*

Proof. For a frame $\mathsf{F} = (W, R)$, we have

$$\mathbf{L}(\mathsf{F}) \curlyvee \mathbf{L}(\mathsf{F}) = \bigcap_{a,b \in W} \mathbf{L}(\mathsf{F}^a \curlyvee \mathsf{F}^b) = \bigcap_{aR^=b} \mathbf{L}(\mathsf{G}^a \curlyvee \mathsf{F}^b).$$

Due to (7), $\mathbf{L}(\mathsf{F}) \curlyvee \mathbf{L}(\mathsf{F}) = \mathbf{L}_{\langle \overline{B} \vee \overline{E} \rangle}(\mathsf{F})$, that proves (i). Now (ii) and (iii) follow from Theorems 5.9, 5.14. □

Corollary 6.3 *The logics $\mathbf{L}_{\langle \overline{B} \vee \overline{E} \rangle}(\mathbb{R}, <)$ and $\mathbf{L}_{\langle \overline{B} \vee \overline{E} \rangle}(\mathbb{Q}, <)$ are undecidable and lack the finite model property.*

7 Further results and open questions

The main results of the paper are stated in Theorems 5.9 and 5.14. At the same time, the method of proof is presented in Lemmas 5.6, 5.7 and 5.13. Basing on this method, many other results on products can be transferred to the unimodal case. In particular, various not recursively enumerable $\curlyvee$-products and fragments of HS can be constructed using Theorems 3 and 4 from [7] and Lemma 5.7.

There are many questions about logical properties of $\curlyvee$-products. Let us formulate some of them.

As it was shown in Section 5, some special axioms appear from $\curlyvee$-products of transitive locally-one component frames. However, no complete axiomatizations for logics of this kind are known.

The logic $\mathrm{K4} \curlyvee \mathrm{K4}$ is of special interest. We know that $\Diamond^3 p \to \Diamond^2 p \in \mathrm{K4} \curlyvee \mathrm{K4}$. Does it have the finite model property? Is it decidable? Is it equal to the logic $\mathrm{K} + \Diamond^3 p \to \Diamond^2 p$? Note that the finite model property (and, apparently, the decidability) of the latter logic is a long-standing open problem.

Another question was asked by one of anonymous referees: to give an example of decidable logic $\mathrm{L}_1 \curlyvee \mathrm{L}_2$ with undecidable $\mathrm{L}_1 \times \mathrm{L}_2$. Theorem 4.4 now gives the answer in the non-transitive case: if L_1 is an undecidable logic, $\mathrm{L}_2 = \mathrm{T}$, then $\mathrm{L}_1 \times \mathrm{L}_2$ is undecidable, and $\mathrm{L}_1 \curlyvee \mathrm{L}_2 = \mathrm{T}$. At the same time, the author was unsuccessful in finding such an example for transitive L_1 and L_2.

8 Acknowledgements

I would like to thank Valentin Shehtman, Stanislav Kikot and Andrey Kudinov for useful discussions.

Also, I would like to thank three anonymous referees for their essential remarks on the earlier version of the paper.

The work on this paper was supported by Poncelet Laboratory (UMI 2615 of CNRS and Independent University of Moscow) and by RFBR grant 06-01-72555.

References

[1] Blackburn, P., M. de Rijke and Y. Venema, "Modal Logic," Cambridge University Press, 2001.

[2] Bresolin, D., V. Goranko, A. Montanari and G. Sciavicco, *Propositional interval neighborhood logics: Expressiveness, decidability, and undecidable extensions*, Ann. Pure Appl. Logic **161** (2009), pp. 289–304.

[3] Bresolin, D., D. Monica, V. Goranko, A. Montanari and G. Sciavicco, *Decidable and undecidable fragments of Halpern and Shoham's interval temporal logic: Towards a complete classification*, in: *LPAR '08: Proceedings of the 15th International Conference on Logic for Programming, Artificial Intelligence, and Reasoning* (2008), pp. 590–604.

[4] Bresolin, D., D. D. Monica, V. Goranko, A. Montanari and G. Sciavicco, *Undecidability of the logic of overlap relation over discrete linear orderings*, Electronic Notes in Theoretical Computer Science **262** (2010), pp. 65 – 81, proceedings of the 6th Workshop on Methods for Modalities (M4M-6 2009).

[5] Gabbay, D., A. Kurucz, F. Wolter and M. Zakharyaschev, "Many-dimensional modal logics: theory and applications," Elsevier, 2003.

[6] Gabbay, D., V. Shehtman and D. Skvortsov, "Quantification in Nonclassical Logic," Elsevier, 2009.

[7] Gabelaia, D., A. Kurucz, F. Wolter and M. Zakharyaschev, *Products of 'transitive' modal logics*, Journal of Symbolic Logic **70** (2005), pp. 993–1021.

[8] Halpern, J. Y. and Y. Shoham, *A propositional modal logic of time intervals*, Journal of the ACM **38** (1996), pp. 279–292.

[9] Kudinov, A., *Topological modal logics with difference modality*, in: G. Governatori, I. M. Hodkinson and Y. Venema, editors, *Advances in Modal Logic* (2006), pp. 319–332.

[10] Marx, M. and Y. Venema, "Multi-dimensional modal logic," Kluwer Academic Publishers, 1997.

[11] Rabinovich, A., *On compositional method and its limitations*, Technical report, University of Edinburgh (2001), eDI-INF-RR-0035.

[12] Shapirovsky, I. and V. Shehtman, *Modal logics of regions and Minkowski spacetime*, Journal of Logic and Computation **15** (2005), pp. 559–574.

[13] Shehtman, V., *Derived sets in Euclidean spaces and modal logic*, Technical report, University of Amsterdam (1990).

A Remark on Propositional Kripke Frames Sound for Intuitionistic Logic

Dmitrij Skvortsov

All-Russian Institute of Scientific and Technical Information, VINITI
Russian Academy of Science
Usievicha, 20, Moscow, Russia, 125190

Abstract

Usually, in the Kripke semantics for intuitionistic propositional logic (or for superintuitionistic logics) partially ordered frames are used. Why? In this paper we propose an intrinsically intuitionistic motivation for that. Namely, we show that every Kripke frame (with an arbitrary accessibility relation), whose set of valid formulas is a superintuitionistic logic, is logically equivalent to a partially ordered Kripke frame.

Keywords: Intuitionistic propositional logic, intermediate logics, Kripke semantics, logical soundness.

While considering the Kripke semantics for intuitionistic propositional logic (or for superintuitionistic logics), only partially ordered (p.o.) Kripke frames are usually used. It is well known that quasi-ordered (i.e., reflexive and transitive) frames can be accepted as well, but this is in essence just the same. In fact, the quotient (modulo the usual equivalence) of a quasi-ordered Kripke frame is a p.o. frame, and intuitionistic formulas 'do not notice' this transformation. On the other hand, it is known (cf. e.g. [5], [4]) that some non-quasi-ordered Kripke frames are sound for intuitionistic logic as well (via the usual definition of intuitionistic validity).

So the conventional restriction of all frames to partial orderings seems to be slightly ad hoc. The most common motivation [1] appeals to modal logic and to Gödel – Tarski

[1] An anonymous referee reminded an informal motivation for using partial orderings in intuitionistic semantics: possible worlds represent knowledge and the accessibility corresponds to acquiring knowledge. The inclusion of sets is definitely a partial ordering, but this informal motivation does not seem convincing to us by different reasons. E.g., the generally accepted 'monotonicity condition' (knowledge grows in future) is controversial, because it ignores delusions and mistakes. Nevertheless, the monotonicity is almost necessary for intuitionistic semantics, because it corresponds to an intuitionistic axiom (cf. [4] or Lemma 2.1(2) in Section 2.1).

translation of intuitionistic logic in logic **S4** (recall that Kripke frames sound for **S4** are exactly the quasi-ordered ones). But this motivation seems rather external for intuitionistic logic, and so it is not quite convincing (all the more, **S4** is not the weakest normal modal logic, in which intuitionistic logic can be embedded via the Gödel – Tarski translation, see [1], [2], and [3]).

One can try to find a more intrinsically intuitionistic argument to explain, why namely quasi-orderings (or equivalently, partial orderings) are in some sense immanent for a sound interpretation of intuitionistic, or of superintuitionistic logics. Here we propose such a motivation.

Namely, we show that every Kripke frame whose set of valid formulas is a superintuitionistic logic, is logically equivalent to a partially ordered frame.

Now let us turn to exact definitions.

1 Intuitionistic sound Kripke frames

We consider *superintuitionistic propositional logics* understood in the usual way, as sets of formulas containing all axioms of *intuitionistic* (or *Heyting*) *propositional logic* **H** and closed under modus ponens and the substitution rule. So **H** is the smallest superintuitionistic logic. It is well known that all consistent superintuitionistic logics are included in *classical logic* $\mathbf{C} = (\mathbf{H} + p\vee\neg p)$; these logics are also called *intermediate*.

For convenience, in addition to connectives $\&, \vee, \rightarrow$, we also use constants $\bot$ (the falsity) and $\top$ (the truth). We use the standard abbreviations $\neg A = (A \rightarrow \bot)$ and $(A\leftrightarrow B) = (A\rightarrow B)\&(B\rightarrow A)$.

Let Var be the set of variables and Im be the set of all implications.

Remark Obviously, one can eliminate the constant $\top$ by replacing it with $(\bot \rightarrow \bot)$ or with $(A \rightarrow A)$ for any formula A. Usually one can also eliminate $\bot$ by using the independent connective $\neg$; then $\bot$ is defined as $A\&\neg A$ or $\neg(A\rightarrow A)$ for an arbitrary A. However, this is not quite semantically adequate, as we will see later (cf. e.g. Example 2 in Section 1.3). On the other hand, our considerations can be easily transferred to the language with the basic connective $\neg$ instead of $\bot$ (with minor modifications[2]).

1.1 We consider propositional *Kripke frames*; such a frame F is a non-empty set W with an arbitrary binary relation R on W. We write $u\in F$ for $u\in W$.

A *valuation* (more precisely, an *intuitionistic valuation*) in a Kripke frame $F = (W, R)$ is a forcing relation $u \vDash A$ between points $u\in F$ and formulas A, satisfying the usual intuitionistic clauses (for all $u\in F$):

$$u \vDash (B\&C) \Leftrightarrow (u \vDash B)\,\&\,(u \vDash C), \quad u \vDash (B \vee C) \Leftrightarrow (u \vDash B) \vee (u \vDash C),$$

$$u \vDash (B\rightarrow C) \Leftrightarrow \forall v\,[\,uRv\ \&\ v \vDash B \Rightarrow v \vDash C\,], \quad u \nvDash \bot, \quad u \vDash \top,$$

[2] E.g. one can use $\top\rightarrow p$ at some point, where we use $\top\rightarrow\bot$, etc.

and the following condition (for variables $p \in \mathrm{Var}$):[3]

$$\text{(Atomic heredity)} \qquad \forall u, v \in F\, [\, uRv \;\&\; u \vDash p \;\Rightarrow\; v \vDash p\,].$$

Clearly, $u \vDash \neg B \Leftrightarrow \forall v \in R(u)\, [\, v \nvDash B\,]$ and $u \vDash (A \leftrightarrow B) \Leftrightarrow \forall v \in R(u)\, [\, v \vDash A \Leftrightarrow v \vDash B\,]$.

A formula A is said to be *true* under a valuation in F if $u \vDash A$ for any $u \in F$. A formula A is *valid* in a Kripke frame F if it is true under any valuation in F. Let $\mathbf{L}(F)$ denote the set of all formulas valid in F.

We call a frame F *intuitionistic sound* (or **H**-*sound*, for short) if $\mathbf{L}(F)$ is a superintuitionistic (or equivalently, an intermediate) logic. We say that F is *weakly* **H**-*sound* (or *w***H**-*sound*) if all intuitionistic theorems are valid in F, i.e., if $\mathbf{H} \subseteq \mathbf{L}(F)$.

Later on we will see that a w**H**-sound frame is **H**-sound iff $\mathbf{L}(F)$ is closed under modus ponens. We suppose that Kripke frames violating modus ponens are rather unsatisfactory for intuitionistic logic (even if they validate **H**).[4]

Note that $\mathbf{L}(F)$ is closed under modus ponens iff

$$(\top \to A) \in \mathbf{L}(F) \;\Rightarrow\; A \in \mathbf{L}(F) \quad \text{for any formula } A. \qquad \text{(MP)}$$

In fact, (MP) implies that $\quad B,\, (B \to A) \in \mathbf{L}(F) \;\Rightarrow\; A \in \mathbf{L}(F)$,
because if $B \in \mathbf{L}(F)$, then $\forall u (u \vDash B)$ and so $\forall u [\, u \vDash (B \to A) \;\Leftrightarrow\; u \vDash (\top \to A)\,]$ for every valuation in F.

By the way, the converse implication $\quad A \in \mathbf{L}(F) \;\Rightarrow\; (\top \to A) \in \mathbf{L}(F)$
obviously holds in every Kripke frame F; so the implication in the condition (MP) actually means the equivalence.

By the semantics generated by a class $\mathcal{K}$ of frames we mean the class of logics $\mathcal{S}(\mathcal{K}) = \{\, \mathbf{L}(F) \mid F \in \mathcal{K}\,\}$. So the class of **H**-sound Kripke frames generates the maximal possible Kripke semantics for superintuitionistic logics (with the usual definition of intuitionistic forcing).

The following simple technical lemma will be unexpectedly useful further on:

Lemma 1.1 *For any formula A there exists a formula*

$$A' = \mathop{\&}_i \bigvee_j A_{ij}, \quad \text{where } \forall i, j\, [\, A_{i,j} \in \mathrm{Im} \cup \mathrm{Var}\,] \qquad (*)$$

(or $A' = \bot$)[5] *such that:*

(1) $\mathbf{H} \vdash (A \leftrightarrow A')$;

(2) $u \vDash A \;\Leftrightarrow\; u \vDash A'$ *for every valuation in a frame*[6] *F and $u \in F$.*

So $\quad A \in \mathbf{L}(F) \;\Leftrightarrow\; A' \in \mathbf{L}(F) \quad$ for every frame F.

---

[3] Here we use the terminology from [5], [4].

[4] If one prefers, another motivation for the validity of modus ponens is related to the notion of strong soundness mentioned in Section 2.5.

[5] Note that $\top$ can be presented by implication $(\bot \to \bot)$, but $\bot$ is not semantically equivalent to $(\top \to \bot)$, as we will see later on (e.g. in a one-element irreflexive frame, cf. Example 2 in Section 1.3).

[6] Here we in general do not suppose that F is (weakly) **H**-sound.

Proof. Note that the formula A is a &,∨-combination of its subformulas from $\mathrm{Im} \cup \mathrm{Var} \cup \{\bot, \top\}$. We transform it into an equivalent conjunctive normal form (∗). Now (1) is obvious. And (2) holds, because the 'inner' connectives & and ∨ in a formula correspond, by the definition of valuation, to the 'external' conjunction and disjunction satisfying the distributivity laws etc. □

1.2 Let R^+ and R^* be the transitive closure of R (in a frame $F = (W, R)$) and the corresponding quasi-ordering, i.e.,

$$uR^+v \Leftrightarrow \exists n>0\,(u\,R^n\,v), \qquad uR^*v \Leftrightarrow (u=v) \vee (uR^+v).$$

For $u \in F$ define the *cone* $F^u = (W^u, R|W^u)$, where $W^u = R^*(u) = \{u\} \cup R^+(u)$. A set $W' \subseteq W$ is *open* if it is upward closed, i.e., $\forall u \in W' \forall v \in R(u)\,(v \in W')$,[7] or equivalently, iff $\forall u \in W'\,(W^u \subseteq W')$ (in other words, $W' = \bigcup(W^u : u \in W')$). Naturally, an open set W' gives rise to an *open subframe* $F' = (W', R|W')$ of F. Clearly, the restriction of a valuation in F to a cone F^u or to any open subframe is a valuation again (the inductive clauses are preserved obviously). So we conclude that

$$\mathbf{L}(F) \subseteq \mathbf{L}(F') \quad \text{for any open subframe } F' \text{ of } F,$$

and $$\mathbf{L}(F) = \bigcap(\mathbf{L}(F^u) : u \in F). \qquad \text{Thus}$$

(1) $(F \textit{ is } w\mathbf{H}\textit{-sound}) \Leftrightarrow \forall u \in F\,(F^u \textit{ is } w\mathbf{H}\textit{-sound})$;

(2) $\forall u \in F\,(F^u \textit{ is } \mathbf{H}\textit{-sound}) \Rightarrow (F \textit{ is } \mathbf{H}\textit{-sound})$,

the converse to (2) in general does not hold, as we shall see later on.

It is well known that all p.o. frames are **H**-sound; moreover, quasi-ordered frames are **H**-sound as well. Actually, for a quasi-ordered Kripke frame its quotient modulo the equivalence relation $(u \equiv v) \Leftrightarrow (uRv) \,\&\, (vRu)$ is a p.o. frame $S[F] = (F/\equiv)$ (the *skeleton* of F). Clearly, F and $S[F]$ have in essence the same valuations (modulo $\equiv$); thus $\mathbf{L}(F) = \mathbf{L}(S[F])$ in this case.

Now we define the *skeleton* $S[F]$ for an arbitrary frame $F = (W, R)$ as the skeleton of the associated quasi-ordered frame $F^* = (W, R^*)$. In other words, $S[F]$ is the quotient $(W/\equiv)$ modulo the equivalence relation $\equiv$ given by the following three equivalent conditions:

$$(u \equiv v) \text{ iff } [(u=v) \vee (uR^+v \,\&\, vR^+u)] \text{ iff } (v \in F^u \,\&\, u \in F^v) \text{ iff } (F^u = F^v),$$

partially ordered by the relation

$$(u/\equiv)\,R_S\,(v/\equiv) \Leftrightarrow (uR^*v) \Leftrightarrow (v \in F^u).$$

Clearly $(u \equiv v) \Rightarrow (u \vDash p \Leftrightarrow v \vDash p)$ for a valuation in F,
so the following valuation in $S[F]$ is well-defined (and satisfies the atomic heredity):

$$(u/\equiv) \vDash_S p \Leftrightarrow u \vDash p \qquad \text{for variables } p.$$

Therefore there exists the natural one-to-one correspondence between valuations in F and in $S[F]$. However, in general we cannot guarantee that

$$(u/\equiv) \vDash_S A \Leftrightarrow u \vDash A \tag{S}$$

[7] Obviously, the cone W^u is the least open set containing u.

for non-atomic A, and the equality $\mathbf{L}(F) = \mathbf{L}(S[F])$ in general does not hold. Namely, definitely $\mathbf{L}(F) \neq \mathbf{L}(S[F])$ for every frame F that is not **H**-sound. Also there exist **H**-sound frames such that $\mathbf{L}(F) \neq \mathbf{L}(S[F])$; we will give an example in Section 2.4 (Appendix).

1.3 It is known that there exist **H**-sound not quasi-ordered Kripke frames. Let us begin with some useful and instructive examples.

Example 1 Let F be a two-element frame $W = \{v_1, v_2\}$ with irreflexive and non-transitive relation $R = \{\langle v_1, v_2\rangle, \langle v_2, v_1\rangle\}$, see Figure 1.

Clearly its skeleton $S[F]$ is a one-element reflexive frame, and one can easily check the equivalence (S) by induction on the complexity of A. So we obtain that $\mathbf{L}(F) = \mathbf{L}(S[F]) = \mathbf{C}$, and thus F is **H**-sound.

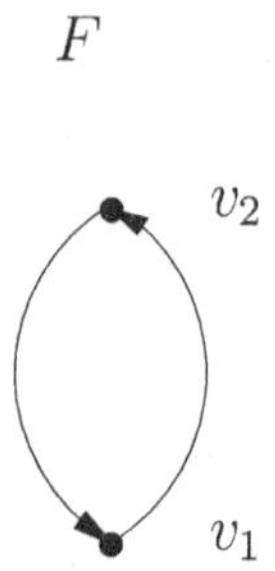

Fig. 1.

Similarly, one can take an arbitrary p.o. frame F' and replace some of its points u by cycles $u_1 \, R \, u_2 \, R \ldots R \, u_n \, R \, u_1$; these cycles can be chosen either reflexive or non-reflexive, transitive or non-transitive; and anyway we obtain a frame F such that $S[F] = F'$, $\mathbf{L}(F) = \mathbf{L}(F')$.

Example 2 Let $\Pi_0 = \{\pi_0\}$ be irreflexive one-element frame (with empty R), see Figure 2.

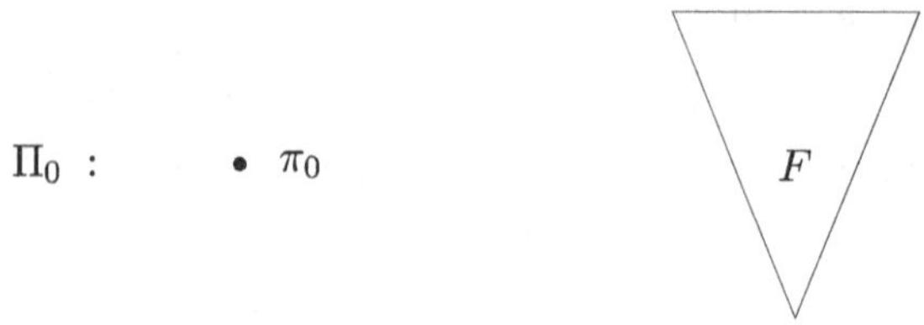

Fig. 2.

Clearly all implications are valid in Π_0, i.e., $\mathrm{Im} \subset \mathbf{L}(\Pi_0)$.[8]

[8] So $\neg(A \to A), (A \to A)\&\neg(A \to A) \in \mathbf{L}(\Pi_0)$, and both these formulas are not semantically equivalent

Lemma 1.2 *(1) Let A' be a formula (*) from Lemma 1.1. Then:*

$$A' \in \mathbf{L}(\Pi_0) \Leftrightarrow \forall i \, \exists j \, (A_{ij} \in \mathrm{Im}).$$

(2) $\mathbf{C} \subset \mathbf{L}(\Pi_0)$.

Proof. The 'if' part of (1) is obvious. Also, a formula $\bigvee_j p_j$ with variables p_j clearly does not belong to $\mathbf{L}(\Pi_0)$ and is not classically valid. Hence the 'only if' part of (1) follows as well. And if $A' \in \mathbf{C}$, then $A' \in \mathbf{L}(\Pi_0)$ by (1). □

Thus the frame Π_0 is w**H**-sound. However it is not **H**-sound; in fact, modus ponens fails in $\mathbf{L}(\Pi_0)$, since $(\top \to \bot) \in \mathbf{L}(\Pi_0)$ and $\bot \notin \mathbf{L}(\Pi_0)$.

On the other hand, for any partially ordered, and moreover, for any **H**-sound frame F the disjoint union (F, Π_0) of F and Π_0 (see Figure 2) is **H**-sound, since

$$\mathbf{L}(F, \Pi_0) = \mathbf{L}(F) \cap \mathbf{L}(\Pi_0) = \mathbf{L}(F)$$

(recall that here $\mathbf{L}(F) \subseteq \mathbf{C} \subset \mathbf{L}(\Pi_0)$).

Example 3 Let F be a p.o. frame. Take a frame $(\Pi_0 + F)$ obtained by adding a minimal irreflexive point π_0 to F (see Figure 3). Now, to extend relation R from F to $(\Pi_0 + F)$, we have to define (in an arbitrary way) a non-empty set $R(\pi_0) \subseteq F$ (what does π_0 'see'?).[9]

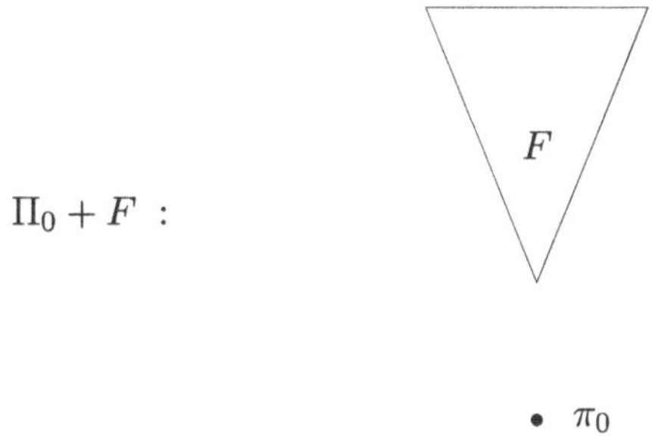

Fig. 3.

One can put, say, $R(\pi_0) = F$; then R is transitive on $(\Pi_0 + F)$. And moreover, obviously, R is transitive iff $R(\pi_0)$ is an open subset of F.

Clearly, for any choice of $R(\pi_0)$, all implications valid in F are valid in $(\Pi_0 + F)$ as well, i.e., $\mathrm{Im} \cap \mathbf{L}(\Pi_0 + F) = \mathrm{Im} \cap \mathbf{L}(F)$. So we obtain:

Lemma 1.3 $\mathbf{H} \subseteq \mathbf{L}(\Pi_0 + F)$, *i.e.,* $(\Pi_0 + F)$ *is* w**H***-sound for any p.o. frame F (and for any choice of $R(\pi_0)$).*

---

to $\bot$ in this frame (cf. Remark at the beginning of Section 1). Moreover, one can easily see that there does not exist a &,∨,→,¬-formula equivalent to $\bot$; in fact, any formula of this kind is true at π_0 if all variables are true, while $\bot$ is false (cf. also Lemma 1.1).

[9] If $R(\pi_0)$ is empty, then we have a disjoint union (F, Π_0).

Proof. One can show that $\mathbf{H} \vdash A \Rightarrow A \in \mathbf{L}(\Pi_0 + F)$,
by induction on the complexity of A (not by induction on its proof!); for the case $A = A_1 \vee A_2$ use the disjunction property of **H**. □

Thus, if $\mathbf{L}(F) = \mathbf{H}$, then $\mathbf{L}(\Pi_0 + F) = \mathbf{H}$, and so $(\Pi_0 + F)$ is **H**-sound.

Similarly, if the logic $\mathbf{L}(F)$ has the disjunction property, then one can show that $\mathbf{L}(\Pi_0 + F) = \mathbf{L}(F)$. On the other hand, let F be a one-element reflexive frame, then $(\Pi_0 + F)$ is not **H**-sound; in fact, modus ponens fails, since

$$(\top \to p \vee \neg p) \in \mathbf{L}(\Pi_0 + F) \quad \text{and} \quad p \vee \neg p \notin \mathbf{L}(\Pi_0 + F).$$

Similarly, one can show that $(\Pi_0 + F)$ is not **H**-sound for any rooted finite F, and moreover, for any rooted F of finite height (if $\pi_0 R u_0$, u_0 being the root of F). [10]

The considered examples show us that there exist very simple and small (actually, $\leq$3-element) **H**-sound frames, which are: (a) non-transitive and non-reflexive, (b) transitive and non-reflexive, (c) non-transitive and reflexive. Also there exist frames of kinds (a) and (b) that are w**H**-sound, but not **H**-sound (on the other hand, every reflexive w**H**-sound frame is **H**-sound, cf. Proposition 1 in Section 2.3).

In Section 2, basing on the presented examples, we shall describe the classes of w**H**-sound and **H**-sound frames, and establish Reducibility Theorem:

Theorem *For every* **H***-sound Kripke frame F there exists a partially ordered frame F' such that* $\mathbf{L}(F) = \mathbf{L}(F')$.

This statement shows that the semantics of all **H**-sound Kripke frames equals the usual semantics of partially ordered Kripke frames. In other words, non-transitive or non-reflexive frames give nothing new for superintuitionistic logics (if we deal with the usual definition of intuitionistic forcing).

2 Description of intuitionistic sound frames

In this section we describe the classes of **H**-sound and w**H**-sound frames. As the main result, we obtain Reducibility Theorem.

2.1 Let $F = (W, R)$ be a Kripke frame. We call $u \in W$ a *parasite* if $\neg\exists w\,(wRu)$; in other words, parasites are minimal irreflexive points 'stuck' to the frame from below. A parasite u is *isolated* if $R(u) = \varnothing$, i.e., if its cone $F^u = \{u\}$ is isomorphic to Π_0 (see Example 2). Let $\Pi[F]$ and $\Pi_0[F]$ be the sets of all parasites and of isolated parasites from F, respectively. The *essential part* of F is $E[F] = F \setminus \Pi[F]$; obviously, $E[F]$ is an open subframe of F.

If $E[F] = \varnothing$, then $F = \Pi_0[F]$ and $\mathbf{L}(F) = \mathbf{L}(\Pi_0)$, see Example 2.

A frame F is called *co-serial* if $\forall u \in F\, \exists w \in F\,(wRu)$, i.e., iff $\Pi[F] = \varnothing$ (or equivalently, $E[F] = F$).

[10] All mentioned claims will be proved later, in Section 2.4 (Appendix).

Now we recall (and slightly reformulate) some notions introduced in [5]. A Kripke frame $F = (W, R)$ is called *weakly reflexive* if $\forall u \in F\,(uR^{+}u)$ and *weakly transitive* if $\forall u, v \in F\,[\,(uR^2 v) \Rightarrow \exists w\,(uRw \equiv v)\,]$. A frame F is *weakly quasi-ordered* if it is weakly reflexive and weakly transitive. Clearly, reflexivity, transitivity, and quasi-ordering imply the corresponding weak properties. Every weakly reflexive frame F is co-serial. Also note that a transitive frame is weakly reflexive iff it is reflexive (since $R^{+} = R$ in a transitive frame).

Remark By the way, we can give a 'uniform' presentation of these notions.
Put $[\Omega_n]$: $\forall u, v \in F\,[\,(uR^n v) \Rightarrow \exists w\,(uRw \equiv v)\,]$ for $n \geq 0$.

Clearly $[\Omega_2]$ is the definition of weak transitivity. Moreover,

(I) F is weakly transitive iff $\forall n > 0\,[\Omega_n]$, i.e.,

$\forall u, v \in F\,[\,(uR^{+}v) \Rightarrow \exists w\,(uRw \equiv v)\,]$.

In fact, we can establish $[\Omega_n]$ by induction on n. The case $n = 1$ is obvious; take $w = v$. The induction step is straightforward, as well. Namely, if $uR^{n-1}u'R^2 v$, then we find w' such that $u'Rw' \equiv v$,and by induction hypothesis $uR^n w'$ implies $\exists w\,[uRw \equiv w'(\equiv v)]$.

Also we have

(II) F is weakly reflexive iff $[\Omega_0]$, i.e.,

$\forall u \in F\,\exists w\,(uRw \equiv u)$.

In fact, if $uR^{+}u$, i.e., $uRwR^n u$ for some w and $n \geq 0$, then $w \equiv u$.

Therefore,

(III) F is weakly quasi-ordered iff $\forall n \geq 0\,[\Omega_n]$, i.e.,

$\forall u, v \in F\,[\,(uR^{*}v) \Rightarrow \exists w\,(uRw \equiv v)\,]$.

Recall that R is transitive iff $R^{+} = R$ and R is a quasi-ordering iff $R^{*} = R$. So we see that a frame F is weakly reflexive, weakly transitive, or weakly quasi-ordered iff the composition of R and $\equiv$ is reflexive, transitive, or quasi-ordered, respectively.[11] In other words, weak reflexivity means 'reflexivity up to equivalence' (and similarly for weak transitivity).

Let A be a formula. A valuation $\models$ in a frame F is called *A-hereditary* if

$$\forall u, v \in F\,[\,uRv \,\&\, u \models A \;\Rightarrow\; v \models A\,], \qquad (A\text{-heredity})$$

or equivalently, $\forall u \in F\,[\,(u \models A) \Rightarrow \forall v \in F^u\,(v \models A)\,]$.
By definition, all our valuations are p-hereditary for variables p. A frame F is called *A-hereditary* if all valuations in F are A-hereditary; F is called **H**-*hereditary* if it is A-hereditary for all formulas A.

Lemma 2.1 *Let $F = (W, R)$ be a Kripke frame.*

(1) $((\top \to p) \to p) \in \mathbf{L}(F)$ iff $E[F]$ is weakly reflexive.

(2) $(A \to (\top \to A)) \in \mathbf{L}(F)$ iff $E[F]$ is A-hereditary (for a formula A).

(3) The following conditions are equivalent:

[11] By the way, (I) implies that $(\equiv) \circ R \circ (\equiv)$ equals $R \circ (\equiv)$ etc. (in a weakly transitive frame).

(i) F is **H**-*hereditary;*

(ii) F is $(p \to q)$-*hereditary;*

(iii) F is weakly transitive.

Therefore, *if F is w***H**-*sound, then* $E[F]$ *is weakly quasi-ordered.*

Note that only ($\Rightarrow$)-parts of (1) and (2) are used for the latter statement; actually, the converse statements are only to complete the picture.[12]

Proof. (1) ($\Rightarrow$). Let wRu and $\neg(uR^{+}u)$. Take a valuation in F such that $v \vDash p \Leftrightarrow uR^{+}v$. Then $u \vDash (\top \to p)$ and $u \nvDash p$, thus $w \nvDash (\top \to p) \to p$.

($\Leftarrow$). Let $E[F]$ be weakly reflexive, and wRu, $u \vDash (\top \to p)$ (for a valuation in F). Then $v \vDash p$ for any $v \in R(u)$, and thus by (Atomic heredity), for any $v \in R^{+}(u)$. Also $uR^{+}u$ since $u \in E[F]$, and hence $u \vDash p$.

(2) ($\Rightarrow$). Let $u \in E[F]$, uRv, and $u \vDash A$; also let wRu and $w \vDash (A \to (\top \to A))$ (for a valuation in F). Then $u \vDash (\top \to A)$, and so $v \vDash A$.

($\Leftarrow$). Let wRu and $u \vDash A$. Then $u \in E[F]$, and so by A-heredity it follows that $\forall v \in R(u)\,(v \vDash A)$, i.e., $u \vDash (\top \to A)$.

(3) (ii) $\Rightarrow$ (iii). Let $uRv'Rv$. Consider a valuation in F such that

$$w \vDash p \Leftrightarrow w \in F^{v} \qquad \text{and} \qquad w \nvDash q \Leftrightarrow v \in F^{w}.$$

Then $v \vDash p$, $v \nvDash q$, thus $v' \nvDash (p \to q)$, and so, by $(p \to q)$-heredity, $u \nvDash (p \to q)$. Hence there exists $w \in R(u)$ such that $w \vDash p$ and $w \nvDash q$, i.e., $w \equiv v$.

(iii) $\Rightarrow$ (i). Let F be weakly transitive. We establish A-heredity by induction on the complexity of A. Clearly, it is sufficient to consider the induction step for $A = (A_1 \to A_2)$.

Let uRv', $v' \nvDash A$, i.e., $v \vDash A_1$ and $v \nvDash A_2$ for some $v \in R(v')$. Then by the weak transitivity, we have $w \in R(u)$ such that $w \equiv v$, i.e., $w \in F^{v}$ and $v \in F^{w}$. Hence by A_1-heredity and A_2-heredity, $w \vDash A_1$ and $w \nvDash A_2$. Thus $u \nvDash A$. □

2.2 We see that weak transitivity expresses heredity of valuations (for all formulas). Similarly, weak reflexivity is related to another property of valuations introduced in [5] and called converse heredity.[13] Namely, a frame F is *conversely A-hereditary* if the following condition holds (for any valuation in F):

$$\forall u \in F\, [\forall v \in R(u)\,(v \vDash A) \Rightarrow u \vDash A]. \qquad \text{(converse } A\text{-heredity)}$$

A frame F is *conversely* **H**-*hereditary* if this property holds for all formulas.

Clearly, converse A-heredity means that

$$[\, u \vDash (\top \to A) \Rightarrow u \vDash A, \quad \text{for every valuation} \vDash \text{and point } u\,].$$

Therefore: $((\top \to A) \to A) \in \mathbf{L}(F) \Leftrightarrow$ ($E[F]$ *is conversely A-hereditary*).[14]

---

[12] The statement (3) actually reformulates Proposition 15 from [5]; we repeat its short proof here, to make our exposition self-contained and to reveal that $(p \to q)$-heredity is sufficient for the weak transitivity. Similarly, (2) almost reformulates Proposition 9 from [4] (note that in [4] the heredity in $E[F]$ is called the conditional heredity in F); here we give a simple proof omitted in [4].

[13] We will not use this notion, but we mention it here to explain the sense of the notions we consider.

[14] Cf. Proposition 10 in [4]; note that there converse heredity in $E[F]$ is called conditional converse heredity in F.

Moreover, for any frame F:

F is weakly reflexive iff *F is conversely p-hereditary for variables p* (adapt the proof of Lemma 2.1(1),[15] or see Proposition 21 in [5]).

However, *weak reflexivity is not sufficient for converse* **H***-heredity.*

Namely, take e.g. a weakly reflexive frame F shown at Figure 4, a formula $A = \neg\neg p$, and a valuation F such that $u \vDash p \Leftrightarrow u = v_3$. Then $u \vDash \neg p \Leftrightarrow u = v_1$, hence $v_2 \nvDash A$ and $\forall u \in R(v_2)(=\{v_1, v_3\})\,[u \vDash A]$.

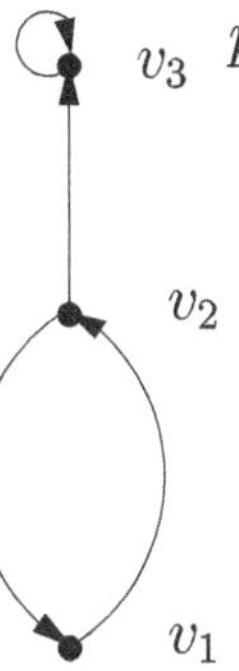

Fig. 4.

On the other hand, weak reflexivity together with weak transitivity imply converse **H**-heredity; in other words,

every weakly quasi-ordered frame is conversely **H***-hereditary.*

In fact,[16] let $\forall v \in R(u)\,(v \vDash A)$. Since $uR^{+}u$ (by weak reflexivity), take $v \in R(u)$ such that $vR^{n}u$ for some $n \geq 0$ (i.e., $u \in F^{v}$) . Then $v \vDash A$, and by A-heredity (see Lemma 2.1(3)) $u \vDash A$.

Therefore, we obtain:

Claim *For a frame F the following conditions are equivalent:*

(i) *F is* **H***-hereditary and conversely* **H***-hereditary;*

(ii) *F is $(p \to q)$-hereditary and conversely p-hereditary;*

(iii) *F is weakly quasi-ordered.*

Remark[17] By the way, Lemma 2.1(2) makes a hint, how one can try to modify the notion of intuitionistically acceptable valuations. Namely, one can admit *weak valuations*

[15] Lemma 2.1(1) readily gives the same equivalence for $E[F]$ (and so for any co-serial frame F). However the argument actually goes through for an arbitrary F as well.

[16] The statement actually follows from [5, Propositions 21 and 19], but in an indirect way, involving some complicated and refined notions, which relate to rudimentary Kripke models; so we give a simple and straightforward direct argument here.

[17] The reader may skip this remark and turn to Section 2.3 that contains the proof of Reducibility Theorem.

satisfying (Atomic heredity) only on $E[F]$, but not necessarily on F. In fact, this form of heredity actually corresponds to the well-known intuitionistic axiom $p \to (q \to p)$. However this modification gives nothing new, because the following statement holds:

(+) *for any*[18] *Kripke frame* F, *the set of formulas true under all weak valuations equals our* $\mathbf{L}(F)$, *defined via traditional valuations from Section 1.1.*

Proof. In fact, it is sufficient to show that every formula refuted under a weak valuation, does not belong to $\mathbf{L}(F)$, i.e., it is refuted by a valuation defined in Section 1.1. This statement easily follows from Lemma 1.1. Namely, let a weak valuation $\models$ in F satisfying (Atomic heredity) on $E[F]$, be given. Take a valuation $\models'$ in F equal to $\models$ on $E[F]$ and such that $u \not\models' p$ for all $u \in \Pi[F]$ and all variables p. Clearly, the valuation $\models'$ satisfies (Atomic heredity) on F. Now

$$u \models A \Leftrightarrow u \models' A \quad \text{for all } u \in E[F] \text{ and all formulas } A.$$

Also $u \models A \Leftrightarrow u \models' A$ for $u \in \Pi[F]$ and $A \in \text{Im}$. Hence we conclude that for every $u \in \Pi[F]$ and for every formula A' of the form (*) (from Lemma 1.1):

$$u \not\models A' \quad \text{implies} \quad u \not\models' A'.$$

Therefore: $u \not\models A \Rightarrow u \not\models' A$ for every A and $u \in F$. □

Actually, the statement (+) together with Lemma 2.1(2) allows to show (in the standard way) that

the set $\mathbf{L}(F)$ *is substitution closed for every* $w\mathbf{H}$*-sound frame* F.[19]

Therefore: F *is* $\mathbf{H}$*-sound* iff

(F *is* $w\mathbf{H}$*-sound and* $\mathbf{L}(F)$ *is closed under modus ponens*).

Later on we will obtain another proof of the latter equivalence (see Proposition 3).

2.3. Clearly, the frames considered in Example 1 are weakly quasi-ordered; so this example makes a hint, how to reduce an arbitrary weakly quasi-ordered frame to a quasi-ordered one. Namely, a weakly reflexive frame $F = (W, R)$ gives rise to a quasi-ordered frame $F^* = (W, R^*) = (W, R^+)$ (recall that R^+ is reflexive here, i.e., $R^+ = R^*$).

Lemma 2.2 *Let* $F = (W, R)$ *be a weakly quasi-ordered frame and* $F^* = (W, R^*)$ *be the corresponding quasi-ordered frame. Then a forcing relation* $\models$ *is a valuation in* F *iff it is a valuation in* F^*.

Hence $\mathbf{L}(F) = \mathbf{L}(F^*)$, and so this set is an intermediate logic. Therefore,

all weakly quasi-ordered frames are $\mathbf{H}$*-sound.*

Note that $S[F] = S[F^*]$, thus $\mathbf{L}(F) = \mathbf{L}(S[F])$ for a weakly quasi-ordered F.

Proof. [20] Note that a valuation is uniquely determined for <u>all</u> formulas by its restriction

---

[18] not necessarily (weakly) $\mathbf{H}$-sound

[19] In fact, if $[B/q]A \notin \mathbf{L}(F)$, i.e., $v \not\models [B/q]A$ for a valuation $\models$ in F, then $v \not\models' A$ for a weak valuation $\models'$ such that $u \models' q \Leftrightarrow u \models B$; by Lemma 2.1(2), this $\models'$ satisfies (Atomic heredity) on $E[F]$. Thus $A \notin \mathbf{L}(F)$ by (+).

[20] Cf. the proof of Proposition 6 from [5].

to Var, by applying the usual inductive clauses, see Section 1.1. Thus, it is sufficient to establish the 'only if' part.[21]

Clearly, the atomic heredity in F and in F^* is just the same. So we have to check the inductive clause for implication, i.e., to show that for any valuation in F and for all formulas B, C the following equivalence holds:

$$\forall v \in R(u)\,[v \vDash B \Rightarrow v \vDash C] \Leftrightarrow \forall v \in R^+(u)\,[v \vDash B \Rightarrow v \vDash C].$$

In fact, let uR^+v, $v \vDash B$, $v \nvDash C$. Then $uR^n v' Rv$ for some v' and $n \geq 0$ (i.e., $v' \in F^u$). Then $v' \nvDash (B \to C)$ in F, and by **H**-heredity of F (see Lemma 2.1(3)), $u \nvDash (B \to C)$ in F, i.e., there exists $v'' \in R(u)$ such that $v'' \vDash B$, $v'' \nvDash C$. □

Recall that a frame F is called co-serial if $E[F] = F$; all weakly reflexive frames are co-serial. So by applying Lemmas 2.1 and 2.2, we readily obtain the following description of **H**-sound frames without parasites:

Proposition 1 *For a frame F the following conditions are equivalent:*[22]

(1) *F is co-serial and* **H***-sound;*

(2) *F is co-serial and w***H***-sound;*

(3) *F is co-serial and* $(\top \to p) \to p$, $(p \to q) \to (\top \to (p \to q)) \in \mathbf{L}(F)$;

(4) *F is weakly quasi-ordered.*

Thus in particular,

*all transitive co-serial (w)***H***-sound frames are quasi-ordered.*

Now let us consider the general case.

Proposition 2 *For a frame F the following conditions are equivalent:*

(1) *F is w***H***-sound;*

(2) $(\top \to p) \to p$, $(p \to q) \to (\top \to (p \to q)) \in \mathbf{L}(F)$;

(3) *$E[F]$ is weakly quasi-ordered.*[23]

Proof. (2) ⇒ (3) follows from Lemma 2.1.

(3) ⇒ (1). If $E[F] = \varnothing$, i.e., $F = \Pi_0[F]$, then $\mathbf{L}(F) = \mathbf{L}(\Pi_0) \supset \mathbf{C}$, by Lemma 1.2. If $E[F]$ is a non-empty weakly quasi-ordered frame, then $\mathbf{L}(E[F])$ is an intermediate

By the way, that proposition actually states that $\mathbf{L}(F^*) \subseteq \mathbf{L}_r(F)$ for an arbitrary frame F, where $\mathbf{L}_r(F)$ is the set of formulas true under all rudimentary Kripke models in F (recall that in [5] a valuation is called *rudimentary* if it satisfies A-heredity and converse A-heredity for all formulas A). One can show that $\mathbf{L}(F) = \mathbf{L}_r(F) = \mathbf{L}(F^*)$ for weakly quasi-ordered frames F (because all valuations in these frames are rudimentary, cf. Claim in Section 2.2). However, this argument is rather detour, and we prefer a simple direct argument to prove Lemma 2.2 without using the notion of rudimentary Kripke models etc.

[21] Then the 'if' part readily follows as well. In fact, for a valuation $\vDash$ in F^* take its restriction to Var and prolong it to a valuation in F. Now, by the 'only if' part, this is a valuation in F^*, and so it equals the original valuation $\vDash$.

[22] In fact, Lemma 2.1 gives (3) ⇒ (4) and Lemma 2.2 gives (4) ⇒ (1).

[23] For a reader familiar with [4] note that these conditions actually mean that all valuations in F form conditionally rudimentary Kripke models; therefore, this proposition indirectly follows from [4, Proposition 12 etc.].

logic and $\mathbf{H} \subseteq \mathbf{L}(F\backslash\Pi_0[F]) \subseteq \mathbf{L}(E[F]) \subseteq \mathbf{C}$, cf. Lemma 1.3.[24] Now, if $\Pi_0[F] \neq \varnothing$, then $\mathbf{L}(F) = \mathbf{L}(F\backslash\Pi_0[F]) \cap \mathbf{L}(\Pi_0) = \mathbf{L}(F\backslash\Pi_0[F])$, since $\mathbf{C} \subset \mathbf{L}(\Pi_0)$. □

Therefore,

to describe *the semantics of (weakly)* **H**-*sound Kripke frames* it is sufficient to consider only *(weakly)* **H**-*sound frames without isolated parasites.*

Note that Proposition 2 implies in particular that

*a transitive frame F is w***H***-sound* iff *$E[F]$ is quasi-ordered.*

Proposition 3 *For a frame F the following conditions are equivalent:*

(1) *F is* **H**-*sound;*

(2) *F is w***H***-sound and $\mathbf{L}(F)$ is closed under modus ponens;*

(3) *F is w***H***-sound and $\mathbf{L}(F) = \mathbf{L}(E[F])$.*

Proof. (3) ⇒ (1) follows from Proposition 2 (the implication (1) ⇒ (3)) and Proposition 1 (the implication (4) ⇒ (1)).

(2) ⇒ (3). If $A \in \mathbf{L}(E[F])$, then $(\top \to A) \in \mathbf{L}(F)$. Thus $A \in \mathbf{L}(F)$ by (MP). □

Therefore we have established Reducibility Theorem stated at the end of Section 1. In fact, if a frame F is **H**-sound, then $E[F]$ is weakly quasi-ordered and $\mathbf{L}(F) = \mathbf{L}(E[F]) = \mathbf{L}(F')$ for a partially ordered (p.o.) frame $F' = S[E[F]]$.

2.4 APPENDIX

A description of **H**-soundness given in Proposition 3, unlike Propositions 1 and 2, is slightly implicit. Namely, it involves a vague condition

$$\mathbf{L}(F) = \mathbf{L}(E[F]). \qquad (\lambda)$$

By Lemma 1.1, the condition (λ) means that

$$A' \in \mathbf{L}(E[F]) \Rightarrow A' \in \mathbf{L}(F) \quad \text{for any } A' = \bigvee_j A_j,\ A_j \in \mathrm{Im} \cup \mathrm{Var}.$$

Open Problem *Try to find a more explicit description of* **H**-*soundness; in other words, represent in a more 'convenient' form the condition (λ) for frames F with weakly quasi-ordered essential part $E[F]$.*

We suppose, Example 3 (from Section 1.3) makes a hint that this problem may not have a satisfactory solution. However to conclude our considerations we mention here some straightforward approaches to the problem (and explain that they do not give a general description). So the whole problem seems to be too hopeless and does not worth serious efforts.

[24] Note that $\mathbf{L}(F\backslash\Pi_0[F]) = \bigcap(\mathbf{L}(\{w\} \cup E[F]) : w \in \Pi[F]\backslash\Pi_0[F])$ and every $(\{w\} \cup E[F])$ is a frame of the kind $(\Pi_0 + E[F])$ described in Example 3.

First, it is sufficient to consider only frames described in Example 3, i.e.:

$$\text{frames } F \text{ with empty } \Pi_0(F), \text{ one-element } \Pi(F)=\{\pi_0\}, \text{ and p.o. } E[F]. \qquad (\varphi)$$

In fact, for a frame F with a weakly quasi-ordered $E[F]$ we consider a frame $\tilde{F} = S[E[F]] \cup \Pi[F]$, in which $\Pi[\tilde{F}] = \Pi[F]$, $E[\tilde{F}] = S[E[F]]$, and

$$w\tilde{R}(u/\equiv) \Leftrightarrow \exists u' \equiv u\ (wRu') \qquad \text{for } w \in \Pi[F],\ u \in E[F].$$

There exists a natural one-to-one correspondence between valuations in $\tilde{F}$ and in F (cf. the end of Section 1.2); namely, the correspondent valuations satisfy the condition (S) on $E[F]$ and coincide on $\Pi[F]$ (recall that $E[F]$ is **H**-hereditary, so forcing for all formulas in $E[F]$ does not distinguish $\equiv$-equivalent points). Hence $\mathbf{L}(F) = \mathbf{L}(\tilde{F})$.

Now[25]

$$\mathbf{L}(\tilde{F}) = \bigcap (\mathbf{L}(\tilde{F}_w) : w \in \Pi[F]),$$

where $\tilde{F}_w = E[\tilde{F}] \cup \{w\}$ for $w \in \Pi[F]$ is an open subframe of $\tilde{F}$.

Thus $\quad \mathbf{L}(\tilde{F}) = \mathbf{L}(E[\tilde{F}]) \Leftrightarrow \forall w \in \Pi[F]\,(\mathbf{L}(\tilde{F}_w) = \mathbf{L}(E[\tilde{F}]))$,
since $\mathbf{L}(\tilde{F}_w) \subseteq \mathbf{L}(E[\tilde{F}])$ for any $w \in \Pi[F]$.

Therefore we conclude that

*F is **H**-sound* iff *all $\tilde{F}_w$ are **H**-sound.*

Clearly, every $\tilde{F}_w$ is a frame of the form (φ).

Hence now we consider only frames F of the form (φ).

Recall that a frame F of this form can be described by a non-empty[26] subset $R(\pi_0) \subseteq E[F]$, cf. Example 3.

We say that $R(\pi_0) (\subseteq E[F])$ *violates* a formula $A' = \bigvee_j (B_j \to C_j)$ if there exists a valuation in $E[F]$ such that $\quad \forall j\, \exists u \in R(\pi_0)\, [\, u \vDash B_j,\ u \nvDash C_j\,]$.

Clearly, $R(\pi_0)$ violates A' iff

there exists a valuation in F such that $\pi_0 \nvDash A' \vee \bigvee_k p_k$

for an arbitrary (perhaps, empty) list of variables $(p_k : k)$.[27]

Thus, if $(A' \vee \bigvee_k p_k) \in \mathbf{L}(E[F]) \backslash \mathbf{L}(F)$, then $R(\pi_0)$ violates A'. Hence we obtain:

Lemma 2.3 *Let F be a frame of the form (φ). Then:*
$\mathbf{L}(F) = \mathbf{L}(E[F])$ *(i.e., F is **H**-sound)* iff
[*for any formula $A = A' \vee A''$, where $A' = \bigvee_j (B_j \to C_j)$, $A'' = \bigvee_k p_k$:*

$$\textit{if } R(\pi_0) \textit{ violates } A', \quad \textit{then } A \notin \mathbf{L}(E[F])\,]. \qquad (\lambda^*)$$

Actually in (λ^*) we may assume that A'' is the disjunction of all variables occurring in A' (in fact, if (λ^*) holds for this A'', then it readily holds for an arbitrary A'' as well).

---

[25] Here we suppose that $\Pi[F] \neq \varnothing$, since otherwise $F = E[F]$ is definitely **H**-sound.

[26] If $R(\pi_0) = \varnothing$, then $\mathbf{L}(F) = \mathbf{L}(E[F])$, cf. Example 2, and F is **H**-sound again.

[27] For the 'only if' part one can prolong a given valuation in $E[F]$ (actually refuting A' at π_0) to a valuation in F; namely, put $\pi_0 \nvDash p$ for all variables p.

Clearly, if $R(\pi_0)$ violates A', then $(B_j \to C_j) \notin \mathbf{L}(E[F])$ for any j. Therefore if $\mathbf{L}(E[F])$ has the disjunction property, then (λ^*) holds, and so F is $\mathbf{H}$-sound.

Also note that F is $\mathbf{H}$-sound e.g. if $E[F]$ is an infinite chain (recall that its logic $[\mathbf{H} + (p \to q) \vee (q \to p)]$ lacks the disjunction property). In fact, if $R(\pi_0)$ violates A', then $A' \notin \mathbf{L}(E[F])$. Hence A' is falsified in any sufficiently large ($\geq n$-element for some n) cone in $E[F]$. Now, take $u, v \in F$ such that uRv, $u \neq v$, and $A' \notin \mathbf{L}(F^v)$, then $A = (A' \vee A'') \notin \mathbf{L}(F^u)$, since all variables can be refuted at u.

Another sufficient condition for $\mathbf{H}$-soundness is in terms of $R(\pi_0)$ rather than $E[F]$. We say that a set $R(\pi_0) \subseteq E[F]$ is *co-directed* if for any finite $X \subseteq R(\pi_0)$ there exists $v_X \in E[F] \backslash X$ such that $\forall u \in X\, (v_X R u)$. Clearly, if $R(\pi_0)$ is co-directed and $R(\pi_0)$ violates A', then $v_X \not\models A = A' \vee A''$ for some X and a valuation $\models$ in $E[F]$. Thus any F with a co-directed $R(\pi_0)$ satisfies (λ^*), and so it is $\mathbf{H}$-sound.

Now we are ready to describe the $\mathbf{H}$-soundness for a natural particular case.

Proposition 4 *Let F be a frame of the form (φ) with a rooted $E[F]$. Then:*

$$F \textit{ is } \mathbf{H}\textit{-sound} \quad \textit{iff} \quad [\,u_0 \notin R(\pi_0)\,] \textit{ or } [\,\mathbf{L}(E[F]) = \mathbf{L}(E[F] \backslash \{u_0\})\,],$$

where u_0 is the root of $E[F]$.

Proof. If $u_0 \notin R(\pi_0)$, then $R(\pi_0)$ is co-directed (take $v_X = u_0$ for any X).

Now let $\mathbf{L}(E[F]) = \mathbf{L}(E[F] \backslash \{u_0\})$. If $R(\pi_0)$ violates A', then $u_0 \not\models A'$ for a valuation in $E[F]$, i.e., $A' \notin \mathbf{L}(E[F])$. Thus there exists a valuation $\models'$ in $E[F] \backslash \{u_0\}$ such that $u \not\models{}' A'$ for some $u \neq u_0$. Then $u_0 \not\models{}' A = A' \vee A''$ in $E[F]$, since we may assume that all variables are false at u_0. Hence $A \notin \mathbf{L}(E[F])$.

Finally let $u_0 \in R(\pi_0)$ and $B \in \mathbf{L}(E[F] \backslash \{u_0\}) \setminus \mathbf{L}(E[F])$. Consider a variable p non-occurring in B, put $A' = (p \to B)$, $A = p \vee A'$. Then $A \in \mathbf{L}(E[F])$ and $R(\pi_0)$ violates A' with a valuation such that $u_0 \not\models B$, $u_0 \models p$. □

Remark By the way, one can rewrite the latter condition (in Proposition 4) in a more 'syntactic' form, using δ-operation of Hosoi – Ono. Namely,

$$[\,\mathbf{L}(E[F]) = \mathbf{L}(E[F] \backslash \{u_0\})\,] \text{ iff } [\,A \in \mathbf{L}(E[F]) \Leftrightarrow \delta A \in \mathbf{L}(E[F]) \textit{ for every formula } A\,],$$

where $\delta A = p \vee (p \to A)$ for a variable p non-occurring in A.[28]

In fact, recall that clearly: $\delta A \in \mathbf{L}(E[F])$ iff $A \in \mathbf{L}(E[F] \backslash \{u_0\})$.

The proposition implies that F is not $\mathbf{H}$-sound if $E[F]$ is a rooted frame of finite height (and its root belongs to $R(\pi_0)$). In fact, here frames $E[F]$ and $E[F] \backslash \{u_0\}$ have different heights, so their logics are distinguishable by a well-known axiom. A similar argument is applicable if, say, $E[F] \backslash X$ is linearly ordered, where X is a non-empty (and non-linear!) downward closed set of finite height (and in many other cases).

But note that we cannot reduce a general case to the case of rooted $E[F]$. For example, let $E[F]$ be a disjoint union of finite cones, say, Jaśkowski's trees, or finite chains, or finite binary trees (and $R(\pi_0) = E[F]$). Then F is $\mathbf{H}$-sound, since $\mathbf{L}(E[F]) =$

[28] Note also that the implication $(\Rightarrow)$ is obvious, so one can read $(\Leftarrow)$ for $(\Leftrightarrow)$ here.

H (or Dummett's logic, or Gabbay – de Jongh's logic of binary trees, [29] respectively), but all its (non-open!) subframes $F' \cup \{\pi_0\}$ for cones F' in $E[F]$ are not **H**-sound.

So we do not hope to find a general description of **H**-sound frames essentially less cumbersome than the technical condition (λ^*). And such a description seems useless, because we have proved that all these frames are actually reducible to partially ordered ones.

Example Proposition 4 gives us an **H**-sound frame such that $\mathbf{L}(F) \neq \mathbf{L}(S[F])$, mentioned in Section 1.2. Namely, let F be a frame of the form (φ) such that $E[F]$ is the set of all proper subsets of a 3-element set I (partially ordered by inclusion), see Figure 5, and $R(\pi_0)$ is the set of all 2-element subsets of I. Then F is **H**-sound, since the root $\varnothing$ of $E[F]$ does not belong to $R(\pi_0)$. Now, π_0 becomes a reflexive minimal point in $S[F]$, and its cone is the tree $T_{2,3}$ of height 2 and branching 3. Thus $\mathbf{L}(S[F]) = \mathbf{L}(E[F]) \cap \mathbf{L}(T_{2,3})$, and one can easily show that the Kreisel – Putnam's formula

$$K = (\neg p \to q \vee r) \to (\neg p \to q) \vee (\neg p \to r)$$

belongs to $\mathbf{L}(E[F]) \backslash \mathbf{L}(S[F]) = \mathbf{L}(F) \backslash \mathbf{L}(S[F])$.

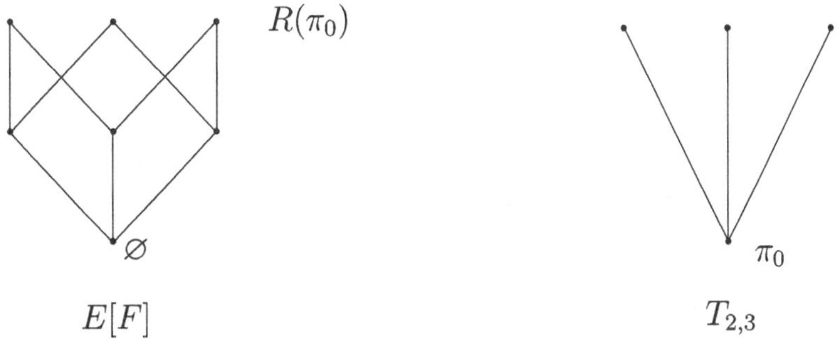

Fig. 5.

2.5 Additional Remark

Anonymous referees proposed to consider a stronger version of intuitionistic soundness, cf. e.g. [5, Proposition 8]. [30]

Recall the well-known notions of *intuitionistic logical consequence*:

$$\Gamma \vdash_{\mathbf{H}} A \text{ iff } [\mathbf{H} \vdash (\&\Gamma_0 \to A) \text{ for a finite } \Gamma_0 \subseteq \Gamma]$$

and the *semantic consequence* (in a frame F):

$$\Gamma \vDash_F A \text{ iff } [\forall u(u \vDash \Gamma) \Rightarrow \forall u(u \vDash A), \text{ for any valuation } \vDash \text{ in } F]$$

(here $u \vDash \Gamma$ for a set Γ of formulas means that $\forall B \in \Gamma (u \vDash B)$).

[29] Recall that it has the disjunction property.

[30] We suppose that this version of soundness is too strong. However, its description is quite simple and readily follows from our previous considerations, so we present it here.

Say that a frame F is *strongly* **H**-*sound* (or *s***H**-*sound*) if

$$\Gamma \vdash_{\mathbf{H}} A \Rightarrow \Gamma \vDash_F A \qquad \text{for all } \Gamma \text{ and } A. \qquad (\sigma)$$

Obviously, $\top \vdash_{\mathbf{H}} A \Leftrightarrow \mathbf{H} \vdash A$ and $\top \vDash_F A \Leftrightarrow A \in \mathbf{L}(F)$. So

(I) *Every s***H**-*sound frame F is w***H**-*sound.*

Also, clearly: if $\Gamma \vDash_F A$, then $(\Gamma \subseteq \mathbf{L}(F) \Rightarrow A \in \mathbf{L}(F))$.

Hence every s**H**-sound frame satisfies modus ponens, i.e., the condition (MP) from Section 1.1, because $(\top \to A) \vdash_{\mathbf{H}} A$. Therefore by Proposition 3 we obtain

(II) *Every s***H**-*sound frame F is* **H**-*sound.*

It is well known that all p.o. (and thus, all quasi-ordered) frames are s**H**-sound.

Finally, we obtain the following description of s**H**-soundness:

Claim *For a frame F the following conditions are equivalent:*

(1) *F is s***H**-*sound, i.e., it satisfies (σ) (with all Γ and A);*

(2) *the condition (σ) holds with all one-element Γ (and all A);*

(3) *F is co-serial and w***H**-*sound;*

(4) *F is weakly quasi-ordered.*

By the way, this claim gives another, slightly indirect, proof of (II), cf. Proposition 1.[31]

Proof. (2) $\Rightarrow$ (3). First, F is w**H**-sound by (I). Now suppose F is not co-serial, i.e, $\Pi[F] \neq \varnothing$. Take a valuation such that $u \vDash p \Leftrightarrow u \in E[F]$. Then we have: $\forall u \in F\,(u \vDash (\top \to p))$ and $u \nvDash p$ for $u \in \Pi[F]$, so $(\top \to p) \nvDash_F p$, while $(\top \to p) \vdash_{\mathbf{H}} p$.

(4) $\Rightarrow$ (1). By Lemma 2.2, the frames F and F^* have the same valuations, so the s**H**-soundness of F readily follows from the s**H**-soundness of F^*.

Now Proposition 1 gives (3) $\Rightarrow$ (4), and concludes the proof. □

So we see that the natural counterpart of Reducibility Theorem for s**H**-soundness holds as well, namely:

*For every s***H**-*sound Kripke frame F there exists a partially ordered frame F' such that*

$$\Gamma \vDash_F A \quad \text{iff} \quad \Gamma \vDash_{F'} A.$$

Put $F' = S[F](= S[F^*])$.

By the way, we can also mention another, modified version of semantical consequence, called *local* (cf. e.g. (**) after Proposition 8 in [5]):

$$\Gamma \vDash'_F A \text{ iff } [\, u \vDash \Gamma \Rightarrow u \vDash A, \text{ for any } u \in F \text{ and for any valuation } \vDash \text{ in } F\,].$$

Naturally, a frame F is *s′***H**-*sound* if

$$\Gamma \vdash_{\mathbf{H}} A \Rightarrow \Gamma \vDash'_F A \qquad \text{for all } \Gamma \text{ and } A. \qquad (\sigma')$$

[31] Note that our proof of the claim uses (I) and does not use (II).

Clearly, $\Gamma \models'_F A$ implies $\Gamma \models_F A$. The converse implication in general does not hold. In fact, obviously, $A \models_F (\top \to A)$ for every frame F (and formula A). On the other hand, $A \models'_F (\top \to A)$ iff F is A-hereditary (see Section 2.1). Therefore, by Lemma 2.1(3), $\models_F$ does not imply $\models'_F$ for every F that is not weakly transitive (e.g. for the frame shown at Figure 4, or for 3-element non-transitive chains, reflexive or non-reflexive). By the way, one can show that $\models_F$ does not imply $\models'_F$ for p.o. frames as well, but we do not know so small and simple counterexamples.[32]

Nevertheless,

*a frame is s′**H**-sound* iff *it is s**H**-sound.*

In fact, an s′**H**-sound frame F is obviously s**H**-sound. On the other hand, if F is s**H**-sound, i.e., weakly quasi-ordered, then (σ') holds, again by Lemma 2.2, since it definitely holds for a quasi-ordered frame F^*.

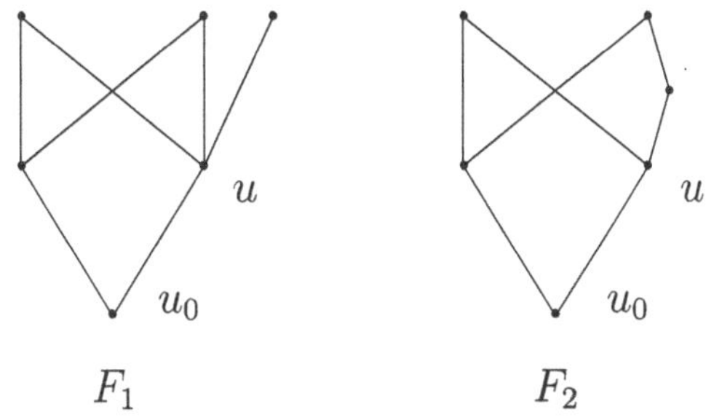

Fig. 6.

Acknowledgements

The author would like to thank the anonymous referees for useful comments and remarks. The author is also grateful to K. Došen, whose papers [5] and [4] inspired the observations presented in this paper.

---

[32] Two simplest p.o. frames that we know are shown at Figure 6; they contain 6 points. Also let us present formulas A_k, B_k such that $B_k \models_{F_k} A_k$ and $B_k \not\models'_{F_k} A_k$ for $k=1,2$. Namely,

$$A_1 = \bigvee_{i=0}^{2} \neg C_i, \quad B_1 = \mathop{\&}_i (\neg C_i \to \bigvee_{j \neq i} C_j), \quad \text{where} \quad C_0 = p\&q,\ C_1 = p\&\neg q,\ C_2 = \neg p\&q.$$

and $A_2 = \neg p \vee \neg\neg p$, $B_2 = ((\neg\neg p \to p) \to p \vee \neg p)$.
In fact, one can easily construct valuations $\models_k$ in F_k such that $u \models_k B_k$, $u \not\models_k A_k$.
On the other hand, $u_0 \models B_k \Rightarrow u_0 \models A_k$ for any valuation in F_k.
Note that F_2 is the 6-element cone in Nishimura's ladder and $(B_2 \to A_2)$ is the well-known Scott's formula.

References

[1] Chagrov, A., *On boundaries of the set of modal companions of intuitionistic logic*, in: *Non-Classical Logics and their Applications*, Institute of Philosophy of the USSR Academy of Sciences, Moscow, 1989, pp. 74–81, (In Russian).

[2] Chagrov, A. and M. Zakharyashchev, *Modal companions of intermediate propositional logics*, Studia Logica **51, No. 1** (1991), pp. 49–82.

[3] Došen, K., *Normal modal logics in which the Heyting propositional calculus can be embedded*, in: P. P. Petkov, editor, *Mathematical Logic*, Plenum Press, New York (1990), pp. 281–291.

[4] Došen, K., *Rudimentary Beth models and conditionally rudimentary Kripke models for the Heyting propositional calculus*, Journal of Logic and Computation **1** (1991), pp. 613–634.

[5] Došen, K., *Rudimentary Kripke models for the intuitionistic propositional calculus*, Annals of Pure and Applied Logic **62** (1993), pp. 21–49.

Bi-approximation Semantics for Substructural Logic at Work

Tomoyuki Suzuki

Department of Computer Science, University of Leicester
Leicester, LE1 7RH, United Kingdom
tomoyuki.suzuki@mcs.le.ac.uk

Abstract

In this paper, we introduce bi-approximation semantics, a two-sorted relational semantics, via the canonical extension of lattice expansions. To characterise Ghilardi and Meloni's parallel computation, we introduce doppelgänger valuations which allow us to evaluate sequents and not only formulae. Moreover, by introducing the bi-directional approximation and bases, we track down a connection to Kripke-type semantics for distributive substructural logics through a relationship between basis and the existential quantifier. Based on the framework, we give a possible interpretation of the two sorts, and prove soundness via bi-approximation and completeness via an algebraic representation theorem plus invariance of validity along a back-and-force correspondences.

Keywords: substructural logic, relational semantics, canonicity

1 Introduction

What is a natural relational semantics for substructural logic or resource sensitive logics? Unlike Kripke semantics for modal logic, we can find several types of relational semantics for substructural logic based on their philosophy or on their mathematical frameworks. For example, the study of relational semantics for distributive substructural logics has led to an operational semantics for relevant implication [20]. In [17], a ternary relational semantics, a.k.a. Routley-Meyer semantics, has been introduced by a different interpretation of relevant implication. For distributive substructural logics, we can also find other relational semantics, see e.g. [16]. Reasoning about relational-type semantics for non-distributive substructural logics, one encounters the interpretation problem of disjunction, namely how to avoid the distributivity of *conjunction* and *disjunction*. For orthologic, one can solve the problem on Goldblatt frames [12], by introducing a non-standard interpretation of disjunction. With Dedekind-MacNeille frames and the closure operator interpretation in [13] and [14], one can also solve the problem by using a closure operator. Generalized Kripke frames [7], which are introduced by characterising the intermediate level of canonical extensions of lattice expansions (see e.g. [5] or [8]), provide another semantics in which one can avoid the disjunction problem by a Galois connection.

The aim of the current paper is to propose another possible relational-type semantics for substructural logic. To achieve our goal, we introduce a two-sorted relational-type semantics, called *bi-approximation semantics*, and describe Ghilardi and Meloni's parallel computation [11], see also [19]. Our framework is closely related to the works [13], [14] and [7]. On the other hand, bi-approximation semantics has novel aspects: *bi-directional approximation*, *bases*, and *doppelgänger valuations* which allow us to evaluate *sequents* (Section 3). Based on our setting, we will come across one possible interpretation of the two sorts, *premises* and *conclusions*, and discover a relationship to Kripke-type semantics for distributive substructural logics through *bases* and *the existential quantifier* (Section 4). Furthermore, the connection between *bases* and *the existential quantifier* provides an *effective evaluation of sequents* in bi-approximation semantics, which is useful to prove the soundness theorem (Theorem 6.1). In Section 5, we prove the representation theorem of FL-algebras via p-frames, which is used to show the completeness theorem in Section 6.

2 Substructural logic

In this paper, we denote propositional variables by $p, q, r, p_1, \ldots$, the set of all propositional variables by Φ, and $\mathbf{t}$ and $\mathbf{f}$ are logical constants representing *true* and *false*, respectively. As logical connectives, we use *disjunction* $\vee$, *conjunction* $\wedge$, *fusion (multiplication)* $\circ$, *implications (residuals)* $\rightarrow$ and $\leftarrow$. Formulae of substructural logic are denoted by ϕ, ψ, $\phi_1, \ldots$ and $\psi_1, \ldots$, and the set of all formulae is denoted by Λ. The following BNF generates formulae of substructural logic.

$$\phi ::= p \mid \mathbf{t} \mid \mathbf{f} \mid \phi \vee \phi \mid \phi \wedge \phi \mid \phi \circ \phi \mid \phi \rightarrow \phi \mid \phi \leftarrow \phi$$

$\Gamma, \Delta, \Sigma, \Pi$ are (possibly empty) finite lists of formulae, and φ is a list of at most one formula. Then, we call $\Gamma \mapsto \varphi$ a *sequent.*

Gentzen's sequent system for substructural logic Let ϕ, ψ be arbitrary formulae, $\Gamma, \Delta, \Sigma, \Pi$ arbitrary (possibly empty) finite lists of formulae, φ a list of at most one formula: see e.g. [15]. The sequent system FL is the following.

Initial sequents :

$$\phi \mapsto \phi \qquad\qquad \mapsto \mathbf{t} \qquad\qquad \mathbf{f} \mapsto$$

Cut rule :

$$\frac{\Gamma \mapsto \phi \qquad \Sigma, \phi, \Pi \mapsto \varphi}{\Sigma, \Gamma, \Pi \mapsto \varphi}\ (\text{cut})$$

Rules for constants :

$$\frac{\Gamma, \Delta \mapsto \varphi}{\Gamma, \mathbf{t}, \Delta \mapsto \varphi} \text{ (tw)} \qquad \frac{\Gamma \mapsto}{\Gamma \mapsto \mathbf{f}} \text{ (fw)}$$

Rules for logical connectives :

$$\frac{\Gamma, \phi, \Delta \mapsto \varphi \qquad \Gamma, \psi, \Delta \mapsto \varphi}{\Gamma, \phi \vee \psi, \Delta \mapsto \varphi} (\vee \mapsto)$$

$$\frac{\Gamma \mapsto \phi}{\Gamma \mapsto \phi \vee \psi} (\mapsto \vee_1) \qquad \frac{\Gamma \mapsto \psi}{\Gamma \mapsto \phi \vee \psi} (\mapsto \vee_2)$$

$$\frac{\Gamma, \phi, \Delta \mapsto \varphi}{\Gamma, \phi \wedge \psi, \Delta \mapsto \varphi} (\wedge_1 \mapsto) \qquad \frac{\Gamma, \psi, \Delta \mapsto \varphi}{\Gamma, \phi \wedge \psi, \Delta \mapsto \varphi} (\wedge_2 \mapsto)$$

$$\frac{\Gamma \mapsto \phi \qquad \Gamma \mapsto \psi}{\Gamma \mapsto \phi \wedge \psi} (\mapsto \wedge)$$

$$\frac{\Gamma, \phi, \psi, \Delta \mapsto \varphi}{\Gamma, \phi \circ \psi, \Delta \mapsto \varphi} (\circ \mapsto) \qquad \frac{\Gamma \mapsto \phi \qquad \Sigma \mapsto \psi}{\Gamma, \Sigma \mapsto \phi \circ \psi} (\mapsto \circ)$$

$$\frac{\Gamma \mapsto \phi \qquad \Sigma, \psi, \Pi \mapsto \varphi}{\Sigma, \Gamma, \phi \to \psi, \Pi \mapsto \varphi} (\to\mapsto) \qquad \frac{\phi, \Gamma \mapsto \psi}{\Gamma \mapsto \phi \to \psi} (\mapsto\to)$$

$$\frac{\Gamma \mapsto \phi \qquad \Sigma, \psi, \Pi \mapsto \varphi}{\Sigma, \psi \leftarrow \phi, \Gamma, \Pi \mapsto \varphi} (\leftarrow\mapsto) \qquad \frac{\Gamma, \phi \mapsto \psi}{\Gamma \mapsto \psi \leftarrow \phi} (\mapsto\leftarrow)$$

In the sequent system FL, a formula ϕ is *provable in FL* if the sequent $\mapsto \phi$ is derivable in FL. The substructural logic **FL** is the set of all provable formulae in FL.

Proposition 2.1 *For all formulae ϕ and ψ, we have*

(i) *ϕ is provable if and only if $\mathbf{t} \mapsto \phi$ is derivable,*

(ii) *$\phi \mapsto \psi$ is derivable if and only if $\phi \to \psi$ is provable in FL if and only if $\psi \leftarrow \phi$ is provable in FL,*

(iii) *$\phi_1, \ldots, \phi_n \mapsto \varphi$ is derivable in FL if and only if $\phi_1 \circ \cdots \circ \phi_n \mapsto \varphi$ is derivable in FL.*

The algebraic counterparts of substructural logic **FL** are known as FL-algebras [6].

Definition 2.2 [FL-algebra] An 8-tuple $\mathbb{A} = \langle A, \vee, \wedge, *, \backslash, /, 1, 0 \rangle$ is a *FL-algebra*, if $\langle A, \vee, \wedge \rangle$ is a lattice, $\langle A, *, 1 \rangle$ is a monoid, 0 is a constant in A, and for all $a, b, c \in A$,

$$a * b \leq c \iff b \leq a \backslash c \iff a \leq c/b.$$

By Proposition 2.1, we sometimes state that **FL** is the set of all sequents derivable in FL. On FL-algebras, each sequent $\phi_1, \ldots, \phi_n \mapsto \varphi$ is interpreted as an inequality $\phi_1 * \cdots * \phi_n \leq \varphi$.

3 Bi-approximation semantics

In this section, we firstly introduce a *polarity*, see [1] or [21], which is the foundation of *bi-approximation semantics.*

Polarity and bi-directional approximation

Definition 3.1 [Polarity] A triple $\langle X, Y, B\rangle$ is a *polarity,* if X and Y are non-empty sets, and B a binary relation on $X \times Y$, i.e. $B \subseteq X \times Y$.

Given a polarity $\langle X, Y, B\rangle$, we induce a preorder $\leq_B$ on $X \cup Y$ as follows, see [7]: for all $x_1, x_2 \in X$ and all $y_1, y_2 \in Y$, we let

(i) $x_1 \leq_B x_2 \iff$ for each $y \in Y$, $x_2 B y$ implies $x_1 B y$,

(ii) $y_1 \leq_B y_2 \iff$ for each $x \in X$, $x B y_1$ implies $x B y_2$,

(iii) $x_1 \leq_B y_1 \iff x_1 B y_1$,

(iv) $y_1 \leq_B x_1 \iff$ for each $x' \in X$ and each $y' \in Y$, $x' B y'$ if $x' B y_1$ and $x_1 B y'$.

Hereinafter, we sometimes omit the subscript $_B$ from the induced preorder $\leq_B$, and refer to the triple $\langle X, Y, \leq\rangle$ as the polarity. That is, a polarity $\langle X, Y, \leq\rangle$ is a preordered set $\langle X \cup Y, \leq\rangle$.

Next, we introduce two approximation functions for polarities. Let $\langle X, Y, \leq\rangle$ be a polarity, $\wp(X)$ the poset of all subsets of X ordered by inclusion $\subseteq$, and $\wp(Y)^\partial$ the poset of all subsets of Y ordered by reverse-inclusion $\supseteq$. Two functions $\lambda : \wp(X) \to \wp(Y)^\partial$ and $\upsilon : \wp(Y)^\partial \to \wp(X)$ are defined as follows: for each $\mathfrak{X} \in \wp(X)$ and each $\mathfrak{Y} \in \wp(Y)^\partial$,

(i) $\lambda(\mathfrak{X}) := \{y \in Y \mid \forall x \in \mathfrak{X}.\ x \leq y\}$,

(ii) $\upsilon(\mathfrak{Y}) := \{x \in X \mid \forall y \in \mathfrak{Y}.\ x \leq y\}$.

The functions λ and υ form a *Galois connection*, i.e. $\lambda \dashv \upsilon$. Hence, the images $\lambda[\wp(X)]$ and $\upsilon[\wp(Y)^\partial]$ are isomorphic. Hereafter, we denote the image $\lambda[\wp(X)]$ by $\mathbb{U}$ and the image $\upsilon[\wp(Y)^\partial]$ by $\mathbb{D}$. We mention that the images are *the Dedekind-MacNeille completion* of the quotient poset of $\langle X, Y, \leq\rangle$ with respect to the equivalence relation associated with $\leq$, see [1] or [4]. We call each element in $\mathbb{D}$ a *Galois stable X-set* and refer to each Galois stable X-set by adding the superscript $_{-}^{\downarrow}$, e.g. $\alpha^{\downarrow}$. We call each element in $\mathbb{U}$ a *Galois stable Y-set* and refer to each Galois stable Y-set by adding the subscript $_{-\uparrow}$, e.g. $\alpha_{\uparrow}$. Since every Galois stable X-set is an image of some (not necessarily unique) subset of Y, and every Galois stable Y-set is an image of some (not necessarily unique) subset of X, we introduce the following terminology.

Definition 3.2 [Approximation and basis] Let $\mathfrak{X} \in \wp(X)$, $\mathfrak{Y} \in \wp(Y)^\partial$, $\alpha^{\downarrow} \in \mathbb{D}$ and $\beta_{\uparrow} \in \mathbb{U}$. An element $\alpha^{\downarrow}$ is *approximated from above by* $\mathfrak{Y}$ and $\mathfrak{Y}$ is a *(Y-)basis of* α, if $\alpha^{\downarrow} = \upsilon(\mathfrak{Y})$. An element $\beta_{\uparrow}$ is *approximated from below by* $\mathfrak{X}$ and $\mathfrak{X}$ is a *(X-)basis of* β, if $\beta_{\uparrow} = \lambda(\mathfrak{X})$.

Later, we will construct two isomorphic FL-algebras on $\mathbb{D}$ and $\mathbb{U}$: see Section 5. Namely, we will take the abstract algebra whose underlying poset is isomorphic to both $\mathbb{D}$ and $\mathbb{U}$. Then, we can see every point α as $\alpha^{\downarrow}$ and as $\alpha_{\uparrow}$. In other words, every point in an abstract algebra is approximated from both *above* and *below*. The main concept of *bi-approximation semantics* is to keep the two directions of approximation: see e.g. Proposition 4.7.

Bi-approximation model Based on a polarity, we introduce *bi-approximation semantics* for substructural logic.

Definition 3.3 [P-frame for substructural logic] A *p-frame for substructural logic, p-frame* for short, is a 8-tuple $\mathbb{F} = \langle X, Y, \leq, R, O_X, O_Y, N_X, N_Y \rangle$, where $\langle X, Y, \leq \rangle$ is a polarity, $R \subseteq X \times X \times Y$ a ternary relation, O_X a non-empty Galois stable X-set, N_X a Galois stable X-set, O_Y and N_Y Galois stable Y-sets, and $\mathbb{F}$ satisfies

R-order: for all $x, x' \in X$, $x' \leq x$ if and only if
$$\exists o \in O_X.[\forall y \in Y.[R(x, o, y) \Rightarrow x' \leq y] \text{ or } \forall y \in Y.[R(o, x, y) \Rightarrow x' \leq y]],$$

R-identity: for each $x \in X$, $[\exists o_2 \in O_X, \forall y \in Y.[R(x, o_2, y) \Rightarrow x \leq y]$
$$\text{and } \exists o_1 \in O_X, \forall y \in Y.[R(o_1, x, y) \Rightarrow x \leq y]],$$

R-transitivity: for all $x_1, x_1', x_2, x_2' \in X$ and $y, y' \in Y$,
$$x_1' \leq x_1, x_2' \leq x_2, y \leq y' \text{ and } R(x_1, x_2, y) \Rightarrow R(x_1', x_2', y'),$$

R-associativity: for all $x_1, x_2, x_3, x \in X$,
$$\exists x' \in X.[\forall y \in Y.(R(x_1, x', y) \Rightarrow x \leq y) \text{ and } \forall y' \in Y.(R(x_2, x_3, y') \Rightarrow x' \leq y')]$$
$$\text{if and only if}$$
$$\exists x'' \in X.[\forall y \in Y.(R(x'', x_3, y) \Rightarrow x \leq y) \text{ and } \forall y'' \in Y.(R(x_1, x_2, y'') \Rightarrow x'' \leq y'')],$$

O-isom: $O_X = \upsilon(O_Y)$ and $O_Y = \lambda(O_X)$,

N-isom: $N_X = \upsilon(N_Y)$ and $N_Y = \lambda(N_X)$,

$\circ$-tightness: for all $x_1, x_2 \in X$, there exists $x \in X$ such that
$$\forall y \in Y.[R(x_1, x_2, y) \text{ if and only if } x \leq y],$$

$\rightarrow$-tightness: for each $x_1 \in X$ and each $y \in Y$, there exists $y_2 \in Y$ such that
$$\forall x_2 \in X.[R(x_1, x_2, y) \text{ if and only if } x_2 \leq y_2],$$

$\leftarrow$-tightness: for each $x_2 \in X$ and each $y \in Y$, there exists $y_1 \in Y$ such that
$$\forall x_1 \in X.[R(x_1, x_2, y) \text{ if and only if } x_1 \leq y_1].$$

A p-frame $\mathbb{F} = \langle X, Y, \leq, R, O_X, O_Y, N_X, N_Y \rangle$ is intuitively explained as follows: the Galois stable sets O_X, O_Y, N_X and N_Y define the worlds where we assume **t**, conclude **t**, assume **f** and conclude **f**. The conditions O-isom and N-isom guarantee that every $x \in X$ where we assume the formula **t** (**f**), if and only if every $y \in Y$ where we conclude the formula **t** (**f**) have the consequence relation $x \leq y$. The ternary relation R is another consequence relation which allows us to reason about logical consequences between two premises and one conclusion. The R-order condition says that the induced relation on X, $x' \leq x$ is also obtained by the ternary consequence relation R. The tightness conditions guarantee that the ternary consequence relation R respects $\leq$.

Remark 3.4 In Definition 3.3 one may feel that the conditions R-order, R-identity and R-associativity look too complicated. However, we reformulate them in Remark 4.3.

Our framework is similar to generalized Kripke frames in [7]. However, we do not assume neither Separation axioms nor Reduced axioms, hence p-frames may not be RS-frames. Our current purpose is to characterise Ghilardi and Meloni's parallel computation [11], see also [19]. The most distinct points are how to evaluate formulae on bi-approximation semantics, i.e. the valuation on two-sorted frames by introducing *doppelgänger valuation*, and how to interpret the satisfaction relation $\Vdash$ on each sort, X and Y.

Definition 3.5 [Doppelgänger valuation] Let $\mathbb{F}$ be a p-frame. A pair $V = \langle V^{\downarrow}, V_{\uparrow} \rangle$ of two functions $V^{\downarrow} : \Phi \to \mathbb{D}$ and $V_{\uparrow} : \Phi \to \mathbb{U}$ is a *doppelgänger valuation*, if $V^{\downarrow}(p)$ and $V_{\uparrow}(p)$ coincide for every propositional variable $p \in \Phi$. That is, $V^{\downarrow}(p) = \upsilon(V_{\uparrow}(p))$ and $V_{\uparrow}(p) = \lambda(V^{\downarrow}(p))$ for each propositional variable $p \in \Phi$.

Given a p-frame $\mathbb{F}$ and a doppelgänger valuation V, we call the pair $\mathbb{M} = \langle \mathbb{F}, V \rangle$ a *bi-approximation model.* On a bi-approximation model $\mathbb{M} = \langle \mathbb{F}, V \rangle$, we inductively define a *satisfaction relation* $\Vdash$ as follows: for each $x \in X$,

X-1: $\mathbb{M}, x \Vdash p \iff x \in V^{\downarrow}(p)$ for each $p \in \Phi$,

X-2: $\mathbb{M}, x \Vdash \mathbf{t} \iff x \in O_X$,

X-3: $\mathbb{M}, x \Vdash \mathbf{f} \iff x \in N_X$,

X-4: $\mathbb{M}, x \Vdash \phi \vee \psi \iff \forall y \in Y. \; [\mathbb{M}, y \Vdash \phi \vee \psi \Rightarrow x \leq y]$,

X-5: $\mathbb{M}, x \Vdash \phi \wedge \psi \iff \mathbb{M}, x \Vdash \phi$ and $\mathbb{M}, x \Vdash \psi$,

X-6: $\mathbb{M}, x \Vdash \phi \circ \psi \iff \forall y \in Y. \; [\mathbb{M}, y \Vdash \phi \circ \psi \Rightarrow x \leq y]$,

X-7: $\mathbb{M}, x \Vdash \phi \to \psi \iff \forall x' \in X, y \in Y. \; [\mathbb{M}, x' \Vdash \phi$ and $\mathbb{M}, y \Vdash \psi \Rightarrow R(x', x, y)]$,

X-8: $\mathbb{M}, x \Vdash \psi \leftarrow \phi \iff \forall x' \in X, y \in Y. \; [\mathbb{M}, x' \Vdash \phi$ and $\mathbb{M}, y \Vdash \psi \Rightarrow R(x, x', y)]$.

For each $y \in Y$,

Y-1: $\mathbb{M}, y \Vdash p \iff y \in V_{\uparrow}(p)$ for each $p \in \Phi$,

Y-2: $\mathbb{M}, y \Vdash \mathbf{t} \iff y \in O_Y$,

Y-3: $\mathbb{M}, y \Vdash \mathbf{f} \iff y \in N_Y$,

Y-4: $\mathbb{M}, y \Vdash \phi \vee \psi \iff \mathbb{M}, y \Vdash \phi$ and $\mathbb{M}, y \Vdash \psi$,

Y-5: $\mathbb{M}, y \Vdash \phi \wedge \psi \iff \forall x \in X. \; [\mathbb{M}, x \Vdash \phi \wedge \psi \Rightarrow x \leq y]$,

Y-6: $\mathbb{M}, y \Vdash \phi \circ \psi \iff \forall x_1, x_2 \in X. \; [\mathbb{M}, x_1 \Vdash \phi$ and $\mathbb{M}, x_2 \Vdash \psi \Rightarrow R(x_1, x_2, y)]$,

Y-7: $\mathbb{M}, y \Vdash \phi \to \psi \iff \forall x \in X. \; [\mathbb{M}, x \Vdash \phi \to \psi \Rightarrow x \leq y]$,

Y-8: $\mathbb{M}, y \Vdash \psi \leftarrow \phi \iff, \forall x \in X. \; [\mathbb{M}, x \Vdash \psi \leftarrow \phi \Rightarrow x \leq y]$.

In bi-approximation models, the satisfaction relation $\Vdash$ has two distinct interpretations depending on the domains X and Y. On X, we comprehend $\mathbb{M}, x \Vdash \phi$ as the formula ϕ is *assumed at* x, and on Y, $\mathbb{M}, y \Vdash \phi$ as the formula ϕ is *concluded at* y. Moreover, we also define $\mathbb{F}, x \Vdash \phi$ and $\mathbb{F}, y \Vdash \phi$ as usual: for every doppelgänger valuation V, we have $\mathbb{F}, V, x \Vdash \phi$ and $\mathbb{F}, V, y \Vdash \phi$, respectively.

An interpretation of the two-sorted semantics To reason about resource sensitive logics, we make a clear distinction between premises and conclusions, and evaluate logical consequences as relations between premises and conclusions. On p-frames, we think about X as a set of *premise worlds* where we evaluate only premises, and about Y as a set of *conclusion worlds* where we evaluate just conclusions. One may feel that the satisfaction relation $\mathbb{M}, y \Vdash \phi$, which says "the formula ϕ is *concluded* at the conclusion world y", is the same with "the formula ϕ is *true* at y." However, these two concepts are not the same. This is because, even if we conclude a formula ϕ at y, we cannot logically judge whether the formula is true or not. For example, if we conclude a formula ϕ meaning "tomorrow is Sunday" at a conclusion world y, we do not have any clue to justify that the formula is a fact. In other words, we may explain $\mathbb{M}, y \Vdash \phi$ as someone is just claiming "ϕ should be concluded" without any reason. Of course, we cannot consider it as logical reasoning. Only when we also have a reasonable premise like "today is Saturday" or "tomorrow is Sunday," we can justify that the logical consequence is *true*. More precisely, only when we have a pair of a premise and a conclusion, we can justify the logical consequence.

Formally the concept of *truth* of logical consequences on bi-approximation models is defined as follows. To reason about truth on bi-approximation models, it is necessary to extend the satisfaction relation $\Vdash \subseteq (X \times \Lambda) \cup (Y \times \Lambda)$ to a relation between $X \times Y$ and pairs of two formulae $\Lambda \times \Lambda$, or *sequents*. For our purpose, we fix the interpretation between sequents and pairs of two formulae. Given a sequent $\phi_1, \ldots, \phi_n \mapsto \varphi$, we translate it to $(\phi_1 \circ \cdots \circ \phi_n, \varphi)$. If $n = 0$, the left-hand side is empty and we write $(\mathbf{t}, \varphi)$. If the right-hand side is empty, we write $(\phi_1 \circ \ldots \circ \phi_n, \mathbf{f})$. But, whenever it is not confusing, we do not make any distinction between sequents and pairs of two formulae. So, both are called just *sequents* and are denoted by $\Gamma \mapsto \varphi$.

Definition 3.6 [Truth] Let $\mathbb{M} = \langle \mathbb{F}, V \rangle$ be a bi-approximation model and $\Gamma \mapsto \varphi$ a sequent. We let

(i) $\mathbb{M}, (x, y) \Vdash \Gamma \mapsto \varphi \iff x \leq y$ whenever $\mathbb{M}, x \Vdash \Gamma$ and $\mathbb{M}, y \Vdash \varphi$,

(ii) $\mathbb{F}, (x, y) \Vdash \Gamma \mapsto \varphi \iff \langle \mathbb{F}, V \rangle, (x, y) \Vdash \Gamma \mapsto \varphi$ for each doppelgänger valuation V,

(iii) $\mathbb{M} \Vdash \Gamma \mapsto \varphi \iff \mathbb{M}, (x, y) \Vdash \Gamma \mapsto \varphi$ for all $x \in X$ and $y \in Y$,

(iv) $\mathbb{F} \Vdash \Gamma \mapsto \varphi \iff \langle \mathbb{F}, V \rangle, (x, y) \Vdash \Gamma \mapsto \varphi$ for all $x \in X$ and $y \in Y$, and every doppelgänger valuation V.

We interpret $\mathbb{M}, (x, y) \Vdash \Gamma \mapsto \varphi$ as the sequent $\Gamma \mapsto \varphi$ is *true at the pair* (x, y), and $\mathbb{F} \Vdash \Gamma \mapsto \varphi$ as the sequent $\Gamma \mapsto \varphi$ is *valid on* $\mathbb{F}$.

Remark 3.7 Unlike what happens in the setting of the normal Kripke semantics, in bi-approximation models we reason about *sequents* but not *formulae*, in general. But, thanks to Proposition 2.1, this distinction is not critical when we consider substructural logic.

Hereinafter, we sometimes write $(x, y) \Vdash \phi \mapsto \psi$ instead of $\mathbb{M}, (x, y) \Vdash \phi \mapsto \psi$.

External reasoning and internal reasoning on p-frames Before we show preliminary results for bi-approximation semantics, we explain how to evaluate premises, conclusions and logical consequences on p-frames.

Recall the satisfaction relation $\Vdash$ in (X-1) - (X-8) and (Y-1) - (Y-8). We notice that there are two types of reasoning: *internal* and *external*. Namely, there is the reasoning on X, e.g. (X-4), or on Y, e.g. (Y-5), and there is the reasoning given by the relation $\leq$ or R between X and Y, e.g. (X-4) or (Y-6). Intuitively speaking, the internal reasoning derives a premise from premises, or a conclusion from conclusions, e.g. we assume $\phi \wedge \psi$ at x if and only if we assume ϕ and ψ at x (X-5). On the other hand, the external reasoning evaluates logical consequences. That is, we describe a premise world by conclusion worlds, and vise versa. For example, a conclusion world y where we conclude $\phi \wedge \psi$ is described by all premise worlds where we assume ϕ and ψ (Y-5). We also say that the conclusion world y is *approximated* by the corresponding premise worlds. Analogously, e.g. (X-4), a premise world is *approximated* by the corresponding conclusion worlds. See also Proposition 3.9. This is what we call *bi-approximation* in our framework.

Whereas the external reasoning is fundamental in bi-approximation models, we also have the internal reasoning as well. One may feel that the internal reasoning (Y-4) is far from our intuition. However, we can also explain it as follows. Recall the sequent calculus LK. In LK, we consider a sequent as a pair of a finite list of premises and a finite list of conclusions, $\phi_1, \ldots, \phi_m \Rightarrow \psi_1, \ldots, \psi_n$. The intuitive interpretation of this sequent is "if we assume *all* premises $\phi_1, \ldots, \phi_m$ then we conclude *one of these* conclusions $\psi_1, \ldots, \psi_n$." In other words, premises are *compulsory* and conclusions are *elective*. Therefore, it is natural to consider (Y-4) as "ϕ and ψ are possible conclusions at y if and only if $\phi \vee \psi$ is a possible conclusion at y."

Preliminary results for bi-approximation semantics In this paragraph, we show basic properties on bi-approximation semantics. The following proposition corresponds to Hereditary property in Kripke semantics for intuitionistic logic, e.g. [3]. But, it is two-sorted in our case.

Proposition 3.8 (Hereditary) *Let $\mathbb{M}$ be a bi-approximation model and ϕ a formula. For all elements $x, x' \in X$ and $y, y' \in Y$, we have*

(i) *if $x' \leq x$ and ϕ is assumed at x, $x \Vdash \phi$, then it is also assumed at x', $x' \Vdash \phi$,*

(ii) *if $y \leq y'$ and ϕ is concluded at y, $y \Vdash \phi$, then it is also concluded at y', $y' \Vdash \phi$.*

Proposition 3.9 *For each bi-approximation model $\mathbb{M}$, each $x \in X$, each $y \in Y$, and every formula ϕ, if $\mathbb{M}, x \Vdash \phi$ and $\mathbb{M}, y \Vdash \phi$, then $x \leq y$. Furthermore, we have*

(i) $\mathbb{M}, x \Vdash \phi \iff$ *for every $y \in Y$. if $\mathbb{M}, y \Vdash \phi$ then $x \leq y$,*

(ii) $\mathbb{M}, y \Vdash \phi \iff$ *for every $x \in X$. if $\mathbb{M}, x \Vdash \phi$ then $x \leq y$.*

Remark 3.10 Proposition 3.9 tells us initial sequents $\phi \Rightarrow \phi$ is valid on every p-frame $\mathbb{F}$. Intuitively, if ϕ is assumed at x, then it should be concluded everywhere in Y above x. Conversely, if ϕ is concluded at y, then it should be assumed everywhere in X below y.

As a corollary of Proposition 3.9, we obtain the following.

Corollary 3.11 *For every p-frame $\mathbb{F}$, each doppelgänger valuation V is naturally extended from the set of all propositional variables Φ to the set of all formulae Λ, i.e. for each formula ϕ, we let*

(i) $\tilde{V}^{\downarrow}(\phi) := \{x \in X \mid \mathbb{F}, V, x \Vdash \phi\}$,

(ii) $\tilde{V}_{\uparrow}(\phi) := \{y \in Y \mid \mathbb{F}, V, y \Vdash \phi\}$.

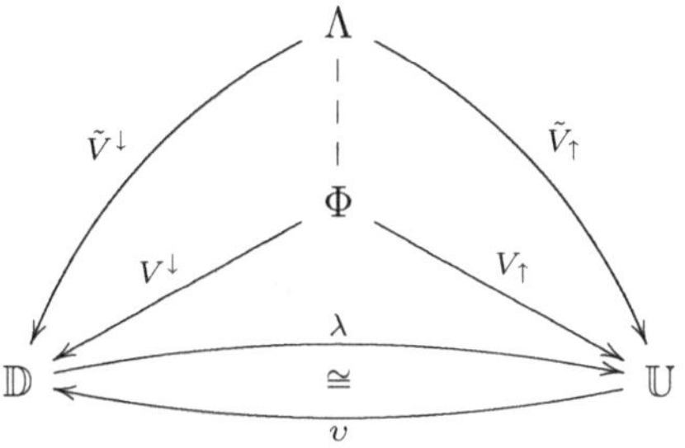

4 Bi-approximation, bases and the existential quantifier

In Kripke semantics, we have a simple interpretation of modal operators $\Diamond$ and $\Box$ as follows: for each Kripke model $\mathbb{M}$ and each possible world w, we let

(i) $\mathbb{M}, w \Vdash \Diamond\phi \iff \exists v \in W$ such that $R(w, v)$ and $\mathbb{M}, v \Vdash \phi$,

(ii) $\mathbb{M}, w \Vdash \Box\phi \iff \forall v \in W$. if $R(w, v)$ then $\mathbb{M}, v \Vdash \phi$,

whereas, in bi-approximation models, all logical connectives are interpreted uniformly with *conjunction*, *implication* and *universal quantifier* $\forall$. For example, if we introduce $\Diamond$ on bi-approximation semantics, it is interpreted as follows:

(iii) $\mathbb{M}, x \Vdash \Diamond\phi \iff \forall y \in Y$. if $\mathbb{M}, y \Vdash \Diamond\phi$ then $x \leq y$,

(iv) $\mathbb{M}, y \Vdash \Diamond\phi \iff \forall x \in X$. if $\mathbb{M}, x \Vdash \phi$ then $R(x, y)$,

where R is a binary relation on $X \times Y$. This is because it is essential to set up our interpretation to return Galois stable sets. Note that item (iv) gives the definition of $\Diamond$ on $\mathbb{U}$, and item (iii) copies the same value to $\mathbb{D}$: see also Section 5. As we saw in Corollary 3.11, this setting allows us to assign the corresponding Galois stable X-set and Y-set for every formula between $\mathbb{D}$ and $\mathbb{U}$. On the other hand, to evaluate any formula on bi-approximation models, we encounter the universal quantifier $\forall$ and an implication in each step, which generates considerable complexity.

However, in this section, we will show that we can reduce the complexity in specific cases by introducing *auxiliary relations* for R. In other words, some logical connectives are translated into other simpler conditions with *the existential quantifier*, which may not be equivalent to the original conditions anymore. Through these simpler conditions, we will find the relationship between relational semantics and bi-approximation semantics. Furthermore, we will also unearth a connection among bi-approximation, bases and the existential quantifier.

Definition 4.1 [Auxiliary relations] For every bi-approximation model $\mathbb{M}$ and the ternary relation $R \subseteq X \times X \times Y$, we let the following three ternary relations $R^\circ \subseteq X \times X \times X$, $R^\rightarrow \subseteq X \times Y \times Y$ and $R^\leftarrow \subseteq Y \times X \times Y$:

(i) $R^\circ(x_1, x_2, x) \iff$ for every $y \in Y$. if $R(x_1, x_2, y)$ then $x \leq y$,

(ii) $R^\rightarrow(x_1, y_2, y) \iff$ for every $x_2 \in X$. if $R(x_1, x_2, y)$ then $x_2 \leq y_2$,

(iii) $R^\leftarrow(y_1, x_2, y) \iff$ for every $x_1 \in X$. if $R(x_1, x_2, y)$ then $x_1 \leq y_1$.

Note that R° is related to $R^\downarrow$ in [7], but we also introduce $R^\rightarrow$ and $R^\leftarrow$ to show Theorem 6.1. Thanks to the tightness conditions in p-frames, see Definition 3.3, we obtain the following.

Lemma 4.2 *For every bi-approximation model* $\mathbb{M}$ *and the ternary relation* $R \subseteq X \times X \times Y$,

(i) $R(x_1, x_2, y) \iff$ *for every* $x \in X$*. if* $R^\circ(x_1, x_2, x)$ *then* $x \leq y$,

(ii) $R(x_1, x_2, y) \iff$ *for every* $y_2 \in Y$*. if* $R^\rightarrow(x_1, y_2, y)$ *then* $x_2 \leq y_2$,

(iii) $R(x_1, x_2, y) \iff$ *for every* $y_1 \in Y$*. if* $R^\leftarrow(y_1, x_2, y)$ *then* $x_1 \leq y_1$.

Remark 4.3 By Definition 4.1, we can reformulate R-order, R-identity and R-associativity in Definition 3.3 as follows:

R-order: for all $x, x' \in X$, $x' \leq x \iff \exists o \in O_X.\ [R^\circ(x, o, x')$ or $R^\circ(o, x, x')]$,

R-identity: for every $x \in X$. $[\exists o_2 \in O_X.\ R^\circ(x, o_2, x)$ and $\exists o_1 \in O_X.\ R^\circ(o_1, x, x)]$,

R-associativity: for all $x_1, x_2, x_3, x \in X$.

$$\exists x' \in X.\ [R^\circ(x_1, x', x) \text{ and } R^\circ(x_2, x_3, x')] \iff \exists x'' \in X.\ [R^\circ(x'', x_3, x) \text{ and } R^\circ(x_1, x_2, x'')].$$

We note that similar conditions for R-order, R-identity and R-associativity can be found in a relational semantics for distributive substructural logics, e.g. [18, Definition 6].[1] Thanks to the auxiliary relations R°, $R^\rightarrow$ and $R^\leftarrow$, we obtain other interpretations of formulae on bi-approximation semantics.

Theorem 4.4 *For every bi-approximation model* $\mathbb{M}$ *and all formulae* ϕ, ψ, *we have*

(i) $y \Vdash \phi \circ \psi \iff \forall x_1 \in X, y_2 \in Y.$ *if* $x_1 \Vdash \phi$ *and* $R^\rightarrow(x_1, y_2, y)$ *then* $y_2 \Vdash \psi$,

(ii) $y \Vdash \phi \circ \psi \iff \forall x_2 \in X, y_1 \in Y.$ *if* $x_2 \Vdash \psi$ *and* $R^\leftarrow(y_1, x_2, y)$ *then* $y_1 \Vdash \phi$,

(iii) $x_2 \Vdash \phi \rightarrow \psi \iff \forall x_1, x \in X.$ *if* $x_1 \Vdash \phi$ *and* $R^\circ(x_1, x_2, x)$ *then* $x \Vdash \psi$,

(iv) $x_2 \Vdash \phi \rightarrow \psi \iff \forall y_1, y \in Y.$ *if* $y \Vdash \psi$ *and* $R^\leftarrow(y_1, x_2, y)$ *then* $y_1 \Vdash \phi$,

(v) $x_1 \Vdash \psi \leftarrow \phi \iff \forall x_2, x \in X.$ *if* $x_2 \Vdash \phi$ *and* $R^\circ(x_1, x_2, x)$ *then* $x \Vdash \psi$,

(vi) $x_1 \Vdash \psi \leftarrow \phi \iff \forall y_2, y \in Y.$ *if* $y \Vdash \psi$ *and* $R^\rightarrow(x_1, y_2, y)$ *then* $y_2 \Vdash \phi$,

(vii) $x \Vdash \phi \circ \psi \Longleftarrow \exists x_1, x_2 \in X$ *such that* $x_1 \Vdash \phi$, $x_2 \Vdash \psi$ *and* $R^\circ(x_1, x_2, x)$,

(viii) $y_2 \Vdash \phi \rightarrow \psi \Longleftarrow \exists x_1 \in X$, $\exists y \in Y$ *such that* $x_1 \Vdash \phi$, $y \Vdash \psi$ *and* $R^\rightarrow(x_1, y_2, y)$,

[1] The order of the ternary relation is different. That is, $R^\circ(x_1, x_2, x)$ in this paper is the same with $R_\circ(x, x_1, x_2)$ in [18].

(ix) $y_1 \Vdash \psi \leftarrow \phi \Longleftarrow \exists x_2 \in X, \exists y \in Y$ *such that* $x_2 \Vdash \phi$, $y \Vdash \psi$ *and* $R^{\leftarrow}(y_1, x_2, y)$.

In Theorem 4.4, item (iii) and item (v) correspond to the normal interpretations in Kripke semantics. The same results for item (iii) and item (v) are obtained by generalized Kripke frames [7]. Moreover, item (vii) looks similar to the interpretation on ternary-relational semantics of distributive substructural logics. Item (vii) must be closely related to the discussion in [7, p.264]. However, unlike what happens in the setting of generalized Kripke frames, the conditions of item (vii), item (viii) and item (ix) are *more beneficial* to evaluate formulae in our framework. More precisely, the auxiliary relations R°, $R^{\rightarrow}$ and $R^{\leftarrow}$ provide bases of $V(\phi \circ \psi)$, $V(\phi \rightarrow \psi)$ and $V(\psi \leftarrow \phi)$: see Theorem 4.6 and Proposition 4.7.

Related to Theorem 4.4, we also obtain the following results for $\vee$ and $\wedge$.

Theorem 4.5 *Let* $\mathbb{M}$ *be an arbitrary bi-approximation model,* ϕ, ψ *be all formulae. For each* $x \in X$ *and each* $y \in Y$,

(i) $\mathbb{M}, x \Vdash \phi \vee \psi \Longleftarrow \mathbb{M}, x \Vdash \phi$ *or* $\mathbb{M}, x \Vdash \psi$,

(ii) $\mathbb{M}, y \Vdash \phi \wedge \psi \Longleftarrow \mathbb{M}, y \Vdash \phi$ *or* $\mathbb{M}, y \Vdash \psi$.

Items (vii)-(ix) in Theorem 4.4 and Theorem 4.5 indicate that, when we reason about formulae with the existential quantifier and disjunction, we may not accumulate all worlds in X (in Y) where the formulae are assumed (concluded). However, as we will see below, we can still collect *essential worlds* in X (in Y) to gather *all worlds* in Y (in X) where the formulae are concluded (assumed): see Theorem 4.6. Hereinafter, to discuss the connection between the existential quantifier and the bi-approximation clearly, we introduce an auxiliary relation $\Vdash_{\mathfrak{bs}}$ of $\Vdash$ as follows (the subscript $_{-\mathfrak{bs}}$ refers to *bases*, see Theorem 4.6):

(i) $x \Vdash_{\mathfrak{bs}} \phi \vee \psi \iff x \Vdash_{\mathfrak{bs}} \phi$ or $x \Vdash_{\mathfrak{bs}} \psi$,

(ii) $y \Vdash_{\mathfrak{bs}} \phi \wedge \psi \iff y \Vdash_{\mathfrak{bs}} \phi$ or $y \Vdash_{\mathfrak{bs}} \psi$,

(iii) $x \Vdash_{\mathfrak{bs}} \phi \circ \psi \iff \exists x_1, x_2 \in X$ s.t. $x_1 \Vdash_{\mathfrak{bs}} \phi$, $x_2 \Vdash_{\mathfrak{bs}} \psi$ and $R^{\circ}(x_1, x_2, x)$,

(iv) $y_2 \Vdash_{\mathfrak{bs}} \phi \rightarrow \psi \iff \exists x_1 \in X, \exists y \in Y$ s.t. $x_1 \Vdash_{\mathfrak{bs}} \phi$, $y \Vdash_{\mathfrak{bs}} \psi$ and $R^{\rightarrow}(x_1, y_2, y)$,

(v) $y_1 \Vdash_{\mathfrak{bs}} \psi \leftarrow \phi \iff \exists x_2 \in X, \exists y \in Y$ s.t. $x_2 \Vdash_{\mathfrak{bs}} \phi$, $y \Vdash_{\mathfrak{bs}} \psi$ and $R^{\leftarrow}(y_1, x_2, y)$.

(vi) $x \Vdash_{\mathfrak{bs}} \phi \iff x \Vdash \phi$, whenever ϕ is a propositional variable or a constant, or the outermost connective of ϕ is either $\wedge$, $\rightarrow$ or $\leftarrow$,

(vii) $y \Vdash_{\mathfrak{bs}} \psi \iff y \Vdash \psi$, whenever ψ is a propositional variable or a constant, or the outermost connective of ψ is either $\vee$ or $\circ$.

By parallel induction, we obtain the following straightforwardly. For every formula ϕ, each $x \in X$ and each $y \in Y$, we have

(i) if $x \Vdash_{\mathfrak{bs}} \phi$, then $x \Vdash \phi$,

(ii) if $y \Vdash_{\mathfrak{bs}} \phi$, then $y \Vdash \phi$.

Furthermore, we also obtain the following.

Theorem 4.6 *Let* $\mathbb{M}$ *be an arbitrary bi-approximation model and* ϕ *each formula. Then, we have the following (recall* $\tilde{V}$ *in Corollary 3.11 and basis in Definition 3.2):*

(i) *the set* $\{x \in X \mid \mathbb{M}, x \Vdash_{\mathfrak{bs}} \phi\}$ *is a basis of* $\tilde{V}_{\uparrow}(\phi)$*,*

(ii) *the set* $\{y \in Y \mid \mathbb{M}, y \Vdash_{\mathfrak{bs}} \phi\}$ *is a basis of* $\tilde{V}^{\downarrow}(\phi)$*.*

Proof. Parallel induction. Base cases are trivial.

(i) $\upsilon(\{y \in Y \mid y \Vdash_{\mathfrak{bs}} \phi \wedge \psi\}) = \tilde{V}^{\downarrow}(\phi \wedge \psi)$. ($\subseteq$). For each x, suppose that $x \leq y$, if $y \Vdash_{\mathfrak{bs}} \phi$ or $y \Vdash_{\mathfrak{bs}} \psi$ for every y. It is equivalent to both $x \leq y$ if $y \Vdash_{\mathfrak{bs}} \phi$ and $x \leq y$ if $y \Vdash_{\mathfrak{bs}} \psi$. By induction hypothesis, we have $x \Vdash \phi$ and $x \Vdash \psi$, hence $x \Vdash \phi \wedge \psi$. ($\supseteq$). trivial.

(ii) $\lambda(\{x \in X \mid x \Vdash_{\mathfrak{bs}} \phi \circ \psi\}) = \tilde{V}_{\uparrow}(\phi \circ \psi)$. For each $y \in Y$, by Theorem 4.4,

$$\begin{aligned} y \Vdash \phi \circ \psi &\iff \forall x_1, x_2.[x_1 \Vdash_{\mathfrak{bs}} \phi, x_2 \Vdash_{\mathfrak{bs}} \psi \Rightarrow R(x_1, x_2, y)] \\ &\iff \forall x_1, x_2, y_2.[x_1 \Vdash_{\mathfrak{bs}} \phi, R^{\rightarrow}(x_1, y_2, y) \Rightarrow (x_2 \Vdash_{\mathfrak{bs}} \psi \Rightarrow x_2 \leq y_2)] \\ &\iff \forall x_1, x_2' \in X.[x_1 \Vdash_{\mathfrak{bs}} \phi, x_2' \Vdash \psi \Rightarrow R(x_1, x_2', y)]. \end{aligned}$$

Note that $x_2 \Vdash_{\mathfrak{bs}} \psi$ changes to $x_2' \Vdash \psi$. Repeat the same replacement for x_1.

The other cases are analogous. □

Theorem 4.6 tells us that, in bi-approximation semantics, bases are (partly) inductively characterised by the existential quantifier and disjunction: see also ⊍-terms, ⋒-terms, pseudo-⊍-terms and pseudo-⋒-terms in [19]. Moreover, this property works beneficially together with the following proposition: see Remark 6.2.

Proposition 4.7 *Let* $\mathbb{M}$ *be an arbitrary bi-approximation model and* ϕ, ψ *all formulae. Then, we have*

$$\mathbb{M} \Vdash \phi \mapsto \psi \iff \forall x \in X, \forall y \in Y. \textit{ if } \mathbb{M}, x \Vdash_{\mathfrak{bs}} \phi \textit{ and } \mathbb{M}, y \Vdash_{\mathfrak{bs}} \psi, \textit{ then } x \leq y.$$

5 The Representation theorem

In this section, to prove Theorem 6.3, we show that FL-algebras can be represented by p-frames. By analogy to the situation in modal logic (e.g. [2]), we will show that the dual frames of FL-algebras are p-frames and the dual algebras of p-frames are FL-algebras. Moreover, the validity relations between p-frames and FL-algebras are also proved as in the case of modal logic: see Theorem 5.3 and Theorem 5.5.

Dual algebra of p-frame For each p-frame $\mathbb{F}$, we construct two isomorphic FL-algebras *in parallel* based on the isomorphic posets $\mathbb{D}$ and $\mathbb{U}$. Namely, we define the operations $\vee$, $\wedge$, $*$, $\backslash$ and $/$, and the constants 1 and 0 on both $\mathbb{D}$ and $\mathbb{U}$, as they are isomorphic FL-algebras, i.e. $\langle \mathbb{D}, \vee, \wedge, *, \backslash, /, 1, 0\rangle \cong \langle \mathbb{U}, \vee, \wedge, *, \backslash, /, 1, 0\rangle$.

Since $\mathbb{D}$ and $\mathbb{U}$ are isomorphic through the Galois connection $\lambda \dashv \upsilon$, we have two natural ways to define each operation, in general. That is, an operation on $\mathbb{U}$, *approximated from below*, and take the copy to the other side via $\upsilon : \mathbb{U} \to \mathbb{D}$. Or, an operation on $\mathbb{D}$, *approximated from above*, and take the copy to the other side via $\lambda : \mathbb{D} \to \mathbb{U}$. We define additive operations $\vee$ and $*$ are defined on $\mathbb{U}$, approximated from below, and multiplicative operations $\wedge$, $\backslash$ and $/$ are defined on $\mathbb{D}$, approximated from above. Otherwise, we cannot prove the residuality (see [9] and [10]).

For each p-frame $\mathbb{F} = \langle X, Y, \leq, R, O_X, O_Y, N_X, N_Y \rangle$, we define $\vee$, $\wedge$, $*$, $\backslash$ and $/$ are defined as follows: on $\mathbb{D}$, for all $\alpha^{\downarrow}, \beta^{\downarrow} \in \mathbb{D}$,

$\mathbb{D}$-1: $\alpha^{\downarrow} \vee \beta^{\downarrow} := \upsilon(\alpha_{\uparrow} \vee \beta_{\uparrow})$,

$\mathbb{D}$-2: $\alpha^{\downarrow} \wedge \beta^{\downarrow} := \alpha^{\downarrow} \cap \beta^{\downarrow}$,

$\mathbb{D}$-3: $\alpha^{\downarrow} * \beta^{\downarrow} := \upsilon(\alpha_{\uparrow} * \beta_{\uparrow})$,

$\mathbb{D}$-4: $\alpha^{\downarrow} \backslash \beta^{\downarrow} := \{x_2 \in X \mid \forall x_1 \in \alpha^{\downarrow}, \forall y \in \beta_{\uparrow}.\ R(x_1, x_2, y)\}$,

$\mathbb{D}$-5: $\beta^{\downarrow} / \alpha^{\downarrow} := \{x_1 \in X \mid \forall x_2 \in \alpha^{\downarrow}, \forall y \in \beta_{\uparrow}.\ R(x_1, x_2, y)\}$.

On $\mathbb{U}$, for all $\alpha_{\uparrow}, \beta_{\uparrow} \in \mathbb{U}$,

$\mathbb{U}$-1: $\alpha_{\uparrow} \vee \beta_{\uparrow} := \alpha_{\uparrow} \cap \beta_{\uparrow}$,

$\mathbb{U}$-2: $\alpha_{\uparrow} \wedge \beta_{\uparrow} := \lambda(\alpha^{\downarrow} \wedge \beta^{\downarrow})$,

$\mathbb{U}$-3: $\alpha_{\uparrow} * \beta_{\uparrow} := \{y \in Y \mid \forall x_1 \in \alpha^{\downarrow}, \forall x_2 \in \beta^{\downarrow}.\ R(x_1, x_2, y)\}$,

$\mathbb{U}$-4: $\alpha_{\uparrow} \backslash \beta_{\uparrow} := \lambda(\alpha^{\downarrow} \backslash \beta^{\downarrow})$,

$\mathbb{U}$-5: $\beta_{\uparrow} / \alpha_{\uparrow} := \lambda(\beta^{\downarrow} / \alpha^{\downarrow})$.

Based on these operations, we can show the following.

Theorem 5.1 *Both* $\langle \mathbb{D}, \vee, \wedge, *, \backslash, /, O_X, N_X \rangle$ *and* $\langle \mathbb{U}, \vee, \wedge, *, \backslash, /, O_Y, N_Y \rangle$ *are FL-algebras, and they are isomorphic.*

By Theorem 5.1, we naturally define the dual FL-algebras of p-frames.

Definition 5.2 [Dual algebra] Let $\mathbb{F}$ be a p-frame. *The dual algebra of* $\mathbb{F}$ is an algebra $\mathbb{F}^+ = \langle A, \vee, \wedge, *, \backslash, /, 1, 0 \rangle$ which is isomorphic to $\langle \mathbb{D}, \vee, \wedge, *, \backslash, /, O_X, N_X \rangle$ and $\langle \mathbb{U}, \vee, \wedge, *, \backslash, /, O_Y, N_Y \rangle$.

Along with the definition of dual algebras, we obtain the equivalence of validity, as usual.

Theorem 5.3 *For every p-frame* $\mathbb{F}$ *and each sequent* $\Gamma \mapsto \varphi$, *the sequent* $\Gamma \mapsto \varphi$ *is valid on* $\mathbb{F}$ *if and only if it is valid on the dual algebra* $\mathbb{F}^+$.

$$\mathbb{F} \Vdash \Gamma \mapsto \varphi \iff \mathbb{F}^+ \models \Gamma \leq \varphi$$

Dual frame of FL-algebras Here we construct the dual frames of FL-algebras. We mention that the dual frame corresponds to *the intermediate level* introduced in [11] but see also [5] and [19].

Let $\mathbb{A} = \langle A, \vee, \wedge, *, \backslash, /, 1, 0 \rangle$ be a FL-algebra. The set of all filters and the set of all ideals are denoted by $\mathcal{F}$ and $\mathcal{I}$. On $\mathcal{F} \cup \mathcal{I}$, we define a binary relation $\sqsubseteq$ as follows: for all filters $F, F_1, F_2 \in \mathcal{F}$ and all ideals $I, I_1, I_2 \in \mathcal{I}$,

(i) $F_1 \sqsubseteq F_2 \iff F_1 \supseteq F_2$,

(ii) $F \sqsubseteq I \iff F \cap I \neq \emptyset$,

(iii) $I \sqsubseteq F \iff \forall a \in I, \forall b \in F.\ a \leq b$,

(iv) $I_1 \sqsubseteq I_2 \iff I_1 \subseteq I_2$.

Next, on the triple $\langle \mathcal{F}, \mathcal{I}, \sqsubseteq \rangle$, we build a ternary relation R, and subsets $O_{\mathcal{F}}$, $O_{\mathcal{I}}$, $N_{\mathcal{F}}$ and $N_{\mathcal{I}}$ as follows: for all $F_1, F_2 \in \mathcal{F}$ and each $I \in \mathcal{I}$,

(i) $R(F_1, F_2, I) \iff F_1 * F_2 \sqsubseteq I$,
where $F_1 * F_2 := \{a \in A \mid \exists f_1 \in F_1, \exists f_2 \in F_2.\ f_1 * f_2 \leq a\}$,

(ii) $O_{\mathcal{F}}$ is the set of all filters containing 1,

(iii) $O_{\mathcal{I}}$ is the set of all ideals containing 1,

(iv) $N_{\mathcal{F}}$ is the set of all filters containing 0,

(v) $N_{\mathcal{I}}$ is the set of all ideals containing 0.

Then, the 8-tuple $\mathbb{A}_+ = \langle \mathcal{F}, \mathcal{I}, \sqsubseteq, R, O_{\mathcal{F}}, O_{\mathcal{I}}, N_{\mathcal{F}}, N_{\mathcal{I}} \rangle$ is *the dual frame of* $\mathbb{A}$. To prove the following theorems, we here mention that, for all $F, F_1, F_2 \in \mathcal{F}$ and each $I, I_1, I_2 \in \mathcal{I}$,

(i) $F_1 * F_2$ is a filter,

(ii) $F \backslash I := \{a \in A \mid \exists f \in F, \exists i \in I.\ a \leq f \backslash i\}$ is an ideal,

(iii) $I / F := \{a \in A \mid \exists i \in I, \exists f \in F.\ a \leq i / f\}$ is an ideal.

(iv) $R^{\circ}(F_1, F_2, F) \iff F \sqsubseteq F_1 * F_2$,

(v) $R^{\rightarrow}(F_1, I_2, I) \iff F_1 \backslash I \sqsubseteq I_2$,

(vi) $R^{\leftarrow}(I_1, F_2, I) \iff I / F_2 \sqsubseteq I_1$.

Then, we can prove the following.

Theorem 5.4 *For any FL-algebra* $\mathbb{A}$*, the dual frame* $\mathbb{A}_+$ *is a p-frame.*

We prove the validity relationship between FL-algebras and the dual p-frames.

Theorem 5.5 *Let* $\mathbb{A}$ *be every FL-algebra and* $\Gamma \mapsto \varphi$ *each sequent. If the sequent is valid on the dual frame* $\mathbb{A}_+$*, it is also valid on the original FL-algebra* $\mathbb{A}$.

$$\mathbb{A} \models \Gamma \leq \varphi \Longleftarrow \mathbb{A}_+ \Vdash \Gamma \mapsto \varphi$$

6 Soundness and Completeness

In this section, we will show that p-frames are a sound and complete semantics for the substructural logic **FL**. Unlike what happens in the setting of relational semantics for distributive substructural logics, soundness is not straightforward. This is because bi-approximation models evaluate formulae through *the Galois connection* $\lambda \dashv \upsilon$. To avoid this complex argument, we can use the relationship between the *bi-approximation* and the *bases*: recall Proposition 4.7.

Theorem 6.1 (Soundness) *Let* $\Gamma \mapsto \varphi$ *be an arbitrary sequent. If the sequent* $\Gamma \mapsto \varphi$ *is derivable in FL, it is valid on every p-frame* $\mathbb{F}$.

Proof. Let $\mathbb{F}$ be an arbitrary p-frame and V an arbitrary doppelgänger valuation on $\mathbb{F}$. On the bi-approximation model $\mathbb{M} = \langle \mathbb{F}, V \rangle$, all initial sequents are true, by Proposition

3.9. Note that we use Proposition 4.7 to prove the inductive steps. We mention that (fw) and ($\circ \Rightarrow$) are trivial, and ($\Rightarrow \vee_2$), ($\wedge_2 \Rightarrow$), ($\Rightarrow\rightarrow$) and ($\leftarrow\Rightarrow$) are analogous to ($\Rightarrow \vee_1$), ($\wedge_1 \Rightarrow$), ($\Rightarrow\leftarrow$) and ($\rightarrow\Rightarrow$), respectively.

(cut): For arbitrary $x \in X$ and $y \in Y$, let $\mathbb{M}, x \Vdash_{\mathfrak{bs}} \Sigma \circ \Gamma \circ \Pi$ and $\mathbb{M}, y \Vdash_{\mathfrak{bs}} \varphi$. Then, there exist $x_1, x_2, x_3, x' \in X$ such that $x_1 \Vdash_{\mathfrak{bs}} \Sigma$, $x_2 \Vdash_{\mathfrak{bs}} \Gamma$, $x_3 \Vdash_{\mathfrak{bs}} \Pi$, $R^\circ(x_1, x', x)$ and $R^\circ(x_2, x_3, x')$. By induction hypothesis, $\Gamma \Rightarrow \phi$ is true on $\mathbb{M}$. By Proposition 3.9, we obtain that $x_2 \Vdash \phi$, hence $x \Vdash \Sigma \circ \phi \circ \Pi$. Again, by induction hypothesis, $\mathbb{M} \Vdash \Sigma \circ \phi \circ \Pi \Rightarrow \varphi$, which concludes $x \leq y$.

(tw): For arbitrary $x \in X$ and $y \in Y$, let $\mathbb{M}, x \Vdash_{\mathfrak{bs}} \Gamma \circ \mathbf{t} \circ \Delta$ and $y \Vdash_{\mathfrak{bs}} \varphi$. Then, there exist $x_1, x_2, x_3, x' \in X$ such that $x_1 \Vdash_{\mathfrak{bs}} \Gamma$, $x_2 \Vdash_{\mathfrak{bs}} \mathbf{t}$, $x_3 \Vdash_{\mathfrak{bs}} \Delta$, $R^\circ(x_1, x', x)$ and $R^\circ(x_2, x_3, x')$. Because $x_2 \in O_X$ and $R^\circ(x_2, x_3, x')$, we obtain $x' \leq x_3$ by R-order. By Hereditary (Proposition 3.8), we also have $x' \Vdash \Delta$, hence $x \Vdash \Gamma \circ \Delta$ holds. Finally, by induction hypothesis, $\mathbb{M} \Vdash \Gamma \circ \Delta \Rightarrow \varphi$. Therefore, $x \leq y$.

($\Rightarrow \circ$): For arbitrary $x \in X$ and $y \in Y$, let $x \Vdash_{\mathfrak{bs}} \Gamma \circ \Sigma$ and $y \Vdash \phi \circ \psi$. Then, there exist $x_1, x_2 \in X$ such that $x_1 \Vdash_{\mathfrak{bs}} \Gamma$, $x_2 \Vdash_{\mathfrak{bs}} \Sigma$ and $R^\circ(x_1, x_2, x)$. By inductive hypothesis, we have $\mathbb{M} \Vdash \Gamma \Rightarrow \phi$ and $\mathbb{M} \Vdash \Sigma \Rightarrow \psi$. We obtain $x_1 \Vdash \phi$ and $x_2 \Vdash \psi$. By definition, since $y \Vdash \phi \circ \psi$, $R(x_1, x_2, y)$ holds. Because of Definition 4.1, we conclude $x \leq y$.

($\rightarrow\Rightarrow$): For arbitrary $x \in X$ and $y \in Y$, let $x \Vdash_{\mathfrak{bs}} \Sigma \circ \Gamma \circ (\phi \rightarrow \psi) \circ \Pi$ and $y \Vdash_{\mathfrak{bs}} \varphi$. By inductive hypothesis, we have $\mathbb{M} \Vdash \Sigma \circ \psi \circ \Pi \Rightarrow \varphi$, hence $y \Vdash \Sigma \circ \psi \circ \Pi$. Moreover, there exist $x_1, x_2, x_3, x_4, x', x'' \in X$ such that $x_1 \Vdash_{\mathfrak{bs}} \Sigma$, $x_2 \Vdash_{\mathfrak{bs}} \Gamma$, $x_3 \Vdash_{\mathfrak{bs}} \phi \rightarrow \psi$, $x_4 \Vdash_{\mathfrak{bs}} \Pi$, $R^\circ(x_2, x_3, x')$, $R^\circ(x', x_4, x'')$ and $R^\circ(x_1, x'', x)$. By inductive hypothesis, $\mathbb{M} \Vdash \Gamma \Rightarrow \phi$, hence $x_2 \Vdash \phi$. Furthermore, because $x_2 \Vdash \phi$ and $x_3 \Vdash \phi \rightarrow \psi$, we have that, for each $x''' \in X$, if $R^\circ(x_2, x_3, x''')$ holds, then $x''' \Vdash \psi$ (Theorem 4.4). Because of $R^\circ(x_2, x_3, x')$, we obtain $x' \Vdash \psi$. Hence, we derive $x \Vdash \Sigma \circ \psi \circ \Pi$. Therefore, $x \leq y$. □

Remark 6.2 We mention that, in the proof of Theorem 6.1, we effectively use the bi-approximation, bases and the existential quantifier, i.e. Theorem 4.4, Theorem 4.6 and Proposition 4.7, to *stay away from taking the Galois connection.*

Theorem 6.3 (Completeness) *Let $\Gamma \Rightarrow \varphi$ be an arbitrary sequent. If the sequent $\Gamma \Rightarrow \varphi$ is valid on every p-frame $\mathbb{F}$, then it is derivable in FL.*

Proof. Let $\mathbb{L}$ be Lindenbaum-Tarski algebra of substructural logic **FL**. If $\Gamma \Rightarrow \varphi$ is not derivable in FL, then $\Gamma \Rightarrow \varphi$ is not valid on $\mathbb{L}$. By Theorem 5.4, the dual frame $\mathbb{L}_+$ of $\mathbb{L}$ is a p-frame. Furthermore, by theorem 5.5, the sequent $\Gamma \Rightarrow \varphi$ is not valid on $\mathbb{L}_+$. □

Therefore, combined with the canonicity results in [19], we obtain the following.

Theorem 6.4 *Let Ω be a set of sequents which have consistent variable occurrence (see [19]). A substructural logic extended by Ω is complete with respect to a class of p-frames.*

7 Conclusion

We introduced bi-approximation semantics to describe Ghilardi and Meloni's parallel computation of the canonical extension of lattice expansions. Unlike what happens in the setting of standard relational semantics, like Kripke semantics or Routley-Meyer semantics, bi-approximation semantics is two-sorted. However, we claim that this is a natural framework for the study of logic, because logic is *a priori* two-sorted: premises and conclusions. In other words, logic is the study of a consequence relation.

From this point of view, bi-approximation semantics is a reasonable relational-type semantics for lattice-based logics. This framework could be valuable when we think about resource sensitive logics, since we explicitly distinguish premises from conclusions. Even over distributive lattice-based logics like intuitionistic logic, our two-sorted semantics may be worthwhile. For example, the first-order definability for intuitionistic modal logic on Kripke semantics is still open (see the footnote in [11, p.2]), whereas the first-order definability on bi-approximation semantics is effectively solved (in preparation).

References

[1] Birkhoff, G., "Lattice theory," American Mathematical Society Colloquium Publications **XXV**, American Mathematical Society, New York, 1948, revised edition.

[2] Blackburn, P., M. de Rijke and Y. Venema, "Modal logic," Cambridge Tracts in Theoretical Computer Science **53**, Cambridge University Press, Cambridge, 2002.

[3] Chagrov, A. and M. Zakharyaschev, "Modal logic," Oxford Logic Guides **35**, Oxford Science Publications, New York, 1997.

[4] Davey, B. and H. Priestley, "Introduction to Lattices and Order," Cambridge University Press, Cambridge, 2002, 2nd edition.

[5] Dunn, M., M. Gehrke and A. Palmigiano, *Canonical extensions and relational completeness of some substructural logics*, The Journal of Symbolic Logic **70** (2005), pp. 713–740.

[6] Galatos, N., P. Jipsen, T. Kowalski and H. Ono, "Residuated lattices: an algebraic glimpse at substructural logics," Studies in Logics and the Foundation of Mathematics **151**, Elsevier, Amsterdam, 2007.

[7] Gehrke, M., *Generalized Kripke frames*, Studia Logica **84** (2006), pp. 241–275.

[8] Gehrke, M. and J. Harding, *Bounded lattice expansions*, Journal of Algebra **239** (2001), pp. 345–371.

[9] Gehrke, M. and H. Priestley, *Non-canonicity of MV-algebras*, Houston Journal of Mathematics **28** (2002), p. 449.

[10] Gehrke, M. and H. Priestley, *Canonical extensions of double quasioperator algebras: an algebraic perspective on duality for certain algebras with binary operators*, Journal of Pure and Applied Algebra **209** (2007), pp. 269–290.

[11] Ghilardi, S. and G. Meloni, *Constructive canonicity in non-classical logics*, Annals of Pure and Applied Logic **86** (1997), pp. 1–32.

[12] Goldblatt, R., *Semantic analysis of orthologic*, Journal of Philosophical Logic **3** (1974), pp. 19–35.

[13] Hartonas, C., *Duality for lattice-ordered algebras and normal algebraizable logics*, Studia Logica **58** (1997), pp. 403–450.

[14] Hartonas, C. and J. M. Dunn, *Stone duality for lattices*, Algebra Universalis **37** (1997), pp. 391–401.

[15] Ono, H., *Substructural logics and residuated lattices - an introduction*, in: V. F. Hendricks and J. Malinowski, editors, *50 Years of Studia Logica: Trends in Logic*, Kluwer Academic Publishers, Dordrecht, 2003 pp. 193–228.

[16] Restall, G., "An Introduction to Substructural Logics," Routledge, London, 2000.

[17] Routley, R., V. Plumwood, R. K. Meyer and R. T. Brady, "Relevant logics and their rivals. Part 1. The basic philosophical and semantical theory," Ridgeview Publishing Company, Atascadero, 1982.

[18] Suzuki, T., *A relational semantics for distributive substructural logics and the topological characterization of the descriptive frames*, CALCO-jnr 2007 Report No.367, Department of Informatics, University of Bergen (2008).
URL `http://www.ii.uib.no/publikasjoner/texrap/pdf/2008-367.pdf`

[19] Suzuki, T., *Canonicity results of substructural and lattice-based logics*, The Review of Symbolic Logic **3** (2010).

[20] Urquhart, A., "The semantics of entailment," Ph.D. thesis, University of Pittsburgh (1972).

[21] Wright, F. B., *Polarity and duality*, Pacific Journal of Mathematics **10** (1960), pp. 723–730.

A Appendix of proofs

Proof. [Proposition 3.8] Parallel induction. Base cases are straightforward, since every Galois stable X-set is a downset and every Galois stable Y-set is an upset.

Inductive steps: $\vee$: Assume $y \Vdash \phi \vee \psi$. By definition, $y \Vdash \phi$ and $y \Vdash \psi$. By induction hypothesis, we obtain $y' \Vdash \phi$ and $y' \Vdash \psi$, hence $y' \Vdash \phi \vee \psi$. Suppose $x \Vdash \phi \vee \psi$. For each $y \Vdash \phi \vee \psi$, we have $x \leq y$. Because of $x' \leq x$, we obtain $x' \leq y$, hence $x' \Vdash \phi \vee \psi$.

$\wedge$: Assume $x \Vdash \phi \wedge \psi$. By definition, $x \Vdash \phi$ and $x \Vdash \psi$. By induction hypothesis, we obtain $x' \Vdash \phi$ and $x' \Vdash \psi$, hence $x' \Vdash \phi \wedge \psi$. Suppose $y \Vdash \phi \wedge \psi$. For each $x \Vdash \phi \wedge \psi$, we have $x \leq y$. Because of $y \leq y'$, we obtain $x \leq y'$, hence $y' \Vdash \phi \wedge \psi$.

$\circ$: Assume $y \Vdash \phi \circ \psi$. If $x_1 \Vdash \phi$ and $x_2 \Vdash \psi$, then we have $R(x_1, x_2, y)$. Since $y \leq y'$, by R-transitivity, we obtain $R(x_1, x_2, y')$, hence $y' \Vdash \phi \circ \psi$. Suppose $x \Vdash \phi \circ \psi$. For each $y \Vdash \phi \circ \psi$, we have $x \leq y$. Because of $x' \leq x$, we obtain $x' \leq y$, hence $x' \Vdash \phi \circ \psi$.

$\rightarrow$: Assume $x \Vdash \phi \rightarrow \psi$. For each $x_1 \Vdash \phi$ and each $y \Vdash \psi$, we have $R(x_1, x, y)$. By R-transitivity, we have $R(x_1, x', y)$, hence $x' \Vdash \phi \rightarrow \psi$. Suppose $y \Vdash \phi \rightarrow \psi$. For each $x \Vdash \phi \rightarrow \psi$, we have $x \leq y$. Since $y \leq y'$, we obtain $x \leq y'$, hence $y' \Vdash \phi \rightarrow \psi$.

$\leftarrow$: Assume $x \Vdash \psi \leftarrow \phi$. For each $x_2 \Vdash \phi$ and each $y \Vdash \psi$, we have $R(x, x_2, y)$. By R-transitivity, we have $R(x', x_2, y)$, hence $x' \Vdash \psi \leftarrow \phi$. Suppose $y \Vdash \psi \leftarrow \phi$. For each $x \Vdash \psi \leftarrow \phi$, we have $x \leq y$. Because of $y \leq y'$, we obtain $x \leq y'$, hence $y' \Vdash \psi \leftarrow \phi$.

□

Proof. [Lemma 4.2] Item (i). ($\Rightarrow$). Suppose $R^{\circ}(x_1, x_2, x)$, i.e. if $R(x_1, x_2, y')$ then $x \leq y'$ for every $y' \in Y$. By assumption, we obtain $R(x_1, x_2, y)$, which derives $x \leq y$.

($\Leftarrow$). Contraposition. Namely, we claim that there exists $x \in X$ such that $R^{\circ}(x_1, x_2, x)$ and $x \not\leq y$, under the assumption that $R(x_1, x_2, y)$ does not hold. Suppose that $R(x_1, x_2, y)$ does not hold. By $\circ$-tightness, there exists $x \in X$ such that, $R^{\circ}(x_1, x_2, x)$, and, for each $y' \in Y$, if $x \leq y'$, then $R(x_1, x_2, y')$. Since $R(x_1, x_2, y)$ does not hold, we have $x \not\leq y$. Item (ii) and item (iii) are analogous to item (i). □

Proof. [Theorem 4.4] Items (i) - (v) are analogous to item (vi). And, item (viii) and item (ix) are analogous to item (vii).

(vi) By Proposition 3.9, Definition 4.1 and Lemma 4.2, we can prove as follows.

$$\begin{aligned} x_1 \Vdash \psi \leftarrow \phi &\iff \forall x_2 \in X, \forall y \in Y.[x_2 \Vdash \phi, y \Vdash \psi \Rightarrow R(x_1, x_2, y)] \\ &\iff \forall x_2 \in X, \forall y_2, y \in Y.[x_2 \Vdash \phi, y \Vdash \psi, R^{\rightarrow}(x_1, y_2, y) \Rightarrow x_2 \leq y_2] \\ &\iff \forall y_2, y \in Y.[y \Vdash \psi, R^{\rightarrow}(x_1, y_2, y) \Rightarrow y_2 \Vdash \phi] \end{aligned}$$

(vii) Suppose that there exist $x_1, x_2 \in X$ such that $x_1 \Vdash \phi$, $x_2 \Vdash \psi$ and $R^{\circ}(x_1, x_2, x)$. We claim that every element $y \in Y$ at which $\phi \circ \psi$ is concluded is above x. If $y \Vdash \phi \circ \psi$ holds, then, by definition, $R(x_1, x_2, y)$ holds. By Definition 4.1, we also obtain that $x \leq y'$, whenever $R(x_1, x_2, y')$ holds for every $y' \in Y$. Hence, $x \leq y$ holds, which derives $x \Vdash \phi \circ \psi$.

□

Proof. [Theorem 4.5] (i). Suppose that $x \Vdash \phi$ or $x \Vdash \psi$. For an arbitrary $y \in Y$, if $y \Vdash \phi \vee \psi$, by definition, $y \Vdash \phi$ and $y \Vdash \psi$. By Proposition 3.9, $x \Vdash \phi$ or $x \Vdash \psi$, either way, $x \leq y$ holds. Therefore, $x \Vdash \phi \vee \psi$. Item (ii) is analogous to item (i). □

Proof. [Proposition 4.7] ($\Rightarrow$). Since $x \Vdash \phi$ ($y \Vdash \psi$) whenever $x \Vdash_{\mathfrak{bs}} \phi$ ($y \Vdash_{\mathfrak{bs}} \psi$), this is trivial. ($\Leftarrow$). Let x be an arbitrary element where ϕ is premised, y an arbitrary element where ψ is concluded. By our assumption, for an arbitrary $x_B \Vdash_{\mathfrak{bs}} \phi$, we have $x_B \leq y_B$ for every $y_B \Vdash_{\mathfrak{bs}} \psi$. By Theorem 4.6, we obtain $x_B \Vdash \psi$, hence $x_B \leq y$ (Proposition 3.9). As x_B is arbitrary, by Theorem 4.6, $y \Vdash \phi$ also holds. Therefore, $x \leq y$ (Proposition 3.9). □

Proof. [Theorem 5.1] Firstly, we need to check well-definedness of each operation. Namely, it is necessary to show that every value returns a Galois stable set. The copying parts are trivial, hence we need to check the following definition parts.

$\vee$: We claim that $\alpha_\uparrow \cap \beta_\uparrow = \lambda(\alpha^\downarrow \cup \beta^\downarrow)$. ($\subseteq$). For each $y \in \alpha_\uparrow \cap \beta_\uparrow$, since $y \in \alpha_\uparrow$ and $y \in \beta_\uparrow$, $x \leq y$ for each $x \in \alpha^\downarrow \cup \beta^\downarrow$. ($\supseteq$). If $y \in \lambda(\alpha^\downarrow \cup \beta^\downarrow)$, for arbitrary $x_a \in \alpha^\downarrow$ and $x_b \in \beta^\downarrow$, we have $x_a \leq y$ and $x_b \leq y$, hence $y \in \alpha_\uparrow$ and $y \in \beta_\uparrow$.

$\wedge$: We claim that $\alpha^\downarrow \cap \beta^\downarrow = \upsilon(\alpha_\uparrow \cup \beta_\uparrow)$. ($\subseteq$). For each $x \in \alpha^\downarrow \cap \beta^\downarrow$, since $x \in \alpha^\downarrow$ and $x \in \beta^\downarrow$, $x \leq y$ for each $y \in \alpha_\uparrow \cup \beta_\uparrow$. ($\supseteq$). If $x \in \upsilon(\alpha_\uparrow \cup \beta_\uparrow)$, for arbitrary $y_a \in \alpha_\uparrow$ and $y_b \in \beta_\uparrow$, we have $x \leq y_a$ and $x \leq y_b$, hence $x \in \alpha^\downarrow$ and $x \in \beta^\downarrow$.

$*$: We claim that $\alpha_\uparrow * \beta_\uparrow = \lambda(\{x \in X \mid x_1 \in \alpha^\downarrow, x_2 \in \beta^\downarrow, R^\circ(x_1, x_2, x)\})$.

$$\begin{aligned}\alpha_\uparrow * \beta_\uparrow &= \{y \in Y \mid \forall x_1 \in \alpha^\downarrow, \forall x_2 \in \beta^\downarrow, R(x_1, x_2, y)\}\\ &= \{y \in Y \mid \forall x \in X, \forall x_1 \in \alpha^\downarrow, \forall x_2 \in \beta^\downarrow, R^\circ(x_1, x_2, x) \Rightarrow x \leq y\}\\ &= \lambda(\{x \in X \mid x_1 \in \alpha^\downarrow, x_2 \in \beta^\downarrow, R^\circ(x_1, x_2, x)\})\end{aligned}$$

$\backslash$: We claim that $\alpha^\downarrow \backslash \beta^\downarrow = \upsilon(\{y_2 \in Y \mid x_1 \in \alpha^\downarrow, y \in \beta_\uparrow, R^\rightarrow(x_1, y_2, y)\})$.

$$\begin{aligned}\alpha^\downarrow \backslash \beta^\downarrow &= \{x_2 \in X \mid \forall x_1 \in \alpha^\downarrow, \forall y \in \beta_\uparrow, R(x_1, x_2, y)\}\\ &= \{x_2 \in X \mid \forall y_2 \in Y, \forall x_1 \in \alpha^\downarrow, \forall y \in \beta_\uparrow, R^\rightarrow(x_1, y_2, y) \Rightarrow x_2 \leq y_2\}\\ &= \upsilon(\{y_2 \in Y \mid x_1 \in \alpha^\downarrow, y \in \beta_\uparrow, R^\rightarrow(x_1, y_2, y)\})\end{aligned}$$

$/$ is analogous to $\backslash$.

Therefore, all operations are well-defined. Furthermore, these two algebras are isomorphic by definition. Next, we prove they are FL-algebras.

$\langle \mathbb{D}, \vee, \wedge \rangle$ and $\langle \mathbb{U}, \vee, \wedge \rangle$ are lattices. For all α, β, γ, we claim that [2]

$$\alpha \leq \gamma \text{ and } \beta \leq \gamma \iff \alpha \vee \beta \leq \gamma, \tag{A.1}$$

$$\gamma \leq \alpha \text{ and } \gamma \leq \beta \iff \gamma \leq \alpha \wedge \beta. \tag{A.2}$$

($\Rightarrow$) of the condition (A.1). For each $y \in \gamma_\uparrow$, since $\alpha_\uparrow \supseteq \gamma_\uparrow$ and $\beta_\uparrow \supseteq \gamma_\uparrow$, we have $y \in \alpha_\uparrow$ and $y \in \beta_\uparrow$, hence $y \in \alpha_\uparrow \cap \beta_\uparrow$. ($\Leftarrow$) of the condition (A.1). For each $y \in \gamma_\uparrow$, since $\alpha_\uparrow \cap \beta_\uparrow \supseteq \gamma_\uparrow$, we obtain $y \in \alpha_\uparrow$ and $y \in \beta_\uparrow$. The condition (A.2) is analogous.

[2] Recall that the order $\leq$ is $\subseteq$ on $\mathbb{D}$ and $\supseteq$ on $\mathbb{U}$.

$\langle \mathbb{D}, *, O_X \rangle$ and $\langle \mathbb{U}, *, O_Y \rangle$ are monoids. For all α, β, γ, we claim that

$$\alpha * (\beta * \gamma) = (\alpha * \beta) * \gamma, \tag{A.3}$$

$$\alpha * O = \alpha = O * \alpha, \tag{A.4}$$

where O is either O_X or O_Y depending on the domain. The condition (A.3). Let y be an arbitrary element in $\alpha_\uparrow * (\beta_\uparrow * \gamma_\uparrow)$. By Theorem 4.6, for all $x_1, x_2, x_3, x', x \in X$, if $x_1 \in \alpha^\downarrow$, $x_2 \in \beta^\downarrow$, $x_3 \in \gamma^\downarrow$, $R^\circ(x_1, x', x)$ and $R^\circ(x_2, x_3, x')$, then $x \leq y$ holds. By R-associativity (see Remark 4.3), the condition is equivalent to that, for each element x, for all $x'' \in X$, if $x_1 \in \alpha^\downarrow$, $x_2 \in \beta^\downarrow$, $x_3 \in \gamma^\downarrow$, $R^\circ(x'', x_3, x)$ and $R^\circ(x_1, x_2, x'')$, then $x \leq y$, which concludes $y \in (\alpha_\uparrow * \beta_\uparrow) * \gamma_\uparrow$.

The left equality of the condition (A.4). ($\subseteq$). For each $x_1 \in \alpha^\downarrow$, by R-identity, there exists $o_2 \in O_X$ such that $R^\circ(x_1, o_2, x_1)$. By definition, for every $y' \in Y$, if $R(x_1, o_2, y')$, then $x_1 \leq y'$ holds. Now, for every $y \in \alpha_\uparrow * O_Y$, by definition, $R(x_1, o_2, y)$ holds, hence $x_1 \leq y$. Since x_1 is arbitrary in $\alpha^\downarrow$, which derives $y \in \alpha_\uparrow$. ($\supseteq$). For arbitrary $x \in \alpha^\downarrow$ and $o \in O_X$, by $\circ$-tightness, there exists $x' \in X$ such that $R^\circ(x, o, x')$ and $x' \leq y' \Rightarrow R(x, o, y')$ for each $y' \in Y$. For every $y \in \alpha_\uparrow$, we have $x \leq y$, because of $x \in \alpha^\downarrow$. Furthermore, by R-order, $x' \leq x$ holds. Since $\leq$ is transitive, we obtain $x' \leq y$, hence $R(x, o, y)$. The right equality of the condition (A.4) is analogous.

Finally, we will show the residuality: for all α, β, γ,

$$\alpha * \beta \leq \gamma \iff \beta \leq \alpha \backslash \gamma \iff \alpha \leq \gamma / \beta. \tag{A.5}$$

($\Rightarrow$) of the first equivalence in the condition (A.5). Let x_2 be an arbitrary element in $\beta^\downarrow$. For arbitrary $x_1 \in \alpha^\downarrow$ and $y \in \gamma_\uparrow$, since $\alpha_\uparrow * \beta_\uparrow \supseteq \gamma_\uparrow$, we have $R(x_1, x_2, y)$. Hence, $x_2 \in \alpha^\downarrow \backslash \gamma^\downarrow$. ($\Leftarrow$) of the first equivalence in the condition (A.5). Let y be an arbitrary element in $\gamma_\uparrow$. For arbitrary $x_1 \in \alpha^\downarrow$ and $x_2 \in \beta^\downarrow$, since $\beta^\downarrow \subseteq \alpha^\downarrow \backslash, \gamma^\downarrow$, we obtain $R(x_1, x_2, y)$. Hence, $y \in \alpha_\uparrow * \beta_\uparrow$. The other equivalence is analogous. □

Proof. [Theorem 5.4] By definition, $\langle \mathcal{F}, \mathcal{I}, \sqsubseteq \rangle$ is a polarity.

R-order: Let F, F' be arbitrary filters. Suppose $F' \sqsubseteq F$. Since $F = F * \uparrow 1$, we obtain $F' \sqsubseteq F * \uparrow 1$. Conversely, if $F' \sqsubseteq F * O$ or $F' \sqsubseteq O * F$ for some $O \in O_{\mathcal{F}}$, because $1 \in O$, we obtain $F * O \sqsubseteq F$ or $O * F \sqsubseteq F$, hence $F' \sqsubseteq F$.

R-identity: Let $\uparrow 1$ be the principal filter generated by 1. For each filter F, we have $F * \uparrow 1 = \uparrow 1 * F = F$, hence $R^\circ(F, \uparrow 1, F)$ and $R^\circ(\uparrow 1, F, F)$.

R-transitivity: For all $F_1, F_1', F_2, F_2' \in \mathcal{F}$ and all $I, I' \in \mathcal{I}$, if $F_1' \sqsubseteq F_1$, $F_2' \sqsubseteq F_2$, $I \sqsubseteq I'$ and $F_1 * F_2 \sqsubseteq I$, then there exist $f_1 \in F_1$, $f_2 \in F_2$ and $i \in I$ such that $f_1 * f_2 \leq i$. Since $f_1 \in F_1'$, $f_2 \in F_2'$ and $i \in I'$, we also have $F_1' * F_2' \sqsubseteq I'$.

R-associativity: For all $F_1, F_2, F_3 \in \mathcal{F}$, we have $F_1 * (F_2 * F_3) = (F_1 * F_2) * F_3$, by the associativity of $*$ on $\mathbb{A}$. If $F \sqsubseteq F_1 * F'$ and $F' \sqsubseteq F_2 * F_3$, we obtain $F \sqsubseteq F_1 * (F_2 * F_3) = (F_1 * F_2) * F_3$. Let $F'' = F_1 * F_2$. Then, $F \sqsubseteq F'' * F_3$ and $F'' \sqsubseteq F_1 * F_2$ hold.

O-isom (N-isom): For each $F \in O_{\mathcal{F}}$ ($N_{\mathcal{F}}$) and each $I \in O_{\mathcal{I}}$ ($N_{\mathcal{I}}$), they have 1 (0) in common.

$\circ$-tightness: For all $F_1, F_2 \in \mathcal{F}$, it is trivially true that $R(F_1, F_2, I)$ if and only if $F_1 * F_2 \sqsubseteq I$ for every $I \in \mathcal{I}$. The other is analogous.

$\rightarrow$-tightness: For each $F_1 \in \mathcal{F}$ and each $I \in \mathcal{I}$, by definition, for each $F_2 \in \mathcal{F}$, $R(F_1, F_2, I)$ if and only if $F_2 \sqsubseteq F_1 \backslash I$.

$\leftarrow$-tightness: For each $F_2 \in \mathcal{F}$ and each $I \in \mathcal{I}$, by definition, for each $F_1 \in \mathcal{F}$, $R(F_1, F_2, I)$ if and only if $F_1 \sqsubseteq I / F_2$.

□

Proof. [Theorem 5.5] Let $f : \Phi \to \mathbb{A}$ be an arbitrary assignment. We also denote the normally extended assignment $f : \Lambda \to \mathbb{A}$ by f. Then, we define a doppelgänger valuation V based on f as follows: for each proposition $p \in \Phi$,

(i) $V^{\downarrow}(p) := \{F \in \mathcal{F} \mid f(p) \in F\} = \upsilon(\{\downarrow f(p)\})$,

(ii) $V_{\uparrow}(p) := \{I \in \mathcal{I} \mid f(p) \in I\} = \lambda(\{\uparrow f(p)\})$.

We claim that, for each filter F, each ideal I and each formula ϕ, $f(\phi) \in F \iff \mathbb{A}_+, V, F \Vdash \phi$ and $f(\phi) \in I \iff \mathbb{A}_+, V, I \Vdash \phi$. Base cases are trivial. Inductive steps. For each filter $F \in \mathcal{F}$ and each ideal $I \in \mathcal{I}$,

$\vee$: Suppose that $f(\phi) \vee f(\psi) = f(\phi \vee \psi) \in I$. It is equivalent to $f(\phi) \in I$ and $f(\psi) \in I$. By induction hypothesis, it is also equivalent to $I \Vdash \phi$ and $I \Vdash \psi$, which, by definition, $I \Vdash \phi \vee \psi$.

If $f(\phi \vee \psi) \in F$, then F has non-empty intersection with all ideals containing $f(\phi \vee \psi)$. We obtain $F \Vdash \phi \vee \psi$, because every ideal I satisfying $I \Vdash \phi \vee \psi$ contains $f(\phi \vee \psi)$. Conversely, if $F \Vdash \phi \vee \psi$, then it must have non-empty intersection with $\downarrow f(\phi \vee \psi)$ as well. Therefore, $f(\phi \vee \psi) \in F$.

$\wedge$: Suppose that $f(\phi) \wedge f(\psi) = f(\phi \wedge \psi) \in F$. It is equivalent to $f(\phi) \in F$ and $f(\psi) \in F$. By induction hypothesis, it is also equivalent to $F \Vdash \phi$ and $F \Vdash \psi$, which $F \Vdash \phi \wedge \psi$ by definition.

If $f(\phi \wedge \psi) \in I$, then I has non-empty intersection with all filters containing $f(\phi \wedge \psi)$. We obtain $I \Vdash \phi \wedge \psi$, because every filter F satisfying $F \Vdash \phi \wedge \psi$ contains $f(\phi \wedge \psi)$. Conversely, if $I \Vdash \phi \wedge \psi$, then it must have non-empty intersection with $\uparrow f(\phi \wedge \psi)$ as well. Therefore, $f(\phi \wedge \psi) \in I$.

$\circ$: Suppose that $f(\phi) * f(\psi) = f(\phi \circ \psi) \in I$. For arbitrary $F_1, F_2 \in \mathcal{F}$, if $F_1 \Vdash \phi$ and $F_2 \Vdash \psi$, by induction hypothesis, $f(\phi) \in F_1$ and $f(\psi) \in F_2$, hence $f(\phi) * f(\psi) \in F_1 * F_2$, which derives $F_1 * F_2 \sqsubseteq I$, i.e. $R(F_1, F_2, I)$. Conversely, assume that $I \Vdash \phi \circ \psi$. By definition, for arbitrary $F_1 \Vdash \phi$ and $F_2 \Vdash \psi$, $F_1 * F_2 \sqsubseteq I$ holds. Then, $\uparrow f(\phi) * \uparrow f(\psi) \sqsubseteq I$ must hold, hence $f(\phi \circ \psi) \in I$.

If $f(\phi \circ \psi) \in F$, then F has non-empty intersection with all ideals containing $f(\phi \circ \psi)$. Since every ideal I satisfying $I \Vdash \phi \circ \psi$ contains $f(\phi \circ \psi)$, we have $F \Vdash \phi \circ \psi$. Conversely, if $F \Vdash \phi \circ \psi$, then it must have non-empty intersection with $\downarrow f(\phi \circ \psi)$ as well. Therefore, $f(\phi \circ \psi) \in F$.

$\rightarrow$: Suppose that $f(\phi) \backslash f(\psi) = f(\phi \to \psi) \in F$. For arbitrary $F' \in \mathcal{F}$ and $I \in \mathcal{I}$, if $F' \Vdash \phi$ and $I \Vdash \psi$, by induction hypothesis, $f(\phi) \in F'$ and $f(\psi) \in I$, hence $f(\phi) \backslash f(\psi) \in F' \backslash I$. By the residuality on $\mathbb{A}$, we obtain $F' * F \sqsubseteq I$, i.e. $R(F', F, I)$

holds. Conversely, assume that $F \Vdash \phi \to \psi$. By definition, for arbitrary $F' \Vdash \phi$ and $I \Vdash \psi$, we have $F' * F \sqsubseteq I$. Then, $\uparrow f(\phi) * F \sqsubseteq \downarrow f(\psi)$ must hold as well. Therefore, there exists $x \in F$ such that $x \leq f(\phi) \backslash f(\psi) = f(\phi \to \psi)$, hence $f(\phi \to \psi) \in F$.

If $f(\phi \to \psi) \in I$, then I has non-empty intersection with all filters containing $f(\phi \to \psi)$. Since every filter F satisfying $F \Vdash \phi \to \psi$ contains $f(\phi \to \psi)$, we have $I \Vdash \phi \to \psi$. Conversely, if $I \Vdash \phi \to \psi$, then it must have non-empty intersection with $\uparrow f(\phi \to \psi)$ as well. Therefore, $f(\phi \to \psi) \in I$.

$\leftarrow$**:** Suppose that $f(\psi)/f(\phi) = f(\psi \leftarrow \phi) \in F$. For arbitrary $F' \in \mathcal{F}$ and $I \in \mathcal{I}$, if $F' \Vdash \phi$ and $I \Vdash \psi$, by induction hypothesis, $f(\phi) \in F'$ and $f(\psi) \in I$, hence $f(\psi)/f(\phi) \in I/F'$. By the residuality on $\mathbb{A}$, we obtain $F * F' \sqsubseteq I$, i.e. $R(F, F', I)$ holds. Conversely, assume that $F \Vdash \psi \leftarrow \phi$. By definition, for arbitrary $F' \Vdash \phi$ and $I \Vdash \psi$, we have $F * F' \sqsubseteq I$. Then, $F * \uparrow f(\phi) \sqsubseteq \downarrow f(\psi)$ must hold as well. Therefore, there exists $x \in F$ such that $x \leq f(\psi)/f(\phi) = f(\psi \leftarrow \phi)$, hence $f(\psi \leftarrow \phi) \in F$.

If $f(\psi \leftarrow \phi) \in I$, then I has non-empty intersection with all filters containing $f(\psi \leftarrow \phi)$. Since every filter F satisfying $F \Vdash \psi \leftarrow \phi$ contains $f(\psi \leftarrow \phi)$, we have $I \Vdash \psi \leftarrow \phi$. Conversely, if $I \Vdash \psi \leftarrow \phi$, then it must have non-empty intersection with $\uparrow f(\psi \leftarrow \phi)$ as well. Therefore, $f(\psi \leftarrow \phi) \in F$.

Finally, we finish up the proof. Assume $\Gamma \mapsto \varphi$ is not valid on $\mathbb{A}$. Then, there exists an assignment $f : \Phi \to \mathbb{A}$ such that $f(\Gamma) \not\leq f(\varphi)$. We have that $\uparrow f(\Gamma) \in \mathcal{F}$ and $\downarrow f(\varphi) \in \mathcal{I}$. Moreover, we also have $\mathbb{A}_+, V, \uparrow f(\Gamma) \Vdash \Gamma$ and $\mathbb{A}_+, V, \downarrow f(\varphi) \Vdash \varphi$. However, since $f(\Gamma) \not\leq f(\varphi)$, $\uparrow f(\Gamma) \not\sqsubseteq \downarrow f(\varphi)$. Therefore, $\mathbb{A}_+ \not\Vdash \Gamma \mapsto \varphi$. □

Proof. [The other cases of Theorem 6.1]

$(\vee \mapsto)$**:** For arbitrary $x \in X$ and $y \in Y$, let $x \Vdash_{\mathfrak{bs}} \Gamma \circ (\phi \vee \psi) \circ \Delta$ and $y \Vdash_{\mathfrak{bs}} \varphi$. By inductive hypothesis, we have $\mathbb{M} \Vdash \Gamma \circ \phi \circ \Delta \mapsto \varphi$ and $\mathbb{M} \Vdash \Gamma \circ \psi \circ \Delta \mapsto \varphi$. So, we obtain $y \Vdash \Gamma \circ \phi \circ \Delta$ and $y \Vdash \Gamma \circ \psi \circ \Delta$. With repeating Definition 4.1 and Lemma 4.2, we obtain the following:

$$y \Vdash \Gamma \circ \phi \circ \Delta \iff \forall y', y_2 \in Y, \forall x_1, x_3 \in X.$$
$$x_1 \Vdash \Gamma, x_3 \Vdash \Delta, R^{\leftarrow}(y_2, x_3, y'), R^{\rightarrow}(x_1, y', y) \Rightarrow y_2 \Vdash \phi,$$
$$y \Vdash \Gamma \circ \psi \circ \Delta \iff \forall y', y_2 \in Y, \forall x_1, x_3 \in X.$$
$$x_1 \Vdash \Gamma, x_3 \Vdash \Delta, R^{\leftarrow}(y_2, x_3, y'), R^{\rightarrow}(x_1, y', y) \Rightarrow y_2 \Vdash \psi,$$
$$y \Vdash \Gamma \circ (\phi \vee \psi) \circ \Delta \iff \forall y', y_2 \in Y, \forall x_1, x_3 \in X.$$
$$x_1 \Vdash \Gamma, x_3 \Vdash \Delta, R^{\leftarrow}(y_2, x_3, y'), R^{\rightarrow}(x_1, y', y) \Rightarrow y_2 \Vdash \phi \vee \psi.$$

Therefore, we obtain $y \Vdash \Gamma \circ (\phi \vee \psi) \circ \Delta$, hence $x \leq y$.

$(\mapsto \vee_1)$**:** For arbitrary $x \in X$ and $y \in Y$, let $x \Vdash_{\mathfrak{bs}} \Gamma$ and $y \Vdash \phi \vee \psi$. By definition, $y \Vdash \phi$. By induction hypothesis, we have $\mathbb{M} \Vdash \Gamma \mapsto \phi$, hence $x \leq y$.

$(\wedge_1 \mapsto)$**:** For arbitrary $x \in X$ and $y \in Y$, let $x \Vdash_{\mathfrak{bs}} \Gamma \circ (\phi \wedge \psi) \circ \Delta$ and $y \Vdash_{\mathfrak{bs}} \varphi$. Then, there exist $x_1, x_2, x_3, x' \in X$ such that $x_1 \Vdash_{\mathfrak{bs}} \Gamma$, $x_2 \Vdash \phi \wedge \psi$, $x_3 \Vdash_{\mathfrak{bs}} \Delta$, $R^{\circ}(x_1, x', x)$ and $R^{\circ}(x_2, x_3, x')$. By definition, we also have $x_2 \Vdash \phi$, hence $x \Vdash \Gamma \circ \phi \circ \Delta$. By induction hypothesis, $\mathbb{M} \Vdash \Gamma \circ \phi \circ \Delta \mapsto \varphi$, hence $x \leq y$.

$(\mapsto \wedge)$**:** For arbitrary $x \in X$ and $y \in Y$, let $x \Vdash_{\mathfrak{bs}} \Gamma$ and $y \Vdash_{\mathfrak{bs}} \phi \wedge \psi$. By inductive hypothesis, we have $\mathbb{M} \Vdash \Gamma \mapsto \phi$ and $\mathbb{M} \Vdash \Gamma \mapsto \psi$. Therefore, we obtain that $x \Vdash \phi$ and $x \Vdash \psi$, which derives $x \Vdash \phi \wedge \psi$. Then, $x \leq y$.

($\Rightarrow\leftarrow$): For arbitrary $x \in X$ and $y \in Y$, let $x \Vdash_{\mathfrak{bs}} \Gamma$ and $y \Vdash_{\mathfrak{bs}} \psi \leftarrow \phi$. Then, there exist $x_2 \in X$ and $y' \in Y$ such that $x_2 \Vdash_{\mathfrak{bs}} \phi$, $y' \Vdash_{\mathfrak{bs}} \psi$ and $R^{\leftarrow}(y, x_2, y')$. By induction hypothesis, we have $\mathbb{M} \Vdash \Gamma \circ \phi \Rightarrow \psi$, hence $y' \Vdash \Gamma \circ \phi$. By Theorem 4.4, for every $y'' \in Y$, if $R^{\leftarrow}(y'', x_2, y')$, then $y'' \Vdash \Gamma$. Finally, since $x \Vdash \Gamma$, we conclude $x \leq y$.

□

Logics of Space with Connectedness Predicates: Complete Axiomatizations

Tinko Tinchev

Sofia University "St. Kliment Ohridski"
5 James Bourchier blvd
1164 Sofia, Bulgaria, e-mail tinko@fmi.uni-sofia.bg

Dimiter Vakarelov

Sofia University "St. Kliment Ohridski"
5 James Bourchier blvd
1164 Sofia, Bulgaria, e-mail dvak@fmi.uni-sofia.bg

Abstract

In this paper we present a complete quantifier-free axiomatization of several logics on region-based theory of space based on contact relation and connectedness predicates. We prove completeness theorems for the logics in question with respect to three different semantics: algebraic – with respect to several important classes of contact algebras, topological – based on the contact algebras over various classes of topological spaces, and relational semantics with respect to Kripke frames with reflexive and symmetric relations.

Keywords: Spatial logics, mereotopology, contact algebras, connectedness, representation theorems, completeness theorems.

Introduction

This paper is in the field of region-based theory of space (RBTS). The origin of this theory goes back to Whitehead [34] and de Laguna [10] and consists of a radical reconstruction of the classical Euclidean approach to the theory of space, putting on the base of the new approach the more realistic primitive notions, like region as an abstraction of physical body, and some intuitive relations between regions, like part-of, overlap and contact. In this way geometry has been based on mereology – the theory of parts and wholes [30]. While at the beginning this new approach was only on the center of attention of some philosophers and philosophically oriented logicians and mathematicians, now Whitehead's ideas on RBTS flourished and found applications in some areas of computer science: qualitative spatial reasoning (QSR), knowledge representation, geographical information systems, formal ontologies in information systems, image processing, natural language semantics etc. The reason is that the languages of RBTS are quite simple and suitable for description of some qualitative spatial features and properties of space bodies. Recent surveys are [1,4,27,32], surveys concerning various applications are [8,9] (see also special issues of Fundamenta Informaticæ [16] and the Journal of Applied Non-classical Logics [2]). One of the most popular systems among the community of QSR-researchers is the system of Region Connection Calculus (RCC) introduced in [28]. RCC attracted quite intensive research in the field of region-based theory of space and related spatial logics, both on its applied and mathematical aspects. An algebraic reformulation of RCC as a Boolean algebra with an additional relation C called *connection* was presented in [31] (in the subsequent literature *connection* has been renamed with the more convenient name *contact*). Now a common name for various similar systems is the notion of *contact algebra*, the simplest one introduced in [11]. The elements of a given contact algebra are formal counterparts of *regions* and by means of the Boolean operations one can define new regions by means of given ones. Standard models of contact algebras are the algebras of regular closed subsets over some topological spaces with contact aCb holding if the regions a and b share a common point. The relationship of a class of contact algebras and the corresponding class of topological spaces is studied in the topological representation theory of that class, which states that each algebra of the class is representable (in some definite sense) as a contact algebra of regular-closed subsets of a topological space. Representation theory for contact algebras corresponding to RCC was given for the first time in [13], representation theory for contact algebras corresponding to various important classes of topological spaces, was given in [11]. Let us note that contact algebras have also non-topological models based on the notion of adjacency space formalizing some discrete versions of region-based theory of space (see [17], [12], [3]). Note that adjacency spaces can be identified with the standard notion of reflexive graph.

In the present paper we study several spatial logics related to RCC system, containing the connectedness predicates $c(a)$ and $c^{\leq n}(a)$. In topological models the predicate $c(a)$ says that the region a is connected (in a topological sense) and $c^{\leq n}(a)$ says that the region a has at most n connected components. These predicates were studied for the first time in [25,26] (see also [32]). Recently a quite intensive investigation of spatial logics containing $c(a)$ and $c^{\leq n}(a)$ with respect to their expressiveness and complexity has been done in [19] – [23]. In some sense we continue the study started in [19] – [23], but with respect to the question of complete quantifier-free axiomatizations of some of the logics considered in [19]– [23]. Namely we are interested in logics based on the language of contact algebras extended with predicates c and $c^{\leq n}$. Let us note that in contact algebras considered as first-order theories, the predicates c and $c^{\leq n}$ are definable, for instance for c we have the following equivalence:

(#) $c(a) \leftrightarrow (\forall b, d)(b \neq 0 \wedge d \neq 0 \wedge a = b + d \rightarrow bCd)$.

The problem is that the logics considered in [19]– [23] are quantifier-free, and we also want to obtain quantifier-free axiomatizations. The implication from left to the right in (#) is a universal sentence and is equivalent to the following quantifier-free formula, which can be taken as an axiom:

$c(a) \wedge b \neq 0 \wedge d \neq 0 \wedge a = b + d \rightarrow bCd$

The converse implication of (#) is not, however, a universal formula, and we will simulate it axiomatically by a special finitary rule of inference. Quantifier-free axiomatizations of spatial logics based on contact algebras, and some additional finitary rules of inference imitating reasoning with quantifiers were studied for the first time in [3]. In this paper we continue this study in the presence of the new predicates $c^{\leq n}$ (in fact c is just $c^{\leq 1}$).

We introduce also a simplified language in which we replace the predicates $c^{\leq n}$, $(n = 1, 2, \ldots)$, by corresponding sets of nominals $C^{\leq n}$, denoting regions satisfying $c^{\leq n}$.

We introduce three kinds of semantics of the languages in consideration: algebraic – based on contact algebras, topological – corresponding to the main type of point-based models of space, and a Kripke style semantics based on the notion of *adjacency space*. Let us note that the Kripke semantics is a new one for the considered logics and gives a graph sense of the connectedness predicates, for instance $c(a)$ means that a is a connected (in a graph-theoretic sense) set of points. We present axiomatic systems strongly complete in several important classes of topological spaces including all topological spaces, all connected topological spaces, all spaces related to RCC system, all (connected) compact Hausdorff spaces. This makes possible to transfer some of the complexity results obtained in [19] to some new classes of topological spaces. Completeness theorems are based on a special representation theory of contact algebras with predicates $c^{\leq n}$ in topological spaces and separately, the representation theory in adjacency spaces. We show also that with respect to weak completeness some of the additional rules of the logics in question can be eliminated, which implies collapsing of some of the logics. Using a new filtration techniques applicable to axiomatic systems with additional rules of inference, we prove that each of the considered logics is complete in a corresponding class of finite models, from which we derive their decidability.

The rest of the paper is organized as follows. In Section 1 we remind some facts for contact algebras. Section 2 is devoted to contact algebras with predicates c and $c^{\leq n}$ and their representation theory both in topological spaces and in adjacency spaces. In Section 3 we introduce two kinds of spatial logics: one based on contact algebras and predicates c and $c^{\leq n}$, and another one, in which predicates $c^{\leq n}$ are replaced by sets $C^{\leq n}$ of nominals denoting regions satisfying $c^{\leq n}$. We prove here several completeness theorems. Section 4 is for some concluding remarks and future plans.

1 Contact algebras

Definition 1.1 Following [11], by a *contact algebra* we mean any system $\underline{B} = (B, C) = (B, 0, 1, \cdot, +, *, C)$, where $(B, 0, 1, \cdot, +, *)$ is a nondegenerate Boolean algebra, $*$ denotes the Boolean complement, and C is a binary relation in B, called a *contact*, such that

(C1) if xCy, then $x, y \neq 0$,

(C2) $xC(y + z)$ if and only if xCy or xCz,

(C3) if xCy, then yCx,

(C4) if $x \cdot y \neq 0$, then xCy.

$\underline{B}$ is a complete contact algebra if it is a complete Boolean algebra.

Elements of B are called *regions*. The complement of C is denoted by $\overline{C}$. The relation $\ll$ of *nontangential inclusion* is defined as follows: $x \ll y$ if and only if $x\overline{C}y^*$.

Axioms (C2) and (C3) imply the monotonicity of C with respect to $\leq$:

(Mono) if aCb and $a \leq a'$ and $b \leq b'$, then $a'Cb'$.

We consider the standard definitions of subalgebra, isomorphism, embedding, etc. (cf. [7, Ch. V]). A contact subalgebra $\underline{B_1}$ of $\underline{B_2}$ is said to be *dense* if

(Dense) $(\forall a_2 \in B_2)(a_2 \neq 0 \rightarrow (\exists a_1 \in B_1)(a_1 \neq 0 \text{ and } a_1 \leq a_2))$

If h embeds $\underline{B}$ as a dense contact subalgebra, then h is called *dense embedding.*

Consider contact algebras satisfying some of the following axioms:

(Con) if $a \neq 0$ and $a \neq 1$, then aCa^* *connectedness*

(Ext) if $a \neq 1$, then $\exists b \neq 0$ such that $a\overline{C}b$ *extensionality*

(Nor) if $a \ll b$, then $\exists d$ such that $a \ll d \ll b$ *normality*

A contact algebra satisfying axiom (Con) ((Ext) or (Nor)) is said to be *connected* (*extensional* or *normal*).

Contact algebras satisfying axioms (Con) and (Ext) were introduced in [31] under the name *Boolean contact algebras* and were considered as an equivalent formulation of the system RCC [28]. It is proved in [31] that (Ext) is equivalent (under axioms (C1)–(C4)) to each of the following axioms:

(Ext$'$) $a \leq b$ if and only if $(\forall d \in B)(aCd \rightarrow bCd)$,

(Ext$''$) $a = b$ if and only if $(\forall d \in B)(aCd \leftrightarrow bCd)$,

(Ext$'''$) $(\forall b \neq 0)(\exists a \neq 0)(a \ll b)$.

Note that axiom (Con) is equivalent to the axiom

(Con$'$) if $a \neq 0$, $b \neq 0$, and $a + b = 1$, then aCb.

Similarly, (Nor) is equivalent to the axiom

(Nor$'$) if $a\overline{C}b$, then $(\exists a'b')(a\overline{C}a'$ and $b\overline{C}b'$ and $a' + b' = 1)$.

Example 1.2 (1) *Topological contact algebras.* Let X be a topological space with operations of closure $Cl(a)$ and interior $Int(a)$. A subset a of X is *regular closed* if $a = Cl(Int(a))$. The set of all regular closed subsets of X is denoted by $RC(X)$. As is known, the regular closed sets with operations $a + b = a \cup b$, $a \cdot b = Cl(Int(a \cap b))$, $a^* = Cl(X \setminus a) = Cl(-a)$, $0 = \varnothing$, and $1 = X$ form a Boolean algebra. Moreover, if we consider the infinite join operation $\sum_{i \in I} a_i = Cl(\bigcup_{i \in I} a_i)$, then the Boolean algebra $RC(X)$ is complete. The contact is defined as follows: $a\, C_X\, b$ if and only if $a \cap b \neq \varnothing$. It satisfies axioms (C1)–(C4) and consequently $RC(X)$ is a contact algebra. All the algebras of the kind $RC(X)$ and all their subalgebras are called *topological contact algebras.*

(2) *Discrete contact algebras.* Let (W, R) be a relational system with a symmetric and

reflexive binary relation R on W. Following Galton [17] we call such systems *adjacency spaces* and the relation R – *adjacency relation*. For $a, b \subseteq W$ define a contact aC_Rb iff $(\exists x \in a)(\exists y \in b)(xRy)$. It can be seen that the Boolean algebra $CA(W, R)$ of all subsets of W with the above relational contact is a contact algebra. Since this is a non-topological example of contact algebras we call $CA(W, R)$ and its subalgebras *discrete contact algebras*. Regions in $CA(W, R)$ are arbitrary subsets of W.

Lemma 1.3 ([12]) *Let $CA(W, R)$ be discrete contact algebra over the adjacency space (W, R). Then:*

(i) $CA(W, R)$ satisfies (Nor) iff R is an equivalence relation.

(ii) $CA(W, R)$ satisfies (Con) iff (W, R), considered as a graph, is a connected graph, i.e. for any two points $x, y \in W$ there is an R-path connecting x and y.

We recall in Appendix A some topological notions used later on.

Theorem 1.4 (Topological representation of contact algebras) *Let $\underline{B} = (B, C)$ be a contact algebra. Then:*

(I) (i) There exists a compact semiregular T_0-space X and a dense embedding h of $\underline{B}$ in the contact algebra of regular closed sets $RC(X)$. Moreover,

(i1) $\underline{B}$ satisfies (Nor) *iff X is κ-normal.*

(i2) $\underline{B}$ satisfies (Con) *iff X is connected.*

(ii) If $\underline{B}$ satisfies axiom (Ext)*, then there exists a compact weakly regular T_1-space X and a dense embedding h of $\underline{B}$ in the contact algebra of regular closed sets $RC(X)$. Moreover,*

(ii1) $\underline{B}$ satisfies (Nor) *iff X is κ-normal.*

(ii2) $\underline{B}$ satisfies (Con) *iff X is connected.*

(iii) If $\underline{B}$ satisfies both axioms (Ext) *and* (Nor)*, then there exists a compact Hausdorff space (X, τ) and a dense embedding h of $\underline{B}$ in the contact algebra of regular closed sets $RC(X)$. Moreover,*

(iii1) $\underline{B}$ satisfies (Con) *iff X is connected.*

(II) If $\underline{B}$ is a complete contact algebra then in all of the above cases h becomes an isomorphism between (B, C) and $(RC(X), C_X)$. In all cases the set $\{h(a) : a \in B\}$ is a base for the closed sets of X.

Remark 1.5 Different parts of Theorem 1.4 have been proved by different authors. In the present form the theorem is taken from [32]. The case (iii) was proved for the first time in [33]. The case (i)+(i1) and (i)+(i1)+(II) was proved in [11, Sec. 5]. The case (ii)+ (ii1) covers RCC system [28]. This case, without compactness, was proved for the first time in [13], and the case with compactness – in [11, Sec. 5]. The fact that in all cases we have compact representation is important, because it will be used in the next section in the representation theory of contact algebras with connectedness predicates.

Definition 1.6 By Theorem 1.4 for each class Σ of contact algebras determined by some of the axioms (Con), (Ext) and (Nor) there exists a class $Top(\Sigma)$ of topological spaces in which the members of Σ are representable. We call the spaces from $Top(\Sigma)$ *corresponding* to the algebras from Σ.

Theorem 1.7 (Discrete representation of contact algebras [12]) *Let $\underline{B} = (B, C)$ be a contact algebra. Then there exists an adjacency space (W, R) and an embedding h into the contact algebra $CA(W, R)$. Moreover, $\underline{B} = (B, C)$ satisfies* (Nor) *iff R is an equivalence relation. If $\underline{B}$ is finite then (W, R) is also finite and h becomes an isomorphism between $\underline{B}$ and $CA(W, R)$.*

2 Contact algebras with connectedness predicates

We will use the notations of connectedness predicates as in [19,26] – $c(a)$ and $c^{\leq n}(a)$. In topological spaces $c(a)$ means that the region a is connected in a topological sense and $c^{\leq n}(a)$ – that a is a sum of at most n ($n \geq 1$) connected components. Obviously $c(a)$ iff $c^{\leq 1}(a)$. The following lemma is well known for c (see Lemma 4.1 (iii)) and can easily be proved for $c^{\leq n}$ by induction on n.

Lemma 2.1 *Let X be a topological space and $a \in RC(X)$. Then:*
(i) $c(a)$ iff $(\forall b_0 \neq \varnothing, b_1 \neq \varnothing \in RC(X))(a = b_0 \cup b_1 \rightarrow b_0 \cap b_1 \neq \varnothing)$,
(ii) $c^{\leq 1}(a)$ iff $c(a)$,
$c^{\leq (n+1)}(a)$ iff $(\forall b \neq \varnothing, d \neq \varnothing \in RC(X))(a = b \cup d \rightarrow c^{\leq n}(b) \vee b \cap d \neq \varnothing)$,
(iii) $c^{\leq n}(a)$ iff $(\forall b_0 \neq \varnothing \ldots b_n \neq \varnothing \in RC(X))(a = b_0 \cup \ldots \cup b_n \rightarrow (\exists i, j : 1 \leq i < j \leq n)(b_i \cap b_j \neq \varnothing))$.

Note that Lemma 2.1 (ii) presents an inductive definition for the predicate $c^{\leq n}$ and (iii) – a direct definition for each $n \geq 1$. This suggests to adopt the following abstract definition of predicates $c^{\leq n}$ in arbitrary contact algebras.

Definition 2.2 Let $\underline{B}$ be a contact algebra. We define $c^{\leq n}(a)$ for an arbitrary $a \in B$, $n = 1, 2, \ldots$ by induction as follows:
(i) $c(a) \leftrightarrow_{def} (\forall b, d \in B)(b \neq 0 \wedge d \neq 0 \wedge a = b + d \rightarrow bCd)$,
(ii) $c^{\leq 1}(a) \leftrightarrow_{def} c(a)$,
(iii) $c^{\leq (n+1)}(a) \leftrightarrow_{def} (\forall b, d \in B)(b \neq 0 \wedge d \neq 0 \wedge a = b + d \rightarrow c^{\leq n}(b) \vee bCd)$.
We denote by $C^{\leq n}(B)$ the set of all regions a such that $c^{\leq n}(a)$.

Note that the sets $C^{\leq n}(B)$ are non-empty, because we always have $c^{\leq n}(0)$.

Lemma 2.3 *The following equivalence is true for any $a \in B$:*
$c^{\leq n}(a)$ iff $(\forall b_0 \ldots b_n \in B)(b_0 \neq 0 \wedge \ldots \wedge b_n \neq 0 \wedge a = b_0 + \cdots + b_n \rightarrow (\exists i, j : 0 \leq i < j \leq n)(b_i C b_j))$.

Proof. The proof proceeds by induction on n and the definition of $c^{\leq n}$. □

Now we will see that the abstract definition of connectedness predicates $c^{\leq n}$ in contact algebras $CA(W, R)$ over adjacency spaces coincides with the standard graph-theoretic connectedness (note that each adjacency space (W, R) can be considered as a graph in a standard way).

In order to characterize the predicates $c^{\leq n}$ in discrete contact algebras (see Example 1.2 (2)) we will use the following notations. Let (W, R) be an adjacency space and

$a \subseteq W$. We denote by R_a the restriction of R on a, and R_a^* will denote the reflexive and transitive closure of R_a. Since R is a symmetric and reflexive relation on W, then R_a is the same on the set a and in this case R_a^* is an equivalence relation on a. Having in mind the above remarks, (W, R) is connected in a graph-theoretic sense (path-connected) if for each $x, y \in W$ we have xR^*y. Since for any $a \subseteq W$ the system (a, R_a) is also an adjacency space, a is connected if for each $x, y \in a$ we have xR_a^*y. Since R_a^* is an equivalence relation then a is path-connected iff R_a^* is the universal relation on a. In general R_a^* divides a into equivalence classes called connected components of a. Having the notion of a connected component, $c(a)$ means that a is itself a connected component and $c^{\leq n}(a)$ means that a is a sum of at most n connected components.

Lemma 2.4 *Let (W, R) be adjacency space and $a \subseteq W$. Then:*
(i) $c(a)$ iff $(\forall x, y \in a)(xR_a^*y)$.
(ii) $c^{\leq n}(a)$ iff $(\forall x_0 \ldots x_n \in a)(\exists i, j : 0 \leq i < j \leq n)(x_i R_a^* x_j)$.

Proof. See the appendix B. □

In the next theorem we deal with topological representation theory of contact algebras with connectedness predicates $c^{\leq n}$. Although $c^{\leq n}$ is definable in contact algebras, if we put this predicate among the signature of contact algebra, it changes the notion of embedding – now every embedding must preserve also the new predicate.

Theorem 2.5 *Let $\underline{B} = (B, C)$ be a contact algebra, X be a semiregular and compact topological space and h be an embedding of (B, C) into the contact algebra $RC(X)$ such that the set $\{h(b) : b \in B\}$ forms a base for the closed sets in X. Then for any $a \in B$: $c^{\leq n}(a)$ holds in $\underline{B}$ iff $c^{\leq n}(h(a))$ holds in $RC(X)$.*

Proof. See the Appendix C. □

Let us recall that all embeddings h in Theorem 1.4 are in compact spaces X such that the set $\{h(a) : a \in \underline{B}\}$ forms a base for closed subsets of X. So the assumptions of Theorem 2.5 are fulfilled and hence h preserves the connectedness predicates $c^{\leq n}$. So we have:

Corollary 2.6 *All embeddings from Theorem 1.4 preserve the predicates $c^{\leq n}$.*

In the next theorem we deal with discrete representation of finite contact algebras with connectedness predicates.

Theorem 2.7 *Let $\underline{B} = (B, C)$ be a finite contact algebra. Then there exists a finite adjacency space (W, R) and an isomorphism h between $\underline{B}$ and $CA(W, R)$ such that for every $a \in B$ the following equivalence is true: $c^{\leq n}(a)$ holds in $\underline{B}$ iff $c^{\leq n}(h(a))$ holds in $CA(W, R)$. Moreover,*
(i) $\underline{B}$ satisfies (Con) iff the graph (W, R) is connected;
(ii) $\underline{B}$ satisfies (Nor) iff the relation R is an equivalence relation.

Proof. The theorem is a direct corollary of Theorem 1.7 and Lemma 2.4. □

3 Spatial logics with connectedness predicates

We will introduce in this section two kinds of spatial logics. The first kind is based on the language of contact algebras extended with connectedness predicates $c^{\leq n}$. The second kind is also based on the language of contact algebras, but instead of the predicates $c^{\leq n}$, we consider for each natural number $n \geq 1$ a special set of nominals denoting regions with the property $c^{\leq n}$.

Definition 3.1 (RCC-like logics) The minimal logic without connectedness predicates and nominals based on the class of all contact algebras, was studied in [3] under the name "Propositional Weak RCC" and denoted by PWRCC. The adjective "Propositional" is used because all logics are quantifier-free, i.e. propositional. We considered in [3] several other logics based on the same language corresponding to the classes of contact algebras satisfying some or all of the axioms (Con), (Ext) and (Nor). We denote these logics putting indices Con, Ext and Nor to the abbreviation PWRCC. The logic $\text{PWRCC}_{Con,Ext}$ is denoted also by PRCC, because it corresponds to the class of all connected and extensional contact algebras, the algebraic equivalent of the RCC system. By the same reason the system $\text{PWRCC}_{Con,Ext,Nor}$ is denoted by PRCC_{Nor}.

In [3] axiomatization of all of these logics was given (see also Appendix D) and strong and weak completeness theorems were proved with respect to topological semantics, corresponding to contact algebras over some classes of topological spaces (namely the topological spaces in which the corresponding contact algebras are representable), and weak completeness with respect to Kripke semantics – corresponding to the contact algebras over some classes of adjacency spaces. It was shown in [3] that with respect to the weak completeness theorem all RCC-like logics collapse into two systems – PWRCC and PWRCC_{Con}. The collapsing classes can be seen in the following diagram.

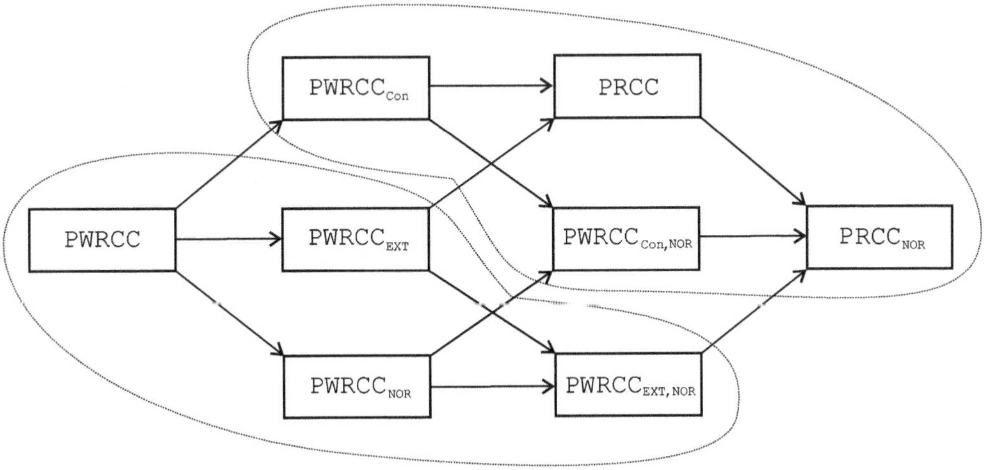

Fig. 1. Propositional RCC type logics

The systems PWRCC and PWRCC_{Con} are equivalent to the systems BRCC-8 introduced by Wolter and Zakharyschev in [35] and interpreted in all topological spaces and in all connected topological spaces. These two systems extended with predicates c and $c^{\leq n}$ are among the important logics studied in [19] – [23].

In this paper we will study all of the logics in the above diagram extended first with the predicates $c^{\leq n}$, and second extended with nominals denoting some regions satisfying $c^{\leq n}$. We will denote the corresponding logics as follows.

Definition 3.2 (Logics with predicates $c^{\leq n}$ and nominals) If $\mathbf{L}$ is the name of one of the above RCC-like logics, its extension with the predicates $c^{\leq n}$, $n \geq 1$, will be denoted by $\mathbf{L}^C$, while extensions with nominals for the regions satisfying $c^{\leq n}$, $n \geq 1$, will be denoted by $\mathbf{L}^{NomC}$. The logics $\mathbf{L}^C$ and $\mathbf{L}^{NomC}$ have the same inclusion diagram as their counterparts $\mathbf{L}$ from Fig 1.

We will present a complete axiomatization of all of these logics, considered in the corresponding classes of contact algebras and prove strong and weak completeness theorems with respect to algebraic, topological and Kripke semantics.

3.1 Syntax

We consider two languages denoted by $L^C = L(\leq, C, c^{\leq n})$ and $L^{NomC} = L(\leq, C, NomC^{\leq n})$. The language $L(\leq, C, c^{\leq n})$ consists of:

- a denumerable set Var of Boolean variables,
- Boolean operations: $+$ (Boolean sum), $\cdot$ (Boolean meet), * (Boolean complement), 0,1 (Boolean constants).
- Relational symbols: $\leq$ (part-of), C (contact), $c^{\leq n}$ for each natural number $n \geq 1$.
- Standard propositional connectives: $\neg, \wedge, \vee, \Rightarrow, \Leftrightarrow$ and the propositional constants $\bot, \top$.
- parentheses: (,).

The set of Boolean terms is defined in a standard way from Boolean variables and constants by means of Boolean operations.

Atomic formulas are of the type: $a \leq b$, aCb, $c^{\leq n}(a)$, $\bot$, $\top$, where a, b are Boolean terms. Formulas are defined from atomic formulas by using propositional connectives in a standard way.

Abbreviations: $a = b =_{def} (a \leq b) \wedge (b \leq a)$, $a \neq b =_{def} \neg(a = b)$, $a\overline{C}b =_{def} \neg(aCb)$, $a \ll b =_{def} a\overline{C}b^*$.

The restriction of the language $L(\leq, C, c^{\leq n})$ only for $n = 1$ is denoted by $L^c = L(\leq, C, c)$. Let us note that in [19] the above languages are denoted correspondingly L^C by $\mathcal{C}cc$ and L^c by $\mathcal{C}c$.

The language $L(\leq, C, NomC^{\leq n})$ differs from the language $L(\leq, C, c^{\leq n})$ as follows: instead of predicate symbols $c^{\leq n}$, it has for each natural number $n \geq 1$ a denumerable set $NomC^{\leq n}$ of new propositional letters called nominals ($NomC^{\leq 1}$ is denoted also by $NomC$). If $a \in NomC^{\leq n}$ then the number n is attached to the nominal a and called its characteristic number. Terms now are defined by Boolean variables, Boolean constants and nominals by Boolean operations. Atomic formulas are only of the form $a \leq b$, aCb, $\bot$ and $\top$ where a, b are Boolean terms. The difference between Boolean variables and nominals is that variables can be substituted by arbitrary terms, while nominals can be substituted only by nominals with the same characteristic number.

3.2 Semantics

Both languages are interpreted in contact algebras as follows. Let $\underline{B} = (B, C)$ be a contact algebra with definable predicates $c^{\leq n}$. A mapping $v : Var \to B$ is called a valuation. Valuations are extended homomorphically to arbitrary Boolean terms in a standard way. A pair $M = (\underline{B}, v)$ is called a model. We define a truth of a formula α in a model $(\underline{B}, v)$, denoted by $(\underline{B}, v) \models \alpha$, inductively as follows:

$(\underline{B}, v) \models a \leq b$ iff $v(a) \leq v(b)$,
$(\underline{B}, v) \models aCb$ iff $v(a)Cv(b)$
$(\underline{B}, v) \models c^{\leq n}(a)$ iff $c^{\leq n}(v(a))$

The interpretation in the language $L(\leq, C, NomC^{\leq n})$ differs as follows. Any valuation v maps Boolean variables into elements of B and if $p \in NomC^{\leq n}$ then $v(p) \in C^{\leq n}(B)$. So nominals from $NomC^{\leq n}$ are mapped into regions satisfying $c^{\leq n}$. The interpretation of formulas is the same.

A formula α (in both languages) is true in the contact algebra (B, C) if for all valuations v we have $(B, C, v) \models \alpha$. If Σ is a class of contact algebras then α is true in Σ if it is true in each member of Σ. The set $\mathbb{L}(\Sigma)$ of all formulas (from the given language) true in Σ is called the logic of Σ. $M = (\underline{B}, v)$ is a model of a set of formulas A if $(\underline{B}, v) \models \alpha$ for every $\alpha \in A$. This the algebraic semantics of both languages and the models in the form $M = (\underline{B}, v)$ are called algebraic models.

If we consider interpretations in contact algebras of the form $RC(X)$ from a given topological space – this is the topological semantics, and if we consider only interpretations in contact algebras of the form $CA(W, R)$ over some adjacency space (W, R) – this is the Kripke (relational) semantics of the presented languages.

It is shown in [19,21] that the language with predicates $c^{\leq n}$ is equally expressive with the language having only the predicate c, but the translations need exponentially more new variables. Consequently our languages $L(\leq, C, c^{\leq n})$ and $L(\leq, C, c)$ also have equal expressive power.

Although the languages with nominas do not contain the predicates $c^{\leq n}$, we can simulate them by the corresponding nominals from $NomC^{\leq n}$ and some new variables in a way similar to the proof of the above result given in [19,21], showing in this way that the languages with nominals are equally expresive with the languages with predicates $c^{\leq n}$. For instance formulas from the language $L(\leq, C, c)$ can be reduced to formulas from the language $L(\leq, C, NomC)$ as follows. Let α be a formula from $L(\leq, C, c)$, Each positive occurrence of predicates of the form $c(a)$ have to be replaced by $a = q$ with a fresh nominal $q \in NomC$. Each negative occurrence of formulas of the form $c(a)$ have to be replaced by the formula $(a = q_1 + q_2) \wedge (q_1 \neq 0) \wedge (q_2 \neq 0) \wedge q_1 \overline{C} q_2$ with fresh variables q_1, q_2. In this way we obtain a formula α' in the language $L(\leq, C, NomC)$ equally satisfiable with the formula α. So, the advantage of logics with the predicates $c^{\leq n}$ is that they allow to speak in a shorter way about connectedness, while the advantage of logics with nominals is that they are simpler, and as we shall see in the next section, they do not require in their axiomatization additional rules.

3.3 *Axiomatizations and completeness theorems*

Let $\mathbf{L}$ be any of the logics from Definition 3.1. The axiomatization of $\mathbf{L}^C$ (see Definition 3.2) can be obtained from the axiomatization of $\mathbf{L}$ given in [3] (see Appendix D) by adding one additional axiom denoted by (Ax $c^{\leq n}$) for each natural number $n \geq 1$, and by adding one additional rule denoted by (Rule $c^{\leq n}$) for each $n \geq 1$:

$$\text{(Ax } c^{\leq n}\text{)} \quad c^{\leq n}(a) \wedge \bigwedge_{i=0}^{n} p_i \neq 0 \wedge a = \sum_{i=0}^{n} p_i \Rightarrow \bigvee_{0\leq i<j\leq n} p_i C p_j$$

$$\text{(Rule } c^{\leq n}\text{)} \quad \frac{\alpha \wedge \bigwedge_{i=0}^{n} p_i \neq 0 \wedge a = \sum_{i=0}^{n} p_i \Rightarrow \bigvee_{0\leq i<j\leq n} p_i C p_j}{\alpha \Rightarrow c^{\leq n}(a)},$$

where α is a formula and $p_0, \ldots, p_n$ are different Boolean variables not occurring in the term a and the formula α (the parameters of the application of the rule).

The axiomatization of $\mathbf{L}^{NomC}$ can be obtained by adding to the axiomatization of $\mathbf{L}$ an additional axiom denoted by (Ax Nom$C^{\leq n}$) for all $n \geq 1$ and for all nominals $a \in C^{\leq n}$:

$$\text{(Ax Nom}C^{\leq n}\text{)} \quad \bigwedge_{i=0}^{n} p_i \neq 0 \wedge a = \sum_{i=0}^{n} p_i \Rightarrow \bigvee_{0\leq i<j\leq n} p_i C p_j.$$

Theorem 3.3 (Strong Completeness theorem) *Let* $\mathbf{L}$ *be any logic from Definition 3.1 and* $\mathbb{L}$ *be any of the logics* $\mathbf{L}^C$ *and* $\mathbf{L}^{NomC}$. *Denote by* $\Sigma(\mathbf{L})$ *the class of contact algebras corresponding to* $\mathbf{L}$ *and by* $Top(\Sigma(\mathbf{L}))$ *the class of topological spaces corresponding (by Definition 1.6) to* $\Sigma(\mathbf{L})$ *in which the members of* $\Sigma(\mathbf{L})$ *are representable. Then the following conditions are equivalent for any set of formulas* A *in the language of* $\mathbb{L}$*:*

(i) A *is consistent in* $\mathbb{L}$,

(ii) A *has an algebraic model in* $\Sigma(\mathbf{L})$,

(iii) A *has a topological model in* $Top(\Sigma(\mathbf{L}))$.

Proof. See Appendix E. □

Theorem 3.4 (Weak completeness theorem including finite models) *Let* $\mathbf{L}$ *be any logic from Definition 3.1 and* $\mathbb{L}$ *be any of the logics* $\mathbf{L}^C$ *and* $\mathbf{L}^{NomC}$. *Then the following conditions are equivalent for any formula* α *of the language of* $\mathbb{L}$*:*

(i) α *is a theorem of* $\mathbb{L}$.

(ii) α *is true in all contact algebras from* $\Sigma(\mathbf{L})$,

(iii) α *is true in all contact algebras* $RC(X)$ *from* $Top(\Sigma(\mathbf{L}))$.

(iv) α *is true in all finite contact algebras in which all theorems of* $\mathbf{L}$ *are true.*

(v) α *is true in all contact algebras* $CA(W, R)$ *over all finite adjacency spaces* (W, R) *in which all theorems of* $\mathbf{L}$ *are true.*

Proof. The equivalence of (i), (ii) and (iii) follow from Theorem 3.3. The equivalence of (iv) and (v) is a corollary of Theorem 2.7. Obviously (ii) implies (iv). For the implication (iv)→(i) suppose that α is not theorem of $\mathbf{L}$. Then there is a canonical model (B, C, v) of $\mathbf{L}$ determined by a maximal consistent set Γ (see the proof of Theorem 3.3) which falsifies α. Now by a special filtration construction we will define a finite model (B', C', v')

which also falsifies α. Let $p_1, \ldots, p_k$ be the Boolean variables occurring in α. Let β_i, $i = 1, \ldots, \beta_l$ be the sequence of all sub-formulas of α not belonging to Γ which have a form of the conclusion of some of the special rules. By the properties of Γ for each β_i there exists a finite sequence of Boolean variables making the premise of the corresponding rule not belonging to Γ. For instance if the conclusion $\beta_i = \gamma \Rightarrow c^{\leq n}(a)$ of the rule (Rule $c^{\leq n}$) does not belong to Γ then the premise $\gamma \wedge \bigwedge_{i=0}^{n} r_i \neq 0 \wedge a = \sum_{i=0}^{n} r_i \Rightarrow \bigvee_{0 \leq i < j \leq n} r_i C r_j$ also does not belong to Γ for some Boolean variables $r_0, \ldots, r_n$. Now let $q_1, \ldots, q_m$ be the sequence of all such variables determined by the formulas β_i. Consider the finite Boolean subalgebra B' of B generated by the elements $|p_1|, \ldots, |p_k|, |q_1|, \ldots, |q_m|$ of B. It is itself a contact subalgebra of (B, C) with the same contact C. Let v' be the canonical valuation v but restricted to the variables $p_1, \ldots, p_k, q_1, \ldots, q_m$. It is easy to see that for all Boolean terms a from $p_1, \ldots, p_k, q_1, \ldots, q_m$ we have $v(a) = v'(a)$. The following statement is a kind of *Filtration Lemma*:

Filtration Lemma. The following equivalence is true for all subformulas γ of α: $(B, C, v) \models \gamma$ iff $(B', C, v') \models \gamma$.

The proof goes by induction on the construction of γ. The nontrivial part is for the atomic $\gamma = c^{\leq n}(a)$.

$(\rightarrow)$ Suppose $(B, C, v) \models c^{\leq n}(a)$, and for the sake of contradiction that $(B', C, v') \not\models c^{\leq n}(a)$, i.e. $c^{\leq n}(v'(a))$ is not true in (B', C). Then for some Boolean terms $b_0, \ldots, b_n$ from $p_1, \ldots, p_k, q_1, \ldots, q_m$ we have: $v'(a) = v'(b_0) + \ldots + v'(b_n)$, $v'(b_i) \neq 0$ for $i = 0, \ldots, n$ and for all i, j, $0 \leq i < j \leq n$, we have $v'(b_i)\overline{C}v'(b_j)$. Since $B' \subseteq B$ and $v'(b_i) = v(b_i)$ for all $i = 0, \ldots, n$, we obtain that $c^{\leq n}(v(a))$ does not hold in (B, C, v), which contradicts the assumption.

$(\leftarrow)$. We will reason by contraposition. Suppose that $(B, C, v) \not\models c^{\leq n}(a)$. By the properties of the canonical model this implies that $c^{\leq n}(a) \notin \Gamma$. Then by the properties of the maximal consistent set Γ, the formula $\bigwedge_{i=0}^{n} r_i \neq 0 \wedge a = \sum_{i=0}^{n} r_i \Rightarrow \bigvee_{0 \leq i < j \leq n} r_i C r_j$ also does not belong to Γ for some Boolean variables $r_0, \ldots, r_n$, such that (by the above construction) $|r_0|, \ldots, |r_n|$ are from the generators of B'. This implies that $v'(a) = \sum_{i=0}^{n} v'(r_i)$, $v'(r_i) \neq 0$ for $i = 0 \ldots, n$ and for all i, j, $0 \leq i < j \leq n$, we have $v'(r_i)\overline{C}v'(r_j)$. All this shows that $(B', C, v') \not\models c^{\leq n}(a)$, which finishes the proof of the Filtration lemma.

By the Filtration lemma we obtain that α is falsified in the finite Boolean algebra (B', C).

Now we will show that all theorems of **L** are true in (B', C). To this end we first show that all theorems of **L** are true in the canonical algebra (B, C), which would imply that the same is true for (B', C). This follows from the following observation. By the properties of the canonical model (B, C, v) we have that for any formula β: $(B, C, v) \models \beta$ iff $\beta \in \Gamma$. So, if β is a theorem of **L**, then $\beta \in \Gamma$ and hence $(B, C, v) \models \beta$. Let w be an arbitrary valuation in B. Then w defines a substitution on Boolean variables Sub_w, which can be applied to arbitrary formulas. It is easy to see that the following holds: $(B, C, w) \models \beta$ iff $(B, C, v) \models Sub_w(\beta)$ iff $Sub_w(\beta) \in \Gamma$. But if β is a theorem then $Sub_w(\beta)$ is also a theorem which implies that $(B, C, w) \models \beta$. □

As a consequence of Theorem 3.4 we obtain:

Corollary 3.5 *(i) The logics $PWRCC^C$ and $PWRCC^{NomC}$ are complete in the class of all contact algebras $CA(W, R)$ over arbitrary (finite) adjacency spaces.*

(ii) The logics $PWRCC^C_{Con}$ and $PWRCC^{NomC}_{Con}$ are complete in the class of all contact algebras $CA(W, R)$ over arbitrary (finite) connected adjacency spaces.

Theorem 3.6 (Admisisibility of the rules (EXT) and (NOR)) *(i) The rules* (EXT) *and* (NOR) *are admissible in the logics $PWRCC^C$ and $PWRCC^{NomC}$.*

(ii) The rules (EXT) *and* (NOR) *are admissible in the logics $PWRCC^C_{Con}$ and $PWRCC^{NomC}_{Con}$.*

Proof. We proved in [3] that the rules (EXT) and (NOR) (see Appendix D) are admissible for the logics PWRCC and PWRCC^{Nor}. Inspecting the proof in [3] it can be seen that the same construction can be used for the proof of the logics PWRCC^C and PWRCC^{NomC}. The proof uses Corollary 3.5. So we invite the reader to consult [3]. □

Corollary 3.7 (Elimination of the rules (EXT) and (NOR)) *(i) The logics $PWRCC^C$, $PWRCC^C_{Ext}$, $PWRCC^C_{Nor}$, $PWRCC^C_{Ext,Nor}$ have equal sets of theorems. The logic $PWRCC^C$ is weakly complete in all classes of models in which $PWRCC^C_{Ext}$, $PWRCC^C_{Nor}$, $PWRCC^C_{Ext,Nor}$ are strongly complete.*

(ii) The logics $PWRCC^{NomC}$, $PWRCC^{NomC}_{Ext}$, $PWRCC^{NomC}_{Nor}$, $PWRCC^{NomC}_{Ext,Nor}$ have equal sets of theorems. The logic $PWRCC^{NomC}$ is weakly complete in all classes of models in which $PWRCC^{NomC}_{Ext}$, $PWRCC^{NomC}_{Nor}$, $PWRCC^{NomC}_{Ext,Nor}$ are strongly complete.

(iii) The logics $PWRCC^C_{Con}$, $PWRCC^C_{Ext,Con}$, $PWRCC^C_{Nor,Con}$, $PWRCC^C_{Ext,Nor,Con}$ have equal sets of theorems. The logic $PWRCC^C_{Con}$ is weakly complete in all classes of models in which $PWRCC^C_{Ext,Con}$, $PWRCC^C_{Nor,Con}$, $PWRCC^C_{Ext,Nor,Con}$ are strongly complete.

(iv) The logics $PWRCC^{NomC}_{Con}$, $PWRCC^{NomC}_{Ext,Con}$, $PWRCC^{NomC}_{Nor,Con}$, $PWRCC^{NomC}_{Ext,Nor,Con}$ have equal sets of theorems. The logic $PWRCC^{NomC}_{Con}$ is complete in all classes of models in which $PWRCC^{NomC}_{Ext,Con}$, $PWRCC^{NomC}_{Nor,Con}$, $PWRCC^{NomC}_{Ext,Nor,Con}$ are strongly complete.

(v) All mentioned logics have finite model property, are finitely axiomatizable and hence are decidable.

Let us note that with respect to the weak completeness Corollary 3.5 and Corollary 3.7 show that we have only four interesting logics: PWRCC^C, PWRCC$^C_{Con}$, PWRCC^{NomC} and PWRCC$^{NomC}_{Con}$, which do not contain the rules (Ext) and (Nor). Note that PWRCC^C and PWRCC$^C_{Con}$ are just the logics BRCC extended with the predicates $c^{\leq n}$ studied in [19] – [21] considered respectively in the classes of all topological spaces and all connected topological spaces. The Corollary 3.7 in fact shows that these rules are admissible for these logics, which implies that we may apply the Strong completeness theorem 3.3 to obtain weak completeness for these logics for the classes of models considered for the logics having these rules. For instance Corollary 3.7 says for the logic PWRCC$^C_{Con}$ that it is complete in: (1) all connected topological spaces, (2) all compact connected spaces, (3) all connected weakly regular spaces (spaces for the system RCC), (4) all connected weakly regular and compact T_1 spaces, (5) all compact Hausdorff spaces, (6) all models over connected adjacency spaces (W, R). It is shown in [19] that satisfiability for formulas from the language with predicate c is ExpTime complete in the class of all connected topological spaces and NExpTime complete in the same class for formulas from the language containing the predicates $c^{\leq n}$. Now this result can be transferred for all mentioned classes of spaces. Similar transfers are true also for the other three logics. Another remark for these logics is the following open question: is it possible to eliminate the rule (Rule $c^{\leq n}$) replacing it only with some set of axioms. The logics with nominals did not contain this rule and in a sense can be considered as a result of elimination of this rule (Rule $c^{\leq n}$).

4 Concluding remarks

In this paper we have presented complete axiomatizations of several natural spatial logics based on the language of contact algebras with connectedness predicates c and $c^{\leq n}$. Some of these logics were studied semantically with respect to their complexity in [19] – [21]. We proved for the introduced logics strong and weak completeness theorems for various important classes of topological spaces and also completeness theorems with respect to some non-topological spaces – adjacency spaces, which are used for certain discrete models of space. This implies that all these classes have the same complexity of satisfiability and allows to transfer some known results from [19] – [21]. The semantics in adjacency spaces can be considered also as a kind of Kripke semantics over reflexive and symmetric Kripke frames (W, R). This semantics allows to translate our logics in the modal logics over KTB + universal modality, which shows that not only S4, but also KTB has also a spatial meaning (see about this discussion [3, Sec. 1]). In [19] – [21] several other languages containing the predicates c and $c^{\leq n}$ are considered and the problem for their complete axiomatization with respect to the intended semantics remains open. We postpone this for our plans for the future.

Acknowledgements. We would like to thank Philippe Balbiani for the fruitful discussions on the topic of the paper. The work of one of the anonymous reviewers with very sharp recommendations and suggestions on how to improve the quality of presentation is much appreciated.

This research is supported by the project DID02/32/2009 of Bulgarian NSF, *Theories of space and time: algebraic, topological and logical approaches.*

References

[1] M. Aiello, I. Pratt-Hartmann and J. van Benthem (eds.), *Handbook of spatial logics*, Springer, 2007.

[2] Ph. Balbiani (Ed.), *Special Issue on Spatial Reasoning*, J. Appl. Non-Classical Logics **12** (2002), No. 3–4.

[3] Ph. Balbiani, T. Tinchev and D. Vakarelov, *Modal logics for region-based theory of space*, Fundamenta Informaticæ, vol. 81 (1–3), 2007, 29-82.

[4] B. Bennett and I. Düntsch, *Axioms, Algebras and Topology.* In: Logic of Space, M. Aiello, I. Pratt, and J. van Benthem (Eds.), Springer, 2007.

[5] L. Biacino and G. Gerla, *Connection structures*, Notre Dame J. Formal Logic **32** (1991), 242–247.

[6] L. Biacino and G. Gerla, *Connection structures: Grzegorczyk's and Whitehead's definition of point*, Notre Dame J. Formal Logic **37** (1996), 431–439.

[7] P. M. Cohn, Universal Algebra. Harper & Row, 1965.

[8] A. Cohn and S. Hazarika. Qualitative spatial representation and reasoning: An overview. *Fuandamenta informaticae* 46 (2001), 1–20.

[9] A. Cohn and J. Renz. Qualitative spatial representation and reasoning. In: F. van Hermelen, V. Lifschitz and B. Porter (Eds.) *Handbook of Knowledge Representation*, Elsevier, 2008, 551-596.

[10] T. de Laguna, *Point, line and surface as sets of solids*, J. Philos. **19** (1922), 449–461.

[11] G. Dimov and Vakarelov, D. Contact Algebras and Region-based Theory of Space. A proximity approach. I and II. *Fundamenta Informaticæ,* Vol. 74 (2-3) (2006) 209–249, 251–282.

[12] I. Düntsch and D. Vakarelov. Region-based theory of discrete spaces: A proximity approach. In: Nadif, M., Napoli, A., SanJuan, E., and Sigayret, A. EDS, *Proceedings of Fourth International Conference Journées de l'informatique Messine*, 123-129, Metz, France, 2003. Journal version in: *Annals of Mathematics and Artificial Intelligence*, vol. 49, No 1-4, 2007, 5-14.

[13] I. Düntsch and M. Winter. A representation theorem for Boolean contact algebras. *Theoretical Computer Science (B)*, 347, (2005), 498-512.

[14] M. Egenhofer, R. Franzosa, Point-set topological spatial relations, *Int. J. Geogr. Inform. Systems* **5** (1991), 161–174.

[15] R. Engelking, *General Topology*, PWN, Warszawa, 1977.

[16] I. Düntsch (Ed.), Special issue on Qualitative Spatial Reasoning, *Fundam. Inform.* **46** (2001).

[17] A. Galton, *Qualitative Spatial Change* Oxford Univ. Press, 2000.

[18] K. Kuratowski. Topology, vol I. Academic Press, New York and London, 1966.

[19] R. Kontchakov, I. Pratt-Hartmann, F. Wolter and M. Zakharyaschev. Topology, connectedness, and modal logic. In C. Areces and R. Goldblatt, editors, Advances in Modal Logic, vol. 7, pp. 151-176. College Publications, London, 2008

[20] R. Kontchakov, I. Pratt-Hartmann, F. Wolter and M. Zakharyaschev. On the computational complexity of spatial logics with connectedness constraints. In I. Cervesato, H. Veith and A. Voronkov, editors, Proceedings of LPAR 2008 (Doha, Qatar, November 22-27, 2008), pp. 574-589, LNAI, vol. 5330, Springer 2008

[21] R. Kontchakov, I. Pratt-Hartmann and M. Zakharyaschev. Topological logics over Euclidean spaces. In Proceedings of Topology, Algebra and Categories in Logic, TACL 2009 (Amsterdam, July 7-11, 2009).

[22] R. Kontchakov, I. Pratt-Hartmann, F. Wolter and M. Zakharyaschev. Spatial logics with connectedness predicates. Submitted

[23] R. Kontchakov, I. Pratt-Hartmann and M. Zakharyaschev. Interpreting Topological Logics over Euclidean Spaces. Submitted.

[24] C. Lutz and F. Wolter, *Modal logics for topological relations*, Logical Meth. Computer Sci. (2006).

[25] I. Pratt-Hartmann, Empiricism and Racionalizm in Region-based Theories of Space, Fundamenta Informaticæ, 45 (2001) 159-186.

[26] I. Pratt-Hartmann, A topological constraint language with component counting. J. of Applied Non-classical Logics, 12 (2002), 441-467.

[27] I. Pratt-Hartmann, *First-order region-based theories of space*, In: Logic of Space, M. Aiello, I. Pratt and J. van Benthem (Eds.), Springer, 2007.

[28] Randell, D. A., Cui, Z.and Cohn, A. G. A spatial logic based on regions and connection. In: B. Nebel, W. Swartout, C. Rich (EDS.) *Proceedings of the 3rd International Conference Knowledge Representation and Reasoning*, Morgan Kaufmann, Los Allos, CA, pp. 165–176, 1992.

[29] E. Shchepin, *Real-valued functions and spaces close to normal*, Siberian Math. J. **13** (1972), 820–830.

[30] P. Simons, *PARTS. A Study in Ontology*, Oxford, Clarendon Press, 1987.

[31] J. Stell, *Boolean connection algebras: A new approach to the Region Connection Calculus*, Artif. Intell. **122** (2000), 111–136.

[32] D. Vakarelov. Region-Based Theory of space: Algebras of Regions, Representation Theory, and Logics. In: Dov Gabbay et al. (Eds.) *Mathematical Problems from Applied Logic II. Logics for the XXIst Century*, 267-348. Springer, 2007.

[33] H. de Vries, *Compact Spaces and Compactifications*, Van Gorcum, 1962

[34] A. N. Whitehead, *Process and Reality*, New York, MacMillan, 1929.

[35] F. Wolter. and M. Zakharyaschev, *Spatial representation and reasoning in RCC-8 with Boolean region terms*, In: Proceedings of the 14th European Conference on Artificial Intelligence (ECAI 2000), Horn W. (Ed.), IOS Press, pp. 244–248.

Appendix A. Some topological notions

A topological space X is said to be

• *semiregular* if it has a base $\mathbb{B}$ of regular closed sets; namely, every closed set is the intersection of elements of $\mathbb{B}$,

• *normal* if every pair of closed disjoint sets can be separated by a pair of open sets,

• *κ-normal* (cf. [29]) if every pair of regular closed disjoint sets can be separated by a pair of open sets,

• *weakly regular* (cf. [13]) if it is semiregular and for every nonempty open set a there exits a nonempty open set b such that $\mathrm{Cl}(b) \subseteq a$,

• *connected* if it cannot be represented as the sum of two disjoint nonempty open sets (if $a \subseteq X$, then a is connected if it is connected in the subspace topology),

• a T_0-*space* if for every two different points $x \neq y$ there exists an open set that contains one of them and does not contain the other,

• a T_1-*space* if every one-point set $\{x\}$ is a closed set,

• a *Hausdorff space* (or a T_2-*space*) if every two different points can be separated by a pair of disjoint open sets,

• a *compact space* if it satisfies the following condition: if $\{A_i : i \in I\}$ is a nonempty family of closed sets of X such that for every finite nonempty subset $J \subseteq I$ we have $\bigcap\{A_i : i \in J\} \neq \varnothing$, then $\bigcap\{A_i : i \in I\} \neq \varnothing$.

Lemma 4.1 *The following assertions hold:*

(i) Let X be semiregular. Then X is weakly regular if and only if $\mathrm{RC}(X)$ *satisfies* (Ext) *[13].*

(ii) X is κ-normal if and only if $\mathrm{RC}(X)$ *satisfies* (Nor) *[13].*

(iii) X is connected if and only if $\mathrm{RC}(X)$ *satisfies axiom* (Con) *[5,13].*

(iv) If X is a compact Hausdorff space, then $\mathrm{RC}(X)$ *satisfies* (Ext) *and* (Nor) *[33].*

(iv) If X is a normal Hausdorff space, then $\mathrm{RC}(X)$ *satisfies* (Nor) *[6].*

Appendix B. Proof of Lemma 2.4

Proof. (i) Let $a \subseteq W$. Then (a, R_a) is an adjacency space. Then $CA(a, R_a)$ satisfies axiom (Con') iff $(\forall b, d \subseteq a)(b \neq \varnothing \wedge d \neq \varnothing \wedge a = b \cup d \rightarrow bC_R d)$ iff $c(a)$ iff (by Lemma 1.3 (ii)) a is path-connected, i.e. $(\forall x, y \in a)(xR_a^* y)$.

(ii) We will use the inductive definition of $c^{\leq n}$ and proceed by induction on n. The case $n = 1$ (the base of induction) is just (i). So suppose that the statement is true for n and proceed for $n + 1$. We have to prove that the following two conditions are equivalent:

(I) $(\forall b, d \subseteq a)(b \neq \varnothing \wedge d \neq \varnothing \wedge a = b \cup d \rightarrow c^{\leq n}(b) \vee bC_R d)$,

(II) $(\forall x_0, \ldots, x_n, x_{n+1} \in a)(\exists i, j : 0 \leq i < j \leq n + 1)(x_i R_a^* x_j)$

(I) $\rightarrow$ *(II)* Suppose (I) and for the sake of contradiction that (II) does not hold, i.e. $(\exists x_0, \ldots, x_n, x_{n+1} \in a)(\forall i, j : 0 \leq i < j \leq n + 1)(x_i \overline{R_a^*} x_j)$. Denote by $|x_i|$ the R_a^*-equivalence class generated by x_i and let $b = \bigcup_{i=0}^{n} |x_i|$, $d = a \setminus b$. Obviously $b \neq \varnothing$, $d \neq \varnothing$ ($x_{n+1} \notin b$ and hence is in d) and $a = b \cup d$. Since for all i, j, $0 \leq i < j \leq n$,

$x_i\overline{R^*}_a x_j$, $x_i \in b$, then by the inductive hypothesis we have $\neg c^{\leq n}(b)$. So by (I) we obtain $bC_R d$, so there exist $y \in b$ and $z \in d$ such that yRz. Since $b = \bigcup_{i=0}^{n} |x_i|$, then there exists $|x_i| \subseteq b$ such that $y \in |x_i|$, hence $x_i R_a^* y$ and by $yR_a z$ we get $x_i R_a^* z$. This implies that $z \in |x_i|$, so $z \in b$, which is impossible ($z \in d$, hence $z \notin b$).

$(II) \rightarrow (I)$ Suppose (II) and in order to obtain a contradiction that (I) does not hold, i.e. there exist $b \neq \varnothing$, $d \neq \varnothing$, $a = b \cup d$, $\neg c^{\leq n}(b)$ and $b\overline{C_R}d$. The last condition implies $b \cap d = \varnothing$. Applying the inductive condition to $\neg c^{\leq n}(b)$, we obtain that there exist $x_0, \ldots, x_n \in b$ such that for all i and j, $0 \leq i < j \leq n$, $x_i\overline{R_b^*}x_j$, so $x_i\overline{R_a^*}x_j$. Since $d \neq \varnothing$, there exists $x_{n+1} \in d$. Applying (I) to the sequence $x_0, \ldots x_n, x_{n+1}$ we obtain that there are i and j, $i \neq j$, such that $x_i R_a^* x_j$. The only possibility is $j = n+1$ and $i \leq n$. So $(\exists m)(\exists y_0, \ldots, y_m \in a)(y_0 = x_i \wedge y_m = x_{n+1} \wedge (\forall k < m)(y_k R y_{k+1}))$. Since $y_0 \in b$ and $y_m \in d$ and $b \cap d = \varnothing$, then there exist $y_k \in b$ and $y_{k+1} \in d$, so by $y_k R y_{k+1}$ we obtain $bC_R d$ – a contradiction. □

Appendix C. Proof of Theorem 2.5

Proof. We will use the equivalent definition of $c^{\leq n}$ from Lemma 2.3.

($\rightarrow$) Suppose $c^{\leq n}(a)$ holds in $\underline{B}$ and for the sake of contradiction that $c^{\leq n}(h(a))$ does not hold in $RC(X)$. Then there are $P_i \in RC(X)$, $i = 0, \ldots, n$ such that:

(1) $P_i \neq \varnothing$, $i = 0, \ldots, n$,

(2) $h(a) = \bigcup_{i=0}^{n} P_i$,

(3) for all i and j, $0 \leq i < j \leq n$, we have $P_i \cap P_j = \varnothing$.

(4) Since all P_i are closed sets each is an intersection of elements from the base $\{h(q) : q \in B\}$, so for every P_i there is a $A_i \subseteq B$ such that $P_i = \bigcap_{p \in A_i} h(p)$. From (2) we have $P_i \subseteq h(a)$ so we may assume that $a \in A_i$ for all i, $i = 0, \ldots, n$.

Now let $i \neq j$, $0 \leq i, j \leq n$ be fixed. Then from (3) and (4) we obtain

(5) $(\bigcap_{p \in A_i} h(p)) \cap (\bigcap_{p \in A_j} h(p)) = \varnothing$.

(6) Applying compactness to (5), there exist a finite subset $A_i^j \subseteq A_i$ and a finite subset $A_j^i \subseteq A_j$ such that

(7) $(\bigcap_{p \in A_i^j} h(p)) \cap (\bigcap_{p \in A_j^i} h(p)) = \varnothing$. Without loss of generality we may assume that $a \in A_i^j$ and $a \in A_j^i$.

(8) Define $A_i' = \bigcup_{j=0, j \neq i}^{n} A_i^j$ for all $i = 0, \ldots, n$. Obviously A_i' is a finite subset of A_i containing a.

(9) It follows from (4), (6), (7) and (8) that $P_i \subseteq \bigcap_{p \in A_i'} h(p) \subseteq \bigcap_{p \in A_i^j} h(p) \subseteq h(a)$, $0 \leq i, j \leq n$ and $i \neq j$.

Now from (9) and (3) we get:

(10) $(\bigcap_{p \in A_i'} h(p)) \cap (\bigcap_{p \in A_j'} h(p)) = \varnothing$, for all $i, j = 0, \ldots, n$ and $i \neq j$.

(11) It follows from (9) that $P_i \subseteq h(p)$ for all $p \in A_i'$, $i, j = 0, \ldots, n$. Since $P_i \in RC(X)$, then by the Boolean product of $RC(X)$ we obtain

(12) $P_i \subseteq \prod_{p \in A_i'} h(p) = h(\prod_{p \in A_i'} p) = h(p_i)$, where $p_i = \prod_{p \in A_i'} p$, $i = 0, \ldots, n$.

(13) It follows from (12) that $p_i \leq p$ for all $p \in A_i'$ especially $p_i \leq a$, because $a \in A_i'$. From here we get

$h(p_i) \subseteq h(p)$ for all $p \in A'_i$ and hence

(14) $h(p_i) \subseteq \bigcap_{p \in A_i} h(p)$, $i = 0, \ldots, n$.

(15) Now from (14) and (10) we get $h(p_i) \cap h(p_j) = \varnothing$ for all i and j, $i \neq j$, $i, j = 0, \ldots, n$, which implies

(16) $p_i \overline{C} p_j$ for all i and j, $i \neq j$, $i, j = 0, \ldots, n$.

(17) It follows from (1) and (12) that $h(p_i) \neq \varnothing$, hence $p_i \neq 0$, $i = 0, \ldots, n$.

(18) Since by (13) $P_i \subseteq h(p_i) \subseteq h(a)$ then $\bigcup_{i=0}^n P_i \subseteq \bigcup_{i=0}^n h(p_i) \subseteq h(a)$. From here we get $\bigcup_{i=0}^n P_i \subseteq h(\bigcup_{i=0}^n p_i) \subseteq h(a)$.

(19) From (18) and (2) we get $h(a) \subseteq h(p_0 + \cdots + p_n) \subseteq h(a)$ which implies

(20) $a = p_0 + \cdots + p_n$.

Now (16), (17) and (20) imply $\neg c^{\leq n}(a)$ in $\underline{B}$, which contradicts the assumption.

($\leftarrow$) Suppose the $c^{\leq n}(h(a))$ holds in $RC(X)$ and for the sake of contradiction that $c^{\leq n}(a)$ does not hold in $\underline{B}$. Then there are $p_0, \ldots, p_n \in B$ such that

(21) $p_i \neq 0$, $i = 0, \ldots, n$, $a = p_0 + \cdots + p_n$, and $p_i \overline{C} p_j$, for all i and j, $i \neq j$, $i, j = 0, \ldots, n$.

¿From (21) we get

(22) $h(p_i) \neq \varnothing$, $i = 0, \ldots, n$, $h(a) = h(p_0) \cup \ldots \cup h(p_n)$, and $h(p_i) \cap h(p_j) = \varnothing$, for all i and j, $i \neq j$, $i, j = 0, \ldots, n$.

But (22) implies that $c^{\leq n}(h(a))$ does not hold in $RC(X)$. □

Appendix D. Axiomatizations of the RCC-like logics from Definition 3.1

Axioms of PWRCC (see [3, Sec. 6])

I. Axiom schemes of the classical propositional logics.

II. Axioms of Boolean algebra based on $\leq$.

III. Axioms for the contact relation C.

Since all predicate axioms of Boolean algebra and contact algebra are universal sentences, they can be written in our language.

Rules of Inference Modus ponens (MP) $\dfrac{\alpha, \alpha \Rightarrow \beta}{\beta}$

The axiomatizations of the other logics from Definition 3.1 we remind the following two rules from [3]:

For an analog of the first-order axiom (Ext) we introduce the *rule of extensionality*

(EXT) $\dfrac{\alpha \Rightarrow (p = 0 \vee aCp)}{\alpha \Rightarrow (a = 1)}$, where p is a Boolean variable that does not occur in a and α.

For an analog of axiom (Nor) we introduce the following *rule of normality*:

(NOR) $\dfrac{\alpha \Rightarrow (aCp \vee p^* Cb)}{\alpha \Rightarrow aCb}$, where p is a Boolean variable that does not occur in a, b, and α.

If $\mathbf{L}$ is any logic from definition 3.1, then its axiomatization can be obtained from the axiomatization of PWRCC as follows:

• If one wants to axiomatize valid formulas in the class of all connected contact algebras – add the axiom (Con) $a \neq 0 \wedge a \neq 1 \Rightarrow aCa^*$.

• If one wants to axiomatize valid formulas in the class of all contact algebras satisfying the axiom (Ext) – add the rule (EXT).

• If one wants to axiomatize valid formulas in the class of all contact algebras satisfying the axiom (Nor) – add the rule (NOR).

Appendix E. Proof of Theorem 3.3

Proof. We will consider only the case $\mathbb{L} = \mathbf{L}^C$, the other case can be treated similarly. Note that the equivalence (ii)↔(iii) is a corollary from the topological representation Theorem 1.4. The implication (ii)→(i) is obvious and for the implication (i)→(ii) we will use a kind of canonical model construction. This construction is a variant of the Henkin proof of the completeness theorem for the first-order logic adapted for the logics of the considered kind with additional rules. This construction is described in [3, Sec. 7] (see also [32, Sec. 3.3]), so we refer the reader to consult for the details the above references. The main idea is shortly the following.

Each consistent set A can be extended into a maximal consistent set Γ with some special properties depending on the rules of the logic:

(1) Γ contains all theorems of the logic and is closed under the rule modus ponens,

(2) If the conclusion $\alpha \Rightarrow c^{\leq n}(a)$ of the rule (Rule $c^{\leq n}$) does not belong to Γ then the premise $\alpha \wedge \bigwedge_{i=0}^{n} p_i \neq 0 \wedge a = \sum_{i=0}^{n} p_i \Rightarrow \bigvee_{0\leq i<j\leq n} p_i C p_j$ also does not belong to Γ for some parameters $p_0, \ldots, p_n$. Similar conditions are formulated for the other rules.

Then, using Γ, one can construct in a canonical way a contact algebra (B, C) as follows: define in the set of Boolean terms $a \equiv b$ iff $a = b \in \Gamma$. It can be proved that this is a congruence relation with respect to the Boolean operations which makes possible to define a Boolean algebra over the classes $|a|$ modulo this congruence. We define $|a|C|b|$ iff $aCb \in \Gamma$. The axioms of contact guarantee that (B, C) is a contact algebra. Moreover the above properties of Γ and the additional axioms and rules of the logic guarantee that the obtained contact algebra belongs to the class $\Sigma(\mathbf{L})$. For instance the axiom (Ax $c^{\leq n}$) guarantee the implication

$c^{\leq n}(|a|)$ implies $(\forall |p_0|, \ldots, |p_n| \in B)(|p_0| \neq |0| \wedge \ldots \wedge |p_n| \neq |0| \wedge |a| = |p_0| + \cdots + |p_n| \rightarrow (\exists i, j : 0 \leq i < j \leq n)(|p_i|C|p_j|))$.

The converse implication is guarantied by the property (2) of the set Γ, which shows that the definition of $c^{\leq n}(|a|)$ is fulfilled in (B, C).

By means of Γ one can define a canonical valuation v in B as follows: $v(p) = |p|$ and finally we need to prove the truth lemma saying that $(B, C, v) \models \alpha$ iff $\alpha \in \Gamma$. Then this shows that (B, C, v) is a model of Γ and hence a model of A. □

A Simple Semantics for Aristotelian Apodeictic Syllogistics

Sara L. Uckelman and Spencer Johnston

Institute for Logic, Language, and Computation
PO Box 94242
1090 GE Amsterdam, The Netherlands
S.L.Uckelman@uva.nl; spencer_johnston_8603@yahoo.com

Abstract

We give a simple definition of validity for syllogisms involving necessary and assertoric premises which validates all and only the Aristotelian apodeictic syllogisms.

Keywords: Aristotle, modal syllogistic, semantics, two Barbaras

1 The problem

The first systematic study of reasoning and inference in the West was done by Aristotle. However, while his assertoric theory of syllogistic reasoning is provably sound and complete for the class of models validating the inferences in the traditional square of opposition [5, p. 100], his modal syllogistic, developed in chapters 3 and 8–22 of the *Prior Analytics* [1], has the rather dubious honor of being one of the most difficult to understand logical systems in history. Starting with some of his own students, many have considered Aristotle's modal syllogistic to be anywhere from confused to simply wrong [7, ch. 1]. In support of these claims, many critics point to what is called the "two Barbaras problem", that is, Aristotle's treatment of syllogisms of the form LXL Barbara and XLL Barbara.[1] According to Aristotle, arguments of the form

Necessarily A belongs to all B.
B belongs to all C.
Therefore, necessarily A belongs to all C.

[1] See §2 for an explanation of the notation. Throughout this paper we make use of the traditional medieval mnemonic names of syllogisms [13, p. 21].

are valid, while arguments of the form

A belongs to all B.
Necessarily B belongs to all C.
Therefore, necessarily A belongs to all C.

are invalid [1, 30a15–30a33]. Many people have found this position to be inconsistent. [2] Aristotle's student Theophrastus argued that both syllogisms are invalid [7, p. 15], since nothing should follow when one premise is necessary and the other assertoric. Łukasiewicz, whose views on Aristotle's modal syllogistic [5] have been extremely influential on modern approaches to the system, has argued that both syllogisms are valid [7, p. 15]. Łukasiewicz says that "Aristotle's modal syllogistic is almost incomprehensible because of its many faults and inconsistencies" and "modern logicians have not as yet been able to construct a universally acceptable system of modal logic which would yield a solid basis for... Aristotle's work" [5, p. 132]. One of the "faults and inconsistencies" is Aristotle's acceptance of LXL Barbara while rejecting of XLL Barbara. Later attempts have been made to give a consistent interpretation of Aristotle's modal syllogistic. McCall [7] gave a syntactic theory which coincides exactly with the apodeictic fragment of the Aristotelian theory (the fragment containing just the necessity and assertoric modal operators). More recently, Johnson [3,4], Thomason [15], and Malink [6] have given semantics corresponding to McCall's syntax, showing that Aristotle's apodeictic fragment is consistent, if, given the complexity of their semantic models, rather unintuitive.

We offer a new approach to the apodeictic fragment of Aristotelian syllogistics, which provides a clear and simple definition of validity that validates all and only those apodeictic syllogisms accepted by Aristotle. First, in §2 we define the notation we use in this paper. Previous attempts at giving syntactic and semantic characterization of the modal syllogistic are considered in §§4,5. The definition of validity that we give provides a formalization of the philosophical interpretation of Aristotle's apodeictic syllogistic given by Rescher in [8], and refined by McCall in [7]; we discuss this interpretation in §3, and then give our new formalism in §6. In §7 we show it is adequate for the pure necessary/assertoric fragment, and discuss the problems we have faced extending this formalism to the fragment which also contains the possibility operator. We conclude with some comments about future work in §8.

2 Notation

Syllogistics is a term logic, so we fix a set TERM of basic terms, and let capital letters $A, B, C \ldots$ range over TERM. (Assertoric) categorical propositions are formed from

[2] And some people disagree that there is even a separate *modal* syllogistics at all [10].

1st	2nd	3rd
A — B	B — A	A — B
B — C	B — C	C — B
A — C	A — C	A — C

Figure 1. The Three Figures

copulae a, e, i, o and terms as follows:

AaB	'A belongs to all B'	$\leftrightarrow$ 'All B are A'	(universal affirmative)
AeB	'A belongs to no B'	$\leftrightarrow$ 'No B is A'	(universal negative)
AiB	'A belongs to some B'	$\leftrightarrow$ 'Some B is A'	(particular affirmative)
AoB	'A does not belong to some B'	$\leftrightarrow$ 'Some B is not A'	(particular negative)

The term preceding the copula is called the predicate term and the term succeeding it is called the subject term. We follow McCall and use L, X, and M to denote the necessary, assertoric, and possible modes, respectively.[3] Hence, if φ is an assertoric categorical proposition, $L\varphi$, $X\varphi$, and $M\varphi$ are modal categorical propositions. (Note that the "assertoric" mode is not any different from the ordinary propositional mode. We will often designate assertoric propositions without the X.) Categorical propositions, both assertoric and modal, can be combined to form syllogisms.

Definition 2.1 A triple $\mathcal{S} = \langle M, m, c\rangle$, where M, m, and c are categorical propositions, is a *syllogism* if M, m, and c contain exactly three distinct terms, of which the predicate of c (called the major term) appears in M and the subject of c (called the minor term) appears in m, and M and m share a term (called the middle term) which is not present in c.

We call M the major premise, m the minor, and c the conclusion. The three ways that major, minor, and middle terms in the premises can be arranged are called figures (see Figure 1). A figure with three copulae added is called a 'mood'; by, e.g., 'LLL Barbara' we mean the mood Barbara with each of the premises prefaced with mode L.

3 Rescher's interpretation

A supposed drawback of Aristotle's modal syllogistic according to Łukasiewicz is that it "does not have any useful application to scientific problems" [5, p. 181]. In contrast with this conclusion, Rescher believes not only that the modal syllogistic can be given a consistent interpretation, but that, in fact, this interpretation is based on Aristotle's theory of scientific knowledge and inference. Rescher describes attempts such those of Łukasiewicz and Becker [2] as "blind alleys, as regards the possibility of interpreting Aristotle's discussion as it stands, without introducing numerous 'corrections'" [8,

[3] We omit from discussion the mode Q 'contingent'.

p. 165]. He argues that the problem of these formalisms was that they force an incorrect interpretation of the *Prior Analytics.* To address this, Rescher develops a non-formal account of the *Prior Analytics* which stresses the scientific nature of the various modal deductions. He argues that:

> The key to Aristotle's theory lies, I am convinced, in viewing the theory of modal syllogisms of the *Analytica Priora* in the light of the theory of scientific reasoning of the *Analytica Posteriora* [8, p. 170].

On his analysis, the major premise is treated as a general scientific principle or rule and the minor premise as a specific instance of the general rule [8, p. 171]. Further,

> [A] rule that is necessarily (say) applicable to all of a group will be necessarily applicable to any sub-group, pretty much regardless of how this sub-group is constituted. On this view, the necessary properties of a genus must necessarily characterize even a contingently differentiated species. If all elms are necessarily deciduous, and all trees in my yard are elms, then all trees in my yard are necessarily deciduous (even though it is not necessary that the trees in my yard be elms) [8, p. 172].

This interpretation allows him to make a principled distinction between LXL Barbara and XLL Barbara, since in the first case, the general rule is necessary, and the particular instance falls under that necessary rule. The conclusion that results should then be necessary. However, if the general rule is only assertoric, then the conclusion shouldn't be necessary, since for Aristotle, the assertoric generally does not entail the necessary.

McCall rightly points out that this interpretation only works for the first-figure syllogisms with mixed necessary and assertoric premises. In the case of second and third figure syllogisms, such as XLL Camestres, the minor premise is the general rule, and the major premise is the special case. Further, attempting to reduce the validity of these other figures to that of the first figure is problematic, not least because one would have to justify the conversion rules used in the reduction. As an alternative, drawing inspiration from the medieval doctrine of distribution, McCall points out that, with two exceptions, the general rule is the premise in which the middle term is distributed, and in a valid syllogism the special case can be "upgraded" to the modality of the general rule. A term is distributed in a proposition if "it *actually* denotes or refers to, in that premiss, the whole of the class of entities which it is *capable* of denoting" [7, p. 25]. In AaB, B is distributed; in AeB, both terms are distributed; in AiB, neither term is distributed; in AoB, A is distributed. The two restrictions are the following: (1) general rules cannot be particular and (2) special cases cannot be negative [7, p. 26]. The first exception allows us to rule out XLL Baroco while the second exception allows us to avoid XLL Felapton and XLL Bocardo, which are not accepted as valid by Aristotle [1, 31a1–31a18, 31a14–31a33].

The models that we introduce in §6 take seriously this suggestion of Rescher that we understand modal syllogisms as making a statement about the relationship between a general scientific law and a special case falling under that law. We will give a precise definition of what counts as a special case, and make explicit how to "upgrade" the modality of the special case to that the general rule. Thus, we will be able to show that if we accept Rescher's interpretation of the modal syllogistic, a consistent theory of syllogistic reasoning can be extracted from Aristotle's works.

4 Syntactic characterizations of the apodeictic fragment

McCall then used his "completion" (as he calls it) of Rescher's interpretation as the basis for developing a syntactic system characterizing Aristotle's apodeictic fragment of the syllogistic. It is based on the rules of conversion and the perfect syllogisms that Aristotle defined for the apodeictic syllogistic in the *Prior Analytics*. McCall shows that from propositional logic plus an axiomatization of the assertoric syllogistic supplemented with six modal axioms, and four laws of modal conversion and subordination, it is possible to deduce all of the valid apodeictic syllogisms and reject all of the ones that are invalid according to Aristotle [7, §14]. The six modal axioms are LXL Barbara, LXL Cesare, LXL Darii, LXL Ferio, LLL Baroco, LLL Bocardo, and the conversion and subordination rules are:

- from $LAiB$ infer $LBiA$
- from $LAaB$ infer AaB
- from $LAiB$ infer AiB
- from $LAoB$ infer AoB

McCall made no attempt to give a semantic grounding for his syntactic theory.

Rescher, along with Parks, later developed his interpretation into a proof-theoretic account which simplifies McCall's approach [9], but which only deals with the L-X fragment (whereas McCall's syntactic theory can be extended to the L-X-M fragment). At the heart of their account is the following observation:

> The leading idea of our proposal is that given syllogistic terms α and β it is possible to define yet another term $[\alpha\beta]$ to represent the β-species of α... they are those α's which must be β's relative to the hypothesis that they are α's (by conditional or relative necessity) [9, p. 678–679].

This idea is based on Aristotle's notion of ekthesis, which allows for deriving universal propositions from particular ones, and which Aristotle uses to give proofs of the oblique moods LLL Baroco and LLL Bocardo [9, §3]. (For more information on ekthesis and its role in Aristotelian syllogistic proofs, see [12]). This observation allows us to move from "A belongs to all B" to "all Bs, given that they are As, are necessarily A, with relative necessity, given that they are in fact Bs." This notion of relative necessity plays a key role in development of Rescher and Park's system, which has just four conversion rules together with the perfect assertoric and wholly apodeictic syllogisms as axioms. The four conversion rules are as follows:

$$\vdash AaB \Rightarrow \vdash L[BA]aB$$

$$\vdash AiB \Rightarrow \vdash L[BA]iB$$

$$\vdash LAaB \Rightarrow \vdash LAa[CB]$$

$$\vdash LAeB \Rightarrow \vdash LAe[CB]$$

The complex term $[AB]$ is read 'A-conditioned-by-B' or 'A's which are B'. These rules can be understood as follows:

- If A belongs to all B, then being B's which are A necessarily belongs to all B.
- If A belongs to some B, then being a B which is A necessarily belongs to some B.
- If A necessarily belongs to all B, then A necessarily belongs to all those C which are B.
- If A necessarily does not belong to any B, then A necessarily belongs to no C which is a B.

Rescher and Parks prove the consistency of their theory only in an indirect fashion (by reducing the apodeictic syllogistic to the assertoric one, which was proved consistent in [11]).

5 Previous semantic attempts

Later authors have attempted to build semantics for McCall's or an equivalent axiomatization; three rigorous approaches are those of Johnson [3], Thomason [15], and [6]. While these semantics are adequate in so far as they validate all of McCall's (and hence Aristotle's) theses, and reject those that should be rejected, they are not very appealing on grounds of both aesthetics and explanatory value. The systems are very complicated and could be labeled *ad hoc* because they are not motivated beyond being adequate to characterize (McCall's version) of Aristotle's theory.

5.1 Johnson's model

The semantics given by Johnson in [3] are adequate to prove the completeness of the apodeictic fragment of McCall's formalization.

Definition 5.1 A *Johnson-syllogistic model* is a quintuple

$$\mathfrak{M}^{\mathrm{J}} = \langle W, V^e, V^a, V_c^e, V_c^a \rangle,$$

where W is a set and the V_j^i are functions from TERM to 2^W meeting the following conditions:

(i) $V(A) := V^e(A) \cup V^a(A)$

(ii) $V^e(A) \neq \emptyset$

(iii) For each A, $V_k^j(A) \cap V_n^m(A) = \emptyset$ iff either $j \neq m$ or $k \neq n$; and for each A, $V^e(A) \cup V^a(A) \cup V_c^e(A) \cup V_c^a(A) = W$.

(iv) If $V(C) \subset V_c^e(B)$ and $V(A) \subset V(B)$ then $V(A) \subset V_c^e(C)$.

(v) If $V(B) \subset V^e(C)$ and $V(A) \cap V(B) \neq \emptyset$ then $V^e(A) \cap V^e(C) \neq \emptyset$.

(vi) If $V(B) \subset V_c^e(C)$ and $V(A) \cap V(B) \neq \emptyset$ then $V^e(A) \cap V_c^e(C) \neq \emptyset$.

(vii) If $V(C) \subset V^e(B)$ and $V^e(A) \cap V_c^e(B) \neq \emptyset$ then $V^e(A) \cap V_c^e(C) \neq \emptyset$.

We think of $V^e(A)$ as the set of things which are essentially A, $V^a(A)$ as the things which are accidentally A, $V_c^e(A)$ is the set of things essentially non-A, and $V_c^a(A)$ is the set of things accidentally non-A.

The truth conditions for categorical propositions are as expected:

Definition 5.2

$$\mathfrak{M}^{\mathrm{J}} \models AaB \text{ iff } V(B) \subset V(A).$$ [4]

$$\mathfrak{M}^{\mathrm{J}} \models AiB \text{ iff } V(B) \cap V(A) \neq \emptyset.$$

$$\mathfrak{M}^{\mathrm{J}} \models AeB \text{ iff } \mathfrak{M}^{\mathrm{J}} \not\models AiB.$$

$$\mathfrak{M}^{\mathrm{J}} \models AoB \text{ iff } \mathfrak{M}^{\mathrm{J}} \not\models AaB.$$

$$\mathfrak{M}^{\mathrm{J}} \models LAaB \text{ iff } V(B) \subset V^e(A).$$

$$\mathfrak{M}^{\mathrm{J}} \models LAeB \text{ iff } V(B) \subset V_c^e(A).$$

$$\mathfrak{M}^{\mathrm{J}} \models LAiB \text{ iff } V^e(B) \cap V^e(A) \neq \emptyset.$$

$$\mathfrak{M}^{\mathrm{J}} \models LAoB \text{ iff } V^e(B) \cap V_c^e(A) \neq \emptyset.$$

Thom criticizes these semantics in [14], and Johnson responded to Thom's objections in [4]. The revised system of [4] was intended to (a) allow that general terms may designate a property such that no object necessarily has this property (thus giving up (ii) above), (b) require that if some object has the property designated by a general term necessarily, then any object which has this property has it necessarily, and (c) be "intuitively graspable" [4, p. 171]. The system goes beyond Aristotelian modal logic by allowing singular sentences (that is, sentences involving constants instead of terms), but it is more restricted than McCall's syntax in that it does not account for M propositions. The semantics are substitutionally based. Thirteen conditions for an acceptable valuation function are given in §3, thus it is by no means clear that Johnson has succeeded with his goal (c) in the new semantics.

5.2 *Thomason's models*

Thomason feels that Johnson's semantics "is in some respects contrived" [15, p. 111], and offers a proposal of his own. Thomason finds fault with Johnson's semantics in that "the interpretations are explicitly required to satisfy Axioms 6–9 [LXL Cesare, Darii, and Ferio, and LLL Baroco] of L-X-M" [15, p. 112], and he introduces models which do away with this requirement.

[4] Note that this definition does not entail existential import, whereas Aristotle's definitions in the Square of Opposition do.

Definition 5.3 A *Thomason-syllogistic model* is a quintuple

$$\mathfrak{M}^{\mathrm{T}} = \langle W, \mathrm{Ext}, \mathrm{Ext}^{+}, \mathrm{Ext}^{-}, V \rangle,$$

where the Exts are functions assigning subsets of W to each term satisfying $\mathrm{Ext}^{+} \subseteq \mathrm{Ext}$, $\mathrm{Ext}^{+} \neq \emptyset$, $\mathrm{Ext}^{-} \cap \mathrm{Ext} = \emptyset$, and V is an ordinary two-valued valuation function.

The functions $\mathrm{Ext}(x)$, $\mathrm{Ext}^{+}(x)$ and $\mathrm{Ext}^{-}(x)$ should be understood as picking out that which is x, is x necessarily, and is necessarily not x respectively. The truth conditions for the assertoric propositions are the same as in Johnson's semantics (so they also do not satisfy existential import), while those for the modal propositions are defined as follows:

Definition 5.4

$$\begin{aligned}
V(LAaB) = T &\text{ iff } \mathrm{Ext}(A) \subset \mathrm{Ext}^{+}(B)\\
V(LAeB) = T &\text{ iff } \mathrm{Ext}(A) \subset \mathrm{Ext}^{-}(B)\\
V(LAiB) = T &\text{ iff } \mathrm{Ext}^{+}(A) \cap \mathrm{Ext}^{+}(B) \neq \emptyset\\
V(LAoB) = T &\text{ iff } \mathrm{Ext}^{+}(A) \cap \mathrm{Ext}^{-}(B) \neq \emptyset
\end{aligned}$$

Validity and consequence are defined on these models in the expected way. Then, the consequences of Axioms 6–9 on this class of models correspond exactly to the theorems of Johnson's axiomatization, which in turn corresponds exactly to Aristotle's theory [15, p. 120]. Since these models require the truth of Axioms 6–9 to be built into the interpretation function, Thomason does not find them adequate, and instead offers two further classes of models, which satisfy all the requirements previous outlined and additionally

(i) $\mathrm{Ext}(x) \cap \mathrm{Ext}(y) \neq \emptyset \Rightarrow \mathrm{Ext}(x) \cap \mathrm{Ext}^{+}(y) \neq \emptyset$

(ii) Both (i) and $\mathrm{Ext}(x) \subseteq \mathrm{Ext}^{-}(y) \Rightarrow \mathrm{Ext}(y) \subseteq \mathrm{Ext}^{-}(x)$ and $\mathrm{Ext}(x) \subseteq \mathrm{Ext}^{+}(y) \rightarrow \mathrm{Ext}^{-}(y) \subseteq \mathrm{Ext}^{-}(x)$.

Aristotle's theory of the apodeictic syllogistic coincides with the set of consequences of LLL Baroco and the conversion rule $LAeB \Rightarrow LBeA$ on the second class of models [15, p. 122] and with the set of validities of the third class of models [15, p.124]. Thus, if we build extra structure into the interpretation of the terms, we are able to recover Aristotelian syllogistics without further assumptions. However, it is not clear where the justification for this extra structure comes in, other than that its addition makes the system work. It would be preferable to have a justification which is less *ad hoc* and more grounded in Aristotelian philosophy.

5.3 Malink's models

A rather different approach is taken by Malink in [6]. Malink appeals to Aristotle's discussion of types of predication in the *Topics* for the philosophical grounding of his interpretation, and bases his reconstruction of the modal syllogistic on what he calls 'predicable-based modal copula' [6, p. 97]. In the *Topics*, there are four different types of predicables: genus with (a) *differentia*, (b) definition, (c) *proprium*, or (d) accident. These four types of predicables are based on two basic relations, essential predication ($\mathbf{E}ab$) and accidental predication (Υab). Malink characterizes the behavior of these two basic relations via the axiomatic system $\mathcal{A}$, consisting of seven definitions and five axioms, and which "is not intended to give an exhaustive description of Aristotelian predicable-semantics, but to capture only those aspects of it which are relevant for the formal proofs of modal syllogistic" [6, p. 98]. These definitions and axioms are interpreted in graphically-representable structures made up of the following elements:

•	substance term, Σa
∘	nonsubstance term, $\neg\Sigma a$
———	substantial essential predication, $\mathbf{E}ab$
……	merely accidental predication, Υab
– – –	non-substantial essential predication, $\tilde{\mathbf{E}}ab$

The predicative relations between terms are represented by downward paths in the diagrams, with the conventions that it is assumed that all substance terms are **E** predicated of themselves, and all nonsubstance terms are $\tilde{\mathbf{E}}$ predicated of themselves, and when both types of predication coincide, only **E** predication is drawn.

In Malink's system, assertoric, necessary, (merely) possible, and contingent categorical claims are formalized as follows:

$XAaB$ Υab

$XAeB$ $\forall z(\Upsilon bz \to \neg\Upsilon az)$

$XAiB$ $\exists z(\Upsilon bz \wedge \Upsilon az)$

$XAoB$ $\neg\Upsilon ab$

$LAaB$ $\widehat{\mathbf{E}}ab$

$LAoB$ $\mathbf{K}ab$

$LAiB$ $\exists z((\Upsilon bz \wedge \widehat{\mathbf{E}}az) \vee (\Upsilon az \wedge \widehat{\mathbf{E}}bz))$

$LAoB$ $\exists z(\Upsilon bz \wedge \mathbf{K}az) \vee \exists xv(\widehat{\mathbf{E}}bz \wedge \widehat{\mathbf{E}}av \wedge \forall u(\Upsilon au \wedge \widehat{\Sigma}u \to \mathbf{K}zu))$

$MAaB$ $\forall z(\Upsilon bz \to \bar{\Pi}az)$

$MAeB$ $\forall z(\Upsilon bz \to \neg\bar{\mathbf{E}}az) \wedge \forall z(\Upsilon \to \neg\bar{\mathbf{E}}bz)$

$MAiB$ $\exists z(\Upsilon bz \wedge \bar{\Pi}az)$

$MAoB$ $\neg\bar{\mathbf{E}}ab$

$$QAa/eB \ \forall z(\Upsilon bz \to \Pi az)$$

$$QAi/oB \ \ \Pi ab$$

where $\Sigma a := \exists z \mathbf{E} za$, $\mathbf{K}ab := \Sigma a \wedge \Sigma b \wedge \neg \exists z(\Upsilon az \wedge \Upsilon bz)$, $\Pi ab := \neg(\Sigma a \wedge \Sigma b) \wedge \neg \mathbf{E}ab \wedge \neg \mathbf{E}ba \wedge ((\Sigma a \vee \Sigma b) \to \exists z(\Upsilon az \wedge \Upsilon bz))$, $\bar{\Pi}ab := \Pi ab \vee \Upsilon ab$, $\widehat{\mathbf{E}}ab := \mathbf{E}ab \vee \tilde{\mathbf{E}}ab$, $\widehat{\Sigma}ab := \exists z \mathbf{E}za$, $\bar{\mathbf{E}}ab := \mathbf{E}ab \vee (\Sigma a \wedge \Upsilon ab)$.

With this formalization, Malink is able to validate not only the apodeictic fragment but he can also makes sense of the merely possible and the contingent fragments, making his approach an improvement on both McCall's syntax as well as the models of Johnson and Thomason, which only work for the apodeictic fragments. However, this short overview of some of the aspects of Malink's reconstruction should be enough to demonstrate its extreme complexity, and there are other drawbacks with this approach which we discuss in the next section.

5.4 Discussion and critique

While these three types of models are semantically adequate in that their proofs are sound and their systems correspond to (a fragment of) Aristotelian syllogistic, there are a number of issues of their formalisms that we want to highlight. First, the semantics do not really explain what is going on in Aristotle's system. Each of the models introduces a primitive distinction between essential and nonessential predications. In Johnson, the quadripartite interpretation functions correspond to the notions of necessarily belonging, contingently belonging, contingently not belonging and necessarily not belonging. Thomason simplifies this to the tripartite distinction between what is necessary, what is necessarily not, and what is neither. Malink reduces this one more step, and distinguishes essential predication and accidental predication. While building these distinctions into the truth conditions and/or syntax is entirely adequate to capture Aristotle's notion of necessity and contingency, doing so reduces what explanatory power the models might have otherwise had.

Furthermore, none of the authors discussed how their semantics correlate with the or make sense of the new interpretation of Aristotle given by Rescher and discussed by McCall. Given that Rescher's interpretation gives a philosophical grounding for why Aristotle's modal syllogistic validates the syllogisms that it does, it is unfortunate that when Johnson, Thomason, and Malink develop their semantics, none of them discuss this philosophical grounding.

6 A new approach

Our new approach to the apodeictic syllogistic is based on making formal the "upgrade" criterion that McCall gives. Our models are standard models for quantified modal logic:

Definition 6.1 A *simple syllogistic model* is a tuple $\mathfrak{M}^{\mathrm{S}} = \langle W, D, R, O, V \rangle$ where W is a set (of possible worlds); D is a set (of objects); $R \subseteq W \times W$ is reflexive, transitive, and symmetric; for $w \in W, O(w) \subseteq D$ is the set of objects existing in w; and for $A \in \text{TERM}, V(A) \subseteq D$ is the set of objects in the extension of a term A.

$V(A)$ is extended naturally to $V'(A,w) = V(A) \cap O(w)$. The truth conditions for the assertoric propositions are as expected:

Definition 6.2

$$\begin{aligned}
\mathfrak{M}^{\mathrm{S}}, w \models AaB &\text{ iff } V'(B,w) \neq \emptyset \text{ and } V'(B,w) \subseteq V'(A,w).\\
\mathfrak{M}^{\mathrm{S}}, w \models AiB &\text{ iff } V'(A,w) \cap V'(B,w) \neq \emptyset.\\
\mathfrak{M}^{\mathrm{S}}, w \models AeB &\text{ iff } V'(A,w) \cap V'(B,w) = \emptyset.\\
\mathfrak{M}^{\mathrm{S}}, w \models AoB &\text{ iff } V'(B,w) = \emptyset \text{ or } V'(A,w) \not\subseteq V'(B,w).
\end{aligned}$$

The first conjunct ensures that our models satisfy existential import, which Aristotle accepted. Since R is an equivalence relation on W, the modalities L and M are the usual S5 modalities. What is novel in our semantics is the definition of the validity of a syllogism, which is given via the concept of model update:

Definition 6.3 For a model $\mathfrak{M}^{\mathrm{S}}$ and formula φ, the *update of* $\mathfrak{M}^{\mathrm{S}}$ *by* φ is the model $\mathfrak{M}^{\mathrm{S}} \restriction \varphi = \langle W \restriction \varphi, D, R \restriction \varphi, O \restriction \varphi, V \restriction \varphi \rangle$ where $W \restriction \varphi = \{w \in W : \mathfrak{M}^{\mathrm{S}}, w \models \varphi\}$; D is unchanged; and $R \restriction \varphi$, $O \restriction \varphi$, and $V \restriction \varphi$ are the restrictions of the original relations and functions to $W \restriction \varphi$.

Definition 6.4 A premise in a syllogism $\mathcal{S}$ is a *general rule* if (1) the middle term is distributed (cf. Def. 2.1 and §3) and (2) it is not particular.

A premise in a syllogism is a *special case* (1) if the other premise is a general rule and (2) only if it is not negative.

Definition 6.5 A syllogism $\mathcal{S}$ with special case s is valid for any simple model $\mathfrak{M}^{\mathrm{S}}$ and $w \in W$ iff (i) $\mathfrak{M}^{\mathrm{S}}, w \models M$ and (ii) $\mathfrak{M}^{\mathrm{S}}, w \models m$ imply (iii) $\mathfrak{M}^{\mathrm{S}} \restriction s, w \models c$.

The process of model update corresponds to the idea of "upgrading" the special case argued for by Rescher and McCall. When the general rule is necessary, we can consider only those worlds where the special case is in fact true, for if it is false then we do not care whether the conclusion is true or false, and when we restrict our attention in this fashion, we are able to draw necessary conclusions.

Note that on this definition, a syllogism $\mathcal{S}$ can be valid at world W in a simple model $\mathfrak{M}^{\mathrm{S}}$ even if the premises are true at w and the conclusion false. Thus, the definition of validity that we introduce is radically different from standard notions of validity, but this change in approach is justified by Aristotle's use syllogistics in scientific reasoning. If we either required that the conclusion already be true at w in $\mathfrak{M}^{\mathrm{S}}$, then we would collapse into the same problems that earlier attempts to formalize the modal syllogistic have, or defined validity so that syllogisms were only valid in $\mathcal{M}^{\mathrm{S}}$, then we would never have any valid modal syllogisms in the "real world".

As an example of how this system works, consider the problem of the two Barbaras. LXL Barbara is of the form $\mathcal{S} = \langle LAaB, BaC, LAaC \rangle$. To prove that this syllogism is

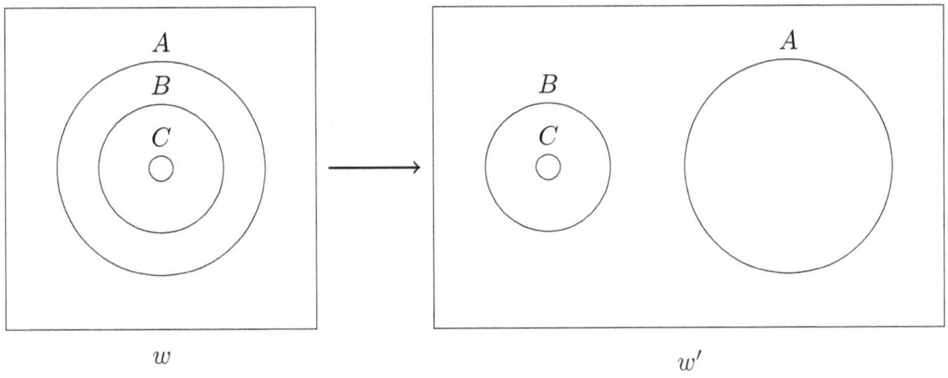

Figure 2. Countermodel for XLL Barbara.

valid, suppose that (i) $\mathfrak{M}^{\mathrm{S}}, w \vDash LAaB$ and (ii) $\mathfrak{M}^{\mathrm{S}}, w \vDash BaC$; we need to show that (iii) $\mathfrak{M}^{\mathrm{S}} \restriction BaC, w \vDash LAaC$. By the definition of model restriction, $\mathfrak{M}^{\mathrm{S}} \restriction BaC, w \vDash LBaC$ since the only worlds remaining are those worlds where BaC is true. From assumption (i), we also have that $\mathfrak{M}^{\mathrm{S}} \restriction BaC, w \vDash LAaB$, and we show that LLL Barbara is valid in the next section. Our proof here is similar in spirit to Aristotle's, as he also reduces the case of LXL Barbara to LLL Barbara by using a type of conditional or relative necessity (cf. [9, §2]). In contrast, XLL Barbara ($= \langle AaB, LBaC, LAaC \rangle$) is not valid, as the counterexample in Figure 2 shows. It is straightforward to show that any syllogism derivable in McCall's L-X fragment [7, Table 7] is validated on these semantics and that any syllogism rejected has a countermodel on our semantics.

7 Adequacy and limitations of the semantics

In this section we prove the adequacy of the semantics introduced in the previous section. It is obvious that all of the purely assertoric syllogisms are valid on our semantics. Further, note that every valid second- or third-figure assertoric syllogism can be derived from one of the four perfect first-figure assertoric syllogism by means of simple conversion (from AeB infer BeA and vice versa, and from AiB infer BiA and vice versa), accidental conversion (from AaB infer AiB, and from AeB infer AoB), and *reductio ad absurdum* or contraposition (interchange the contradictory of the conclusion with either the contradictory of the major premise or the minor premise). In the assertoric syllogistic, contraposition is required only for the argument from XXX Barbara to XXX Bocardo and XXX Baroco. In the modal syllogistic, neither LXL nor XLL Bocardo or Baroco are valid; therefore, for the L-X fragment we do not need to consider proof by modal contraposition.

To prove the soundness of the semantics with respect to the L-X fragment of Aristotelian syllogistic, we first prove that the four perfect LLL syllogisms are valid:

Proposition 7.1 *LLL Barbara is valid on our semantics.*

Proof Take an arbitrary model $\mathfrak{M}^{\mathrm{S}}$ and assume that (i) $\mathfrak{M}^{\mathrm{S}}, w \vDash LAaB$ and (ii) $\mathfrak{M}^{\mathrm{S}}, w \vDash LBac$. It suffices to show that $\mathfrak{M}^{\mathrm{S}} \restriction LBaC, w \vDash LAaC$. From the definition of model restriction, we have that $\mathfrak{M}^{\mathrm{S}} \restriction LBaC, w \vDash LBaC$, which is equivalent to for all w', if wRw' then $\mathfrak{M}^{\mathrm{S}} \restriction LBaC, w' \vDash BaC$, that is, $V'(C,w') \neq \emptyset$ and $V'(C,w') \subseteq V'(B,w')$. Further, from (i) we have w', if wRw' then $V'(B,w') \neq \emptyset$ and $V'(B,w') \subseteq V'(A,w')$. Now, take an arbitrary world v such that wRv, and we have that $V'(B,v) \neq \emptyset$, $V'(B,v) \subseteq V'(A,v)$, $V'(C,v) \neq \emptyset$, and finally $V'(C,v) \subseteq V'(B,v)$. Since subset inclusion is transitive, $V'(C,v) \subseteq V'(A,v)$, and hence $\mathfrak{M}^{\mathrm{S}} \restriction LBaC, v \vDash AaC$. Since v was arbitrary, we have that this holds for all w' such that wRw', which is to $\mathfrak{M}^{\mathrm{S}} \restriction LBaA, w \vDash LAaC$. □

Proposition 7.2 *LLL Celarent is valid on our semantics.*

Proof Take an arbitrary model $\mathfrak{M}^{\mathrm{S}}$ and assume that (i) $\mathfrak{M}^{\mathrm{S}}, w \vDash LAeB$ and (ii) $\mathfrak{M}^{\mathrm{S}}, w \vDash LBaC$. Then it suffices to show that $\mathfrak{M}^{\mathrm{S}} \restriction LBaC, w \vDash LAeC$. From the definition of restriction, we have that $\mathfrak{M}^{\mathrm{S}} \restriction LBaC, w \vDash LBaC$, which is equivalent to for all w' if wRw' then $V'(C,w') \neq \emptyset$ and $V'(C,w') \subseteq V'(B,w')$. Further, from (i) we have for all w', if wRw' then $V'(A,w') \cap V'(B,w') = \emptyset$). Now, take an arbitrary world v such that wRv, then we have that (1) $V'(A,v) \cap V'(B,v) = \emptyset$) and that (2) $V'(C,v) \neq \emptyset$ and $V'(C,v) \subseteq V'(B,v)$. Now take an arbitrary $x \in D$ such that $x \in V'(C,v)$. By (2) we have $x \in V'(B,v)$ since $V'(C,v) \subseteq V'(B,v)$. Now, by (1) we have $x \notin V'(A,v)$. Since x was arbitrary, it follows that $V'(C,v) \cap V'(A,v) = \emptyset$. Hence $\mathfrak{M}^{\mathrm{S}} \restriction LBaC, v \vDash AeC$. Since v was arbitrary, it follows that for any w' such that wRw', $\mathfrak{M}^{\mathrm{S}} \restriction LBaC, w' \vDash AeC$ and so $\mathfrak{M}^{\mathrm{S}} \restriction LBaC, w \vDash LAeC$. □

Proposition 7.3 *LLL Darii is valid on our semantics.*

Proof Take an arbitrary model $\mathfrak{M}^{\mathrm{S}}$ and assume that (i) $\mathfrak{M}^{\mathrm{S}}, w \vDash LAaB$ and (ii) $\mathfrak{M}^{\mathrm{S}}, w \vDash LBiC$. It suffices to show that $\mathfrak{M}^{\mathrm{S}} \restriction LBiC, w \vDash LAiC$. From the definition of restriction, we have that $\mathfrak{M}^{\mathrm{S}} \restriction LBiC, w \vDash LBiC$, which is equivalent to for all w', if wRw' then $V'(B,w') \cap V'(C,w') \neq \emptyset$. Further, from (i) we have for all w', if wRw' then $V'(B,w') \neq \emptyset$ and $V'(B,w') \subseteq V'(A,w')$. Now, take an arbitrary world v such that wRv, and we have that (1) $V'(B,v) \neq \emptyset$ and $V'(B,v) \subseteq V'(A,v)$ and (2) $V'(B,v) \cap V'(C,v) \neq \emptyset$. Thus, we have that $\exists x \in V'(B,v) \cap V'(C,v)$. Call this element y, then since $y \in V'(B,v)$, it follows that $y \in V'(A,v)$. Since $y \in V'(C,v)$, then $V'(C,v) \cap V'(A,v) \neq \emptyset$, from which it follows that $\mathfrak{M}^{\mathrm{S}} \restriction LBiC, v \vDash AiC$. Now since v was arbitrarily chosen, it follows that for all w' such that wRw', $\mathfrak{M}^{\mathrm{S}} \restriction LBiC, w' \vDash AiC$ and so $\mathfrak{M}^{\mathrm{S}} \restriction LBiC, w \vDash LAiC$. □

Proposition 7.4 *LLL Ferio is valid on our semantics.*

Proof Take an arbitrary model $\mathfrak{M}^{\mathrm{S}}$ be assume that (i) $\mathfrak{M}^{\mathrm{S}}, w \vDash LAeB$ and (ii) $\mathfrak{M}^{\mathrm{S}}, w \vDash LBiC$. It suffices to show that $\mathfrak{M}^{\mathrm{S}} \restriction LBiC, w \vDash LAoC$. From the definition of restriction, we have that $\mathfrak{M}^{\mathrm{S}} \restriction LBiC, w \vDash LAiC$. This and (i) are equivalent to for all w' such that wRw', (1) $V'(B,w') \cap V'(C,w') \neq \emptyset$ and (2) $V'(A,w') \cap V'(B,w') = \emptyset$. Now, take an arbitrary world v such that wRv. By (1), since $V'(B,v) \cap V'(C,v) \neq \emptyset$, it follows that $V'(C,v) \neq \emptyset$. Further, since $y \in V'(B,v)$ by (2) it follows that $y \notin V'(A,v)$. Hence, $V'(A,v) \nsubseteq V'(C,v)$, and thus $\mathfrak{M}^{\mathrm{S}} \restriction LBiC, w' \vDash AoC$. Now since v was arbitrary, it follows that for all w', wRw' implies that $\mathfrak{M}^{\mathrm{S}} \restriction LBiC, w' \vDash AoC$, as required. □

The soundness of the necessitated forms of simple and accidental conversion follow directly from their validity in their assertoric forms. The validity of the XLL and LXL syllogisms corresponds directly to the non-modalization of the premise other than one in which the middle term is distributed[5]; when we update with the special case, it becomes necessary, and thus the result corresponds to an LLL syllogism which can, if required, be converted to a first figure one.

However, when we attempt to extend these semantics to the L-X-M fragment, a number of problems emerge. First, we lose the close connection between the modality of the special case and the validity of the syllogism. We cannot retain the original definition of validity, since it makes valid a number of M-X syllogisms that would not have been accepted by Aristotle[6], for example MXM Barbara and MXM Celarent. This follows because when we update with the special case (the minor premise), $\mathfrak{M}^{S} \upharpoonright m, w \models m$ for all $w \in W$; then either the world which made the major premise true is still in the model, in which case the conclusion must also be true, or the conclusion is false in the updated model, but we have then falsified the major premise. This will be the case for any syllogism where the special case is non-modal; the update procedure will always promote an assertoric premise to a necessary one. A similar problem occurs when the special case is modal; since R is an equivalence relation, model reduction by a modal formula does not change the model, and thus we can always create a counter-model where the non-modalized form of the minor premise is true at a world other than the world where the major premises is true, and false elsewhere. Then the minor premise is possible, as required, but the conclusion is falsifiable.

The cause of both these problems is rooted in the same fact, namely, that our semantics does not preserve the validity of modal *reductio ad absurdum* (contraposition) when at least one premise is possible, rather than assertoric or necessary. While it may be possible to give a counterexample for every invalid syllogism in the L-X fragment, there is no straightforward way of converting this into a counterexample for the contraposed syllogism. The failure of contraposition in our semantics stems from the fact that there is no correlation between the premise that is modal and the premise that distributes the middle term in a syllogism and its contraposed form. For example, in MXM Camestres, the major premise is modalized, and the middle term is distributed in the minor; in its contraposition, XLL Ferison, the minor premise is modalized and the middle term is distributed in the major. On the other hand, in XMM Camestres, the minor is modalized and contains the distributed middle, whereas in its contraposition, XLL Darii, the minor premise remains modalized, but the middle term is distributed in the major premise. Thus, contraposition breaks the close association between modality and distribution of the middle that is seen in the L-X fragment of Aristotle's modal syllogistic. At this point, we have not seen a way to generalize our definition of validity

---

5 The rather convoluted description "premise other than one in which the middle term is distributed" is a result of the fact that in Darapti, the middle term is distributed in *both* premises. Thus, either premise can serve as the general rule to the other's special case, which is reflected by the fact that XLL Darapti and LXL Darapti are both valid, and this is the only mood where both the XLL and LXL versions are.

6 Aristotle doesn't explicitly discuss syllogisms with pure possibility (as opposed to contingency) premises; however, given his acceptance of proof by *reductio ad absurdum*, it is easy reconstruct which X-M syllogisms would have to be valid given the validity of the L-X fragment.

to validate contraposition, so that we can extend our results to the L-X-M fragment.

8 Conclusion

We have provided a semantics which validates the axiomatization for the L-X fragments of Aristotle's modal syllogistics proposed in [7]. These semantics, which crucially rely on a model update process, are much simpler than those found in previous literature, e.g., [3,4,6,15]. They take seriously Rescher's proposal in [8] that a modal syllogism should be interpreted as making a claim about a specific case of a general scientific principle. This emphasis on the status of the special case, the premise other than one where the middle term is distributed, gives rise to a type of relative or conditional necessity, which is expressed in our system by the model update process. This independent motivation for the use of the dynamic upgrade of the premise which is special case means that our system is not *ad hoc*, but instead has good philosophical grounding.

We have shown that a new definition of validity based on model update provides a sound semantics for the L-X fragment but we have also shown that it is not straightforward to extend this definition to a larger fragment. We hope to investigate such an extension in future work. Another question that we hope to answer in future work involves relating the semantics we gave for the L-X fragment to standard modern logical theories. Model updates such as the one that we have proposed, where the truth of a formula at the evaluating world is required before the update can proceed, correspond to truthful public announcements, à la the Public Announcement fragment of epistemic logic [16]. Thus, one natural open question is precisely what fragment of dynamic modal logic this fragment of Aristotle's syllogistic corresponds to.

9 Acknowledgments

The first author was funded by the NWO project "Dialogical Foundations of Semantics" (DiFoS) in the ESF EuroCoRes programme LogICCC (LogICCC-FP004; DN 231-80-002; CN 2008/08314/GW), and would like to thank Martijn Buisman for invaluable assistance in the preparation of an earlier version of this paper. Both authors would like to thank the anonymous referees for their comments and suggestions, which helped us clarify some of our arguments.

References

[1] Aristotle, *The Prior Analytics*, in: J. Barnes, editor, *The Complete Works of Aristotle*, InteLex, 1992 Online at `http://pm.nlx.com/xtf/view?docId=aristotle/aristotle.xml`.

[2] Becker, A., "Die Aristotelische Theorie der Möglichkeitsschlüsse," Ph.D. thesis, Münster (1933).

[3] Johnson, F., *Models for modal syllogisms*, Notre Dame Journal of Formal Logic **30** (1989), pp. 271–284.

[4] Johnson, F., *Modal ecthesis*, History and Philosophy of Logic **14** (1993), pp. 171–182.

[5] Łukasiewicz, J., "Aristotle's Syllogistic: From the Standpoint of Modern Formal Logic," Clarendon, 1957.

[6] Malink, M., *A reconstruction of Aristotle's modal syllogistic*, History and Philosophy of Logic **27** (2006), pp. 95–141.

[7] McCall, S., "Aristotle's Modal Syllogisms," North-Holland Publishing Company, 1963.

[8] Rescher, N., *Aristotle's theory of modal syllogisms and its interpretation*, in: M. Bunge, editor, *The Critical Approach to Science and Philosophy*, Collier-MacMillan Limited, 1964 pp. 152–177.

[9] Rescher, N. and Z. Parks, *A new approach to Aristotle's apodeictic syllogisms*, The Review of Metaphysics **24** (1971), pp. 678–689.

[10] Rini, A. A., *Is there a modal syllogistic?*, Notre Dame Journal of Formal Logic **39** (1998), pp. 544–572.

[11] Shepherdson, J. C., *On the interpretation of Aristotelian syllogistic*, Journal of Symbolic Logic **21** (1956), pp. 137–147.

[12] Smith, R., *What is Aristoelian ecthesis?*, History and Philosophy of Logic **3** (1982), pp. 113–127.

[13] Spade, P. V., *Thoughts, words, and things: An introduction to late medieval logic and semantic theory* (2002), preprint, `http://pvspade.com/Logic/docs/thoughts1_1a.pdf`.

[14] Thom, P., *The two Barbaras*, History and Philosophy of Logic **12** (1991), pp. 135–149.

[15] Thomason, S. K., *Semantic analysis of the modal syllogistic*, Journal of Philosophical Logic **22** (1993), pp. 111–128.

[16] van Ditmarsch, H., W. van der Hoek and B. Kooi, "Dynamic epistemic logic," Synthese Library Series **337**, Springer, 2007.

A Complete Proof System for a Dynamic Epistemic Logic Based upon Finite π-Calculus Processes

Eric Ufferman [1]

Department of Mathematics, School of Science, National University of Mexico (UNAM)
Circuito exterior, Ciudad Universitaria. C.P. 04510
México D.F., México

Pedro Arturo Góngora [1]

Postgraduate Program in Computer Science, National University of Mexico (UNAM)
Circuito exterior, Ciudad Universitaria. C.P. 04510
México D.F., México

Francisco Hernández-Quiroz [1]

Department of Mathematics, School of Science, National University of Mexico (UNAM)
Circuito exterior, Ciudad Universitaria. C.P. 04510
México D.F., México

Abstract

The pi-calculus process algebra describes the interaction of concurrent and communicating processes. In this paper we present the syntax and semantics of a dynamic epistemic logic for multi-agent systems, where the epistemic actions are finite processes in the pi-calculus. We then extend the language to include actions from a specified set of action structures. We define a proof system for the extended language, and prove the completeness of the proof system. Thus any valid formula in the original language without action structures can be proved in the proof system for the extended language.

Keywords: Dynamic Epistemic Logic, pi-Calculus, Completeness

[1] The authors received support for this paper from the grant PAPIIT IN109010.

1 Background

1.1 Epistemic Logic

Epistemic logic is a branch of modal logic concerned with reasoning about the knowledge of agents. Given a countable set of proposition symbols $\mathcal{P}$, and a finite set of agents $\mathcal{A}$, formulas of the basic epistemic $\mathcal{L}_E$ logic are formed using the usual connectives and formation rules for formulas of propositional logic, with the additional proviso that for any formula φ, and agent $a \in \mathcal{A}$, $K_a\varphi$ is also a formula. The formula can be read either as "Agent a *knows* φ", or "Agent a *believes* φ", depending on the interpretation.

Dynamic epistemic logics model situations in which external events or actions can modify the knowledge or beliefs of agents. Examples of dynamic epistemic logics include the Logic of Public Announcements introduced in [6], the Action Model Logic introduced in [9], and the logic $\mathcal{L}_{E\pi}$ introduced in [2]. The Logic of Public Announcements is capable of modeling the knowledge of agents after a public announcement that a formula in the language holds. The Action Model Logic is a generalization of the first that is capable of modeling a broad variety of events involving communication, including public and private messages. The logic $\mathcal{L}_{E\pi}$ is capable of modeling knowledge of agents after the execution of a process in a modified version of the π-calculus, in which agents are allowed to send private messages – which are formulas in the logic – to each other.

Here, we will present a dynamic epistemic logic $\mathcal{L}^*_{E\pi}$, which is similar to $\mathcal{L}_{E\pi}$, but involves only the use of finite processes in the modified π-calculus. Due to the private nature of the communication modeled, agents may have misconceptions about epistemic states of other agents. Therefore, the interpretation of $K_a\varphi$ as "Agent a believes φ" is more appropriate for our logic.

1.2 The π-calculus

The π-calculus [5] is a process algebra used to model concurrent computations that may exchange names during computation. The names may represent information or communication links. The ability to exchange communication links makes the π-calculus useful for modeling mobile processes.

In this paper we will work with a basic subset of the π-calculus defined below.

Definition 1.1 [π-Calculus Syntax] Let $\mathcal{N} = \{x, y, \ldots\}$ be a denumerable set of *name* symbols. The *set of all processes of π-calculus* is the least set generated by the grammar:

$$P ::= \mathbf{0} \mid \pi.P \mid P \mid P \mid (\boldsymbol{\nu} x)\, P \mid\, !\pi.P$$
$$\pi ::= \overline{x}y \mid x(z) \mid \tau$$

where $x, y \in \mathcal{N}$ and $\tau, \boldsymbol{\nu} \notin \mathcal{N}$.

Processes are represented by P and prefixes by π. An important aspect of a process is the set of actions it is capable of performing. The process $\mathbf{0}$ is the null process, which is incapable of action. Processes of the form $\pi.P$ are called *guarded*, which means that they must realize the action represented by the prefix π before proceeding. A process $\overline{x}y.P$

is capable of sending message y along channel x and then proceeding as P. The process $x(y).P$ is capable of receiving a name z along channel x and proceeding as $P\{z/y\}$ – P with z substituted for every free occurrence (see below) of y. A process $\tau.P$ can proceed as P after some internal action without any interaction with its environment. The process $P_1|P_2$ represents processes P_1 and P_2 acting in parallel. The notation $(\boldsymbol{\nu}x)\ P$ signifies that the name x is restricted, and cannot be used in communication between P and its environment, but may be used for communication among subprocesses of P. Finally, a process $!P$ is capable of self-replication, and may proceed as $!P|P$.

We often omit null processes from process expressions and may write $\overline{x}y$ in place of $\overline{x}y.\mathbf{0}$, for example.

The variable y is bound if it is the scope of a receiving prefix $x(y)$, or operator $(\boldsymbol{\nu}y)$. All other occurrences of variables in prefixes are considered to be free.

The *transition semantics* of processes describes how processes can be reduced. We write $P \to Q$ if the process P can be reduced to Q in a single step. The basic reductions are $\tau.P \to P$ and $\overline{x}y|x(z).P \to P\{y/z\}$. The operation of parallel composition is considered to be associative and commutative, so parentheses may be omitted, and when multiple parallel components appear, multiple reductions may be possible.

Processes P and Q that can mimic any of each other's transitions are said to be *bisimilar*, written $P \sim Q$. We refrain from giving the full definition of bisimilarity here, but note that if a process P has no possible reductions, then $P \sim \mathbf{0}$.

We give some basic examples of reduction below. For more detail about the π-calculus, we refer the reader to [8].

1.3 Examples of Reductions of Processes

Example 1.2 Let

$$P \stackrel{\text{def}}{=} \overline{x}y|\overline{x}w|\tau.x(z).Q$$

Then P has only one immediate transition, as the rightmost process is "guarded" by the τ prefix. We have:

$$P \to P' \stackrel{\text{def}}{=} \overline{x}y|\overline{x}w|x(z).Q$$

Now there are two components capable of sending a message along channel x, but only process capable of receiving the message. So P' may proceed as either of two distinct processes in a nondeterministic manner.

$$P' \to \overline{x}w|Q\{y/z\}$$

and

$$P' \to \overline{x}y|Q\{w/z\}$$

Example 1.3 Now let

$$P \stackrel{\text{def}}{=} \overline{x}y|(\boldsymbol{\nu}x)\ (\overline{x}w|x(z).Q)$$

Here, the appearance of the operator $(\boldsymbol{\nu}x)$ means that the components within its scope may use the name x only to communicate with each other. So despite the fact that the symbol x also appears in the first component, it does not represent the same channel that is referred to by x in the second two components. Therefore, there is actually only one transition of P:

$$P \rightarrow \overline{x}y|(\boldsymbol{\nu}x)\ Q\{w/z\}$$

Example 1.4 Our final example shows the phenomenon of scope extrusion of a ν-operator. Given the process

$$P \stackrel{\text{def}}{=} ((\boldsymbol{\nu}y)\ \overline{x}(y).Q)|x(z).Q'$$

then we have the transition

$$P \rightarrow (\boldsymbol{\nu}y)\ (Q|Q'\{y/z\})$$

provided y does not appear free in Q'. Before the transition, the scope of the operator $(\boldsymbol{\nu}y)$ is the process Q, but after it is expanded to include the entire parallel composition. The idea is that by sending the restricted name y, process Q has established a private communication link with process Q'. This privacy persists after such a transition, even if the process P is put in parallel composition with other processes that use the name y.

2 Syntax of the logic $\mathcal{L}^*_{E\pi}$

In this section we define the syntax of the logic $\mathcal{L}^*_{E\pi}$, which is a dynamic epistemic logic for modeling knowledge updates using *finite* processes of the π-calculus. The logic $\mathcal{L}_{E\pi}$ defined in [2] - which allowed the use of both finite and infinite processes of the π-calculus - is an extension of $\mathcal{L}^*_{E\pi}$.

Definition 2.1 [Dynamic Epistemic Logic with Finite Processes Syntax] Let $\mathcal{P}$ be a denumerable set of atomic proposition symbols, $\mathcal{A}$ a finite set of agents, and $\mathcal{N}$ a denumerable set of name symbols. The *set* $\mathcal{L}^*_{E\pi}$ *of all formulas of Dynamic Epistemic Logic with Finite Processes* is the least set generated by the grammar:

$$\begin{aligned}
\varphi &::= p \mid \neg\varphi \mid (\varphi \wedge \varphi) \mid K_a\varphi \mid [P]\varphi \\
\psi &::= p \mid \neg\psi \mid (\psi \wedge \psi) \mid K_a\psi \\
P &::= \mathbf{0} \mid \pi.P \mid P \mid P \mid (\boldsymbol{\nu}x)\ P \\
\pi &::= \overline{x}_a y \mid \overline{x}_a\psi \mid x_a(z) \mid \tau
\end{aligned}$$

where $p \in \mathcal{P}$, $a \in \mathcal{A}$, $x, y \in \mathcal{N}$ and $\tau, \boldsymbol{\nu} \notin \mathcal{N}$.

We call the processes represented by P in the above grammar $\mathcal{L}^*_{E\pi}$ *processes.* These differ from standard π-calculus process in two fundamental ways. First, every sending and receiving action is associated with an agent a – the agent responsible for sending or receiving the name or message. Second, not only names but also purely epistemic formulas are allowed to be sent and received by processes.

We note that the processes are the same as those used in $\mathcal{L}_{E\pi}$ with the exception that the replication operation is not allowed in their formation, and that only purely epistemic formulas are allowed to be sent by agents.

Note: A small technical issue arises here when discussing transitions of $\mathcal{L}^*_{E\pi}$ processes. When we write $P \to Q$ we assume both processes are $\mathcal{L}^*_{E\pi}$ processes. So, for example, the following

$$\overline{x}\varphi|x(y).\overline{y}z \to \overline{\varphi}z$$

is *not* an acceptable reduction, as the process on the right has a formula where a name should be, and therefore is not well-formed.

Otherwise, reductions for $\mathcal{L}^*_{E\pi}$ processes are defined using the same rules as for processes in the regular π-calculus.

3 Semantics of Epistemic Logics

Here we give the semantics of the static portion of $\mathcal{L}^*_{E\pi}$, which coincides with the standard definition of semantics for basic epistemic logic $\mathcal{L}_E$. We will later extend the definition to the dynamic part of $\mathcal{L}^*_{E\pi}$. Formulas are interpreted in Kripke models.

Definition 3.1 A Kripke model $\mathfrak{M}$ is a tuple $\mathfrak{M} = \langle \mathcal{W}, \{R_i\}_{i\in\mathcal{A}}, V\rangle$, where:

(i) $\mathcal{W} = \{w_1, w_2, \ldots\}$ is a countable set of possible worlds (also called states).

(ii) $R_i \subseteq \mathcal{W} \times \mathcal{W}$ is an accessibility relation between worlds for each agent $i \in \mathcal{A}$.

(iii) $V : \mathcal{W} \times \mathcal{P} \to \{T, F\}$ is a valuation function that assigns a truth value to each propositional symbol in each possible world.

Given a model $\mathfrak{M}$ and a world $w \in \mathcal{W}$, we call the pair $(\mathfrak{M}, w)$ a pointed Kripke model.

Any sentence in $\mathcal{L}^*_{E\pi}$ can be evaluated in any pointed Kripke model. For purely epistemic formulas, the evaluation is just given by the standard Kripke semantics.

Definition 3.2 [Satisfaction in $\mathcal{L}^*_{E\pi}$ - static part.] We define the satisfaction relation $\models$ pointed Kripke models $\mathfrak{M}, w$ and (purely epistemic) formulas in $\mathcal{L}^*_{E\pi}$ as follows

(i) $\mathfrak{M}, w \models p$ iff $V(w, p) = T$

(ii) $\mathfrak{M}, w \models \neg\varphi$ iff $\mathfrak{M}, w \not\models \varphi$

(iii) $\mathfrak{M}, w \models (\varphi \wedge \psi)$ iff $\mathfrak{M}, w \models \varphi$ and $\mathfrak{M}, w \models \psi$

(iv) $\mathfrak{M}, w \models K_a\varphi$ iff for all $w' \in \mathcal{W}$ if $(w, w') \in R_a$ then $\mathfrak{M}, w' \models \varphi$

We defer the "dynamic" portion of the definition, (ie. the case $\mathfrak{M}, w \models [P]\varphi$) to Section 5.1.

4 Action Model Logic

To introduce our proof system for the logic $\mathcal{L}^*_{E\pi}$, we will use ideas from the Action Model Logic introduced in [1].

Definition 4.1 Let $\mathcal{L}$ be any epistemic language over a set of agents $\mathcal{A}$ and propositional symbols $\mathcal{P}$. An action model $\mathfrak{A}$ is a tuple $\langle E, \{\to_a\}_{a\in\mathcal{A}}, PRE\rangle$, where E is a set of events,

$\rightarrow_a$ is an accessibility relation on $E \times E$ for each $a \in \mathcal{A}$, and $PRE : E \rightarrow \mathcal{L}$ is a function assigning to each action point a precondition, which is a formula in the language $\mathcal{L}$.

We may write $(e, e') \in \rightarrow_a$ as $e \rightarrow_a e'$.

An action α is a pointed action model $(\mathfrak{A}, e)$ where $e \in E$. We may blur the distinction between actions and the associated point in the action model, and write $PRE(\alpha)$ for $PRE(e)$. If $\alpha = (\mathfrak{A}, e)$ and $\alpha' = (\mathfrak{A}, e')$ are actions, and $e \rightarrow_a e'$, we write $\alpha \rightarrow_a \alpha'$. Note that the actions must be pointed models for the same action structure for this relation to hold.

The idea is that in a Kripke model an action may occur, which requires updating the model. The resulting model is the composition of the model and the action, defined below.

Definition 4.2 Given a model and an action structure:

$$\mathfrak{M} = \langle \mathcal{W}, \{R_i\}_{i \in \mathcal{A}}, V \rangle$$

$$\mathfrak{A} = \langle E, \{\rightarrow_a\}_{a \in \mathcal{A}}, PRE \rangle$$

and distinguished elements $w_0 \in \mathcal{W}$, $e_0 \in E$, the product $(\mathfrak{M}, w_0) \otimes (\mathfrak{A}, e_0)$ of pointed models is defined iff $\mathfrak{M}, w_0 \models PRE(e_0)$. When defined,

$$(\mathfrak{M}, w_0) \otimes (\mathfrak{A}, e_0) = (\mathfrak{M}', (w_0, e_0))$$

Where $\mathfrak{M}' = \langle \mathcal{W}', \{R'_a\}_{a \in \mathcal{A}}, V' \rangle$ such that:

(i) $\mathcal{W}' = \{(w, e) \mid w \in \mathcal{W}, e \in E \text{ and } \mathfrak{M}, w \models PRE(e)\}$

(ii) $((w, e), (w', e')) \in R'_a$ iff $(w, w') \in R_a$ and $(e, e') \in \rightarrow_a$

(iii) $V'((w, e), p) = V(w, p)$ for all $(w, e) \in \mathcal{W}'$ and $p \in \mathcal{P}$

So an action is possible in a given world if that world meets the action's precondition. If multiple actions are possible in a given world w, then the world "splits" in the updated model. That is, we get a world of the form (w, e) for each action e that is possible in world w. If an agent considers w' to be possible given that the actual world is w, and also considers e' to be possible given action e, then he considers the pair (w', e') possible in the updated model, given actual world (w, e). Finally, the actions do not change the actual facts of a world w, so w and (w, e) will share a truth valuation, regardless of e.

4.1 Syntax and Semantics of Action Model Logic

We now introduce syntax of the action model logic $\mathcal{L}_{Act}$ as follows:

$$\varphi ::= p \mid \neg\varphi \mid \varphi \wedge \varphi \mid K_a\varphi \mid [\alpha]\varphi$$

We let α range over the set of all actions (up to isomorphism), with preconditions that are $\mathcal{L}_{Act}$ formulas already constructed at a previous stage of the inductive hierarchy. The semantics can then be defined as in Definition 3.2. We need only show how to handle the induction in the case of formulas like $[\alpha]\varphi$, where $\alpha = (\mathfrak{A}, e)$:

$$\mathfrak{M}, w \models [\alpha]\varphi \text{ iff } \mathfrak{M}, w \models PRE(e) \text{ implies } (\mathfrak{M}, w) \otimes (\mathfrak{A}, e) \models \varphi$$

We abbreviate $\neg[\alpha]\neg\varphi$ as $\langle\alpha\rangle\varphi$.

4.2 Composition of Action Models

We may also define an operation of composition on pairs of action structures.

Definition 4.3 Let $\mathfrak{A} = \langle E, \{\rightarrow_a\}, PRE\rangle$ and $\mathfrak{A}' = \langle E', \{\rightarrow'_a\}, PRE'\rangle$ be actions, and let $e_0 \in E$, $e'_0 \in E'$. We define $(\mathfrak{A}, e_0) \circ (\mathfrak{A}', e'_0) = (\mathfrak{A}'', (e_0, e'_0))$, where $\mathfrak{A}'' = \langle E'', \{\rightarrow''_a\}, PRE''\rangle$ such that:

(i) $E'' = E \times E'$

(ii) $((d, d'), (e, e')) \in\rightarrow''_a$ iff $(d, e) \in\rightarrow_a$ and $(d', e') \in\rightarrow'_a$

(iii) $PRE((e, e')) = \langle\mathfrak{A}, e\rangle PRE'(e')$

The following indicates that the effect of executing a composition of actions is the same as that of executing the actions in succession:

Proposition 4.4 *[1] Let $\mathfrak{M}, w$ be a pointed Kripke model, α and β be actions and φ a formula of $\mathcal{L}_{Act}$. Then*

$$\mathfrak{M}, w \models [\alpha][\beta]\varphi \Leftrightarrow \mathfrak{M}, w \models [\alpha \circ \beta]\varphi$$

5 The extended language $\mathcal{L}^+_{E\pi}$

We define an extension $\mathcal{L}^+_{E\pi}$ of the language $\mathcal{L}^*_{E\pi}$, by adding actions to the language. We will define a complete proof system for $\mathcal{L}^+_{E\pi}$. Because $\mathcal{L}^+_{E\pi}$ is an extension of $\mathcal{L}^*_{E\pi}$, every valid formula of $\mathcal{L}^*_{E\pi}$ will be a theorem in the proof system.

We first specify the set of actions that we will add to the language. The *basic actions* consist of the following two types:

(i) The trivial action $\tau = (\mathfrak{A}, t)$, where $\mathfrak{A} = \langle\{t\}, \{\rightarrow_a\}, PRE\rangle$, such that $\rightarrow_a = \{(t, t)\}$ for all $a \in \mathcal{A}$ and $PRE(t) = \top$.

(ii) The communication actions $\alpha^\varphi_{i,j} = (\mathfrak{A}^\varphi_{i,j}, e)$, representing a message φ sent on some channel x by agent i and received by agent j. Here

$$\mathfrak{A}^\varphi_{i,j} = \langle\{e, t\}, \{\rightarrow_a\}, PRE\rangle,$$

such that $\rightarrow_j = \rightarrow_i = \{(e, e), (t, t)\}$, $\rightarrow_a = \{(e, t), (t, t)\}$ for all $a \neq i, j$, $PRE(e) = K_i\varphi$ and $PRE(t) = \top$.

The motivation for including these types of actions is that they mimic the sorts of epistemic updates that correspond to transitions of processes in our modified π-calculus. The communication actions represent a message φ being sent from i to j, unbeknownst to the other agents. The precondition $K_i\varphi$ indicates that agent i may only send a message she believes to be true, ie., agents may not attempt to lie or deceive. We need trivial actions because processes may have transitions in which no information whatsoever is exchanged. Updating with trivial actions has no semantic effect.

The set *Act* of $\mathcal{L}^{+}_{E\pi}$ *actions* is the closure of the set of basic actions under composition. We are now ready to formally define the language $\mathcal{L}^{+}_{E\pi}$.

Definition 5.1 [Dynamic Epistemic Logic with Finite Processes and Actions Syntax] Let $\mathcal{P}$ be a denumerable set of atomic proposition symbols, $\mathcal{A}$ a finite set of agents, and $\mathcal{N}$ a denumerable set of name symbols. The *set* $\mathcal{L}^{+}_{E\pi}$ *of all formulas of Dynamic Epistemic Logic with Finite Processes and Actions* is the least set generated by the grammar:

$$\begin{aligned} \varphi &::= p \mid \neg\varphi \mid (\varphi \wedge \varphi) \mid K_a\varphi \mid [P]\varphi \mid [\alpha]\varphi \\ \psi &::= p \mid \neg\psi \mid (\psi \wedge \psi) \mid K_a\psi \\ P &::= \mathbf{0} \mid \pi.P \mid P \mid P \mid (\boldsymbol{\nu} x)\, P \\ \pi &::= \overline{x}_a y \mid \overline{x}_a \psi \mid x_a(z) \mid \tau \end{aligned}$$

where $p \in \mathcal{P}$, $a \in \mathcal{A}$, $x, y \in \mathcal{N}$, $\alpha \in Act$ and $\tau, \boldsymbol{\nu} \notin \mathcal{N}$.

Essentially, we have taken the language $\mathcal{L}^{*}_{E\pi}$ and actions from the set *Act* to the syntax. The semantics for the "new" formulas of the type $[\alpha]\varphi$ is defined in the same way as formulas of that type in $\mathcal{L}_{Act}$ (see Section 4.1).

5.1 *A translation function for* $\mathcal{L}^{+}_{E\pi}$

We wish to define a translation function $t : \mathcal{L}^{+}_{E\pi} \to \mathcal{L}_E$. We need some preliminary definitions. The first allows us to distinguish reductions of processes in which some information is acquired by an agent from those in which no information (but possibly a name) is acquired by any agent.

Definition 5.2 [Reduction-action Types] To every reduction of a process, we associate a basic action as follows:

(i) If a process $P \to P'$, such that the reduction used is either a τ-reduction, or the transmission of a *name* from one component to another, then we write $P \rightsquigarrow_{\tau} P'$.

(ii) If $P \to P'$, such that the reduction is the reception by agent j of a *formula* φ sent by agent i, then we write $P \rightsquigarrow_{\alpha^{\varphi}_{i,j}} P'$.

Next we define a one-step translation for formulas of shape $[P]\varphi$.

Definition 5.3 Let $\psi = [P]\varphi$. We define $s(\psi)$ as follows:

(i) $s([P]\varphi) = \varphi$ if $P \sim \mathbf{0}$

(ii) Otherwise, suppose that $P \rightsquigarrow_{\alpha_i} P_i$, where $1 \leq i \leq k$, and each α_i is a basic action. We assume that if $P \to Q$ then $Q = P_i$ for some $1 \leq i \leq k$, and that there are no repetitions among the P_i. Then we write:

$$s([P]\varphi) = \bigwedge_{1 \leq i \leq k} [\alpha_i][P_i]\varphi$$

We may now define the full translation function t:

Definition 5.4 The translation function $t : \mathcal{L}^+_{E\pi} \rightarrow \mathcal{L}_E$ is defined inductively according the following rules:

(i) $t(p) = p$

(ii) $t(\neg\varphi) = \neg t(\varphi)$

(iii) $t(\varphi \wedge \psi) = t(\varphi) \wedge t(\psi)$

(iv) $t(K_a\varphi) = K_a t(\varphi)$

(v) $t([\alpha]p) = PRE(\alpha) \rightarrow p$

(vi) $t([\alpha]\neg\varphi) = PRE(\alpha) \rightarrow \neg[\alpha]\varphi$

(vii) $t([\alpha](\varphi \wedge \psi)) = t([\alpha]\varphi) \wedge t([\alpha]\psi)$

(viii) $t([\alpha]K_a\varphi) = t(PRE(\alpha) \rightarrow \bigwedge_{\{\alpha' | \alpha \rightarrow_a \alpha'\}} K_a[\alpha']\varphi)$

(ix) $t([\alpha][\beta]\varphi) = t([\alpha \circ \beta]\varphi)$

(x) $t([P]\varphi) = t(s([P]\varphi))$

(xi) $t([\alpha][P]\varphi) = t([\alpha]s([P]\varphi))$

It is not obvious that $t(\varphi)$ is defined for all $\varphi \in \mathcal{L}^+_{E\pi}$, because in some cases $t(\varphi)$ is defined in terms of formulas that are not subformulas of φ. We will later define a well-ordered complexity measure on formulas of $\mathcal{L}^+_{E\pi}$, and show in Lemma 5.9 that $t(\varphi)$ is always defined in terms of t applied to formulas of strictly smaller complexity. As a consequence, $t(\varphi)$ is defined for every $\varphi \in \mathcal{L}^+_{E\pi}$.

We may now also give the missing case of Definition 3.2:

Definition 5.5 [Satisfaction in $\mathcal{L}^*_{E\pi}$ - dynamic part.]

If φ is an $\mathcal{L}^*_{E\pi}$ formula, and P is an $\mathcal{L}^*_{E\pi}$ process, then

$$\mathfrak{M}, w \models [P]\varphi \text{ iff } \mathfrak{M}, w \models s([P]\varphi).$$

Although $s([P]\varphi)$ is not a subformula of $[P]\varphi$ in general, the induction is still well-founded, as a consequence of Lemma 5.9.

5.2 The Proof System

We are now ready to define a proof system for the extended language $\mathcal{L}^+_{E\pi}$. The system is an extension of the proof system for Action Model Logic given in [1], which is itself an extension of the standard complete proof system **K** for modal logic (see, for example, [7]), with axioms added corresponding to each case in our translation function. Its axioms and inference rules are given in the following table:

All instantiations of proposition tautologies	
K-normality	$K_a(\varphi \to \psi) \to (K_a\varphi \to K_a\psi)$
atomic permanence	$[\alpha]p \leftrightarrow (PRE(\alpha) \to p)$
action and negation	$[\alpha]\neg\varphi \leftrightarrow (PRE(\alpha) \to \neg[\alpha]\varphi)$
action and conjunction	$[\alpha](\varphi \wedge \psi) \leftrightarrow ([\alpha]\varphi \wedge [\alpha]\psi)$
action and knowledge	$[\alpha]K_a\varphi \leftrightarrow (PRE(\alpha) \to \bigwedge_{\{\alpha' \mid \alpha \to_a \alpha'\}} K_i[\alpha']\varphi)$
composition of actions	$[\alpha][\beta]\varphi \leftrightarrow [\alpha \circ \beta]\varphi$
processes	$[P]\varphi \leftrightarrow s([P]\varphi)$
	$[\alpha][P]\varphi \leftrightarrow [\alpha]s([P]\varphi)$
modus ponens	From φ and $\varphi \to \psi$ infer ψ
necessitation of K_a	From φ infer $K_a\varphi$

The soundness of all axioms not involving processes follows from the soundness of the proof system given in [1]. The soundness of the axioms involving processes is immediate from our alternate definition of semantics of formulas of type $[P]\varphi$. Hence:

Proposition 5.6 *The proof system is sound.*

The following lemma will be the key to our completeness proof:

Lemma 5.7 *For every $\varphi \in \mathcal{L}^+_{E\pi}$, $t(\varphi)$ is defined, and moreover $\vdash \varphi \leftrightarrow t(\varphi)$*

In order to prove the lemma, we first define a *complexity function* on formulas of $\mathcal{L}^+_{E\pi}$.

Definition 5.8 For a finite process P, let $l(P)$ be the length of the longest possible reduction of P. The complexity function $c : \mathcal{L}^+_{E\pi} \cup Act \to \mathbb{N}^2$, is defined recursively as follows, where if $c(\varphi) = (x, y)$, we write $c_1(\varphi) = x$, and $c_2(\varphi) = y$.

(i) $c(p) = (0, 1)$

(ii) $c(\neg\varphi) = (c_1(\varphi), c_2(\varphi) + 1)$

(iii) $c(K_a\varphi) = (c_1(\varphi), c_2(\varphi) + 1)$

(iv) $c(\varphi \wedge \psi) = (\max\{c_1(\varphi), c_1(\psi\}, \max\{c_2(\varphi), c_2(\psi\} + 1)$

(v) $c([\alpha]\varphi) = (c_1(\varphi), (4 + c_2(\alpha))c_2(\varphi))$

(vi) $c([P]\varphi) = (c_1(\varphi) + l(P) + 1, c_2(\varphi))$

(vii) $c(\alpha) = (0, \max\{c_2(PRE(d)) \mid d \in \mathfrak{A}\})$ where $\alpha = (\mathfrak{A}, e)$

The following lemma will allow us to see that the semantics of $\mathcal{L}^+_{E\pi}$ is well-founded for formulas of type $[P]\varphi$, and also to prove the key Lemma 5.7.

Lemma 5.9 *Let $<_L$ represent the lexicographic order on $\mathbb{N}^2$. Then the following hold:*

(i) $c(\psi) \leq_L c(\varphi)$ *if ψ is a subformula of φ.*

(ii) $c(PRE(\alpha) \rightarrow p) <_L c([\alpha]p)$

(iii) $c(PRE(\alpha) \rightarrow \neg[\alpha]\varphi) <_L c([\alpha]\neg\varphi)$

(iv) $c([\alpha](\varphi \wedge \psi)) <_L c([\alpha]\varphi \wedge [\alpha]\psi)$

(v) $c(PRE(\alpha) \rightarrow K_a[\beta]\varphi) <_L c([\alpha]K_a\varphi)$ *for all β such that $\alpha \rightarrow_a \beta$*

(vi) $c([\alpha \circ \beta]\varphi) <_L c([\alpha][\beta]\varphi)$

(vii) $c(s([P]\varphi)) <_L c([P]\varphi)$

(viii) $c([\alpha]s([P]\varphi)) <_L c([\alpha][P]\varphi)$

Proof. Item (i) is clear from the definition of c. Items (ii)-(vi) are proved in a similar manner. In (iii), for example, we have that $c_1(PRE(\alpha) \rightarrow \neg[\alpha]\varphi) = c_1([\alpha]\neg\varphi)$, but that $c_2(PRE(\alpha) \rightarrow \neg[\alpha]\varphi) < c_2([\alpha]\neg\varphi)$. For details on the latter claim, see [9]. Analogous remarks hold for the others.

For (vii), we claim $c_1(s([P]\varphi)) < c_1([P]\varphi)$. In the case where $P \sim \mathbf{0}$, $[P]\varphi = \varphi$, so the claim follows immediately from (i). If $P \nsim \mathbf{0}$, then $s([P]\varphi) = \bigwedge_{1\leq i\leq k} [\alpha_i][P_i]\varphi$, where $P \rightsquigarrow P_i$ for each $1 \leq i \leq k$. But if $P \rightsquigarrow P'$, then $l(P') < l(P)$. It follows that $c_1([\alpha_i][P_i]\varphi) < c_1[P]\varphi$ for all i. But c_1 evaluated on a conjunction is the maximum of c_1 of any of the conjuncts, so the claim holds, and (vii) follows immediately. An analogous statement holds for (viii) . □

Proof of Lemma 5.7. By induction on the complexity of formulas. If $c(\varphi) = (0, 1)$, then φ is some propositional symbol p, and therefore $t(p) = p$. By propositional tautology, $\vdash p \leftrightarrow t(p)$. Now suppose that $\vdash \psi \leftrightarrow t(\psi)$ for all ψ such that $c(\psi) <_L c(\varphi)$.

The rest of the proof is handled in various cases. For example, suppose $\varphi = \neg\psi$. Then by Lemma 5.9 and the induction hypothesis, we have $\vdash \psi \leftrightarrow t(\psi)$. By tautology, we have $\vdash \neg\psi \leftrightarrow \neg t(\psi)$, which by definition of t, is the same as $\vdash \varphi \leftrightarrow t(\varphi)$. The other cases are handled similarly.

Lemma 5.10 *For all $\varphi \in \mathcal{L}^+_{E\pi}$, $t(\varphi) \in \mathcal{L}_E$.*

The proof is an easy induction over complexity of formulas. We may now give a short proof of completeness.

Theorem 5.11 *The proof system is complete. Ie., if $\models \varphi$ then $\vdash \varphi$.*

Proof. Suppose $\models \varphi$. Then, by soundness and the fact that $\vdash \varphi \leftrightarrow t(\varphi)$, we have $\models t(\varphi)$. By Lemma 5.10, $t(\varphi)$ is a formula of $\mathcal{L}_E$ (ie., contains no processes or actions). Since our proof system is an extension of the standard complete proof system **K** for $\mathcal{L}_E$, we have that $\vdash t(\varphi)$. This, together with $\vdash \varphi \leftrightarrow t(\varphi)$, yields the desired conclusion $\vdash \varphi$. □

6 Conclusions and Future Work

Following the ideas of [2], we have presented an attempt to integrate the approach of the π-calculus – which allows us to reason about communicating systems with mobility – with that of dynamic epistemic logic – which concerns updating the knowledge or beliefs of agents that are able to communicate. Having defined a dynamic epistemic language whose actions are finite processes in a modified version of the π-calculus, and extending that language with action structures as in [1] representing private messages between pairs of agents, we were able to define a proof system for the extended language and prove its completeness.

Although the above shows that our language is subsumed by the Action Model Logic of [1], there are reasons that the logic $\mathcal{L}^*_{E\pi}$ may be preferable in many applications. In cases where many sequences of actions are possible and can be captured by a single process, $\mathcal{L}^*_{E\pi}$ will gave a much more natural and succinct may of representing the sequences. Furthermore, the possible actions are inherent in the language, and do not need to be defined whenever they appear in a formula. The language is also highly adaptable; there is no reason it has to be restricted to two-party honest communication (as was done here for simplicity's sake). For example, if we wanted to model public or group communication rather than private communication, we need only subscript prefixes with sets of agents, and adjust the actions we use accordingly. Also, the semantics of $\mathcal{L}^*_{E\pi}$ is readily extendible to handle infinite processes (see [2]), which is not the case for Action Model Logic.

There are some obvious directions for future work. We aim to develop a proof system for a logic where the restriction to finite processes is dropped, and also where the formulas that can be sent are not limited to purely epistemic formulas. We do not expect that the present techniques will be applicable to the language with infinite processes, as our system makes essential use of the fact that any sequence of reductions of a process terminates. The introduction of fixed-point operators may be useful in attacking this problem. We also aim to develop a type system for the language so that we need not restrict the allowable transitions of processes. In [2], a slightly different version of the logic $\mathcal{L}^*_{E\pi}$ is used to proof the correctness of preservation of anonymity in the Dining Cryptographer's Algorithm. We believe that additional applications of the logic may be found in the areas of cryptography and security.

References

[1] Baltag, A., L. Moss and S. Solecki, *The logic of public annnouncements, common knowledge and private suspicions*, Technical report, CWI (1999).

[2] Góngora, P., E. Ufferman and F. Hernández-Quiroz, *Formal semantics of a dynamic epistemic logic for describing knowledge properties of π-calculus processes*, in: *Proceedings of Computational Logic in Multi-Agent Systems XI (to appear)*, LNCS (2010).

[3] Mardare, R., *Decidable extensions of hennessy-milner logic*, in: *Proceedings of International Conference on Formal Methods for Networked and Distributed Systems (FORTE 2006)*, LNCS 4229 (2006).

[4] Mardare, R., *Observing distributed computation. a dynamic-epistemic approach*, in: *Proceedings of the second Conference on Algebra and Coalgebra in Computer Science (CALCO2007)*, LNCS 4624 (2007).

[5] Milner, R., J. Parrow and J. Walker, *A calculus of mobile processes i and ii*, Information and Computation **100** (1992), pp. 1–77.

[6] Plaza, J., *Logics of public communications*, in: *Proceedings of 4th International Symposium on Methodologies for Intelligent Systems*, 1989, pp. 201–216.

[7] Popkorn, S., "First Steps in Modal Logic," Cambridge University Press, 2008.

[8] Sangiorgi, D. and D. Walker, "The Pi-Calculus A Theory of Mobile Processes," Cambridge University Press, 2003.

[9] van Ditmarsch, H., W. van der Hoek and B. Kooi, "Dynamic Epistemic Logic," Springer, 2007.

Proofs, Disproofs, and Their Duals

Heinrich Wansing

Dresden University of Technology
Institute of Philosophy
Germany
Heinrich.Wansing@tu-dresden.de

Abstract

Bi-intuitionistic logic, also known as Heyting-Brouwer logic or subtractive logic, is extended in various ways by a strong negation connective used to express commitments arising from denials. These logics have been introduced and investigated in [48]. In the present paper, an inferentialist semantics in terms of proofs, disproofs, and their duals is developed. Whereas the Brouwer-Heyting-Kolmogorov interpretation of intuitionistic logic uses just the notion of proof as primitive, and López-Escobar's inferentialist interpretation of Nelson's logics with strong negation utilizes only the notions of proof and disproof as primitive, the inferentialist interpretation of bi-intuitionistic logic with strong negation employs the four notions of proofs, disproofs, dual proofs, and dual disproofs as primitive concepts.

Keywords: proofs, disproofs, dual proofs, dual disproofs, proof-theoretic semantics, constructive logic, connexive logic, constructive negation, constructive implication, constructive co-implication.

1 Introduction

1.1 Inferential status and speech acts

It seems to be an accepted view that assertion and denial are particularly important speech acts in the context of a use-based, inferentialist account of linguistic meaning. In particular, the idea is that the rules of use that determine the meaning of linguistic expressions provide a basis for warranted assertions and denials. In order to make an assertion, it is enough to seriously utter a sentence, for example the sentence 'Mary is beautiful'. In order to obtain an absolutely clear case of denying that Mary is beautiful, the contrary predicate 'is ugly' may be used, i.e., the sentence 'Mary is ugly' may be seriously uttered. Instead of replacing the adjective 'beautiful' by another, contrary item from the lexicon, namely the adjective 'ugly', one may employ a suitable unary negation connective, $\sim$ 'it is definitely false that'. It is definitely false that Mary is beautiful if and only if Mary is ugly. The "only if" may not be clear. If it is definitely false that Mary is beautiful, then Mary not just fails to be beautiful, but she is ugly. It is then denied that Mary is beautiful by seriously uttering the negated sentence '$\sim$ Mary is beautiful'. This move is supported by the existence of more systematically and regularly connected pairs of contrary predicates in the lexicon: 'sane' versus 'insane', 'believes' versus 'disbelieves', 'desirable' versus 'undesirable', etc. The prefixes 'in', 'dis' and 'un' suggest the introduction of the negation connective $\sim$, so that a denial of a sentence A may be represented as an assertion of $\sim A$.

Negation can be iterated. Is a denial of $\sim A$ an assertion of A? Can denying be iterated? Can asserting be iterated? It seems plausible to assume that a speaker may assert that Mary is beautiful not only by seriously uttering the sentence 'Mary is beautiful', but also by seriously uttering the sentence 'I assert that Mary is beautiful'. Similarly, a speaker may deny that Mary is beautiful not just by seriously uttering the sentences 'Mary is ugly', but also by seriously uttering the sentence 'I deny that Mary is beautiful'. Clearly, first- and other-person asserting-that-ascriptions and denying-that-ascriptions *may* be iterated. A sentence such as 'I deny that I deny that Mary is beautiful' is perfectly grammatical, though perhaps difficult to parse. Seriously uttering this sentence amounts to performing the same speech act as uttering the perhaps more idiomatic sentence 'I deny that Mary is ugly'. A clear case of denying that Mary is ugly is seriously uttering the sentence 'Mary is beautiful'. A denial of $\sim A$ thus seems to be an assertion of A, and recall that an assertion of $\sim A$ was introduced as a denial of A.

The notions of assertion and denial stand in a close relation to the notions of proof and disproof, respectively. If I assert that A, then I commit myself to be ready to prove A, and if I deny that A, then I commit myself to be ready to disprove A. Assertion and denial are basic speech acts which are insensitive to the complexity and composition of the asserted or denied sentence A. No matter how complex A may be and no matter how A is composed, in order to assert or deny A, it is enough to seriously utter the sentence A or its strong negation $\sim A$. A (canonical) proof or a (canonical) disproof of a sentence A, however, *is* sensitive to the complexity and composition of A. A canonical proof of a conjunction $(A \wedge B)$, for example, requires a proof of A and a proof of B, whereas a canonical disproof of $(A \vee B)$ requires a disproof of A and a disproof of B.

If we look only at proofs and disproofs of elementary sentences representable by atomic formulas of a propositional or first-order language, then proofs and disproofs often are basic acts. We can take up an example provided by A. Grzegorczyk [13]. Suppose that l is a yellow lemon. We may prove that l is yellow just by drawing visual attention to l, and we may disprove that l is red again just by drawing visual attention to l. The falsification of the proposition that l is red is as direct as the verification of the proposition that l is yellow. Neither would we attempt to disprove that l is red by leading the assumption that l is red to an absurdity, nor would we attempt to prove that l is yellow by leading the assumption that l fails to be yellow to an absurdity. We would, under normal circumstances at least, just point to the colour of l. It might be objected that the provability of an elementary sentence such as 'l is yellow' requires a theory and that in verifying by eye that l is yellow, we do not just see that l is yellow but *infer* that l is yellow from a theory based on our visual experience. But then in falsifying by eye that l is red, we still do not seem to lead the assumption that l is red to an absurdity. If in the verification case we directly *infer* from (a theory based on) our visual experience that l is yellow, then in the falsification case it seems that we directly infer from (a theory based on) our visual experience that l is definitely not red. Therefore, *if* disproving by eye that l is red is conceived of as an inference of the proposition that l is definitely not red, then this "definitely not" is not a so-called negation as inconsistency. In other words 'l is definitely not red' is not to be understood as 'l implies absurdity', cf. [9,42].

What is absurdity? A sentence expresses absurdity, the absurd proposition, if in every model, the sentence fails to be true. If we consider possible worlds models, a sentence expresses absurdity, if the sentence fails to be true at every possible world in every model. Possible worlds are often conceived of as classical models satisfying the principle of bivalence. But they may also be conceived of as information states that may or may not support the truth or the falsity of propositions. If absence of truth is distinguished from falsity, so that the principle of bivalence is violated, a sentence may express absurdity without being false in every model or false at every state in every model. A sentence thus expresses absurdity if it is never true, and, *in general*, a reduction to absurdity is a reduction to non-truth. Of course, we may then also consider reductions to non-falsity. A sentence expresses non-falsity, if it is never false.

If an act of assertion commits a speaker to be ready to prove the asserted proposition, and an act of denial commits a speaker to be ready to disprove the denied proposition, one may wonder what kind of action is such that it commits a speaker to be ready to reduce the assumption that a certain proposition A is true to absurdity (or, more generally, to non-truth) and what kind of action is such it commits a speaker to be ready to reduce the assumption that A is definitely false to non-truth. It seems that the *first* kind of commitment comes with asserting that nothing supports the truth of A. If I assert that nothing supports the truth of the sentence 'Person b stabbed person c', I am committed to be ready to show that any piece of information (in particular any piece of information that seems to establish the truth of the assumption that b stabbed c) fails to establish the truth of the assumption that b stabbed c. If there is a witness who claims to have seen that b stabbed c, for example, I may point out that the witness used to be extremely unreliable on previous occasions. Proceeding in this way, I may try to show that there is no conclusive evidence in favour of 'b stabbed c'.

What makes it difficult, perhaps, to see the difference between disproofs and reductions to absurdity is that one might hold that every direct falsification of A also reduces the assumption that A to absurdity. If I present a group of very reliable witnesses who confirm that b was not at the crime scene, this may be viewed as a direct falsification of 'b stabbed c', in addition leading the assumption that b stabbed c to absurdity. But, firstly, this does not show that there is no difference between disproving and reducing to absurdity, and, secondly, note that information may be contradictory. Someone else might present another group of highly reliable witnesses who claim that they saw that b stabbed c, so that the available testimony both supports the truth and supports the falsity of 'b stabbed c'. Thus, it is not at all clear that disprovability always implies reducibility to non-truth. Indeed, the implication may fail.

The *second* kind of commitment appears to come with asserting that no information supports the falsity of assumption A. If I assert that no information supports the falsity of 'b stabbed c', I am committed to be ready to show that the assumption that b definitely did not stab c leads to absurdity. Again, I might try to point to certain facts that are incompatible with the assumption under consideration, although they do not prove that b stabbed c. I might, for example, point out that b's fingerprints can be found on the dagger that has been removed from c's corpse.

The view that the denial of a sentence s can be profitably analyzed as the assertion of a suitable negation of s is contentious. According to Greg Restall [31]

> [d]enial is not to be analysed as the assertion of a negation,

whereas Bryson Brown [3, p. 646] explains that he has

> a modest proposal: negation is *denial* in the object language.

Graham Priest [27, p. 105] concedes that the uttering of a negated sentence sometimes may be interpreted as a denial but holds that "asserting a negation (in the Fregean sense) is not necessarily a denial." Priest regards rejection as the linguistic expression of denial and takes rejecting something to be putting a bar on accepting it. "When justified, it is so because there is evidence against the claim: positive grounds for keeping it out of one's beliefs" [27, p. 103]. This exclusion from belief is stronger than agnosticism (absence of belief) but, as it seems, weaker than disbelief. Timothy Williamson [51, p. 10] explains that "we can regard assertion as the verbal counterpart of judgement and judegement as the occurrent form of belief". The association of assertions with proofs and denials with disproofs takes the negative judgement of denial as the occurrent form of disbelief and not as the occurrent form of refusal from belief. [1]

In this paper I would like to discuss logics in which it is important to distinguish between provability, disprovability, and their duals. The term 'duality' has several meanings even in mathematics. In one usage the concept of duality is related to order reversal. In this sense, the dual of provability is reducibility to non-truth. The dual of disprovability is reducibility to non-falsity. The picture summarized in Table 1 emerges.

	inferential status	related speech act
$\varnothing \vdash A$	A is provable	to assert that A
$\varnothing \vdash {\sim}A$	A is disprovable	to deny that A
$A \vdash \varnothing$	A is reducible to non-truth	to assert that no information supports the truth of A
${\sim}A \vdash \varnothing$	A is reducible to non-falsity	to assert that no information supports the falsity of A

Table 1
Speech acts and the inferential status of propositions.

[1] I intend to discuss the relation between assertion, denial, and negation in more detail in a separate paper.

1.2 Inferential relations and logical operations

If A is provable, then it is warranted to assert that A, if A is disprovable, then it is warranted to deny that A, if A is reducible to non-truth, then it is warranted to assert that no information supports the truth of A, and if A is reducible to non-falsity, then it is warranted to assert that no information supports the falsity of A.

The above considerations on the inferential status of a sentence A can be generalized to proofs from a finite set of sentences assumed to true $A_1, \ldots, A_n$ and reductions from a finite set of sentences $A_1, \ldots, A_n$ assumed not to be true. If the expression 'assumptions' is reserved for sentences assumed to be true, there seems to be a semantic gap in English and other natural languages, as there is no idiomatic term for sentences assumed not to be true. Let us agree to call sentences assumed not to be true *counterassumptions*. Sentences assumed to be false may be called *rejections* (or repudiations), so that sentences assumed not to be false might be called *counterrejections* (counterrepudiations). Table 2 lists eight different kinds of inferential relations.

	inferential relation
$A_1, \ldots, A_n \vdash A$	A is provable from assumptions $A_1, \ldots, A_n$
$A_1, \ldots, A_n \vdash {\sim}A$	A is disprovable from assumptions $A_1, \ldots, A_n$
$A \vdash A_1, \ldots, A_n$	A is reducible to absurdity from counterassumptions $A_1, \ldots, A_n$
${\sim}A \vdash A_1, \ldots, A_n$	A is reducible to non-falsity from counterassumptions $A_1, \ldots, A_n$
${\sim}A_1, \ldots, {\sim}A_n \vdash A$	A is provable from rejections $A_1, \ldots, A_n$
${\sim}A_1, \ldots, {\sim}A_n \vdash {\sim}A$	A is disprovable from rejections $A_1, \ldots, A_n$
$A \vdash {\sim}A_1, \ldots, {\sim}A_n$	A is reducible to absurdity from counterrejections $A_1, \ldots, A_n$
${\sim}A \vdash {\sim}A_1, \ldots, {\sim}A_n$	A is reducible to non-falsity from counterrejections $A_1, \ldots, A_n$

Table 2
Inferential relations.

If we want to reduce the inferential relation between the sentences $A_1, \ldots, A_n$ and the sentence A to the inferential status of a single formula, we may use suitable connectives: conjunction $\wedge$, disjunction $\vee$, implication $\rightarrow$, and the less well-known co-implication $\prec$, see Table 3.

We thereby arrive at the following vocabulary: $\{\wedge, \vee, \rightarrow, \prec, \sim\}$. Whereas $\wedge, \vee, \rightarrow$, and $\prec$ may be seen to emerge from the reduction of inferential relations to inferential status stated in Table 3, $\sim$ reflects the distinction between provability and disprovability. Conjunction $\wedge$ combines formulas in antecedent position, i.e., on the left of $\vdash$, and disjunction combines formulas in succedent position, i.e., on the right of $\vdash$. Implication

inferential relation	inferential status
$A_1, \ldots, A_n \vdash A$	$\varnothing \vdash (A_1 \wedge \ldots \wedge A_n) \rightarrow A$
$A_1, \ldots, A_n \vdash {\sim}A$	$\varnothing \vdash (A_1 \wedge \ldots \wedge A_n) \rightarrow {\sim}A$
$A \vdash A_1, \ldots, A_n$	$A \prec (A_1 \vee \ldots \vee A_n) \vdash \varnothing$
${\sim}A \vdash A_1, \ldots, A_n$	${\sim}A \prec (A_1 \vee \ldots \vee A_n) \vdash \varnothing$
${\sim}A_1, \ldots, {\sim}A_n \vdash A$	$\varnothing \vdash ({\sim}A_1 \wedge \ldots \wedge {\sim}A_n) \rightarrow A$
${\sim}A_1, \ldots, {\sim}A_n \vdash {\sim}A$	$\varnothing \vdash ({\sim}A_1 \wedge \ldots \wedge {\sim}A_n) \rightarrow {\sim}A$
$A \vdash {\sim}A_1, \ldots, {\sim}A_n$	$A \prec ({\sim}A_1 \vee \ldots \vee {\sim}A_n) \vdash \varnothing$
${\sim}A \vdash {\sim}A_1, \ldots, {\sim}A_n$	${\sim}A \prec ({\sim}A_1 \vee \ldots \vee {\sim}A_n) \vdash \varnothing$

Table 3
From inferential relations to inferential status.

is a vehicle for registering formulas that appear in antecedent position in succedent position, and co-implication is a vehicle for registering formulas that appear in succedent position in antecedent position. We read $A \prec B$ as "B co-implies A" or as "A excludes B". Whereas implication is the residuum of conjunction, co-implication is the residuum of disjunction:

$$(A \wedge B) \vdash C \text{ iff } A \vdash (B \rightarrow C) \text{ iff } B \vdash (A \rightarrow C),$$

$$C \vdash (A \vee B) \text{ iff } (C \prec A) \vdash B \text{ iff } (C \prec B) \vdash A.$$

The strong negation $\sim$ is a *primitive* negation. Other kinds of negation connectives are *definable* in the presence of $\rightarrow$ and $\prec$. Let p be a certain propositional letter. Then we define non-falsity as follows: $\top := (p \rightarrow p)$, and non-truth in this way: $\bot := (p \prec p)$. We can then introduce two negation connectives:

$$-A := (\top \prec A) \text{ (co-negation), and}$$

$$\neg A := (A \rightarrow \bot) \text{ (intuitionistic negation).}$$

Other defined connectives of HB are equivalence, $\leftrightarrow$, and co-equivalence, $\succ\!\!\prec$, which are defined as follows:

$$A \equiv B := (A \rightarrow B) \wedge (B \rightarrow A); \quad A \succ\!\!\prec B := (A \prec B) \vee (B \prec A).$$

The connectives $\wedge, \vee, \rightarrow$, and $\prec$ are the primitive connectives of bi-intuitionistic logic BiInt, also known as Heyting-Brouwer logic HB or as subtractive logic, see [5,6,10,11,12,16,28,29,30,33,52]. Extensions of HB by strong negation $\sim$ have been introduced and investigated in [48], see also [18]. Logics with strong negation and intuitionistic implication have been introduced by David Nelson in the late 1940s and subsequently have been investigated by many researchers, see, for example, [1,8,14,15,17,19,21,22,23,25,35,37,38,39,40,42,44,46].

2 Syntax and relational semantics of HB

The propositional language $\mathcal{L}'$ of HB is defined in Backus–Naur form as follows:

$$\text{atomic formulas: } p \in \mathit{Atom}$$
$$\text{formulas: } A \in \mathit{Form(Atom)}$$
$$A ::= p \mid (A \wedge A) \mid (A \vee A) \mid (A \rightarrow A) \mid (A \prec A).$$

It is well-known that intuitionistic propositional logic is faithfully embeddable into the modal logic S4 (= KT4), the logic of necessity and possibility on reflexive and transitive frames. The relational frame semantics of HB is simple and reveals that HB can be faithfully embedded into temporal S4 (= $\mathsf{K}_t\mathsf{T4}$).

Definition 2.1 A frame is a pre-order $\langle I, \leq \rangle$. Intuitively, I is a non-empty set of information states, and $\leq$ is a reflexive transitive binary relation of possible expansion of states on I.

Instead of $w \leq w'$, we also write $w' \geq w$.

Definition 2.2 An HB-model is a structure $\langle I, \leq, v^+ \rangle$, where $\langle I, \leq \rangle$ is a frame and v^+ is a function that maps every $p \in \mathit{Atom}$ to a subset of I (namely the states that support the truth of p). It is assumed that v^+ satisfies the following persistence (or heredity) condition for atoms:

$$\text{if } w \leq w', \text{ then } w \in v^+(p) \text{ implies } w' \in v^+(p).$$

The relation $\mathcal{M}, w \models^+ A$ ('state w supports the truth of $\mathcal{L}'$-formula A in model $\mathcal{M}$') is inductively defined as follows:

$$\begin{array}{ll}
\mathcal{M}, w \models^+ p & \text{iff } w \in v^+(p) \\
\mathcal{M}, w \models^+ (A \wedge B) & \text{iff } \mathcal{M}, w \models^+ A \text{ and } \mathcal{M}, w \models^+ B \\
\mathcal{M}, w \models^+ (A \vee B) & \text{iff } \mathcal{M}, w \models^+ A \text{ or } \mathcal{M}, w \models^+ B \\
\mathcal{M}, w \models^+ (A \rightarrow B) & \text{iff for every } w' \geq w : \mathcal{M}, w' \not\models^+ A \text{ or } \mathcal{M}, w' \models^+ B \\
\mathcal{M}, w \models^+ (A \prec B) & \text{iff there exists } w' \leq w : \mathcal{M}, w' \models^+ A \text{ and } \mathcal{M}, w' \not\models^+ B
\end{array}$$

where $\mathcal{M}, w \not\models^+ A$ is the classical negation of $\mathcal{M}, w \models^+ A$.

For intuitionistic negation and co-negation one obtains the following support of truth conditions:

$$\mathcal{M}, w \models^+ \neg A \text{ iff for every } w' \geq w,\ \mathcal{M}, w' \not\models^+ A;$$

$$\mathcal{M}, w \models^+ -A \text{ iff there exists } w' \leq w \text{ and } \mathcal{M}, w' \not\models^+ A.$$

Proposition 2.3 *For every $\mathcal{L}'$-formula A, HB-model $\langle I, \leq, v^+ \rangle$, and w, $w' \in I$:*

$$\textit{if } w \leq w',\ \textit{then } \mathcal{M}, w \models^+ A \textit{ implies } \mathcal{M}, w' \models^+ A.$$

Definition 2.4 HB is the set of all $\mathcal{L}'$-formulas A such that for every HB-model $\langle I, \leq, v^+ \rangle$, and $w \in I$: $\mathcal{M}, w \models^+ A$.

3 Extensions of HB by strong negation

The propositional language $\mathcal{L}$ is defined in Backus–Naur form as follows:

atomic formulas: $p \in Atom$

formulas: $A \in Form(Atom)$

$$A ::= p \mid {\sim}A \mid (A \wedge A) \mid (A \vee A) \mid (A \rightarrow A) \mid (A \prec A).$$

Definition 3.1 A model is a structure $\langle I, \leq, v^+, v^- \rangle$, where $\langle I, \leq \rangle$ is a frame. Moreover, v^+ and v^- are functions that map every $p \in Atom$ to a subset of I (namely the states that support the truth of p and the falsity of p, respectively. The functions v^+ and v^- satisfy the following persistence conditions for atoms: if $w \leq w'$, then $w \in v^+(p)$ implies $w' \in v^+(p)$; if $w \leq w'$, then $w \in v^-(p)$ implies $w' \in v^-(p)$. The relations $\mathcal{M}, w \models^+ A$ ('state w supports the truth of $\mathcal{L}$-formula A in model $\mathcal{M}$') and $\mathcal{M}, w \models^- A$ ('state w supports the falsity of $\mathcal{L}$-formula A in model $\mathcal{M}$') are inductively defined as follows:

$$\mathcal{M}, w \models^+ p \quad \text{iff } w \in v^+(p)$$

$$\mathcal{M}, w \models^- p \quad \text{iff } w \in v^-(p)$$

$$\mathcal{M}, w \models^+ {\sim}A \quad \text{iff } \mathcal{M}, w \models^- A$$

$$\mathcal{M}, w \models^- {\sim}A \quad \text{iff } \mathcal{M}, w \models^+ A$$

$$\mathcal{M}, w \models^+ (A \wedge B) \quad \text{iff } \mathcal{M}, w \models^+ A \text{ and } \mathcal{M}, w \models^+ B$$

$$\mathcal{M}, w \models^- (A \wedge B) \quad \text{iff } \mathcal{M}, w \models^- A \text{ or } \mathcal{M}, w \models^- B$$

$$\mathcal{M}, w \models^+ (A \vee B) \quad \text{iff } \mathcal{M}, w \models^+ A \text{ or } \mathcal{M}, w \models^+ B$$

$$\mathcal{M}, w \models^- (A \vee B) \quad \text{iff } \mathcal{M}, w \models^- A \text{ and } \mathcal{M}, w \models^- B$$

$$\mathcal{M}, w \models^+ (A \rightarrow B) \text{ iff for every } w' \geq w : \mathcal{M}, w' \not\models^+ A \text{ or } \mathcal{M}, w' \models^+ B$$

$$\mathcal{M}, w \models^+ (A \prec B) \text{ iff there exists } w' \leq w : \mathcal{M}, w' \models^+ A \text{ and } \mathcal{M}, w' \not\models^+ B.$$

In Table 4, a number of natural support of falsity conditions for strongly negated implications and co-implications are listed. For each choice of pairs of conditions, support of falsity is persistent for arbitrary formulas.

cI_1	$\mathcal{M}, w \models^- (A \to B)$	iff	$\mathcal{M}, w \models^+ A$ and $\mathcal{M}, w \models^- B$
cI_2	$\mathcal{M}, w \models^- (A \to B)$	iff	for every $w' \geq w : \mathcal{M}, w' \not\models^+ A$ or $\mathcal{M}, w' \models^- B$
cI_3	$\mathcal{M}, w \models^- (A \to B)$	iff	there is $w' \leq w : \mathcal{M}, w' \models^+ A$ and $\mathcal{M}, w' \not\models^+ B$
cI_4	$\mathcal{M}, w \models^- (A \to B)$	iff	there is $w' \leq w : \mathcal{M}, w' \not\models^- A$ and $\mathcal{M}, w' \models^- B$
cC_1	$\mathcal{M}, w \models^- (A \prec B)$	iff	$\mathcal{M}, w \models^- A$ or $\mathcal{M}, w \models^+ B$
cC_2	$\mathcal{M}, w \models^- (A \prec B)$	iff	there is $w' \leq w : \mathcal{M}, w' \models^- A$ and $\mathcal{M}, w' \not\models^+ B$
cC_3	$\mathcal{M}, w \models^- (A \prec B)$	iff	for every $w' \geq w : \mathcal{M}, w' \not\models^+ A$ or $\mathcal{M}, w' \models^+ B$
cC_4	$\mathcal{M}, w \models^- (A \prec B)$	iff	for every $w' \geq w : \mathcal{M}, w' \models^- A$ or $\mathcal{M}, w' \not\models^- B$

Table 4
Support of falsity conditions for implications and co-implications

Proposition 3.2 *For every $\mathcal{L}$-formula A, model $\langle I, \leq, v^+, v^- \rangle$, and w, $w' \in I$: if $w \leq w'$ then $w \models^+ A$ implies $w' \models^+ A$; if $w \leq w'$, then $w \models^- A$ implies $w' \models^- A$.*

The different support of falsity conditions for implications and co-implications listed in Table 4 result in *sixteen* systems of constructive propositional logic with strong negation that extend HB. Valid equivalences characteristic of these logics are stated in Table 5. The logics in the language $\mathcal{L}$ that differ from each other only with respect to validating a certain pair of these equivalences (one from the I-equivalences and one from the C-equivalences) are referred to as systems (I_i, C_j), $i, j \in \{1, 2, 3, 4\}$.[2]

I_1	$\sim(A \to B) \leftrightarrow (A \wedge \sim B)$	negated implication, classical reading
I_2	$\sim(A \to B) \leftrightarrow (A \to \sim B)$	negated implication, connexive reading
I_3	$\sim(A \to B) \leftrightarrow (A \prec B)$	negated implication as co-implication
I_4	$\sim(A \to B) \leftrightarrow (\sim B \prec \sim A)$	negated implication as contraposed co-impl.
C_1	$\sim(A \prec B) \leftrightarrow (\sim A \vee B)$	negated co-implication, classical reading
C_2	$\sim(A \prec B) \leftrightarrow (\sim A \prec B)$	negated co-implication, connexive reading
C_3	$\sim(A \prec B) \leftrightarrow (A \to B)$	negated co-implication as implication
C_4	$\sim(A \prec B) \leftrightarrow (\sim B \to \sim A)$	negated co-implication as contraposed impl.

Table 5
Constructively negated implications and co-implications

[2] In the sequel I will sometimes omit the qualification $i, j \in \{1, 2, 3, 4\}$.

Definition 3.3 The logics (I_i, C_j) are defined as the triples $(\mathcal{L}, \models^+_{I_i,C_j}, \models^-_{I_i,C_j})$, where the entailment relations $\models^+_{I_i,C_j}$, $\models^-_{I_i,C_j} \subseteq \mathcal{P}(\mathcal{L}) \times \mathcal{P}(\mathcal{L})$ are defined as follows:
$\Delta \models^+_{I_i,C_j} \Gamma$ iff for every model $\mathcal{M} = \langle I, \leq, v^+, v^- \rangle$ defined with clauses cI_i and cC_j and every $w \in I$, if $\mathcal{M}, w \models^+ A$ for every $A \in \Delta$, then $\mathcal{M}, w \models^+ B$ for some $B \in \Gamma$, and
$\Delta \models -_{I_i,C_j}\Gamma$ iff for every model $\mathcal{M} = \langle I, \leq, v^+, v^- \rangle$ defined with clauses cI_i and cC_j and every $w \in I$, if $\mathcal{M}, w \models^- A$ for every $A \in \Gamma$, then $\mathcal{M}, w \models^- B$ for some $B \in \Delta$.
For singleton sets $\{A\}$ and $\{B\}$, we write $A \models^+_{I_i,C_j} B$ ($A \models^-_{I_i,C_j} B$) instead of $\{A\} \models^+_{I_i,C_j} \{B\}$ ($\{A\} \models^-_{I_i,C_j} \{B\}$).

This definition of a logic as comprising *two* entailment relations instead of just one is unusual but not at all unnatural, see, for instance, [34,49,50]. The set of all constructively false sentences is not the complement of the set of all constructively true sentences, and we can make the following observation.

Proposition 3.4 *If* $(I_i, C_j) \neq (I_4, C_4)$, *then* $\models^+_{I_i,C_j} \neq \models^-_{I_i,C_j}$.

We do not require that for atomic formulas p, $v^+(p) \cap v^-(p) = \varnothing$. Therefore, the logics under consideration are *paraconsistent*. Neither is it the case that for any formula B, $\{p, {\sim}p\} \models^+_{I_i,C_j} B$ nor is it the case that $B \models^-_{I_i,C_j} \{p, {\sim}p\}$.[3]

The next observation on negation normal forms is used in the proof of the completeness result in Section 5. A formula is in *negation normal form* (nnf) if it contains $\sim$ only in front of atoms. The following translations ρ_{I_i,C_j} send every formula A to a formula in nnf, where $p \in Atom$ and $\odot \in \{\vee, \wedge, \rightarrow, \prec\!\!\!\!-\}$:

$$
\begin{aligned}
\rho_{I_i,C_j}(p) &= p\\
\rho_{I_i,C_j}({\sim}p) &= {\sim}p\\
\rho_{I_i,C_j}({\sim}{\sim} A) &= \rho_{I_i,C_j}(A)\\
\rho_{I_i,C_j}(A \odot B) &= \rho_{I_i,C_j}(A) \odot \rho_{I_i,C_j}(B)\\
\rho_{I_i,C_j}({\sim}(A \vee B)) &= \rho_{I_i,C_j}({\sim}A) \wedge \rho_{I_i,C_j}({\sim}B)\\
\rho_{I_i,C_j}({\sim}(A \wedge B)) &= \rho_{I_i,C_j}({\sim}A) \vee \rho_{I_i,C_j}({\sim}B)\\
\rho_{I_1,C_j}({\sim}(A \rightarrow B)) &= \rho_{I_1,C_j}(A) \wedge \rho_{I_1,C_j}({\sim}B)\\
\rho_{I_2,C_j}({\sim}(A \rightarrow B)) &= \rho_{I_2,C_j}(A) \rightarrow \rho_{I_2,C_j}({\sim}B)\\
\rho_{I_3,C_j}({\sim}(A \rightarrow B)) &= \rho_{I_3,C_j}(A) \prec\!\!\!\!- \rho_{I_3,C_j}(B)\\
\rho_{I_4,C_j}({\sim}(A \rightarrow B)) &= \rho_{I_4,C_j}({\sim}B) \prec\!\!\!\!- \rho_{I_4,C_j}({\sim}A)\\
\rho_{I_i,C_1}({\sim}(A \prec\!\!\!\!- B)) &= \rho_{I_i,C_1}({\sim}A) \vee \rho_{I_i,C_1}(B)\\
\rho_{I_i,C_2}({\sim}(A \prec\!\!\!\!- B)) &= \rho_{I_i,C_2}({\sim}A) \prec\!\!\!\!- \rho_{I_i,C_2}(B)\\
\rho_{I_i,C_3}({\sim}(A \prec\!\!\!\!- B)) &= \rho_{I_i,C_3}(A) \rightarrow \rho_{I_i,C_3}(B)\\
\rho_{I_i,C_4}({\sim}(A \prec\!\!\!\!- B)) &= \rho_{I_i,C_4}({\sim}B) \rightarrow \rho_{I_i,C_4}({\sim}A)
\end{aligned}
$$

[3] Co-negation is, of course, also a paraconsistent negation, see [4,36], whereas intuitionistic negation is paracomplete, i.e., does not validate the law of excluded middle.

Lemma 3.5 *For every formula A, $\rho_{I_i,C_j}(A)$ is in negation normal form and $A \models^{+}_{I_i,C_j} \rho_{I_i,C_j}(A)$, $\rho_{I_i,C_j}(A) \models^{+}_{I_i,C_j} A$, $A \models^{-}_{I_i,C_j} \rho_{I_i,C_j}(A)$, $\rho_{I_i,C_j}(A) \models^{-}_{I_i,C_j} A$.*

4 Inferentialist (proof-theoretic) interpretation

The plan now is to interpret the connectives of $\mathcal{L}$ in the style of the Brouwer-Heyting-Kolmogorov (BHK) interpretation of the intuitionistic connectives in terms of canonical proofs, see, for example, [7, p. 154]. It is well-known that David Nelson's constructive logics with strong negation admit of a sound interpretation in terms of both proofs and disproofs, see [20,40]. We will supplement the BHK interpretation by interpretations in terms of canonical disproofs, canonical reductions to absurdity (alias non-truth), and canonical reductions to non-falsity. That is, we will define the notions of canonical proofs, disproofs, dual proofs, and dual disproofs of complex $\mathcal{L}$-formulas by simultaneous induction. To make sure that the interpretation is correct for the logics (I_i, C_j), we will make the following assumptions:

(i) for no $\mathcal{L}$-formula A there exists both a proof and a dual proof of A;

(ii) for no $\mathcal{L}$-formula A there exists both a disproof and a dual disproof of A;

(iii) every $\mathcal{L}$-formula A either has a proof or dual proof;

(iv) every $\mathcal{L}$-formula A either has a disproof or dual disproof.

Note that we do not need clauses for the constants $\bot$ and $\top$ and the negation operations $\neg$ and $-$, because in $\mathcal{L}$ these connectives are definable. We also assume that the conditions under which an entity is a canonical proof, disproof, dual proof, or dual disproof of an atomic sentence depend on the appropriate and relevant social practice and are not a matter of logic.

4.1 Canonical proofs

We first consider the inductive definition of the notion of a canonical proof of a compound $\mathcal{L}$-formula.

- A canonical proof of a strongly negated formula $\sim A$ is a canonical disproof of A.
- A canonical proof of a conjunction $(A \wedge B)$ is a pair (π_1, π_2) consisting of a canonical proof π_1 of A and a canonical proof π_2 of B.
- A canonical proof of a disjunction $(A \vee B)$ is a pair (i, π) such that $i = 0$ and π is a canonical proof of A or $i = 1$ and π is a canonical proof of B.
- A canonical proof of an implication $(A \rightarrow B)$ is a construction that transforms any canonical proof of A into a canonical proof of B.
- A canonical proof of a co-implication $(A \prec B)$ is a pair (π_1, π_2), where π_1 is a canonical proof of A and π_2 is a canonical dual proof of B. (This pair is a canonical dual proof of $(A \rightarrow B)$).

4.2 Canonical disproofs

- A canonical disproof of a strongly negated formula $\sim A$ is a canonical proof of A.
- A canonical disproof of a conjunction $(A \wedge B)$ is a pair (i, π) such that $i = 0$ and π is a canonical disproof of A or $i = 1$ and π is a canonical disproof of B.
- A canonical disproof of a disjunction $(A \vee B)$ is a pair (π_1, π_2) consisting of a canonical disproof π_1 of A and a canonical disproof π_2 of B.
- A canonical disproof of an implication $(A \to B)$ in
  (I_1C_j) is a pair (π_1, π_2) consisting of a canonical proof π_1 of A and a canonical disproof π_2 of B.
  (I_2C_j) is a construction that transforms any canonical proof of A into a canonical disproof of B.
  (I_3C_j) is a pair (π_1, π_2), where π_1 is a canonical proof of A and π_2 is a canonical dual proof of B. (This pair is a canonical dual proof of $(A \to B)$).
  (I_4C_j) is a pair (π_1, π_2), where π_1 is a canonical disproof of B and π_2 is a canonical dual disproof of A.
- A canonical disproof of a co-implication $(A \prec B)$ in
  (I_iC_1) is a pair (i, π) such that $i = 0$ and π is a canonical disproof of A or $i = 1$ and π is a canonical proof of B.
  (I_iC_2) is a pair (π_1, π_2), where π_1 is a canonical disproof of A and π_2 is a canonical dual proof of B. (This pair is a canonical dual proof of $(\sim A \to B)$).
  (I_iC_3) is a construction that transforms any canonical proof of A into a canonical proof of B.
  (I_iC_4) is a construction that transforms any canonical disproof of B into a canonical disproof of A.

4.3 Canonical reductions to non-truth (canonical dual proofs)

- A canonical reduction to non-truth of a strongly negated formula $\sim A$ is canonical dual disproof of A.
- A canonical reduction to non-truth of a conjunction $(A \wedge B)$ is a pair (i, π) such that $i = 0$ and π is a canonical dual proof of A or $i = 1$ and π is a canonical dual proof of B.
- A canonical reduction to non-truth of a disjunction $(A \vee B)$ is a pair (π_1, π_2) consisting of a dual proof π_1 of A and a dual proof π_2 of B.
- A canonical reduction to non-truth of an implication $(A \to B)$ is a pair (π_1, π_2), where π_1 is a canonical proof of A and π_2 is a canonical dual proof of B. (This pair is a canonical proof of $(A \prec B)$).
- A canonical reduction to non-truth of a co-implication $(A \prec B)$ is a construction that transforms any dual proof of B into a dual proof of A.

4.4 *Canonical reductions to non-falsity (canonical dual disproofs)*

- A canonical reduction to non-falsity of a strongly negated formula $\sim A$ is a canonical dual proof of A.
- A canonical reduction to non-falsity of a conjunction $(A \wedge B)$ is a pair (π_1, π_2) consisting of a dual disproof π_1 of A and a dual disproof π_2 of B.
- A canonical reduction to non-falsity of a disjunction $(A \vee B)$ is a pair (i, π) such that $i = 0$ and π is a canonical dual disproof of A or $i = 1$ and π is a canonical dual disproof of B.
- A canonical reduction to non-falsity of an implication $(A \rightarrow B)$ in
  $(I_1 C_j)$ is a pair (i, π) such that $i = 0$ and π is a canonical dual proof of A or $i = 1$ and π is a canonical dual disproof of B.
  $(I_2 C_j)$ is a pair (π_1, π_s), where pi_1 is a canonical proof of A and π_2 is a canonical dual disproof of B.
  $(I_3 C_j)$ is a construction that transforms any canonical dual proof of B into a canonical dual proof of A. (This pair is a canonical dual proof of $(A \prec B)$).
  $(I_4 C_j)$ is a construction that transforms any canonical dual disproof of A into a canonical dual disproof of B.
- A canonical reduction to non-falsity of a co-implication $(A \prec B)$ in
  $(I_i C_1)$ is a pair (π_1, π_2), where π_1 is a canonical dual disproof of A and π_2 is a canonical dual proof of B.
  $(I_i C_2)$ is a construction that transforms any canonical dual proof of B into a canonical dual disproof of A. (This construction is a canonical dual proof of $(\sim A \prec B)$).
  $(I_i C_3)$ is a pair (π_1, π_2), where π_1 is a canonical proof of A and π_2 is a canonical dual proof of B. (This pair is a canonical dual proof of $(A \rightarrow B)$).
  $(I_i C_4)$ is a pair (π_1, π_2), where π_1 is a canonical disproof of B and π_2 is a canonical dual disproof of A.

In order to show by induction on the construction of inferences that the logics (I_i, C_j) are sound with respect to the above BHK-style interpretation in terms of proofs, disproofs, and their duals, we need proof systems for the semantically defined logics (I_i, C_j). We consider the display calculi defined in [48].

5 Display calculi

The structural proof theory of bi-intuitionistic logic is confronted with a number of problems, which are described, for example, in [5,10,48]. The sequent calculus for Heyting-Brouwer logic in [6] uses single-conclusion sequents but imposes a 'singleton on the left' constraint on the left introduction rule for co-implication (and a 'singleton on the right' constraint on the right introduction rule for implication). This asymmetric sequent calculus does not enjoy cut-elimination. Also the sequent calculus for HB in [28] does not allow cut-elimination. These problems can be overcome in display logic and in other types of sequent calculi that differ from ordinary Gentzen systems, see [5,10,11,12,48]. We will use the display sequent calculi $\delta(I_i, C_j)$ for the logics (I_i, C_j)

developed [48] and therefore briefly rehearse the presentation of $\delta(I_i, C_j)$.

The set of structures (or Gentzen terms) is defined as follows:

$$\text{formulas: } A \in \mathit{Form(Atom)}$$

$$\text{structures } X \in \mathit{Struc(Form)}$$

$$X ::= A \mid \mathbf{I} \mid (X \circ X) \mid (X \bullet X).$$

The intended interpretation of the connective $\circ$ as conjunction in antecedent position and as implication in succedent position and of $\bullet$ as co-implication in antecedent position and as disjunction in succedent position justifies certain 'display postulates' (*dp*) (we omit outer brackets, each column states two structural inference rules):

$$\frac{\dfrac{Y \vdash X \circ Z}{X \circ Y \vdash Z}}{X \vdash Y \circ Z} \qquad \frac{\dfrac{X \vdash Y \circ Z}{X \circ Y \vdash Z}}{Y \vdash X \circ Z} \qquad \frac{\dfrac{X \bullet Z \vdash Y}{X \vdash Y \bullet Z}}{X \bullet Y \vdash Z} \qquad \frac{\dfrac{X \bullet Y \vdash Z}{X \vdash Y \bullet Z}}{X \bullet Z \vdash Y}$$

Moreover, the interpretation of $\mathbf{I}$ as the empty structure suggests the following structural inference rules:

$$\frac{\dfrac{X \circ \mathbf{I} \vdash Y}{X \vdash Y}}{\mathbf{I} \circ X \vdash Y} \qquad \frac{\dfrac{\mathbf{I} \circ X \vdash Y}{X \vdash Y}}{X \circ \mathbf{I} \vdash Y} \qquad \frac{\dfrac{X \vdash Y \bullet \mathbf{I}}{X \vdash Y}}{X \vdash \mathbf{I} \bullet Y} \qquad \frac{\dfrac{X \vdash \mathbf{I} \bullet Y}{X \vdash Y}}{X \vdash Y \bullet \mathbf{I}}$$

In addition there are various 'logical' structural rules:

$$\frac{}{p \vdash p}\ (id) \qquad \frac{}{{\sim}p \vdash {\sim}p}\ (id{\sim}) \qquad \frac{X \vdash A \quad A \vdash Y}{X \vdash Y}\ (cut)$$

and versions of the familiar structural rules from standard Gentzen systems for classical logic, monotonicity (weakening), exchange (permutation), and contraction, together with associativity, see Table 6.

$$\frac{X \vdash Y}{X \vdash Y \bullet Z}\ (rm) \qquad \frac{X \vdash Y}{X \circ Z \vdash Y}\ (lm)$$

$$\frac{X \vdash Y \bullet Z}{X \vdash Z \bullet Y}\ (re) \qquad \frac{X \circ Z \vdash Y}{Z \circ X \vdash Y}\ (le)$$

$$\frac{X \vdash Y \bullet Y}{X \vdash Y}\ (rc) \qquad \frac{X \circ X \vdash Y}{X \vdash Y}\ (lc)$$

$$\frac{X \vdash (Y \bullet Z) \bullet X'}{X \vdash Y \bullet (Z \bullet X')}\ (ra) \qquad \frac{(X \circ Y) \circ Z \vdash X'}{X \circ (Y \circ Z) \vdash X'}\ (la)$$

Table 6
Structural sequent rules

$$\frac{X \vdash A \quad Y \vdash B}{X \circ Y \vdash (A \wedge B)}\ (\vdash \wedge) \qquad \frac{A \circ B \vdash X}{(A \wedge B) \vdash X}\ (\wedge \vdash)$$

$$\frac{X \vdash A \bullet B}{X \vdash (A \vee B)}\ (\vdash \vee) \qquad \frac{A \vdash X \quad B \vdash Y}{(A \vee B) \vdash X \bullet Y}\ (\vee \vdash)$$

$$\frac{X \vdash A \circ B}{X \vdash (A \rightarrow B)}\ (\vdash \rightarrow) \qquad \frac{X \vdash A \quad B \vdash Y}{(A \rightarrow B) \vdash X \circ Y}\ (\rightarrow \vdash)$$

$$\frac{X \vdash B \quad A \vdash Y}{X \bullet Y \vdash B \prec A}\ (\vdash \prec) \qquad \frac{B \bullet A \vdash X}{B \prec A \vdash X}\ (\prec \vdash)$$

$$\frac{X \vdash \sim A \bullet \sim B}{X \vdash \sim(A \wedge B)}\ (\vdash \sim\wedge) \qquad \frac{\sim A \vdash X \quad \sim B \vdash Y}{\sim(A \wedge B) \vdash X \bullet Y}\ (\sim\wedge \vdash)$$

$$\frac{X \vdash \sim A \quad Y \vdash \sim B}{X \circ Y \vdash \sim(A \vee B)}\ (\vdash \sim\vee) \qquad \frac{\sim A \circ \sim B \vdash X}{\sim(A \vee B) \vdash X}\ (\sim\vee \vdash)$$

$$\frac{X \vdash A}{X \vdash \sim\sim A}\ (\vdash \sim\sim) \qquad \frac{A \vdash X}{\sim\sim A \vdash X}\ (\sim\sim \vdash)$$

Table 7
Introduction rules shared by all logics (I_i, C_j)

The display sequent calculi $\delta(I_i, C_j)$, $i, j \in \{1, 2, 3, 4\}$, for the constructive logics (I_i, C_j) share these rules and the introduction rules stated in Table 7. The particular display calculus $\delta(I_i, C_j)$ then is the proof system obtained by adding the rules rI_i and rC_j from Table 8.

A derivation of a sequent s from a set of sequents $\{\mathsf{s}_1, \ldots, \mathsf{s}_n\}$ in $\delta(I_i, C_j)$ is defined as a tree with root s such that every leaf is an instantiation of (id), $(id\sim)$, or a sequent from $\{\mathsf{s}_1, \ldots, \mathsf{s}_n\}$, and every other node is obtained by an application of one of the remaining rules. A proof of a sequent s in $\delta(I_i, C_j)$ is a derivation of s from $\varnothing$. Sequents s and s' are said to be interderivable iff s is derivable from $\{\mathsf{s}'\}$ and s' is derivable from $\{\mathsf{s}\}$.

Two sequents s and s' are said to be structurally equivalent if they are interderivable by means of display postulates only. It is characteristic for display calculi that any substructure of a given sequent s may be displayed as the entire antecedent or succedent of a structurally equivalent sequent s'.

If $\mathsf{s} = X \vdash Y$ is a sequent, then the displayed occurrence of X (Y) is an antecedent (succedent) part of s. If an occurrence of $(Z \circ W)$ is an antecedent part of s, then the displayed occurrences of Z and W are antecedent parts of s. If an occurrence of $(Z \bullet W)$ is an antecedent part of s, then the displayed occurrence of Z (W) is an antecedent (succedent) part of s. If an occurrence of $(Z \circ W)$ is a succedent part of s, then the displayed occurrence of Z (W) is an antecedent (succedent) part of s. If an occurrence

rI_1	$\dfrac{X \vdash A \quad Y \vdash {\sim}B}{X \circ Y \vdash {\sim}(A \to B)}$	$\dfrac{A \circ {\sim}B \vdash X}{{\sim}(A \to B) \vdash X}$
rI_2	$\dfrac{X \vdash A \circ {\sim}B}{X \vdash {\sim}(A \to B)}$	$\dfrac{X \vdash A \quad {\sim}B \vdash Y}{{\sim}(A \to B) \vdash X \circ Y}$
rI_3	$\dfrac{X \vdash A \quad B \vdash Y}{X \bullet Y \vdash {\sim}(A \to B)}$	$\dfrac{A \bullet B \vdash X}{{\sim}(A \to B) \vdash X}$
rI_4	$\dfrac{X \vdash {\sim}B \quad {\sim}A \vdash Y}{X \bullet Y \vdash {\sim}(A \to B)}$	$\dfrac{{\sim}B \bullet {\sim}A \vdash X}{{\sim}(A \to B) \vdash X}$
rC_1	$\dfrac{X \vdash {\sim}A \bullet B}{X \vdash {\sim}(A \mathrel{-\!\!\prec} B)}$	$\dfrac{{\sim}A \vdash X \quad B \vdash Y}{{\sim}(A \mathrel{-\!\!\prec} B) \vdash X \bullet Y}$
rC_2	$\dfrac{X \vdash {\sim}A \quad B \vdash Y}{X \bullet Y \vdash {\sim}(A \mathrel{-\!\!\prec} B)}$	$\dfrac{{\sim}A \bullet B \vdash X}{{\sim}(A \mathrel{-\!\!\prec} B) \vdash X}$
rC_3	$\dfrac{X \vdash A \circ B}{X \vdash {\sim}(A \mathrel{-\!\!\prec} B)}$	$\dfrac{Y \vdash A \quad B \vdash X}{{\sim}(A \mathrel{-\!\!\prec} B) \vdash Y \circ X}$
rC_4	$\dfrac{X \vdash {\sim}B \circ {\sim}A}{X \vdash {\sim}(A \mathrel{-\!\!\prec} B)}$	$\dfrac{Y \vdash {\sim}B \quad {\sim}A \vdash X}{{\sim}(A \mathrel{-\!\!\prec} B) \vdash Y \circ X}$

Table 8
Sequent rules for negated implications and co-implications

of $(Z \bullet W)$ is a succedent part of s, then the displayed occurrences of Z and W are succedent parts of s.

Theorem 5.1 (cf. (Belnap 1982)) *For every sequent* s *and every antecedent (succedent) part* X *of* s*, there exists a sequent* s' *structurally equivalent to* s *such that* X *is the entire antecedent (succedent) of* s'.

Proposition 5.2 *For every* $\mathcal{L}$*-formula* A *and every calculus* $\delta(I_i, C_j)$*,* $A \vdash A$ *is provable.*

One can define translations τ_1 and τ_2 from structures into formulas such that these translations make explicit the intuitive, context-sensitive interpretation of the structural connectives: translation τ_1 translates structures which are antecedent parts of a sequent, whereas τ_2 translates structures which are succedent parts of a sequent.

Definition 5.3 The translations τ_1 and τ_2 from structures into formulas are inductively defined as follows, where A is a formula and p is a certain atom:

$$\begin{array}{rlrl} \tau_1(A) &= A & \tau_2(A) &= A \\ \tau_1(\mathbf{I}) &= p \to p & \tau_2(\mathbf{I}) &= p \prec p \\ \tau_1(X \circ Y) &= \tau_1(X) \wedge \tau_1(Y) & \tau_2(X \circ Y) &= \tau_1(X) \to \tau_2(Y) \\ \tau_1(X \bullet Y) &= \tau_1(X) \prec \tau_2(Y) & \tau_2(X \bullet Y) &= \tau_2(X) \vee \tau_2(Y) \end{array}$$

Theorem 5.4 (Soundness) *(1) If the $X \vdash Y$ is provable in $\delta(I_i, C_j)$, then $\tau_1(X) \models^+_{I_i,C_j} \tau_2(Y)$. (2) If $X \vdash Y$ is provable in $\delta(I_i, C_j)$, then $\sim\tau_2(Y) \models^-_{I_i,C_j} \sim\tau_1(X)$.*

The language $\mathcal{L}^*$ results from $\mathcal{L}$ by adding for every atomic formula p a new atom p^*. If A is an $\mathcal{L}$-formula, $(A)^*$ is the result of replacing every strongly negated atom $\sim p$ in A by p^*.

Lemma 5.5 *For every $\mathcal{L}$-formula A, if $\varnothing \models^+_{I_i,C_j} A$, then $(\rho_{I_i,C_j}(A))^*$ is valid in* HB.

Lemma 5.6 *For every $\sim$-free $\mathcal{L}$-formula A, if A is provable in* HB, *then $\mathbf{I} \vdash A$ is provable in $\delta(I_i, C_j)$ without using any sequent rules for strongly negated formulas.*

Lemma 5.7 *For every $\mathcal{L}$-formula A, $A \vdash \rho_{I_i,C_j}(A)$ and $\rho_{I_i,C_j}(A) \vdash A$ are provable in $\delta(I_i, C_j)$.*

Lemma 5.8 *Every sequent $X \vdash \tau_1(X)$ and $\tau_2(X) \vdash X$ is provable in $\delta(I_i, C_j)$, for all $i, j \in \{1, 2, 3, 4\}$.*

Theorem 5.9 (Completeness) *(1) If $\rho_{I_i,C_j}(\tau_1(X)) \models^+_{I_i,C_j} \rho_{I_i,C_j}(\tau_2(Y))$, then $X \vdash Y$ is provable in $\delta(I_i, C_j)$. (2) If $\rho_{I_i,C_j}(\sim\tau_2(Y)) \models^-_{I_i,C_j} \rho_{I_i,C_j}(\sim\tau_1(X))$, then $X \vdash Y$ is provable in $\delta(I_i, C_j)$.*

Let $\delta(I_i, C_j)^+$ denote the result of dropping all sequent rules exhibiting $\sim$ from $\delta(I_i, C_j)$.

Theorem 5.10 *If $X \vdash Y$ is provable in system $\delta(I_i, C_j)$, then $(\rho_{Ii,C_j}(\tau_1(X)))^* \vdash (\rho_{Ii,C_j}(\tau_2(Y)))^*$ is provable in $\delta(I_i, C_j)^+$ without any applications of (cut).*

6 Correctness of the the logics (I_i, C_j) with respect to the inferentialist semantics

We show that if a sequent is provable in $\delta(I_i, C_j)$, then there exists a certain construction made up from proofs, disproofs, and their duals that transforms any proof of the antecedent of the sequent into a proof of its succedent.[4]

Theorem 6.1 *Let $i, j \in \{1, 2, 3, 4\}$. If $X \vdash Y$ is provable in $\delta(I_i, C_j)$, then*

(i) *there exists a construction π such that $\pi(\pi')$ is a canonical proof of $\tau_2(Y)$ whenever π' is a canonical proof of $\tau_1(X)$.*

[4] This result may give rise to a four-sorted typed λ-calculus.

(ii) *there exists a construction* π *such that* $\pi(\pi')$ *is a canonical dual proof of* $\tau_1(X)$ *whenever* π' *is a canonical dual proof of* $\tau_2(Y)$.

Proof. By simultaneous induction on derivations in $\delta(I_i, C_j)$.
(i): We first consider the display postulates. The first display postulates for $\circ$ are:

$$\frac{\dfrac{Y \vdash X \circ Z}{X \circ Y \vdash Z}}{X \vdash Y \circ Z}$$

Suppose, by the induction hypothesis for (i), that there exists a construction π that transforms any canonical proof of $\tau_1(Y)$ into a canonical proof of $\tau_2(X \circ Z)$ $(= \tau_1(X) \to \tau_2(Z))$, i.e., into a construction that transforms any canonical proof of $\tau_1(X)$ into a canonical proof of $\tau_2(Z)$. Let π' be any canonical proof of $\tau_1(X \circ Y)$ $(= \tau_1(X) \wedge \tau_1(Y))$. The proof π' is a pair (π'_1, π'_2), where π'_1 is a canonical proof of $\tau_1(X)$ and π'_2 is canonical proof of $\tau_1(Y)$. Then $(\pi(\pi'_2))(\pi'_1)$ is a proof[5] of $\tau_2(Z)$.
Suppose next, by the induction hypothesis for (i), that there is a construction π that transforms any proof of $\tau_1(X \circ Y)$ $(= \tau_1(X) \wedge \tau_1(Y))$ into a proof of $\tau_2(Z)$. Let π' be any proof of $\tau_1(X)$. Then $\pi^*(\pi') = \pi((\pi', \ ))$ is a construction that transforms any proof of $\tau_1(Y)$ into a proof of $\tau_2(Z)$.
The second pair of display postulates for $\circ$ is dealt with similarly.
The first pair of display postulates for $\bullet$ is:

$$\frac{\dfrac{X \bullet Z \vdash Y}{X \vdash Y \bullet Z}}{X \bullet Y \vdash Z}$$

Suppose, by the induction hypothesis for (i), that there exists a construction π that transforms any proof of $\tau_1(X \bullet Z)$, $(= \tau_1(X) \prec\!\!\!- \tau_2(Z))$, i.e., any pair (π_1, π_2), where π_1 is a proof of $\tau_1(X)$ and π_2 is a dual proof of $\tau_2(Z)$, into a proof of $\tau_2(Y)$. Let π' be any proof of $\tau_1(X)$. There either is a dual proof of $\tau_2(Z)$ or there is not. If there is such a dual proof, let π'' be a fixed dual proof of $\tau_2(Z)$. Then $(0, \pi((\pi', \pi'')))$ is a proof of $(\tau_2(Y) \vee \tau_2(Z))$. If there does not exist any dual proof of $\tau_2(Z)$, then there exists a proof of $\tau_2(Z)$. Let π''' be such a proof. Then $(1, \pi''')$ is a proof of $(\tau_2(Y) \vee \tau_2(Z))$.
Suppose now, by the induction hypothesis for (i), that there is a construction π that transforms any proof of $\tau_1(X)$ into a proof of $\tau_2(Y \bullet Z)$ $(= \tau_2(Y) \vee \tau_2(Z))$. Let π' be any proof of $\tau_1(X \bullet Y)$ $(= \tau_1(X) \prec\!\!\!- \tau_2(Y))$, i.e., any pair (π_1, π_2), where π_1 is a proof of $\tau_1(X)$ and π_2 is a dual proof of $\tau_2(Y)$. Since $\tau_2(Y)$ has no proof, $\pi(\pi_1)$ is a proof of $\tau_2(Z)$.
The second pair of display postulates for $\bullet$ is dealt with similarly.
The case of the logical structural rules is simple; the axiomatic sequents are dealt with by the identity function and (*cut*) by functional application. The case of the other structural sequent rules from Table 6 is quite obvious.
We present here, by way of example, just the cases of three introduction rules.

[5] In the sequel I will often omit the expression 'canonical'.

$(\vdash \prec)$:

$$\frac{X \vdash B \quad A \vdash Y}{X \bullet Y \vdash B \prec A}$$

Suppose, by the induction hypothesis for (i), that there is a construction π that transforms any proof of $\tau_1(X)$ into a proof of $\tau_2(B)$ and, by the induction hypothesis for (ii), that there is a construction π' that transforms any dual proof of $\tau_2(Y)$ into a dual proof of $\tau_1(A)$. Let π^* be a proof of $\tau_1(X \bullet Y)$ $(= \tau_1(X) \prec \tau_2(Y))$. Then π^* is a pair (π_1, π_2), where π_1 is a proof of $\tau_1(X)$ and π_2 is a dual proof of $\tau_2(Y)$. Therefore, the pair $(\pi(\pi_1), \pi'(\pi_2))$ is proof of $(B \prec A)$.
rI_4, first rule:

$$\frac{X \vdash \sim B \quad \sim A \vdash Y}{X \bullet Y \vdash \sim (A \rightarrow B)}$$

Suppose, by the induction hypothesis for (i), that π is a construction that transforms any proof $\tau_1(X)$ into a proof of $\sim B$, and, by the induction hypothesis for (ii), that π' is a construction that transforms any dual proof of $\tau_2(Y)$ into a dual proof of $\sim A$, i.e., into a dual disproof of A. Let π^* be a proof of $\tau_1(X \bullet Y)$ $(= \tau_1(x) \prec \tau_2(Y))$. Then π^* is a pair (π_1, π_2), where π_1 is a proof of $\tau_1(X)$ and π_2 is a dual proof of $\tau_2(Y)$. A proof of $\sim (A \rightarrow B)$ in (I_4, C_j) is a disproof of $(A \rightarrow B)$ in (I_4, C_j), which is a pair (π_1', π_2'), where π_1' is a proof of $\sim B$, and π_2' is a dual proof of $\sim A$. Note that $(\pi(\pi_1), \pi'(\pi_2))$ is such a pair.
rC_1, second rule:

$$\frac{\sim A \vdash X \quad B \vdash Y}{\sim (A \prec B) \vdash X \bullet Y}$$

Suppose, by the induction hypothesis for (i), that π' is a construction that transforms any proof of B into a proof of $\tau_2(Y)$ and that π'' is a construction that transforms any proof $\sim A$ into a proof of $\tau_2(X)$. Let π^* be a proof of $\sim (A \prec B)$ in (I_i, C_1), i.e., a disproof of $(A \prec B)$ in (I_i, C_1). Then π^* is a pair (i, π) such that $i = 0$ and π is a disproof of A or $i = 1$ and π is a proof of B. But then either $(0, \pi''(\pi))$ or $(1, \pi'(\pi))$ is a proof of $\tau_2(X \bullet Y)$.
(ii): We present here just the case of the second pair of display postulates for $\circ$:

$$\frac{\dfrac{X \vdash Y \circ Z}{X \circ Y \vdash Z}}{Y \vdash X \circ Z}$$

Suppose, by the induction hypothesis for (ii), that π is a construction that transforms any dual proof of $\tau_1(Y) \rightarrow \tau_2(Z)$ (i.e., any pair (π_1, π_2), where π_1 is a proof of $\tau_1(Y)$ and π_2 is dual proof of $\tau_2(Z)$) into a dual proof of $\tau_1(X)$. Either there is a proof of $\tau_1(Y)$ or not. If there is a proof of $\tau_1(Y)$, let π'' be such a proof. Then $(0, \pi(\pi'', \pi'))$ is a dual proof of $\tau_1(X \circ Y)$. If there is no proof of $\tau_1(Y)$, then there is a dual proof of $\tau_1(Y)$. Let π''' be such a dual proof. Then $(1, \pi''')$ is a dual proof of $\tau_1(X \circ Y)$.
Now suppose that, by the induction hypothesis for (ii), there exists a construction π that transforms any dual proof of $\tau_2(Z)$ into a dual proof of $\tau_1(x) \wedge \tau_1(Y)$. Let π' be any dual proof of $\tau_1(X) \rightarrow \tau_2(Z)$, i.e., a pair (π_1, π_2), where π_1 is a proof of $\tau_1(X)$ and

π_2 is a dual proof of $\tau_2(Z)$. Then there is no dual proof of $\tau_1(X)$ and $(0, \pi(\pi_2))$ is a dual proof of $\tau_1(Y)$. □

Corollary 6.2 *Let $i, j \in \{1,2,3,4\}$. If $X \vdash Y$ is provable in $\delta(I_i, C_j)$, then*

(i) *there exists a construction π such that $\pi(\pi')$ is a canonical disproof of $\tau_2(Y)$ whenever π' is a canonical disproof of $\tau_1(X)$.*

(ii) *there exists a construction π such that $\pi(\pi')$ is a canonical dual disproof of $\tau_1(X)$ whenever π' is a canonical dual disproof of $\tau_2(Y)$.*

Proof. Every canonical disproof of A is a canonical proof of $\sim A$ and every canonical dual disproof of A is a canonical dual proof of $\sim A$. □

The following claims follow from Theorem 6.1 and Corollary 6.2.

Theorem 6.3 *Let $i, j \in \{1,2,3,4\}$.*

(i) *If $\mathbf{I} \vdash A$ is provable in $\delta(I_i, C_j)$, then there exists a construction π which is a proof of A.*

(ii) *If $A \vdash \mathbf{I}$ is provable in $\delta(I_i, C_j)$, then there exists a construction π which is a dual proof of A.*

(iii) *If $\mathbf{I} \vdash \sim A$ is provable in $\delta(I_i, C_j)$, then there exists a construction π which is a disproof of A.*

(iv) *If $\sim A \vdash \mathbf{I}$ is provable in $\delta(I_i, C_j)$, then there exists a construction π which is a dual disproof of A.*

Proof. Note that any canonical proof of $\tau_1(\mathbf{I}) = (p \rightarrow p)$ and any canonical dual proof of $\tau_2(\mathbf{I}) = (p \prec p)$ is the identity function. □

Example 6.4 The sequent $\mathbf{I} \vdash q \vee -q$ is provable in the logics (I_i, C_j), and it can easily be seen that there exists a construction that is a (canonical) proof of $q \vee ((p \rightarrow p) \prec q)$. A proof of $q \vee ((p \rightarrow p) \prec q)$ is a pair (i, π), where $i = 0$ and π is a proof of q, or $i = 1$ and π is a proof of $((p \rightarrow p) \prec q)$. Now, π is a proof of $((p \rightarrow p) \prec q)$ iff π is a pair (π_1, π_2), where π_1 is a proof of $(p \rightarrow p)$ and π_2 is a dual proof of q. Since the identity function is a proof of $(p \rightarrow p)$ and since every $\mathcal{L}$-formula either has a proof or a dual proof, there exists a proof of $q \vee ((p \rightarrow p) \prec q)$.

Example 6.5 There exists a construction that is a proof of $\sim (p \rightarrow q) \rightarrow (p \prec q)$ in the logics (I_3, C_j). A proof of $\sim (p \rightarrow q) \rightarrow (p \prec q)$ in (I_3, C_j) is a construction that transforms any proof of $\sim (p \rightarrow q)$ into a proof of $(p \prec q)$. A proof of $\sim (p \rightarrow q)$ in (I_3, C_j) is a disproof of $(p \rightarrow q)$, which is a pair (π_1, π_2), where π_1 is a proof of p and π_2 is a dual proof of q. But this pair is a proof of $(p \prec q)$ in (I_3, C_j), so that the identity function is a proof of $\sim (p \rightarrow q) \rightarrow (p \prec q)$ in (I_3, C_j).

7 Summary

We have considered sixteen extensions of propositional Brouwer-Heyting logic by strong negation. Each of these logics (I_i, C_j) $(i, j \in \{1,2,3,4\})$ turned out to be correct with respect to an extended BHK-style inferentialist interpretation. The interpretation makes use of four primitive notions, namely the notions of proof, disproof, dual proof, and dual disproof. This correctness result supports the view that the logics (I_i, C_j) are indeed constructive propositional logics.[6] The findings of this paper can be summarized as in a Table 9.[7]

(propositional) logic	*soundness with respect to an interpretation*
intuitionistic logic	in terms of proofs
Nelson's logics	in terms of proofs and disproofs
dual intuitionistic logic	in terms of dual proofs
bi-intuitionistic logic	in terms of proofs and dual proofs
bi-intuitionistic logic extended by strong negation	in terms of proof, disproofs, and their duals

Table 9
Summary

References

[1] A. Almukdad and D. Nelson, Constructible Falsity and Inexact Predicates, *Journal of Symbolic Logic*, 49 (1984), 231–233.

[2] N.D. Belnap, Display Logic, *Journal of Philosophical Logic* 11 (1982), 375–417. Reprinted with small modifications as §62 of A.R. Anderson, N.D. Belnap, and J.M. Dunn, *Entailment: the logic of relevance and necessity*. Vol. 2, Princeton University Press, Princeton, 1992.

[3] B. Brown, On Paraconsistency, in. L. Goble (ed.), *A Companion to Philosophical Logic*, Blackwell Publishers, Oxford, 2002, 628–650.

[4] A. Brunner and W. Carnielli, Anti-intuitionism and Paraconsistency, *Journal of Applied Logic* 3 (2005), 161-184.

[5] L. Buisman and R. Goré, A Cut-Free Sequent Calculus for Bi-intuitionistic Logic, in: N. Olivetti (ed.), *TABLEAUX 2007*, Springer Lecture Notes in AI 4548, Springer-Verlag, Berlin, 2007, 90–106.

[6] T. Crolard, Subtractive Logic, *Theoretical Computer Science* 254 (2001), 151–185.

[7] D. van Dalen, *Logic and Structure*, Springer-Verlag, Berlin, 2004.

[8] J.M. Dunn, Partiality and its Dual, *Studia Logica* 66 (2000), 5–40.

[9] D. Gabbay, What is Negation in a System, in: F. Drake and J. Truss (eds.), *Logic Colloquium '86*, Elsevier, Amsterdam, 95–112.

---

[6] Earlier versions of this paper have been presented at the Foundations of Logical Consequence Workshop II: The Logic of Denial, St Andrews, October 2009, and the Workshop on Truth, Falsity, and Negation, Dresden, April 2010. I would like to thank the audience of these events, in particular Peter Schroeder-Heister, for helpful comments and Stephen Read for the invitation to St Andrews. Moreover, I would like to thank the organizers of *AiML 2010* for inviting me to Moscow.

[7] In intuitionistic logic, $\bot$ is primitive and has no proof; in dual intuitionistic logic, $\top$ is primitive and has no dual proof.

[10] R. Goré, Dual Intuitionistic Logic Revisited, in: Roy Dyckhoff (ed.), *TABLEAUX 2000*, Springer Lecture Notes in AI 1847, Springer-Verlag, Berlin, 2000, 252–267.

[11] R. Goré and L. Postniece, Combining Derivations and Refutations for Cut-free Completeness in Bi-intuitionistic Logic, *Journal of Logic and Computation* 20 (2010), 233–260.

[12] R. Goré, L. Postniece, and A. Tiu, Cut-elimination and Proof-search for Bi-intuitionistic Logic using Nested Sequents, in: C. Areces and R. Goldblatt (eds.), *Advances in Modal Logic. Vol. 7*, College Publications, London, 2008, 43–66.

[13] A. Grzegorczyk. A Philosophically Plausible Formal Interpretation of Intuitionistic Logic, *Indagationes Mathematicae* 26 (1964), 596-601.

[14] Y. Gurevich, Intuitionistic Logic with Strong Negation, *Studia Logica* 36 (1977), 49–59.

[15] N. Kamide, A Canonical Model Construction for Substructural Logics with Strong Negation, *Reports on Mathematical Logic* 36 (2002), 95–116.

[16] N. Kamide, A Note on Dual-intuitionistic Logic, *Mathematical Logic Quarterly* 49 (2003), 519–524.

[17] N. Kamide, Phase Semantics and Petri Net Interpretation for Resource-Sensitive Strong Negation, *Journal of Logic, Language and Information* 15 (2006), 371–401.

[18] N. Kamide and H. Wansing, Symmetric and Dual Paraconsistent Logics, 2008, to appear in: *Logic and Logical Philosophy.*

[19] M. Kracht, On Extensions of Intermediate Logics by Strong Negation, *Journal of Philosophical Logic* 27 (1998), 49–73.

[20] E.G.K. López-Escobar, Refutability and Elementary Number Theory, *Indigationes Mathematicae* 34 (1972), 362–374.

[21] D. Nelson, Constructible Falsity, *Journal of Symbolic Logic* 14 (1949), 16–26.

[22] S. Odintsov, Algebraic Semantics for Paraconsistent Nelson's Logic, *Journal of Logic and Computation* 13 (2003), 453–468.

[23] S. Odintsov, *Constructive Negations and Paraconsistency*, Springer-Verlag, Dordrecht, 2008.

[24] S. Odintsov and H. Wansing, Inconsistency-tolerant Description Logic. Motivation and Basic Systems, in: V. Hendricks and J. Malinowski (eds.), *Trends in Logic. 50 Years of Studia Logica*, Kluwer Academic Publishers, Dordrecht, 2003, 301–335.

[25] S. Odintsov and H. Wansing, Constructive Predicate Logic and Constructive Modal Logic. Formal Duality versus Semantical Duality, in: V. Hendricks et al. (eds.), *First-Order Logic Revisited*, Logos Verlag, Berlin, 2004, 269–286.

[26] S. Odintsov and H. Wansing, Inconsistency-tolerant Description Logic. Part II: Tableau Algorithms, *Journal of Applied Logic* 6 (2008), 343–360.

[27] G. Priest, *Doubt Truth to Be a Liar*, Oxford University Press, Oxford, 2006.

[28] C. Rauszer, A Formalization of the Propositional Calculus of H-B Logic, *Studia Logica* 33 (1974), 23–34.

[29] C. Rauszer, Applications of Kripke Models to Heyting-Brouwer Logic, *Studia Logica* 36 (1977), 61–72.

[30] C. Rauszer, An Algebraic and Kripke-style Approach to a certain Extension of Intuitionistic Logic, *Dissertationes Mathematicae* 167, Institute of Mathematics, Polish Academy of Sciences, Warsaw 1980, 62 pp.

[31] G. Restall, Multiple Conclusions, in: P. Hájek, L. Valdes-Villanueva, and D. Westerstahl (eds.), *Logic Methodology and Philosophy of Science. Proceedings of the Twelfth International Congress*, King's College Publications, London, 2005, 189–205.

[32] R. Routley, Semantical Analyses of Propositional Systems of Fitch and Nelson, *Studia Logica*, 33 (1974), 283–298.

[33] P. Schroeder-Heister, Schluß und Umkehrschluß Ein Beitrag zur Definitionstheorie, to appear in: C.F. Gethmann (ed.), *Lebenswelt und Wissenschaft. Akten des XXI. Deutschen Kongresses für Philosophie 2009. Deutsches Jahrbuch Philosophie*, Vol. 3, 2010.

[34] Y. Shramko und H. Wansing, Some Useful 16-valued Logics: How a Computer Network Should Think, *Journal of Philosophical Logic* 34 (2005), 121–153.

[35] R. Thomason, A Semantical Study of Constructive Falsity, *Zeitschrift für mathematische Logik und Grundlagen der Mathematik* 15 (1969), 247–257.

[36] I. Urbas, Dual-Intuitionistic Logic, *Notre Dame Journal of Formal Logic* 37 (1996), 440–451.

[37] D. Vakarelov, Notes on N-lattices and Constructive Logic with Strong Negation, *Studia Logica* 36 (1977), 109–125.

[38] D. Vakarelov, Constructive Negation on the Basis of Weaker Versions of Intuitionistic Negation, *Studia Logica* 80 (2005), 393–430.

[39] D. Vakarelov, Non-classical Negation in the Works of Helena Rasiowa and Their Impact on the Theory of Negation, *Studia Logica* (84), 2006, 105–127.

[40] H. Wansing, *The Logic of Information Structures*, Springer Lecture Notes in AI 681, Springer-Verlag, Berlin, 1993.

[41] H. Wansing, *Displaying Modal Logic*, Kluwer Academic Publishers, Dordrecht, 1998.

[42] H. Wansing, Negation, in: L. Goble (ed.), *The Blackwell Guide to Philosophical Logic*, Basil Blackwell Publishers, Cambridge/MA., 2001, 415–436.

[43] H. Wansing, Connexive Modal Logic, in: R. Schmidt et al. (eds.), *Advances in Modal Logic. Volume 5*, King's College Publications, London, 2005, 376–383 (http://www.aiml.net/volumes/volume5/).

[44] H. Wansing, On the Negation of Action Types: Constructive Concurrent PDL, in: P. Hájek, L. Valdes-Villanueva, and D. Westerstahl (eds.), *Logic Methodology and Philosophy of Science. Proceedings of the Twelfth International Congress*, King's College Publications, London, 2005, 207–225.

[45] H. Wansing, Connexive Logic, The Stanford Encyclopedia of Philosophy (Winter 2006 Edition), Edward N. Zalta (ed.), URL = <http://plato.stanford.edu/archives/win2006/entries/logic-connexive/>.

[46] H. Wansing, Logical Connectives for Constructive Modal Logic, *Synthese* 150 (2006), 459–482.

[47] H. Wansing, A Note on Negation in Categorial Grammar, *Logic Journal of the IGPL* 15 (2007), 271–286.

[48] H. Wansing, Constructive Negation, Implication, and Co-implication, *Journal of Applied Non-Classical Logics* 18 (2008), 341-364.

[49] H. Wansing und Y. Shramko, Harmonious Many-valued Propositional Logics and the Logic of Computer Networks, in: C. Dégremont, L. Keiff and H. Rückert (ed.), *Dialogues, Logics and Other Strange Things. Essays in Honour of Shahid Rahman*, College Publications, London, 2008, 491-516.

[50] H. Wansing und Y. Shramko, Suszko's Thesis, Inferential Many-valuedness, and the Notion of a Logical System, *Studia Logica* 88 (2008), 405–429, 89 (2008), 147.

[51] T. Williamson, *Knowledge and Its Limits*, Oxford University Press, Oxford, 2000.

[52] F. Wolter, On Logics with Coimplication, *Journal of Philosophical Logic* 27 (1998), 353–387.

www.ingramcontent.com/pod-product-compliance
Lightning Source LLC
Chambersburg PA
CBHW060312230426
43663CB00009B/1679